Silicate Science

VOLUME II

SILICATE SCIENCE

VOLUME I. SILICATE STRUCTURES

SECTION A. Silicate Crystal Structures

SECTION B. Clay Minerals; Structures

SECTION C. Silicate Dispersoids

VOLUME II. GLASSES, ENAMELS, SLAGS

SECTION A. Properties and Constitution of Silicate Glasses

SECTION B. Industrial Glass and Enamels

SECTION C. Industrial Slags

VOLUME III. DRY SILICATE SYSTEMS

SECTION A. Dry Silicate Equilibria; Fusion and Polymorphism

SECTION B. Dry Silicate Systems; Fusion and Polymorphism

VOLUME IV. HYDROTHERMAL SILICATE SYSTEMS

SECTION A. Silicate Systems with Volatiles

SECTION B. Dehydration Behavior of Silicate Hydrates: Zeolites and Related Materials

Appendix

VOLUME V. CERAMICS AND HYDRAULIC BINDERS

SECTION A. Solid-State Reactions and Their Uses

SECTION B. Reactions in Ceramic Bodies

SECTION C. Portland Cements and Related Hydraulic Binders

SILICATE SCIENCE

BY

WILHELM EITEL

INSTITUTE FOR SILICATE RESEARCH
UNIVERSITY OF TOLEDO
TOLEDO, OHIO

VOLUME II

GLASSES, ENAMELS, SLAGS

1965

ACADEMIC PRESS New York San Francisco London
A Subsidiary of Harcourt Brace Jovanovich, Publishers

ACADEMIC PRESS INC.
111 Fifth Avenue, New York, New York 10003

United Kingdom Edition published by
ACADEMIC PRESS, INC. (LONDON) LTD.
24/28 Oval Road, London NW1

Library of Congress Catalog Card Number: 63-16981

PRINTED IN THE UNITED STATES OF AMERICA.
79 80 81 82 83 84 9 8 7 6 5 4 3 2

General Preface

The evolution of Silicate Science has tremendously accelerated in the decade since the publication of "The Physical Chemistry of the Silicates" by the University of Chicago Press in 1954. One is tempted to speak, in the language of our newspapers, of an "explosive" advance in this field of scientific and technological investigation.

The author has decided, in complete accord with the publishers of the 1954 book, and with the new publishers, Academic Press, New York, to restrict this revision exclusively to the progress made in the period between 1952 and 1962.

The new title "Silicate Science" expresses the more extensive fields of investigation covered by the present treatise. The author is very grateful to the publishers for having agreed to issue this new version, which required the more comprehensive form of a series of five encyclopedic volumes.

In this treatise, references cited from "Physical Chemistry of the Silicates" are indicated only by naming the authors and the year of publication. The paragraph system of subdivision, which was welcomed by many readers of the 1954 book, is also used for the new treatise and facilitates cross-references and index quotations. However, the former reference system using numerical keys for periodicals is abandoned. In the footnotes, periodicals are cited by use of the abbreviations now universally known from the "List of Periodicals, Key to Library Files and Other Information," published by *Chemical Abstracts* under the auspices of the American Chemical Society, Washington, D. C., 1961.

The present treatise would not have been possible without the kind help extended by many friends and colleagues all over the world who sent the author their reprints and other publications. This wide correspondence gave the author hope that this treatise will be well received.

With the greatest appreciation the author wishes to express his obligation to Dr. W. S. Carlson, President of the University of Toledo, and the Board of Directors, who granted him multiple facilities for independent work. The National Science Foundation, Washington, D. C., supported his work over the years 1960 to 1962 by important subsidies.

Particular gratitude is owed to Mrs. M. M. Gillham, M. A., and her staff, who helped him master the abundant material through search of the literature and acquisition of rare publications, not to speak of the hospitality the author enjoyed in the office and reading rooms of this University Library.

It would require too much space to give a complete list of those Publishers who granted permission for reproduction of figures. Their names will be cited in special acknowledgments in the individual volumes. The author, however, wishes to express his most sincere appreciation for the spirit of cooperation he met everywhere.

The author would like to include in this general acknowledgment his cordial gratitude to his indefatigable secreatry, Miss Arleen Altaffer, who worked with him assiduously and with the greatest accuracy for more than three years, in preparation of the voluminous manuscripts, indexes, and correspondence.

The author also wishes to express his thanks and appreciation to Col. G. Wilfrid Hibbert, B. A., for editorial assistance, comments, and advice. Furthermore, he offers words of personal gratitude to Professor G. E. Pankratz, M. S., M. E., University of Toledo, and Mr. Marvin Johns, B. E., who prepared most of the figures.

Toledo, Ohio W. E.
May 1964

Preface to Volume II

In this second volume the modern principles of the physics and chemistry of glass and its constitution (Section A. II of the 1954 book) are combined with the basic facts of glass technology (Section E. I of the previous book). This combination removes many inconveniences and simplifies considerably the use of cross references in both sections. The reader will now find collected in one volume all the necessary information on glass formation, properties, and theory.

In Section A the reader will find that tremendous progress has been made in the theory of glass constitution and that a reconciliation has been achieved between previously sharply contrasting hypotheses of structural characteristics. Among modern methods of glass investigation that are of basic scientific and technological interest, infrared absorption and radioactive tracer methods have become increasingly important.

Section C, treating of metallurgical slags, illustrates the great progress made in the application of silicate melt equilibria of systems containing iron oxides following the basic investigations of E. F. Osborn and A. Muan. New studies of the role of phosphoric acid in the melts of pig iron slags are equally important, as seen from the investigations of G. Trömel and his collaborators.

W. E.

Toledo, Ohio
September 1964

Acknowledgments

The organizations listed below kindly granted permission to reproduce figures taken from their copyrighted publications:

Akademische Verlagsgesellschaft mbH, Frankfurt a. M., Germany
American Association for the Advancement of Science, Washington, D. C.
American Ceramic Society, Columbus, Ohio
American Chemical Society, Washington, D. C.
American Institute of Mining and Metallurgical Engineers, Inc., New York, New York
American Institute of Physics, New York, New York
Asahi Glass Company Ltd., Yokohama, Japan
Bergbau-Berufsgenossenschaft Hauptverwaltung, Bochum, Germany
The British Ceramic Society, Stoke-On-Trent, England
Chemical Society of Japan, Tokyo, Japan
Clarendon Press, Oxford, England
Gauthier-Villars Imprimeur-Librairie, Paris, France
Gebrüder Borntraeger, Berlin, Germany
Hermann Hübener Verlag, Goslar, Germany
Industrie Chimique Belge, Brussels, Belgium
Institut du Verre, Paris, France
Institute of Vitreous Enamellers, Ripley (Derby), England
The Iron & Steel Institute, London, England
Le Journal de physique, et le radium, Paris, France
Macmillan & Co., Ltd., London, England
Max Planck-Institut für Silikatforschung, Würzburg, Germany
Einar Munksgaard, Copenhagen, Denmark
Österreichisch-Amerikanische Magnesit A. G., Radenthein/Carinthia, Austria
Ogden Publishing Company, New York, New York
Osaka Industrial Research Institute, Osaka, Japan
Pennsylvania State University, School of Mineral Industries, University Park, Pennsylvania
Revue d'optique, théorique et instrumentale, Paris, France
Society of Glass Technology, Sheffield, England
Springer-Verlag, Berlin, Göttingen, Heidelberg, Germany
U. S. Bureau of Standards, Washington, D. C.
University of Chicago Press, Chicago, Illinois
VEB Verlag für Bauwesen, Berlin, Germany
Verlag Chemie G.m.b.H., Weinheim/Bergstr., Germany
Verlag der Deutschen Glastechnischen Gesellschaft, Frankfurt a. M., Germany
Verlag Stahleisen m.b.H., Düsseldorf, Germany
Vulkan-Verlag, Essen, Germany

Dedicated

to the Memory of

HERMANN SALMANG

Contents

Section A

PROPERTIES AND CONSTITUTION OF SILICATE GLASSES

Section B

INDUSTRIAL GLASS AND ENAMELS

Section C

INDUSTRIAL SLAGS

Section A

Properties and Constitution of Silicate Glasses

PHYSICO CHEMICAL-PROPERTIES OF MELTS AND GLASSES

1. The analogies existing between typical inorganic glasses, classical silicate and phosphate glasses as characterized in their atomic structure by anionic chains, annular units, and frameworks of highly complex arrangements, which are the principal reason for impediments in their crystallization tendencies, are discussed from a general viewpoint by E. Thilo.[1] For many such compounds, dissolution in a solvent first loosens the structural (valence) bonds and crystallization may be nucleated from the solution phase by a reconstitution of the valence bonds. The chief advantage of such a theoretical discussion of the glassy state in crystallization tendencies over the structural theories of W. H. Zachariasen (1932) and A. Smekal (1951) is its equal applicability to inorganic *and* organic molecular complexes, say equally for "elemental glasses" of sulfur, selenium, tellurium, or sulfides, oxides, and so on, as well as for polysaccharides, plastomers, and proteins. In addition, for inorganic glasses the specific nature of so-called mineralizers to promote crystallization is evident, such as the use of water for the reconstitution of a crystalline phase of lower-molecular units derived from a polymeric "glass." Similar effects can be exerted by small cations with a high polarization strength. For example, lithium ions reduce the activation energy of the transition glass $\rightarrow$ crystalline phase, either directly, or over aging processes, which are a gradual approximation to the crystalline state as that of the higher stability in comparison with the metastable, "quenched" glassy melts.

2. Starting from the optical properties, especially from the refractive indices of binary oxide glasses, and of elements in the groups III to V of the Periodic System, A. Winter[2] derived from the number, L, of the oxygen atoms in compounds $A_K L_L$,

[1] *Silikat Tech.*, **6**, 1955, 278-280.

[2] *Verres et réfractaires*, **9**, 1955, 147-156; *Compt. rend*, **240**, 1955, 73-75; **242**, 1956, 2931-2933; 3057-3059; *Ibid.*, **244**, 1957, 750-752; *J. Am. Ceram. Soc.*, **40**, 1957, 54-58; **41**, 1958, 464-466.

in the ratio $(n_{A_K O_L} - n_O)/L$, as a function of the atomic and molecular weight of the elements A, or of their oxides, a straight line from which an extrapolation can be made to the refractive index of pure idealized "oxygen glass" ($n_O = 1.22$). Such calculations may be extended to sulfur, selenium, and tellurium as elements of the VIth Group, in combination with those of the groups III to V. Basic correlations are in this way found concerning the position of the elements in the Periodic System determining their refractive indices, n, and dispersion characteristics,

$$\nu = \left(\frac{n_D - 1}{n_F - n_C}\right):$$

$$n_{SiO_2} < n_{GeO_2} < n_{SnO_2} \qquad \nu_{SiO_2} > \nu_{GeO_2} > \nu_{PbO_2}$$

$$n_{PbO} < n_{PbS} < n_{PbTe} \qquad \nu_{Al_2O_3} < \nu_{SiO_2} < \nu_{P_2O_5}$$

$$n_{In_2O_3} > n_{SnO_2} < n_{Sb_2O_3}$$

A. Winter's calculations are particularly important because they give information on many interesting but experimentally not directly accessible glass types, such as oxygen (see above), or other elements of "self-vitrifying" character. In Fig. A. 1

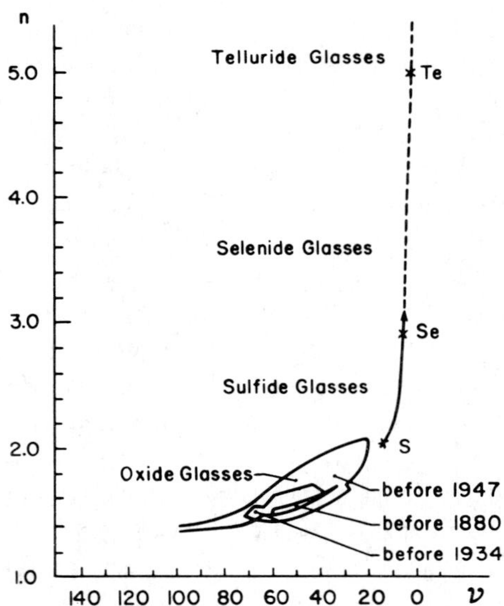

FIG. A. 1. Refractive indices of elemental and diatomic glasses, as a function of their optical dispersion. (Winter-Klein).

The definition of "glass" used by A. Winter is that conventionally given by Committee C-14 of the American Society of Testing Materials, namely "*an inorganic product of fusion which has cooled to a rigid condition without crystallization.*" Parallel to this goes the definition based on the viscosity $\eta = 10^{15}$ poises, as the limit temperature at which glass behaves as a solid material. This corresponds to about the viscosity (rigidity) of aluminum metal at room temperature.

are plotted the refractive indices, n, of such elemental glasses as a function of the dispersion, in relation to the indices of binary oxides, sulfides, selenides, and tellurides (with $PbTe_2$ $n = 5.2$ as the maximum value observed). The elemental glasses of sulfur, selenium, and tellurium have the indices $n = 1.998$, 2.9, and about 5.0, respectively. Winter further discussed glass formation of the elements of the VIIth Group in combination with those of Groups II to IV, or with a transition element. None of the halogen elements are able to act as a self-vitrifying material. Many of the binary compounds forming predicted glasses could be prepared for the first time. Under no circumstances, however, are the elements of Group I able to cooperate in the formation of a vitreous framework structure. As a first rule for the formation of more complex glasses, the capacity to form glasses increases with the number of group elements to the left of group VI (for a given period). Complex glass types can be systematically derived by the mixture of diatomic vitrifiable compounds, say composed of SiO_2 or B_2O_3 with As_2S_5 or As_2Se_5, or by a mixture of a vitrifiable binary compound with a nonvitrifiable diatomic compound, as SiO_2 and Na_2O, in this case up to a limit concentration of about 60 weight per cent Na_2O.

3. The glass of *arsenic trioxide* was studied by I. N. Stranski, H. Böttcher, K. Plieth, and E. Reuber-Kürbs[3] in its particularly interesting structural relations to the crystalline modifications, claudetite, and arsenolite (cf. A. ¶ 221). The analogy with claudetite is decidedly greater than that with arsenolite. Corresponding *arsenic sulfide glasses* were observed by G. C. Amstutz[4] as a natural mineral occurrence, as in vugs of sphalerite ore bodies of Cerro de Pasco, Peru, in reniform aggregates, or tubules, transparent with dark-red color, and a relatively high reflectivity (30 per cent). Artificial glasses of the same type were described in their infrared transmittance properties by several authors. Such glasses have a high interest for the optical industry, just as selenide and telluride (chalcogenide) glasses must have (cf. A. ¶ 389). Beyond that, the arsenic sulfide glass from Cerro de Pasco is scientifically most remarkable because of an exceptionally strong enrichment in the isotope S^{32} with regard to S^{34} (in the ratio 22.96 : 1, whereas normally it is 22.23 to 22.31).

4. As a complex oxide-sulfide glass formation we mention here only that in the system K_2O–Sb_2O_3–Sb_2S_3 which was studied by B. W. King and G. D. Kelly[5] because

[3] *Z. anorg. u. allgem. Chem.*, **266**, 1951, 302-312; **280**, 1955, 205-210; see also K. Plieth, *Proc. Intern. Symposium on Reactivity of Solids, Gothenburg*, 1952, Vol. II, 1952/54, 761-770.

[4] The composition of such glasses is given by Amstutz in formulas like $As_{50-60}S_{35-45}Pb_{3-8}Tl_{\pm 1}$, or $As_{23.1\pm .7}S_{51.4\pm .25}Pb_{20.5\pm .45}Tl_{.04}$. Cf. *Bull. Geol. Soc. Am.*, **68**, 1957, 1696. See also W. A. Fraser and D. M. Fraser, *J. Opt. Soc. Am.*, **49**, 1959, 497. Important discussions on the covalent bonding in chalcogenide glasses were given recently by N. A. Goryunova and B. T. Kolomiets, *Structure of Glass. Proc. All-Union Conf. Glassy State, 3rd, Leningrad (English Translation)* **3**, 1959/60, 58-63 (see also A. ¶ 389*).

[5] *J. Am. Ceram. Soc.*, **41**, 1958, 367-371. The analogy to the mineral *kermesite*, $2Sb_2S_3.Sb_2O_3$, (of orthorhombic symmetry) is remarkable.

of its infrared transmittance properties. Melts of potassium pyroantimonate and Sb_2S_3 have an annealing and softening temperature of 150° and 230°C., respectively. They are stable to moisture of the air atmosphere and have a remarkably high transmittance in the range from 3.5 to 7 μ (Fig. A. 2).

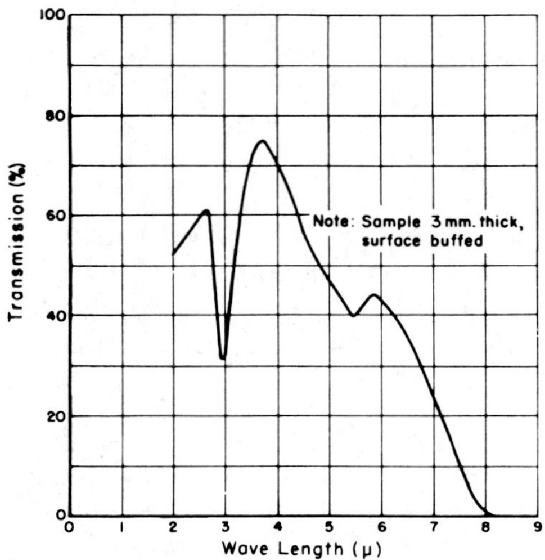

FIG. A. 2. Infrared transmittance curve of a potassium pyroantimnate-antimony sulfide glass. (King and Kelly).

5. Other than silicate glasses based on complex oxide compounds were generally discussed by H. M. Heaton and H. Moore,[6] also in respect to their infrared transmittance characteristics over the range of wavelengths between 1 and 15 μ. Main absorption bands typical of all oxide constituents are observed near and beyond 6 μ, often overlapped by the H_2O bands causing in many cases a virtual "cut-off" either at 3, or at 6 μ, for glass samples thicker than 2 mm. The replacement of oxygen by sulfur in some cases was recommended for an improvement of infrared transmittance by W. A. Fraser and J. E. Brugger,[7] as in the comparison of the optical properties of calcium aluminate or calcium gallate glasses, with corresponding thioaluminate and thiogallate glass compositions. From the rich experimental material given in the extensive work of Heaton and Moore one may summarize that binary glasses can be prepared in the systems Al_2O_3–BaO, Sb_2O_3–PbO, As_2O_3–K_2O,

[6] *J. Soc. Glass Technol.*, **41**, 1957, 3-28 T; 29-71 T (on the water content of glasses and its effects on infrared transmittance phenomena); 72-85 T (on the factors which determine the possibilities of glass formation).

[7] *J. Opt. Soc. Am.*, **49**, 1959, 497; see also P. L. Baynton, H. Rawson, and J. E. Stanworth, *Nature*, **179**, 1957, 434-435, and quite recently, H. H. Blau, *Glastech. Ber.*, **32**, K, 1959, VII. 6-10.

$-PbO$, $-Sb_2O_3$, $-MoO_3$, $-SeO_2$, $-V_2O_5$, further TeO_2–K_2O, $-BaO$, $-Al_2O_3$, $-Sb_2O_3$, $-Bi_2O_3$, $-MoO_3$, $-WoO_3$, and $-TiO_2$, the latter forming glasses in the presence of some silica (5 to 10 per cent). The general discussions of Heaton and Moore on the factors of glass-forming systems are particularly valuable for a systematic development of new types of glasses, in certain limits chiefly determined by adequate ratios of anion/cation for the framework-forming cation of definite coordination with oxygen. The tabulations given by these authors also include polarizability effects of the anions O^{2-}, S^{2-}, Se^{2-}, and Te^{2-}, and the coordination types of these anions with different cations. Corresponding considerations, with a complete tabulation of known and of predicted glass compositions on the basis of the anion/cation relationships, were also developed by M. A. Bezborodov,[8] with particularly illustrative "checkerboard charts" of possible complex oxide systems (up to eight components), with SiO_2 as the constant framework-building ingredient, and noting the principal properties observed, or to be expected. Bezborodov also included short discussions of corresponding germanate and phosphate glass systems and effects of the introduction of fluorine, or of the complex anion SiF_6^{2-}.

6. In connection with general discussions on glass-forming oxide systems, previously given descriptions of *zirconia*, ZrO_2, glasses as quenched from the melts prepared in a solar furnace (cf. W. M. Cohn, 1931) had to be revised. N. Köppen[9] reexamined such alleged glass samples and identified in these the so-called "stabilized" crystalline cubic ZrO_2, contaminated by 2 to 3 per cent MgO, with some baddeleyite (monoclinic), but *no* glass phase at all could be detected. The exterior aspect of the translucent and even transparent fusion products in the solar furnace may even be more fallacious if fine ZrO_2 hairs or filaments (wools) are "spun," which, however, are not a real glass but only polycrystalline fiberlike ceramic products as described by C. G. Harman.[10] Also alumina forms such "wools" of nonglassy fiber aggregates.

7. Another very interesting nonsilicate glass group are the *tellurite* glasses studied by J. E. Stanworth and J. A. James,[11] of the systems TeO_2–BaO and TeO_2–PbO, further TeO_2–B_2O_3, and TeO_2–MoO_3, or $-WoO_3$, which can be melted in gold crucibles. Corresponding glasses with CuO, MnO_2, and V_2O_5 are entirely opaque and black in the visible range, but have a transmittance for short-wave infrared radiation over the range from 0.8 to 5.0 μ. On the basis of the trioxides of *molybdenum*, *wolfram*, and *uranium*, B. L. Baynton, H. Rawson, an J. E. Stanworth[12] observed glasses in

[8] *Steklo i Keram.*, **15**, 1958 (11) 7-12; *Silikat Tech.*, **10**, 1959, 54-57.

[9] *Ber. deut. keram. Ges.*, **35**, 1958, 313-316. On zirconia and its modifications see B. C. Weber and M. A. Schwartz, *Ibid.*, **34**, 1957, 391-396.

[10] *Ceram. Ind.*, **73**, 1959 (1) July, p. 74.

[11] *J. Soc. Glass Technol.*, **38**, 1954, 421-424, 425-435.

[12] *Nature*, **178**, 1956, 910-911.

the systems Li_2O–MoO_3, K_2O–MoO_3, and BaO–MoO_3, of amber color, and of a high crystallization tendency, in a very narrow range of concentrations (about only 5 per cent) for the glass forming compositions. In the system MoO_3–P_2O_5, however, all mixtures with more than 11 per cent P_2O_5 gave glasses of a rather high stability and of black color. In the wolframate type systems the glass formation is restricted to some colorless compositions near 80 to 82 per cent WoO_3, with a high tendency to crystallize. Also here the system WoO_3–P_2O_5 gives above 11 per cent P_2O_5, much better durable black glasses. UO_3 combined with P_2O_5 yields rather readily glasses with acidic oxides. The same authors[13] previously described *vanadate* glasses with P_2O_5, GeO_2, TeO_2 (see above), or with As_2O_3, but also with BaO or PbO. All these glasses have in common a complete absorption of the visible light, whereas the transmittance in the short-wave infrared is rather distinct. In addition such vanadate glasses are semiconductors, as is V_2O_5, although the pure oxide does not form a simple glass.[14]

8. Among all nonsilicate glasses those of the *phosphate* group are the best investigated compositions. The physical and chemical properties of this group were extensively discussed by E. Kordes, W. Vogel, and R. Feterowsky,[15] with a clear distinction of normally built phosphate glasses of the systems P_2O_5–CaO, –BaO, –CdO, and –PbO, and the irregularly built glasses of the systems P_2O_5–BeO, –MgO, and –ZnO. These distinctions will be evident when we discuss the structural details (A. ¶ 220). Of much more general chemical interest are the alkali metaphosphate glasses which have attained great technical importance as modern detergents ("Calgon") and water softeners because of their solubility in water and their capacity for dissolving calcium oleate at room temperature, on which especially W. Dewald and H. Schmidt[16] gave a general report. I. Schulz and W. Hinz[17] described the analogy existing between the constitution problems of phosphate and silicate glasses (see more details in A. ¶ 380, and Vol. IV, Section A), based on previous theoretical ideas of G. Hägg (1935) on group formations in the structure of vitreous states. The soluble phosphate glasses have the great advantage over silicate glasses in that simple paper-chromatographic methods can be applied in determination of the sizes of the group complexes formed. In this way the investigations of J. R. van Wazer (1950), and of A.

[13] *Nature*, **173**, 1954, 1030-1032.

[14] *J. Electrochem. Soc.*, **104**, 1957, 237-240.

[15] *Z. Elektrochem.*, **57**, 1953, 282-288. A most comprehensive investigation on possibilities of developing complex *aluminophosphate glasses* was made by Z. M. Syritskaya, *Structure of Glass. Proc. All-Union Conf. Glassy State, 3rd, Leningrad (English Translation)*, 1959/60, 295-298 (cf. more in details in A. ¶ 388* and B. ¶ 123*).

[16] *Angew. Chem.*, **65**, 1953, 78-81.

[17] *Silikat Tech.*, **6**, 1955, 235-241; *Glastech. Ber.*, **29**, 1956, 319-323. On G. Hägg's theory of glass constitution see *J. Chem. Phys.*, **3**, 1935, 42-49.

F. R. Westman, J. Crowther, and P. A. Gartaganis,[18] and those of H. Grunze[19] are particularly significant. The fundamental importance of the molecular units for glass formation is also in contrast with the commonly and exclusively discussed interpretation of glasses as infinite framework structures (cf. A. ¶ 205 ff., 305, etc.).

9. The use of chromatographic methods for continuous control of the polymerization degree of metaphosphate units proved to be an equally reliable way as say in organic complex compounds, or as X-ray and electron diffraction methods are for inorganic materials. The discussions, of I. Schulz and W. Hinz mentioned above, on these questions are chiefly based on binary metaphosphate glasses with divalent cations, and their aqueous solutions, containing chain or ring structural elements. Their studies further demonstrate the premises of glass formation and spontaneous crystallization, namely if stable di-(pyro-)phosphate anions are formed. Glasses are only observed in quenched melts of such pyrophosphates as contain highly polarizable cations with a high field strength, or Cd^{2+}, Zn^{2+}, Pb^{2+}, or Bi^{3+}. The paper-chromatographic methods show in these special cases involving aqueous solutions of the glass melts a relative maximum in diphosphate anions (69.7 per cent for lead, 37.8 per cent for bismuth pyrophosphate glasses), and "disproportionate" amounts of other anions, such as mono-, tri-, tetra-, penta- or hexa-phosphates. This disproportion does not alone determine the glass-formation tendency, but also the difference of the energy contents of the crystalline and the glassy phases. Thus, zinc and cadmium pyrophosphates are, in spite of their higher disproportion, *not* better than lead pyrophosphate as glass formers, but are not as good. The heat of devitrification is say, for $2ZnO.P_2O_5$ much larger than that of $2PbO.P_2O_5$.

10. The important *optical* phosphate glasses were studied by W. Vogel[20] in their interesting devitrification products (cf. B. ¶ 208), especially in the appearance of the regular aluminum metaphosphate, $Al(PO_3)_3$, and concerning the influence of a substitution of Al^{3+} by Zn^{2+}, with a considerable improvement of the homogeneity and chemical durability of the glass products.

[18] *J. Am. Ceram. Soc.*, **37**, 1954, 420-427; **40**, 1957, 293-299. See more recently on the constitution of *mixed-alkali* phosphate glasses: M. Krishna Murthy, M. J. Smith, and A. E. R. Westman, *Ibid.* **44**, 1961, 97-105, 475-480; **45**, 1962, 401-407.

[19] *Silikat Tech.*, **7**, 1956, 134-138. On the chromatographic methods used see H. Grunze and E. Thilo, "Die Papierchromatographie der kondensierten Phosphate," Akademie-Verlag Berlin, 1954; *Sitzber. deut. Akad. Wiss. Berlin, Kl. Math. u. allgem. Naturwiss.*, 1953, No. 5, 26 pp. On the thermal characteristics of phosphate glasses in combination with paper chromatography see also V. Gottardi and G. Bonatti, *Vetro Silicati*, **3**, 1959, (14) 31-38.

[20] *Silikat Tech.*, **6**, 1955, 510-517.

VISCOSITY OF FUSED SILICATES

11. For accurate determination of the viscosity of molten glasses, slags, and the like, the standard of the *absolute viscosity* of pure water at 20°C. was reexamined and redetermined by J. F. Swindells, J. R. Coe and F. B. Godfrey,[21] using the elementary method of capillary flow phenomenon. The standard $\eta^{20}_{H_2O}$ is = 0.010019 poises, ± 0.000.003 (in the place of the previous primary standard value = 0.01005 poises), for the calibration of all conventional viscosimeter instruments.

12. The simple *outflow-viscosimetry* principle was applied by R. L. Tiede[22] in a platinum-bushing (one-orifice or nozzle) device, for the range of 100 to 450 poises, as is usually applied in glass-fiber spinning processes. The construction of the one-hole bushing is seen in the cross-section Fig. A. 3. Although this method for the spe-

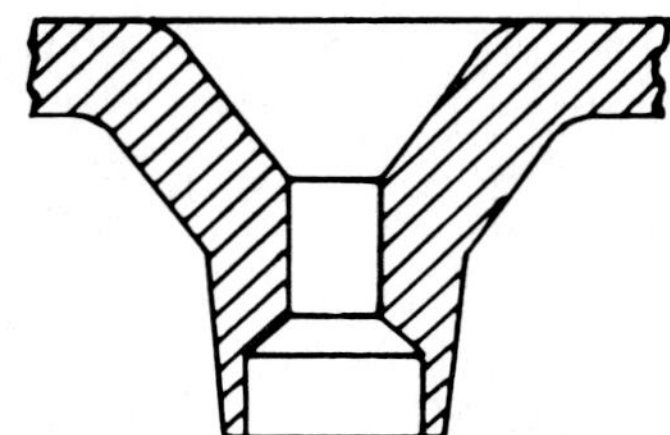

Fig. A. 3. Cross section of the one-hole platinum bushing (without the insulating refractory parts) for viscosimetry at high temperatures. (Tiede).

cial purpose is an excellent routine control of filament-spinning operations, it is subjected to the inconvenience that it does not furnish absolute viscosity data, and that it is difficult to increase the precision of the measurements by correcting for density, radiation emission, wetting and surface tension on the bushing metal, and so on. The density, however, can be calculated from the weight output of the spinning process as a function of time. Weight delivered is proportional to the square of density. Because of the complicated shape of the bushing orifice a direct and rigorous application of Poiseuille's formula is not possible. The chief advantage of the method is, on the other hand, the ease with which rapid determinations can be made of differences in viscosities for glasses of similar chemical composition.

13. For the simple *sphere penetration* (falling sphere) method on the basis of Stokes's equation, J. Schurz[23] demonstrated the accurate calculation of the fundamental rheological factors τ_m (shear stress), and the average gradient of the rate of the

[21] *J. Research Natl. Bur. Standards*, **48**, 1952, 1-31.

[22] *J. Am. Ceram. Soc.*, **38**, 1955, 183-186.

[23] *Naturwissenschaften*, **42**, 1955, 339-340.

sphere dislocation, $(dv/dR)_m = v(R - r)$, over the length of the free fall-way, l, $= v \cdot t$, as a function of time:

$$\tau_{ni} = \eta \cdot l/[t(R - r)] = \frac{\Delta \varrho \cdot gr^2}{4.5\ (R - r)}$$

(R the radius of the fall tube, r that of the sphere). The ways proposed by Schurz are particularly important if structural effects (viscosity anomalies by non-Newtonian flow) are safely detected by varying the radius of the falling sphere. It is only necessary that the flow pattern around the sphere be not disturbed. Use of the equations is therefore not possible for the Höppler flow viscosimeter. Improvements of the latter instrument to eliminate thixotropic and other effects by application of ultrasound vibrational treatment were proposed by R. Dallendörfer[24] and are valuable in determination of the viscosity of barium titanate suspensions or for the study of polymerization processes of organic agents as "models" of glasses (cf. Vol. I, C. ¶ 339).

14. The most generally used instrument for measurement of viscosity in silicate melts is actually the rotation viscosimeter which was universally applied by J. S. Machin, Tin Boo Yee, and D. L. Hanna[25] in their fundamental studies of viscosity as a function of composition in the quaternary system $CaO–MgO–Al_2O_3–SiO_2$. The basic knowledge of viscosity as a function of temperature for blast furnace slags is nowadays completely developed in a systematic collection of isothermal and isokom curves plotted on the triangular basis of the system $CaO–Al_2O_3–SiO_2$, with sections of constant MgO contents of 10 and 20 per cent, or of the system $CaO–MgO–Al_2O_3$, with constant contents in silica (35, 40, 45, 50, 60, and 65 per cent), or of the system $CaO–MgO–SiO_2$, with constant contents of 10 and 20 per cent alumina, or of the system $MgO–Al_2O_3–SiO_2$ (with constant contents of 20, 30, and 40 per cent CaO), as given in the diagrams of Machin *et al.*

15. In their extensive studies on the optimum concentration conditions of blast furnace slags in the system $CaO–MgO–Al_2O_3–SiO_2$, E. F. Osborn, R. C. De Vries, K. H. Gee, and H. M. Kraner[26] (cf. Vol. III, B) also discuss the premises of a minimum viscosity for a given temperature of slag flow. It is an important metallurgical fact that the systematic compositions for minimum viscosity and of maximum desulfurization effects practically nearly coincide, in agreement with the data given in the work of Machin *et al.*, especially in the important range from 1350° to 1500°C., with isokoms nearly parallel to lines of constant CaO contents in the diagrams. In this respect, alumina and silica appear to be approximately equivalent with their influence on viscosity (on a weight per cent basis). MgO is slightly more effective than is CaO in lowering the viscosity. A recent critical discussion of previous data

[24] *Silikat Tech.*, **6**, 1955, 519-524.
[25] *J. Am. Ceram. Soc.*, **28**, 1945, 310-316; **31**, 1948, 200-204; **35**, 1952, 322-325; **37**, 1954, 177-186.
[26] *J. Metals*, **6**, 1954 (1) 3-15, 33-45.

of viscosities especially of the systems $CaO–Al_2O_3–SiO_2$, and $CaO–MgO–Al_2O_3–SiO_2$, by E. T. Turkdogan and T. M. Bills[27] concludes that a series of older data must be abandoned as being erroneous, whereas the more recent measurements may be used with confidence. Although alumina and silica exert in glass melts similar effects on viscosity, the silica equivalent of alumina, N_A, depends very much on the concentration in alumina and even on the ratio Al_2O_3/CaO. Also the activation energy for the viscous flow varies with the temperature. The work done by J. O'M. Bockris and other authors[28] (see below) emphasizes particularly the role of structural changes in the silicate anion groups as a function of temperature, ruling the flow mechanisms, especially above 1350°C. Average viscosity versus composition relationships can be found only if viscosity is plotted versus $N_{SiO_2} + N_A$, i.e. as a function of

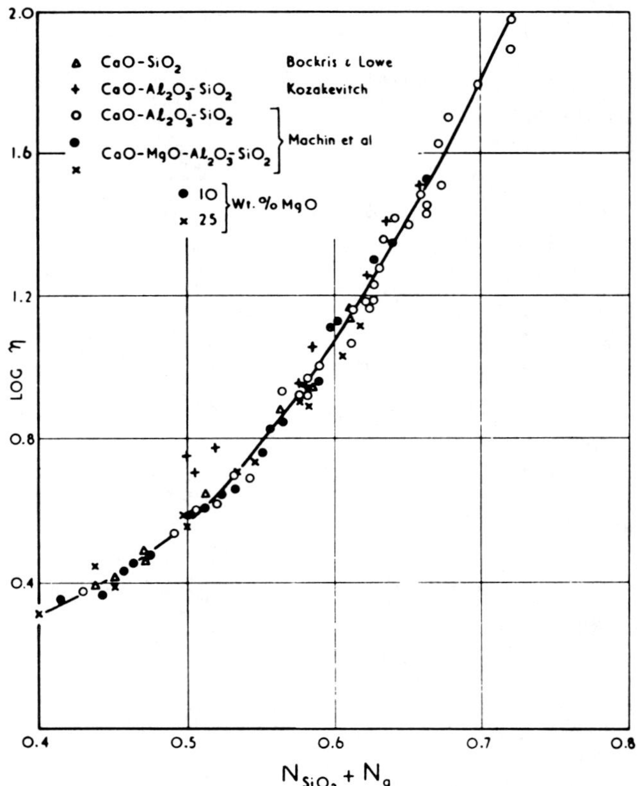

FIG. A. 4. Variation of viscosity with composition of $CaO—MgO—Al_2O_3—SiO_2$ melts at 1500°C. (Turkdogan and Bills).

[27] *Am. Ceram. Soc. Bull.*, **39**, 1960, 682-687.

[28] *Discussions Faraday Soc.*, 1948, No. 4, 65-81; *Trans. Faraday Soc.*, **48**, 1952, 75-91; 423-435.

the molecular amount of silica plus the silica-equivalent of alumina, N_A, defined by $N_A = N_{SiO_2\ bin.} - N_{SiO_2\ tern.}$, and corresponding relations are valid for log η, and the activation energies (Figs. A. 4 and 5).

16. The great importance of viscosity for the flow of metallurgical slags encouraged measurements of this parameter, such as the early work by A. Feild (1916) up to 1600°C., and later by I. P. Semik (1938, 1941),[29] especially as a function of the com-

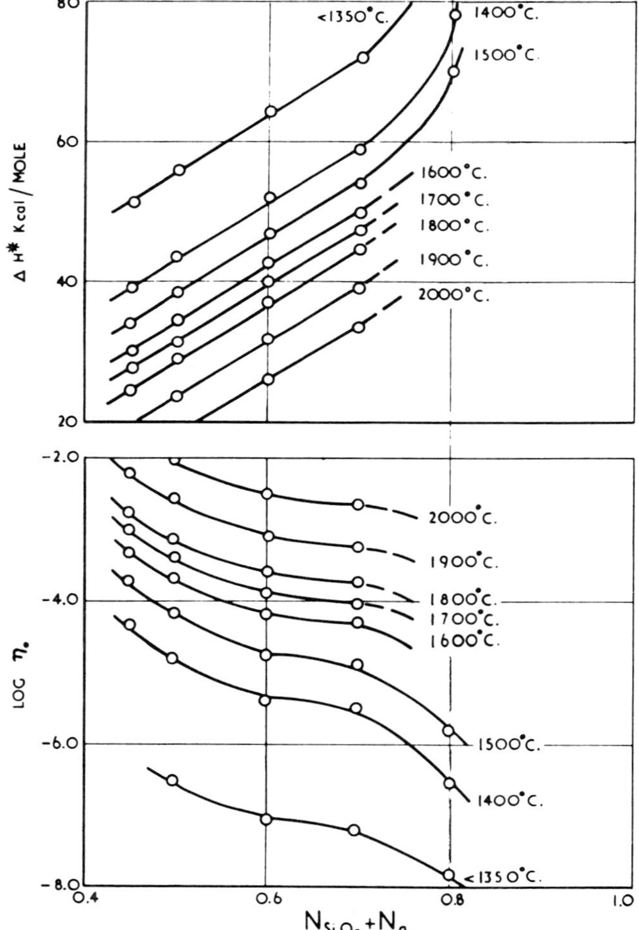

FIG. A. 5. Variation of the activation energy, ΔH^*, of viscous flow, and of the viscosity, log η, as a function of temperature and composition. (Turkdogan and Bills).

[29] *Sovet. Met.*, 10, 1938, (2) 22-34; *Izvest. Akad. Nauk S.S.S.R.*, Otdel. Tekhn. Nauk 1941 (4) 55-66.

position considering the contents of blast furnace slags in MnO, CaS, and the like, or titanium slags produced from titanomagnetite ore metallurgy (cf. C. ¶ 25 ff.). Also P. Kozakevitch[30] measured the viscosity of acidic metallurgical slags by a rotation viscosimeter (with coaxial cylindric rotors), partly with the unusually high crystallization temperatures of 1600° to 1700°C., but with a considerable decrease in viscosity if calcium or magnesium fluoride is added in small amounts. Difficulties in handling the acidic slags (with, say, 60 per cent silica) by the relatively high viscosity at operation temperatures (1450° to 1300°C.) are evident also from investigations of V. I. Loginov,[31] with a detailed discussion of the "length" of such slags, more specifically of the steepness of the slopes of the curves for log η as a function of temperature (cf. C. ¶ 67).

17. For viscosity measurements at temperatures above 1600°C. the use of a rotation instrument with molybdenum as the construction material becomes indispensable. I. A. Bulavin[32] developed such a viscosimeter working *in vacuo* (Fig. A. 6)

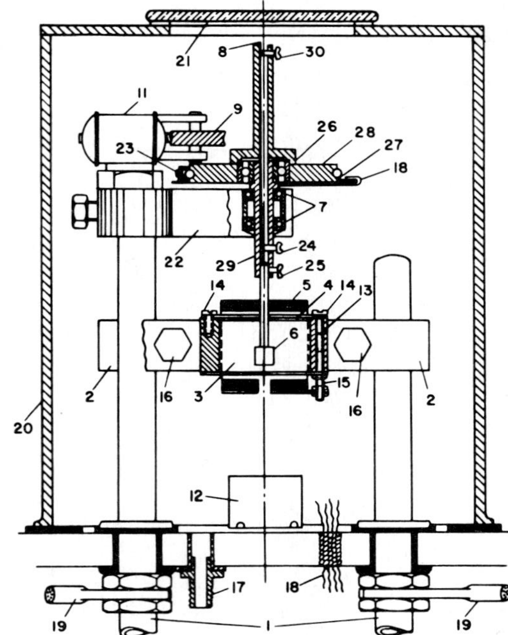

3 cylindrical crucible
4 heating element
5 radiation screens
6 cylindrical stirrer
11 electromotor (12 r. p. m.)
23 gear
28 wheel
29 spindle axis

Fig. A. 6. Vacuum molybdenum rotation viscosimeter for high temperatures. (Bulavin).

[30] *Rev. mét.*, **51**, 1954, 569-587.
[31] "Investigations on the Blast Furnace Process," Acad. Sci. U.S.S.R., Moscow 1957, 148-166.
[32] *Doklady Akad. Nauk S.S.S.R.*, **130**, 1960, 133-136.
[33] *Zavodskaya Lab.*, **4**, 1950, 813-818.

with molybdenum heater, crucibles, and stirrers, useful up to 2500°C., as was demonstrated for melts of the system $CaO–BaO–Al_2O_3–SiO_2$ in the aluminous portions (constant 20 per cent Al_2O_3). At 1600°C. such compositions may become surprisingly low in viscosity, as with only 10 poises. Also M. P. Volarovich and O. I. Yatsunskaya[33] developed high-temperature viscosimeters for slag melts, but based on the principle of damped pendulum oscillations and using the equation of Fawsitt for systems of a known inertia moment of the swinging periodic system. Such an instrument has the advantage of great simplicity and rapidity of measurements (cf. Fig. A. 7).

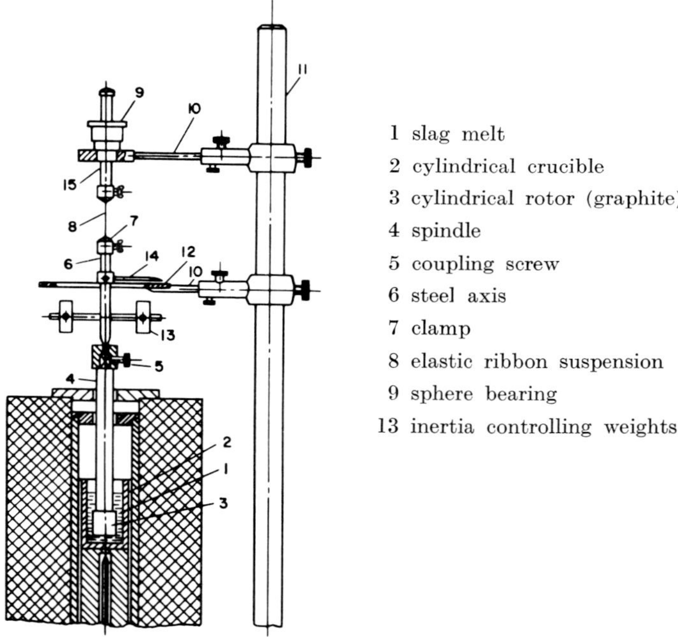

Fig. A. 7. Damped torsion oscillation viscosimeter. (Volarovich and Yatsunskaya).

18. The *theory* of rotation viscosimetry was revised by R. Schwaben and H. Umstätter,[34] with a new mathematical formulation by Bessel functions for angular velocity of flow, and torque conditions for a *conical* type of the rotating body with an apex angle of 70,5°, or spherical and cylindrical rotation bodies as the limit conditions, especially with the purpose to avoid the much disturbing effects from the walls or from the bottom surface of the melt containers. A rather ideal form in this respect is offered by a conical trunk (frustum) shape of rotors.

[34] *Konstrukt. Maschinen App. u. Gerätebau*, **5**, 1953, (3) 79-95.

19. To eliminate all the more or less uncontrolled wall effects and other factors in common viscosimetry, A. Dietzel and R. Brückner[35] undertook to develop an *absolute* viscosimeter of the rotator type. These authors also discuss the influence of the different shapes for rotors on wall and other effects, namely the hemispherical, half-ellipsoidal, paraboloidal, and hemispherical + cylindrical forms (cf. Fig. A. 8).

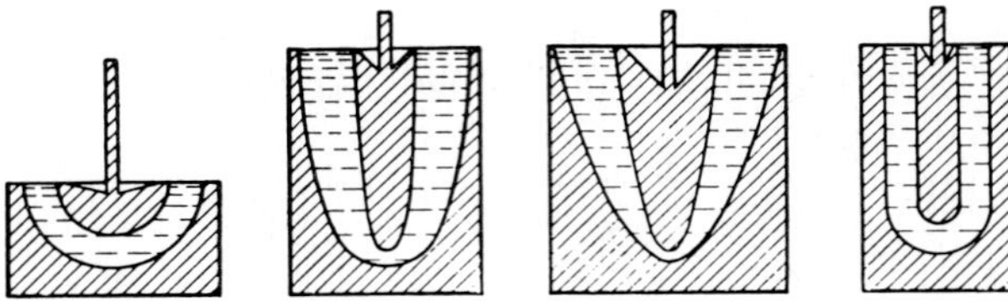

FIG. A. 8. Variation of the shapes of rotation stirrers for absolute viscosimetry. (Dietzel and Brückner).

The last one is the most appropriate shape that was adopted to best approximate the Navier-Stokes formulas of hydrodynamics and for a strictly laminar-circular flow of a Newtonian liquid. The rotating parts were constructed of platinum/20 per cent rhodium-platinum alloy; the damped oscillations of the rotor were recordered and evaluated for the logarithmic decrement, ln X, to determine the viscosity by the equation $\eta = K/\pi \cdot \sqrt{d\Theta} \cdot \ln X$ (d the direction force, Θ the moment of inertia). The thermal expansion effects were compensated in the steel wire suspension system by a precision adjustment with an observation tube to control the surface and the meniscus of the liquid. Dietzel and Brückner used the "absolute viscosimeter" over the range from 10^1 to 10^7 poises, with an average error of $\pm$ 1.5 per cent in the range from 10^1 to 10^3, $\pm$ 2 to 3 per cent between 10^3 and 10^5, and $\pm$ 3 to 4 per cent between 10^5 and 10^7 poises. The apparatus was calibrated with the melt of anhydrous B_2O_3, between 320° and 1300°C. ($\eta = 10^1$ to 10^5 poises), and an extrapolation formula of the Tammann-Fulcher type

$$\log \eta = 0.296 \pm 1327.5\ (T - 435) \pm 0.019$$

20. Of two modern viscosimeters, we mention further that constructed by A. G. F. Dingwall and H. Moore,[36] characterized by the rotation of the *exterior* cylindrical melt container (Fig. A. 9), with Globar heating resistors, a very accurate adjustment mechanism for the coaxial movements of the cylinders, and a special torsion head (Fig. A. 10). Much care was also given to accurate determination of the transmission of heat radiation through iron-containing glasses (Fe replacing Si up to 8 per cent).

[35] *Glastech. Ber.*, **28**, 1955, 455-467.

[36] *J. Soc. Glass Technol.*, **37**, 1953, 316-372 T. The calculations were based on the H. Eyring – K. J. Laidler theory of viscous flow (1941).

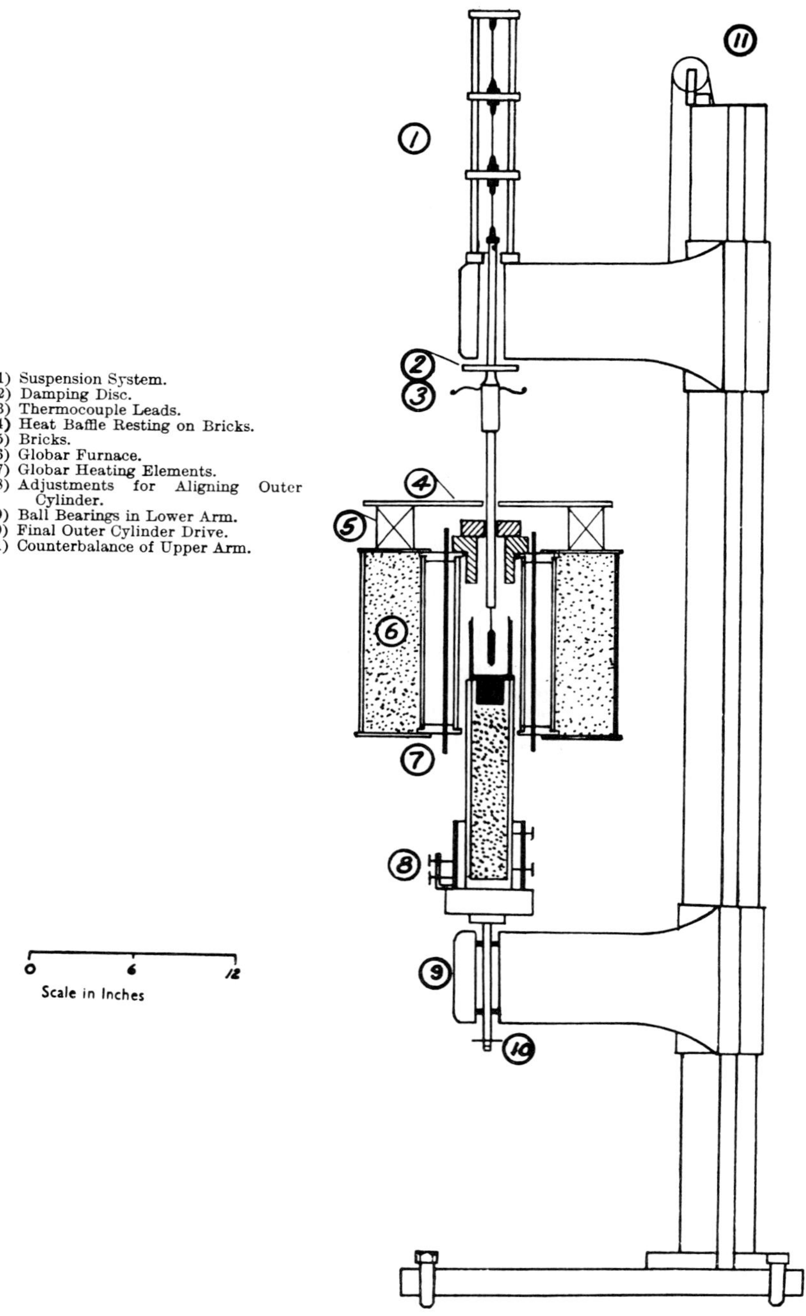

Fig. A. 9. Concentric cylinder viscosimeter, exterior cylindrical platinum crucible rotating, for high temperatures. (Dingwall and Moore).

The instrument was used for high-precision studies of the effects of different cations in their sizes and field strengths on viscosity as a function of temperature, and further on the structure-building effects of the oxides R_2O_3 and RO_2. The special conditions introduced by iron as FeO, Fe_2O_3, or Fe_3O_4 confirm the fact that Fe^{2+} colors blue, Fe_3O_4 gray in a colloidal suspension dispersed through the glass melt, and

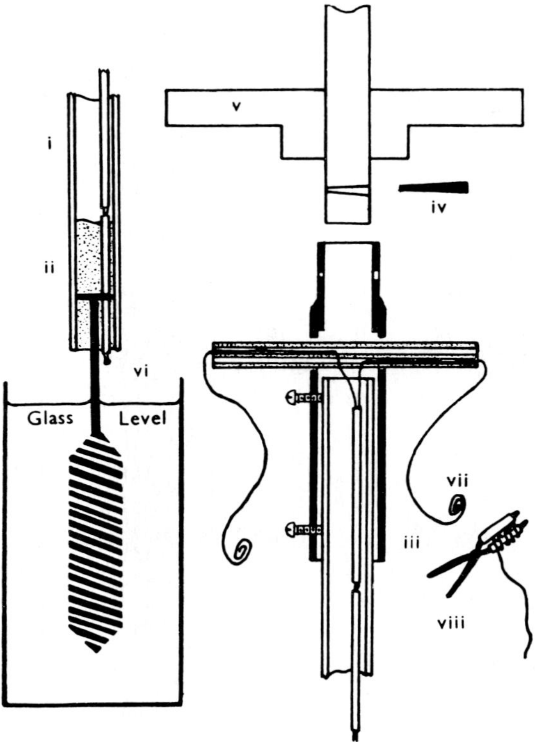

FIG. A. 10. Special torsion head for the suspension of the rotor in high-temperature viscosimeter. (Dingwall and Moore).

Fe^{3+} either amber or brown, structure-building units similar to $[SiO_4]$ making the glass practically colorless in the visible light. For a viscosity up to 13,000 poises, R. L. Tiede[37] recently constructed a routine instrument for rapid measurements (15-minute periods for every viscosity-*vs*-temperature point on the curve). Particularly remarkable in this construction of a viscosimeter is the well-calculated shape of the platinum heating element with 8 fins (Fig. A. 11) for an optimum heat flow to maintain constant temperature conditions of the liquid in the heating chamber.

[37] *J. Am. Ceram. Soc.*, **42**, 1959, 537-541.

The spindle is of a combined cylindrical-conical shape as seen in Fig. A. 12. The instrument is calibrated with a standard glass with an excellent fit of the viscosity data as well as with increasing and decreasing temperatures.

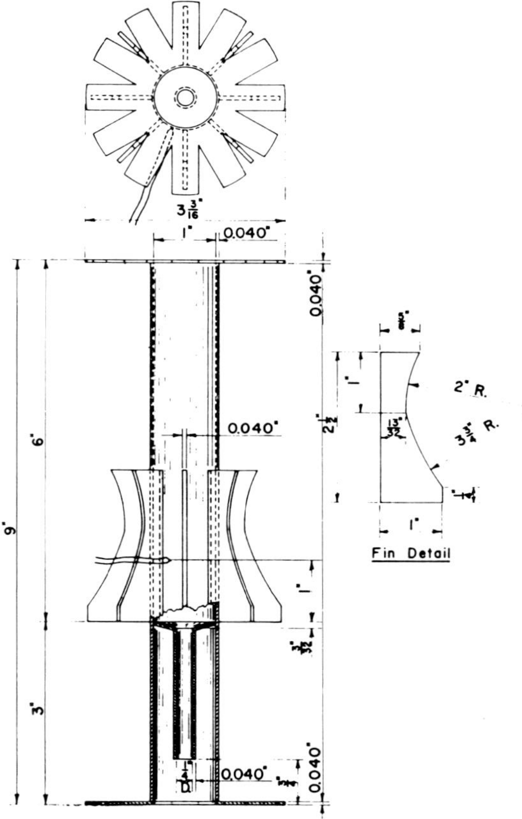

FIG. A. 11. Heating element for rotation viscosimeter cylinders. (Tiede).

21. The often-discussed problems of whether chemical compounds (silicates) appearing in the constitution of glass melts can be indicated and identified in the curves representing viscosity as a function of the composition were interpreted anew by D. K. Belashchenko.[38] He found for the kinetic viscosity ($\nu = \eta/\varrho$) and the electrical resistance of molten alloys of the systems Cd–Sb and Cd–Cu by a combined torsion-oscillation viscosimetric and electrodynamic method (observing the logarithmic decrements of free oscillation and the angle of twist deviation in a magnetic

[38] *Doklady Akad. Nauk S.S.S.R.*, **117**, 1957, 98-101; *Zhur. Fiz. Khim.*, **31**, 1957, 2269-2276.

field) distinct maxima on the curves for ν, indicating the composition of the intermetallic compound CdSb which is also indicated in the phase equilibrium diagram, whereas the maxima of the isotherms are shifted from 50 to 40 atomic per cent Sb in the direction towards the metastable compound Cd_3Sb_2. The resistance curves show for 600° and 700°C. the maxima at 50 per cent antimony. Belashchenko concludes that the atomic packing is the same in the crystalline and the molten CdSb

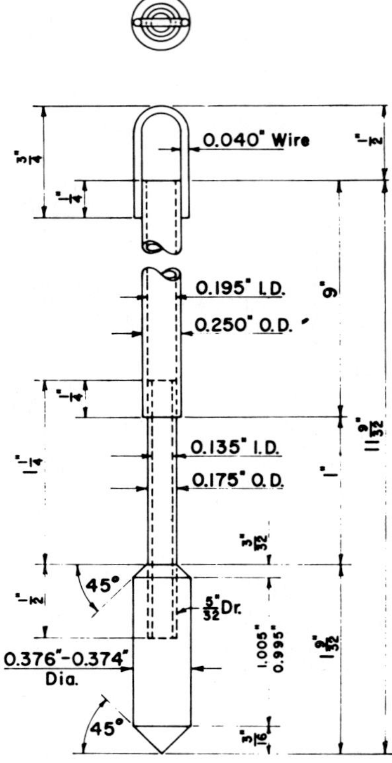

FIG. A. 12. Design and dimensions of platinum spindle as stirrer in rotation viscosimeter. (Tiede).

phases, although with a better approximation to the compound Cd_3Sb_2 at lower temperatures. Distinct hystereses appear in the viscosity curves on heating and cooling cycles. Typical undercooling was observed for alloys of the systems Tl–Pb and Tl–Bi. In the latter case anomalies become evident in viscosity and resistance between 350° and 400°C. indicating metastable configurations. In the system Cd–Cu the anomalous maximum in viscosity and resistance is near the composition Cd_8Cu_5 (δ-phase).

22. It is particularly interesting that recently E. Eipeltauer and A. More[39] determined in the simple system K_2O–SiO_2 by precision measurements of the viscosity (using the Hänlein-Endell sphere-extraction method) very distinct humps in the curves of $\log \eta$ versus the composition (on the isotherms for 1100° and 1200°C.), near 28, 36, and 44 weight per cent K_2O corresponding to molecules of $K_2O.2SiO_2$, $K_2O.3SiO_2$, and $K_2O.4SiO_2$, in the liquid melt. This is sensitively indicated if the parameters b, a, and T_0 in the Vogel-Tammann-Fulcher equation, $\log \eta = a + b/(T - T_0)$, are plotted as a function of the K_2O contents (Fig. A. 13). The sharp discontinuities

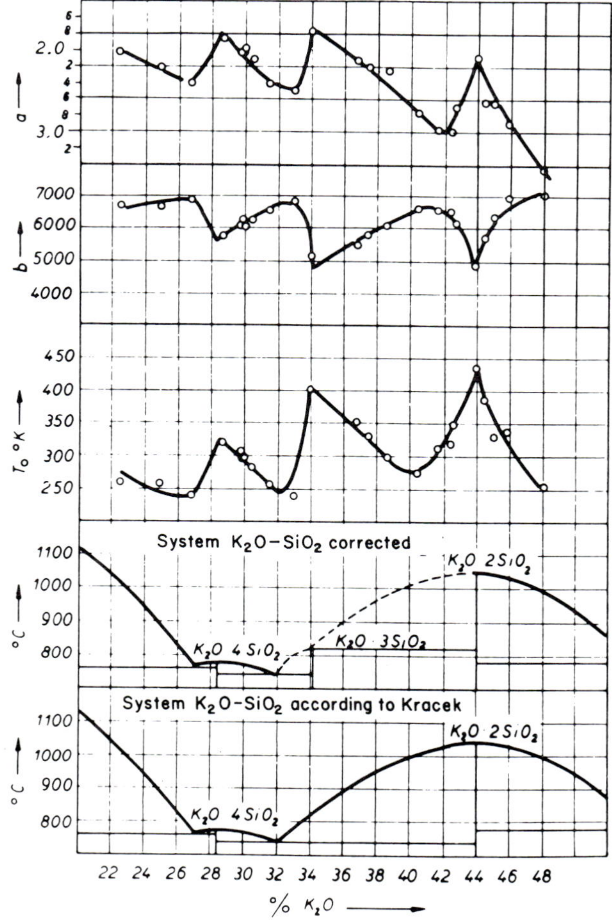

FIG. A. 13. Parameters, a, b, and T_0, in Tammann-Fulcher equation for the viscosity of potassium silicate glasses as a function of the contents in K_2O. (Eipeltauer and More).

[39] *Radex Rundschau*, 1960, 230-238. On the system Na_2O–SiO_2, see E. Eipeltauer and G. Jangg, *Kolloid-Z.*, **142**, 1955, 77-84; **143**, 1955, 83-92.

thesis that both properties, viscosity and electrical resistance, are determined by the concentration in more or less weak bonds in the glass structure. The constancy of the ratio between the temperature coefficients for both is in every case striking. Figure A. 15 shows how well the previous observations of M. P. Volarovich and D. M. Tolstoĭ (1930, 1934) are confirmed. They concern the log η and resistance as a function of the soda content (in weight per cent) in the borate glasses, in agreement with the hypothetical free-space and weak bonds concept.

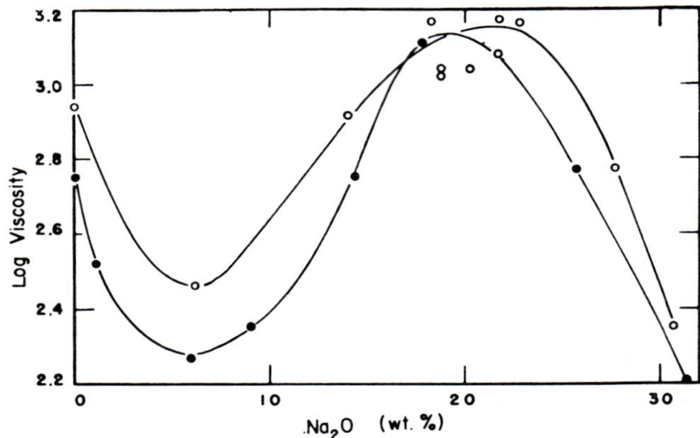

FIG. A. 15. Comparison of the viscosity data of Volarovich and Tolstoĭ (1930), and of Shartsis, Capps, and Spinner (1953), for 700°C., as a function of alkali concentration.

VISCOSITY IN THE SOFTENING AND TRANSFORMATION RANGES

25. The general tendency to give characteristic *fixed points* on the *viscosity scale* for glasses of widely variable compositions, but in corresponding states of glass structure and physical constitution, namely at corresponding temperature conditions for these states, was the reason that a new definition and reexamination of the previous proposals was made. H. R. Lillie[46] who had given in 1931 a nearly generally adopted scale of such fixed points for viscosities, corresponding to a strain point, an annealing, and a softening point (at $\eta = 4.0 \times 10^{14}$, 2.5×10^{13}, and 4.5×10^{7} poises, respectively), revised these data by new precision measurements. His new recommendations are:

for the strain point: $10^{14.50}$ poises, } during a constant cooling rate of
for the annealing point: $10^{13.00}$ poises, } 4°C./minute.

[46] *J. Am. Ceram. Soc.*, 37, 1954, 111-117.

These determinations were based on the fiber elongation method. The revised definitions raise the annealing points of all glasses by 2° to 10°C. They lower the strain points for a high-expansion glass by 3° to 7°C., whereas they are raised for low-expansion glasses by about 5°C.

26. From purely practical viewpoints, and essaying to devise a rapid routine method for observing the "mobility" behavior of a given glass as a function of temperature, H. Jebsen-Marwedel[47] developed the definition of a "*point of mobilization*" ("Mobilpunkt") by observing rectangular glass samples under dark-field illumination in the heating microscope in their first rounding on corners and edges (analogous to the older "deformation point," or "flow point"; cf. F. Weidert and G. Berndt, 1920). The surface tension overcomes the viscosity of the glass in this temperature point and the mobilization is observed by appearance of a bright point of reflection in the corners of the samples, with a reproducibility better than by 5°C. at a heating rate of 5°C./minute.

27. The determination of a conventional "*sag point*" as a viscosity reference point was recommended by S. Spinner, G. W. Cleek, and E. H. Hamilton,[48] and is defined as the temperature at which a glass fiber of 0.5 to 0.8 mm. in diameter, horizontally supported at 1.25 cm. intervals will start sagging under its own weight in 25 $\pm$ 5 minutes.

28. Impressed by the practical needs of having viscosity fixed points in the *working range* of industrial glasses, A. Dietzel and R. Brückner[49] more recently defined a "*penetration point*" for $\eta = 10^{4.22}$ poises by that temperature at which a metal rod (of platinum-20 per cent rhodium alloy) of 0.5 mm. in thickness, and 200 mm. in length (weight 0.746 gm.) with hemispherical endings, penetrates under its own weight into the glass specimen to a depth of 2.00 cm., following the simple relation $\eta = C \cdot t/l^2$. One may also consider a definition of another "conventional penetration point" for $\eta = 10^{4.00}$ poises. Both definitions must be based on absolute viscosimeter measurements (see above). The chief practical advantage of the Dietzel-Brückner penetration test, however, is the fact that, together with the annealing and softening points in Lillie's definition, one has a possibility to apply, with a good approximation, the entire $\log \eta$-temperature variation curve for a given glass in the range from $\log \eta = 2$ to 13.00 by the Vogel-Tammann-Fulcher equation. A series of examples are given by Dietzel and Brückner which make evident the practical usefulness of the penetration test in this respect. The average accuracy of the calculated viscosity data within $\pm$ 5 to 10 per cent for many orienting experiments is satisfactory enough.

[47] *Glastech. Ber.*, **27**, 1954, 172-173.

[48] *J. Research Natl. Bur. Standards*, **59**, 1957, 227-231. Specifically on the sag point see Ch. Hirayama, *J. Am. Ceram. Soc.* **45**, 1962, 113-115.

[49] *Glastech. Ber.*, **30**, 1957, 73-79.

29. Micheline Prod'homme[50] used for the determination of viscosity in a range including the transformation range ($\eta = 10^9$ to 10^{17} poises) the well-known Chevenard dilatometer. The mirror system for the self-recording mechanism limits the lowest viscosity values available, whereas the highest load applied to the glass filament was chosen as 1 kg. Prod'homme offers valuable data not only for common glasses of the Crown type, but also of As_2S_3, As_2Se_3, and glycerin glass, the conventional transformation temperatures of which are 120°, 75°, and — 90°C., respectively. For the viscosity in the transformation range a lower limit of $10^{12.6}$, and an upper limit of $10^{13.8}$ poises is given, in good agreement with the observed anomalies of the refractive index, the thermal expansion, the specific heat, and other characteristics, as a function of temperature. The lower the transformation range, the steeper the curve for the change of log η versus temperature. At lowest temperatures the viscosity curve becomes concave to the temperature axis and reaches a final amount of log $\eta = 19$ or 20.

30. Modern uses of molten glasses as lubricants in metallurgical shaping (extrusion) processes (cf. B. ¶ 102 f.) bring about particular problems of the viscous flow behavior of the glassy material under very rapid dislocation conditions for which not the normal viscosity, but an "*instantaneous*" viscosity (η*) parameter is to be taken into account. R. Guy[51] defined and measured such instantaneous viscosities which are always relatively lower if the quenching is very rapid. Fine-spun commercial fiber glass is in a rapidly chilled state and it is possible to determine the transition from instantaneous to normal viscous flow conditions under stress over a spectrum of variable heating and cooling rates. In the determination of the standard softening point of a glass for $\eta = 10^{7.6}$ according to Littleton, one observes the instantaneous viscosity effect by an apparent lowering of the softening temperature under increased rates of chilling by more than 50°C. During the measurements the instantaneous effects are, of course, masked by the (slower) relaxation processes. The observed reductions in the softening points are lower than the really existing effects. For an instantaneous viscosity $\eta^* = 10^{6.5}$ poises the corresponding time is in the order of magnitude of several seconds. For "soft" glasses the difference of the rupture temperature of freshly prepared and of annealed filaments will be relatively small. The "harder" the glass samples, the more pronounced is the difference.

31. The classical *fiber elongation method* was used by V. A. Pospelov and K. S. Evstrop'ev[52] for the range of $\eta = 10^8$ to 10^{15} poises especially for sodium-lead sil-

[50] *Verres et réfractaires*, **10**, 1956, 208-214.

[51] *Silicates inds.*, **21**, 1956, 69-73. The experimental method used by Guy is an ingenious cathetometric observation of the rupture process with, or without, changes in the length of heat-treated filaments, either quenched or heated with great speed, by a double system of annealing and temperature-shock furnace applied to the glass sample.

[52] "Physical-Chemical Properties of the Ternary System Na_2O–PbO–SiO_2," I. V. Grebenshchikov, ed., Acad. Sci. U.S.S.R., Section Chem. Sci., Moscow 1949, 70-82.

icate glasses. As an interpolation formula valid in the temperature interval from 400° to 1400°C. the equation $\log \eta = A + B/T \cdot e^{a/T}$ is used. For an approximation in the range from 0.1 to 10^4 to 10^{15} poises one may apply an equation of the type $\log \eta = A + \alpha/T^2$. For the partial system $Na_2O.2SiO_2$–$PbO.SiO_2$, as a function of concentration, and up to 700°C. it was sufficient to write: $\log \eta = a + b \cdot C$ in which C is the molecular percentage of the silicate components. The viscosity as a function of temperature and composition in the range from 550° to 600°C. shows two straight lines intersecting at the composition of the ternary compound $Na_2O.2PbO.4SiO_2$ and at lower temperatures indicating discontinuities near the binary compositions $PbO.SiO_2$ and $Na_2O.2SiO_2$.

32. For plant routine tests, L. Brichard and E. Plumat[53] modified the Littleton fiber elongation method in maintaining the constant temperature of 600° ± 0.1°C. by an electronic (Thyratron) device, observing the strip-shaped samples (10 × 155 mm.) of window glass by micrometric examination, with two fine marks applied at 5 mm. distances. This relatively short specimen form greatly facilitates uniformity of the temperature distribution required. The reproducibility of the method is excellent (± 3.5 per cent). A substitution of Na_2O by SiO_2 by only 0.1 per cent in the glass composition is immediately indicated in the viscosity data.

33. By the fiber elongation method, M. Watanabe and R. Koyama[54] were able to study in the important range from 10^8 to 10^{13} poises, effects of load, temperature, and time on the *rates of elongation*, with the elaborate apparatus shown in Fig. A. 16. The time-dependent characteristics of viscosity which gives lower values in the early period of the measurements is influenced by the load applied, and this effect is the more pronounced the lower the load. Watanabe and Koyama used a borosilicate glass of the "Nonex" type with a relatively short relaxation time for $\eta = 10^8$ to 10^{11} poises; the higher the temperatures the shorter the relaxation period. Important is the fact that the relation between the reciprocal of viscosity and time corresponds to that between a distribution function and time (Fig. A. 17). The long relaxation time disappears at higher temperatures, the short relaxation time when the glass is given a suitable heat treatment. The full explanation of the correlation of temperature, load, time, and viscosity by using the theories of R. W. Douglas and A. Q. Tool (cf. C. ¶¶ 73, 160), and the so-called fictive temperatures is given later. Generally, the flow characteristic of the glass corresponds to that of a Newtonian liquid and only under extremely large shear stress deviations of the non-Newtonian type

[53] *Verres et réfractaires*, **9**, 1955, 69-81. Particularly interesting is the proposal to construct the furnace and the elongation device of transparent silica glass to observe details of elastic aftereffects and other changes directly during the heating and cooling processes.

[54] *J. Soc. Glass Technol.*, **41**, 1957, 137-156. The rheological methods used by Watanabe and Koyama correspond to those used for *creep* phenomena of high-polymers, with the theoretical application of Voigt-Maxwell and similar dashpot-model aspects.

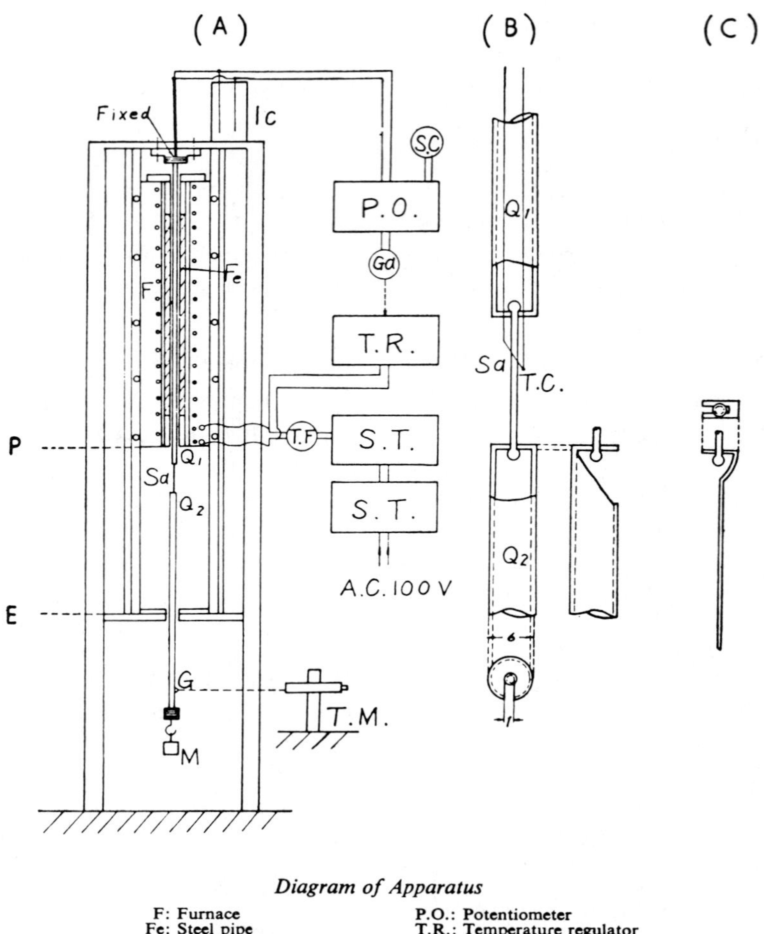

Diagram of Apparatus

F: Furnace	P.O.: Potentiometer
Fe: Steel pipe	T.R.: Temperature regulator
Q_1: Fused quartz tube	S.T.: Voltage stabiliser
Q_2: Fused quartz tube	T.C.: Thermocouple
Sa: Specimen	Ga: Galvanometer
M: Added load	S.C.: Standard cell
T.M.: Travelling microscope	T.F.: Transformer

FIG. A. 16. Fiber elongation apparatus. (Watanabe and Koyama).

become evident as pointed out by H. R. Lillie (1931). Anomalous viscosity effects as described by W. B. Pietenpol (1940) were not observed at all.

34. A. Winter-Klein[55] observed that change of viscosity with temperature can be widely variable in the transformation range dependent on the specific nature of different glass forming systems. For binary, ternary, and complex glasses she studied

[55] *Verres et réfractaires*, 13, 1959, 293-310.

the temperature correction coefficients, defined as temperature differences, ΔT, required to maintain a selected viscosity value for the different compositions, relative to standard glasses, such as Na_2O, CaO, $6SiO_2$, or CaO, Al_2O_3, $6SiO_2$. It is generally not possible to extrapolate the viscosity *vs* concentration curves of the glasses, if changes in the structure, which complicate the viscosity-temperature functional relations, are omitted. Curves are given which show the different degrees of steepness of $(d\eta/dT)$ for the glasses Na_2O–CaO–SiO_2, PbO–SiO_2, borosilicate Crown, selenium, glycerin; all glasses for the range from 10^8 to 10^{15} poises. The attempts to find a simple additivity for the viscosity of complex glass compositions and attempts to establish the "increments" for the oxide constituents are at best only approximations, and are not very reliable.

35. For measurements of *high* viscosities under a strong deformation and constant stress H. J. Oel[56] developed an extension apparatus for glass samples, based on the

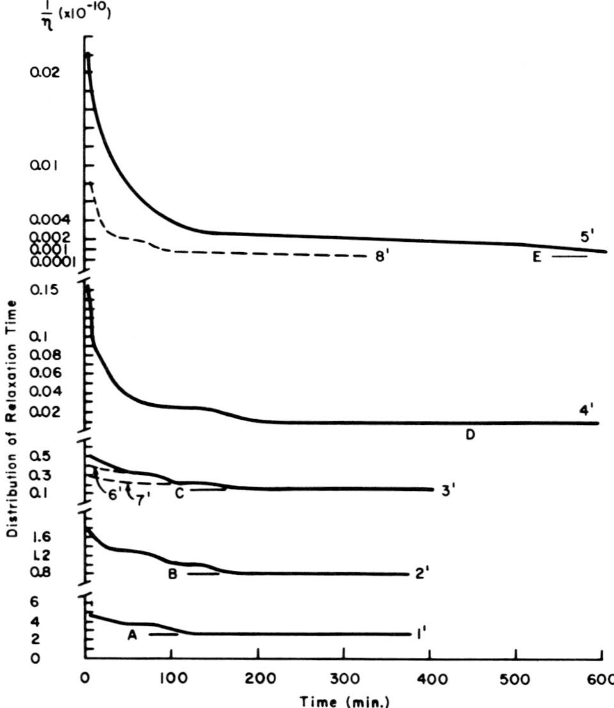

Fig. A. 17. Distribution of relaxation (retardation) times for a borosilicate glass, at various temperatures, after different heat treatments. (Watanabe and Koyama).

[56] *Glastech. Ber.*, **33**, 1960, 219-224. A highly valuable introduction into modern aspects of glass rheology see by H. F. Mark and Sh. M. A. Atlas, *Science*, **138**, 1962, 413-416.

principle of an acting force which is systematically reduced in the measure of the decrease in cross section during elongation. A program wheel device (cf. Fig. A. 18) controls the change in load. The samples have the shape of chain links to maintain a uniform decrease in cross section during the experiment. The length dimension of

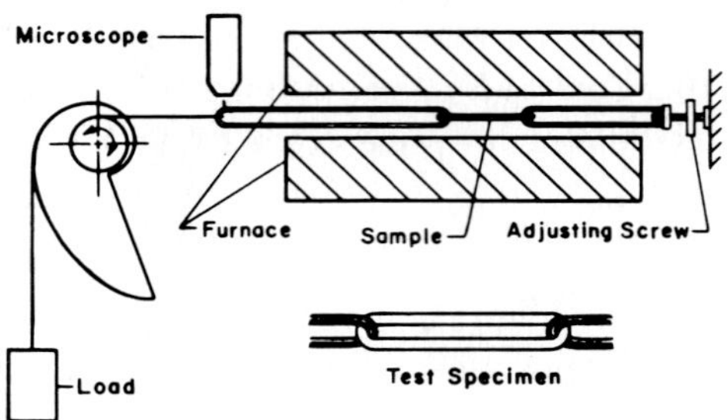

Fig. A. 18. Principle sketch of a special viscosimeter for a uniform decrease of cross section of the specimen, and shape of the specimen. (Oel).

the samples is sometimes increased by more than 200 per cent under such an arrangement of constant stress deformation. For a viscosity above 10^9 considerable deviations occur from the behavior of a normal Newtonian flow. The viscosity is increased during the deformation and reaches a final value only after a definite time period. The functional relation of viscosity to the percentage of elongation, l/l_0, shows for Jena glass 16 III impressively an increase of η during the deformation (Fig. A. 19) below 690°C. Above this temperature, however, the glass has a constant viscosity and is a normal Newtonian liquid. Probably, structural changes are responsible for this behavior in the highly viscous state. The measurements of Oel are also of particular significance for interpretation of the experiments of E. D. Lynch and F. V. Tooley[57] who investigated the influence of temperature and deformation forces during the drawing process of glass samples (large-diameter fibers of a sodium-calcium silicate glass) drawn from a cane at a constant rate, using loads from 100 to 8,000 gm. Tensile strengths were determined by transverse loading and differences up to 60 per cent as an increase in strength were resulted from variation of temperature and load during forming. If the surface of fibers was abraded, the same tendency to an increase in strength persisted, although to a somewhat lesser extent (cf. Fig. A. 20).

36. A rich numerical tabulation of viscosities of systematically varied glass compositions was given by L. C. Hoffman, T. A. Kupinski, R. L. Thakur, and W. A, Weyl,[58] for η between 10^8 and 10^{13} poises, between 370° and 700°C. constant, by de-

[57] *J. Am. Ceram. Soc.*, **40**, 1957, 107-112.

[58] *J. Soc. Glass. Technol.*, **36**, 1952, 196-216.

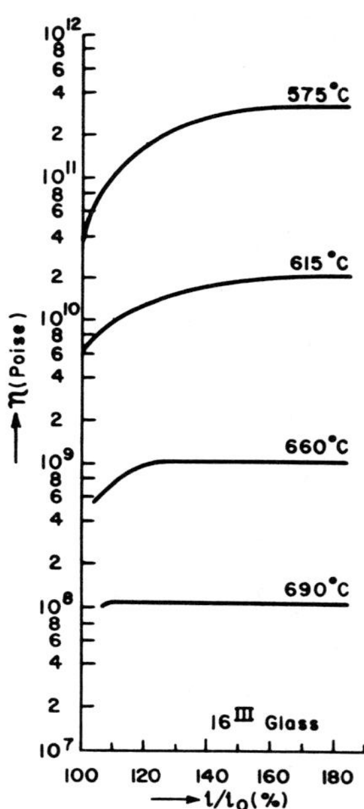

FIG. A. 19. Viscosity of Jena glass 16 III. as a function of the deformation for different temperatures, at a constant stress of 300 gm./cm.[2] (Oel).

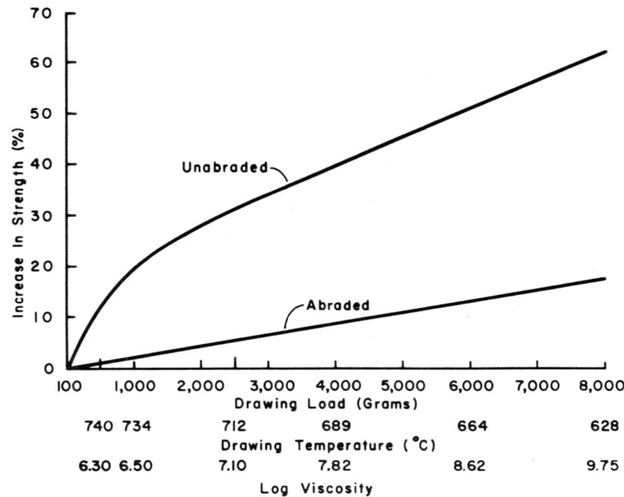

FIG A. 20. Increase in tensile strength (in per cent) of a sodium-calcium silicate glass, due to forming conditions during drawing, for abraded, and not-abraded fiber specimens. (Lynch and Tooley).

termining the time required for a definite fiber elongation under load. Interesting curves are observed in glasses of the composition $Na_2O, MgO, 5SiO_2$, if Mg^{2+} is replaced by Zn^{2+}, Ni^{2+}, Co^{2+}, Fe^{2+}; or in glasses $Na_2O,2MgO,Al_2O_3,5P_2O_5$ by the substitution of K^+ and Li^+ for Na^+, or of Cu^{2+} replacing Mg^+; or in glasses $Na_2O,MgO,5B_2O_3$, by replacement of Mg^{2+} by Cu^{2+}, Co^{2+}; or substitution of Na^+ by K^+ in glasses $Na_2O, MgO,3SiO_2$; $Na_2O,BaO,3SiO_2$; $Na_2O,CuO,3SiO$; $Na_2O,5B_2O_3$; or in glasses $Na_2O, 3SiO_2$ by replacement of Si^{4+} by Ti^{4+}, Th^{4+}, Zr^{4+}, Ge^{4+}, Al^{3+}, Ga^{3+}, In^{3+}, B^{3+}, P^{5+},

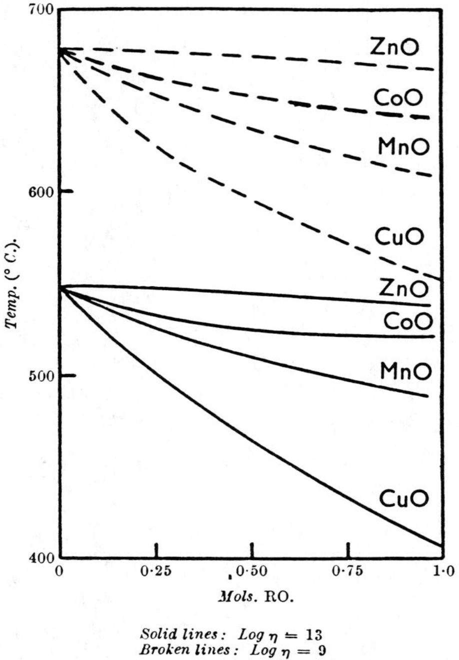

FIG. A. 21. Isokomes for the series of glass compositions Na_2O, $(1-x)$ MgO, xRO, 5 SiO_2, (MgO + RO = 14.3 mol. per cent). (Hoffman, Kupinski, Thakur, and Weyl).

V^{5+},Ta^{5+},Nb^{5+}. Corresponding curves for the constant viscosity = 10^9 and 10^{13} poises and corresponding temperatures as a function of the concentration are given. We show in Fig. A. 21 the curves for the substitutions in glasses of the type Na_2O, $(1-x)MgO,xRO,5SiO_2$, in Fig. A. 22 those for the series $xNa_2O,(1-x)K_2O,RO,3SiO_2$.

37. H. Wondratschek[59] demonstrated the possibility of a simultaneous determination of viscosity, η, and surface tension, σ, of glasses at relatively low tempera-

[59] *Glastech. Ber.*, 32, 1959, 276-278. Wondratschek criticizes and corrects important details in the theoretical deductions of Parikh.

tures, analogous to a previous method recommended by N. M. Parikh.[60] The calculation of both parameters, η and σ, is based on an equation system in which L is the length, $2r$ the diameter of the glass fiber, G the load applied for the extension, E the elongation $= (1/3\eta)\ (GL/\pi r^2 + L^2\varrho/2 - \sigma L/r)$. This calculation makes the vis-

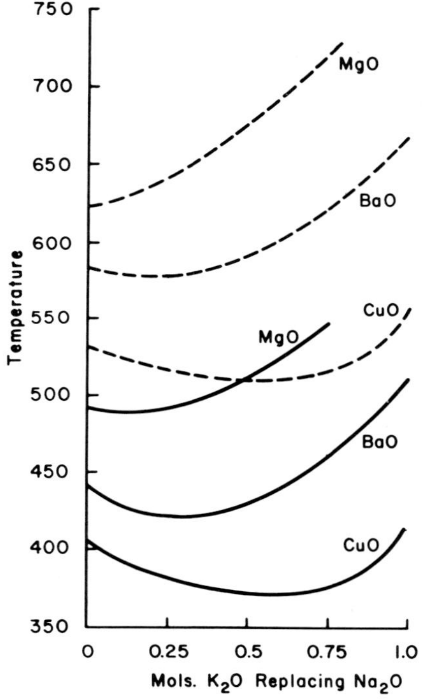

FIG. A. 22. Isokomes for the series $x\text{Na}_2\text{O},(1 - x)\text{K}_2\text{O},\text{RO},3\text{SiO}_2$, for log $\eta = 8$ (broken curves), and $= 13$ (solid lines). (Hoffman, Kupinski, Thakur, and Weyl).

cosity available in a temperature range which previously was not practical by the normal methods. Only very low amounts of material are needed for the fiber samples. For evaluation, two ways are indicated by Wondratschek, both based on the graphic evaluation according to the sketches Figs. A. 23 and 24.

38. Another very interesting way of determining absolute viscosities in the softening range was shown by R. Krause[61] with a *parallel plate plastometer*, as shown in

[60] *J. Am. Ceram. Soc.*, **41**, 1958, 18-22.

[61] *Silikat Tech.*, **11**, 1960, 263-266. The principle of the instrument goes back to J. R. Scott, "Theories and Application of the Parallel Plate Plastometer," 1921, especially constructed for rubber testing. See also in the book of R. Houwink, "Elasticity, Plasticity and Structure of Matter," Harren Press, Washington, D. C., 2nd ed. 1953, 368 pp., especially pp. 19, 199ff.

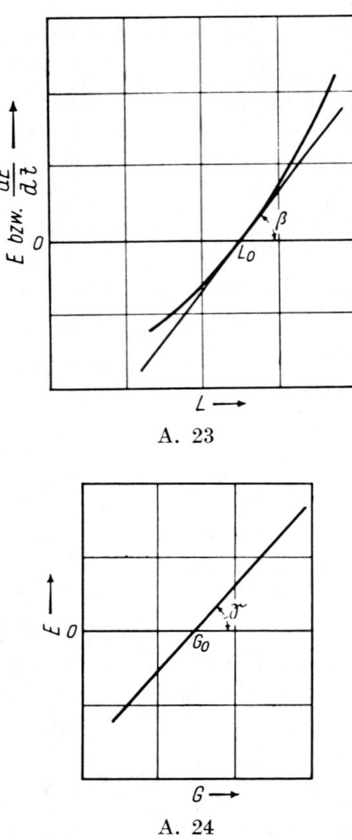

FIGS. A. 23 and 24. Principle sketch for graphical evaluation of experimental curves of fiber elongation under a constant load, G (details in text). (Wondratschek). Two modes are shown.

its working principle by Fig. A. 25, under hydrodynamically exactly defined flow conditions. Glass behaves in this way as a Newtonian liquid, but a plastic flow with a yield point phenomenon is not excluded in some cases. It is not yet possible to judge the appearance of a typical structural viscosity in glass, in which case much higher shearing stresses should be applied than are accessible in the actual plate plastometer constructions, and at temperatures nearer to the transformation range. The full usefulness of the method, however, was made evident for three Jena optical glasses at about 650° ± 1.5°C.

39. G. Tammann and R. Klein[62] previously discussed the limits between the plastic and the elastic behavior of glass as a function of temperature, by a qualitative observation of changes of the phenomena in F. Auerbach's sphere intrusion hardness

[62] *Z. anorg. u. allgem. Chem.*, **191**, 1930, 161-178.

test and in that of the *resilience* of a steel sphere from the polished surface of a given elastic or plastic material. These fundamental ideas on behavior of glass near, and below, the softening range were studied with modern aspects and apparatus possibilities by P. L. Kirby[63] in observing the contact times of impact of a freely falling steel sphere of 8 mm. in diameter onto a slab of Pyrex glass, by an electronic-con-

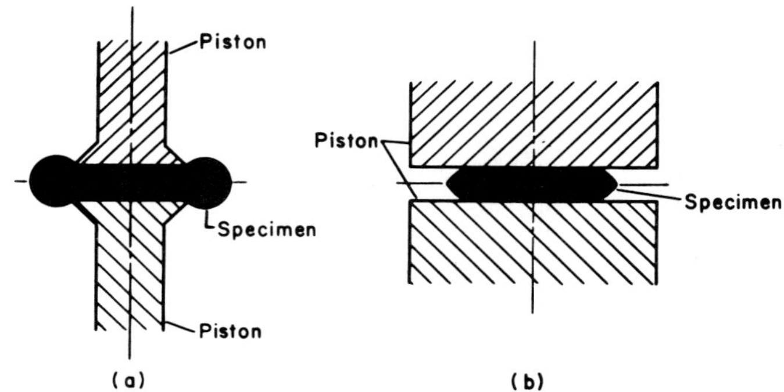

Fig. A. 25. Principle of the action of the parallel plate viscosimeter, in two different experimental versions. (Krause).

trolled (oscilloscopic) proximity meter without any direct connection to the sphere (cf. Fig. A. 26). The time of the impact contact was measured at room temperature and up to 700°C. with only a slight change, about 10 per cent, and an average of 18 μsec. for this time. Above 700°C., however, the absorption of fractional impact energy is rapidly increased, and at 900° to 1000°C. there is no more rebounding at all. These results are of great importance for understanding the instantaneous-elastic, and the anelastic (delayed-elastic after-effect) portion, if the experimental temperature is above the Littleton point with $\eta = 10^{7.6}$ poises. This point can ($\pm$ 1.5°C.) be determined rather well from resilience observations if one measures the loss in kinetic energy of the steel sphere by absorption in the glass after impact. It is $\Delta E/E = f(h_1 - h_2)/h_1$, with heights of the free fall (h_1), and the height of the resilience (h_2). It is $= 2\pi \cdot \tan \delta$ in which $\tan \delta$ is the "internal friction" of the glass for an equivalent periodic oscillation frequency, ν. At temperatures here concerned for glass, the (external) friction effect of air in the free fall, and the anelasticity of the cold steel sphere are negligible. Resilience at the Littleton point is 50 per cent ($\Delta l/l = 0.5$). The temperature gradient of log ($\Delta l/l$) is almost identical for different glasses tested, independent of their chemical composition (see Fig. A. 27).

[63] *Brit. J. Appl. Phys.*, 7, 1956, 227-228; *Verres et réfractaires*, 10, 1956, 201-207. See also the discussions of E. I. Kozlovskaya, *Structure of Glass. Proc. All-Union Conf. Glassy State, 3rd, Leningrad (English Translation)*, 1959/60, 344-347.

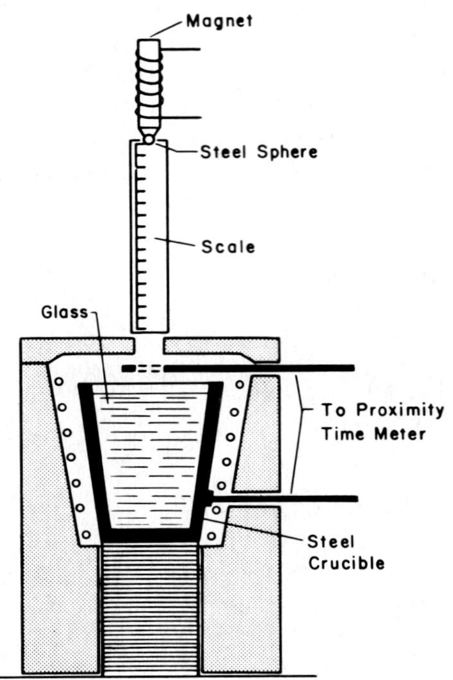

FIG. A. 26. Apparatus for the study of the resilience of a steel sphere from a Pyrex glass block, from 20° to 1000°C. (Kirby).

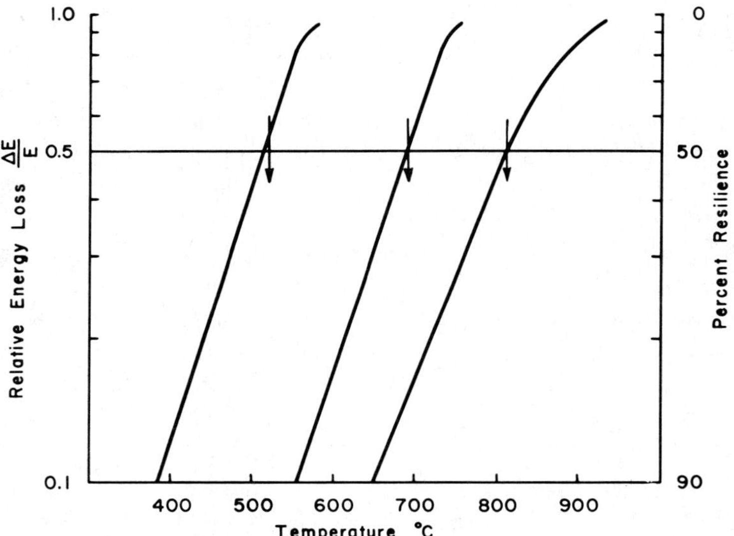

FIG. A. 27. Loss of energy of impact of a steel sphere on a glass surface, for three different glasses, as a function of temperature and resilience. (Kirby).

40. For *very high* degrees of viscosity Micheline Prod'homme[64] recommended the interferometric method of Lord Rayleigh, and the Reiner equation

$$\eta = mg \cdot l^3/12\ bh^3(dl/dt),$$

with l, b, h = length, width, and thickness of the sample, m the mass for the deformation of a given glass sample under a load applied over a long time period (several days for 10^{15}, months for 10^{17} poises). She experimentally studied several optical glasses, anhydrous B_2O_3 glass, and those of As_2O_5, As_2Se_5, and glycerin. By a combination with results of A. Dietzel and R. Brückner (see A. ¶ 28) complete viscosity versus temperature curves were developed for the ranges from 115° to 1269°C. Variation of the annealing point temperatures as a function of the composition is interpreted by M. Prod'homme by the bonding energies of the framework-forming and framework-modifiying cations with oxygen anions, the boric acid anomaly by the change of the coordination $[BO_3] \rightleftharpoons [BO_4]$. In this latter case the bonding mechanism is determined by the weak molecular bonding strengths, and the same reason is given for As_2S_5 and As_2Se_5 with their low annealing temperatures. The activation energy of viscous flow has a characteristic maximum in the transformation range at $\log \eta = 11$. B_2O_3 shows a corresponding maximum activation energy of 100,000 cal./mole, glycerin of 40,000 cal./mole, correcting the previous results of E. Seddon and J. E. Stanworth (1939, 1948). The average difference quotient $\Delta\ (\log \eta)/\Delta T$ in the range from $\eta = 10^{15}$ to 10^{11} is 0.016 for silica glass, 0.041 for B_2O_3 glass, 0.071 for As_2S_5 glass, 0.111 for As_2Se_5 glass, and 0.125 for glycerin glass. This quotient characterizes the different gradients of the viscosity curves in the transformation range.

41. We owe to Aniuta Winter-Klein a series of most fascinating discussions on the correlations of viscosity with other fundamental physicochemical properties of glass materials.[65] Starting from the Arrhenius equation $\eta = A \cdot T \cdot e^{Q/RT}$ for the transformation range and together with the available data on viscosity of silicate glasses, a representative, or idealized, viscosity versus temperature curve can be drafted. A slight lateral shifting parallel to the temperature axis brings two characteristic branches of this curve to coincidence, thus the gradients in the overlapping portions are the same. The transformation range is then characterized by an *accelerated* increase in viscosity with decreasing temperaure, having an inflection point at $\log \eta = 13$ (Fig. A. 28). If one compares this curve with the corresponding one of aluminum

[64] Thèse, Université de Paris, 1960, 41 pp.; *Verres et réfractaires*, **14**, 1960, 193-205; 261-273. On the Reiner equation cf. M. Reiner, "Deformation and Flow," 2nd ed., Wiley (Interscience) Publishing Corp., Inc., New York, 1960, 347 pp. (cf. A. ¶ 268).

[65] *Verres et réfractaires*, **7**, 1953, 217-227; *Compt. rend.*, **241**, 1955, 1281-1284; *Silicates inds.*, **21**, 1956, 157-162; *Verres et réfractaires*, **11**, 1957, 71-88. See also T. Moriya, *Tokyo Kogyo Daigaku Gakuho*, 1951, B, (1), especially pp. 3-5.

metal (Fig. A. 29), one sees the striking difference of the curve types. Also the values Q and A in the Arrhenius equation plotted as a function of temperature, show anomalies in the transformation range ($Q_{max.}$ = 130,000 cal./mole; $A_{min.}$ = — 25)

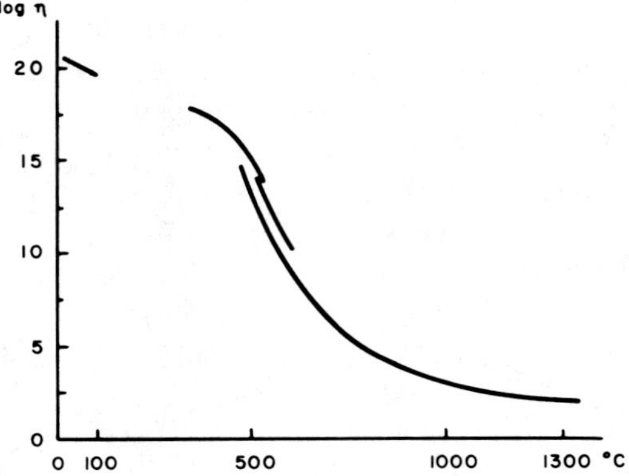

FIG. A. 28. "Idealized" viscosity curve as a function of temperature. (Winter-Klein).

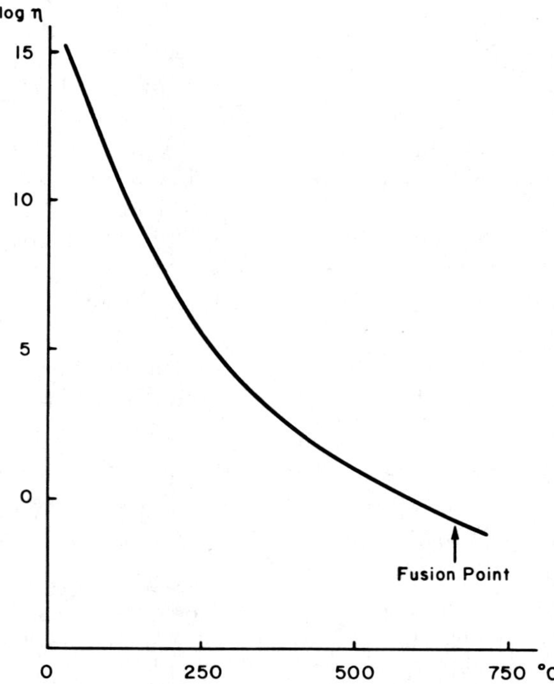

Fig. A. 29. Viscosity Curve as a Function of Temperature for Aluminum Metal. (Winter-Klein).

(Figs. A. 30 and 31). In the transformation range, the viscosity/temperature curve is nearly a straight line, A and Q are independent on temperature, in other words, the glass shows properties which are similar to those of liquids of monoatomic constitution (e.g. metal melts), or with liquids bonded only by van der Waals forces for $\eta = 10^8$ to 10^{15}. The time factor $\tau = f(\eta, T)$ indicates the conditions of "inner equilibria" establishment; τ is of the order of magnitude of 10^{-4} seconds for $\eta = 10^6$; it is 0.4 second for $\eta = 10^8$, 2 seconds for $\eta = 10^{12}$, 20 minutes for $\eta = 10^{13}$, but 14 days for $\eta = 10^{15}$.

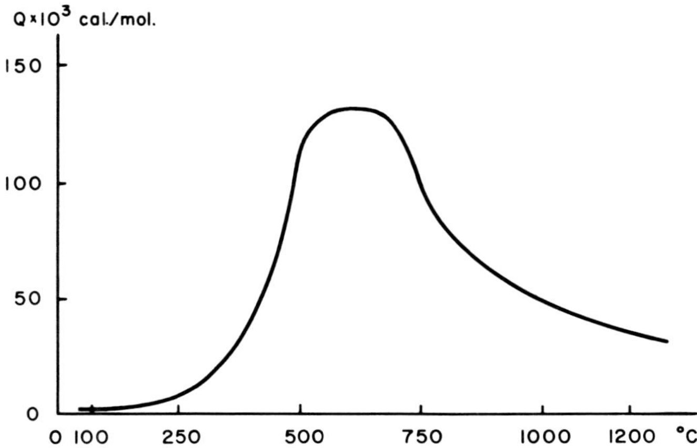

Fig. A. 30. Function $Q = f(T)$ from the Arrhenius equation of viscous flow. (Winter-Klein).

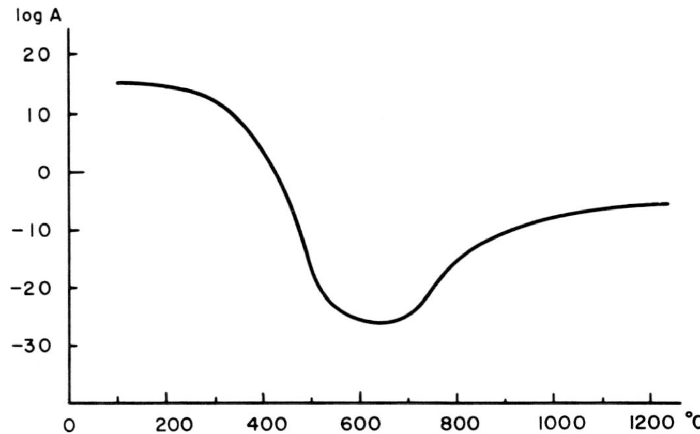

Fig. A. 31. Function $A = f(T)$ from the Arrhenius equation for viscous flow. (Winter-Klein).

42. Winter-Klein especially emphasizes the thorough uniformity of the viscosity/temperature relations in the transformation ranges for all available glasses, independent of their chemical nature, although the position and length of the transformation range may vary from one glass to the other as demonstrated for common sodium-calcium glasses, lead silicate glasses, glassy selenium, glycerin, and others. This is, of course, only a more modern formulation of what G. Tammann had recognized several decades ago in his studies of organic "model glasses," with a deep justification of analogies developed in his extensive investigations. Winter-Klein gives a new quantum-mechanical hypothesis for these analogies by the fact that the *p*-electronic binding in glass-forming materials is chiefly directional and not symmetric. The transformation range and its constant limit viscosities are the most characteristic physicochemical phenomenon in widely varying glass compositions. The solidification goes through steps from the true-liquid state, the undercooled liquid, the transformation range, to the solid state. The transition temperature of the glass-forming true liquid to the undercooled liquid is the analog to the *fusion point*, observed as a minute "knee" on the curve for $\log \eta$ versus $1/T$. See A. ¶ 277 for further discussion of this subject and an emphasis on the analogy of transformation range phenomena with structural conversions, as in alloys, and related problems in high-polymers.

43. In a further discussion of low-temperature behavior of glass in the problematic transition range from the solid-elastic to the plastic state, we must here restrict our considerations to those phenomena which are related to basic physico-chemical and constitution problems and omit the purely mechanical side of the whole, tremendously complex group of phenomena. We go back to the experiments of A. Smekal which in their classical simplicity and conclusiveness demonstrated the possibility of a fractureless deformation of glass under stress. This was shown most impressively in the scratching process of a glass surface which at room temperature showed in the microscopic study (including the phase contrast method) of the traces, typical effects of a plastic dislocation of material.[66] How far the principles of molecular mobility in solids will influence reactivity in the solid state, will be discussed in conjunction with Smekal's theories in Vol. V, A and here we shall only mention that on the general basis given by Smekal, E. Brüche and G. Schimmel[67] more recently developed an elaborate idea of the physical state of a glass surface, say in its response to hardness tests, chiefly of the "microscriber" type. A "soft" surface layer of 60 mμ in thickness was observed in microscope slide glass (10 years old) and up to 300 mμ in different kinds of industrial glass. The electron-microscope study of scratching traces, of a width below 1 μ, produced by a diamond pointer in a micromanipulator, made evident that there are characteristic and different

[66] Cf. A. Smekal, W. Loos, and W. Klemm, *Naturwissenschaften*, **29**, 1941, 688-690, 769-770; *Ibid.*, **30**, 1942, 224-225. For most remarkable report on these crucial experiments and on the evolution of Smekal's theories of the solid state in glass see *Proc. Intern. Symposium on Reactivity of Solids, Gothenburg*, 1952/54, Vol. I, 125-132.

[67] *Glastech. Ber.*, **27**, 1954, 239-247; *Z. angew. Phys.*, **7**, 1955, 378-385.

responses of the glass surface to face pressure exerted in the scratch or pointer effect: (1) an elasticity range (with resilience) observed as long as the face pressure is below 500 kg./cm.^{2}; (2) a plasticity range beyond the flow limit of the glass, with permanent grooves around the scratch or pointer puncture, for a face pressure of about 500 kg./cm.^{2}; (3) a scraping range in which the diamond tool ejects parts of the soft surface material and squeezes laterally the glass material in irregular lumps; walls of the traces may be plastically smoothed, however; (4) if the face pressure is very high, there is no more plastic deformation at all, the glass behaves really brittle, and the strains bring about the formation of characteristic cracks and typical fracturing. Temperature has no marked effects on the third process. Glass plasticity is "athermal" as a gliding mechanism of molecular aggregates. No indications of a "local fusion" are observed which were sometimes presumed as participating in glass plasticity at room temperature.

44. Fr. Kerkhof, R. Seeliger, and W. Westphal[68] described the brittle fracturing by the fact that accurately fitting positive and negative surfaces of the fracture faces must occur, whereas fracturing in plastic materials brings about a partial displacement. In this latter case, the positive and negative surfaces cannot fit each other. The electron-microscopic collodion replica details show full brittleness, as in a sodium fluoride-turbid opal glass, with exactly fitting positive and negative fracture surfaces; whereas polystyrene, Plexiglas, and other plastomers show plastic deformation and nonfitting in the corresponding test.

45. The practical interest in transition phenomena between the elastic and plastic behavior of glass is concentrated on the problem of adequate *annealing* or *stabilization* of shaped glass to remove all of the detrimental inner tensions. For this purpose the classical annealing equation of L. H. Adams and E. Williamson (1920) offered a first approximation, although since that date many critical improvements of the basic theory were necessary. More recently, P. G. Migeotte and H. P. C. Vandecapelle[69] measured the behavior of window glass bars ($50 \times 50 \times 150$ mm.) at rising temperature by a mirror dilatometer (extensometer) under a constant compression load, exerted by a mechanical lever press (Fig. A. 32). The bars were observed between crossed Nicols and birefringence by strains was directly measured. The resulting deformation-time curves were analyzed to characterize the elastic flow properties as a function of temperature and load. In a second series of experiments, relaxations under decreasing loads were measured. The modulus of elasticity which was at room temperature equal to 7,320 kg./cm.^{2}, was decreased with increasing temperature, rapidly near 500°C. The viscosity at 525°C. was equal to $10^{14.37}$ poises (directly

[68] *Glastech. Ber.*, **28**, 1955, 262-264; on tests of organic plastomers see H. Schardin, *Kunststoffe*, **44**, 1954, 48-55.

[69] *Silicates inds.*, **18**, 1953, 181-186; *Verres et réfractaires*, **7**, 1953, 275-280; *Glastech. Ber.*, **27**, 1954, 405-409.

determined by fiber elongation it was equal to $10^{14.35}$). The photoelastic constant C was for a given glass equal to 2.63×10^{-7} cm.^{2}/kg. at room temperature and equal to 2.77×10^{-7} at 500°C. The Maxwell relaxation law $(d\sigma/dt) = -\alpha\,\sigma$ (σ the strain) is not valid unless for longer times, the Adams-Williamson equation $(d\sigma/dt) = -A \cdot B\sigma^2$ was not strictly fulfilled, and no better agreement could be found with

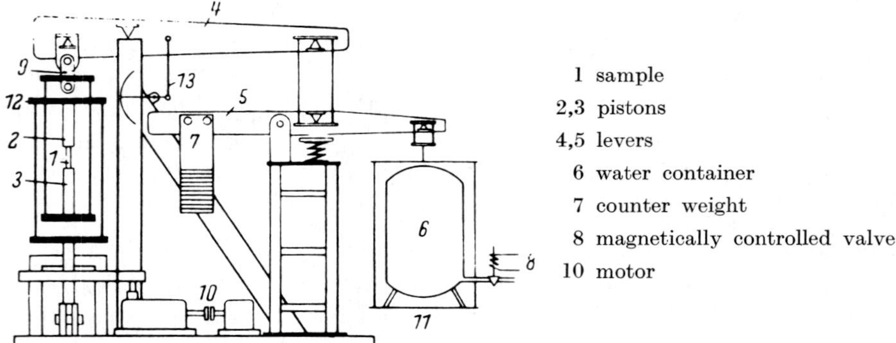

Fig. A. 32. Apparatus for the study of glass flow and relaxation under pressure. (Migeotte and Vandecapelle).

Nutting's equation $\sigma = A \cdot t^n$, which is valid for organic polymers. Migeotte and Vandecapelle tried with some success to introduce the empirical relation for variable temperatures and loads: $\sigma = \sum A_n e^{-a_n t}$.

46. From the purely practical viewpoint, namely, concerning the behavior of optical glass units (large astronomic objective lenses, mirrors, etc.), the urgent problems of so-called "*creep*" and of "*secular annealing*" by a very sluggish disappearance of inner strains is highly important in this connection. L. G. Gherling and F. W. Preston[70] gave very valuable information on "low-temperature annealing" extended over days, months, and years of observation at room temperature and up to 250°C. These observations gave the conclusive results that a glass with ordinary annealing temperature of 500°C. can lose a part of its birefringence at 200°C. although this is only a fraction of the total amount of birefringence. The glass is then in a stable state and does not undergo further changes (cf. A. ¶ 275 f.). From the theoretical side, F. W. Preston[71] made valuable contributions to the mathematical-statistical treatment of problems here concerned by discussing the surprising analogy to biological-genetic problems, such as the random loss of genes in a population (analogous to the extinction of "*polarons*," the structural units causing birefringence in glass).

[70] *J. Am. Ceram. Soc.*, **33**, 1950, 221-222.
[71] *J. Am. Ceram. Soc.*, **36**, 1953, 232-238.

Such a formulation is identical in its final results with the deduction of the Adams-Williamson law, in its integrated form $t/2 = 1/f + 1/f_0$ (f the residual, f_0 the initial stress), corresponding to a rectangular hyperbola. Gherling's results mentioned above would in this meaning indicate that beyond this theoretically plausible simplification, certain "cooperations" or "collaborations" of polaron units are necessarry. If these are not realized the ideal case cannot be valid, as the experiment makes believable, and a residual birefringence remains. An attempt at improving the Adams-Williamson relation to illustrate completely the annealing process cannot omit the structural changes (changes in constitution) of the glass to be discussed in A. ¶ 275. The annealing process is, in other words, not "purely mechanical" or statistical.[72]

47. The practical ways to find an entirely satisfactory solution of the annealing problem for optical quality glass specimens were extensively discussed by H. R. Lillie and H. N. Ritland.[73] The most efficient practical method is a soaking of the glass for a short time near the temperature of the annealing point, and then to cool it at a constant rate to a temperature below the annealing range, and at an increased rate subsequently down to room temperature. For this process Lillie and Ritland gave empirical formulas to develop the complete annealing schedule in terms of the final refractive index and stress conditions tolerable in the glass, especially on the basic assumption of a parabolic stress distribution in which the isostress surfaces are plane and parallel to the surface of the glass sample (a slab). For silica glass, J. O. Isard and R. W. Douglas[74] demonstrated for stress relaxation the elementary Maxwell law $S/S_0 = f\text{ (time)} = e^{-Gt/\eta}$ (G the modulus of rigidity) for a sample entirely homogeneous in stress distribution and heat treatment. There are, however, characteristic deviations from this ideal state since viscosity is commonly changed with time because of inner nonequilibria and cause a delayed elasticity effect (viscoelasticity). In every case, the Adams-Williamson equation for annealing which previously was found as an empirical law, can be derived on this theoretical basis, as well as the approximation given by J. Bailey and D. E. Sharp for release of strains in glass. Also H. N. Ritland[75] based his discussions of stress-time relations in glass during annealing on a simple viscous flow mechanism in which the viscosity linearily increases with time (at constant temperature). Thus he was able to find an excellent agreement between the experimental data of H. R. Lillie (1936) with this hypothesis. It corresponds to the approach of the glass structure towards its equilibrium state based on the "fictive temperature" concept (cf. A. ¶ 270 ff.). L. W. Tilton[76] developed

[72] *J. Soc. Glass Technol.*, **36**, 1952, 287-296.

[73] *J. Am. Ceram. Soc.*, **37**, 1954, 466-473. The importance of the constant cooling rate is emphasized in the deductions of H. N. Ritland, *Ibid.*, **37**, 1954, 370-378 (cf. A. ¶ 272 ff.).

[74] *J. Soc. Glass Technol.*, **39**, 1955, 61-82 T, 83-98 T, on theoretical background. On the Bailey-Sharp equation cf. *J. Am. Ceram. Soc.*, **16**, 1933, 367-379.

[75] *J. Soc. Glass Technol.*, **39**, 1955, 99-112 T.

[76] *J. Soc. Glass Technol.*, **40**, 1956, 338-352.

special ideas on structural rearrangements combined with the annealing mechanism, as a function of composition, with particular emphasis given to the fraction of non-silica particles, their bond character, with steadily sluggish changes in the tetrahedral framework and especially of the angle Si–O–Si (cf. A. ¶ 234).

48. Using a bending device for glass bars, A. F. Van Zee and H. M. Noritake[77] more recently investigated the stress-birefringence behavior of sodium-calcium silicate glasses in the temperature range from 23° to 650°C. The stress-optical coefficient slowly increases with the temperature and reaches a distinct maximum at 570° to 590°C. (Fig. A. 33). In addition, the rates of stress release were measured. The

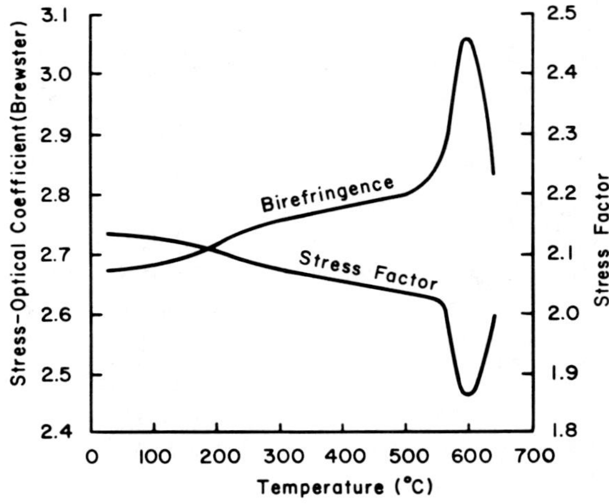

FIG. A. 33. Stress-optical coefficient and stress factor for a plate glass, as a function of temperature. (Van Zee and Noritake).

theoretical analysis of the results suggests cooperation of two antagonistic mechanisms acting in the stress release over the viscosity range from 10^{12} to $10^{14.5}$ poises, one with a period, τ_1, of 0.6 to 10 minutes, the other, τ_2, of 2 to 60 minutes, corresponding to an equation for the stress level $S = C_1e^{-(t/\tau_1)} + C_2e^{-(t/\tau_2)} + C_3$. Whether these differences should mean the presence of two different atomic arrangements ("phases") in the glass structure, with their individual viscosities and rates of stress release, is not easily decided although a possibility of this kind was discussed in greater detail by C.L. Babcock, S. W. Barber, and K. Fajans[78] (cf. A. ¶ 230). At least, the fact of the excellent agreement of the equation given above with the

[77] *J. Am. Ceram. Soc.*, **41**, 1958, 164-175.

[78] *Ind. Eng. Chem.*, **46**, 1954, 161-166. The previous formulation of annealing based on glass density data, as given by E. U. Condon, *Glass Ind.*, **33**, 1952, 307, 322, 323, also contains a dualism of the mechanism in a similar kind.

experiments of stress release for plate glass at 549°C. seen in Fig. A. 34 is highly convincing. The older data of Adams and Williamson do fit much better to the present equation than to their original empirical equation of annealing. The change of the time constants τ_1 and τ_2 is shown in Fig. A. 35.

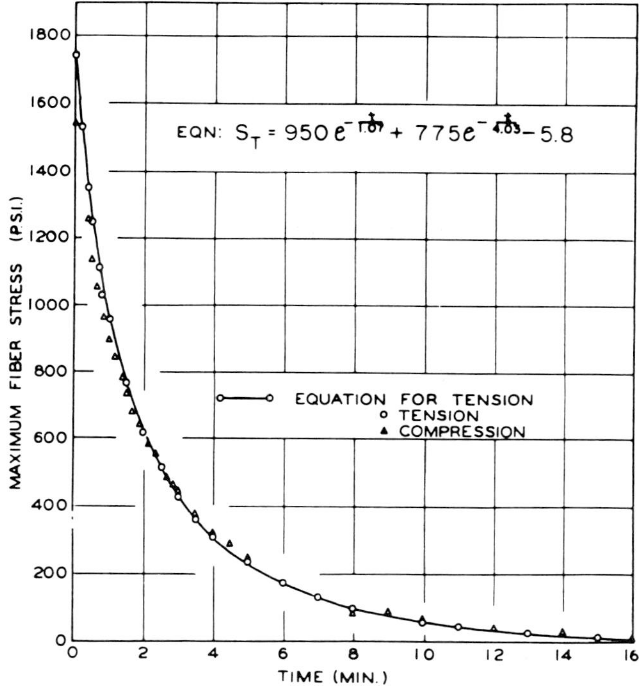

FIG. A. 34. Stress-release curve for a plate glass at constant temperature of 549°C. (Van Zee and Noritake).

49. Based on their extensive discussions of the microheterogeneity of glass structure (cf. A. ¶¶ 245, 340*) Russian authors also explain the particular behavior of glass resulting from inner tensions. V. L. Indenbom[79] describes anomalous birefringence phenomena in borosilicate glass (Nonex type, etc.) which are *not* removed by a normal annealing method. The anisotropy is observed in the axis of cylindrical rods but not in massive blocks which are apparently free from inner tensions. The anomalous "second class" birefringence is also independent of surface forces, and not

[79] *Doklady Akad. Nauk S.S.S.R.*, **89**, 1953, 509-511. On the microheterogeneity concepts see also O. A. Esin and P. V. Gel'd, *Steklo i Keram.*, **11**, 1954 (3) 4-6; *Stroenie Stekla, Inst. Khim. Silikatov Akad. Nauk S.S.S.R., Trudy Soveshchaniya, Leningrad*, 1953 (Publ. 1955), 44-55 (cf. A. ¶ 203).

changed by shaping and working, analogous to similar phenomena known from resins. The dualism of aggregations in the microheterogeneous glasses brings about a splitting of the relaxation curves into two different zones of annealing temperatures. The dilatometric curves then show the interval between the upper limit temperature of the annealing zone and the softening temperature considerably larger than in normal glasses. Indenbom,[80] refuting the previous theoretical methods of glass an-

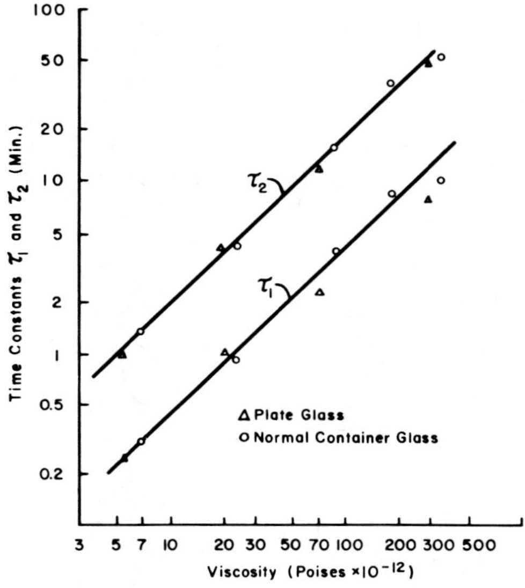

FIG. A. 35. Time constants, τ_1 and τ_2, for release of stress, as a function of viscosity, for a plate glass, and a container glass. (Van Zee and Noritake).

nealing develops a theory of this process in which he defines the annealing point, T_g, as the limit temperature above which the glass is plastic, with an instantaneous plastic deformation in the absence of any strains. If the glass is cooled below T_g, the remaining strain is defined by the equation

$$F = \left[\frac{\alpha E}{1 - \sigma}\right] \cdot T_g \left(1 - \frac{\sin \delta}{\delta}\right)$$

where α is the expansion coefficient, σ the Poisson coefficient, E Young's modulus, δ given by $\delta \cdot \tan \delta = h \cdot a$, with h the coefficient of heat exchange at the surface

[80] *Zhur. Tekh. Fiz.*, **24**, 1954, 925-928; see also I. I. Kitaĭgorodskiĭ and V. L. Indenbom, *Doklady Akad. Nauk S.S.S.R.*, **108**, 1956, 843-845. The theory displayed goes back to G. M. Bartenev, *Ibid.*, **76**, 1951, 227-230. H. D. Weymann, *J. Am. Ceram. Soc.* **45**, 1962, 517-522, presented a thermo-viscoelastic interpretation of the *tempering process* of glass, for optimum mechanical properties, by controlled rapid cooling.

and a the half-thickness of the glass slab. Introducing the relaxation period, τ, at the start of the cooling, one finds:

$$F = \left[\frac{a E}{1 - \sigma}\right] (1 - e^{-\Theta/\omega\tau}) \frac{1}{6 \varkappa}$$

with Θ the temperature of the medium, ω the rate of cooling, $\varkappa$ the thermal conductance of the glass. If τ is much smaller than Θ/ω the annealing is proportional to the cooling rate, independent of the annealing temperature. If τ is much larger than Θ/ω the annealing is independent of the cooling rate, but dependent on temperature.

50. How indispensable the consideration of *structural factors* is in the viscous anomalies in the transformation range, was early emphasized by I. Peychès[81] concerning the chemical heterogeneity assumed to exist in chilled glasses. Such microheterogeneities cannot be detected by common methods since they concern "domains" of relatively very small molecular clusters. The chilled surface of glass fibers or of rapidly blown thin-walled glass bubbles represents the higher degree of depolymerization of a structure which corresponds to the temperature of fused glass, whereas a slow cooling and "fine-annealing" develops the more or less approximate equilibrium state of the fully polymerized state of strainless glass (cf. A. ¶ 340 ff.). Generally spoken, the measurements of many physical properties of glass depend so much on these factors that their results must be misleading and vitiated by an insufficiently controlled thermal history of the glass samples. E. G. Marboe and W. A. Weyl[82] followed such ideas deeper by considering the properties of the ions in glass structures and an atomistic interpretation of the viscous flow in the light of the theories of K. Fajans on "depth action" of ionic interactions and of the screening theory. The behavior of a glass to mechanical forces exerted on the chemical binding mechanism brings about a polarizing action and dipole moments developed which contribute to the separation of inner boundaries in the glass structure with ensuing flow.

51. The *anelasticity* behavior of glass at temperatures below the annealing point in their relation to structural factors was first discovered by C. E. Guye and S. Vasileff (1914) in the logarithmic decrement characteristics of torsional vibrations, and interpreted by H. Rötger and K. Bennewitz (1936) as an *internal friction* phenomenon caused by stress. The experimental series of investigations of J. V. Fitzgerald[83]

[81] *J. Soc. Glass Technol.*, **36**, 1952, 164-180.

[82] *J. Soc. Glass Technol.*, **39**, 1955, 16-36 T. On the theories of K. Fajans see *Chem. Eng. News*, **27**, 1948, 900-904; *Ceram. Age*, **54**, 1949, 288-293.

[83] *J. Soc. Glass Technol.*, **36**, 1952, 90-104 (with K. M. Laing and G. S. Bachman; *J. Am. Ceram. Soc.*, **34**, 1951, 314-321; 339-342; 388-392. The theoretical background is given in the book of C. Zener: "Elasticity and Anelasticity of Metals," Univ. of Chicago Press, 1948; cf. A. J. R. de Bast and P. G. Migeotte, *Silicates inds.*, **21**, 1956, 329-334, and especially E. G. Ri-

have contributed much to a deeper interpretation of the simultaneous phenomena of internal friction, relaxation, elasticity modulus, and recoverable creep, as functions of temperature and time. The basic experiment is the torsional oscillation in a pen-

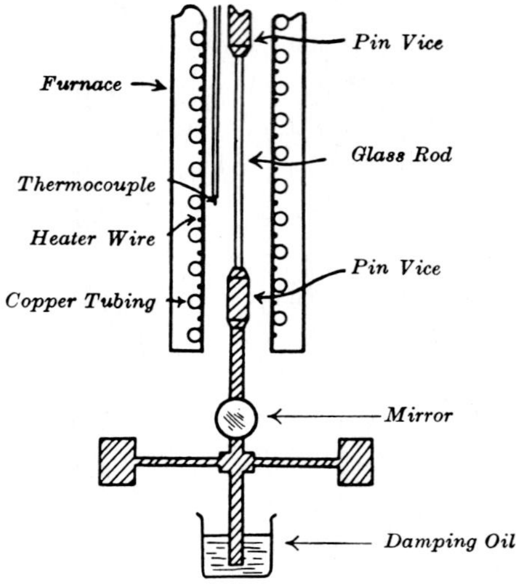

FIG. A. 36. Apparatus for torsional vibrations of rods of a sodium silicate glass. (Fitzgerald).

dulum device as shown by Fig. A. 36, from which a linear relation between the logarithms of the amplitudes, A, with the number, n, of oscillations results, and the internal friction

$$Q^{-1} = \frac{2.30}{n \cdot \pi} \cdot \log \left(\frac{A_o}{A_n} \right)$$

This internal friction is an important function of temperature of a given glass, and different in its characteristic maximum for different states of thermal treatment as shown in Fig. A. 37 for a simple sodium silicate glass (24 Na_2O;76 SiO_2). Of the two characteristic maxima, that at about -43^oC. corresponds to the diffusion of Na^+ cations in the glass, the other is the viscous-elastic after-effect, both with different activation energies (16,000 and 34,600 cal./mole).

52. The so-called *acoustic spectrum* of glass in which internal friction is represented as a function of the frequencies of vibrations exerted (Fig. A. 38) not only shows these two maxima, but a third maximum, for the transverse thermal current, espe-

chardson, "Relaxation Spectrometry," Wiley (Interscience) Publishers, Inc., New York, 1957, 160 pp. On anelasticity of glass see more recently I. Mohyuddin and R. W. Douglas, *Phys. & Chem. Glass*, **1**, 1960, 71-86.

cially characteristic of glass and metals. A typical dispersion of the internal friction peaks as a function of the frequencies is seen in Fig. A. 39 for a chilled tank plate glass. Annealing decreases the peak intensity and shifts the maximum at the same

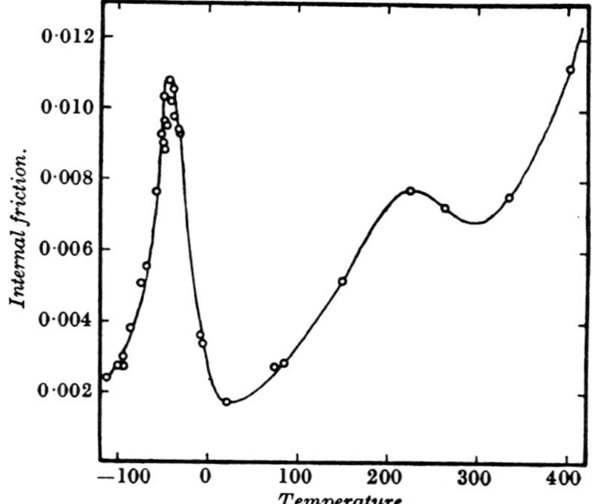

Fig. A. 37. Changes in internal friction with temperature, for a chilled glass 0.24 Na_2O; 0.76 SiO_2. (Fitzgerald).

time to higher temperatures (at constant frequency), or to lower frequencies (at contant temperature). Particularly important are correlations of the electrical strain with the mechanical strain in glass bringing about characteristic thermal effects on the electrical conductance of glass (cf. A. ¶ 158 f.).

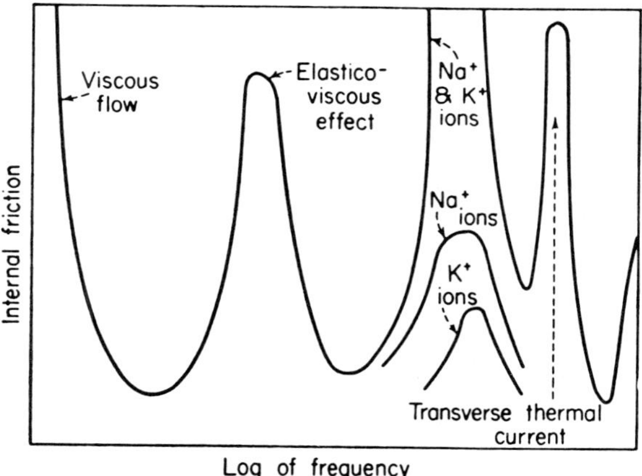

Fig. A. 38. Typical acoustic relaxation spectrum of glass (schematic). (Fitzgerald).

53. The theoretical aspects of internal friction in glass were fundamentally given by P. L. Kirby[84] in the connection between the damping factors for oscillatory and aperiodic stress-time phenomena in general, and relaxation ratios in the acoustic spectrum in particular. This discussion makes possible a thorough analysis of effects of heterogeneities in glass structure (a model for the delayed elasticity effect in Fig.

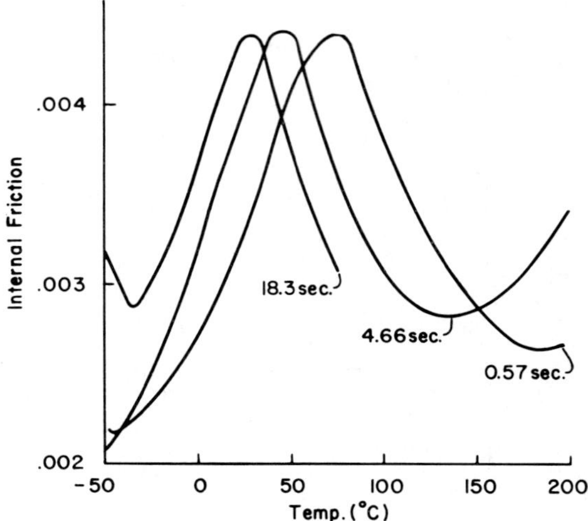

Fig. A. 39. Internal friction peaks in chilled tank plate glass (soda-lime silicate type), by using three different periods of vibration. (Fitzgerald).

A. 40), thermal diffusivity, ionic mobility, and effective temperature configuration as a reason of the surprisingly small differences in time order of elastic after-effects at room temperature and those in the transformation range. In his experiments Kirby used for frequencies of 5 to 20 cycles/second a thin glass rod of 1 mm. in diameter held in a pin chuck mounted in a massive cast iron block (Fig. A. 41) enclosed in an air-tight glass cylinder, transverse oscillations being exerted from an electromagnetic plectrum, with a thin steel plate attracted by the magnets and catching the lower end of the rod. For higher frequencies a stroboscopic flash lamp is used to record vibrations by a photographic or a similar method. Torsional vibrations were applied with an apparatus similar to that of Fitzgerald, only using an inertia disc in the place of a pendulum with removable inertia weights. The recording was entirely automatized with electronic-controlled counting circuits.

[84] *J. Soc. Glass Technol.*, **37**, 1952, 7-26 T; **38**, 1954, 383-420, 548-560; **39**, 1955, 385-393 T. Further physical aspects on anelasticity and viscoelasticity, see P. L. Kirby, *Ibid.*, **41**, 1957, 95-116 T. On the anelastic behavior of glass fibers and their *internal damping in vacuo* and under the action of water treatments, see S. L. Blum, *J. Am. Ceram. Soc.*, **38**, 1955, 205-210.

54. In the discussion of the results of measurements Kirby confirms that there are two processes contributing to internal friction, namely, relaxation of the alkali ions in the glass, and the viscous-elastic behavior of the remainder of the framework in its structure, as is particularly well demonstrated in the dependence of the internal friction on thermal history effects in the specimen. A series of commercial glasses were examined, including Pyrex as a borosilicate type. For such an exceedingly hard borosilicate glass Kirby applied a flexural deformation in the temperature

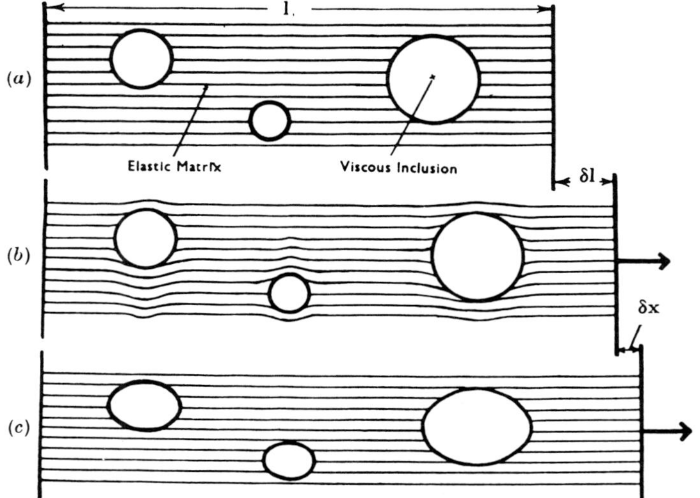

Fig. A. 40. Delayed elasticity effect in a viscoelastic material with a heterogeneity in its structure. (Kirby).

range from 250° to 500°C., with micrometric readings (Fig. A. 42). The typical Na^+ diffusion peak of internal friction in different chilled and annealed Pyrex specimens is shown in Fig. A. 43. These curves may be compared with those given by J. W. Marx and J. M. Sivertsen[85] for the frequency $\nu = 37$ kc./second, with a maximum considerably above 200°C. We further emphasize in this connection the investigations of R. J. Ryder and G. E. Rindone[86] on internal friction phenomena in alkali-alkaline earth silicate systems with the same results in principle as previously discussed by Fitzgerald and Kirby. The two characteristic peaks corresponding to alkali ion dif-

[85] *J. Appl. Phys.*, **24**, 1953, 81-87; Marx and Sivertsen applied *longitudinal* modes of vibration. The same alkali ion diffusion peaks were observed in alkali borate glasses by K.-H. Karsch and E. Jenckel, *Glastech. Ber.*, **34**, 1961, 397-408.

[86] *J. Am. Ceram. Soc.*, **43**, 1960, 662-669; **44**, 1961, 532-540. See also on inner friction in sodium aluminosilicate glasses: D. E. Day and G. E. Rindone, *Ibid.* **45**, 1962, 496-504. Internal friction of progressively *crystallized* glasses was studied by the same authors, *Ibid.* **44**, 1961, 161-167.

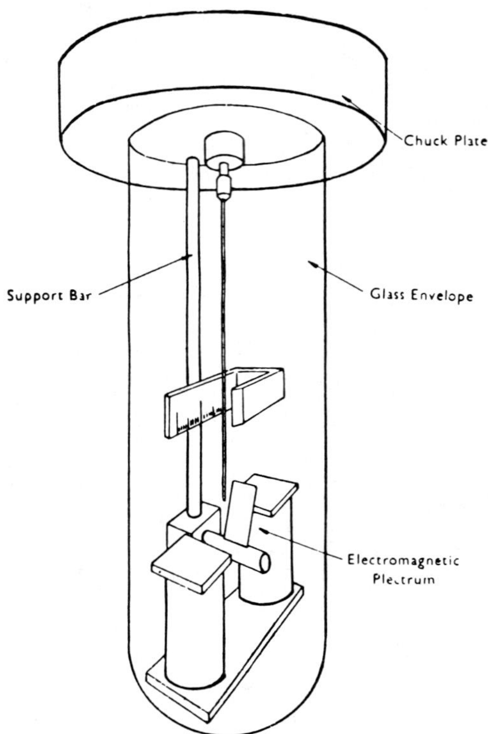

Fig. A. 41. Apparatus used for the determination of the energy losses (tan δ) at intermediate frequencies. The large iron chuck plate is supported on two massive concrete piers. (Rigby).

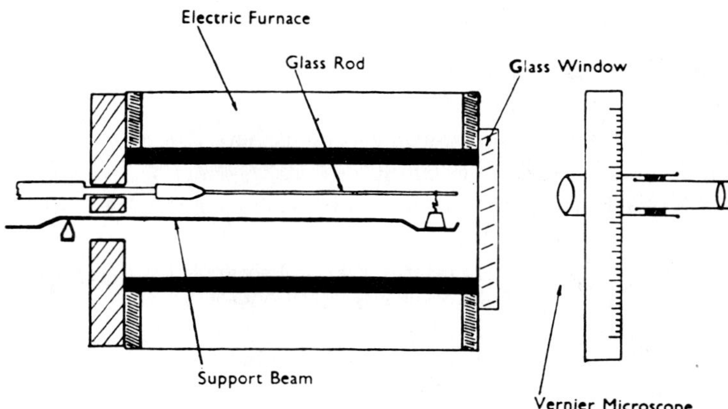

Fig. A. 42. Flexural deformation apparatus for the study of retarded elasticity behavior of a hard glass rod at elevated temperatures. (Kirby).

fusion and viscous-elastic response, however, appear shifted to higher temperatures and have lower relative intensities. In addition, a third peak appeared, increasingly prominent, the higher the contents in alkaline earth oxides. Particularly remarkable, however, is the fact that even alkali-free barium and calcium silicate glasses, and

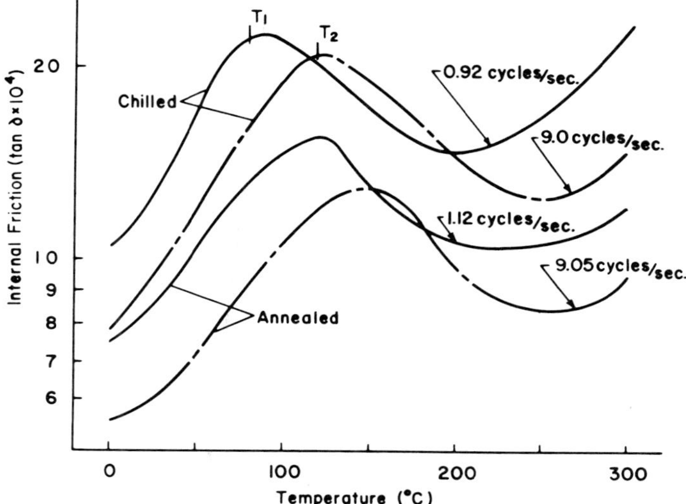

Fig. A. 43. Alkali diffusion peak of internal friction as a function of temperature, for Pyrex borosilicate glass. (Kirby). For different frequencies.

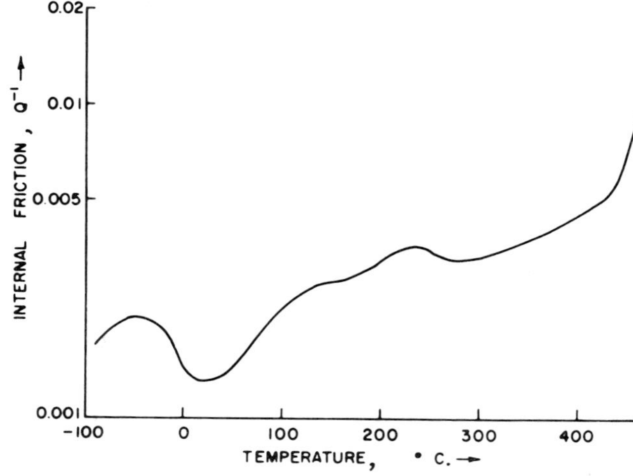

Fig. A. 44. Internal friction curve for a glass 0.25 CaO; 0.75 BaO; 2.00 SiO_2. (Ryder and Rindone).

some phosphate glasses (with BaO, MgO, Al_2O_3) show the same peaks (Figs. A. 44 and 45).

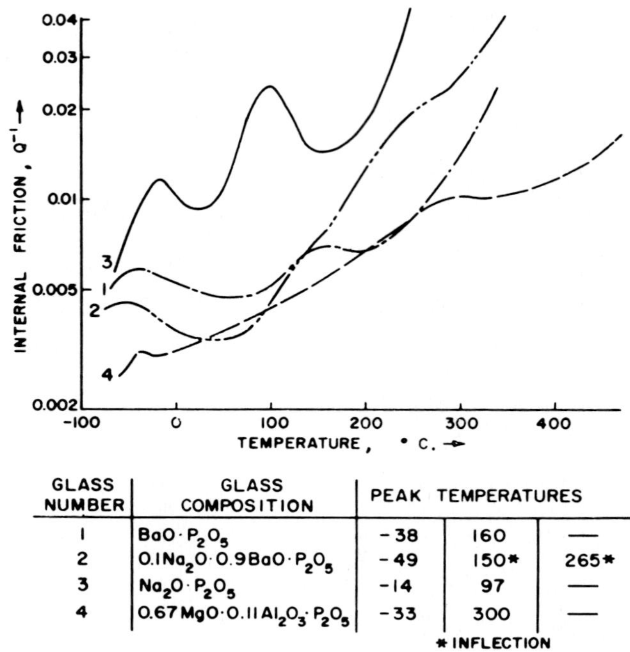

GLASS NUMBER	GLASS COMPOSITION	PEAK TEMPERATURES		
1	$BaO \cdot P_2O_5$	-38	160	—
2	$0.1Na_2O \cdot 0.9BaO \cdot P_2O_5$	-49	150*	265*
3	$Na_2O \cdot P_2O_5$	-14	97	—
4	$0.67MgO \cdot 0.11Al_2O_3 \cdot P_2O_5$	-33	300	—

* INFLECTION

FIG. A. 45. Internal friction curve for some phosphate glasses. (Ryder and Rindone).

55. Advanced knowledge on the elastic relaxation behavior of simple and mixed silicate and borax glasses was theoretically discussed anew by H. Rötger[87] in respect to the binding energies of the structural groups in glass ruling the place exchange of alkali ions, or of neutral oxide complexes, as when R_2O_3 combined with disruptions in the framework structure. Activation energy of the "loose group" exchanges is in full agreement with the calculations from internal friction maxima, in the order of magnitude of 13,000 cal., and that of the "firm groups" of 26,000 cal./mole, whereas the framework binding energy is of the order of magnitude of 100,000 to 150,000 cal./mole. An important application of the "elastic remanence" phenomena is that of the secular *ice point increase* of thermometer glass. The experimental curves show the very high effects of alkali-mixing on the intensity of the peak of resonance for the loosely bound group (Fig. A. 46), the greatest effect if Li^+, Na^+, and K^+ simultaneously participate in the glass composition. If Si^{4+} is replaced by B^{3+} in the

[87] *Glastech. Ber.*, **31**, 1958, 54-60; *Silikat Tech.*, **10**, 1959, 57-62; cf. *Ibid.*, **7**, 1956, 496-502, on prediction of the ice-point shiftings in thermometers.

framework-forming skeleton, resonance effect of the loosely bound groups is little influenced, whereas that of firmly bound groups is shifted to considerably higher temperatures and disappears entirely in pure borax glass.

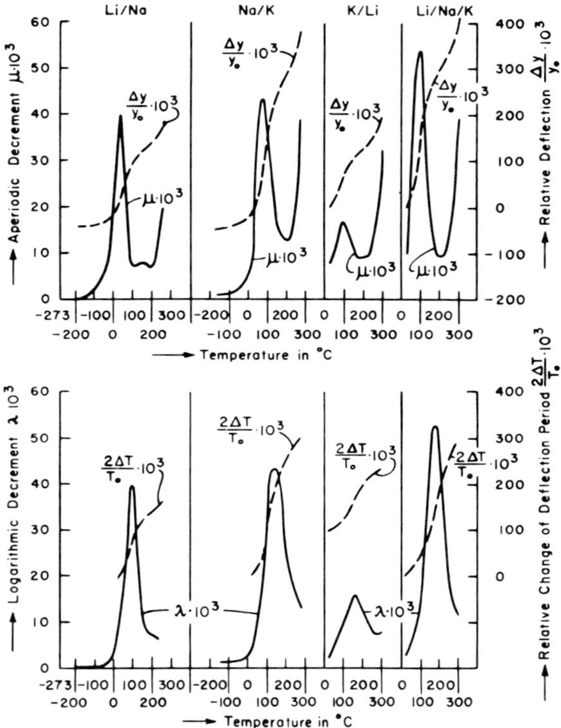

FIG. A. 46. Elastic aftereffects for a relaxation time $\tau = 100$ sec (μ), 0.1 sec (λ), and as a function of temperature, for mixed alkali silicate glasses. (Rötgers).

56. Interesting correlations between inner damping, sound velocity, and the elasticity modulus of glass as a function of temperature were discussed by A. Dietzel and E. Deeg[88] starting from the fact that different silicate glasses show much different sound patterns (Lissajous figures) in oscillation. By using the F. Förster elastometer[89] *in vacuo* the internal damping factor $\eta = \vartheta/\pi$ (ϑ is the logarithmic decrement)

[88] *Glastech. Ber.*, **27**, 1954, 105-116. On mechanical relaxation in the temperture range of metastable equilibrium see M. Coenen and E. M. Amrhein, *Union Scient. Continent. du Verre*, Charleroi, Belg., 1962, 529-550; *Veröffentl. Max Planck-Inst. f. Silikatforschg.* **22**, 1962, 105-126.

[89] *Z. Metallk.*, **29**, 1937, 109-115. Application of this instrument to hard porcelain bodies from the moist-green state to fired and finished products, was shown by E. Deeg, *Silicates inds.*, **21**, 1956, 265-270.

and $c = \sqrt{E/\varrho}$ can be used as characteristic parameters for a given glass. Typical thermal aging and hysteresis after-effects are observed on rods the damping of which was measured immediately after annealing at room temperature and the same after aging for some days (Fig. A. 47). The complete analogy of the temperature

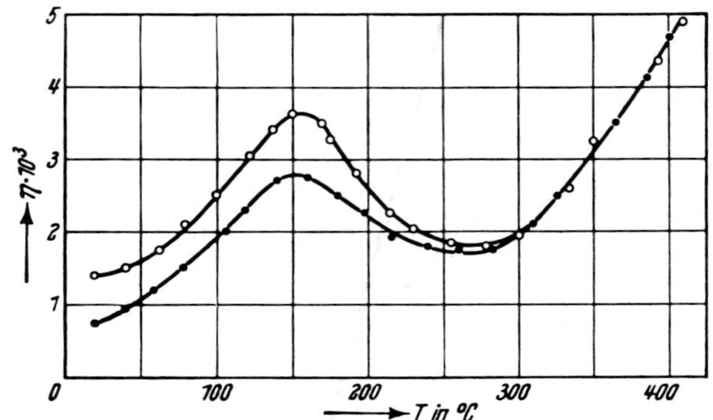

FIG. A. 47. Temperature hysteresis of the damping factor of glass as measured immediately after annealing —○—○—, and aged for one day —●—●—. (Dietzel and Deeg).

function of the inner damping factor, η, with the ice point increase was discussed by E. Deeg[90] and demonstrated for an insufficiently annealed plate glass sample as an aging phenomenon (cf. Fig. A. 48), with the time law $\sqrt{\eta} = A - B \cdot \ln t$. Further, the observation of anomalies in the damping-temperature curve for two

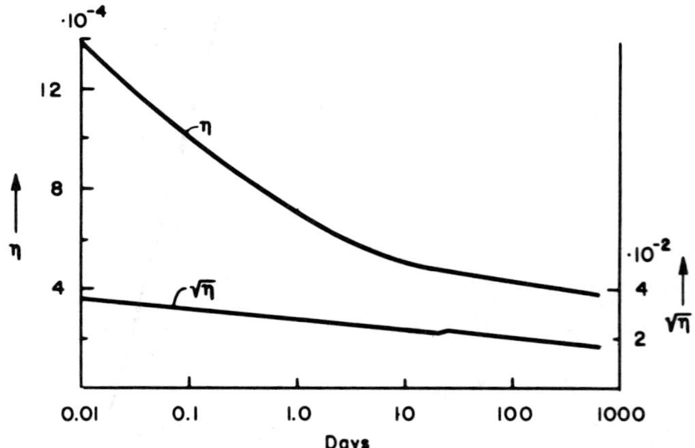

FIG. A. 48. Change of the internal damping of plate glass with time. (Deeg).

[90] *Naturwissenschaften*, 44, 1957, 303, 440.

commercial silica glass brands is important which makes the structural heterogeneity of such products evident (cf. Fig. A. 49).

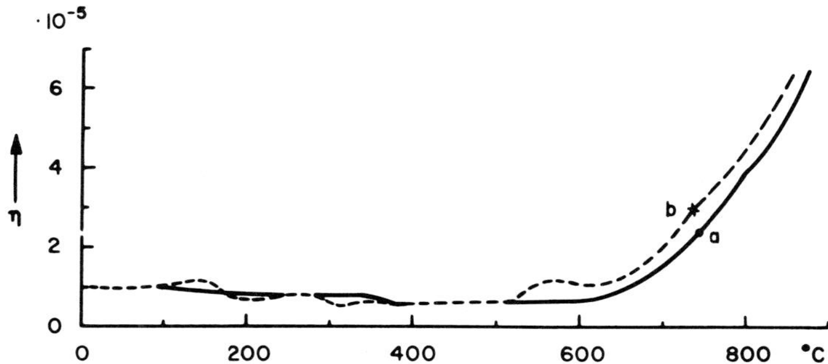

FIG. A. 49. Anomalies of internal damping versus temperature curve of two commercial silica glasses. (Deeg).

57. E. Deeg[91] presented a comprehensive discussion of the entire complexity of the interrelations between the mechanical-acoustic properties of silicate glasses and their structural constitution. Particularly interesting is the fact that an irradiation with soft X-rays influences the "elastic losses" (inner damping) in the measure of the X-ray absorption, with a distinct increase in losses, whereas an irradiation with hard X-rays is nearly without any effect of this kind. Omitting purely acoustical measurements, as with the Freystedt acoustic frequency spectrometer applied to different glasses, it is essential that different anomalies are indicated by higher damping degrees if inner strains remain in the glass structure. Cross-sectional heterogeneities (cords) show specific disturbance effects for specific oscillations. Rich numerical material was made available on temperature coefficients of the speed of sound propagation, of the modulus of elasticity, and of the inner damping (acoustic loss), in a temperature range up to the transformation interval, with particularly high thermal after-effects for mixed alkali silicate glasses (see below). The complicated relaxation effects observed compel abandonment once and for all of the common idealization of the glass structure as a statistically homogeneous framework for the interpretation of those after-effects. It must be assumed that heterogeneities with two or more components ("phases") make the multiplicity of maxima in the curves of elastic losses plausible. Even when only one maximum of this kind appears, as in silica glass, it is very broad and the modulus of elasticity has an anomalous positive temperature coefficient. The theoretical conditions then correspond somewhat to

[91] *Glastech. Ber.*, **31**, 1958, 1-9; 85-93; 124-132; 229-244. On the acoustic frequency spectrometer of E. Freystedt see *Z. tech. Physik.*, **16**, 1935, 533-539. See also A. Dietzel and E. Deeg, *Vetro*, **1**, 1956, 13-25.

K. H. Meyer's and W. Kuhn's model concepts for an entropy of elasticity, such as in rubber with fibrous-crystalline molecules. This is near to the crystallite hypothesis of glass constitution (cf. A. ¶ 202 ff.).

58. Very recently, R. Jagdt[92] investigated the relaxation phenomenon in alkali silicate glasses by measurements of rod vibrations under bending deformation and for a constant frequency $\nu = 1000$ cycles/second from a permanent-dynamic sender system. The measurements extended over the wide temperature interval from — 180° to the transformation range. Glass samples of this type show two pronounced maxima of damping, whereas glasses with mixed alkali oxides have only one maximum but of higher intensity. The temperature coefficients of inner friction and of the modulus of elasticity are not only functions of the concentration in alkalies, but also very sensitive to the thermal history of the samples, namely to their degree of annealing. The curves of relaxation are derived as a function of the frequency in equal states of annealing. The better the annealing the higher the activation energy of the relaxation, which can be calculated from the shifting of the maxima with frequencies for the definite reactions concerning damping effects. The results favor the idea that a place exchange occurs in the structural separation discontinuities for the second maximum of damping. In the equation $\tau = \tau_\infty \cdot e^{-Q/RT}$ the parameter τ_∞ is independent of temperature only at low temperatures. The modulus of elasticity for the maximum observed at higher temperatures is therefore only an approximation under the assumption that τ_∞ would be independent.

59. The problems of the *flow* (*creep*) of glass under extremely high-viscous conditions and of its recovery were discussed by S. Pearson.[93] An ingenious apparatus (Fig. A. 50) was constructed for study of torsional creep in the temperature range from 20° to 80°C. (and a viscosity in the order of magnitude of 10^{20} to 10^{23} poises) for a sodium-calcium silicate glass. The phenomena observed are explained by cooperation in the deformation effects of an instantaneous-elastic strain, a, and a delayed-elastic strain, c, which is entirely reversible, and a Newtonian viscous flow at a constant rate which is not reversible. The ratio of the elastic strains, c/a, is at 20°C. about 0.0025, and the temperature function $(c/a)_T = A \cdot e^{-H/RT}$, with the activation energy $H = 4{,}100$ cal./mole, which is nearly identical with the activation energy of the viscous flow (= 4,300 cal./mole). In this connection, one is reminded of the remarks of G. W. Morey[94] about the creep phenomena in the 150-cm. mirror of the Mt. Wilson Observatory which were carefully measured in the very slight astigmatism caused by the weight being supported on a small area. Morey concludes

[92] *Glastech. Ber.*, **33**, 1960, 10-19. On the moduli of elasticity of glass at elevated temperatures, cf. S. Spinner, *J. Am. Ceram. Soc.* **39**, 1956, 113-118, and specifically for silica glass. *Ibid.* **45**, 1962, 394-397.

[93] *J. Soc. Glass Technol.*, **36**, 1952, 105-114.

[94] *J. Opt. Soc. Am.*, **42**, 1952, 356-857.

that there is however, no evidence of any detectable distortion having occurred during the 25 years that the mirror stood on edge. A distortion of as much as one-fourth of a wavelength of visible light would have been safely detected. Glass therefore does not flow simply under its own weight at room temperature. Strains from insufficient annealing are surely much larger than those produced by the weight of the glass disk. Flow at room temperature under the action of strains by poor anneal-

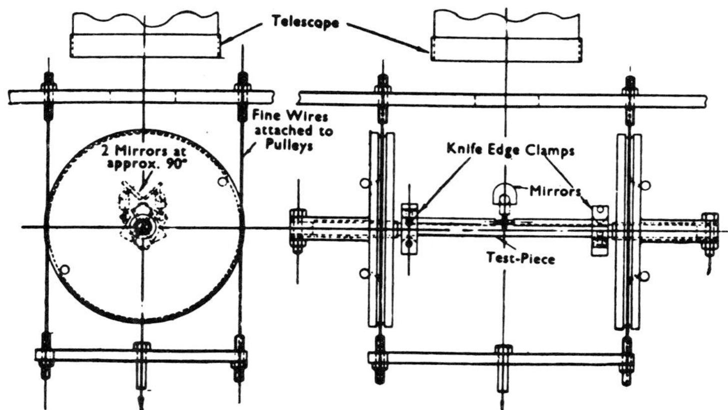

FIG. A. 50. Apparatus for the measurement of creep in torsion. (Pearson).

ing is in every case not excluded. More recently, W. A. Weyl[95] gave interesting opinions on room temperature deformation of glass rods by bending and contradicts the assumption of R. Houwink (1937) that this phenomenon would make a real flow phenomenon evident. Weyl applies his ideas on the polarization of the anions in the surface of glass under stress, which brings about a permanent deformation as a consequence of anelasticity, but not of a viscous flow.

60. The *effects of pressure* on glass structure and, therefore, on its viscous-elastic behavior were discussed by O. L. Anderson[96] in connection with the anelastic response. Pressure changes the amount of skewness, and, therefore, the average bond lengths which trivially determine glass density. Anderson distinguishes two types of pressure response in volume flow in the glass, namely a reversible densification, and an irreversible compaction. With Corning glass 7052 the combined parameters pressure and temperature, were examined, under 3800 and 6600 atm., at 100° to 300°C., and ap-

[95] *Glastech. Ber.*, **30**, 1957, 269-282, especially p. 280; *Central Glass & Ceram. Research Inst. Bull. (India)*, **4**, 1957, 121-129, especially pp. 137-138.

[96] *J. Appl. Phys.*, **27**, 1956, 943-949. See more recently R. Roy and H. M. Cohen, *Nature*, 1961, 798 f.; *J. Am. Ceram. Soc.* **44**, 1961, 523 f.; **45**, 1962, 398 f.; further discussion remarks by C. E. Weir and S. Spinner, *Ibid.*, **45**, 1962, 196.

plied for an exposure period from 15 minutes to 7 days. Densification begins immediately (cf. Fig. A. 51), and in about 5 hours it reaches an asymptotic end value, whereas compaction sets in abruptly after 15 hours and does not show any evidence of leveling off in 5 days. Under the action of extremely high pressures (order of magnitude 10^4 to 10^5 atm.) P. W. Bridgman and I. Šimon[97] were able to demonstrate

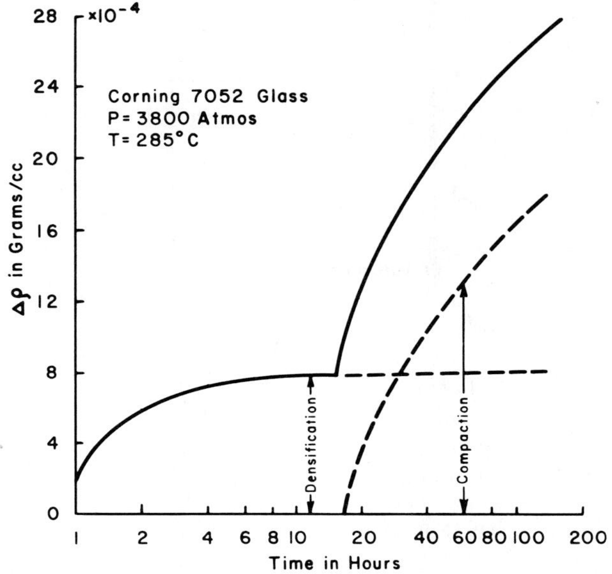

FIG. A. 51. Corning glass 7052 under a pressure of 3,800 atmospheres, and at 285°C.; changes in density, $\Delta\varrho$, as a function of time. (Anderson).

the compaction effect which is most probably connected with a change of the $[SiO_4]$ coordinations into $[SiO_6]$. This effect can be annealed and the density of the compacted glass restored to its original value. The mechanism in the densification of B_2O_3 is basically different from that observed in silica glass. It shows an activation energy in the order of magnitude of 13,400 cal./mole (cf. A. ¶ 228). As a matter of fact, the curves for compaction of silica and B_2O_3 glass (Figs. A. 52 and 53) show in the first case permanent compactions by about 10 per cent above 200,000 atm. The density of SiO_2 glass (2.22 at 1 atm.) is increased to 2.61 under these extreme conditions, say near that of quartz. As soon as alkali ions participate in the glass structure, the compaction is considerably reduced (Fig. A. 54) indicating the polarizability of the alkali ions.

[97] *J. Appl. Phys.*, 24, 1953, 405-413; cf. E. Plumat, *Silicates inds.*, 21, 1956, 447-457.

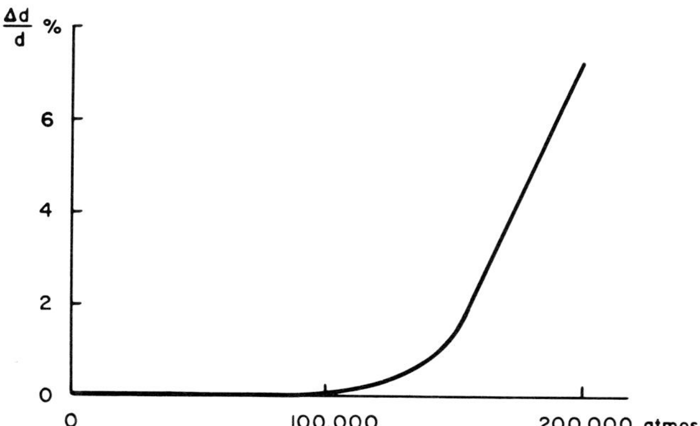

FIG. A. 52. Relative increase in density (in per cent) of silica glass as a function of the pressure applied. (Bridgman and Šimon).

61. How volume and relaxation rates in the transformation range of glass are affected by pressure was investigated by R. D. Maurer.[98] Starting from an empirical differential equation for the annealing rate at atmospheric pressure (introducing the "fictive temperature") Maurer found an analytical expression also for the pressure effects under the assumption that *compressibility* and the thermal expansion

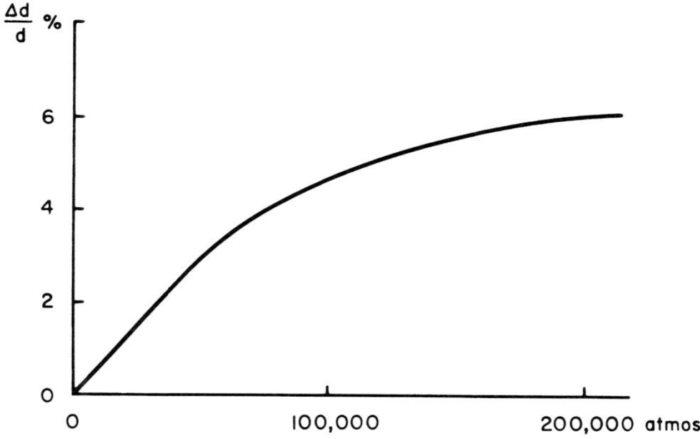

FIG. A. 53. Relative increase in density (in per cent) for B_2O_3 glass, as a function of pressure applied. (Bridgman and Šimon).

coefficient are independent of pressure. From the schematic diagram Fig. A. 55 changes in glass volume are seen as a function of temperature and pressure. It is evident that the volume difference $(\Delta V)_S$ indicated leads to an approximated de-

[98] *J. Am. Ceram. Soc.*, 40, 1957, 211-214.

termination of the compressibility in the transformation range. For a borosilicate (optical glass 517–645) as an example is $\varkappa = 40 \times 10^{-7}$ cm.²/kg. in the "frozen" glass, whereas compressibility of the liquid is split into two factors, namely a solid-like (instantaneously restored), and a liquidlike (sluggishly restored, relaxation-dependent) compression effect. On quenching only the liquidlike compression is frozen. The liquid compressibility is $= 68 \times 10^{-7}$ cm.²/kg.

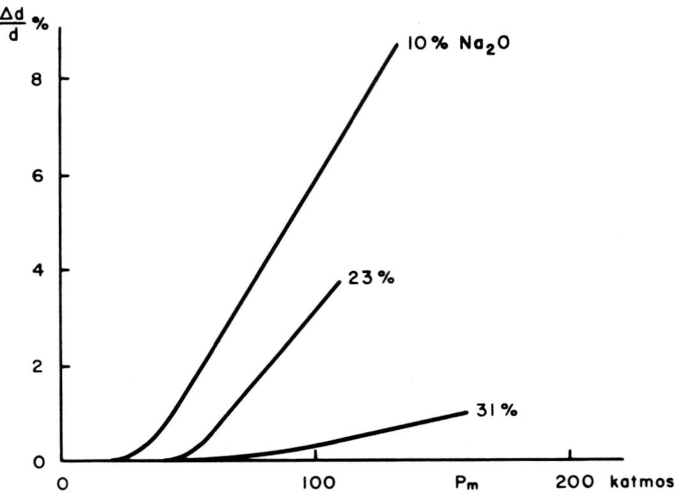

FIG. A. 54. Relative increase in density (in per cent) for sodium silicate glasses with 10; 23; and 31 mol. per cent Na_2O. (Bridgman and Šimon).

62. The diffusion of Na^+ cations even at relatively very low temperatures in the glass structure, indicating a high degree of mobility and internal friction phenomena, was discussed by R. Kamel[99] in the same interpretation encountered before in the investigations of P. L. Kirby *et al.* Kamel observed the corresponding peaks in the elastic spectrum for transverse and torsional frequencies between 20 and 400 cycles/second, with a heat of activation of the ion diffusion of 17,000 to 21,000 cal./mole, slightly variable with the Na_2O content of the glass. If there is less than 4 per cent Na_2O in pure silica glass, the diffusion peak is completely absent.

63. An important improvement of the apparatus for measuring internal friction as a function of temperature by the pendulum method was contributed by K. E. Forry[100] for synthetic sodium silicate glasses in the range of only 1 to 3 cycles/second from — 68° to + 260°C. with a statistical-analytical evaluation of the peak distri-

[99] *J. Appl. Phys.*, **24**, 1953, 1308-1311.
[100] *J. Am. Ceram. Soc.*, **40**, 1957, 90-94.

bution. The heats of activation are distributed in a Gauss curve, showing for high temperatures a much wider distribution than at low temperatures, indicating a cooperative action among Na^+ cations of the glass structure. L. C. Hoffman and W. A. Weyl[101] further gave a survey of effects of glass composition on internal friction data, with tightening effects by the replacement of Na^+ by K^+ or Li^+, and of

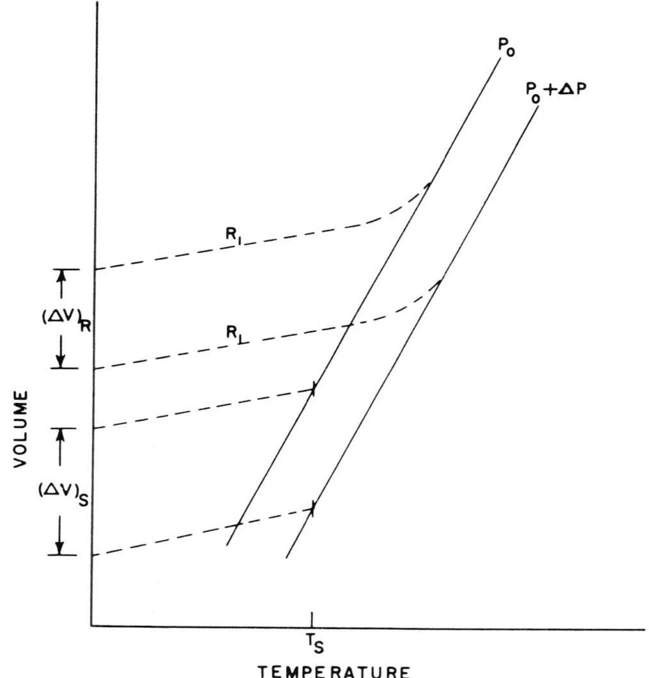

FIG. A. 55. Schematic diagram, showing the changes in glass volume for various temperature-pressure treatments. (Maurer). Solid lines show the liquid equilibrium volume for two different pressures. Upper two dashed lines show the changes in volume during cooling at a rate, R_1, for two pressures, with the resulting difference, $(\Delta V)_R$. The lower dashed lines correspond to the changes in volume for samples quenched from equilibrium at temperature T_S, for two pressures, with the resulting difference $(\Delta V)_S$.

Mg^{2+} by Ba^{2+}, Cu^{2+}, and Pb^{2+} as mentioned already in A. ¶ 36, on the basis of an atomistic interpretation. The concepts of a discontinuous structure of the glass are fully supported. What Weyl repeatedly emphasized in his deductions is a drastic revision of the common static-structural discussion of such effects as viscous response,

[101] *Glass Ind.*, **38,** 1957, 81-84, 104-105; see also E. C. Marboe and W. A. Weyl, *J. Soc. Glass Technol.*, **39,** 1955, 16-36 T, from the viewpoint of K. Fajans' polarization concepts and of screening theory.

and a thorough reexamination under dynamic aspects of a mechanical chemistry of the glassy state. This viewpoint is also as considerable an advance in the same direction as H. Eyring's reaction theories (since 1936) have been in general chemistry. His interpretation was also fully adopted by O. L. Anderson and D. A. Stuart.[102] The application of these theories makes evident the close correlations existing between viscosity and electrical conductivity by the mechanism of cation mobility, which is common to both phenomena as a basic principle (cf. A. ¶ 157).

64. O. L. Anderson and H. E. Bömmel[103] further studied the *ultrasonic wave absorption* at low temperatures and high frequencies (60 kilocycles to 20 megacycles per second) in silica glass which is specific for the vitreous state since quartz crystals do not show an analogous phenomenon. The reason for ultrasonic absorption is a structural relation as a function of temperature and frequency, in confirmation of the previous (low-frequency) results of J. W. Marx and J. M. Sivertsen (for a comparison of the results see Fig. A. 56). The internal friction is independent of the cooling rate, the absorption, therefore, a result of thermal strains.

65. The ultrasound absorption in silica glass is largely due to shear elastic waves. Activation energy of this process is surprisingly small and evidently can not correspond to an atomistic diffusion. It is also too small for a molecular rotation effect, but might be correlated to a change in the bond angles Si–O–Si by about 4° or 5°. The magnitude of activation energy makes the order of relaxation time at lower temperatures equal to the order of the period of sound waves. Transverse vibrations of the oxygen atoms would also explain other low-temperature anomalies shown by silica glass and even account for the very low thermal expansion coefficient at common temperatures, including the negative expansion at low temperatures observed by H. T. Smyth, H. S. Skogen, and W. B. Harsell[104] (A. ¶ 288).

66. Ultrasound absorption of sodium-calcium silicate glasses at *elevated* temperatures was measured by B. L. Steierman, J. C. C. Wu, and J. M. McCormick[105] in the range of frequencies between 0.25 and 1.0 megacycles/second, using a monocrystal

[102] *Ind. Eng. Chem.*, **46**, 1954, 154-160; *J. Am. Ceram. Soc.*, **37**, 1954, 572-580; for theoretical-statistical aspects of viscosity, and nonequilibrium phenomena in glasses, with a discussion of the theories of E. F. Poncelet, see *Verres et réfractaires*, **3**, 1949, 149-160, 289-299; **4**, 1950, 158-171; **5**, 1951, 69-80. See also A. G. F. Dingwall and H. Moore, *J. Soc. Glass Technol.*, **37**, 1953, especially pp. 332 ff.

[103] *J. Am. Ceram. Soc.*, **38**, 1955, 125-131; cf. H. J. McSkimin, *J. Appl. Phys.*, **24**, 1953, 988-997, on measurements at temperatures of 1.6 to 300°K. in silica glass, with a maximum loss at about 30°K. See also on low-temperature internal friction and elasticity effects in silica glass: M. E. Fine, H. Van Duyne, and N. T. Kenney, *Ibid.* **25**, 1954, 402-405 (cf. A. ¶ 287). On the low-temperature "*excess*" specific heat of silica glass see recently A. E. Clark and R. E. Strakna, *Phys. & Chem. Glass*, **3**, 1962, 121-126.

[104] *J. Am. Ceram. Soc.*, **36**, 1953, 327-328.

sapphire rod as the transmission between the sample and the barium metatitanate transducers (cf. Fig. A. 57). Characteristic is a small but uniform increase in absorption of energy up to 425°C. followed by a rapid increase in the range between

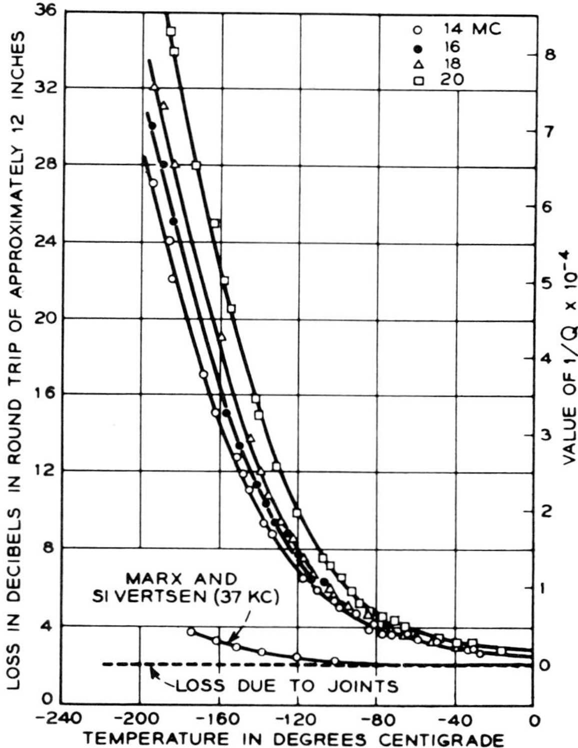

Fig. A. 56. Attenuation of ultrasound waves absorbed in silica glass, as a function of temperature, measured for lower frequencies (Marx and Sivertsen), and for higher frequencies (Mc Skimin).

425° and 815°C. At 815°C. the energy passing through 1 cm. of the sodium-calcium silicate glass was 0.2 per cent of the incident energy. At higher temperatures no decrease in the high ultrasound absorption was observed. No differences in the absorption behavior were noted when the frequencies were changed over the range from 250 to 1,000 kc./second.

67. Most instructive is the application of ultrasound vibrational damping by viscous-elastic materials in a general comparison of glasses, ceramic glazes, or other ceramic products, with organic plastomers, resins, and other materials. W. Roth

[105] *J. Am. Ceram. Soc.*, **38**, 1955, 211-213.

and S. R. Rich[106] constructed for this purpose a viscosimeter with a magnetostrictive transducer which makes possible measurements under continuously variable conditions of external pressure (from a high vacuum to 70 kg./cm.^{2}), at temperatures from — 124° to + 340°C. and frequencies up to 50 kc. In this connection also a hint may be found in the theoretical ideas of G. M. Bartenev[107] on a comparison of structural "vitrification" as in rubber by slow cooling at — 73°C., but at higher temperatures if ultrasound mechanical stresses of increasing frequencies are exerted. Similar ef-

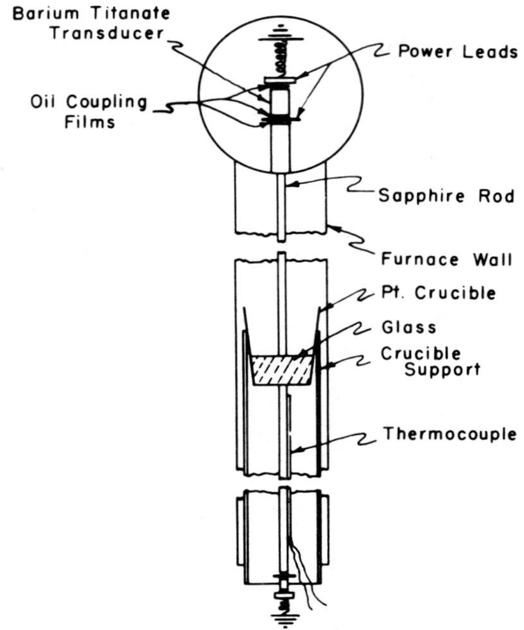

FIG. A. 57. Ultrasound transmission test apparatus. (Steierman, Wu, and McCormick).

[106] *J. Appl. Phys.*, **24**, 1953, 940-950. E. C. Bernhardt, *Ind. Eng. Chem.*, **46**, 1954, 742-746, applied ultrasound absorption methods for study of thermoplastic melts and observed no depolymerizing effects in the temperature and frequency ranges applied, but a rapid rise in temperature especially for the frequency near 1 megacycle/second (cf. Vol. I, C. ¶ 11).

[107] *Doklady Akad. Nauk S.S.S.R.*, **110**, 1956, 805-809, and recently *Structure of Glass. Proc. All-Union Conf. Glassy State, 3rd, Leningrad* (*English Translation*) 1959/60, 125-130. Cf. A. P. Aleksandrov and Yu. S. Lazurkin, *Zhur. Tekh. Fiz.*, **9**, 1939, 1249-1260, 1261-1266, on dynamic methods of studying highly elastic polymers. For high-polymeric methylmethacrylates and similar materials, I. A. Bolotina, *Structure of Glass. Proc. All-Union Conf. Glassy State, 3rd, Leningrad* (*English Translation*), 1959/60, 121-124, discussed interesting correlations between optical activity and vitrification, especially in copolymers with citronellol, glucose. At the critical T_g temperatures there is a sudden change in optical activity, with hysteresis phenomena on cooling and heating.

fects are observed with jellies of low viscosity in all transitions to highly viscous and brittle true glasses. For ordinary conditions (cooling rates of 1° per minute, and $\nu = 1$ cycle per second) both structural and mechanical vitrification coincide, but for intense stresses of short duration in a cooling amorphous material, first the mechanical vitrification effect prevails, followed by structural vitrification. Liquids and "amorphous solids" belong to the structural type of the aggregate state, viscous-flowing, highly elastic and flexible systems belong to the mechanical type. Vitrification under the action of an electrostatic field is analogous to that under mechanical action, with coincident vitrification temperatures (determined by maximum losses) when mechanical and electric fields act on the same structural unit.

68. There are practically important correlations between the viscous-elastic behavior of glass and its mechanical properties, namely those commonly called *grindability* or *abradibility*, and (micro-) *hardness*. J. Gypser[108] studied the grinding wheel response of glasses of widely different chemical compositions and of different soften-

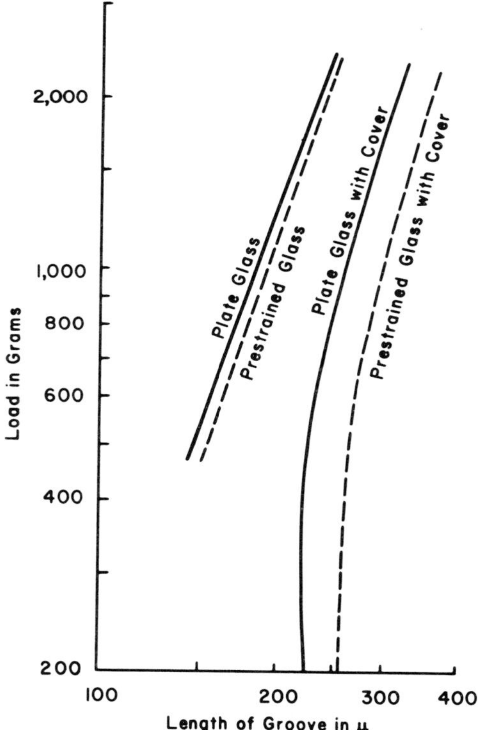

Fig. A. 58. Double-cone diamond pointer microhardness of plate glass, in different states of thermal history. (Grodzinski).

[108] *Silikat Tech.*, **6**, 1955, 372-377.

ing qualities. There is a strong dependence of the abradibility of the speed of the grinding wheels, every glass having an optimum speed under constant experimental conditions. P. Grodzinski[109] used the double-cone shaped diamond pointer (in the place of the Vickers pyramid pointer of the standard microhardness tester) in an extensive critical examination of the optimum conditions for reproducible microhardness determinations, for plate glass and later for a Crown glass, in the quenched and annealed states, under different loads. In the determination of the surface penetration marks, the effect of an *elastic resilience* after unloading (cf. A. ¶ 43) was eliminated by covering the glass before scratching with a layer of soot. Graphs of the lengths of the scratching marks (in microns) as a function of the applied load (in grams) show in this technique considerable differences in both cases (cf. Fig. A. 58). The curves for chilled glass (above 200 gm.) show a *lower* hardness than that of normal plate glass which is also less elastic. Whereas in the pure glass surface only plastic deformation governs the scratching process, the soot-covered surface corresponds to plastic + elastic deformation conditions. If the load is less than 150 gm. the chilled glass sample appears *harder* than the annealed glass. The hardness curves intersect at a characteristic load, between 150 and 200 mg.

VISCOSITY AS A FUNCTION OF TEMPERATURE

69. From the purely practical viewpoint nomograms and tabulations of viscosity-temperature data for sodium-calcium silicate glasses are given in a very convenient form, based on the older data of M. V. Okhotin,[110] and many other authors. V. T. Slavyanskiĭ[111] and M. M. Skornyakov[112] discussed especially the manner in which viscosity-temperature functions can be plotted with the greatest efficiency, and how to bring the viscosity data into line with the actual opinions on Liquidus temperatures in silicate systems, in analogy with the model systems of organic compounds (salol, betol, etc.) for which G. Tammann (1903) first derived the rules of temperature functions of viscosity. B_2O_3 is, in this understanding, *not* a true model of silicate glasses in comparison with other substances because of its entirely anomalous curve (Fig. A. 59). These considerations show impressively that the idea of model analogies must not be interpreted too widely because of fundamental differences existing between real polymorphic effects observed in many of the simple models and the

[109] *Radex Rundschau*, 1953, 3-8; *Glastech. Ber.*, **26**, 1953, 309-310; **28**, 1955, 58; for information on the double-cone pointer tester, see the Bergsman British Patent No. 651,800 (1952).

[110] *Silikat Tech.*, **5**, 1954, 503-504, with two plates of nomograms.

[111] *Stroenie Stekla, Inst. Khim. Silikatov Akad. Nauk S.S.S.R., Trudy Soveshchaniya, Leningrad*, 1953 (Publ. 1955), 251-255.

[112] *Stroenie Stekla, Inst. Khim. Silikatov Akad. Nauk S.S.S.R., Trudy Soveshchaniya, Leningrad*, 1953 (Publ. 1955), 256-257.

structurally much different silicate glasses. Skornyakov also applied simple interpolation formulas like $\log \eta = a + b/T^2$ (as recommended by K. S. Evstrop'ev) for the latter systems, and interpreted the break points observed (see below) as discontinuities caused by the Liquidus equilibria. Again, from the practical viewpoint of the difficulty of routine viscosity determinations of commercial glasses,

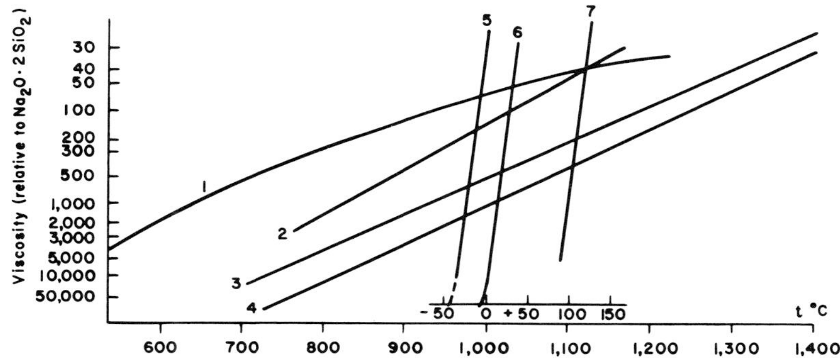

FIG. A. 59. Comparison of the viscosity-temperature correlation of typical silicate glasses (curves 2, 3, 4), organic model glasses (salol, curve 5; betol, curve 6), a fused salt ($NaNO_3$, curve 7), and B_2O_3 glass (curve 1). (Slavyanskiĭ).

A. K. Lyle[113] gave simplified correlations as characteristics of the *working* qualities, with chosen examples, such as introducing the relative machine speed, M, in relation to the annealing point, A, and the softening temperature, S, the strain point, and other parameters, in a relation developed by T. C. Vaughan of the form

$$M = \frac{(S - 450) \cdot 100}{(1 - A) + 70}$$

valid for M between 95 and 130 (95 to 115 for container glass) (Fig. A. 60). Other formulas can be applied for the feeding, or the "gob" temperature, a devitrification index, and such working range being defined as the difference $S - A$.

70. More recently, J. Cornelissen, J. van Leeuwen, and H. I. Waterman[114] determined the *kinematic viscosity*, $\nu = \eta/\varrho$ by an interpolation formula of the classic type $\log \nu = B + A/T \cdot x$, especially for fused alkali silicate glasses in the range from 1000° to 1700°K. It is equally valid for sodium-calcium silicate, borosilicate, and lead silicate glasses, in the range from 455° to 1413°C. A more detailed mathe-

[113] *Ind. Eng. Chem.*, **46**, 1954, 166-170. It is interesting that the first practical interpolation curve of the form $\log \eta = a + b/T$ was developed by O. Reynolds, *Phil. Trans. Roy. Soc. London Ser. A.*, **177**, 1886, 157.

[114] *Génie chim.*, **75**, 1956, 116-121; *Chim. & ind. (Paris)*, **77**, 1957, 69-78; **80**, 1958, 398-401.

matical analysis of viscosity-temperature data for sodium and lead silicate glasses was given by M. Kh. Karapet'yants,[115] and by O. Knapp[116] who confirmed the "penetration point" previously defined by A. Dietzel and R. Brückner (cf. A. ¶ 28), for the purpose of calculating the complete viscosity-temperature characteristics of a given glass on the basis of the Vogel-Tammann-Fulcher equation (see below).

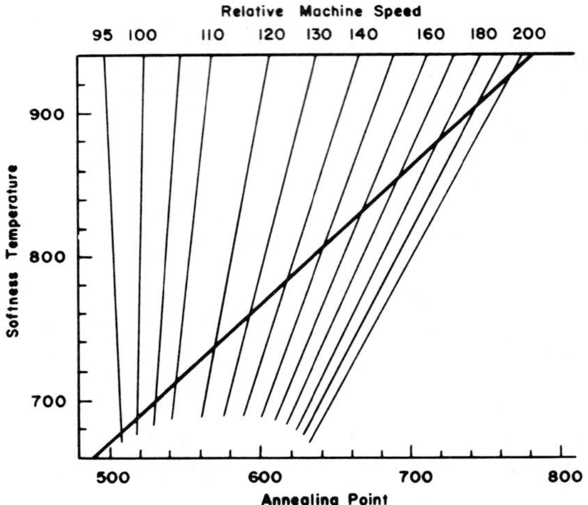

FIG. A. 60. Plot of the annealing points versus softening temperatures of commercial glasses. (Lyle).

Other approximations to linear interpolation equations are those of J. R. Poole (1949): arc sin h $(\log \eta) = A + B/T$, and an equation based on a theory of viscous liquids given by T. A. Litovitz: $\log \eta = A + B/T^2$, with a highly satisfactory approximation to experimental data of previous literature.

71. For sodium silicate glasses, E. Eipeltauer and G. Jangg[117] found, by precision measurements in the concentration range from 22 to 45 per cent Na_2O and for the isotherms from 900° to 1300°C., the Vogel-Tammann-Fulcher equation satisfactory, as the most accurate existing approximation and interpolation formula, especially in the range for $\log \eta = 1$ to 9, and approximately up to $\log \eta = 13$. Highly interesting is the derivation of this formula from the premises of the van der Waals equation of state. A theoretical curve corresponding to Fig. A. 61 is shown expressing

[115] *Steklo i Keram.*, **15**, 1958 (1) 22-25.

[116] *Glastech. Ber.*, **31**, 1958, 94-95; on the theory of T. A. Litovitz cf. *J. Chem. Phys.*, **20**, 1952, 1088-1089, 1980.

[117] *Kolloid-Z.*, **142**, 1955, 77-84; **143**, 1955, 83-92. Se also previous ideas of E. Berger (1932) on concepts of molecular mobility.

the Vogel-Tammann-Fulcher relation as a continuous function, whereas the actual experimental data yield a discontinuous curve, with the "association temperature," T_A, the "Liquidus" temperature, T_2, and the softening temperature, T_F (Fig. A. 62). The validity of this equation is made evident anew by E. Eipeltauer and A. More[118]

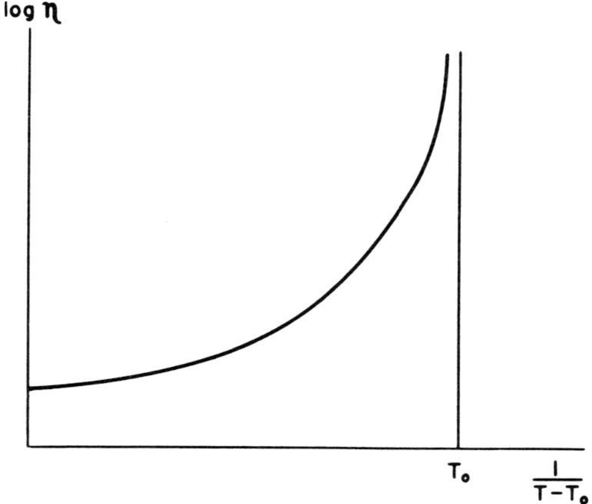

Fig. A. 61. Theoretical viscosity-temperature curve, representing the Vogel-Tammann-Fulcher formula. (Eipeltauer and More).

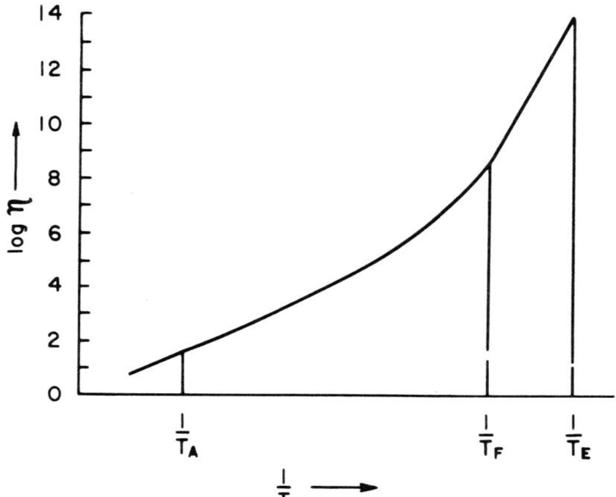

Fig. A. 62. Actual experimental viscosity-temperature curve, for sodium silicate melts and glasses. (Eipeltauer and More).

[118] *Radex Rundschau*, 1960, 292-300. See also the formula developed for mineral lubrication oils by U. Rost, *Erdöl u. Kohle*, 1955, 549-552, 650, 718-722; *Kolloid-Z.*, **142**, 1955, 132-150, which is identical with the Vogel-Tammann-Fulcher equation if the empirical constant $B = T_0$. We further emphasize the importance of these factors in the theoretical basis given by A. Dietzel

also for potassium and lithium silicate glasses. The temperature T_0 in the equation is defined as that temperature at which the molecular mobility discontinuously stops.

72. The repeatedly mentioned basic aspects of the viscosity mechanism as developed by H. Eyring[119] culminates in the conclusion that the activation energy (at constant volume) is composed of two portions, namely the energy, E, needed to form a "hole" into which the particles may move, and an additional energy, ΔE, required to detach the particle from its surroundings and to move it into the hole. At high temperatures the energy expenditure required for the formation of "holes" will be expected to become small whereas at low temperatures the energy will be large because the rupture of individual structural bonds is required to open the holes. In Eyring's formulation the expression for activation energy η of viscous flow in silicate glasses will be written as

$$\eta = A' \cdot e^{E/RT} \cdot e^{\Delta E/RT}$$

for low temperatures, and

$$\eta = A'' \cdot e^{\Delta E/RT}$$

for high temperatures, with A′ and A″ as constants.

73. In another formulation by R. W. Douglas the analogous expressions are in general

$$\eta = AT \cdot e^{B/T} \cdot (1 + Ce^{D/T})$$

which at high temperatures will be

$$\eta = A' \cdot T \cdot e^{B/T} \quad \text{with} \quad A' = A\ (1 + C \cdot e^{D/RT})$$

and at low temperatures

$$\eta = A'' \cdot e^{G/T} \quad \text{with} \quad G = B + D, \quad \text{and} \quad A'' = A \cdot C \cdot T$$

Though developed from different theoretical bases these expressions are entirely analogous to those of Eyring's equations. They postulate that two independent terms are required, one of which becomes predominant at high temperatures, the other predominates at low temperatures. At high temperatures the viscosity of silicate glasses depends on the nature of forces between the constituent atoms, whereas at low temperatures the viscosity is largely dependent on the arrangement of the atoms in the nearly established glass structure.

and R. Brückner, *Glastech. Ber.*, **28**, 1955, 455-467; especially pp. 462 ff.; *Ibid.*, **30**, 1957, 73-79, especially pp. 77-78.

[119] Cf. A. G. F. Dingwall and H. Moore, *J. Soc. Glass Technol.*, **37**, 1953, especially pp. 332 ff.; see further R. W. Douglas, *Ibid.*, **33**, 1949, 138-162.

74. The reality and the physicochemical significance of *break points* on the log η versus temperature lines were early discussed by H. Lechatelier (1924), more recently by J. Herbert,[120] and interpreted by an elementary delineation of the plastic and the fluid states of glass. The intersection point of the two linear branches is given as a temperature, T_2, and by $\log (10^{13}/\eta) = 7.5$, or by $\log \eta = 5.5$, a constant for widely variable glass compositions. Only glasses with high contents in B_2O_3 and Na_2O behave anomalously. The temperature T_2 together with the transformation point is remarkably near the mechanical glass working range and the Littleton point with $\log \eta = 5.0$. Herbert writes for the plasticity range an empirical formula $\log (10^{13}/\eta) = A \cdot \log (T/T_1)$, with T_1 as the temperature for $\log \eta = 13$, and for the fluidity range the linear relation $\log (10^{13}/\eta) = B_1 \log (T/T_1) + C$. T_1 is determined by a simple device in which the starting yield of a cylindrical glass sample to a pointer of nonscaling steel is measured. If the Littleton point ($\log \eta = 5.0$) is known the constant A is simply defined by $A = (\log T_2 - \log T_1)/5.35$, and B is determined by the viscosity measured with the sphere-drawing method. As a good approximation one can use: $\log T_2 = (A \cdot \log T_1 + 7.5)/A$. T_2 practically coincides with the temperature of the maximum rate of devitrification. The anomaly of B_2O_3 glass ($T_2 = 436$°C., with $\log \eta = 8.84$) and of borax glass ($T_2 = 742$°C. with $\log \eta = 11.48$) is explained by Herbert with the reinforcement of bindings of the structure in the transition of $[BO_3]$ groups into $[BO_4]$.

75. E. Plumat[121] critically analyzed common methods for determining viscosity at high and at low temperatures (rotation, and fiber-elongation methods). He developed for the transition range with $\log \eta = 7$ to 9, in which both classical methods cannot be useful, the new viscosimeter discussed in A. ¶ 32. For simple sodium silicate glasses the linear branches of the log η *vs* temperature functions intersect again in distinct break points, T_1 and T_2, the reality of which is in agreement with the experimental data given by C. L. Babcock (1934) and L. Shartsis, S. Spinner, and W. Capps (1952), with discontinuous changes in the activation energies of flow and the "structural coefficient" of viscosity.

76. In a certain contrast with most of the theoretical aspects of changes in activation energy during the transitions from the elastic to the plastic, and to the fluid state of glass, in which ruptures of valence bonds were made responsible for the mechanism of flow R. L. Myuller[122] is of the opinion that there are only structural

[120] *Verres et réfractaires*, **8**, 1954, 74-88.

[121] *Silicates inds.*, **21**, 1956, 391-396, 447-457; the new viscosimeter discussed in this investigation is that of E. Brichard and E. Plumat, *Verres et réfractaires*, **9**, 1955, 69-81.

[122] *Zhur. Priklad. Khim.*, **28**, 1955, 363-371; 1077-1087. On this basis, V. T. Slavyanskiĭ, *Structure of Glass. Proc. All-Union Conf. Glassy State, 3rd, Leningrad* (English Translation), 1959/60, 289-291, calculated the activation energies of viscous flow in different systems, e.g., SiO_2–Na_2O, PbO, and –B_2O_3–Al_2O_3–SiO_2. Whereas Myuller takes only Si–O bonds into account, Slavyanskiĭ considers also the bonds Me–O.

regroupings of atoms, and relative displacements by exchanges of valence-chemical bonds among the atoms. The principal point of Myuller's discussions is the much higher activation energy, E_d, of a rupture in the structure, say of silica or B_2O_3 glass, than that to be expected only for the valence bond exchange in the transformation phenomenon, E_t. The activation energy of fluidity, E_η, should be in the following correlation to the energy of activation of valence vibration, E_ν: $E_\nu < E_t < E_\eta < E_d$. The fluidity of silicate glasses as a function of temperature is then expressed by the equation $\log (1/\eta) = - [E_\eta(T)]/2.30 \cdot RT + B$ (with $B = 1.30$ for B_2O_3 glass, and $= 3$ for silica and lead silicate glass), and more explicitly

$$\log (1/\eta) = - \frac{E_\eta (T)}{2.30\, RT} + \log \left(\frac{\nu\, \delta^2}{nkT\Delta y} \right)$$

with $n =$ the number of valence bonds/cm² cross section. Myuller further defends the opinion that η is chiefly determined by the content of nonionized oxide in the glass. It distinctly depends on the packing density of valence bonds. For the temperatures between 773° and 1673°K. Myuller calculated the thermodynamic data for E_η, S_η, and H_η as a function of temperature in the form

$$\log (1/\eta) = - \frac{\Delta H_\eta (T)}{2.30\, RT} + \frac{\Delta S_\eta (T)}{2.30\, R} + B$$

and confirmed the relation $E_\nu < E_t < E_\eta < E_d$. The characteristic physical states of the glass are then: (1) the low-temperature state of rigidity of the valence bonds; (2) the labile state of valence vibrations and shifting valence bonds, accompanied by regroupings of atoms; (3) the liquid state of maximum disordered vibrations and atomic regroupings by the non-entropy transformation of the bonds.

IMPORTANCE OF VISCOSITY IN NATURAL SILICATE MELTS

77. The theory of volcanic effusions, specifically lava silicate melts, is still rather unsatisfactory in respect to our knowledge of its physicochemical basis. On the influence of *pressure* on viscosity at elevated temperatures we are still dependent on conclusions from the more or less analogous relations for "model systems." We may, however, to a certain degree conclude from experiments of J. H. Heiks and E. Orban[123] that for benzene under its own vapor pressure the variation of $\log \eta$ with the reciprocal of temperature is a straight line up to 180°C., but above this temperature the rate of change increases with increasing temperature and pressure from 10.5 to 50 kg./cm.² (at the critical point, with 288.5°C.). The method used is interesting because in the common sphere-falling tube (of stainless steel) a steel

[123] *J. Phys. Chem.*, **60**, 1956, 1025-1027.

plummet was used into which a radioactive Co^{60} source was inserted as the position indicator. The Reynolds number was in most of the measurements only about 100, in maximum 500, or far enough below the limit of turbulence in the liquid. The timing assembly consisted of commercial electronic counter tubes and circuits for an accurate determination of the time period of the fall of the plummet over a distance of 50 cm. With similar somewhat modified experimental conditions it might be possible to derive the actually lacking information on pressure effects on the viscosity of model silicate systems to be compared with magmatic melts.

78. Nevertheless, there are more recent observations of volcanological interest which may shed new light on the changes of viscosity of lavas during, and after, the paroxysmatic stage of effusions. We mention in this connection a valuable report of F. von Wolff[124] on the great eruption of Mt. Vesuvius of the year 1906, which took place under the hydrodynamic conditions of periodically clogged volcanic channels. A complete calculation of the flowrate and viscosity as well for the silicate liquid as for the gaseous phase was attempted for a constant temperature of 1050°C. The Bosco di Cognoli lava current had an effective viscosity of $\log \eta = 4.94$, the gas phase $\log \eta = -3.47$, whereas the degassed lava at the same temperature had a $\log \eta = 6.70$. Even if these data may appear relatively high for the average volcanic lava streams, the calculations of von Wolff show at least the order of magnitude and the possibility of having a more accurate opinion on conditions ruling in the volcanic throat, such as the hearth temperatures, and other factors which are so important for an understanding of possible assimilation reactions in contrast with the previous mere estimates made of the depths and the pressure-temperature conditions of volcanic hearths. The Vesuvius hearth had, as a best approximation, a temperature of 1300°C. and a gas pressure of 580 kg./cm.^{2}, with a lava load of 1000 m. at the outflow (with a temperature gradient from 1300° to 1050°C., a pressure gradient of 580 to 1 atm., and a caloric energy content of 3.35×10^{10} cal., released to the surface).

79. Volcanic eruptions may be interpreted as a special *magmatic differentiation* phenomenon of liquid melts and gas phases. As differentiations of liquid to liquid phases we usually classify the *liquation* (unmixing) processes analogous to those observed in borosilicate melts of the Vycor type (A. ¶ 239), but now restricted to petrogenic system units. We recall in this respect the classical discussion of F. Yu. Loewinson-Lessing[125] on the variolites of Yalguba, Kareliya, with a distinct splitting

[124] *Fortschr. Mineral.*, **26**, 1950, 131-134. On temperature and orienting viscosity measurements of Hawaiian volcano lavas of basaltic type see also the report of G. A. MacDonald, *Ibid.*, **33**, 1955, 133-140. Average temperature data are around 1100°C. the Aa lavas having a certain mobility in flowing over smooth hill slopes down to 700°C. The viscosity in the lava lake of the Mauna Loa 1952 eruption was 3×10^3 to 1×10^4 poises at 940°C.

[125] *Trudy Mineral. Petrograf. Inst. im. Levinson-Lessinga, Akad. Nauk S.S.S.R.*, 1935 (5) 21-27. See also C. N. Fenner (1938) on immiscibility in magmas, and critical remarks of A. Holmes and Fr. Hecht, *Mineral. Mag.*, **24**, 1936, 408-421, especially p. 419 f.

of a formerly uniform melt phase to two immiscible liquids. Even if there may be other interpretations of this most striking phenomenon, say by quartz xenoliths "transfused" into a volcanic glass, D. S. Belyankin[126] discussed these problems of liquation especially in highly siliceous rock melts, with an application of the synthetic experience in the system Na_2O–SiO_2–H_2O (O. F. Tuttle and J. J. Friedman, 1948). Although the existence of such true liquations cannot be doubted, Belyankin is of the opinion that in nature they are restricted to local phenomena as observed in the variolites and some porphyries, but not of a large-scale petrogenetic and therefore geochemical importance.

80. In the strongly unmixed systems observed in perlites from Arizona, W. D. Keller and E. E. Pickett[127] particularly studied the simultaneous loss of absorbed water which brings about a tremendous increase in viscosity of the highly siliceous melts to form characteristic glass spherules (cf. more in details, Vol. IV, A). In this respect the discussions of W. A. Cassidy[128] on the origin of the tektites in comparison with other interplanetary chondritic and achondritic meteoric stones are also highly interesting. Also in this case a liquid immiscibility phenomenon is the most plausible reason for observed contrasts in the chemical composition ranges of these petrogenic types, with the extremely siliceous tektites on one side, the basic chondrites on the other side. A small planet might have existed with a chondritic core, and achondritic subcrust, and a tektitic crust. This outer layer was above its liquidus temperature, thus completely liquid. Upon disruption from the planet it separated into clouds of droplets adopting the rotational shapes before quenching to tektite glass (cf. Vol. III, B. ¶ 41 f.).

81. The high importance of magmatic differentiation of basic and ultrabasic rock melts under the action of viscosity-temperature conditions is richly discussed in geological and geophysical literature. We mention in this respect only the fascinating descriptions of differentiated *ore magmas* from Japan, by A. Matsubara[129] (see C. ¶ 1). In the mines described cupriferous pyrite was segregated in melts from abyssal reservoirs in three liquid phases of restricted mutual miscibility. The ore bodies have the characteristic rounded margins and smooth slippery surfaces to be expected from such a liquation process, in fluidal textures. In every case the "ore solution" phase must have been of high viscosity, unable to exert any appreciable chemical reaction with the wall rocks and inclusions of rock fragments. This interpretation is in agreement with metallurgical observations of the jellylike, pasty state of similar unmixed mattes, if they are very low in dissolved volatiles, especially water.

[126] *Izvest. Akad. Nauk S.S.S.R.*, Geol. Series 1949, No. 5, 35-39. See also the extensive discussions by V. A. Nikolaev, *Zapiski Vsesoyuz. Mineral. Obshchestva*, **80**, 1951, 3-14.

[127] *Am. J. Sci.*, **252**, 1954, 87-98, especially p. 91.

[128] *Geochim. et Cosmochim. Acta*, **14**, 1958, 304-315.

[129] *J. Geol. Soc. Japan*, **59**, 1953, 79-87.

82. The striking variation in chemical composition of the lavas of the Paricutin volcano, Mexico, over its relatively short eruptive activity was discussed by R. E. Wilcox.[130] The variation is thoroughly gradual and surprisingly smooth, but after all, Wilcox concludes that fractional crystallization alone cannot account for these chemical changes. More fitting to global chemical changes is the assumption of a combined action of fractional crystallization with bulk assimilation of included xenolithic material. The heat of crystallization would not be adequate to accomplish the large amount of assimilation implied by the analytical data observed, whereas sufficient heat would have been furnished by the incremental crystallization of a magma circulating by thermal convection in a cupola above a magma body of basaltic type. The progressive changes in the Paricutin lavas are from 55 per cent silica in olivine-bearing basalt-andesite (1943) to more than 60 per cent in olivine-pyroxene andesite erupted in 1958.

83. The classical problem of *dunites* and of *olivine nodules* in basaltic effusive rocks was extensively investigated by Cl. S. Ross, M. D. Foster, and A. T. Myers[131] (cf. Vol. III, Section B). These formations show a surprisingly uniform character and mineralogical composition. After all there is no doubt that they are derived from deep-seated products, brought up from the profound depths of the peridotite zone of the Eearths' crust, by orogenic forces in dunite bodies, and as single olivine nodules from the same hearths by basaltic magmas but not as their differentiates. This is convincingly demonstrated in the distribution of the major constituents and of characteristic trace elements determined by quantitative spectrometry methods.

84. An example of true fractional crystallization differentiation, however, is that described by E. D. Jackson[132] for an ultramafic zone of the Stillwater Complex, Montana, with distinctly settled accumulations of peridotite minerals (olivine, chromite, orthopyroxene) in euhedral crystals, and interstitial gabbroic material (ortho- and clinopyroxene, plagioclase). A good example of crystal fractionation is also the practically monomineralic plagioclasite described by M. Galli[133] from the Western Alps, originated from pillow diabases, in intimate association with peridotite and gabbro.

85. In a special study of the crystallization of basalts from Schonen (Sweden), H. G. F. Winkler and G. Weitz[134] made extensive observations on the systematic crystal growth of plagioclase in the salbands and in the central portions of a sill as

[130] *Science*, **119**, 1954, 515-516.

[131] *Am. Mineralogist*, **39**, 1954, 693-737.

[132] *Am. Mineralogist*, **40**, 1955, 322-323.

[133] *Periodico Mineral.* (*Rome*), **28**, 1959, 1-81. As a particularity of these plagioclasites, their high content in albite must be emphasized, in contrast with Norwegian anorthosites.

[134] *Geol. Fören. i Stockholm Förh.*, **78**, 1956, 619-641.

a distinct manner of the advancing fractionation, with intermediate transitions from the high-temperature to the low-temperature optical characteristics (cf. Vol. III, Section B). The mineralogical data and chemical analyses make evident that fractional crystallization is in agreement with N. L. Bowen's theories, and especially with the formation of an alkali feldspar-enriched residual melt of granitic composition (with about equal amounts of orthoclase, albite, and quartz). The intrusion temperature of the magmatic sill may have been 1170°C. in maximum, the crystallization of the anorthite-rich plagioclase (60 per cent An) starting at about 1010°C., that of augite at 985°C., of magnetite at 960°C., of feldspar $Ab_{50}An_{50}$ at 885°C., of $Ab_{70}An_{30}$ at 810°C., and of the residual melt below 750°C.

86. E. Tröger[135] tried to connect the experimentally known viscosity values of basaltic rocks (cf. Kani, Volarovich *et al.*) with the petrochemical character of effusive rocks expressed, as in the Niggli parameter *si* (molecular ratio of SiO_2 to the sum of the basic oxides, including alumina × 100). Such calculations which of course only concern the mineralogical composition without the water content (dry crystallization) give a certain orientation on the petrographic province (pacific or atlantic) character of the effusive melts. Sodium-rich rocks in this calculation would have a very high viscosity as dry melts, whereas they are particularly rich in volatiles and therefore crystallize better than basic rocks of the pacific series, evidently from a liquid of higher fluidity.

87. N. N. Ormont[136] described basalts and diabases in so-called "asymmetric phase diagrams" as of the system diopside-albite-anorthite (cf. C. ¶ 8) in their crystallization from fluid liquids, such as basalts from the Rhinelands which are particularly suitable for production of synthetic cast pavement stones. Less suitable or even unsuitable are basalts with primary plagioclase, of high viscosity and forming much glass. The usual grouping of such rocks as raw materials for artificial stones according to their silica contents (in weight per cent) is often entirely misleading. The best-suitable basalts for this purpose are all projected in Bowen's diagram of the system diopside-albite-anorthite in the field of primary diopside (pyroxene). Olivine is always subordinate. In complete agreement with these conclusions are the characteristics given by J. Voldán[137] for Bohemian basalts used for the purposes of "petrurgy" (the technology of rock material castings for pavement stones or chutes of high resistance to abrasion, etc.), and further reference by E. Eipeltauer and F. Thenner[138] to Austrian basalts for the same purposes. A slight addition of magnetite and soda increased the mechanical durability of such cast products, such as of chutes, by a considerable increase in the uniformity of the texture.

[135] *Fortschr. Mineral.*, **26**, 1950, 124-125.

[136] *Vestnik Moskov. Univ. Ser. Fiz. Mat. i Estestven. Nauk*, 1950 (3) 117-133; 1951 (11) 107-121.

[137] *Sklář a keram.*, **5**, 1955, Nos. 1 and 5; *Silikat Tech.*, **7**, 1956, 48-53.

[138] *Radex Rundschau*, 1958, 224-234.

88. Other remarkable attempts to make useful molten vitreous or recrystallized basalts were made by A. V. Abramyan[139] who drew thin rods from the melts and determined their glass-technological properties. Such glasses produced from magmatic melts of higher fluidity show a much higher thermal expansion than do those of higher viscosity. Reducing conditions in the fusion process bring about thermal expansions which are lower than those of glasses molten in a neutral or oxidizing atmosphere. The annealing temperature, T_g, is for basalts of industrial casting quality about 610 to 625°C., the deformation temperature (determined by the method of G. Gehlhoff and M. Thomas, 1926) for glass produced from neutral fusion conditions is 740° to 775°C., for that molten under reducing conditions 720° to 755°C.

89. Arecent attempt to determine directly the viscosity of rock and silicate melts by a dynamometric torsion viscosimeter was made by R. Euler and H. G. F. Winkler,[140]

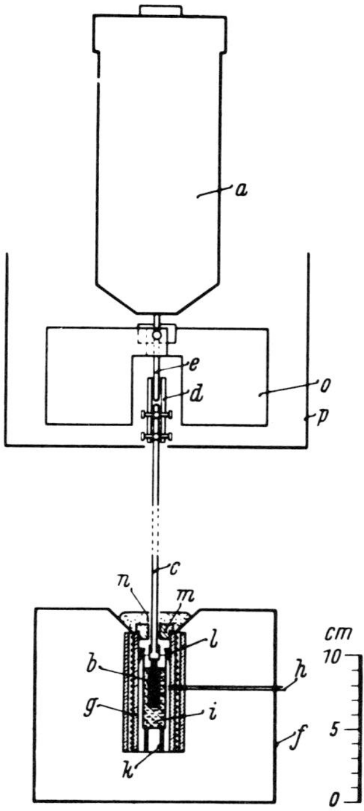

Fig. A. 63. Drage torsion viscosimeter for the determination of the viscosity of molten silicate rocks. (Euler and Winkler). (*a*) motor with dynamometer; (*b*) Pt-20% Rh rotor; (*c*) alumina rod; (*d*) adjustment head; (*e*) axis; (*f*) Pt resistor furnace; (*g*) alumina tube; (*h*) thermocouples; (*i*) Pt crucible; (*k*) alumina tube; (*l*) alumina wedges; (*m*) porcelain plug; (*n*) silica glass wool insulation; (*o*) aluminum sheet vane; (*p*) aluminum sheet pot.

[139] *Izvest. Akad. Nauk Armyan. S.S.R., Fiz. Mat. Estestven. i Tekh. Nauki*, **9**, 1956, (2) 13-21.
[140] *Glastech. Ber.*, **30**, 1957, 325-332.

over the range from 10 to 4,000 poises (cf. Fig. A. 63). All of the rock melts behaved as Newtonian liquids, without any anomalies indicating structural viscosity effects (Fig. A. 64), including kersantites, andesites, basalts, and trachytes, if fused to full homogeneity and without any gas bubbles. The results are in general agreement with the previous determinations by M. P. Volarovich *et al.* (1936), but those of

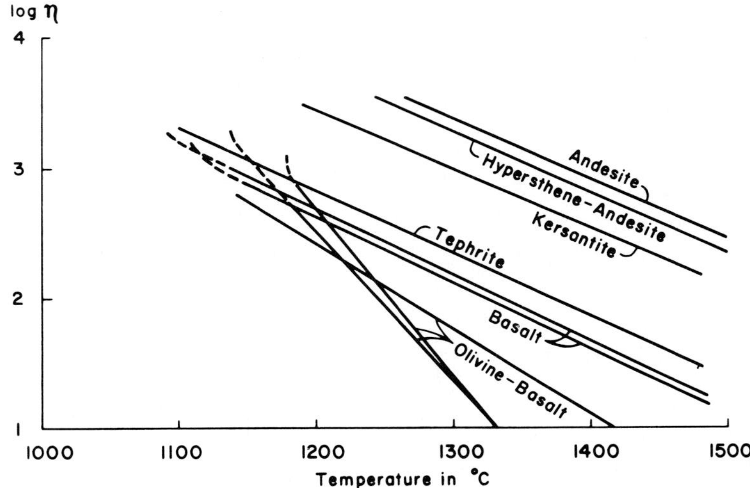

FIG. A. 64. Viscosities as a function of temperature for some characteristic effusive rock melts. (Euler and Winkler).

K. Kani (1934) are often too high by one order of magnitude and the log η versus temperature curves are flatter. The slopes of the lines are steepest for olivine basalt melts, crystallization of magnetite and olivine are immediately indicated by humps on the curves. The activation energies of viscous flow are rather constant = 50,000 cal./mole, in the range of the ratio of the atomic per cents (Si + Al) : O = 0.40 to 0.45, in comparison with 30,000 to 40,000 cal./mole for calcium and sodium silicate melts, with the ratio Si : O = 0.30 to 0.45 (cf. Fig. A. 65). These facts are in agreement with the theory of J. O'M. Bockris and D. C. Lowe[141] on existence of separate aluminosilicate complex anions and not of networks or frameworks in the structure of the melts. If (Si + Al) : O is below 0.40 a wide scattering of activation energies is observed, indicating specific effects of different cations in the melt.

90. In a close relation to the phenomena of viscous flow are also those of *diffusion* and *convection* in magmatic, or industrial silicate melts. Equally under geochemical

[141] *Proc. Roy. Soc.* (*London*), 226 A, 1954, 423-435.

and technological aspects, thermal diffusion plays an important role, as in the homogenization of large glass melts in modern tank furnaces (cf. B. ¶ 43). Omitting here the technical details of the glass flow in such a tank furnace we may mention the studies of H. Shimada[142] on special conditions in an electrically heated model tank which offers much simpler working conditions than do the large gas-heated units of industry. The indicators for the convective material transfer may be either strongly

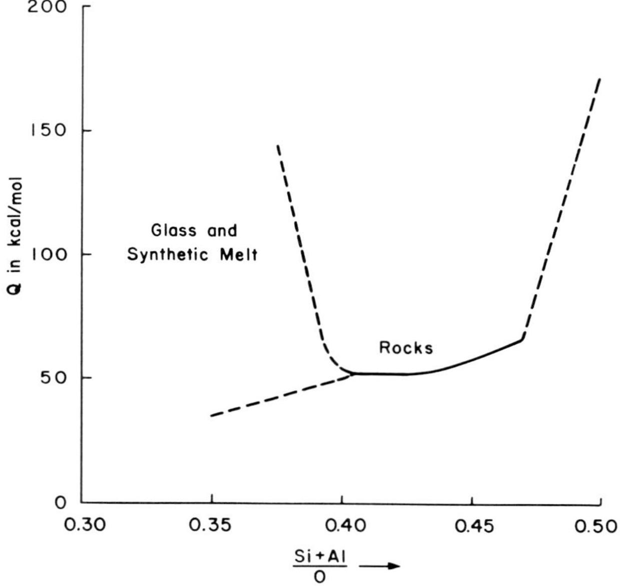

FIG. A. 65. Activation energy, Q, of viscous flow of silicate rock melts, as a function of the ratio (in atomic per cent) (Si + Al) : O. (Euler and Winkler).

colored additions (mostly cobaltum), or radioactive tracers, especially Na^{24}. Shimada's results with Co show coloration effects, especially the whirl formation in the melt. Beyond these qualitative observations he was able to calculate the apparent diffusion coefficient, D', for the change in concentration with time for a specific constituent of the glass, and the true diffusion coefficient by an equation of the general form $dC/dt = D(C + V \text{ grad } C)$ where V is the flow rate of the glass. Although

[142] *Glastech. Ber.*, 27, 1954, 157-159. On physical premises for the use of *radioactive* tracers for such purposes see L. Meyer-Schützmeister, *Naturwissenschaften*, 37, 1950, 501-517, with most valuable tabulations of numerical data. Because of their adequate half-life periods, Na^{24}, K^{42}, Ba^{131}, and As^{76} are recommended for studies with silicate melts. Using the secondary color changes in glass layers by the diffusion of Cu^{+} ions into glass, E. A. Ivanova, *Structure of Glass, Proc. All-Union Conf. Glassy State, 3rd, Leningrad* (English Translation), 1959/60, 241-243, determined the diffusion coefficient as a function of the alkali contents of the glasses.

the true diffusion coefficient cannot be separated from the convectional effects, the apparent coefficient D' could be calculated numerically from another equation of the form $dC/dt = D' \cdot d^2C/dx^2$ for the coordination axis x in the flow direction. The explicit solution of the equation is possible for special boundary conditions with a certain approximation. The best condition is the assumption of a concentration gradient = 0 on the surface of the melt. The apparent diffusion coefficient was then computed for CaO = 1.4×10^{-2} m.^{2}/day, and for ZnO = 1.26×10^{-2} m.^{2}/day, in formal agreement with previous data of I. Sawai (1952) for a similar model tank furnace with $D' = 2.8$ to 7.5×10^{-2} m.^{2}/day (cf. B. ¶ 43*).

91. A very fascinating report of I. Peychès[143] on the use of radioactive tracers for glass tank problems, especially that of combined diffusion and convection in large bodies during homogenization, and others, cannot yet demonstrate a distinct supremacy of such methods, in spite of their great elegance, over the elementary observation of color indicators. Too many difficulties of a most diversified kind are a hindrance for the safe application of radioactive tracer tests of this kind. Not only Na^{24}, but also Sr^{89} with a considerably longer half-life period might be used, but the latter, nevertheless, is better reserved for studies of distribution of glass lubricants in metallurgical extrusion techniques, or that of polishing processes. Peychès labeled, in an interesting plant experiment, 25 kg. of a glass batch with 20 gm. of a Na^{24} containing Na_2CO_3 sample (radiation intensity 30 mc.) and introduced this soda with the batch into a glass furnace. At regular time intervals the activity of samples taken from different spots on the glass surface and from different depths of the glass bath were measured by common counter tubes. Thus details of the distribution of the tracer over the whole glass body were hoped to be determined with accuracy, but evidently, the glass furnace did not homogenize the charge properly enough.

92. That thermal diffusion and the Ludwig-Soret phenomenon can as a matter of fact play an important role in the differentiation of magmatic intrusions and extrusions, and their derivatives, was recently demonstrated by T. F. W. Barth[144] in a study of aplitic veins. The chilled margins of such an intrusion body display a distinctly increased concentration in alkalies and silica, but a lower concentration in Mg, Ca, Fe, and Al than that of the coarse-granular central portion of the same

[143] *Silicates inds.*, **17**, 1952, 241-246; *Glass Ind.*, **35**, 1954, 309-313, 340; see also H. Muth, *Glastech. Ber.*, **27**, 1954, 248-255, and more recently R. Fischer, *Silikat Tech.*, **11**, 1960, 559-561, on a plant experiment with interesting technological results on glass flow to shaping machines. Also K. Evtrop'ev, *Structure of Glass. Proc. All-Union Conf. Glassy State, 3rd, Leningrad* (English Translation), 1959/60, 237-240, studied the diffusion of alkali ions in silicate glasses by radioactive tracer techniques; in mixed Na-K silicate glasses the diffusion coefficient for Na^+, increases sharply with increase in K^+ ions.

[144] *Am. J. Sci.*, Bowen Volume, 1952, 27-36. The experiments of K. Clusius and G. Hückel are described in *Naturwissenschaften*, **26**, 1938, 546; **27**, 1939, 148-149, 486; *Z. physik. Chem. (Leipzig)*, **44**, B, 1939, 397-473.

vein over a distance of less than 1 m. The fractional crystallization should have differentiated the elements mentioned just in the opposite direction. The chilled margins must have congealed *before* the central portion, building up a strong temperature gradient which drives the material in the same way as is observed in the diffusion column used by K. Clusius and G. Hückel (1938, 1939) in their impressive experiments, and corresponding with the theory of W. Wahl (1946). It is also to be expected that the strong separation of ions by thermodiffusion may be accompanied by a distinct shifting of the isotopic composition of oxygen in natural rocks and minerals of the chilled marginal and central portions of a magmatic body.[145] Material transfer in temperature gradients of industrial furnaces, i.e., changes in the composition of refractories in service, were observed on a similar basis by N. Skalla and F. Trojer,[146] exemplified by a silicate migration in chrome-magnesite roof bricks.

SURFACE TENSION PHENOMENA[147]

93. The important problems of the solid state physics involved in the surface energies, σ, of *rigid* materials, crystalline or glassy, were experimentally challenged by V. D. Kuznetsov[148] in studies on mutual abradibility of crystals. The ratio σ_1/σ_2 for two different crystal faces can be determined by the ratio of the volumes of abraded material, V_1/V_2, in other words, by a quantitative determination of the mechanical energy used for displaying the removed layers of the crystal faces, by the equation $V_1/V_2 = \sigma_2/\sigma_1$. As a standard abrasive grain, green silicon carbide powder was used between faces of alkali halide crystals. In the same way, the ratio of surface energy of glass of different compositions is a valuable criterion of the relative surface characteristics of different glasses if one standard crystal face, say (100) of NaCl, is used in the abradibility test, as in the range of 380 to 980 ergs/cm.^{2} for potassium-lead silicate, sodium and sodium-calcium silicate, and borosilicate glasses, the last as the most surface wear-resistant glass type. The abradibility is different if the material is ground in the dry state, or with water, or ethyl alcohol as "lubricants," with different degrees of formation of swollen surface layers. Water decreases

[145] See to this interesting consequence S. R. Silverman, *Geochim. et Cosmochim. Acta*, **2**, 1951, 26-42, and the data given by R. J. Allenby, *Ibid.*, **5**, 1957, 40-48, on the distribution of Si^{28} and Si^{30} in silicate rocks, with a differentiation of the ratio Si^{28}/Si^{30} by up to 1.3 per cent.

[146] *Radex Rundschau*, 1955, 506-517.

[147] As a most useful introduction to the theory and special discussion of surface energy and tension phenomena we recommend the recent book by K. L. Wolf, "Physikalische Chemie der Grenzflächen," Vol. I, 1957, Vol. II, 1959, Springer-Verlag, Berlin, Göttingen, Heidelberg, 262 and 360 pp., respectively.

[148] *Doklady Akad. Nauk S.S.S.R.*, **84**, 1952, 927-930, 1151-1153; **85**, 1952, 761-764; **90**, 1953, 537-540. Theory of the Kuznetsov method is given by M. Born and O. Stern, *Sitzber. preuss. Akad. Wiss. Physik math. Kl.*, **48**, 1919, 901-913.

the relative surface energy by about 14 per cent, ethyl alcohol even by 33 per cent in the case of soft potassium-lead silicate glass.

94. The unequally more important factors ruling surface energy and surface tension of *liquids*, say of glass melts, connected with the fundamental problems of adhesive forces acting by short-range attractions, were critically discussed by D. W. Mitchell, S. P. Mitoff, V. F. Zackay, and J. A. Pask.[149] In the same way, common methods for determination of surface tensions were critically reexamined, making evident the uncertainties of methods based on fiber elongation or drop-weight (including G. Quincke's approximations for very small drops) which, however, may be helpful in many cases as routine methods because of their simplicity for relative comparisons. Better absolute results may be expected from the method of the pendant drop. The methods of the dipping cylinder are accurate, and also that of F. M. Jaeger characterized as the maximum bubble pressure method (with a correction given by N. E. Dorsey (1926), for the hydrostatic flattening of larger bubbles). The bulb method of W. B. Pietenpol and H. H. Scott (1930) on the other hand, is much more complicated because the viscosity of the liquid interferes with expansion of the bulb. The most "popular" method may remain the sessile drop determination. Valuable tabulations were given in the discussions of Mitchell *et al.* for surface tension data of typical glass melts from older literature, with the temperature coefficients of surface tension which, as a rule, are negative. The numerous cases, however, in which they are positive are connected with more or less anomalous constitution effects.

95. The *additivity* of surface tension as a function of the oxide composition and oxide increments is to be discussed later in B. ¶ 114 ff. Beyond his investigations of 1936 on this problem, A. A. Appen, with K. A. Shishov, and S. S. Kayalova[150] gave an excellent collection of new numerical data for the surface tension of silicate glasses, and a discussion of their additivity calculation, with rules for the specific surface tension decreasing effects of MoO_3 and WoO_3, less of V_2O_5 and CrO_3. Appen *et al.* preferred in their measurements the determination of the surface tension by the falling drop method, with corrections as given by W. D. Harkins and F. E. Brown (1919). Appen classified the oxides in silicates concerning their effects on the surface tension of the melts (starting from the partial molecular surface tension, $\sigma_1 = \sigma + (1 + \gamma_1) \cdot d\sigma/d\gamma_1$ with $\gamma_1 =$ the molar fraction of the oxide component) in the following way:

(1) Oxides for which the partial molecular surface tension, σ_1, is almost independent of γ. To this class belong the oxides SiO_2 ($\sigma_1 = 290$), TiO_2 (250), Al_2O_3 (580), BeO (390), MgO (520), CaO (510), SrO (490), BaO (470), ZnO (450), CdO

[149] *Glss Ind.*, **33**, 1952, 453-457, 482, 515-523.

[150] *Zhur. Fiz. Khim.*, **26**, 1952, 1131-1138, 1399-1404; *Silikat Tech.*, **4**, 1953, 104-105. See also A. Petzold, *Ibid.*, **5**, 1954, 11-12.

(430), MnO (490), CoO (430), NiO (400), Li_2O (450), and Na_2O (295), in ergs/cm.[2]. These are not surface-active components.

(2) Oxides with a small dependence of the partial molecular surface tension on γ, it may even be negative. Here belong K_2O, PbO, B_2O_3, Sb_2O_3, and P_2O_5 which all behave as components of intermediate surface activity.

(3) Oxides with a strong dependence of the molecular partial surface tension on γ, of negative sign. Here belong the typical surface-active components like As_2O_5, V_2O_5, WoO_3, MoO_3, CrO_3, and SO_3.

The oxides of classes (2) and (3) do not follow distinct additivity relations.

96. With the *maximum bubble pressure method*, and based on the E. Schrödinger equation (1915) recently L. R. Barrett and A. G. Thomas[151] simultaneously measured the surface tension and density (cf. A. ¶ 185) of melts in the system $CaO–Al_2O_3–SiO_2$. The data fit remarkably well to elementary linear functional relations of the surface tension as well as the density and temperature, without any discontinuities which might indicate the presence of definite molecular associations ("compounds") in the melts. The surface tension increases with both increasing alumina and silica contents. In the composition field studied the temperature coefficients are all positive. With exception of the compositions highest in alumina (30 to 40 per cent), there is a remarkable clustering of the lines intersecting surface tension versus temperature in a narrow range near 1650°C. and surface tension = 465 ergs/cm.[2] (cf. Fig. A. 66). Also L. Merker[152] used the same method (modified by a vertical shifting of the water-cooled platinum capillary under micrometric control) for sodium-calcium silicate glass melts. The rapidly increasing viscosity, however, limits practical application of the maximum bubble pressure method to temperatures above 1200° to 1300°C. The density data were determined in the same series of measurements. The melts must be carefully degassed, but bubbles on the wall of the crucible have no disturbing effects.

97. The method developed by E. W. Washburn and F. E. Libman (1923) which is characterized by determination of the *maximum pull* exerted on a thin-walled platinum cylinder when its lower edge is in contact with the test melt, was recently used again by L. Shartsis, H. F. Shermer, and A. B. Bestul[153] for measurement of the surface tension of alkaline earth borate melts, with a constant content of 3 molecular per cent K_2O. Up to 40 mol. per cent RO, the order is Ca $<$ Sr $<$ Ba, and is inverse to the ionic potentials z/r^2 of the cations with positive temperature coefficients of the surface tension if the RO contents are low. The increase of surface tension

[151] *J. Soc. Glass Technol.*, **43**, 1959, 179-190 T.

[152] *Glastech. Ber.*, **32**, 1959, 501-503.

[153] *J. Am. Ceram. Soc.*, **42**, 1959, 242-249; on correlation of immiscibility with ionic potentials of the cations R^+, cf. L. Shartsis and W. Capps, *Ibid.*, **35**, 1952, 169-172.

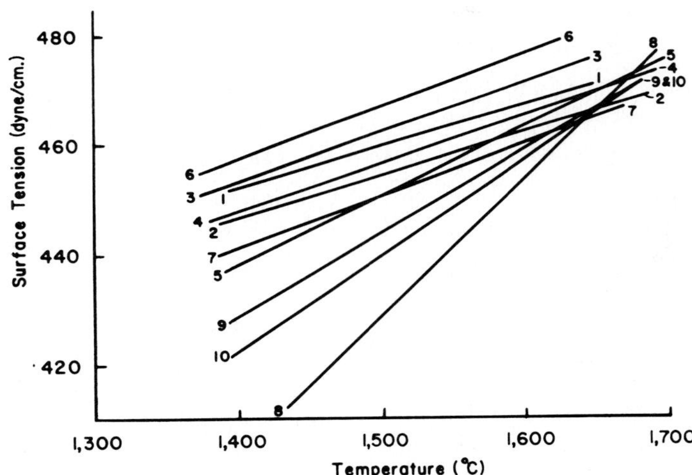

CALCULATED PERCENTAGE COMPOSITIONS OF GLASSES

Glass	CaO	Al_2O_3	SiO_2
1	39	19	42
2	35	10	55
3	34	30	36
4	30	25	45
5	30	10	60
6	29	40	31
7	25	20	55
8	25	10	65
9	23	15	62
10	15	20	65

FIG. A. 66. Surface tension of glasses of the system CaO—Al_2O_3—SiO_2, as a function of temperature. (Barrett and Thomas).

with increasing R^{2+} concentration is relatively slow (Fig. A. 67). Some of these phenomena are observed in systems with regions of liquid phase separation, or at least with a certain tendency to unmixing (cf. A. Dietzel, 1942), in correlation with high ionic potentials. Concerning the application of results of these investigations to the Gibbs adsorption equation for Γ (the difference of surface and body concentrations of a given molecular species), to the change of the surface tension, γ, with the chemical potential, μ, there is the ideal form $-\Gamma = d\gamma/d\mu$ and $-\Gamma/\gamma = 1/RT \cdot d\gamma/dc$, it is evident from Bestul's studies on variants of surface tension with body and surface concentrations[154] that an increase in ionic potential will result in

[154] *J. Am. Ceram. Soc.*, 42, 1959, 236-241, a discussion of the mathematical-theoretical background.

a relative increase in $-\Gamma$, the *depression* of surface ion concentration and a decrease in $d\gamma/dc$. Thus, at appropriately high values of the potentials there occurs a reversal from the direct relation between $d\gamma/dc$, and the ionic potential.

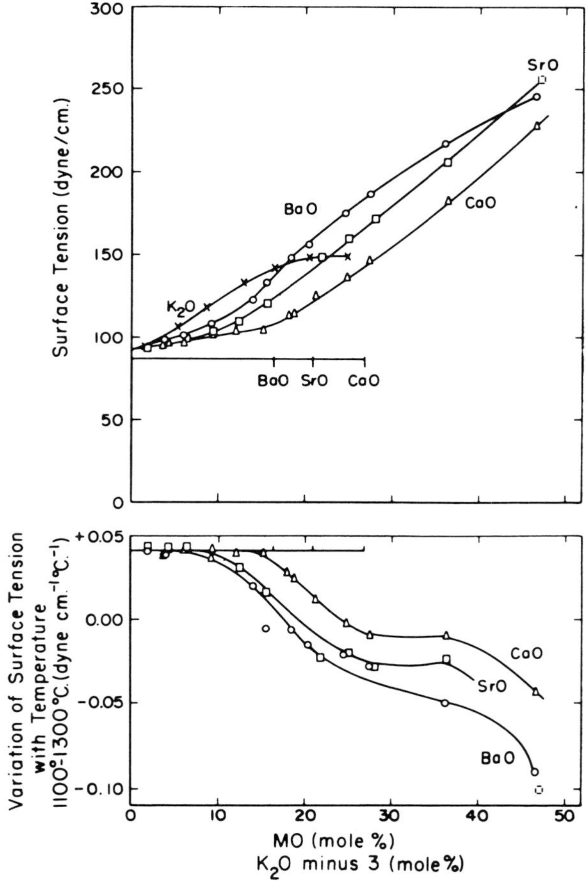

FIG. A. 67. Surface tension at 1100°C. of alkaline earth borate melts (with constant 3 mol. per cent K_2O), and average variation of surface tension with temperature from 1100° to 1300°C. (Shartsis, Shermer, and Bestul).

98. Because of their low ionic potentials, the alkali ions increase the ratio of surface to body concentration for alkaline earth cations above the value observed in the absence of the alkalies. It is in this respect interesting to see that the common opinion that alkali silicate glass melts are "improved" or "stabilized" by the addition of CaO, is just reversed by the concept that these composite glasses can be interpreted as calcium silicate melts to which sufficient alkalies were added to make a homogeneous liquid from what would otherwise be a two-liquid phase system. The alkali present is responsible for the negative temperature coefficient of the surface tension and the normal order of effects on the surface tension of various additions or substitutions of R^{2+} ions in such glasses.

99. An interesting study of W. D. Kingery[155] concerns the surface tension of *molten oxides* (alumina, B_2O_3, GeO_2, P_2O_5, and SiO_2), either by the sessile drop method (see below) or with the *pendant-drop* techniques. Alumina was fused to form a drop hanging on the end of an inert rod in pure helium atmosphere, or B_2O_3, P_2O_5, and SiO_2, by dipping a platinum rod into the melt to "gather" a satisfactory "gob" and air-quenching. The drops were directly observed in their shapes, and a volume increase of 20 per cent in melting alumina occurred, with a surface tension of 690 dynes/cm. (average) at 2050°C., in contrast with 905 ergs/cm.^{2} at 1885°C. in the solid state. For liquid P_2O_5 was found the surprisingly low surface tension of 60 dynes/cm. at 100°C., and 50 dynes/cm. at 400°C., for B_2O_3 at 1000°C. 83 dynes/cm., for GeO_2 at 1150°C. 250 dynes/cm., for SiO_2 at 1800°C. 307 dynes/cm. P_2O_5 has a negative temperature coefficient but B_2O_3, GeO_2, and SiO_2 have a positive temperature coefficient of surface tension. The most probable source of the positive coefficients in these single-components systems is the association into larger effective ion sizes at lower temperatures, and a liquid structure changing at elevated temperatures, not simply a surface orientation effect of the molecules as J. O'M. Bockris[156] assumed.

100. The *sessile drop* method as the most "popular" determination of surface tensions in silicate melts was used in the older studies of W. Steger[157] on the fusibility of ceramic feldspars by the elementary "button test" and the observations of the drop formation in the E. Leitz heating microscope (cf. Vol. III, A. ¶ 4 ff.), especially with measurements of the contact angle as a function of surface tension. On platinum sheet as support was found for one feldspar melt at 1540°C. the not too convincing value of 29.0 for another feldspar melt 51.6 dynes/cm. For ceramic glazes, slags, and others, H. Towers[158] recommended the same method for the important problem of *wetting* on refractory surfaces. The optimum conditions for reduction of corrosion by slag melts must therefore be a high contact angle Θ, associated with a high interfacial free energy between the slag and the refractory, and a low surface energy of the refractory, relative to the slag. On this principle is based the development of the "Vanal" protecting layers applied by A. Staerker (see below).[159] Towers gives

[155] *J. Am. Ceram. Soc.*, **42**, 1959, 6-10.

[156] "Physical Chemistry of Melts," Institution of Mining and Metallurgy, London, 1953, 106 pp., especially p. 42.

[157] *Proc. Intern. Symposium on Reactivity of Solids. Gothenburg*, 1952, Vol. II, 1954, 689-701, especially pp. 698-700.

[158] *Trans. Brit. Ceram. Soc.*, **53**, 1954, 180-202.

[159] *Ber. deut. keram. Ges.*, **34**, 1957, 329-334.

a complete graphic discussion of the Dupré equation $\gamma_{SL} = \gamma_{SA} - \gamma_{LA} \cdot \cos \Theta$, for γ_{SA} and $\gamma_{LA} = 100$, and 600 ergs/cm.², and a simple technique of observing the sessile drop in a furnace. The variation of γ with time is given in Fig. A. 68 in the

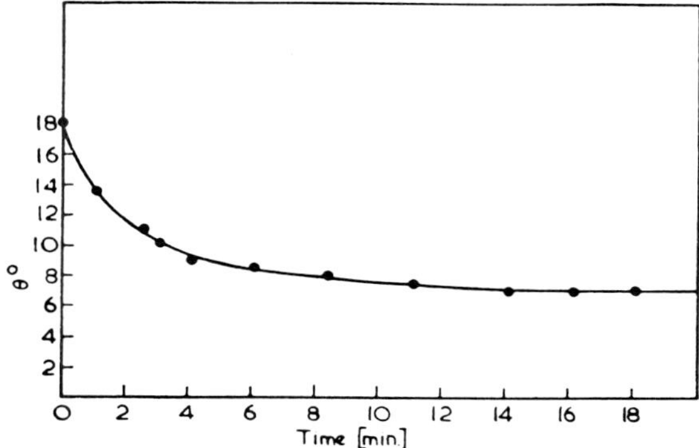

FIG. A. 68. Variation of surface tension with time for a melt composition 30.6 CaO; 37.0 Al_2O_3; 32.4 SiO_2, on polished alumina support, at 1415°C. (Towers).

example of a slag composition with 30.6 CaO, 37.0 Al_2O_3, and 32.4 SiO_2, on polished alumina as support. On graphite the results are influenced by side reactions like $SiO_2 + C \rightarrow SiO + CO$, or $SiO_2 + 3\ C \rightarrow SiC + 3\ CO$, with a distinct influence of the oxygen potential in the surrounding atmosphere on the surface tension.

101. The sessile drop method was used more recently by J. C. Williams and J. W. Nielsen[160] for determination of the surface tension of original and metallized high-in-alumina ceramic surfaces, by molten so-called brazing solders (alloys), as a function of time, at constant temperatures, in terms of contact angles. The measurements were made totally *in vacuo*, or in strictly controlled gas atmospheres, with an inductively heated furnace chamber, modified from the apparatus previously developed by W. D. Kingery and M. Humenik, as shown in Fig. A. 69, with a tubular molybdenum susceptor. Such a type of experimental kiln may be a universally useful apparatus for the sessile drop method. The contact angles, as a function of time, show

[160] *J. Am. Ceram. Soc.*, **42**, 1959, 229-235. On furnace construction of W. D. Kingery and M. Humenik mentioned in the text see *Ibid.*, **37**, 1953, 19-20. The method and samples described cover a remarkably wide field of compositions important in modern electronics, for transistors, quartz crystals, capacitors, and for sealings of metals. An important metallurgical problem is, e.g., the corrosion of alumina by molten aluminum metal which was studied by a similar vacuum furnace method by R. D. Carnahan, T. L. Johnston, and C. H. Li, *Ibid.*, **41**, 1958, 343-347, on sapphire monocrystals, and recrystallized alumina refractories, at 1200°C. and 10^{-4} mm. Hg. The interfacial geometry between molten aluminum and alumina due to dissolution is different for monocrystalline and polycrystalline materials of corundum.

curves of entirely the same type as those given by Towers. Similar studies of K. Semba[161] concerned the sealing of sodium-calcium silicate and of lead silicate glasses to gold and platinum, with their characteristic surface tension versus time curves. Semba remarked that many of the curves for variable temperatures fit to one another, and a "master curve" can be derived by simple shifting along the time axis. The

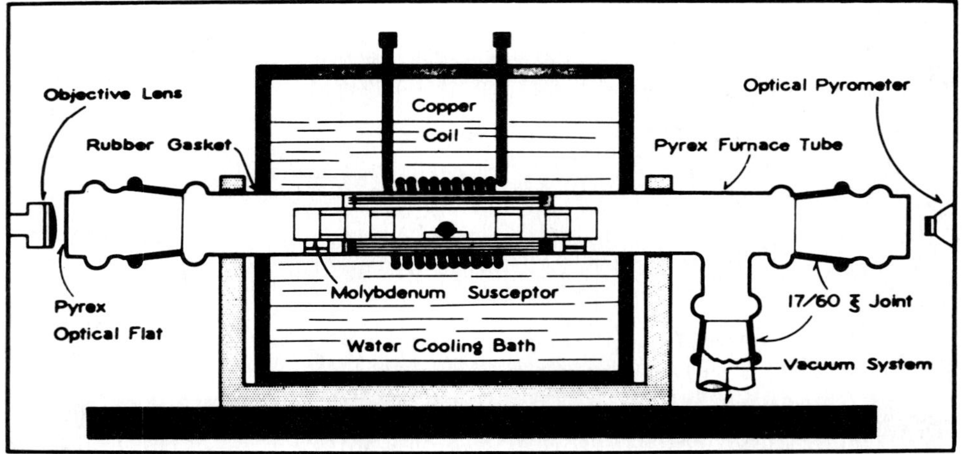

Sessile-drop induction-furnace assembly.

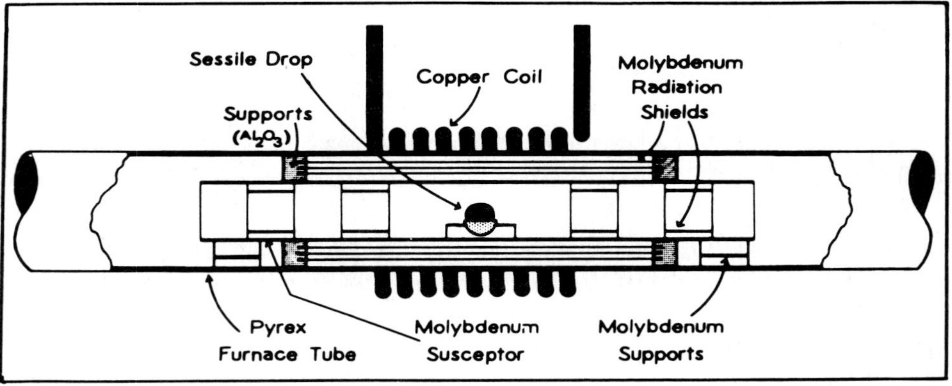

Induction-furnace susceptor and radiation shields.

FIG. A. 69. Inductively heated vacuum furnace for sessile drop method for the determination of contact angles of metal/ceramic systems. (Humenik and Kingery).

[161] *Dainippon Yogyo Kyokai Zasshi*, 66, 1958, 213-218.

amount of this shift, a_T, can be used for deriving the apparent activation energy of the wetting process from the linear plots of a_T as a function of $1/T$, which is about 40,000 to 60,000 cal./mole, or in the same order of magnitude as the activation energy of the viscous flow of the molten glass. The importance of the contact angle Θ for ceramic systems was also emphasized by U. S. Singh.[162] The influence of the gravity field on the shape of the sessile drop is negligible concerning possible deviations from a sphere segment, and the angle Θ to the volume, V, of the drop, and the area, A, of contact is a simple function of the ratios $A/V^{2/3}$ which was tabulated for $\Theta = 1^o$ to 150°, for a rapid determination of Θ from the geometric variables measured in the shadow-casting optical methods mentioned above.

102. An example of the use of the E. Leitz *heating microscope* for such purposes was given by H. Lehmann and U. S. Singh[163] especially in the study of wetting of aluminum silicate refractories by silicate melts combined with typical corrosion effects, and on the favorable action of the "Vanal" protection method (cf. Vol. V, Section B). The evaluation with the formula of A. W. Worthington (1914) is refuted as inaccurate for the calculation of surface tension, the method of F. Bashforth and S. C. Adams (1883) preferred. In most cases, the contact angle is increased with increasing ratios $SiO_2 : Al_2O_3$ in the refractory brick material. These results are particularly important for wetting of refractories in glass tanks by the bath which was investigated by K. Konopicky and H. Woelk,[164] also in the Leitz heating microscope, and with a detailed discussion of sucking effects exerted by more or less large pore cavities in the brick structure of high rugosity. Practical experiments were made with plate or window glass melts, observing the changes in height and contact angle of the initial sphere-segment shaped drops versus time, by measurements from kinematographic projections of the shadows. Most of the refractories showed final contact angles of about 20°, but quartz firebrick (with only 21 per cent Al_2O_3 and a porosity of 30 per cent) had a particularly low wettability. Bricks with a particularly high wettability and an angle Θ of about 7° also have a high permeability (cf. Vol. IV, Section B). Bricks covered with a reaction glaze are rapidly wetted and the glass melt finally spreads over the whole surface (with $\Theta = 0^o$).

103. In connection with the practically premordial importance of wettability problems of *silicates on metals*, especially of the Ist and VIIIth group of the Periodic System, the systematic investigations of J. A. Pask and his collaborators,[165] who

[162] *Glas-Email-Keramo-Tech.*, **10**, 1959, 165-169.

[163] *Ber. deut. keram. Ges.*, **34**, 1957, 353-362, with extensive literature references on p. 354.

[164] *Glastech. Ber.*, **27**, 1954, 409-414.

[165] *J. Am. Ceram. Soc.*, **36**, 1953, 84-89 (with V. F. Zackay, D. W. Mitchell, and St. P. Mitoff); *Am. Ceram. Soc. Bull.*, **33**, 1954, April, Program p. 25 (with A. Kaminski and D. W. Mitchell); *J. Am. Ceram. Soc.*, **40**, 1957, 269-272 (with R. M. Fulrath and St. P. Mitoff); *Ibid.*, **42**, 1959,

preferably used the sessile drop method in their studies, are to be emphasized. They worked either in a high vacuum, or in controlled atmospheres of helium, hydrogen, air, or oxygen, with copper, silver, gold, palladium, platinum, and nickel, wetted by melts of sodium silicates. No reaction between metal and silicate was observed if the surroundings were inert, whereas chemical reactions were observed in air and oxygen, also in hydrogen which then changed the nature of the interfaces between the metal and the glass melt, and favored the spreading out of the sessile drops on the metal surface. In contrast to the previous data given by B. S. Ellefson and N. W. Taylor (1938), gold shows in an oxidizing atmosphere a distinct lowering of the contact angle and better wetting as a reproducible effect. An interesting exceptional case is the system palladium/glass. Although this metal shows definite tendencies to form an oxide (PdO is decomposed at 877°C.) the experimental results do not exhibit a zero contact angle in air or in oxygen. In hydrogen the contact angle is very low, evidently due to the tendency to form a palladium hydride. Also in the systems gold/glass, and platinum/glass, there is an increased wettability in hydrogen atmosphere. Pask *et al.* prefer to discuss these and similar effects under aspects of the polarization properties of the metals, or of their polarizabilities which rule the changes in wettability. Changes in the contact angle with time are also particularly typical in systems of potassium silicate and sodium borate melts at 900°C. in vacuo, on gold, platinum, iron, tantalum, molybdenum, and wolfram.

104. For a deeper understanding of the problems of the glass/metal wetting characteristics in oxidizing atmospheres a special investigation of the influence of *oxygen pressure* was made. It is evident that oxidizing gases primarily affect the surface energy of the metal and then the interfacial energy of the system metal/glass following the basic equation of Dupré-Young. At very low pressures, the surface energy of the metal is high, and only partial wetting is possible. At slightly higher pressures, the surface energy of the metal is reduced but the interfacial energy

102-106 (with M. L. Volpe and R. M. Fulrath). Interesting observations of V. A. Presnov see in *Stroenie Stekla, Inst. Khim. Silikatov Akad. Nauk S.S.S.R., Trudy Soveshchaniya, Leningrad,* 1953 (Pub. 1955), 268-269, who measured contact angles of a glass melt on Pt; MoO_3, MoO_2; MnO_2, MnO; Cu_2O, CuO; in each one of these cases with characteristic wettabilities. The same author discussed in *Structure of Glass. Proc. All-Union Conf. Glassy State, 3rd, Leningrad* (English Translation), 1959/60, 368-370, the bonding mechanism of glass to metal sealings, especially by linking oxygen atoms. Carbiding or nitriding of the metal surface may have similar effects. In continuation of his fundamental studies of glass/metal bonding, the following recent publications of J. A. Pask and his school are important: R. W. Cline, R. M. Fulrath, and J. A. Pask, *J. Am. Ceram. Soc.* **44**, 1961, 423-428, on wettability of iron by $Na_2O.2SiO_2$ melt; L. G. Hagan and S. Fr. Ravitz, *Ibid.* pp. 428 ff., on reactions beween iron and $Na_2O.2SiO_2$; R. B. Adams and J. A. Pask, *Ibid.* pp. 430-433, on wettability of iron by $Na_2O.2SiO_2$ melt containing iron oxide; J. A. Pask and R. M. Fulrath, *Ibid.* **45**, 1962, 592-596, on the nature of wetting and adherence.

remains high, and even poorer wetting results. If the pressure is further increased, the interfacial energy is more rapidly decreased than is the surface energy of the metal, and a complete wetting finally results. Interfacial compounds are in this case not formed. It depends on the oxygen affinity of the metal over which ranges of pressure and temperature the relationships here outlined are observed, as in the systems platinum/glass in oxygen, or iron/glass in oxygen, with a higher affinity of the metal to oxygen. The latter system shows a temperature dependence of the contact angle at lower pressures than does platinum/glass. On the other hand, gold is the most inert metal to oxygen and shows only partial wetting even at atmospheric pressure. As soon as interfacial compounds appear the Dupré-Young relation is not valid anymore. As important data we mention here the interfacial energy of the system gold/$Na_2O.2SiO_2$ melt at 1040°C. = 1230 dynes/cm., either in air on *in vacuo*, based on this equation, with the surface energies of gold = 1370, and of $Na_2O.2SiO_2$ melt = 286 dynes/cm.

105. Further investigations considered the wettability of gold and platinum by sodium disilicate melts at 1000°C. in oxygen, nitrogen, hydrogen, helium, argon, carbon monoxide and dioxide, and water vapor, and in mixtures of different gases, in the pressure range from 10^{-5} to 10^0 mm.Hg in a special furnace apparatus. Gold in its wetting characteristics was not affected by the different atmospheres, but the contact angle of the system platinum/$Na_2O.2SiO_2$ depended very distinctly on the nature and pressure of the surrounding gases, most strikingly in water vapor, CO, hydrogen, and oxygen. Water is easily displaced from the metal by oxygen; the latter gas is evidently adsorbed by platinum. Particularly enigmatic reversible changes in contact angles appeared which became more sluggish with increasing test time, and apparently indicate some change in the glass rather than in the platinum. On the periphery of the sessile drop a more viscous glass is observed, which if it extends over the whole interface would indicate losses in sodium by a limited reaction between both phases. Pask *et al.* conclude from their experiments that generally any gas that is *ad*sorbed only on the metal surface, lowers surface energy and will increase the contact angle. If the gas, however, is *ab*sorbed, it will diffuse to the metal/silicate interface and also lower the interfacial energy, the effect of the combined action depending on the balance of the forces.

106. Concerning the problems of glass/metal *adherence* it must be concluded that the "work of adherence," of the energy, $\Delta F_{\text{separation}} = \gamma_{SA} + \gamma_{LA} - \gamma_{SL}$ expended in separating the two phases, is influenced by gas absorption at the interface and affects the strength of the bonding. Oxygen and hydrogen are such surface-active gases; one might imagine that two platinum/glass systems with the same contact angle might exist but with *different* interfacial tensions. The practically important room-temperature strength of glass to metal sealings would be complicated by existence of strains due to a mismatch in thermal expansion and other factors. We may in this respect emphasize the studies of W. Weiss[166] who used an interesting vacuum bulb furnace as shown in Fig. A. 70, for platinum, molybdenum, and wolfram at 1300°C. The wettability defined as the difference of the surface tension, τ, of the metal and the contact face tension ("affinity"), λ, is equal to the surface tension of

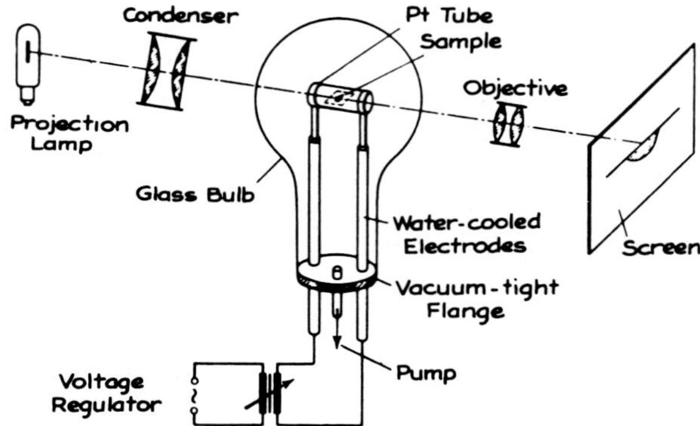

FIG. A. 70. Vacuum bulb furnace for the measurement of contact angles between small glass drops and metals. (Weiss).

the glass, σ, multiplied by $\cos \Theta$. The glasses examined were composed in homologous series with variable R_2O, MgO, CaO, BaO, Al_2O_3, and B_2O_3. *Evaporation* effects were chiefly observed for B_2O_3 at 1300°C. over 3 hours. In general, R_2O and RO in the glasses improve wettability, but there is no general analogy to surface tension. For RO the parameter λ is evidently more important than σ, and also in those glasses in which Na_2O was replaced by K_2O or Li_2O, the observed wettability effect is contrary to what would be expected from the change in σ alone. If SiO_2 is replaced by B_2O_3 the surface tension σ is strongly reduced, the wettability also considerably decreased, with a maximum effect at about 20 molecular per cent B_2O_3. For the adhesion strength, H, of the metal sealed to the experimental glasses a suitable tensile strength device was used (Fig. A. 71). The adhesion is the better the smaller the contact angle, and the higher the surface tension of the glass. The product $\sigma(1 + \cos \Theta)$ often called the specific adhesion energy (in ergs/cm.^{2}) $= \tau + \lambda + \sigma$; it is for platinum as metal in good agreement with the surface tensions determined at 900° to 1000°C. by the Berggren (fiber elongation) method, with those calculated according to A. Dietzel from oxidic increments, and fits very well to adhesion strength H measured at room temperature in the function $H^{1/4} = \sigma(1 + \cos \Theta)$.

107. More recently, Y. Godron[167] confirmed the results and conclusions of Pask

[166] *Glastech. Ber.*, **29**, 1956, 386-392. For sessile drop measurements on platinum and graphite at 650° to 1100°C. in different gas atmospheres with lead containing glasses see also by T. Yamauchi, H. Suzuki, and M. Ozima, *Yogyo Kyokai Shi*, **65**, 1957, 314-320.

[167] *Silicates inds.*, **24**, 1959, 539-549. See also the excellent report given by H. J. Oel, *Ber.*

and Weiss, and extended his discussions on adhesion strengths and wettability to systems of *Cermets*, such as WoC/Co, or TiC/Ni, and TiC/Ni + 10 per cent Mo, further to enamels (cf. B. ¶ 259). In the latter field we especially emphasize the studies of R. M. King and R. L. Cook,[168] with the observation of the changes in the contact angles with time, in dry and moist atmospheres, and the influence of iron

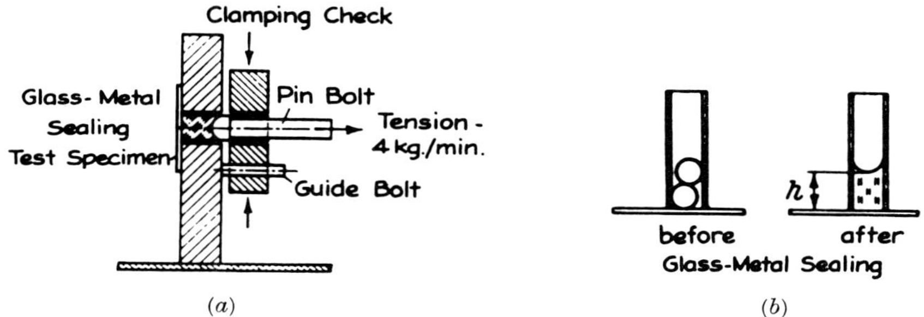

Fig. A. 71. (*a*) Device for the measurement of tensile strength and adherence strength of a metal/glass sealing. (Weiss). (*b*) Platinum sheet, and platinum tube sealed to form the specimen for tensile strength test. (Weiss).

oxides demonstrating how basically important the content of dissolved iron oxides must be for a good wetting of enamelling steel on the contact angle in ground coat enamels. Iron-free ground coat does not wet nonoxidized steel sheet in a dry state, but is wetting in the presence of water (0.5 to 1.6 per cent H_2O are sufficient). A one-coat titania-white enamel wetted with difficulty even in the presence of moisture.

108. Coming back again to the *Cermets* as metal/metal oxide and carbide systems, we have to emphasize the important experimental investigations made first by M. Humenik and W. D. Kingery[169] for systems of oxides with molten metals (iron, nickel, silicon), and later by M. Humenik and N. M. Parikh.[170] If the carbide phase shows a coalescence in the structure giving the impression of carbide particles growing to larger sizes, the wetting easily becomes incomplete, the adhesion strength unsatisfactory. To avoid this highly undesirable effect, as in TiC/Ni Cermets, the addition of molybdenum to the nickel metal phase improves the dispersion of the carbide as well as their hardness and impact resistance. These experiments give the interesting rule that the wetting tendency of metals to carbides, say TiC, is analogous

deut. keram. Ges., **38**, 1961, 258-267, with very valuable bibliography, and an extensive discussion on the multiple factors developing a good wetting and adhesion of glass melts on metals and ceramic high temperature materials (cf. Vol. V, Section B).

[168] *Am. Ceram. Soc. Bull.*, **36**, 1957, 293-296.

[169] *J. Am. Ceram. Soc.*, **37**, 1954, 18-23; *J. Phys. Chem.*, **57**, 1953, 359-363.

[170] *J. Am. Ceram. Soc.*, **39**, 1956, 60-63; **40**, 1957, 315-320.

to that of metals with oxides. In the latter case the wetting tendency is related according to Humenik and Kingery to the free energy of the formation of the oxides. Metals which are strong oxide formers have a greater wetting tendency on oxides. For the systems of nickel with different carbides the contact angle as a function of the free energy, ΔF/gm. atom carbon, in the formation of carbides of different metals, shows an increase in wettability for the systems with carbides of lower stability. In contrast with what might be expected, metals which are strong carbide formers, alloyed with nickel, do not exert a wettability increasing effect, and the contact angle remains high, whereas molybdenum which is a very low carbide former element, decidedly reduces this angle to zero. On the importance of the coalescence concept in combination with the wettability factor for perfection of sintering of solids to dense compacts see Vol. V, Section A.

109. In *metallurgical* experience the observation of *creeping* phenomena as a function of surface energy differences of melts and crucible materials is common and mostly troublesome observation, as in manganese or sulfide containing slag melts (cf. C. ¶ 72). F. P. Glasser[171] in his basic investigation of the system $MnO–SiO_2$ (Vol. III, Section B) described the creeping and overflow of such melts from platinum crucibles, especially for compositions between the MnO–tephroite eutectic, and that of tephroite and rhodonite, with an extremely high fluidity of the liquid phase and a strong wetting action on platinum. Melts containing primary MnO or SiO_2 are more readily contained in the crucibles.

110. In the school of O. A. Esin, the basic importance of surface energy relations to the physical chemistry of *slag melts* in metallurgical processes was fully recognized and extensively investigated. We mention in this respect first the studies of S. I. Popel', O. A. Esin, and P. V. Gel'd[172] on measurements of interfacial tensions, as derived from the height, h, (measured from the top to the equator of the sessile drops) of a metal melt suspended in fused slag and photographed in shadow projections by X-rays. If ϱ_1 and ϱ_2 are the densities of the metal and the slag, respectively, the interfacial tension is $\sigma = \frac{1}{2}(\varrho_1 - \varrho_2) \cdot gh^2$. For mercury drops in water, as a model system, copper at 1300°C. and cast iron in slag at 1400°C. a particular effect of freezing was detected for the metal drops, with heights h smaller by about 10 per cent and causing deviations in σ by 14 to 32 per cent if density data were taken from low temperatures, or 6 to 20 per cent if density of the real melts was considered. If carbon is dissolved in the iron up to 5 per cent, the method of the sessile drops, in a slag of the composition 39 CaO, 26 Al_2O_3 and 36 SiO_2 at 1400° to 1500°C. (on a corundum support in a graphite crucible) shows distinct effects of the carbon con-

[171] *Am. J. Sci.*, **256**, 1958, 398-412, especially p. 410.

[172] *Doklady Akad. Nauk S.S.S.R.*, **74**, 1950, 1097-1100; **83**, 1952, 253-255, 431-434 (with Yu P. Nikitin).

centration in the metal. With increasing carbon contents, surface tension is decreased, i.e., carbon is surface-active at the boundary with the slag. The surface tension values are not much different from those measured against air, because there is no chemical interaction between the metal and the slag.

111. Interesting *electrocapillary effects* were observed in an electrolyte of the composition 14.6 Na_2O, 8.3 CaO, 5.6 Al_2O_3, and 71.5 SiO_2, with iron-carbon alloys of 2.5, 3.0, and 4.0 per cent C, on drops of 14 to 16 mm. in diameter, with a second graphite electrode, and a current density below 50 ma./cm.^{2}, at 1320° to 1380°C. The interfacial tension, σ, is increasing with increasing negative polarization, φ. Cathodic polarization has a larger effect on σ than the anodic polarization, and sodium ions are more surface-active than silicate anions. The curves of σ are steeper for lower carbon contents in the metal. The charge density $E = -\delta\sigma/\delta\varphi$ between 0 and 0.2 volts is 50 and 30 μcoul./cm.^{2}, with 2.5 and 4 per cent, respectively. The adsorption of carbon on the surface of the drop is reduced with increasing negative polarization, or, in other words the positive charge of the double layer on the electrolyte side repels the carbon cations from the metal surface more strongly than it repels the Fe^{2+} ions.

112. P. M. Shurygin and O. A. Esin[173] give the surface tension of gold/air = 1180 ergs/cm.^{2}, of gold/Fe_2SiO_4 = 805 ergs/cm.^{2}, of 97 per cent gold, 3 per cent iron alloy/Fe_2SiO_4 = 720 ergs/cm.^{2}, all at 1300°C. Wetting properties of slags of the electrosteel process were studied by A. I. Kholodov, S. I. Suchil'nikov, and I. P. Malkin.[174] By its capillary (surface) activity oxygen is driven into the surface and the electrostatic double layer previously observed by O. A. Esin, and is characterized by a positive electrostatic charge of the slag and a negative charge of the metal. Sodium which is in its ion more active than Ca^{2+} displaces the latter from the double layer. Therefore the gas contents of steel melts and removal of silicate inclusions (cf. C. ¶ 99 ff.) are influenced by capillary wetting phenomena. The contact angle of an Armco steel melt to refractory material (magnesite) was measured at 1530° to 1580°C. and with oxidizing slag melts at 1450° to 1520°C. by A. M. Levin. For deoxidizing slags, Kholodov *et al.* used an optical method for direct measurement of the wetting angle to cast iron melt (with 4 per cent carbon, 1.5 per cent silicon) and a synthetic slag composition with 5, 15, and 20 per cent CaF_2, or industrial slags with 0, 0.27, and 2.2 per cent CaC_2. The cos Θ measured by the ratio $\gamma_{\text{met. gas}}$: $\gamma_{\text{slag gas}}$ shows for all these slags an increase with temperature; the curves are concave to the temperature axis, with the one exception of a convex curve for slag with 5 per cent CaF_2. As a function of the ratio CaO/SiO_2, the cos Θ decreases linearly

[173] *Doklady Akad. Nauk S.S.S.R.*, **95**, 1954,1043-1045.

[174] *Doklady Akad. Nauk S.S.S.R.*, **101**, 1955, 1093-1096. The investigation of O. A. Esin used by the authors is that in *Trudy Ural. Ind. Inst. im. S. M. Kirova*, **49**, 1954, 87; see also A. M. Levin, *Nauch. Trudy Dnepropetrovskogo Met. Inst.*, 1952, No. 28.

with this ratio increasing from 2.5 to 3.1. As a rule the contact angle Θ varies in the temperature range between 1350° and 1630°, and σ increases with rising temperatures.

113. Unusually low surface tension data are characteristic of *lead silicate* melts ($PbSiO_3$ and Pb_2SiO_4), which are able to dissolve relatively large amounts of lead *sulfate* (up to 33 mol. per cent) as L. Merker and H. Wondratschek[175] observed. By the method of maximum pressure in bubbles, the measurement of surface tension at 900°C. showed an unexpectedly low influence of the SO_3 contents, although in common glasses this oxide is highly surface-active. The concentration coefficient is positive whereas in common glasses it is negative. This apparent contradiction is tentatively explained by the latent tendency of the sulfate containing glass melts to unmix, not by an effect proper to the dissolved SO_3. That, on the other hand, SO_3 has a very pronounced influence on the surface tension of glass melts was established by A. Dietzel and E. Wegner[176] who worked with the bubble pressure method (modified by G. Keppeler and A. Albrecht, 1940), and the fiber elongation method (in the form given by Dietzel, 1949). For a glass of the composition 20 Na_2O, 5 CaO, and 75 SiO_2, the following data (in dynes/cm.^{2}) for a temperature of 900°C. are characteristic:

SO_3—free, on laboratory atmosphere:	307	with 1 per cent SO_3:	264
SO_3—free, on moist air:	301	with sulfide, equivalent to 0,4 per cent SO_3, in air:	301
SO_3—free, on dry air:	311	with sulfide, in a mixture of 75 nitrogen and 25 hydrogen:	314

114. The temperature of evaporation (above 1200°C.) from the glass melt is decisive in eliminating entirely the influence of SO_3, whereas the SO_3 effect is typical of temperatures below those of the fining process of glass in the tank. However, also this effect fades away after a prolonged heat exposure. For each 1 per cent SO_3 content in the glass one may calculate a depression in surface tension by —40 dynes/cm., valid over the working range of industrial glass. Sulfides have no direct influence on surface tension of glass as such, whereas atmospheric moisture noticeably decreases surface tension. For *enamels* the corresponding studies of A. Dietzel and E. Wegner[177] confirmed the apparently strong wettability-increasing effects of sulfides introduced as desulfurization slags (cf. C. ¶ 260). This, however, is not a property to be ascribed to the sulfide anion but to the fact that sulfide containing

[175] *Glastech. Ber.*, **30**, 1957, 473-475; **32**, 1959, 54-58.

[176] *Atti congr. intern. vetro, 3rd Congr. Venice* 1953, 354-363; *Veröffentl. Max-Planck-Inst. Silikatforsch.*, **13**, 1953, 121-130; on the fiber elongation method for determination of surface tension cf. *Mitteilungsbl. deut. keram. Ges.*, **26**, 1949, 15-21.

[177] *Mitt. Ver. deut. Email-fachleute*, **2**, 1954, (3) 13-14; *Veröffentl. Max-Planck-Inst. Silikatforsch.*, **14**, 1954, 204-205.

melts, at enamelling temperatures, are promptly oxidized when in contact with air. The resulting SO_3 content of the oxidized melts acts in the same way as described above for glass melts in the working range. The local reduction of sulfate in the enamel flux by contact with the metallic iron causes the formation of secondary sulfides and even of sulfoferrites which, however, are unequally less surface-active than was the previous sulfate.

115. The *fiber elongation method*, first used by G. Tammann and R. Tampke (1927), and later by I. Sawai (1928), for highly viscous glass melts was applied by M. V. Okhotin and I. G. Bazhbeuk-Melikova[178] to a study of the effects of systematically increased alumina contents on surface tension of sodium-calcium silicate glasses. Alumina considerably increases surface tension. The phenomena of contraction of the filament under the action of surface tension are identical with those previously described by Tammann-Tampke and Sawai. The rather sharp point at which contraction of the filament reverts to extension is characteristic for the calculation of surface tension from the filament's own weight, F, from the lower end to the point of zero deformation, and r, the radius of the filament: $\sigma = F/\pi r$. Okhotin and Bazhbeuk-Melikova[179] studied the influence also of magnesia on surface tension of a commercial glass with 16 Na_2O, 6 CaO, 0 to 5 MgO, 3 Al_2O_3, and 70 to 75 SiO_2, with the same filament extension test. The surface tension decreases with rising temperature in a linear function, but is raised with increasing magnesia contents.

116. A more recent reexamination of the theoretical background of the filament elongation method was given by N. M. Parikh[180] over the temperature range of 500° to 700°C. for commercial sodium-calcium silicate glass, treating a single filament as a cylinder of a highly viscous material. With the stress $\sigma_0 = \gamma/r$ at the equilibrium state between the contracting action of the surface tension, and the pull under the pending filament's own weight, one may develop an accomplished method for the accurate determination of surface tension in the range from $\log \eta = 7.5$ to 15.7 (corresponding to temperatures of 450° and 700°C. respectively). The furnace chamber used by Parikh is shown in Fig. A. 72 a, b, especially constructed for cathetometric control of the filament elongation either *in vacuo*, or in different gas atmospheres, or in dry and moist air. Silica glass has a surface tension of 290 dynes/cm. *in vacuo* and is not noticeably affected in this property by water vapor or other gases. A sodium-calcium silicate glass (Corning 0080, lamp bulb glass) showed *in vacuo* a surface tension of 315 dynes/cm.; $PbO.2SiO_2$ glass 210 dynes/cm. The polarity of the molecules H_2O, SO_2, NH_3, and HCl in the surrounding atmosphere is in a linear functional relation to the depression in surface tension, whereas dry air, hydrogen, or

[178] *Steklo i Keram.*, **9**, 1952 (4) 12-14; (6) 3-4.

[179] *Zhur. Priklad. Khim.*, **26**, 1953, 1320-1322.

[180] *J. Am. Ceram. Soc.*, **41**, 1958, 18-22. CO_2 is a remarkable exception in the regularity shown in Fig. A. 73.

nitrogen as nonpolar gases do not exert any depression effect (cf. Figs. A. 73 and 74). The specific effects of water vapor on the surface tension of sodium-calcium silicate glass according to the Gibbs equation, say at 6 mm. Hg for the partial pressure of

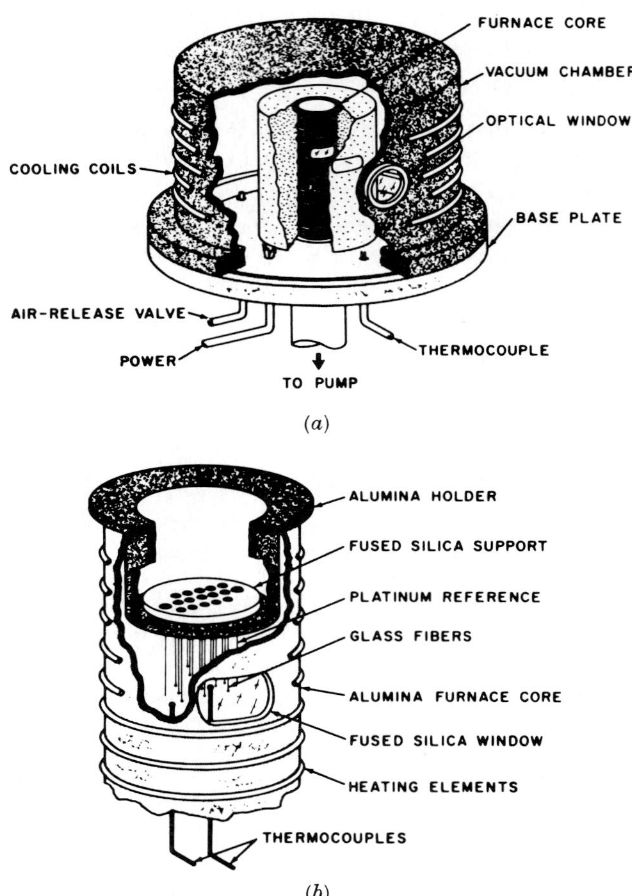

FIG. A. 72. (*a*) Furnace chamber for the observation of filament elongation for the determination of surface tension. (*b*) Details of the furnace core, with holder and support for the glass filaments. (Parikh).

H_2O, cause the formation of an adsorption monolayer of 7.7×10^{14} hydroxyl ions (equivalent to 3.85×10^{14} water molecules) per cm.^{2} surface area, corresponding to a depression of the surface tension by about 50 dynes/cm.

117. H. Wondratschek[181] criticized in Parikh's discussions of his improved fila-

[181] *Glastech. Ber.*, **32**, 1959, 276-278.

ment elongation method that the latter author did not consider the systematic changes in the surface of the volume elements in the theoretical derivation of the basic equations. The fact is that the viscous glass cannot be treated as a solid but is really deformed under the stresses applied. If one introduces the rate of filament

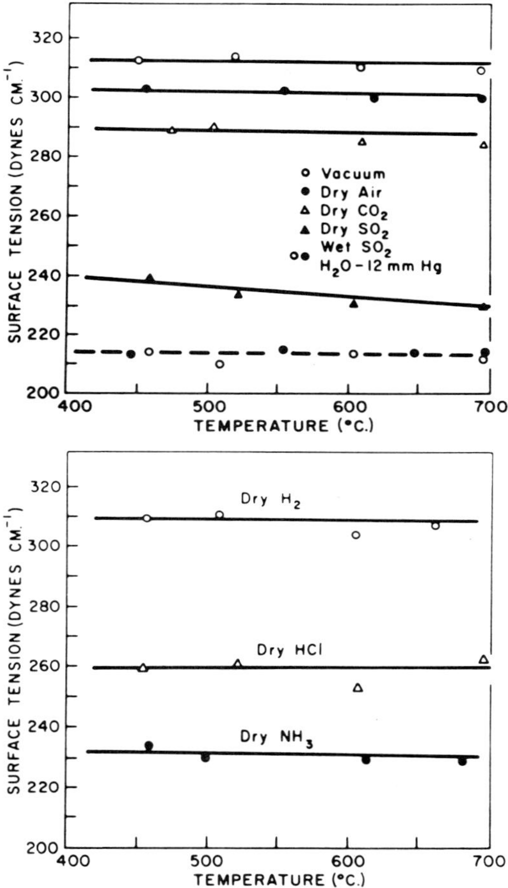

FIG. A. 73. Effect of different gases on the surface tension of a sodium-calcium silicate glass. (Parikh).

elongation, dE/dt, and neglects, on the other hand, the time-function of the changes in the filament radius, the integration over the time brings about an interesting equation for the elongation, E, in which both the surface tension and the viscosity participate:

$$E = \frac{t}{3\,\eta}\left(\frac{GL}{\pi\,r^2} + \frac{L^2\,\varrho}{2} - \frac{\sigma\,L}{r}\right)$$

(L being the length, G the weight of the filament in action, ϱ the density of the glass). This opens a way for a simultaneous determination of both parameters in the experimental series (cf. A. ¶ 37).

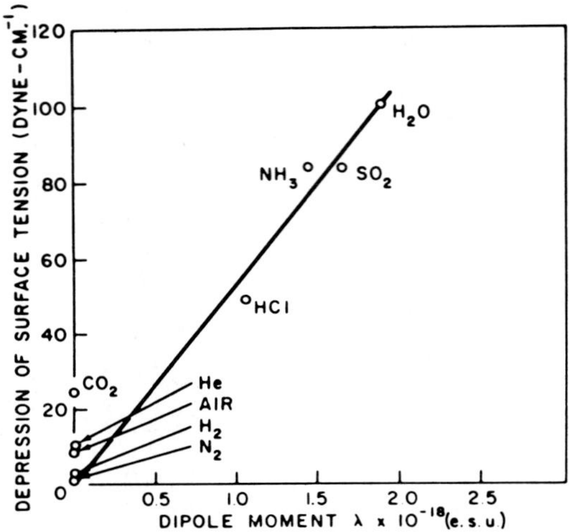

FIG. A. 74. Depression of surface tension of sodium-calcium silicate glass, caused by gases with different dipole moments. (Parikh).

ELECTROLYTIC CONDUCTION IN SILICATES

118. The classical investigations of E. Warburg and F. Tegetmeier (1887) on electrolysis phenomena in the crystallographic *c*-axis of *quartz* were supplemented by L. G. Chentsova and N. E. Vedeneeva[182] in connection with problems of the coloration in smoky quartz (cf. Vol. I, C. ¶ 75), and a study of the displacement of lithium and sodium cations under the action of an electrostatic field at 300° to 500°C. Evidently, these cations migrate in free interstices of the quartz framework structure, with diameters of about 2 Å. but no anions move in this process. Especially H. E. Wenden[183] thoroughly examined the direct current resistivity of quartz crystals as a function of temperature, the intensity of the electric field, the crystallographic orientation, and the time of current flow. The conductivity phenomenon as a whole is rather complex and characterized by time factors and activation energies, anomalous charging and discharging currents with a reversible interfacial polarization, and a relaxation frequency near 0.01 to 0.001/sec. The initial current flow is caused by typical impurity cations. It decays as a result of irreversible ion trapping and

[182] *Trudy Inst. Krist. Akad. Nauk S.S.S.R.*, **7**, 1952, 191-202.
[183] *Am. Mineralogist*, **42**, 1957, 859-888.

electrolytic exclusion of these ions. In the principal direct current flow period the quartz crystal behaves as a typical semiconductor. The room temperature resistance is in the order of magnitude of 10^{13} to 10^{19} ohms, at 500°C. of the order of 10^9 ohms, with an activation energy of 20,000 to 25,000 cal./mole. If the current has flowed for more than 500 hours, a steady state mechanism with an activation energy of 35,000 to 40,000 cal./mole is reached with a room temperature resistance of the order of magnitude 10^{27} ohms, and at 500°C. about 5×10^9 ohms. Quartz does not rigorously obey Ohm's law and has a typical voltage-dependent coefficient of resistivity, varying with temperature and time of current flow. Ionic diffusion in the *equatorial* direction was demonstrated as the reason for a conductance perpendicular to the *c*-axis. The often observed contradictions in the previous literature on the electric resistance, R, of quartz (as shown in the wide scattering of the log R versus $1/T$ lines) are for the major part caused by insufficient study and consideration of the time factors. Only when impurity cations are exhaustively removed, the difference of resistances parallel and perpendicular to the *c*-axis, and also the differences between different samples of natural quartz are of the same order of magnitude as the experimental errors.

119. Concerning the often emphasized tremendous anisotropy of the electrical resistance of quartz parallel and perpendicular to the *c*-axis, the older literature, in neglect of the important time factors mentioned above, had overlooked that the "conductance" observed in relatively short periods of current flow was not a characteristic parameter of the crystal as such but more of the "mobilities" of impurity ions in action. The apparent anisotropy is drastically reduced by a sufficiently long time of current flow to exhaustion of the impurities. Then, the "structure-sensitivity" disappears for the greatest part. There are in the quartz structure *not only* the often discussed "channels" parallel *c* but similar continuous openings, such as channels inclined by 59° to the *c*-axis, nearly perpendicular to the tetrahedral layers of the major rhombohedron planes, which also permit some ionic diffusion. The reality of "defect channels" in the direction of the *c*-axis in which the initial transport of impurity ions takes place was quite recently confirmed by H. H. Pfenninger and F. Laves[184] by an electron microscopic replica examination of the surface of quartz plates parallel (0001) through which gold was electrolyzed. The gold was precipitated as a filling in these channels, made visible as noble-metal "bristles" which are absent on $(10\bar{1}0)$.

120. H. Wondratschek, G. O. Brunner, and F. Laves[185] gave a highly plausible discussion of the mechanism of electrolysis in quartz crystals, based on the fact that

[184] *Naturwissenschaften*, **48**, 1961, 23; cf. H. Pfenninger, Inaug.-Dissert. Univ. Zürich, 1961, 130 pages. In 1928, R. Klemm observed gold filaments oriented in the *c*-axis direction of quartz as deposited by electrolysis between gold electrodes and deposited in the channels.

[185] *Naturwissenschaften*, **47**, 1960, 275; see also *Ibid.*, **46**, 1959, 664; E. Manegold and F. A.

alkali ions transferred to the cathode at 500° to 700°C. in an electrostatic field of 200 to 300 volts, can be replaced by protons from moisture of the surrounding atmosphere. These protons can be identified by intensity measurements of the infrared absorption bands near 3 μ (OH valence vibrations). The identification can be made even more impressive by introducing deuterium ions from D_2O in the surrounding atmosphere. This phenomenon, however, is not only observed from alkalies migrating in channels parallel *c* but also perpendicular to the axis, as H. H. Pfenninger and F. Laves described.[186] It is, therefore, necessary to consider also the transport of H^+ or D^+ cations in the electrolysis of quartz crystals and not to omit the partial pressure of H_2O and D_2O in the surrounding atmosphere or other sources of those ions.

121. The last-mentioned studies of Pfenninger and Laves were made using the *color centers* in the quartz crystals as indicators of material transport of this kind, a phenomenon which was intensely investigated by F. K. Drescher-Kaden and G. Böttcher[187] in the progressive displacement of the precipitation of copper or silver from the migrating ion fronts, "portraying" the area of the anode on the cathode surface and gradually extending into the inner of the crystal in the direction of the *c*-axis channels. The migration of the ions in these starts strictly from the area of the anode, never outward from it (Fig. A. 75).

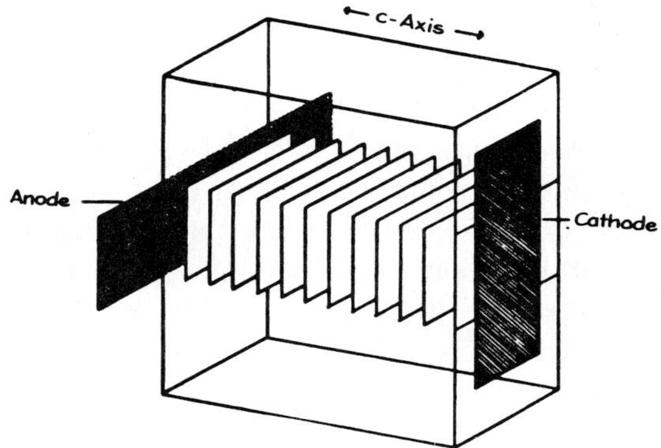

FIG. A. 75. Silver precipitations in layer "portraying" shape and outlines of the anode in the inner of a quartz crystal, after electrolysis in direction of the *c*-axis. (Drescher-Kaden and Böttcher). (*1*) First deposition on the cathode, (*2*) (*3*) (*4*) subsequent precipitations in the crystal, shifting towards the anode.

Schneider, *Z. physik. Chem. (Leipzig)*, **158**, A, 1932, 197-215; and recently H. U. Bambauer, G. O. Brunner, and F. Laves, *Fortschr. Mineral.*, **39**, 1961, 336; *Schweiz. mineral. petrog. Mitt.*, 1962, **41**, 1961, 335-369; **42**, 1962, 220-236.

[186] *Naturwissenschaften*, **47**, 1959, 276.

[187] *Naturwissenschaften*, **42**, 1955, 341-342; more extensively in *Hamburger Beitr. angew. Mineral. u. Kristallphysik*, **2**, 1959, 69-85. Gold behaved much differently than silver in these experiments.

122. Especially on the color phenomena connected with alkali electrolysis in quartz and other silicates (cf. Vol. I, C. ¶ 80), J. Lietz and W. Münchberg[188] made extensive studies, including corresponding phenomena in *silica glass*. When silica glass ("Herasil I") was γ-irradiated with a strong Co^{60} source (intensity 1.5 C. for 66 hours), and then electrolysed at 1000°C. with a direct current potential of 220 to 1000 volts, using purest graphite electrodes in nitrogen atmosphere, a colorless zone developed at the anode which was displaced gradually towards the cathode as the electrolysis proceeded. During irradiation, silicon valence electrons migrate to $[SiO_4]$ tetrahedra with *defects* in oxygen, whereas the electron-deficient sites are located near foreign Al^{3+} ions. Oxygen migrating towards the anode causes disappearance of the color centers. Analogous phenomena are observed in amethyst. The bell-shaped Gauss absorption curves show a band at 4300 Å. and another band at 4130 Å. which was previously not known. In amethyst the band at 2980 Å. is split in a doublet. After exhaustive electrolysis silica glass is no longer discolored by γ-rays; evidently the color centers migrate from the anode to the cathode and are also completely removed thermally at 1000°C. under a potential of 1000 volts and an initial current intensity of several milliamperes, which is however reduced after some time to a few microamperes. The rate of displacement of the sharp boundary between the colorless and the colored zone is 3×10^{-7} cm./sec./volt (cf. Fig. A. 76). The migrating cations

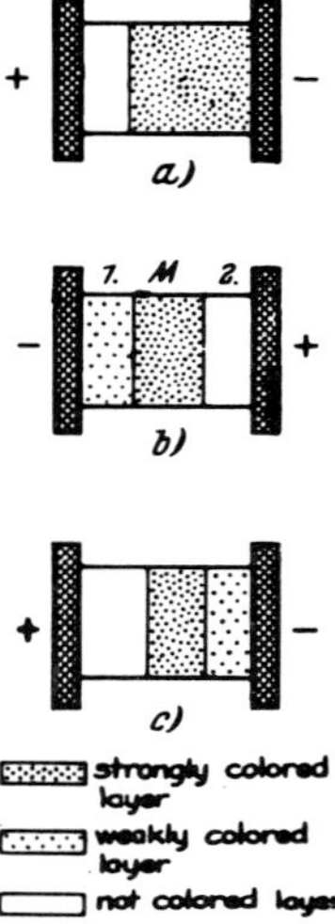

Fig. A. 76. Coloration of commercial silica glass ("Herasil I") after irradiation with γ-rays. (Lietz and Münchberg), (*a*) before the first, (*b*) after the first, (*c*) after the second reversal of the direction of the electrolysis current.

[188] *Glastech. Ber.*, **31**, 1958, 121-124; *Fortschr. Mineral.*, **37**, 1959, 72-73; *Naturwissenschaften*, **44**, 1957, 487-488; **46**, 1959, 67-68 (with R. Hänisch).

are Si^{4+} ions, and at the anode oxygen is evolved. It is therefore established that the discoloration by γ-irradiation is caused by sodium and aluminum cations of impurities. In perfectly pure silica glass no discoloration can occur. In quartz (amethyst and smoky quartz) the electrolytic coloration is stable up to 1000°C. whereas the natural coloration or that artificially produced by γ-irradiation, fades away at 250° to 300°C.

123. H. H. Pfenninger and F. Laves (1960) confirmed the observations of Lietz and Münchberg; by a spectral-analytical comparison of alkali contents of the anode and cathode areas in electrolyzed as well as in non-electrolyzed quartz portions, it was found that the lithium contents near the cathode are by 50 to 1000 times enriched, as was also confirmed by the experiments. Electrolysis brings about brown color centers at room temperature, which are vividly luminescent at the temperature of the electrolysis, but disappear at the cathode. No new centers of this kind are supplied from the anode where the colorless zone builds up. Reversal of current direction dissolves the "clouds" of color centers *in situ*, before a new cloud forms from the previous cathode (the anode), migrating in the reversed direction. That, in addition to small lithium ions, protons also may participate in these experiments is evident by an increase in intensity of the electrolysis current if the process is observed in an atmosphere of 75 per cent nitrogen and 25 per cent hydrogen, with the appearance of the hydroxyl band near 2.95 μ in the infrared absorption spectrum (see above).

124. Electrolysis phenomena in crystalline *aluminosilicates*, namely beryl, sodium aluminosilicates, glasses of corresponding composition and others, caused by ionic migration in a direct current potential field, were studied by V. A. Ioffe.[189] Side-by-side with ionic transfer also some electronic conductivity and polarization phenomena must be considered. These are the same factors which cause the color centers in irradiated crystals and glasses. Particularly characteristic is the *paramagnetic resonance* of aluminum ions connected with structural defects and electron excesses, as in plagioclase and alkali feldspars, synthetic carnegieite, cancrinite, and others, with a maximum for tan δ as a function of temperature at acoustic frequencies ($\nu = 10^3$/sec.) (Fig. A. 77) and a range of anomalous dispersion of the dielectric constant as a function of the frequencies. The resonance absorption phenomenon is related to the $[(Si,Al)O_4]$ tetrahedral units in aluminosilicates with the electron localized near one of the oyxgen anions in configuration

$$\left[\begin{array}{c} \text{O} \\ | \\ \text{O—Al}^{-}\text{—O} \\ | \\ \text{O} \end{array}\right]$$

[189] *Zhur. Fiz. Khim.*, **27**, 1957, 1454-1460; *Silikat Tech.*, **10**, 1959, 599-601. Concerning the rule of W. Meyer and H. Neldel see *Physik. Z.*, **38**, 1937, 1014-1019; *Z. tech. Physik.*, **18**, 1937, 588-593. Valuable previous literature see in the report of O. A. Esin and P. V. Gel'd in *Stroenie Stekla*,

and the presence of empty positions of monovalent and divalent cations in their neighborhood. The electrical properties of crystalline aluminosilicates in the range of acoustic and radio frequencies below 200°C. are chiefly determined by electron transfer at temperatures in which ionic mobility is still very subordinate. As an example there is the conductivity of microcline in different crystallographic directions.

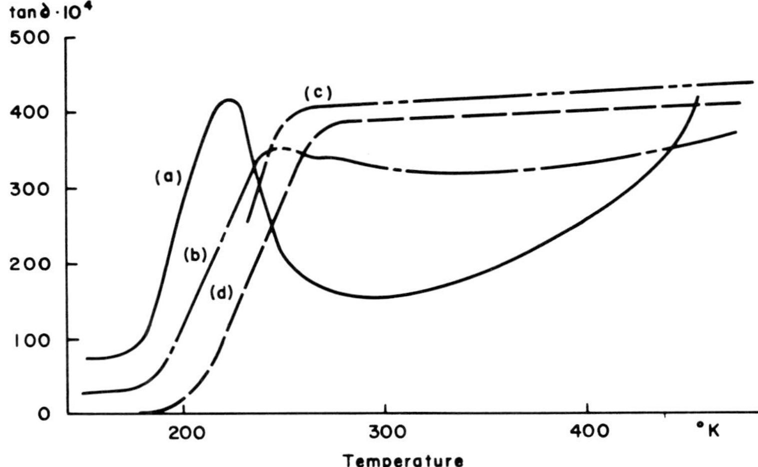

FIG. A. 77. Tan δ as a function of temperature for a plagioclase ($Ab_{75}An_{25}$), (*a*) for the frequency $\nu = 10^3$; (*b*) for $\nu = 8 \times 10^4$; (*c*) for $\nu = 5 \times 10^5$; (*d*) for $\nu = 10^6$ Hertz. (Ioffe).

The curves of the temperature coefficient of conductance show definite breakpoints indicating two different mechanisms with different activation energies (Fig. A. 78). The conductance of a sodium aluminosilicate glass is not so much determined by the contents in sodium as by the number of aluminum atoms relative to silicon atoms, following a rule for semiconductors with a high specific resistance as developed by W. Meyer and H. Neldel: the conductance increases with increasing activation energies, according to the exponential law $\log \varkappa = a + b \cdot U$ (U is the height of the potential barrier separating different equilibrium positions of the ions).

125. The electrolysis in *melts* of the system $CaO–Al_2O_3–SiO_2$ was investigated by O. A. Esin and V. A. Chechulin[190] to examine whether Faraday's law is fulfilled, as for a determination of CO and CO_2 evolved by the oxidation of anodic graphite electrodes at 1370° to 1440°C. The total current output was only 42 to 90 per cent

Inst. Khim. Silikatov Akad. Nauk S.S.S.R., Trudy Soveshchaniya, Leningrad 1953 (Publ. 1955) pp. 44-55, especially pp. 50-55.

[190] *Doklady Akad. Nauk S.S.S.R.*, **113**, 1957, 109-111.

of the theoretical, with a strong prevalence of silicon (38 to 71 per cent of theoretical output) on the cathode side, whereas aluminum is much lower. In calcium aluminate melts even only 24 per cent of the theoretical were found in the cathodic output. That of aluminum improves with increasing alumina contents of the aluminosilicate melt composition. For silicon there is a broad minimum of output (38 per cent) for a melt with 30 to 40 per cent SiO_2 which also shows the lowest viscosity. Increasing temperatures for a constant melt composition decrease the silicon output. Oxidation of silicon may be responsible for these anomalies as indicated by a change in silicon output with varying dimensions of the cathode space (ratio height : diameter), namely by diffusion of oxygen or of trivalent iron ions in the fluid melts into the cathode space, or more probably, a reaction of the type $Si^{4+} + 2\,e^- \rightarrow Si^{2+}$ which causes a typical polarization. The Faraday law is (with a scattering from 90 to 104 per cent) fulfilled for the solution of silica in the anodic melt.

126. J. O'M. Bockris, J. A. Kitchener, S. Ignatowicz, and J. W. Tomlinson[191]

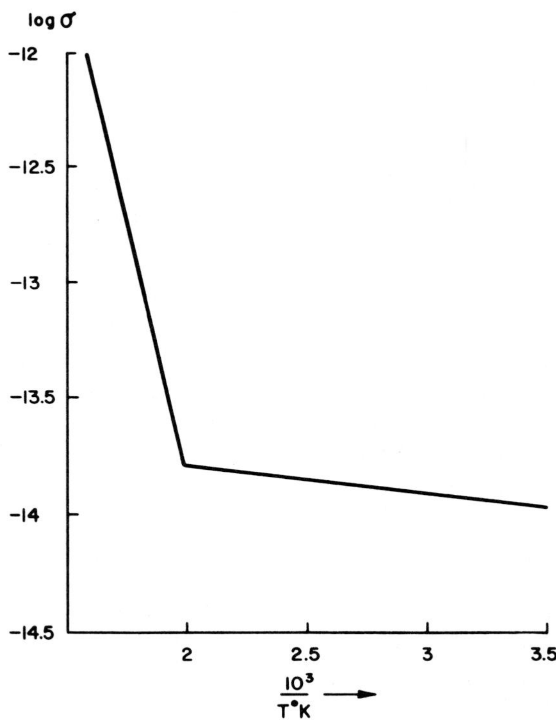

Fig. A. 78. Conductance (log ϰ) of a calcium aluminosilicate glass (composition 1 CaO; 0.3 Al_2O_3; 2 SiO_2), as a function of $1/T$. (Ioffe).

[191] *Trans. Faraday Soc.*, 48, 1952, 75-91. The analogy of the conductance in the fused salt

discussed the high importance of electrolytic conductance of silicate melts in connection with problems of their constitution, specifically the possibility of forming more or less complex anionic groups. Bockris, Kitchener, and A. E. Davies[192] also emphasized essential characteristics of the electricity transport in such systems. Starting from the simple binary systems SiO_2–Li_2O,Na_2O,K_2O; SiO_2–MgO,CaO,SrO, BaO; SiO_2–MnO; SiO_2–Al_2O_3, and SiO_2–TiO_2, further of a few selected ternary compositions of the systems CaO–MnO–SiO_2, and MnO–MnS–SiO_2, at temperatures up to 1800°C., validity of the general exponential law $\varkappa = A \cdot e^{E/RT}$ was established. The conductance increases with increasing concentrations in metal oxides. The activation energy decreases with increasing metal oxide contents, whereas the empirical constant A increases with increasing alkalies, but decreases with the other oxides. The conductivity is prevailingly of the ionic type. The entropy of activation, ΔS^*, for the conduction process and the free energy of activation, ΔG^*, were determined from the theory of absolute reaction rates. The relation of the nature of the ion in the melt to the same composition is described by the assumption of a distortion in the lattice structure, with the coulombic field between the added metal ions and neighboring oxygen anions. When the ratio O/Si is about = 4, the main factor determining the reaction is an "electroviscosity" of the cations in interaction with neighboring anionic groups.

127. The electricity transport in silicate melt electrolysis is interpreted by Bockris, Kitchener, and Davies as being prevailingly of ionic character in the systems SiO_2–Li_2O, SiO_2–CaO, SiO_2–MnO, and SiO_2–PbO, whereas SiO_2–FeO behaves differently. The validity of the Faraday law could be examined by the CO and CO_2 evolution by anodic oxidation of graphite electrodes (see above). Deviations from this law, however, are striking at definite composition in the system CaO–SiO_2 and by anodic oxidation of divalent to trivalent iron ions. Bockris and J. D. Mackenzie[193] established anew the almost completely ionic type of electricity transport in the binary systems mentioned above. Determination of activation energy affords a possibility of distinguishing between cations of different sizes and valencies; higher-valent ions have a

systems here investigated with silicate glass melts was especially emphasized by I. Peychès, *Silicates inds.*, **21**, 1956, 209-218, with valuable thermodynamic-electrochemical data. See also L. Riccobone, M. Fiorani, and L. Oleari, on glass as an electrolyte, *Vetro*, **2**, 1957, (2) 3-9.

[192] *Trans. Faraday Soc.*, **48**, 1952, 536-548. The conductance of alkali-free barium silicate glasses was discussed by K. K. Evstrop'ev and V. A. Khar'yuzov, *Doklady Akad. Nauk S.S.S.R.*, **136**, 1961, 140-142, starting from A. Einstein's equation for the diffusion coefficient and its determination by radioactive tracer method (using Ba^{140}). The Ba^{2+} ions are the electricity carriers in this case, their rates and mobilities are commensurable to those of alkali ions, if present. On stresses in glass caused by nonuniform diffusion exchange of monovalent ions, see S. S. Kistler, *J. Am. Ceram. Soc.* **41**, 1962, 59-68.

[193] *Rev. mét.*, **51**, 1954, 658-664. "*Network liquids*" are also discussed by J. D. Mackenzie in their viscosity-temperature relations, cf. *J. Am. Ceram. Soc.* **44**, 1961, 598-601.

tendency to compete with silicon in binding the oxygen anions. From the discussion of activation energies of the viscous flow of the melts, on the other hand, it is possible to calculate the size and structure of aggregates in the constitution of the melts which are displaced in viscous flow. Bockris *et al.*, therefore, defend the opinion that there is a real framework structure only possible in silica glass and its melts, whereas in all of the binary, ternary, and polynary melts a breakdown of the silica framework to more or less smaller aggregates takes place by interaction with metal oxides.

128. Tr. Bååk[194] studied the validity of the Faraday law for melts of the composition $2FeO.SiO_2$, $2.5\ CoO.SiO_2$, $2.75\ CoO.SiO_2$, and $3.0\ CoO.SiO_2$, with the "moving boundary method" using an electrolytic cell as shown in Fig. A. 79. From den-

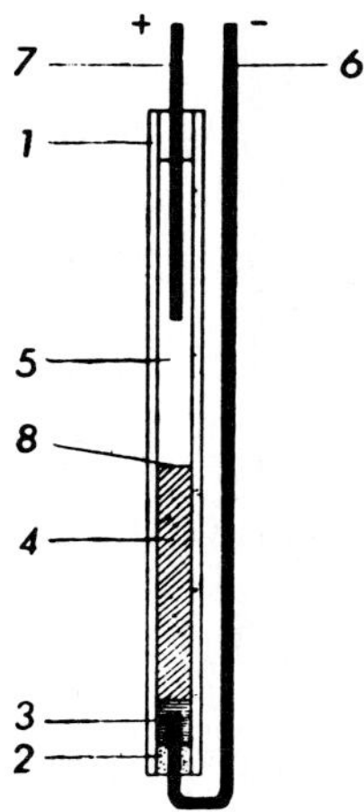

1 alumina tube
2 alumina cement
3 silver cathode
4 bottom electrolyte (dark)
5 top electrolyte (white)
6 mo-wire
7 mo-anode
8 boundary

FIG. A. 79. Electrolysis cell for the determination of cation transference in silicate melts, according to the "moving boundary" method. (Bååk).

sity, d, of the lower electrolyte, radius, r, of the tube cell, and distance, h, which the boundary has travelled, one may calculate the number of equivalents which the known quantity of electricity, f, has deposited, the degree of ionic conductivity,

$$I = \frac{F}{f} \cdot \frac{\pi r^2 \cdot h \cdot d}{E}$$

(E, the equivalent weight = molecular weight/4). For $2FeO.SiO_2$ and $2.5CoO.SiO_2$

[194] *Acta Chem. Scand.*, **8**, 1954, 166-174. On systematic measurements of conductance of molten fayalite, $2FeO.SiO_2$, with additions of $2CaO.SiO_2$, $CaO.SiO_2$, $2MnO.SiO_2$, and other compounds (chiefly oxides and sulfides) of metallurgical importance, over the temperature range from 1300° to 1400°C. and down to the primary crystallizations, see G. Björling, *Trans. Roy. Inst. Technol. Stockholm*, 1952 (52) 32 pp.

melts this degree of ionic conductivity is practically = 1.00. Only the iron and cobalt ions carry the current, the oxygen anions do not participate in the transport, and no oxygen deficiencies appear at the cathode, whereas at the anode oxygen gas is evolved and some silica is deposited. Transference numbers can be given for Co^{2+} in the melt with 2.5CoO. SiO_2. If the concentration in CoO is increased beyond this limit, semiconductance gradually develops, and prevails in the melt of $3CoO.SiO_2$ (I dropping to zero), semiconductivity also being typical of CoO. The ionic conductance in the lower-in-CoO silicate melts is a typical case of the presence of small cations and large anions, with an extensive Frenkel disorder of the structure similar to that in AgCl or $PbCl_2$. In analogy with the usual conditions for solids,[195] the specific conductance is $\varkappa = n \cdot z^2 \cdot e^2 \cdot u \cdot S$, with the *defect factor* S, n the number of cations/mole, $z \cdot e$ the charge, and u the mobility of the migrating particles. The calculation for different silicates (ortho-, meta-, and disilicates with $S = 3$, 2, and 1, respectively) gives an inverse proportion of u to the square of the ionic radius: $u = 1/r^2$, and for a pair of corresponding silicate systems $u_1/u_2 = (r_2/r_1)^2$ or $\varkappa_1 = \varkappa_2 \cdot (n_1/n_2)\,(r_2/r_1)^2 \cdot S_1/S_2$, in a very satisfactory agreement of computed and observed conductance data.

129. Bååk further observed interesting conditions in the electric conductivity of melts in the system CaO–CaF_2[196] at 1500° to 1545°C. (Fig. A. 80). In spite of the very low conductivity of CaO at its fusion temperature 2570°C. ($\varkappa = 0.175$ (ohm · cm.)$^{-1}$ the conductance of the mixtures with CaF_2 shows an increase of this parameter with increasing concentration in CaO and then an abrupt decrease beyond a limit value. At lower percentages of CaO oxygen anions are apparently built into the "liquid lattice" of the fluorine ions, causing anion vacancies due to the different valencies, therefore, an increasing conductance, and a lower density of the melt. This type of a "liquid lattice," however, breaks down causing rapid decrease in conductance and an increase in density, as was observed above a concentration of 0.0735 mole fraction for CaO. The maximum number of vacancies tolerable in the liquid lattice decreases with increasing temperature, that is with greater amplitudes of thermal agitation.

130. Typical *semiconducting glasses* were developed by P. L. Baynton, H. Rawson, and J. E. Stanworth[197] on the basis of alkali and alkaline earth phosphate-vana-

[195] Cf. W. Jost, "Diffusion in Solids, Liquids, and Gases," Academic Press Inc., New York, 1952, 558 pp.; especially p. 180 ff.

[196] *Acta Chem. Scand.*, **9**, 1955, 1406. The analogy to the system KCl–$SrCl_2$ investigated by C. Wagner and P. Hantelmann, *J. Chem. Phys.*, **8**, 1950, 72-74, is striking.

[197] *J. Electrochem. Soc.*, **104**, 1957, 237-240. See V. A. Ioffe, I. V. Patrina, and S. V. Poberovskaya, *Structure of Glass. Proc. All-Union Conf. Glassy State, 3rd, Leningrad* (English Translation), 1959/60, 407-409, on glasses of the systems P_2O_5–V_2O_5 and P_2O_5–V_2O_5–BaO. The conductance depends on the ratio P_2O_5/V_2O_5, not on the contents in BaO.

dates, with V_2O_5 contents of 50 to 87 molecular per cent and a conductance, $\varkappa$, of the order of magnitude of 10^{-5} (ohm. cm.)$^{-1}$ at room temperature. The slope of the log $\varkappa$ versus $1/T$ linear plots corresponds to activation energies of 0.35 to 0.40 e.v. for all glasses investigated. If log $\varkappa$ is plotted as a function of the contents in

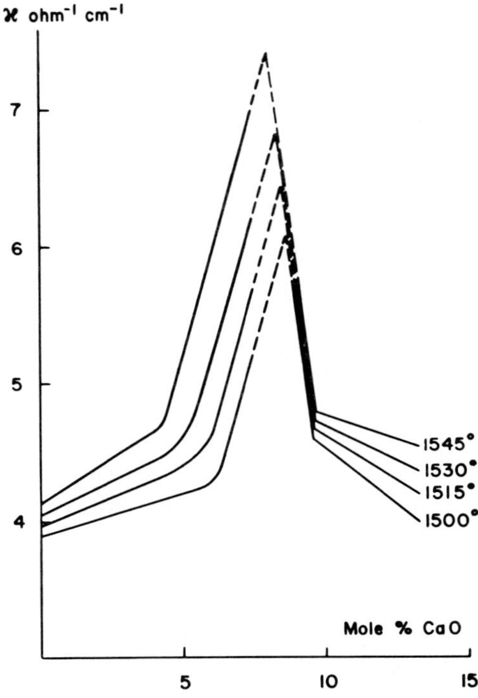

FIG. A. 80. Electric conductivity of melts in the system CaO—CaF_2, at different temperatures. (Bååk).

V_2O_5 in molecular per cent, a straight line is found. Such glasses are to be distinguished from those with a hydrogen-reduced surface layer, such as those containing lead metal precipitated by the reduction method of R. L. Green and K. B. Blodgett (1948) (cf. Vol. I, C. ¶ 86). Because of their particular technological interest H. H. Funk[198] described such surface-conducting glasses on the basis of lead silicates into which B_2O_3 and especially Bi_2O_3 were introduced. Bismuth borate glasses show a rather constant resistance after hydrogen treatment, and anomalously low resistances in pronouncedly low-ohmic types. The temperature coefficients of the resistance of hydrogen-treated lead silicate glasses are much lower than those of the same, but not treated, glasses (cf. B. ¶ 158). Reduced lead silicate glasses do not show any

[198] *Glastech. Ber.*, **31**, 1958, 269-272.

functional relation to voltage and surface resistance up to 10,000 volts. No changes in the resistance are observed in the magnetic field.

131. Very interesting experiments of Kl. Kühne[199] concern the introduction of iron, copper, lead, silver, arsenic, and other elements, into highly siliceous porous glasses (with 98.5 SiO_2, 1.03 B_2O_3, and 0.42 Na_2O) from aqueous salt solutions (concentration 0.02 gm. atom in 50 cm.^{3} H_2O, = 0.4 mole) to a concentration in the glass phase of 20.5×10^{-4} gm. atom/cm.^{3}, and sintering at 1100°C. The specific resistance, ϱ, was measured at 200°C., the ionic and electronic conductance as a function of temperature as well as electric polarization. A minimum resistance is characteristic of the elements with 0.8 to 0.9 Å. radii for the alkalies and alkaline earth ions (in the latter a maximum at 1.0 Å.). Greater anomalies appear in the elements of the side groups of the Periodic System, but those of the VIth Group (Cr, Mo, Wo, U) have a regular increase of ϱ with increasing atom number. The resistance was changed in characteristic curves which show an increase in ϱ with concentration for As,K,Rb, and Cs, a slight decrease for Na,Ba,Al. The ionic and electronic conductivity as a function of temperature for a glass with 99.2 SiO_2, and 0.05 Na_2O, showed only ionic conductivity up to 600°C. Electronic conductivity increases constantly up to 1100°C. but no saturation value could be observed. Silica glass is interpreted by Kühne as an electrolyte in which the framework of [SiO_4] groups is the highly viscous "solvent" while the cations of trace elements (especially sodium) are highly mobile as in aqueous solutions of high degree of dilution. The intensity of the current flowing not only depends on the ion concentration and the impoverishment in ions during electrolysis but also on the viscosity of the solvent, the "inner friction" of the cations, or say the volume effects of their radii and coordination numbers, and on the electrostatic binding forces. The determination of T_{k100} (log $\varrho = 8$) for sodium-calcium silicate and sodium borosilicate glasses was directed by practical viewpoints of optimum electrical properties, as in a special glass with 0.5 Na_2O, 0.5 ZrO_2, 0.2 Al_2O_3, 1 CaO, 2 B_2O_3, and 4 SiO_2, with T_{k100} of 295°C. and a thermal expansion coefficient of 49×10^{-7}.

132. In a melt of the composition Na_2O, CaO, and 4 SiO_2, V. I. Malkin, S. F. Khokhlov, and L. A. Shvartsman[200] made systematic measurements of *transference numbers*, using the radioactive tracers Na^{24} and Ca^{45}, at 1150°C. The difference of the half life periods of these elements (15.1 hours, and 152 days, respectively) makes an accurate discrimination of the activity effects in the Geiger counter devices possible in the cathode as well as the anode spaces of the electrolysis cells. The transference numbers of Na^+ (0.70) is double that for Ca^{2+} (0.35), or the double-charged cation of the same radius, both of noble-gas type. The weights of sodium

[199] *Silikat Tech.*, **6**, 1955, 191-200; **7**, 1956, 451-465.

[200] *Doklady Akad. Nauk S.S.S.R.*, **106**, 1956, 491-493.

and calcium volatilized from the melts during electrolysis and transference measurements are also approximately the same for both elements. No anion mobility was observed in the system at 1150°C. V. I. Malkin and V. V. Pokidyshev[201] determined from similar melts of the systems alkalies-silica, labeled with Na^{24}, K^{42}, Rb^{86}, Cs^{124}, respectively, at 1200°C. the transfer numbers by activity measurements in the cathode and anode spaces of the electrolysis cell. The transference numbers generally increase with increasing radii, or with decreasing intensity of the binding forces between cation and anion in the "liquid lattice." The functional relation is, however, not monotonous and mobilities of the cations are not determined alone by the strengths of the metal-oxygen bondings. There is a geometric factor related to repulsion forces which must be overcome by the migrating cations. The larger the ions, the more these repulsion forces impede the mobilities. Steps as discontinuities indi-

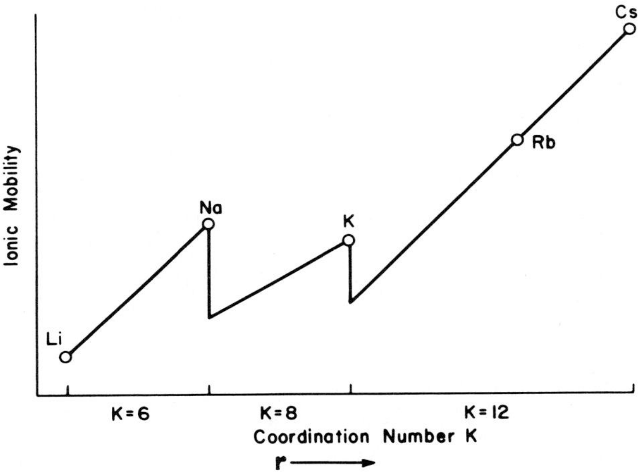

Fig. A. 81. Schematic diagram showing the discontinuous changes in cation mobility of alkalies, as a function of the coordination number with oxygen anions. (Malkin and Pokidyshev).

cated in a schematic graph (Fig. A. 81) of mobilities versus coordination numbers, show a break in the transition from Na^+ (coordination VI/VIII) to K^+ (VIII/XII)), and a corresponding change in density of the melt structure combined with a decrease in conductance of binary silicates with one cation, if a third component with another cation is added.

133. Coming back again to the phenomenon of *mixed conductivities* (of cationic beside electronic type) we emphasize their great importance in the theory of metallurgical *slags*. Ya. I. Ol'shanskiĭ[202] demonstrated such effects in the reaction products

[201] *Doklady Akad. Nauk S.S.S.R.*, **127**, 1959, 1253-1255.
[202] *Voprosy Petrog. i Mineral., Akad. Nauk S.S.S.R.*, **2**, 1953, 211-219.

of copper with H_2S at 500° to 800°C., with distinct pseudomorphs and "brushes" or "whiskers" of chalcocite up to 3 mm. in length, and "inner electrolysis" with diffusion of copper through a layer of the sulfide. The rate of material transfer, M/t, is proportional to the free energy change, ΔF, and the mixed conductance, with a maximum for equal shares of electronic and ionic conductivity. C. Tubandt in his classical experiments (1920) with Ag_2S/AgI neglected the possibility of mixed conductivity, and was therefore in error with his interpretation and assumption of a purely ionic conductance in sulfides at high temperatures. Ol'shanskiĭ sent a direct current of 20 amp.-hours through a cell $Cu/Cu_2S/Cu$, in pure nitrogen, at 350°C. and observed a copper migration into the sulfide, but no weight changes, in other words, there was no ionic conductance in the sulfide above 0.1 per cent of the total conductance of 10^3 $(\text{ohm.cm.})^{-1}$. Analogous conditions are even much more important in the system Fe–FeO–FeS–SiO_2, with a positive temperature coefficient of conductivity in the melts.

134. O. A. Esin and P. M. Shurygin[203] dedicated particular studies to liquid iron silicate slags of the system FeO–Fe_2O_3–SiO_2, using in pure nitrogen a cathode of molten gold which easily dissolves iron, and a separate carbon anode (cf. Fig. A. 82), in an iron-free CaO–MgO–Al_2O_3–silicate slag, to avoid the reaction $Fe^{3+} + e^- \rightarrow Fe^{2+}$, at 1300° to 1400°C. with a current density of 0.2 to 0.5 amp./cm.^{2}. The electrolytic output in per cent of the theory was given as a function of the degree of oxidation of the slags (by the ratio Fe^{3+}/Fe^{2+}). An increase in the acidity of the slag from 6 to 30 per cent SiO_2 brings about a decrease in the output from 82.6 per cent to 8.3 per cent. Slags of the composition FeO–Fe_2O_3–CaO–SiO_2 show much higher outputs than corresponding lime-free slags, even for 30 to 35 per cent silica. The results of Esin and Shurygin are in contradiction to previous opinions of W. A. Fischer and H. vom Ende (1950,1951) that pure iron oxide slags would not be able to show any electrolysis. A slag with 40 per cent (FeO + Fe_2O_3), 20 Al_2O_3, and 40 SiO_2 shows a surprisingly good output (up to 60 per cent) in spite of this high acidity. The very different shapes of the electrolysis curves (cf. Fig. A. 83) indicate the high influence of viscosity on the regeneration of Fe^{3+} ions by convection and diffusion from the furnace atmosphere. The electrical conductivity of the silicate slags is not exclusively ionic, it is in some share electronic. Therefore, a dilution of iron oxides by additions of lime, alumina, and silica causes a reduction of the electronic portion of the conductivity and an increased electrolysis output.

135. The only partly ionic nature of the conductivity in iron silicate slag melts was emphasized by M. T. Simnad, G. Derge, and I. George[204] by measurements in

[203] *Doklady Akad. Nauk S.S.S.R.*, **94**, 1954, 1145-1149. Cf. W. A. Fischer and H. vom Ende, *Arch. Eisenhüttenw.*, **21**, 1950, 217-224; **22**, 1951, 417-423.

[204] *J. Metals*, **6**, 1954; *Trans. Am. Inst. Mining Met. Engrs. Iron Steel Div.*, **200**, 1954, 1386-1390.

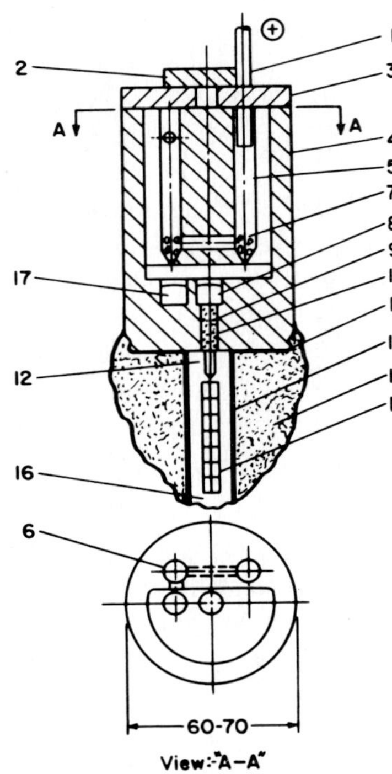

1 carbon anode
2,3 magnesia lids
4 fused magnesia crucible
5 anodic cell
7 magnesite diaphragma
8 gold cathode
10 molybdenum lead
12 coupled molybdenum and nichrome wire.

FIG. A. 82. Electrolysis cell for iron silicate slags. (Esin and Shurygin).

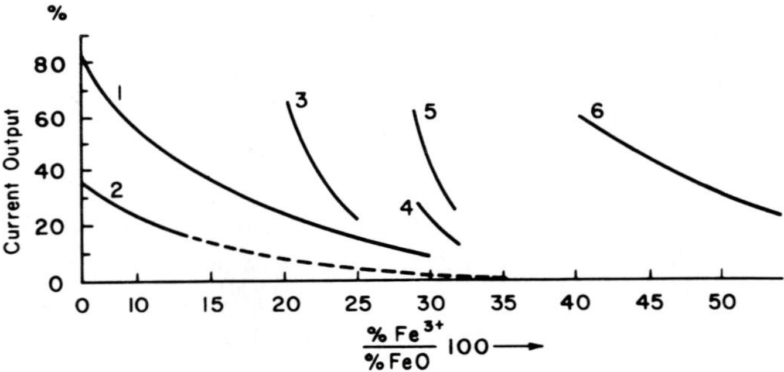

Curve (1): $FeO-Fe_2O_3-SiO_2$ slag;
(2): $FeO-Fe_2O_3$ slag;
(3): slag 45% ($FeO + Fe_2O_3$), 20% CaO, 35% SiO_2;
(4): slag 65% ($FeO + Fe_2O_3$), 15% CaO, 20% SiO_2;
(5): slag 40% ($FeO + Fe_2O_3$), 30% CaO, 30% SiO_2;
(6): slag 40% ($FeO + Fe_2O_3$), 20% Al_2O_3, 40% SiO_2.

FIG. A. 83. Current output curves for iron slag electrolysis, as a function of the oxidation degree in the slag composition. (Esin and Shurygin).

iron crucibles. The conductivity proved to be ionic by only 10 per cent in slags with less than 10 per cent silica, about 90 per cent, however, if silica contents are higher than 34 per cent; in the intermediate range an approximately linear function of composition rules. The remainder of the conductivity is electronic; only the cations Fe^{2+} are electricity carriers in the ionic transport phenomena. The transference number of Fe^{2+} ions in iron silicate melts was determined by O. A. Esin and A. K. Kir'yanov,[205] using the radioactive tracer Fe^{59}. In melts saturated in silica, or in melts of the system $CaO–Al_2O_3–SiO_2$ these transference numbers are about = 1.00 indicating the almost exclusively cationic electricity transport whereas the complex anions do not participate in it. With increasing contents in FeO of lime-free iron silicate slags the transference number systematically decreases as a consequence of increasing electronic conductivity and increasing transport by oxygen anions. In $FeO–Fe_2O_3$ melts free from silica conductivity is prevailingly electronic.

136. O. A. Esin and G. A. Teterin[206] extended the radioactive tracer method using the elements Ca^{45}, Co^{60}, and P^{32} in *phosphate slag* melts of the systems $CaO–P_2O_5$ and $CaO–CoO–P_2O_5$. V. I. Malkin ad L. A. Shvartsman[207] had previously observed that the transfer number of phosphorus in phosphate melts is practically zero,

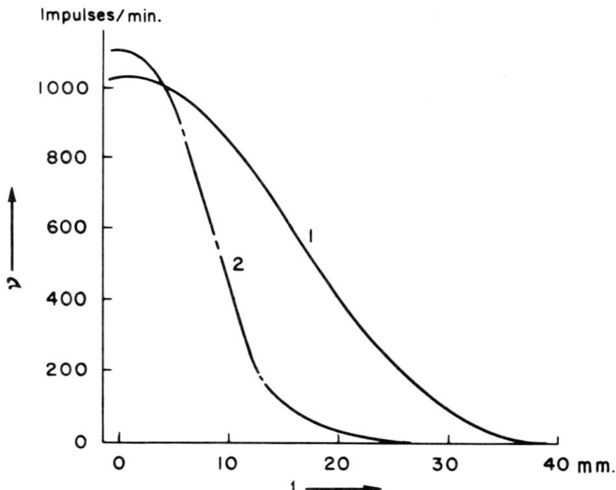

FIG. A. 84. Distribution of radioactive isotope Ca^{45} in the length of the electrolysis cell, (1) under the action of a galvanic potential of 30 volts, (2) without this field, by self-diffusion. (Esin and Teterin).

205 *Izvest. Akad. Nauk S.S.S.R.*, Otdelenie Tekhn. Nauk 1956 (8) 20-27.

206 *Doklady Akad. Nauk S.S.S.R.*, 128, 1959, 567-570.

207 *Doklady Akad. Nauk S.S.S.R.*, 102, 1955, 961-963. On the self-diffusion of alkali ions in silicate melts, using Na^{24} and K^{42} and the determination of alkali ion mobilities, see more recently V. I. Malkin and B. M. Mogutnov, *Ibid.* 141, 1961, 1127-1130.

although the diffusion coefficient of phosphorus is larger than that of calcium in such melts. These facts strictly contradict what is known for silicate melts, their ion transfer (Na^+, Ca^{2+}, Co^{2+}), and the practically negligible diffusion coefficient of silicon. Using a graphite crucible as the cathode, a copper alloy with 10 per cent P as the anode in the electrolysis cell, Esin and Teterin found the transfer numbers of calcium (as Ca^{45}) = 0.8, that of phosphorus (with P^{32}) = 0.3 (theory 0.2). The radioactive P^{32} is almost exclusively found in the *cathodic* portion of the electrolysis cell, it therefore migrates as the cation P^{5+} and not as the PO_4^{3-} anion. Melts of the system $CaO–CoO–P_2O_5$ (in alumina crucibles at 1120°C.) showed the transference number of $Ca^{2+} = 0.5$, that of $Co^{2+} = 0.4$, and of $P^{5+} = 0.2$. The bond Co—O is much more covalent (43 per cent) in comparison with Ca—O (20 per cent). The mobility u_{Ca} was determined by a comparison of the self-diffusion of Ca^{45} in the melt with the transfer number n_{Ca} under a galvanic potential of 30 volts (cf. Fig. A. 84). For the first process is $n_{Ca} = 0.7$, the diffusion coefficient $= 6.5 \times 10^{-5}$ cm.²/sec., and in absolute measures: $u_{Ca} = 0.00008$ cm.²/v.sec., $u_P = 0.000034$ cm.²/v.sec.

CONDUCTANCE AS A FUNCTION OF TEMPERATURE CONSTITUTION EFFECTS

137. Since the first theoretical discussions of temperature effects on the electrical resistance of silicate glasses (Fousseron, 1882) and the derivation of the Rasch-Hinrichsen law (1908) the principal advance was made by J. M. Stevels[208] by application of statistical calculations for the probability of "ion jumps" of mobile Na^+ cations over a potential barrier, and further by an application of Eyring's statistical rate process method (1941) which we repeatedly mentioned in our considerations. D. A. Stuart and O. L. Anderson[209] demonstrated how by this latter method, it is theoretically possible to derive exactly the Rasch-Hinrichsen law for rigid glass, as well as a simple binomial equation of the form $\varrho = A + BT + CT^2$, for the liquid state, both as empirical approximations to a more general understanding of the conductivity mechanism. This general aspect includes the possibility of a nonohmic behavior of silicate glasses under high field strengths as postulated by H. H. Poole (1921), giving a mathematical formulation for these conditions of the type $\log \varkappa = A + B \cdot E$ (E the field strength). Stuart's and Anderson's deductions also include the evident analogies existing between conductivity and fluidity (reciprocal viscosity) to be discussed in A. ¶ 162 ff. The activation energy ΔH^* as calculated from viscosity data in the order of magnitude of 30,000 cal./mole can also be used as a sound basis for the activation energy of the condutivity mechanisms in the thermodynamic derivations.

[208] *J. Soc. Glass Technol.*, **30**, 1946, 31-53 T.

[209] *J. Am. Ceram. Soc.*, **36**, 1953, 27-30.

138. For practical purposes, H. Beyersdorfer[210] presented valuable tabulations of collected data on the resistance of Jena optical glasses as a function of temperature. Especially well shown in these tables are the break points between the linear branches of these functions and the T_{k100} data. The 250° and 350°C. parameters of log ϱ are particularly important industrially. They can be accurately interpolated from the nomograms and be calculated from the Rasch-Hinrichsen law. O. V. Mazurin[211] developed a thorough series of precision measurements of specific resistance and conductance for different alkali-calcium silicate glasses, over the wide range from 100° to 1300°C., or from the rigid to the softening and liquid melt states. The straight lines resulting from the Rasch-Hinrichsen law show a first break point at the critical temperature T_g (Tammann) as a slight discontinuity and a second point with steeper slopes at higher temperatures. The resistances of glasses with *mixed alkalies* are considerably larger than those of glasses with only one alkali oxide (cf. A. ¶ 285 f.).

139. Also J. M. Stevels[212] started from Rasch-Hinrichsen's law and its empirical constants A and B, as a function of the chemical composition. These, however, have no definite physical meaning, and the restriction must be emphasized that the transformation range is the upper limit of the validity of the simple rule expressed in the law. The introduction of *structural* factors makes possible a heuristic rediscussion of these problems from the basis of the framework model hypothesis. If the activation energy of the electric conductivity, φ, is expressed by a formula of the type $\varphi = 0.198 \times 10^{-3} B$, it must be a function of Y, that is of the number of oxygen anions in a structural polyhedron participating in the framework. From the previous measurements of R. L. Myuller (1935) and E. Seddon, E. J. Tippett, and W. E. S. Turner (1922) for sodium and lead silicate glasses it can be demonstrated that a break point is observed in the rapidly decreasing φ/Y curve, with $Y = 3.47$ in the discontinuity point. For lead silicate glasses with $\varphi = 1.1$ eV is Y between 3 to 0 rather constant. Tabulations of the φ data for cations of the alkalies and alkaline earths, of Mn^{2+}, Pb^{2+}, Al^{3+}, and Ti^{4+} are provided. In borate glasses the data for φ are always somewhat lower. The constant A is chiefly influenced by volume changes in the glass framework, the migration rate of the cations, and their mutual impediments (Ca^{2+}, e.g., hinders the mobility of Na^{+} to a certain degree). Stevels cannot recommend any way for computing φ and A from the composition of complex glasses. Simple sodium-potassium silicate glasses show a maximum for φ and a minimum for A which cannot yet be interpreted. The break points for sodium silicate glasses ($Na_2O.SiO_2$) are at $Y = 3.00$, for potassium silicate glass $K_2O,4SiO_2$ at 3.50, for lead silicate glass $PbO.SiO_2$ at 2.00.

[210] *Silikat Tech.*, **5**, 1954, 459-462.

[211] *Trudy Leningrad. Tekhnol. Inst. im. Lensoveta*, 1954 (29) 72-89.

[212] *Silicates inds.*, **22**, 1957, 325-335. On the mechanism of electrical conductance in lead silicate glasses and its dependence on the water content, see G. C. Milnes and J. O. Isard, *Phys. & Chem. Glasses*, **3**, 1962, 157-162.

140. An interesting application of conductometric methods for a study of *phase-unmixing reactions* was demonstrated by C. Kröger and Kl. Lieck.[213] exemplified in the systems $PbO–B_2O_3$ and $PbO–B_2O_3–SiO_2$ (cf. Vol. III, Section B). The straight lines of resistance as a function of $1/T$ show in this case not only anomalies by curvature but considerable deviations from the Rasch-Hinrichsen law for homogeneous melts. Experimentally these conditions are better observed if the electrolysis cell is furnished not only with one pair of electrodes but with two pairs in the melt, one above the other which indicate, in the case of unmixing considerable differences in the resistances of the two prevailing liquid phases. It is not necessary to wait until a complete gravitative settling of the liquids takes place, since the contrasting resistance conditions are well marked in the intermediate stages of an incomplete differentiation of the suspensions. The resistance versus temperature curves are split

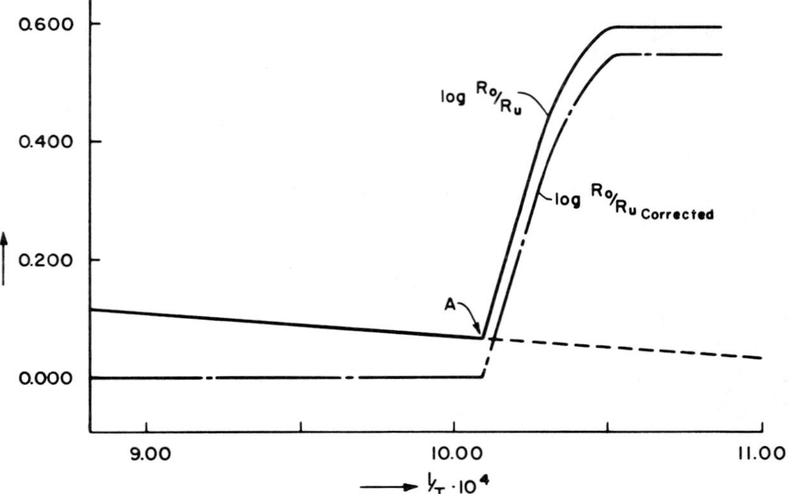

FIG. A. 85. Ratio of resistance in the upper and lower portion of electrolytic cell, used by C. Kröger and Kl. Lieck for discrimination of unmixing reactions, as a function of 1/T.

in two distinct branches (cf. Fig. A. 85) with corrected ratios of the resistance of the upper, and the lower layer, as the ordinates, and the modified Rasch-Hinrichsen equation for this ratio: $\log (\varrho_{upper}/\varrho_{lower}) = B/T + C$.

141. For simple *alkali borate* glasses, L. Shartsis, W. Capps, and S. Spinner[214] measured the conductance and plotted log ϱ as a function of the alkali concentration.

[213] *Z. anorg. u. allgem. Chem.*, **279**, 1955, 300-312.

[214] *J. Am. Ceram. Soc.*, **36**, 1953, 319-326. On density measurements see *Ibid.*, **36**, 1953, 35-43. On alkali, silver and thallium borate melts and their conductance see B. I. Markin, *Stroenie*

The first additions of alkali to B_2O_3 glass cause the most rapid decrease in resistance, followed by a more moderate slope, with increasing concentrations. No boric acid anomaly effect is observed in these curves. Since independently the same authors had also measured the densities of the melts (cf. A. ¶ 183) it was possible to calculate *equivalent conductance* data, or the conductance of an amount of the melt containing one equivalent of alkali ions placed between parallel electrodes 1 cm. apart from each other. These data are most instructive in their course compared with the data from corresponding silicate melts (Fig. A. 86). The equivalent conductances *in*crease

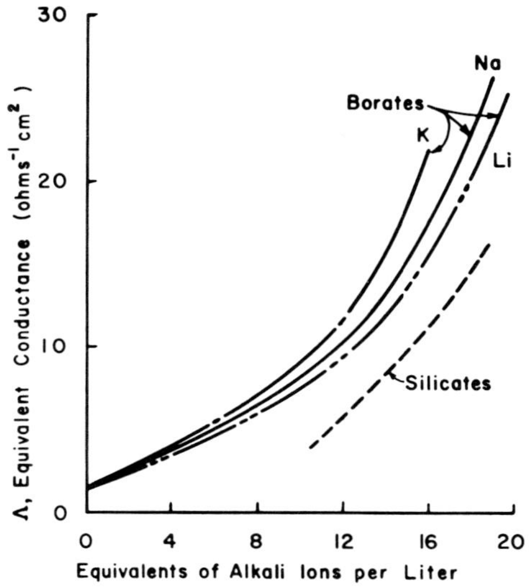

Fig. A. 86. Equivalent conductance of alkali borate and silicate melts at 900°C., as a function of the concentration in alkali ions. (Shartsis, Capps, and Spinner). The data on the same curves include the ions Li^+, Na^+, and K^+, equally.

with the alkali concentrations, in contrast with the well-known behavior of electrolytes in aqueous solutions where they must *de*crease under these conditions. The mechanism of the conduction in borate and silicate melts, therefore, *cannot* be described by the common Arrhenius electrolytic dissociation type but is more similar to that early (in 1805) discussed by Th. von Grotthusz[215] in which conduction takes

Stekla, Inst. Khim. Silikatov Akad. Nauk S.S.S.R., Trudy Soveshschaniya, Leningrad, 1953 (Pub. 1955), 264-266.

[215] On this highly interesting relation to a very old concept in electrochemistry cf. G. Kortüm and J. O'M. Bockris, "Textbook of Electrochemistry," Elsevier Publishing Co., New York, Amsterdam, London, Brussels, Vol. I, 1951, Vol. II, 1951, 882 pp., especially pp. 4 ff., and pp. 178 ff. on the difficulties of applying the Debye-Hückel theory for solvents with low dielectric constants. The strict application of the Grotthusz ideas would postulate that there should be at *very* low

place by means of ion pairs that change charge partners under the influence of an electric field. Since the anions in the present case are practically immobile, the current transport is effected by the cation movement alone. This type of conductance and the apparently unexpected increase of the equivalent conductance with increasing electrolyte concentration is observed in systems of solvents with a low dielectric constant.

142. The assumption of *association* phenomena and appearance of *complex ions* in the melts of fused salts and, therefore, also in silicates, borates, phosphates, and others, was recently especially discussed for a full understanding of their electrolysis behavior by E. R. van Artsdalen.[216] Molten salts are thus interpreted as very concentrated electrolytes (Ch. A. Kraus even speaks of solutions of the "solvent" in the fused salt as typical of the molecular state of such electrolytes) in which a medium with a high dielectric constant (mostly water) participates. The common definition of complex ions (especially anions) must from case to case be reexamined and redefined, experimental methods be modified, as by vapor pressure determinations to detect deviations from the ideal behavior. Only in the minority of particularly favorable cases one may expect a direct identification of such complexes and their more or less complete dissociation in the understanding of the Arrhenius theory of electrolysis. Model complex ions, such as $CdCl_3^-$, are important species which could be interpreted and be discussed in their dissociations. Nevertheless, in the case of silicate melts much more complicated and less evident conditions are to be expected. On the other hand, it is beyond mere speculation that T. Förland[217] derived from phase-equilibrium studies and cryoscopic measurements at high temperatures, namely in the system Na_2SiO_3–Na_2SO_4, the existance of annular complexes of metasilicate anions, such as of four tetrahedral groupings to a complex $[(SiO_3)_4^-]$ anion as the most probable configuration when dissolved in Na_2SO_4 melts (on other rings with 3 and 6 members cf. Vol. IV, Section A). By electrolytic determinations nothing certain can be said in favor or against this interesting consideration.

143. It is, however, not possible to find any indication of the existence of *aluminosilicate* anions from the electrical conductivity of sodium aluminosilicate glasses, as J. O. Isard[218] concluded from measurements of three different composition series.

concentrations in alkali a minimum of equivalent conductance, which, however, was not yet observed in experiments of Shartsis, Capps, and Spinner. On modern aspects of problems of solutions, in analogy with fused salts, and others, in the light of the ideas of Debye and Hückel, Onsager, and Bjerrum, s.a. Ch. A. Kraus, J. *Phys. Chem.*, **58**, 1954, 673-682; **60**, 1956, 129-141, on the "ion-pair concept."

[216] *J. Phys. Chem.*, **60**, 1956, 172-176.

[217] *J. Am. Ceram. Soc.*, **41**, 1958, 524-527.

[218] *J. Soc. Glass Technol.*, **43**, 1959, 113-123 T.

Nevertheless, the activation energy of conductivity shows a distinct minimum in compositions for which the ratio $Na^+/Al^{3+} = 1:1$. It varies as a reciprocal function of the radius of the oxygen anion shell surrounding one Na^+ ion in coordination groups $[NaO_n]$. The size of such groups increases the more $[AlO_4]$ groups are introduced into the framework of the glass structure, with a distinct limit for $Na^+/Al^{3+} = 1:1$. If more aluminum is introduced into the glass composition this cannot participate in the tetrahedral framework but is present in the form of excess $[AlO_6]$ configurations. At the same time, the coordination of the oxygen anions surrounding the central Na^+ ions starts decreasing in size.

144. How *devitrification* influences the resistivity of sodium silicate glasses, with a primary crystallization of $Na_2O.SiO_2$, $Na_2O.2SiO_2$, $Na_2O.2CaO.3SiO_2$, and $Na_2O.Al_2O_3.2SiO_2$ on one hand, and in titania containing glass of $CaO.TiO_2.SiO_2$ (titanite), was made evident by B. L. Joyner and W. C. Bell,[219] over a temperature range from 130° to 550°C. As a rule for the crystallization effects the change of resistivity is inversely proportional to the change in Na_2O contents of the glass phase remaining after the separation of the primary crystal phases mentioned before. There is, therefore, a possibility of foreseeing from the phase equilibrium diagrams of the systems Na_2O–CaO–SiO_2, Na_2O–Al_2O_3–SiO_2, and CaO–TiO_2–SiO_2 how the resistivity will be affected in the course of devitrification. Particularly evident is this change in a sodium aluminosilicate glass (with 5 per cent Al_2O_3, 45 per cent Na_2O) in which, after the primary crystallization the remaining glass is impoverished by not less than 15 per cent alkali, and, therefore, has a definitely higher resistance than the

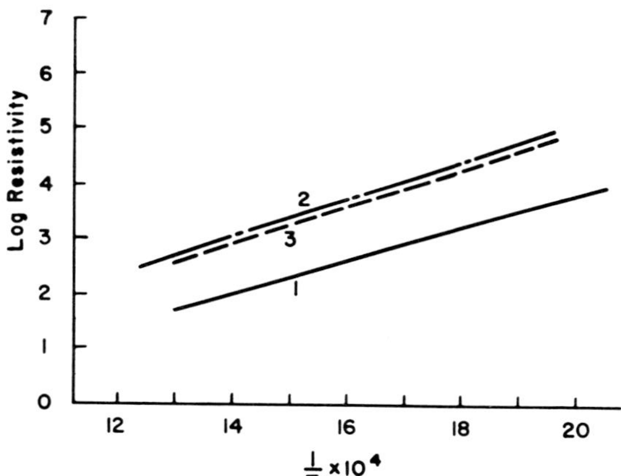

Fig. A. 87. Resistance of sodium aluminosilicate glass (with 45 per cent Na_2O; 5 per cent Al_2O_3), (*1*) in the original state, (*2*) after devitrification; (*3*) of the glass remaining after primary crystallization alone. (Joyner and Bell).

[219] *J. Am. Ceram. Soc.*, **36**, 1953, 263-266.

original glass (cf. Fig. A. 87). If the primary crystallization product does not contain Na_2O (e.g., $CaO.TiO_2.SiO_2$) the remaining glass is enriched in alkali, and resistance, therefore, is lower than in the original glass.

MEASUREMENT OF CONDUCTANCE OF SILICATES. POLARIZATION AND DIELECTRIC LOSSES

145. From experimental and methodological viewpoints of the conductance problems it is interesting that repeatedly in recent times the use of *direct current* in place of the classical Kohlrausch method (with alternating current of different frequencies) is again recommended. Thus, P. L. Kirby[220] describes the advantage of d.c. measurements if disturbing polarization effects were eliminated by an often repeated commutation. For this purpose he developed a relay system (alternator) of high speed for an efficient commutation, with a multivibrator circuit for electron valve tube control, as shown in Fig. A. 88. Also H. E. Taylor[221] worked with d.c. methods, for very high specific resistances from 10^8 to 5×10^{12} ohm.cm., with a Cambridge bench-type battery-operated pH meter even up to 10^{15} ohm.cm., as for

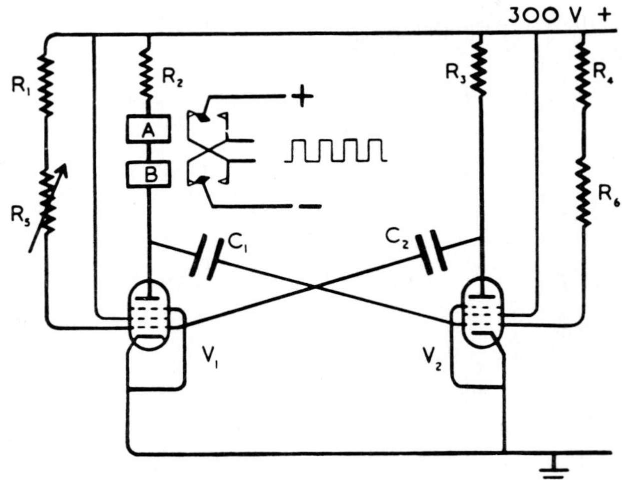

Key to Letters on Figure.

$R_1 = R_4 = 1$ Megohm. $R_6 = 100$K.
$R = 20$K (Kilohms) $C_1 = C_2 = 0{\cdot}05\mu$F.

Fig. A. 88. Multivibrator circuit for high-speed relay alternator. (Kirby).

[220] *J. Soc. Glass. Technol.*, **37**, 1953, 3-6.
[221] *J. Soc. Glass Technol.*, **39**, 1955, 193-204 T.

Pyrex glass examination. Glass discs of 3 mm. thickness are taken as samples, with an electrode assembly as shown in Fig. A. 89, and the circuit used in Fig. A. 90, later with a Ferranti electrometer valve (type BM8A) which operates upon a grid current of about 2×10^{-14} amp., combined with the pH meter.

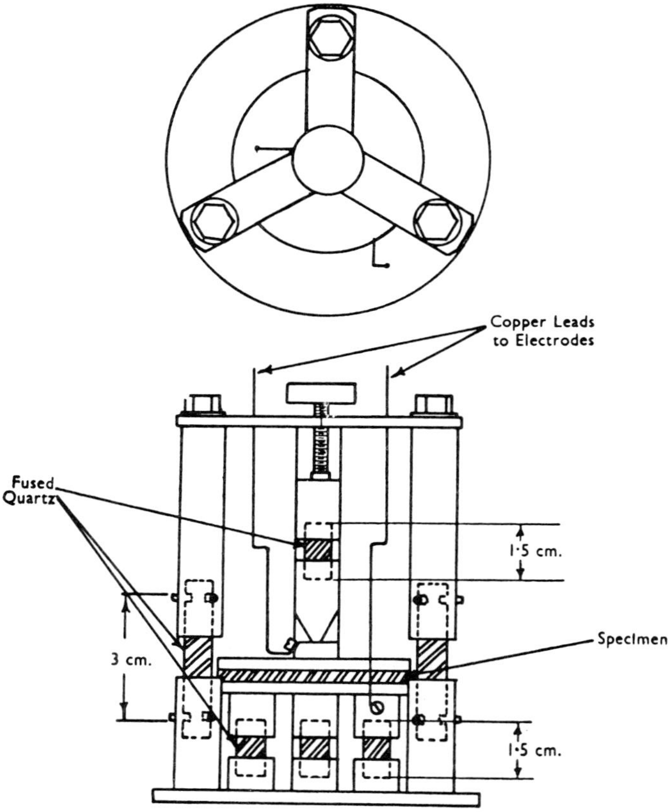

Fig. A. 89. Electrode assembly for measurements of high resistances of glass disc samples. (Taylor).

146. More recently, W. Weiss[222] also recommended the d.c. methods in routine measurements from room temperature to the transformation range, and W. Rothe[223] gave further recommendations for the same purpose in the rapid determination of the Gehlhoff T_{k100} temperature (for log $\varrho = 8$). In such d.c. measurements it is necessary to eliminate temporary or permanent surface conductivity effects of *water*

[222] *Glastech. Ber.*, 31, 1958, 76.
[223] *Silikat Tech.*, 12, 1961, 9-14.

films by the application of a hydrophobic film of silicone and a thermal treatment in high vacuum. By the use of short-time current flow of densities below 10^{-8} amp./cm.^{2} the disturbance by polarization is practically eliminated, especially if the direction of the current is reverted after each measurement. Even an automatation of such a routine method is possible for plant survey purposes. On the other hand, it is remarkable that Rothe prefers for a resistance below 10^6 ohm, the use of an alternating current method with a frequency of 100 or 500 cycles/sec.

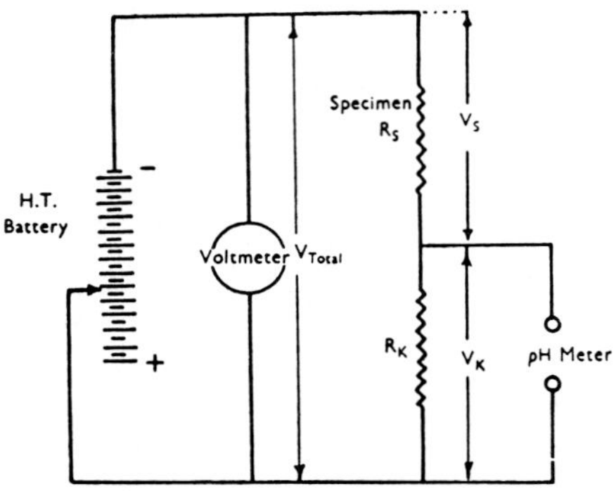

FIG. A. 90. Principle sketch of the *pH*-meter circuit used for the measurements of high resistance. (Taylor).

147. Typical determinations of conductance by the classical *alternating current* method were given for glasses of the system $Na_2O.SiO_2$–$PbO.SiO_2$–SiO_2 by K. S. Evstrop'ev[224] up to 1300°C., using a common Wheatstone bridge. C. Kröger and P. Weisgerber[225] more recently made measurements of conductance of salt melts with a very careful consideration of polarization effects as a function of frequency of alternating potentials on *blank* platinum electrodes. For fused nitrates the additional polarization resistance, ΔR, is proportional to the square root of the cyclic frequency, $\omega = 2\pi\nu$, for the range of $\omega = 1000$ to 4000/sec. which makes an accurate extrapolation to $\omega = \infty$ possible. Such measurements and tabulated data resulting therefrom are of a high value in the calibration of a.c. bridge methods, and also for determination of the conductance of silicate systems. As the Wheatstone bridge a model of the Wien type,[226] is indispensable, as zero indicator a Philips oscillograph GM 5655. A typical nomogram of resistance data found for fused KNO_3 for calibration purposes, with the extrapolation for $\omega = \infty$, is given in Fig. A. 91.

[224] "Physico-Chemical Properties of the Ternary System Sodium Oxide-Lead Oxide-Silica," editor I. V. Grebenshchikov, Moscow-Leningrad, 1949, 83-109.

[225] *Z. physik. Chem.* (*Leipzig*), **5**, 1955, 192-203.

[226] Cf. J. Jones and S. M. Christian, *J. Am. Chem. Soc.*, **57**, 1935, 272-280, and for the platinization of the electrodes pp. 280-284.

148. How considerable the *surface* electrical conductivity in glasses of the system $Na_2O.2SiO_2$–$PbO.SiO_2$ at room temperature or a little above it (up to 70°C.) can become was demonstrated by L. Yu. Kurts.[227] The moisture of the air in the measurement cell was 30 per cent. Freshly fractured surfaces were used as specimens. There is a particularly typical *time factor* in surface conductivity as a function of

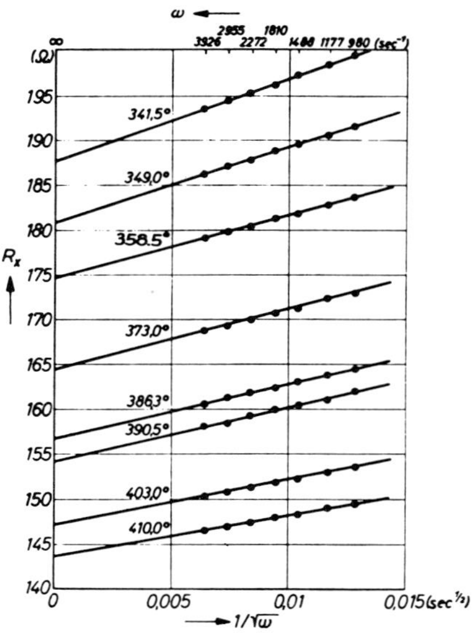

FIG. A. 91. Functional relation between the resistance of KNO_3 melts, and the alternating current frequencies, ω, with extrapolation to $\omega = \infty$. (Kröger and Weisgerber).

chemical composition. For glasses rich in PbO but containing only little $Na_2O.2SiO_2$, the curve of the surface conductance shows a maximum for the chemical compound $Na_2O.2PbO.4SiO_2$. Also the polarization curve, as a function of the chemical composition shows a singular point for this same compound. Polarization effects can become very important in slag melts of metallurgy. L. K. Gavrilov and O. A. Esin[228] gave an example in an electrolyte of 50 per cent CaO, 10 MgO, 10 Al_2O_3, and 30 SiO_2, at 1520°C. After interruption of the electrolysis current decay of the polarization takes place according to a law $\eta = c - d \cdot \ln t$ (t the time between 0.01 sec. and 30 minutes). When the current is switched on again, η is increased in the initial period from 0.02 to 0.2 sec. according to the relation $\ln [\eta'/(\eta' - \eta] = \beta \cdot t$ in which β and η' are empirical constants and η' differs from the value of steady polarization.

[227] "Physico-Chemical Properties of the Ternary System Sodium Oxide-Lead Oxide-Silica," editor I. V. Grebenshchikov, Moscow-Leningrad, 1949, pp. 110-122.

[228] *Zhur. Fiz. Khim.*, **29**, 1955, 566-575, 635-641; *Ibid.*, **30**, 1956, 374-378.

It is probable that for these polarization phenomena, an orientation and deformation of complex silicate anions is the characteristic which determines the rate of polarization.

149. Gavrilov and Esin further give polarization curves for a slag composition 46 CaO, 10 MgO, 43 SiO_2, and of some phosphides, between electrodes of ferrosilicon alloy (containing 23 per cent P) at 1490° to 1610°C. The type of these polarization curves is the same as it is for silicate melts between ferrosilicon electrodes alone. It is presumed that the silicon is present in Si^{2+} ions but it is difficult to reconcile the temperature coefficient of polarization with this interpretation. As a function of temperature, the cathodic and anodic polarization become equal at definite current densities. An increase of the temperature from 1440° to 1620°C. lowered the polarization (measured by commutation methods) by 15 to 30 times. They persisted after switching off the current at high temperatures, say for 30 minutes, and after cooling to room temperature for a day. It is difficult to intepret these facts simply by a slow discharge or a slow diffusion effect. They are thoroughly analogous to what is observed in rigid glasses.

150. Interesting also in this viewpoint is a proposal for improving the insulator functions of Steatite ceramics by a high-temperature treatment with d.c. as recommended by W. Soyck,[229] by pressing discs of this material together at 900°C. under a d.c. field of 1500 v./cm., and cooling in it. The original loss factor (determined for $\nu = 1000$ cycles/sec.) is changed from 10×10^{-4} to 8×10^{-4} on the anode and to 13×10^{-4} on the cathode. The change in specific resistance (measured with a.c.) was slightly higher in the cathode but increased about 12-fold at 800°C. and 6-fold at 1000°C. for the anode. The specific resistance (in a logarithmic scale) as a function of temperature is shown for the range from 700° to 1000°C. in Fig. A. 92. These interesting effects are evidently a consequence of electrolytic ion migration under the electrostatic potential and a "freezing-in" effect which causes the flow of a polarization current if the dielectric material is reheated, as in an accumulator, and is especially characteristic of feldspar containing ceramic bodies. The polarization current cannot be suppressed by applied a.c.

151. The practically important problems involved with *dielectric losses* in glasses in an alternating current potential field, especially as a function of the frequencies applied, were extensively investigated by J. M. Stevels[230] for systematically changed

[229] *Tonind.-Ztg. u. Keram. Rundschau*, **77**, 1953, 234; cf. the previous experiments of "freezing-in" electrolysis effects by K. Schaudinn, *Fachber. Ver. Deut. Elektrotech.*, **10**, 1938, 139-142; *Physik. Ber.*, **20**, 1939, 861; see also K. Backhaus, *Jahrb. Elektrowärme*, 1956, 447-459, especially 449-450.

[230] *Philips Research Repts.*, **5**, 1950, 23-36; **6**, 1951, 34-53; **7**, 1952, 161-68; *Verres et réfractaires*, **4**, 1950, 83-89; **5**, 1951, 4-14; **6**, 1952, 3-7; **7**, 1953, 91-104; *Glass Ind.*, **35**, 1954, 657-662; *Glastech. Ber.*, **26**, 1953, 227-241. See also H. Moore and R. C. De Silva, *J. Soc. Glass Technol.*, **36**,

glass compositions (mostly lead containing silicates and borosilicates, some of them also containing CaF_2). Characteristic at the constant frequency $\nu = 1.5 \times 10^6$ cycles/sec. is the appearance of a minimum for the loss angle, tan $\delta = 5 \times 10^{-4}$ and the power factor (at room temperature) in Na_2O containing glasses. If Na_2O substitutes in a series of magnesia-containing lead silicate glasses (with CaF_2), the shallow

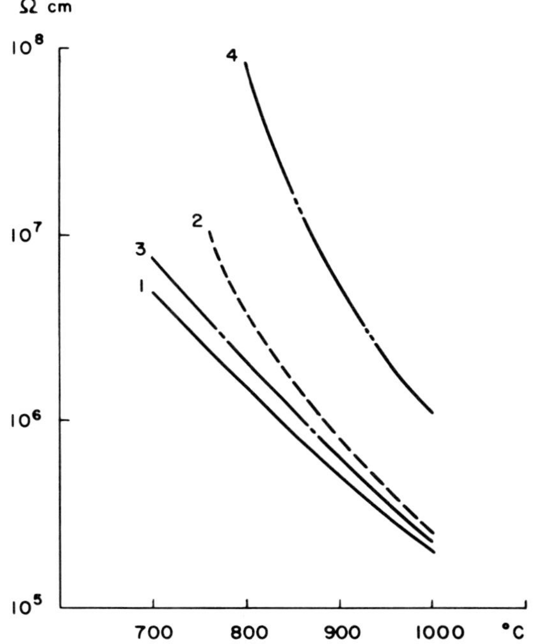

1 original state of a single plate
2 original state of both plates combined
3 plate on cathode side after electrostatic exposure
4 plate on anode side after electrostatic exposure

FIG. A. 92. Specific resistance of Steatite plates before, and after a thermal treatment with a direct current potential of 1500 volts/cm. (Soyck).

minimum at 17×10^{-4} (for 4 per cent Na_2O) is followed by a sharp maximum with 30×10^{-4} (for 16 per cent Na_2O). For potassium containing glasses the curves for tan δ as a function of composition become more complex, and strong maxima are observed (25×10^{-4}) for 6 and 8 per cent K_2O. The influences of molecular substitutions with *mixed alkali* compositions are interpreted by the structural compactness at a distinct ratio Na : K, or of Na : Li, and Li : K.

152. Likewise in systems of alkali-magnesia silicate glasses similar effects occur which indicate in this case the tendency of the Mg^{2+} ions to enter the framework in

1952, 5-55, on the theory of conductivity in its relation to the dielectric constant, the dielectric losses, power factor, and structural bonding in the glass structure (cf. B. ¶ 140), also with a rich experimental data material.

tetrahedral coordination (cf. A. ¶ 308). In this respect Zn^{2+} and Ni^{2+} have similar tendencies, which cause visible changes in color in nickel containing glasses. For alkali-lead silicate glasses a minimum of the power factor is characteristic for frequencies of $\nu = 1.5 \times 10^6$ and 2.4×10^{10} cycles/sec., along the transition line between the fields of "accumulation," with bridging oxygen anions, in the structure (AR), and of "destruction," with non-bridging oxygen anions (DR), as seen in Fig. A. 93. This transition line is the conjugation of SiO_2 with the point 18 per cent R_2O,

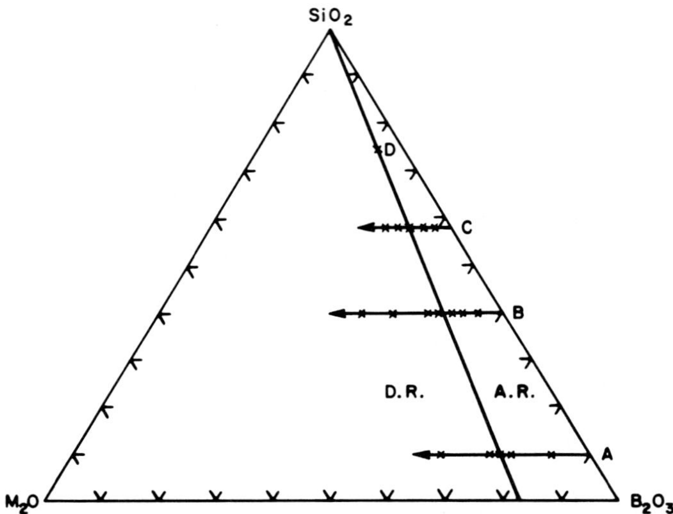

FIG. A. 93. System $R_2O—B_2O_3—SiO_2$, with the transition line dividing the fields of accumulation (*AR*), and destruction (*DR*). (Stevels).

and 82 per cent B_2O_3 in the ternary diagram SiO_2–R_2O–B_2O_3, as a line of minimum cohesion in the structural framework (one of the special cases involved is shown in Fig. A. 94). The minimum for the power factor depends on the silica content; it fades away with increasing SiO_2 and disappears entirely at 65 per cent SiO_2, whereas at 75 per cent it is replaced by a maximum. Probably, such effects are connected with phenomena of phase unmixing which are so characteristic of the Vycor type borosilicate glasses.

153. The theory of the dielectric losses[231] (for low values of the angle δ) demonstrates that the energy consumption observed per unit of time is approximately

[231] On the dispersion of the dielectric constant at very low frequencies, with surprisingly high electrode capacities (strongly dependent on frequency and temperature) see J. M. Stevels and C. van Amerongen, *Philips Research Repts.*, **8**, 1953, 452-470. Dielectric losses of glasses in the system Na_2O–PbO–SiO_2 were determined earlier by A. A. Khar'kov, "Physico-Chemical Proper-

$= \nu \cdot \tan \delta$. The loss angle is also a characteristic function of temperature, and Stevels discussed therefore a three-dimensional model of the $\tan \delta = f(\nu, T)$ relation, in the range from 0° to 600°K. and for $\nu = 10^{-2}$ to 10^{12}, to illustrate the widely different methods required for the measurements of $\tan \delta$, such as with the Klystron or a Magnetron instrument, in liquid nitrogen, hydrogen, and up to high furnace

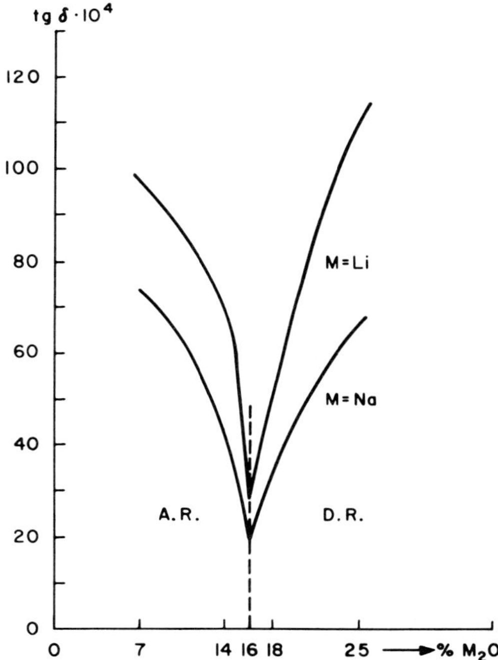

Fig. A. 94. Variation of $\tan \delta$ (for $\nu = 1.5 \times 10^6$ cycles/sec.) as a function of chemical composition, in the series 10 per cent SiO_2 + 90 per cent (R_2O + B_2O_3), for R = Li and Na. (Stevels).

temperatures. The model mentioned before has the shape of a half-moon-like mountain crest, surrounding a valley opening to the frequency axis, with maxima in the low and the high frequency ranges. There is an absolute minimum at $\nu = 10^3$, and 200°K., at room temperature a minimum at 10^6. The complex form of the section for constant temperatures —300°K. and 50°K. (see Fig. A. 96) shows the super-

ties of the Ternary System Na_2O–PbO–SiO_2," editor I. B. Grebenshchikov, Moscow-Leningrad, 1949, 158-168, for the frequencies $\nu = 1.7 \times 10^5$ to 1.7×10^7 cycles/sec., also as a function of temperature, with a minimum corresponding to the compound $Na_2O.2PbO.4SiO_2$. See also recently on dielectric resonance losses in aluminosilicates (natural and synthetic feldspars) V. I. Odelevskiĭ and A. F. Khomylev, *Structure of Glass. Proc. All-Union Conf. Glassy State, 3rd, Leningrad* (English Translation), 1959/60, 251-255, with anomalies caused by contaminations which are assimilated by a high temperature treatment (Fig. A. 95).

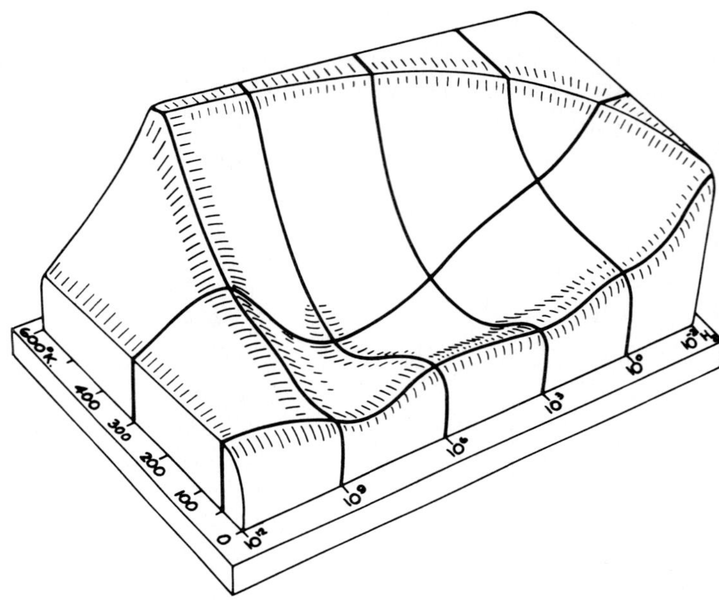

FIG. A. 95. Perspectivic representation of the three-dimensional function frequency (in Hertz units) - temperature - dielectric losses (tan δ). (Stevels).

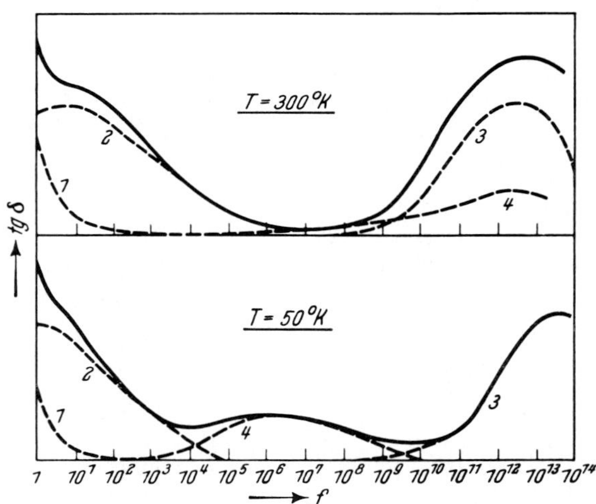

FIG. A. 96. Schematic frequency spectrum of dielectric losses for room temperature and 50°K. Full curve corresponds to total losses, curve (*1*) for conductivity; (*2*) for relaxation; (*3*) for vibration; (*4*) for deformation losses, as components. (Stevels).

position of *conduction losses* (negligible for ν below 50 cycles, of the form $\tan\delta = \varkappa/\omega\varepsilon'\varepsilon_0$) ($\varkappa$ the conductance, ε, the dielectric constant of the glass ε', and of the vacuum ε_0), but with an exponential temperature function, and of *relaxation losses*, with a maximum for circular frequencies $\omega = 2\pi\nu_{max.}$, near $1/\tau$ (τ the relaxation time, $\tan\delta = (\varepsilon'_s - \varepsilon'_\infty)/\varepsilon' \cdot \omega\,\tau/(1 + \omega^2\,\tau^2)$) ($\varepsilon'_s$ the static, ε_∞ the infinite dielectric constant).

154. There is a regular "spectrum" of these relaxation losses at room temperatures from $\nu = 10^{-3}$ to 10^6, making up the principal portion of the low-frequency region of the model "hills" (τ in the order of magnitude of 10^1 to 10^2 sec.). There are further *vibration losses*, important in the region of high frequencies, caused by the resonance of ionic vibrations in the glass structure, with the alternating electromagnetic field. The maximum of oscillation losses is for silicate glasses about 10^{12} in the infrared, for a phosphate glass with heavy structure-modifying cations at $\nu_{max.} = 10^9$. Finally there are *deformation losses* as relaxation losses of larger groups (or chains) in the glass structure which are deformed in the electromagnetic field, with very low τ values (in order of magnitude of 10^{-11} sec.) at room temperature. These last losses become important at very low temperatures, as with $\tau = 10^{-6}$ sec., and $\nu_{max.} = 10^6$. Summary curves of the dielectric losses are given by Stevels for a soft sodium-cal-

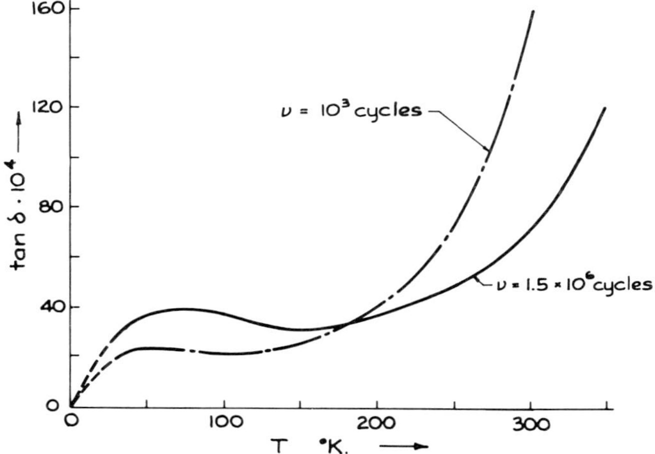

FIG. A. 97. Dielectric losses as a function of temperature for two different frequencies, ν, of a soft sodium-calcium silicate glass. (Stevels).

cium silicate glass, and a hard borosilicate glass (Fig. A. 97 and 98), for $\nu = 10^3$ and 1.5×10^6 cycles/sec., as a function of temperature, with characteristic minima between 40° and 100^oK. From the temperature shift of the maxima the activation energy and the relaxation times can be computed. With Stevels' model, and the rules derived from it, important problems of the systematic development of special glasses for radio technique purposes can be successfully challenged.

155. In many points, the results of Stevels agree with those given by V. A. Ioffe,[232] especially concerning the low-temperature minima of tan δ, say for potassium-magnesium and lead silicate glasses. Ioffe also discussed the structural tightening

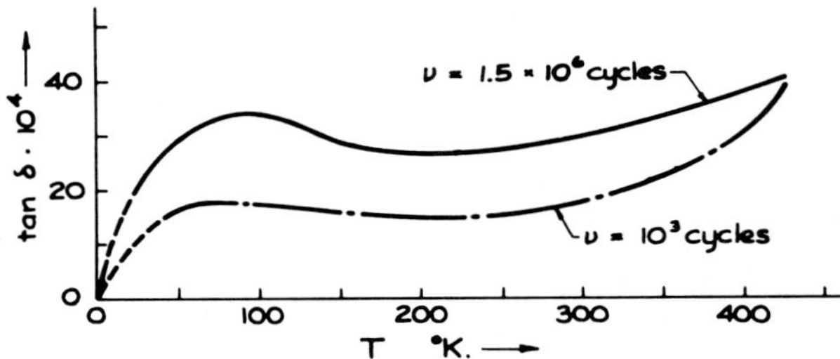

FIG. A. 98. Dielectric losses as a function of temperature, for two different frequencies, ν, for a hard borosilicate glass. (Stevels).

effect of mixed alkali ions (Fig. A. 99). For magnesia-lead silicate glass the activation energy of the vibrational losses were determined to be only 500 cal./mole, and a characteristic frequency $\nu = 10^7$, in other words, very low, whereas for ternary alkali silicate glasses it is about = 1000 cal./mole, and 10^{10} cycles/sec., respectively. For ultrashort waves ($\nu = 8.78 \times 10^8$ cycles) the tan δ values vary rather little with composition and for lead silicate glasses are nearly independent of the composition (Fig. A. 100).

156. Dielectric relaxation was extensively studied by H. E. Taylor,[233] from the theoretical as well as from the experimental basis, for commercial sodium-calcium silicate glasses, and for Pyrex borosilicate, over a range of ν from 100 to 20,000 cycles/sec., from room temperature up to 400°C., using a Schering a.c. bridge, and a special bridge for high power factors. In the latter case experiments were at room temperature from 10^4 up to 10^8 cycles and in close correlation with the d.c. data of conductance of the same glasses. The activation energy of the latter conductance is very similar to the activation energy of dielectric relaxation since in both phenomena the migration of Na^+ ions is the fundamental mechanism. These effects were more

[232] *Stroenie Stekla, Inst. Khim. Silikatov Akad. Nauk S.S.S.R., Trudy Soveshchaniya, Leningrad,* 1953 (Publ. 1955), 258-263.

[233] *J. Soc. Glass Technol.*, **41**, 1957, 350-382; **43**, 1959, 124-146 T. Particularly remarkable are also the measurements of the dielectric constant ε' and $(\varepsilon' - \varepsilon_\infty)$ as a function of temperature, cf. pp. 370-374 of the first publication, and pp. 126-131 of the second. On dielectric properties and ionic diffusion in glasses, see also R. J. Charles, *J. Am. Ceram. Soc.*, **45**, 1962, 105-113.

specifically studied in synthetic sodium silicate and sodium-calcium silicate glasses. A graphical estimate could be made for distribution of the relaxation times involved in these processes. This distribution is very similar for all these glasses and may be interpreted as corresponding to the similarity of the structural principles in the random framework.

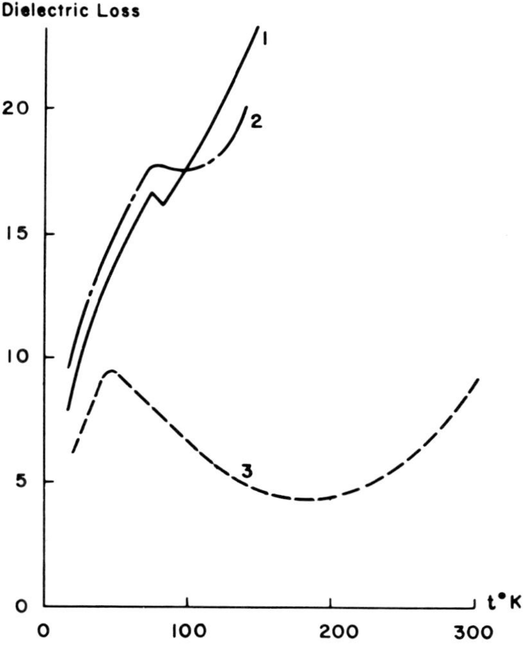

FIG. A. 99. Dielectric losses for alkali silicate glasses (ν constant $= 10^6$ cycles/sec.), (*1*) with 25 per cent Na_2O; (*2*) with 23 per cent K_2O; (*3*) with 12.5 per cent Na_2O; 12.5 per cent K_2O. (Ioffe).

157. The influence of thermal history effects on the dielectric losses of plate glass was studied by D. W. Rinehart,[234] if first heated to 480°C. and slowly cooled to room temperature. The glass was then reheated and quenched from temperatures in the range from 650° down to 90°C. The dielectric losses were measured in the final samples, in comparison with the conductivity phenomena of strained, quickly cooled, and annealed samples, for a constant frequency of 1000 cycles/sec. Characteristic, in the first place, are the irreversible increases in tan δ at quenching temperatures

[234] *J. Am. Ceram. Soc.*, **41**, 1958, 470-475. The hole theory is also particularly important for the explanation of the conductance of inorganic glasses at low temperatures with anomalously increased activation energies of ionic mobility. Cf. Par. B. 155, and Ch. Hirayama and D. Berg, *Am. Ceram. Soc. Bull.*, **40**, 1961, 550-555.

below the transformation range of the glass (540° to 590°C.) as well as within it. They are interpreted as a consequence of the sodium ion distributions, established during cooling among the interstitial holes of the glass structure. Combined with determinations of density, it is possible to conclude that the increases in tan δ ob-

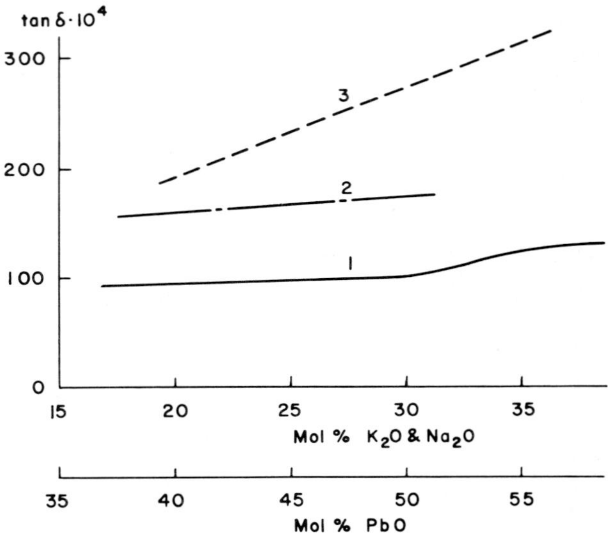

FIG. A. 100. Dielectric losses for ultrashort waves ($\nu = 8.78 \times 10^8$ cycles/sec.), as a function of chemical composition, for alkali and lead silicate glasses. (Ioffe). Curve (*1*) for $PbO—SiO_2$; (*2*) for $K_2O—SiO_2$; (*3*) for $Na_2O—SiO_2$.

served within the transformation range are caused by configuration changes in the silica framework skeleton. How a compaction of the structure influences the tan δ data is seen from the family of curves shown in Fig. A. 101 for tan δ versus quenching temperatures. The configuration state of the glass depends on the thermal history within the transformation range, and influences the specific volume *and* the ionic mobility, whereas the distribution state of the mobile ions observed in the thermal history effects below the transformation range *only* concerns ionic mobility.

158. Concerning the relations existing between the mechanical relaxation of alkali ions and dielectric relaxation phenomena in glasses we wish to avoid repetitions of what was extensively discussed in A. ¶ 53 and 62 f., on the inner friction and damping, as a function of frequency in the mechanical spectrum, especially in the *anelastic* behavior of glass, and refer to the important investigations of J. V. Fitzgerald and P. L. Kirby *et al.* there mentioned.[235]

[235] J. V. Fitzgerald, *J. Soc. Glass Technol.*, **36**, 1952, 90-104 (with V. M. Laing, and G. S. Bach-

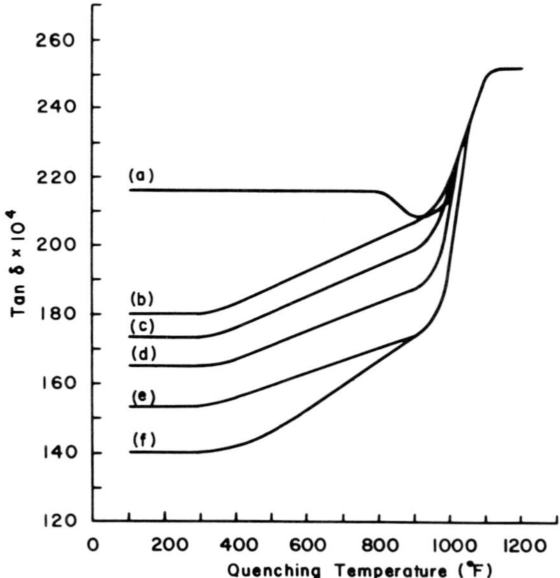

Fig. A. 101. Tan δ versus temperatures (ν constant = 1000 cycles/sec.), for plate glass in different degrees of annealing. (Rinehart). Curve (*a*) for commercial annealing; (*b*) through (*e*) heated at 480°C. for 1 hour, 5 hours, 24 hours, 1 week, respectively, followed by normal cooling. Curve (*f*) for 1 week at 480°C., followed by a prolonged stepwise cooling (in steps of 28°C. every 24 hours, down to room temperature).

159. In their extensive measurements of electrical resistance as a function of temperature (in comparison with the temperature coefficient of thermal esxpansion), S. M. Cox, J. F. Stirling, and P. L. Kirby[236] studied a Pyrex glass type for these properties in the *annealing range*, with the characteristic discontinuity in the point T_g (Tammann), and predicted theoretically the existence of a second discontinuity, called T_0. This was later established in the studies of P. Beyersdorfer[237] for Jena optical glasses, and is interpreted by this author as caused by *stabilization* effects of inner equilibria in the glass constitution and by time effects for the specific resistance at constant temperature. By a graphic method (as shown in Fig. A. 102) the temperature T_0 of stabilization is determined as above T_g and cannot be meas-

man); *J. Am. Ceram. Soc.*, **34**, 1951, 314-321; 339-342; 383-391; P. L. Kirby, *J. Soc. Glass Technol.*, **37**, 1953, 7-26; *Ibid.*, **38**, 1954, 383-420; 548-560 T; *Verres et réfractaires*, **9**, 1955, 197-203, especially also in connection with studies of J. W. Marx and J. M. Sivertsen, *J. Appl. Phys.*, **24**, 1953, 81-87, e.g., on the Na^+ peak at frequency ν = 37,000 cycles/sec., and 300°C.

[236] *J. Soc. Glass. Technol.*, **35**, 1951, 103-135, especially pp. 132-135.

[237] *Silikat Tech.*, **5**, 1954, 459-462. On the previous literature see W. A. Weyl in "Phase Transformations in Solids," edited by R. Smoluchowski, J. E. Mayer, and W. A. Weyl, John Wiley & Sons, Inc., New York 1951, pp. 296-334, especially pp. 322 f.

ured directly because of the deformation of the sample. The specific resistance, ϱ, increases with progressing stabilization. An example is given in the two breakpoints indicated in Fig. A. 103 for an optical glass annealed in common industrial lehrs, whereas quenched samples show only one breakpoint for $T_0 = T_g$.

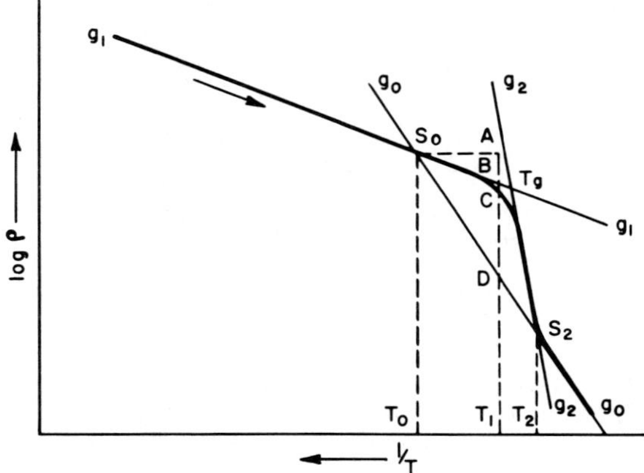

Fig. A. 102. Graphical construction of the points S_0 and T_0, the corresponding temperature, from the function for log ϱ to $1/T$. (Beyersdorfer).

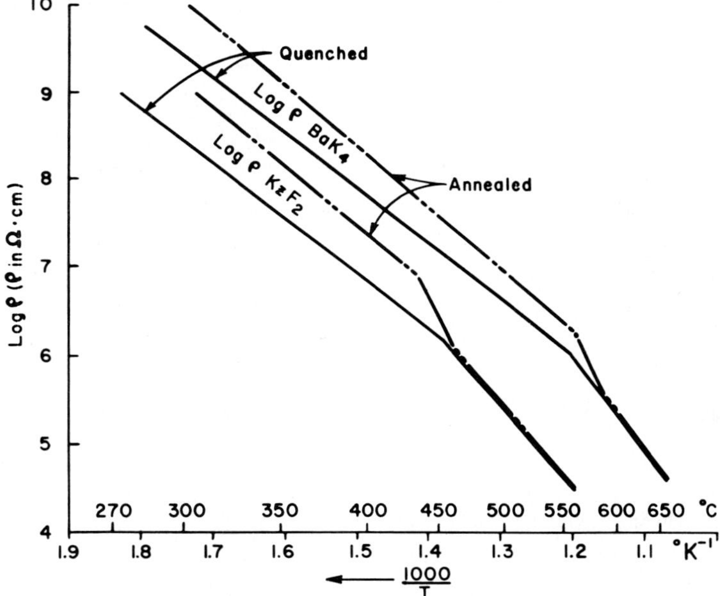

Fig. A. 103. Projection of the functional relation of log ϱ versus $1/T$, for quenched and annealed optical glasses. (Beyersdorfer).

160. More from the theoretical viewpoint (cf. A. ¶ 270 ff.), E. U. Condon[238] discussed the transformation range in its effects on the specific resistance in connection with the so-called *"fictive temperature"* (A. Q. Tool) which characterizes the internal glass structure below the transformation range independently of the thermal history by which the final state is produced. The aspects given by Condon fit very well into a full interpretation of the experimental data of J. T. Littleton and W. L. Wetmore (1936), and of the idea of "frozen-in" structural effects in the glass framework (cf. A. ¶ 265).

161. Finally we may shortly discuss the experiments of V. K. Rohatgi[239] on *photoelectric* effects in borosilicate glasses as quantum actions for different wavelengths *in vacuo*, such as at photon emission from a mercury arc lamp ($\lambda = 2537$ Å.), and a glow discharge in pure argon or neon, from 10^{-1} to 10 mm. Hg pressure. With a potential of 1000 volts/cm. photoelectric current densities of the order of magnitude of 10^{-12} amp./cm.^{2} are evolved for the full light of the mercury lamp focussed by quartz optics. Soda glasses yielded current in proportion to the density of sodium atoms, therefore it is plausible that the photoelectric current of borosilicate glass is derived from sodium contaminations on the surface. Since Na^+ ions next to a cathode cannot leave the glass, electrons enter the glass from the metal cathode with a linear work function to neutralize the ion space charge building up near the cathode. Photons above 3300 Å. do *not* contribute to conductivity in any fashion. The photoelectric yield for $\lambda = 2537$ Å. was determined by a vacuum bolometer.

162. The old problems involved with correlation of electrical *resistance and viscosity* for glasses in their very evident analogies, as previously formulated in the classical P. Walden rule (1906) and later by C. L. Babcock (1934) were extensively studied by J. Yamamoto,[240] and especially for alkali borate melts by L. Shartsis, W. Capps, and S. Spinner[241] over the temperature range from 600° to 1000°C. The important result was deduced that the rate of increase in equivalent conductance with alkali concentration (cf. A. ¶ 141) which is so typical of a liquid medium with a low dielectric constant, was greater in the concentration range in which the viscosity increases than when the viscosity decreases. It is concluded, therefore, that there is no direct controlling relation between viscosity and resistance. In addition, a graphical projection of the experimental resistance data and of $\log \eta$ showed straight lines the slopes of which decrease with increasing alkali contents (Fig. A. 104). The same is valid for optical glasses. Shartsis et al. developed, therefore, a working hypothesis that resistance and viscosity of glass are both related to the concentration of weak bonds in the glass structure. Such an idea can account for the evident constancy of

[238] *Am. J. Phys.*, **22**, 1954, 132-142.
[239] *J. Appl. Phys.*, **28**, 1957, 951-959.
[240] *Yogyo Kyokai Shi*, **60**, 1952, 379-382; **61**, 1953, 108-111, 211-214.
[241] *J. Am. Ceram. Soc.*, **36**, 1953, 319-326.

the relation of the temperature coefficients for the viscosity as well as for resistance. The parameter $B = (d \log \eta / d \log R)$ as a function of the alkali concentration (Fig. A. 105) may indicate the order $Li^+ > Na^+ > K^+$ in the scattering of the observed values of B. They show a tendency to increase with conductance for simple as well as for complex glass compositions (e.g., optical glass types).

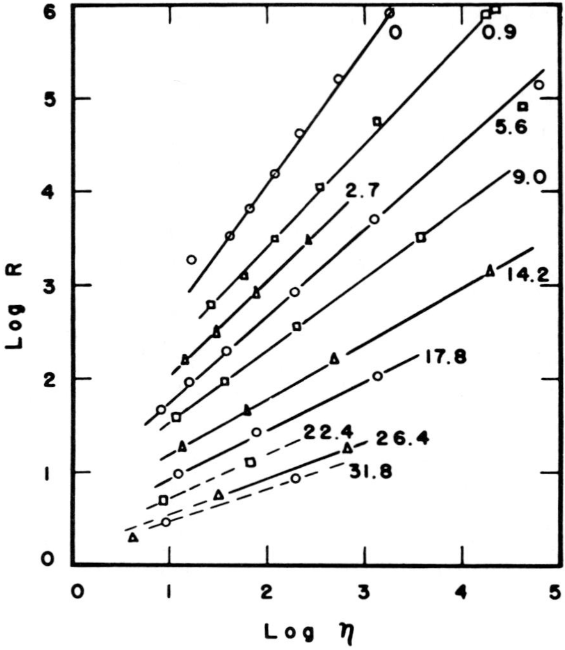

Fig. A. 104. Relation between electrical resistance, R, and log η for melts of sodium borates. (Shartsis, Capps, and Spinner). The numbers indicated on each line are the percentages in Na_2O.

163. More from the practical viewpoint, N. V. Zaimskikh and O. A. Esin[242] determined in a special viscosimeter simultaneously the electric conductance of welding slag melts, in wide limits of composition (SiO_2 from 20 to 52 per cent; MnO from 4.5 to 45 per cent; FeO from 0 to 26 per cent; TiO_2 from 0 to 21 per cent; MgO from 0 to 2.6 per cent; CaO from 0 to 48.1 per cent; Al_2O_3 from 0 to 13 per cent; CaF_2 from 0 to 8.22 per cent), at 1420°C. with a.c. and a Thompson bridge. Highest conductance and lowest viscosity are shown by slags high in FeO and MnO. The reversed behavior is characteristic of slags of the system $CaO–Al_2O_3–SiO_2$, as generally typical of siliceous slags. The opposite curves of viscosity and resistance as a function of temperature are seen also in C. ¶ 60.

[242] *Zhur. Priklad. Khim.*, **26**, 1953, 70-76.

GALVANIC POTENTIALS WITH SILICATE MELT ELECTROLYTES

164. Possibilities for developing reproducible high-temperature reference electrodes (working up to 1300°C.) in fused salts were systematically examined by M. Rey and O. Danner,[243] say of the type $Ag/(K_2SO_4 + Li_2SO_4)$, in the eutectic composition of the salt mixture, the latter representing at the same time a 0.01 molar solution of Ag_2SO_4, useful in glass containers up to 535°C. For the electrode Ag/AgCl the container must be of hard porcelain, working in a pure argon atmosphere. Other sulfate electrolyte electrodes up to 1300°C. (with a lifetime of 10 hours and a reproducibility of ± 100 millivolts) were used for the study of cathodic and anodic polarization in electrolytes at 830°C., or for lead and vanadium silicate glasses at 800° in air, with nichrome or zirconium boride (ZrB_2) electrodes. Platinum electrodes were used at 1200°C. for lead silicate melts. For gold-silver alloys, Thuringian glass was used as an electrolyte (containing silver ions) with a silver reference electrode, by

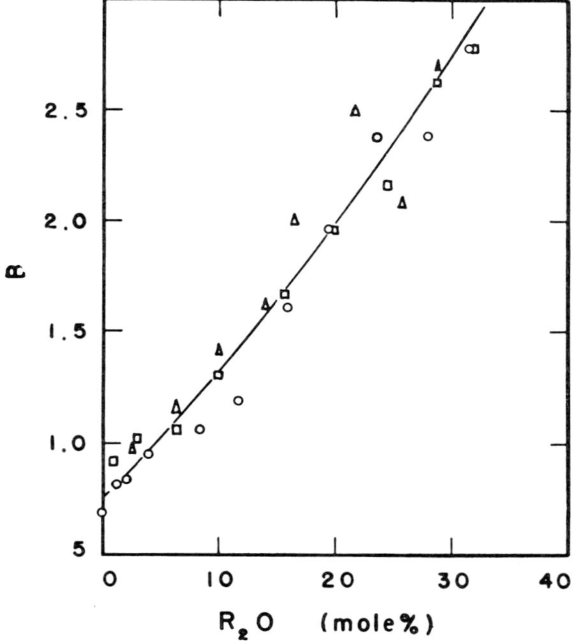

FIG. A. 105. Relation of $B = (d \log \eta / d \log R)$ to concentration in alkali oxide, for alkali borate melts. (Shartsis, Capps, and Spinner).

[243] *Silicates Inds.*, 25, 1960, 19-22.

O. Kubaschewski and O. Hüchler,[244] achieving excellent success in determination of the thermodynamic fundamental parameters ΔF, ΔH, and ΔS for the solution of silver in gold, with a maximum error of only 2 per cent.

165. Glass electrolytes and electrodes combined in galvanic *concentration cells* of the type $Pb/(PbCl_2 + KCl)_{\text{concentration I}}/Glass/(PbCl_2 + KCl)_{\text{concentration II}}/Pb$ were already studied by B. von Lengyel and A. Sammt (1937) with characteristic potential drops of the glass with the liquid fused salt electrolyte across the interface, which decays with time as L. Reed[245] investigated in different glass/salt pairs. A

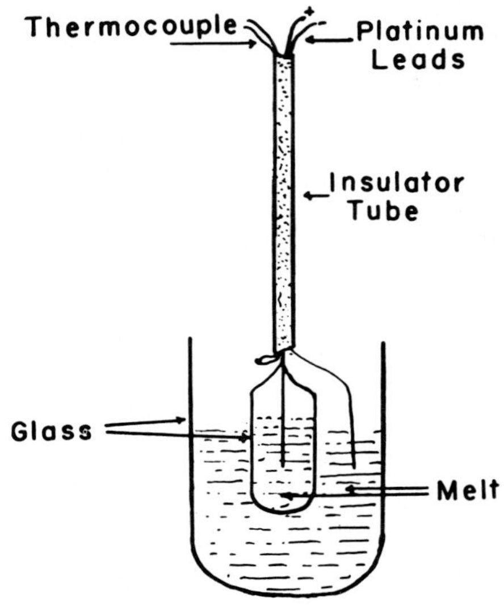

FIG. A. 106. Glass/fused salt cell assembly for the study of electromotive forces in galvanic cells. (Reed). $LiNO_3$ is used as electrolyte melt inside and outside the glass tube system.

qualitative explanation of this phenomenon was found in considering the glass structure with relation to the adsorption and polarizability of the salt ions on the interface. The experimental arrangement shown in Fig. A. 106 (with $LiNO_3$ as fused salt electrolyte in the inner and outer tubes) gave at 430°C. a typical asymmetric potential versus time curve (Fig. A. 107). Similar curves are observed with other combinations of electrolytes and electrodes. Concerning the glass electrode there was a surprisingly small difference between annealed and nonannealed tubes. To understand the mechanism of the rather complex potential drops observed, one must consider the sources of strains by temperature gradients and interdiffusion of ions,

[244] *Z. Elektrochem.*, **52**, 1948, 170-177.
[245] *J. Am. Ceram. Soc.*, **38**, 1955, 131-135.

and also water adsorption and release on the glass surface. When a continuous liquid film is formed over the inner glass surface the maximum potential drop occurs and the highest rate of cation penetration into the glass from the fused salt electrolyte is reached. Excess melt cations are held to the glass surface by adsorption and residual valence forces. These ions tend to move through the glass and to polarize it. As the glass surface becomes increasingly saturated with melt cations, the tendency for cations to be adsorbed decreases and a gradual decay of the potential difference follows between the electrodes. The lower the temperature and the larger the melt cation, the smaller its force field and the lower the rate of decay in adsorption potential difference.

166. In the same way that glass displays electric contact potentials, single crystals and refractories in the sintered state behave at high temperatures, as in the walls of glass tanks in contact with the melt bath. These effects are an important source of corrosion phenomena which were intensely studied by E. Plumat, I. Peychès *et al.* Before we discuss these investigations (cf. A. ¶ 172), we recall the importance of galvanic potentials measured in glass/fused salt (or oxide) systems for metallurgical slags and their constitution, which were especially promoted by O. A. Esin *et al.* He assumed preferably the presence of *complex anions* of the kind which was previously discussed under general aspects given by E. R. van Artsdalen,[246] and also important for the interpretation of the constitution, for example, of the system Na_3AlF_6–Al_2O_3 which is the electrolyte melt of aluminum electrometallurgy.[247] Here anions of the type $Al_3O_2F_x$ are formed, by the addition of AlF_6^{3+} to $(AlO_2F_2)^{3-}$, forming

$$\begin{array}{c} F_2 \\ | \\ F_5Al—AlO—AlF_5 \\ | \\ F_2 \end{array}$$

as was confirmed by cryoscopic measurements. The existence of such complexes with two oxygen atoms in connection with a central Al atom is a phenomenon which is also repeated in complexes with Ti and Si.

167. The specific significance of the electrochemical investigations of O. A. Esin with K. Gavrilov and B. M. Lepinskikh[248] consists in systematic measurements of

[246] *J. Phys. Chem.*, **60**, 1956, 172-176.

[247] Cf. T. Förland, *J. Metals*, **9**, 1957, 1380; T. Förland, H. Storegraven and S. Urnes, *Z. anorg. u. allgem. Chem.*, **279**, 1955, 205-211.

[248] *Doklady Akad. Nauk S.S.S.R.*, **88**, 1953, 713-716, *Silikat Tech.*, **6**, 1955, 10-11. Cf. L. C. Chang and G. Derge, *Am. Inst. Mining Met. Eng. Tech. Publ.*, 1946, No. 2101; *Metals Technol.*, **13**, 1946 (7) 27 pp. Recently, B. M. Lepinskikh, O. A. Esin, and V. I. Musikhin, *Structure of Glass. Proc. All-Union Conf. Glassy State, 3rd, Leningrad* (English Translation), 1959/60, 104-106, gave a positive evidence of the existence of complex anions in melts of the systems Na_2O–SiO_2

galvanic potentials for the discrimination of complex silicate and even aluminosilicate ions in metallurgical slag melts, say those written in the form $[(Si_xO_y)_n]^{3-}$. Typical galvanic cells used for this purpose with electrodes of ferrosilicon and auxiliary

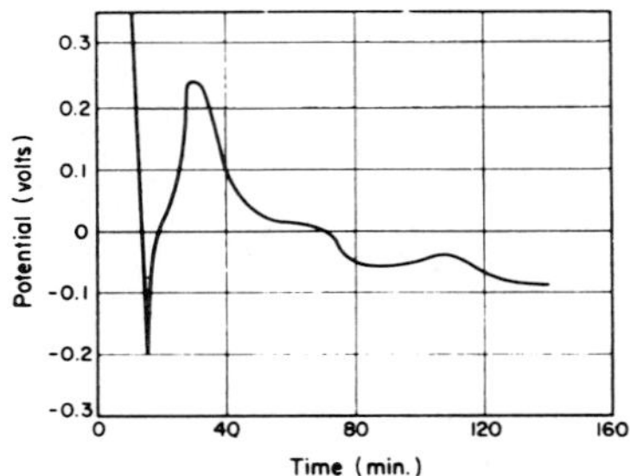

Fig. A. 107. Asymmetric potential development at 430° C. using glass/melt electrolyte contact pairs, with $LiNO_3$ as fused salt. (Reed).

(oxygen controlling) electrodes of magnesia or graphite are shown in the sketch Fig. A. 108. The slag composition was on one side 50 per cent CaO, 38 Al_2O_3, 10 MgO, and 2 SiO_2; in the other half cell the same, but with variations of SiO_2 contents up to 65 per cent. The presence of the auxiliary magnesia or graphite electrodes is indispensable for the determination of the activity of the oxides in the cell composition. A typical curve showing the electromotive force (EMF) at 1470°C. for such

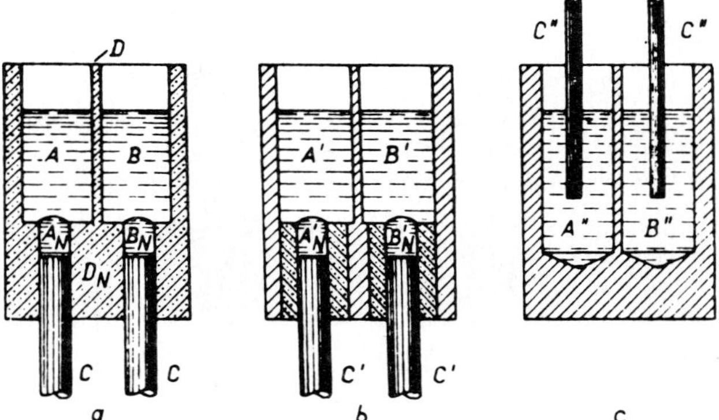

Fig. A. 108. Type of concentration cells for slag melts, (*a*) with magnesia intermediate electrode; (*b*) with graphite and magnesia electrodes; (*c*) only with graphite electrodes. (Eisin, Gavrilov, and Lepinskikh).

and CaO–B_2O_3, by the observation of a distinct anisotropy conductance under flow. The conductance is increased in the direction of flow, and decreased in the transverse direction.

a cell as a function of the silica concentration is given in Fig. A. 109, with distinct breakpoints at 33 per cent and 53 per cent SiO_2, corresponding in the first point to the presence of the anion $(SiO_4)^{4-}$ which is associated with $[(SiO_3)_2^{2-}]$ between the

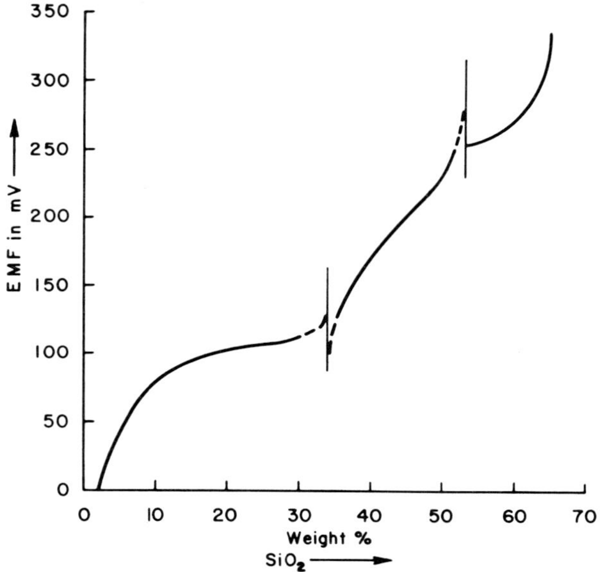

FIG. A. 109. Electromotive force of silicate slag melts with magnesia and graphite electrodes in concentration cells, as a function of silica contents, with steps indicating complex anion formation. (Esin, Gavrilov, and Lepinskikh).

first and the second breakpoint, whereas above 53 per cent silica the complex metasilicate anion alone occurs in the melts. The EMF are formulated

$$E = E_0 + \frac{RT}{4F} \cdot \ln \left(\frac{a_{SiO_4^{4-}}}{a^2_{O^{2-}}} \right) = E_0 + \frac{RT}{4F} \cdot \ln a_{SiO_2}$$

or only dependent on the activity of SiO_2. For concentrations above 53 per cent SiO_2 is

$$E = E_0 + \frac{RT}{4\,nF} \cdot \ln \left(\frac{a^2{}_{(SiO_3)_n^{2-}}}{a^n{}_{(SiO_4)^{4-}}} \right)$$

168. The existence of aluminosilicate complex anions is based on the congruent fusion equilibrium of the aluminates $CaO.Al_2O_3$ and $5CaO.3Al_2O_3$, and of $CaO.Al_2O_3.2SiO_2$, and $2CaO.Al_2O_3.SiO_2$. In this case, Esin and Lepinskikh also used melt electrolytes of quaternary $CaO–MgO–Al_2O_3–SiO_2$ slags, with variable concentrations in alumina, and electrodes of iron-aluminum-carbon alloys, at 1450°C. in a graphite crucible. The EMF plotted as a function of the concentration in alumina shows two discontinuities corresponding to the transition of complex aluminate into aluminosilicate anions. Curves of constant EMF plotted in the system $CaO–Al_2O_3–SiO_2$ (Fig. A. 110) show distinctly the field in which aluminum cations appear, and another field for lime-rich compositions in which aluminum participates in complex

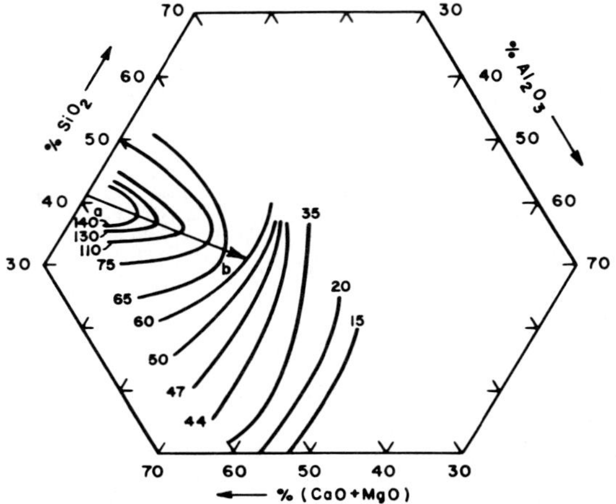

Fig. A. 110. Curves of constant electromotive force, starting from the standard slag composition 40.2 CaO; 8.4 MgO; 50.5 Al_2O_3; 9 SiO_2. (Esin, Gavrilov, and Lepinskikh).

aluminosilicate anions,[249] both fields separated by the line *a–b*. The EMF is formulated

$$E = \frac{RT}{6F} \cdot \ln \left(\frac{a'_{Al_2O_3}}{a''_{Al_2O_3}} \right)$$

I. S. Kaĭnarskiĭ and E. V. Degtyareva[250] made the interesting attempt to apply Esin's conclusions on the existence of different complex silicate anions in silicate melts to a theory of the mineralizing effects exemplified in the important change of silica glass to tridymite (cf. Vol. V, Section B), when the complex anions play an intermediate role in the nucleation of the tridymite framework, between the true framework with Si: O = 1 : 2, and the hexagonal layer-network for Si : O = 1 : 2.5 (cf. Vol. III, A. ¶ 75 f.). Melts with the anions $(Si_2O_5)^{2-}$ with Si : O = 2.50 to 2.75 exert a minor mineralizing effect, and the chain structures in $(SiO_3)^{2-}$ anions have the lowest effect. Aluminosilicate anions, finally, are not favorable for formation and nucleation of tridymite.

[249] *Doklady Akad. Nauk S.S.S.R.*, **91**, 1953, 1187-1190.
[250] *Doklady Akad. Nauk S.S.S.R.*, **99**, 1954, 301-304.

169. Yu. P. Nikitin and O. A. Esin[251] especially studied the *electrocapillary* phenomena in slag/metal systems on contact surfaces with Cu_2S, Ni_3S_2, and other slags in the temperature range from 1350° to 1500°C. (cf. A. ¶ 110 ff.) for a determination of the interfacial tensions, and further V. V. Khlynov and O. A. Esin[252] reported the resulting metal losses by capillary effects in two-liquid systems at 1370°/1500°C. and the migration of suspended droplets in an electrostatic field. It is in this way possible to remove almost completely sulfide inclusions from slags by the application of an exterior field of 5 to 7 volts/cm., with rates of the order of magnitude of 7 to 10×10^{-3} cm./sec. In principle, even more significant are the *exchange currents* flowing between liquid metals and slags as observed during capacity measurements of such systems in a magnesia crucible at 1480° to 1560°C. The capacity of such systems was practically independent of the frequency and low ion concentrations determine the contact potentials. An anomalously high intensity of such exchange currents (880 mamp./cm.^{2}) was observed on contacts of iron with a slag with 28 per cent FeO, 13 CaO, 24 Al_2O_3, and 35 SiO_2. Higher intensities resulted if sulfur was introduced to the metal in contact with a 1.72 per cent FeO and 2.25 per cent CaS containing calcium aluminate slag. One with 36 per cent MnO with molten Mn–C and Mn–Si alloys gave even intensities above 1100 mamp./cm.^{2}. Valence interchanges of the type $R^{2+} \rightleftharpoons R^{3+}$ in the solution phase were important for rapid electron exchange on the contacts.

170. In continuing the previous investigations of L. Ch. Chang and G. Derge[253] on measurements of reversible electromotive forces (EMF) in metal/slag melt systems, especially concerning the systems CaO–SiO_2 and CaO–Al_2O_3–SiO_2 for the slag compositions, with basic thermodynamic calculations of reaction parameters, W. A. Fischer and A. Hoffmann[254] extended the experimental methods of galvanic measurements up to 1500°C. in a systematic study of the system FeO–Al_2O_3 using the cell Pt/Al_2O_3/(Al_2O_3 + FeO)/Pt (cf. Vol. III, B. ¶ 341, and Vol. V, Section A), with the formation of hercynite, FeO.Al_2O_3, as the stable spinel phase, indicated by the typical EMF curve shown in Fig. A. 111, as a function of the concentration in FeO. W. A. Fischer and R. Schäfer[255] also gave important information on the electric potentials in metallurgical systems consisting of iron melts (4 per cent carbon) with the furnace lining of refractory products built up of corundum or magnesia, or with bonding clay-sand ("Klebsand"), and corresponding thermoelectric current potentials

[251] *Doklady Akad. Nauk S.S.S.R.*, **107**, 1956, 847-849; **122**, 1958, 106-108, on kinetics of ion exchanges.

[252] *Doklady Akad. Nauk S.S.S.R.*, **120**, 1958, 134-136; **123**, 1958, 330-332; see also Yu. P. Nikitin and O. A. Esin, *Ibid.*, **116**, 1957, 63-65.

[253] *Am. Inst. Mining Met. Eng. Tech. Publ.*, 1946, No. 2101; *Metals Technol.*, **13**, 1946 (7) 27 pp.

[254] *Arch. Eisenhüttenw.*, **26**, 1955, 43-50; cf. on the method W. A. Fischer and R. Schäfer, *Ibid.*, **24**, 1953, 307-314.

[255] *Arch. Eisenhüttenw.*, **24**, 1953, 105-111.

as a function of time (cf. Fig. A. 112) and of temperature differences (Fig. A. 113). The electromotive forces are, in this case, interpreted by the reduction of silica in the refractory sand to silicon by the carbon content of the iron melt. The considerably higher EMF (up to 1.1 volt), measured in a high-frequency induction furnace are

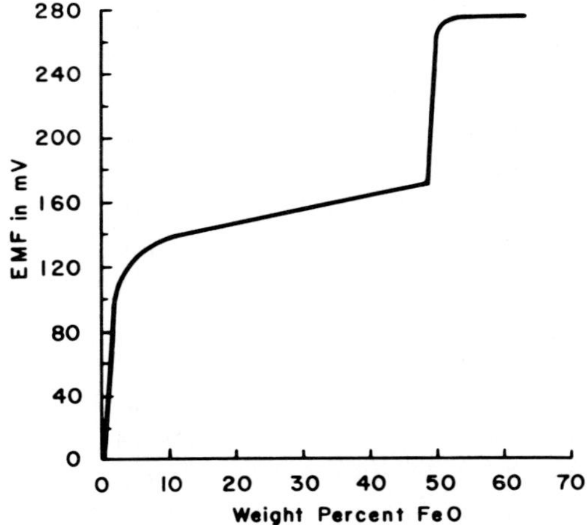

FIG. A. 111. Electromotive force as a function of the concentration in FeO, for the system $Pt/Al_2O_3/(Al_2O_3 + FeO)/Pt$, at 1500°C. (Fischer and Hoffmann).

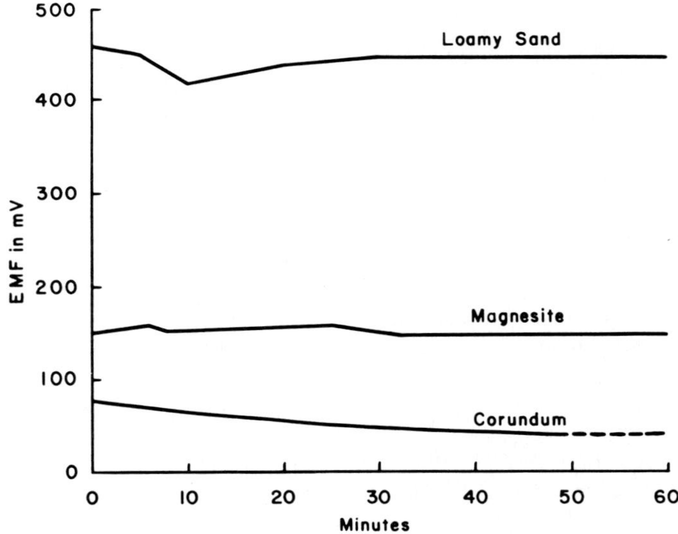

FIG. A. 112. Electromotive force of an iron melt at 1350°C. towards a platinum electrode, as a function of time, with different refractory linings in Tammann furnace. (Fischer and Schäfer).

chiefly caused by thermoelectric currents developing from temperature differences in the crucible. As soon as the carbon content of the iron melt is lowered by refining (oxidation) the EMF is also reduced towards zero. It is remarkable that copper metal melts do not develop any galvanic potentials with corundum or magnesia crucible linings. In the system $Pt/Al_2O_3/(Al_2O_3 + MgO)/Pt$, Fischer and Hoffmann[256] established quite analogous curves of EMF as a function of the concentration in magnesia,

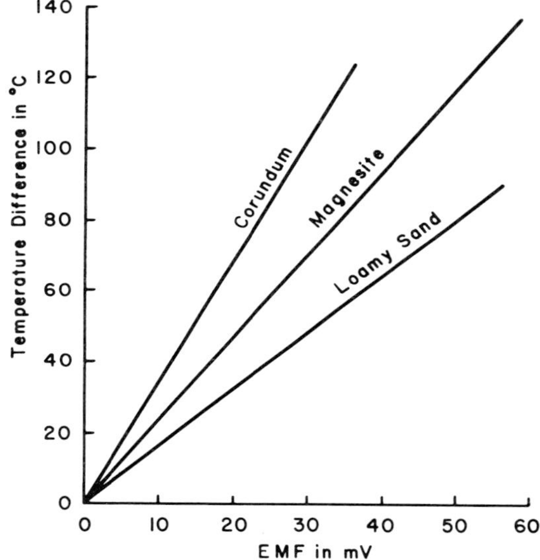

Fig. A. 113. Electromotive force of refractory linings at 1350° to 1500°C., as a function of the temperature difference in the solid oxides between two platinum electrodes. (Fischer and Schäfer).

with a sharp discontinuity indicating the formation of spinel in agreement with the epuilibrium diagram of the system Al_2O_3–MgO and of crystalline solutions between $MgO.Al_2O_3$ and Al_2O_3 in the isometric modification. The maximum alumina content at 1500°C. is 80.5 per cent forming the spinel structure up to 28.3 per cent MgO corresponding to pure $MgO.Al_2O_3$.

171. Corresponding investigations on lead silicate melts as electrolytes in galvanic cells of the type $Pt(O_2)/PbSiO_3/(PbSiO_3 + \text{additional oxides})/Pt(O_2)$, at 800° to 900°C. were made by R. Didtschenko and E. G. Rochow,[257] with the electrolytes in porcelain crucibles and rotating platinum electrodes dipping into the melts. The

[256] *Arch. Eisenhüttenw.*, **26**, 1955, 63-70; a comprehensive report given by W. A. Fischer see in *Silicates inds.*, **20**, 1955, 244-254.

[257] *J. Am. Chem. Soc.*, **76**, 1954, 3291-3294.

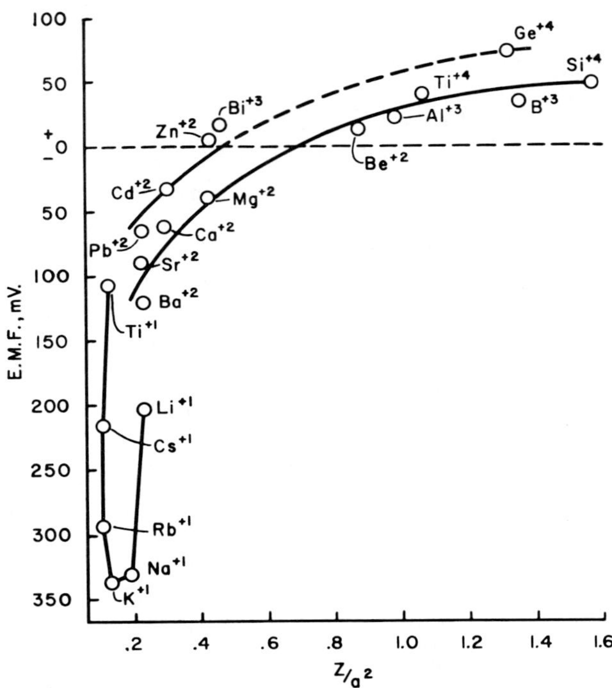

FIG. A. 114. Electromotive force as a function of cationic field strengths, z/a^2, in the oxides added to $PbSiO_3$ melt as the electrolyte, at constant concentration of 0.2 moles/mole $PbSiO_3$, at 900°C. (Didtschenko and Rochow).

oxide additions were constantly 0.2 moles/mole of the solvent ($PbSiO_3$), namely the oxides of the alkali metals, of thallium, the alkaline earth elements, boron, aluminum, silicon, germanium, titanium, zinc, cadmium, and bismuth. For the alkalies, concentrations 0.1 to 0.5 moles/mole $PbSiO_3$ yielded the oxygen anion activities on the electrodes as a function of concentration and basicity of the oxide additions. If the EMF is plotted as a function of the field strengths, z/a^2, of the cations, there is a distinct grouping of three types of cations as shown in Fig. A. 114, or the projection versus a^2/z in Fig. A. 115 from which the relation EMF $= -.050 \cdot a^2/z + 0.079$ gives the value $\ln (O^{2-})_{sol.} = 0.429 \cdot a^2/z - 0.678$ (at 900°C.), as an expression for the activity of the oxygen anions.

172. The planful application of galvanic potentials on contacts between *glass melts and refractories* as the root of troublesome *corrosion* effects in the upper construction portions of glass tanks we owe to P. Le Clerc and I. Peychès[258] as depending on the

[258] *Trans. Intern. Ceram. Congr.*, III, Paris, 1952, 217-222; *Verres et réfractaires*, 7, 1953, 339-345; *Silicates inds.*, 19, 1954, 374-379.

chemical character of both agents, the glass and the refractory body. For a standard glass composition a classification of different refractory oxides can then be developed. Alkali aluminate enrichments were observed near the refractory electrodes, especially those of electrofused alumina blocks (commercial Monofrax) indicating ion migration and corresponding galvanic potentials as soon as the cell is in short-circuit. The

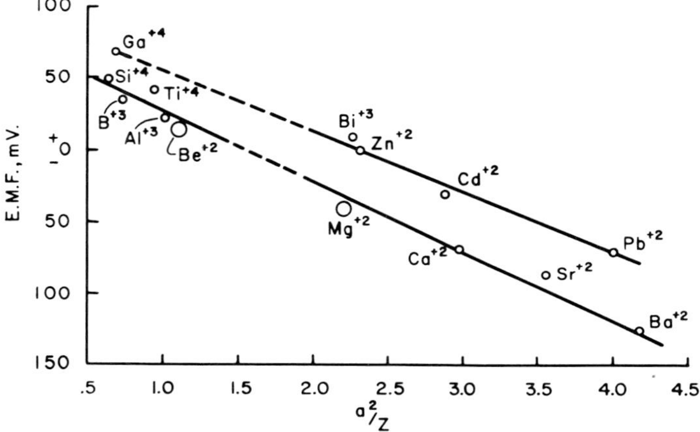

FIG. A. 115. Electromotive force as a function of reciprocal field strengths, a^2/z, for cations in oxides added to $PbSiO_3$ as electrolyte. (Didtschenko and Rochow).

framework-forming elements of the glass behave as electropositive, the framework-modifying elements as electronegative agents in a series of decreasing enthalpies of the formation of the oxides of silicon-aluminum-zirconium-calcium-magnesium, and with the electronegativities decreasing from oxygen to silicon-aluminum-magnesium-

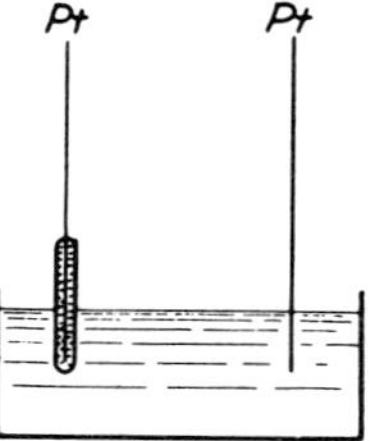

FIG. A. 116. Type of galvanic cell Pt/refractory oxide/glass melt/Pt. (Plumat).

calcium. Cells with Pyrex/sodium silicate glass as electrolytes showed decreasing EMF with increasing silica contents; cells with Pyrex and different alkali and alkaline earth silicates have a distinct maximum of EMF for K^+ silicate systems. By radioautography the electrochemical migration of sodium ions in a block of "β-alumina" was made evident by exposing the whole specimen to neutron irradiation from a pile which makes the sodium radioactive. The radiographs show convincingly the enrichment of highly corrosive alkalies in the cathodes whereas the anodes are not affected. It is practically conclusive that in one and the same glass melt tank blocks of different chemical composition should not be applied in order to avoid internal cell potentials. The alkali ion transfer has a typical *rectifier effect* for alternating currents flowing between two electrodes in glass. The alkali ions exert a specific and most detrimental corrosion particularly on smaller electrodes.

173. Elaborate studies on the electrochemical contact phenomena between glass and oxides at high temperatures were made by E. Plumat[259] who applied to these the hypothesis of "screening" tendencies of highly charged, low-polarizable cations

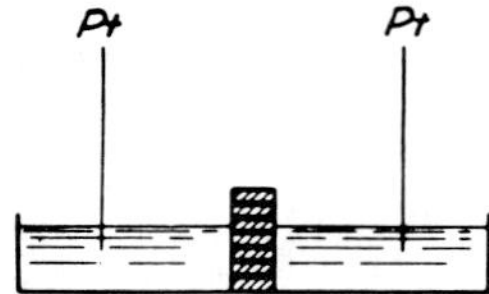

FIG. A. 117. Type of galvanic cells with two glass melts as electrolytes, with refractory diaphragm. (Plumat).

(Si^{4+} and Al^{3+}) to reduce their exterior fields by surrounding electrons and anions. For the electrometric and tube-voltmetric measurements, cells of different types were used, either Pt/crystalline oxide/glass melt/Pt, or Pt/oxide/glass melt/oxide/Pt, or liquid cells Pt/glass melt I/glass melt II/Pt, or solid cells Pt/oxide I/oxide II/Pt, of which Figs. A. 116 to 118 show schematic representations. Potentials on the glass line are shown in Figs. 119 and 120. Among the special results for EMF between glasses the following ones are the most important:

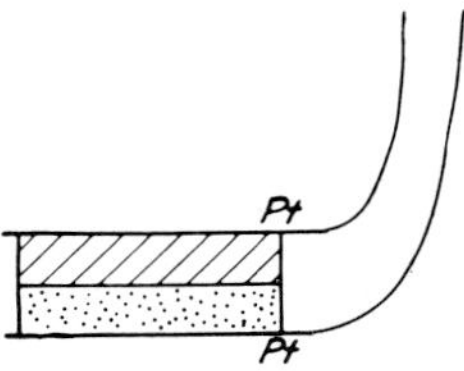

FIG. A. 118. Type of galvanic cells with two glass electrolytes, for glasses of high viscosity. (Plumat).

[259] *Silicates inds.*, **19**, 1954, 141-154; **22**, 1957, 657-667; **23**, 1958, 17-28; on silver glasses cf. W. A. Weyl and G. E. Rindone, *J. Am. Ceram. Soc.*, **33**, 1950, 91-95. On the galvanic corrosion of glass tank refractories see recently L. Leger, M. Boffé, and E. Plumat. *Glass Technol.*, **1**, 1960, 174-179, and F. J. Shonebarger, *Ibid.*, **2**, 1961, 53-59.

(1) *For glasses of the type* $R_2O.4SiO_2$ *and* $R_2O.2SiO_2$ (*at* 400° *to* 800°*C.*): The alkali silicate glasses become increasingly negative with increasing R_2O contents in their potentials. Lithium silicate glasses are more positive than sodium and potassium glasses. Window glass, sodium silicate glass, and fused NaCl have about the same potentials. B_2O_3 is negative toward window glass, borax slightly positive.

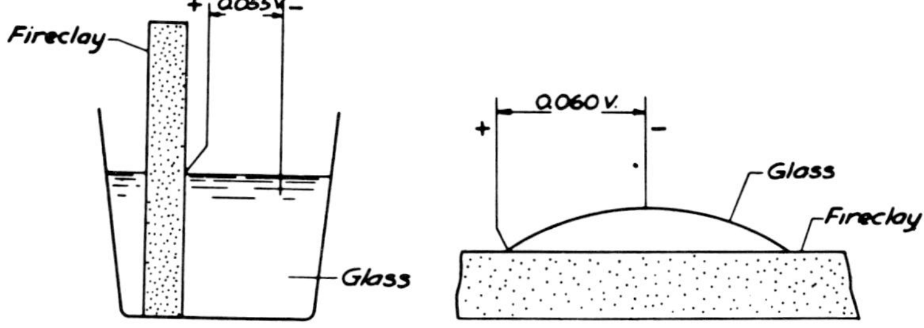

FIGS. A. 119 and 120. Two types of galvanic cells to show the potentials on the glass line. (Plumat).

(2) *Window glass versus oxides at* 1000°*C.*: SiO_2 has the highest potential, alumina is slightly lower. Magnesia gave somewhat inconsistent results because of difficulties in preparing compact electrodes. With increasing ionic radii of metal cations the oxide potentials decrease in the series

BeO	MgO	CaO	BaO	SiO_2	TiO_2	ZrO_2
+ 0.05	— 0.14 to — 0.30	— 0.32	— 0.62	+ 0.13	+ 0.10	+ 0.07 volts

(3) *Mixed oxides and window glass at* 1050°*C.*: Alkalies added to silica decrease the potentials more rapidly than do the alkaline earths; alumina added to silica strongly increases the potentials with a maximum of 0.76 volts, for the ratio $SiO_2 : Al_2O_3 = 1 : 1$. Refractory clays show entirely the same effects. For B_2O_3 added to silica also a maximum is observed, but less distinct than for the alumina-silica system. ZrO_2, TiO_2, ZnO, SnO_2, and Fe_2O_3 do not markedly affect the silica potential. Mixes of Al_2O_3 and B_2O_3, or of MgO and B_2O_3, versus window glass show a distinct maximum of positive character while the pure oxides have negative potentials (Fig. A. 121).

(4) *Pressed pellets of crystalline aluminates and silicates, zirconates, titanates* at 1050°C. show decreasing potentials in the series $Na_2O,6SiO_2$(+ 0.30 v.); $Na_2O,6TiO_2$; Na_2O,Al_2O_3; and Na_2O, $6ZrO_2$(— 0.25 v.); CuO,SiO_2 is positive, $MgO,2SiO_2$; $CaO,2SiO_2$; and $Li_2O.2SiO_2$ are negative.

(5) *Refractory clays versus window glass at* 1000°*C*.: The potentials are strongly affected by the firing temperatures because of changes in composition of the bonding glass; with increasing temperature the potential is reduced. An English

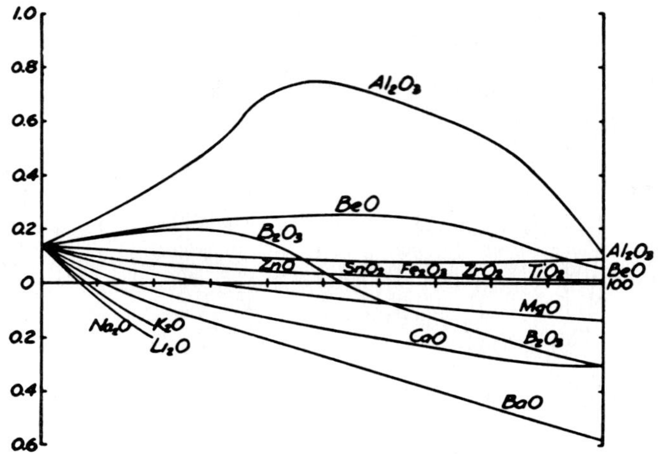

Fig. A. 121. Potentials of galvanic cells of the type Pt/(SiO_2 + oxides)/window glass melt/Pt, at 1000°C. (Plumat).

ballclay fired at 700°C. showed 1.2 volts, but fired at 1400°C. = 0.65 v. The maximum coincides with the highest state of activation in the meta-clay (cf. Vol. V, Section B). If an exterior EMF is applied to such cells a decomposition voltage effect and galvanic polarization are observed (Fig. A. 122).

(6) *Corrosion products of glass melts/refractory materials*: In cells of the type Pt/modified glass/intact glass/Pt the glass corrosion causes a strong increase in glass potentials with increasing percentage of dissolved refractory material. Whereas fireclay shows a potential of + 0.4 to 0.8 v. as soon as corrosion occurs, intermediate modified glass layers are formed with regularly decreasing potentials. If alumina (from commercial Monofrax blocks) is the refractory body, the potential of the pure oxide is only 0.1 v. but that of the intermediate corrosion layer 0.3 to 0.4 v.; it decreases again to that of the pure glass. A temperature gradient and its effect on the EMF of the glass melts are a linear function of increasing temperature difference between the electrodes.

174. Particularly important are the effects of *oxidation and reduction*, as in window glass prepared from a pure carbonate batch, or from a sulfate containing batch (with 40 per cent of the Na_2O introduced as Na_2SO_4 and systematic additions of carbon up to 12 per cent). The higher the carbon addition, the more the potential is changed from positive to negative. Gas evolution on the negative electrode is always observed when the cells are short-circuited, namely first pure oxygen followed by mixtures of oxygen with CO_2,SO_2, and water. Cells with reduced/oxidized glass melts show gas evolution on both electrodes if short-circuited.

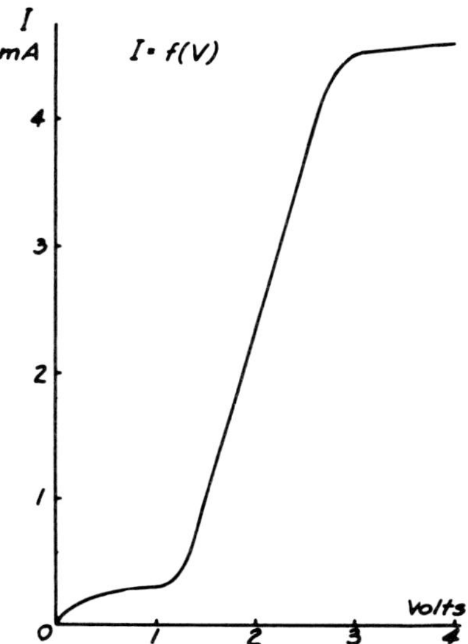

FIG. A. 122. Decomposition voltage for a galvanic cell Pt/window glass melt/fireclay refractory/Pt. (Plumat).

175. Plumat's extensive investigations not only excellently illustrate the alkali migration towards the refractory side in glass tanks as previously observed by Le Clerc and Peychès, they also give a thorough explanation of inner corrosion phenomena imminent to glass tank fusion practice, from the processes in the dog-house to the machines shaping the glass, and are, therefore, of really basic importance. Additional studies of Plumat concern the systematic replacement of alkali cations in simplest R_2O–SiO_2 glasses and for Pyrex glass as electrolytes, influencing the EMF. These decrease in the series Li > Na > K, but increase from K to Rb and Cs. In mixed sodium-potassium silicate glasses there is a definite maximum at 0.5 moles K_2O in $R_2O,3SiO_2$ glass. In alkaline earth silicate glasses, effects of the replacement of R_2O by RO are considerably lower than those caused by one alkali replacing the other. For glasses of the type $Na_2O,xRO,6SiO_2$ the EMF decreases from magnesium to barium, specifically with increasing cation radii and polarizability. Complex phenomena appear with cations of variable valency, namely with Cu,Fe,Mn, but also for B,Al,Cr,Sb,Ti,Mo,Nb. There is a distinct analogy between the silicate potentials here discussed and those of dissolution for the same elements in aqueous solution. The temperature coefficients of the potentials relative to platinum are low for Cu,Ag,Au, and Mo, negative for Wo,Ni,Fe, positive for platinum and rhodium. In cells of the type M/reference glass/$Na_2O,3SiO_2$, or $Na_2O,6SiO_2$ glass/M at 800°C. with M = Fe as electrode material the EMF have particular time effects. It may be that the oxidation by oxygen diffusion plays a role during the dissolution reaction in the glass. Relatively constant EMF (0.102 v.) are observed in cells Pt/glass/Ag at 900°C. in air, if the glass is either Pyrex, or window glass, or $Na_2O,4SiO_2$, or Li_2O, $1.8SiO_2$. For $K_2O,3SiO_2$ glass the potential is 0.180 v. The effects of increasing temperature were studied for the cell Pt/reference glass/$1.6Na_2O,1.13Al_2O_3,6SiO_2$ glass/Pt, with a sharp maximum at 1000° to 1050°C.

176. *Thermogalvanic* cells of the type Pt(hot)/glass/Pt(cold) show a nearly linear increase in EMF with temperature (from 0.040 to 0.055 volts/100°C), until at 600°C. (as the "cold" temperature) a discontinuity appears, probably caused by transformation effects. Between 600° and 1200°C. the EMF are well reproducible if devitrification does not interfere. Considerable oxygen diffusion is characteristic of cells with relatively large potential differences between oxidized and reduced glasses if silver or platinum electrodes are used. Typical *gas reactions*, as by hydrogen surrounding one electrode in silver containing glasses, are additionally remarkable (cf. W. A. Weyl and G. E. Rindone, 1950). If different metals with non-noble gas charcateristics in their ions and with variable valencies are used as electrodes, considerable complications are observed.

177. The particular suitability of electrochemical methods in the determination of the *acidity* and *basicity* of glass melts applying the changes in valencies of colored

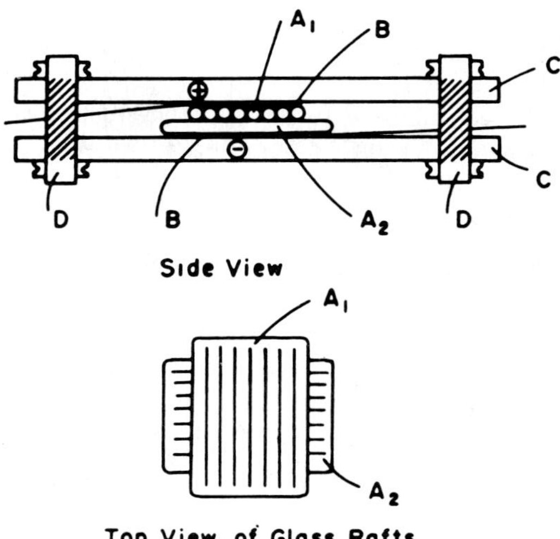

Fig. A. 123. Oxygen electrode for acidity-basicity measurements in glass melts. (Förland and Tashiro).

metal cations as indicators was demonstrated for alkali borate model glasses by A. Dietzel with W. Stegmaier (1940) and with P. Csaki (1940) (cf. B. ¶ 63). In continuation of these studies, T. Förland and M. Tashiro[260] developed an oxygen electrode especially suitable for measurements of acidity-basicity relations in glass melts for which W. A. Weyl (cf. ¶ B. 65 ff.) had developed fundamental new concepts, equally applicable for aqueous solutions, fused salts, silicate or borate glasses, and solids. The principle of Förland's and Tashiro's cell is demonstrated by Fig. A. 123, after the cell type Ag(air)/Glass I/Glass II/Ag(air) with silver foil electrodes and two

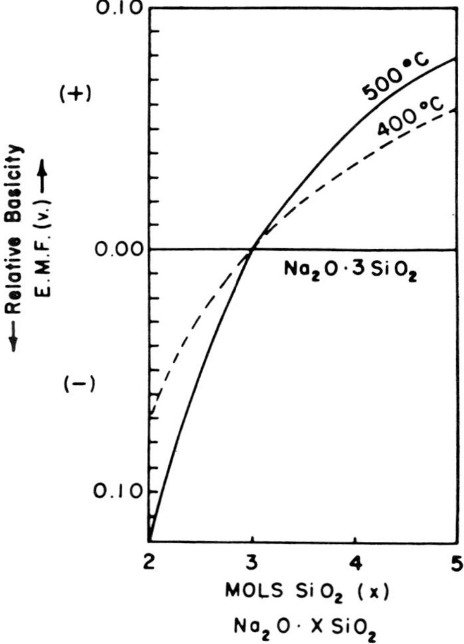

FIG. A. 124. Changes in electromotive force and basicity (relative to standard glass $Na_2O,3SiO_2$ in the series Na_2O—x SiO_2. (Förland and Tashiro).

rapts of glass rods as the electrolytes of different composition, suitable for 300° to 500°C. The direct relation of the EMF measured and the basicity (relative to $Na_2O,3SiO_2$ glass as the standard, EMF = 0) is seen in Fig. A. 124, for glasses of the series Na_2O–SiO_2, and in Fig. A. 125 for the replacement of R_2O by RO, with a considerable reduction in EMF and basicity. If potassium replaces sodium in the alkali silicate melts a definite maximum appears toward positive potentials (Fig. A. 126). G. S. Smith and G. E. Rindone[261] further used concentration cells, of a sim-

[260] *Glass Ind.*, **37**, 1956, 381-385, 399, 402.
[261] *Glass Ind.*, **37**, 1956, 437-443, 454, 456.

ilar type as described above with silver electrodes in air, to study the equivalent replacement of different cations R^+ in $R_2O,3SiO_2$ glasses as the zero standards, with systematic additions of silica, magnesia, and barium oxide. In the opinion of Smith

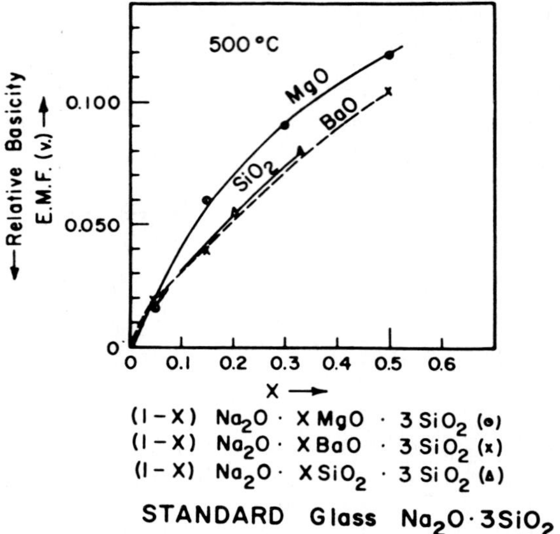

FIG. A. 125. Changes in electromotive force and basicity (relative to standard glass $Na_2O,3SiO_2$ in the series $(1 - x)Na_2O,x\ RO,3SiO_2$. (Förland and Tashiro).

and Rindone, magnesia, in somewhat lower degree BaO, bring about a greater increase in + EMF potentials, i.e., in the acidity of the mixed glass than even silica can do, particularly in lithium silicate glasses. In this case the series is Ba > Mg > Si.

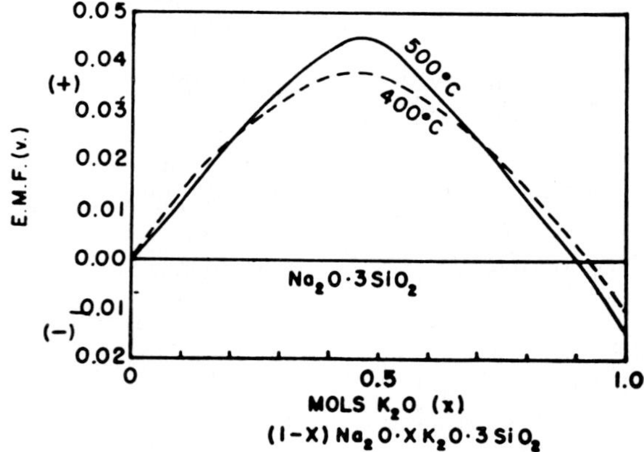

FIG. A. 126. Changes in electromotive forces and of basicity (relative to the standard glass $Na_2O,3SiO_2$), in series $(1 - x)Na_2O,xK_2O,3SiO_2$. (Förland and Tashiro).

G. E. Rindone, D. E. Day, and R. Caporali[262] used the oxygen electrode as described for acidity-basicity studies of sodium and potassium silicate glasses from 300° to 500°C. in which alumina and TiO_2 were introduced to replace the alkalies. This substitution also causes, in this case, a typical increase in relative acidity (positive EMF to the standard $R_2O,3SiO_2$ glass). If alumina is introduced in sodium silicate glasses the first addition of alumina behaves characteristically because $[AlO_4]$ groups are formed which bind nonbridging oxygen anions of the structure into the framework.

178. In a similar way, Ti^{4+} forms $[TiO_6]$ and $[TiO_4]$ groups side-by-side, but there is a characteristic difference in that the tetrahedral group formation is combined with an acidity *de*creasing effect, whereas the octahedral groups *in*crease it. For potassium silicate glasses both aluminum and titanium increase the acidity in a higher degree than in sodium silicate glasses. In the measure as $[AlO_4]$ groups are polymerized in compositions with high aluminum contents, a loosening of the framework occurs (Figs. A. 127 and 128). Particularly instructive also is the projection of the EMF (at. 500°C.) versus the moles of Al^{3+} and Ti^{4+} ions contained in sodium

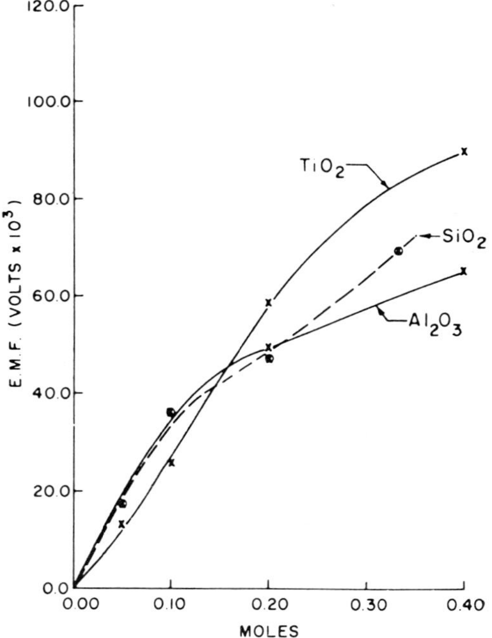

FIG. A. 127. Changes of electromotive forces and of basicity (relative to standard glass $Na_2O,3SiO_2$), in series $(1 - x)Na_2O,xAl_2O_3$, or $xTiO_2,3SiO_2$. (Rindone, Day, and Caporali).

[262] *J. Am. Ceram. Soc.*, **43**, 1960, 571-577.

or potassium silicate glasses (corrected for the differences in the SiO_2/R_2O ratios), in Fig. A. 129.

179. One of the most impressive applications of electrochemical potentials developed in silicate melt electrolytes for technological processes is the theory of *enamel adherence* on steel sheets which was promoted by A. Dietzel (1935 ff.).[263]

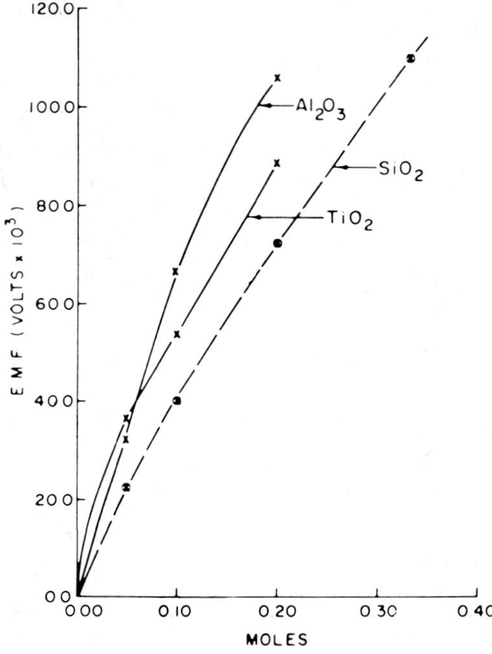

Fig. A. 128. Changes of electromotive forces and of basicity (relative to standard glass $K_2O,3SiO_2$), in series $(1 - x)K_2O,xAl_2O_3$ or $xTiO_2,3SiO_2$. (Rindone, Day, and Caporali).

Although we have to discuss this idea of electrochemical reactions as the origin of the adhesion phenomena concerned, more in detail in B. ¶ 238 ff., especially on the criticisms given to the theory by a different basic understanding of those processes, we may mention here the studies of D. G. Moore *et al.*[264] on the factors of surface roughness for a sufficient "anchoring" of the silicate melt on the metal after firing. This latter process is indispensable for the development of a roughness, as made

[263] A more recent report on this theory see in *Ceram. Age*, **62**, 1953 (6) 17-21, 24, 38f. Previous to A. Dietzel, H. F. Staley, *J. Am. Ceram. Soc.*, **17**, 1934, 163-167, discussed electrolytic reactions in vitreous enamels in their correlation to adherence on steel.

[264] *J. Am. Ceram. Soc.*, **36**, 1953, 410-416 (with J. C. Richmond, H. B. Kirkpatrick, and W. N. Harrison).

evident in metallographic polished sections with the ground coat containing say 0.8 per cent of cobalt oxide. Later D. G. Moore, J. W. Pitts, J. C. Richmond, and W. N. Harrison[265] widely adopted the electrochemical concepts on the adherence reactions, combining them principally with the aspects of anchoring (which were already considered in Dietzel's original ideas), into pits formed by galvanic corrosion on the steel surface by the silicate melt electrolyte in a short-circuited cell. The fact is

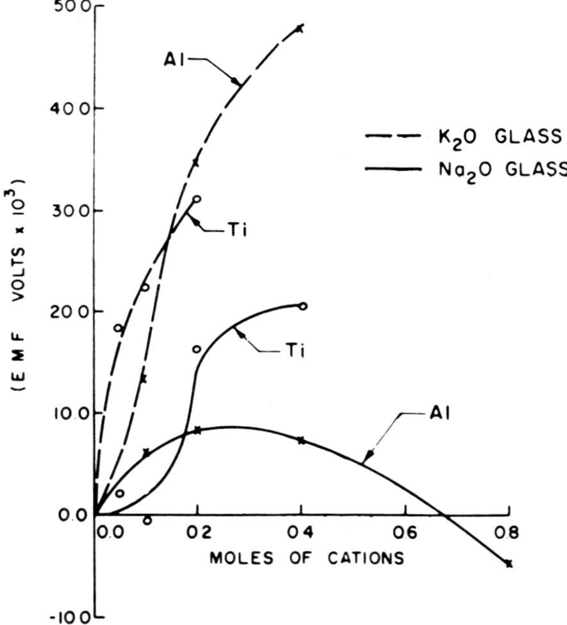

FIG. A. 129. Changes of electromotive forces and basicity for the glasses $Na_2O,3SiO_2$, and $K_2O,3SiO_2$, for varying concentrations in Al^{3+} and Ti^{4+} ions. (Rindone, Day, and Caporali).

emphasized that the corrosion reactions act only in the short period of the enamel firing on the steel sheet surface. There are left, however, by Moore *et al.* a few open questions, such as why copper oxide does not act as an adhesion oxide in the same way as does cobalt oxide, and also why sandblasting of the sheet surface to produce "undercut" adhesion anchoring points cannot result in a similarly intense adhesion as the pits corroded by the galvanic reaction for a correspondingly intense bonding from the silicate melt to the metal.

180. The way in which the technological method of dipping into nickel sulfate solution before the firing process acts so specifically with a cobalt oxide containing

[265] *J. Am. Ceram. Soc.*, 37, 1954, 1-6.

ground coat enamel, was further examined by D. G. Moore, J. W. Pitts, and W. N. Harrison.[266] There are definite optimum conditions for development of the nickel layer deposited on the steel sheet by the dipping process. Much depends on the cleaning method used for preparing the metal surface. The subsequent firing process is important in which the roughening occurs as mentioned above as the consequence of galvanic corrosion. The nickel is deposited selectively on more cathodic fields of the steel surface and forms local galvanic cells functioning immediately after liquefaction of the enamel applied on it, thus wetting the metal. The iron portions of these local cells are then preferably corroded, the nickel-covered fields remaining protected by cathodic effects. In this way the "undercut" structure of the roughened surface is built up and this is the premise of the firmness of the adhesion.

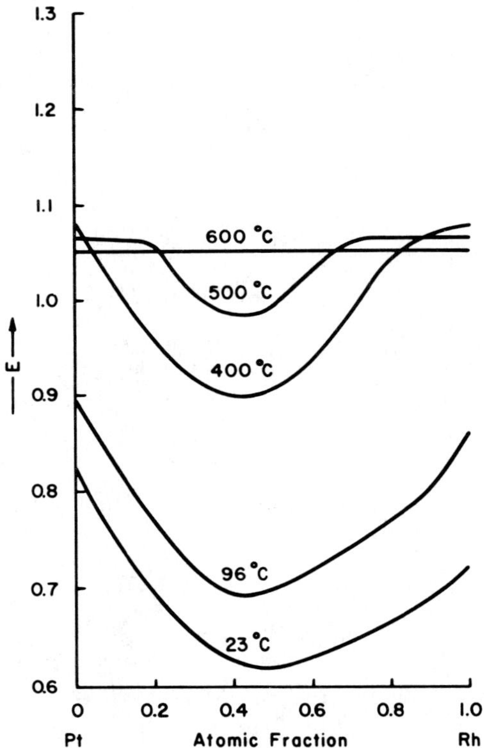

FIG. A. 130. Electromotive force of platinum-rhodium alloys at different temperatures and alloy compositions. (Dietzel and Coenen).

[266] *J. Am. Ceram. Soc.*, **37**, 1954, 363-369. A fortiori, many experimental facts and practical observations were described by R. F. Patrick, E. G. Forest, and G. H. Spencer-Strong, *Ibid.*, **36**, 1953, 305-313.

181. A. Dietzel and M. Coenen[267] answered some of the critical problems involved in the galvanic theory of enamel adhesion on metals by an extensive investigation of the wettability of noble metals (gold, platinum, rhodium, and their alloys) by determining the contact angles (cf. A. ¶ 103 ff.) at 840° and 1000°C. in air and in nitrogen. The potentials on the metal electrode were measured in a 1 N H_2SO_4 solution versus a concentrated calomel electrode (at low temperatures), or a standard hydrogen electrode (for higher temperatures) (Fig. A. 130). Oxygen super-potentials disappear at that temperature at which oxygen atoms begin to permeate the metal by diffusion. This is the same temperature at which the reversible potential of the H_2/O_2 cell is reached (Fig. A. 131). It is, therefore, the oxygen diffusion

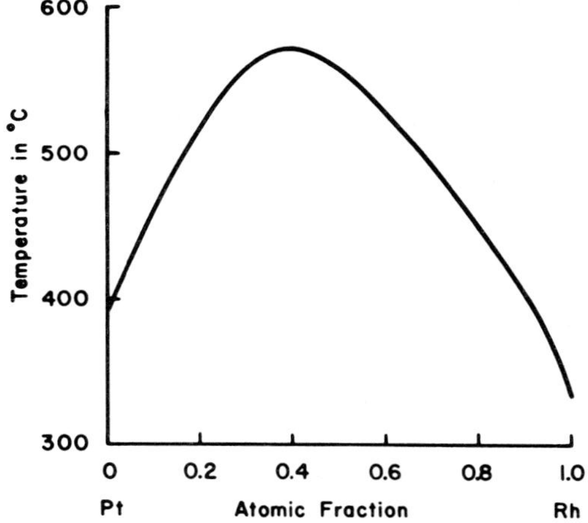

FIG. A. 131. Start of the oxygen diffusion as a function of composition of platinum-rhodium alloys, and of temperature, as well for the reversible potential, as for oxygen diffusion. (Dietzel and Coenen).

into the metal which not only governs the wetting process but also the electrochemical behavior. The adhesion energies are in a linear correlation to the super-potential of the anodic oxygen/metal electrode. These rules are very well followed by platinum-gold and platinum-rhodium alloys, but not by those of platinum and beryllium for which evidently a different electrochemical mechanism is valid, chiefly because of the formation of a layer of beryllium oxide on the active surface. If the current density is low, super-potentials are built up and block out the oxygen diffusion. For pure platinum and platinum-rhodium alloys, the reversible potential of 1.2 volts is reached at 400° and 600°C., respectively. For contacts of copper with the enamel melt no mechanical anchoring by corrosion indentation occurs, but oxygen anions bridge over the metal to the silicate melt phase to explain the adhesion phenomenon of this type.

[267] *Glastech. Ber.*, **32**, 1959, 357-361.

DENSITY MEASUREMENTS, AND CHANGES IN SPECIFIC VOLUME OF SILICATE MELTS

182. Concerning the classical buoyancy method used by A. L. Day, R. B. Sosman, and J. C. Hostetter (1914) a not negligible correction must be made of descriptions given in further publications on this ingenious method.[268] For a better illustration of what is meant we reproduce here in Fig. A. 132 the principle sketch in the original paper of Day, Sosman, and Hostetter which shows that the stem B was *not* determined to hold the upper graphite frame, F, suspended on a long flexible metal wire, but is only to facilitate, by a hook let down from above, the removal of the upper cage and crucible in lifting free from the metal melt before its solidification. This purpose of the stem is quite secondary to the measurements. The inverted graphite crucible, G, and the specimen, S, *float entirely free* in the molten tin bath, and the weights, W, at the bottom are only determined to push them under to the depth set by the contact pin, A, no other attachments being necessary. The container crucible, C, for the tin bath is supported on a column of refractory tubes and screens of hard porcelain or magnesia (not designed). A magnesia ring, E, is to insulate the float crucible G from the cage, while the pointed pin A is to indicate by an electric contact of the upper cross bar of the frame F with the surface of the metal bath when buoyancy for the system float crucible + specimen is established.

183. By a counterbalanced sphere-buoyancy method similar to that developed by G. Heidtkamp and K. Endell (1936), L. Shartsis, W. Capps, and S. Spinner[269] determined the density of molten alkali borates in the temperature range between 600° and 1000°C. Especially characteristic at the higher temperatures is the maximum in density at about 30 molecular per cent of R_2O (Fig. A. 133) connected with minima of mean volume expansivity $(d_1 - d_2)/d_2(T_2 - T_1)$, from 25° to 1000°C. at 10 or 20 per cent R_2O, whereas those in the liquid range increase with increasing alkali contents. Particularly impressive is the relative contraction of molten B_2O_3 by additions of alkalies, as if the glass would act under a high external pressure. In Fig. A. 134 the linear functional relation of the apparent specific volume of the alkalies is shown with the relative concentration in these. The distances of the B^{3+}–B^{3+} ions calculated from the densities are seen in Fig. A. 135. The contraction effect is largest for the system B_2O_3–Li_2O. There is also a large decrease in expansivity in the transformation range. The data of Shartsis, Capps, and Spinner are not di-

[268] We owe these correcting remarks to R. B. Sosmas personally by letter exchange in 1954 with the author of the present book. The description given by F. M. Jaeger in his book "Eine Anleitung zur Ausführung exakter physikochemischer Messungen bei höheren Temperaturen," Groningen, J. B. Wolters' U.M., 1913, 152 pp., especially pp. 125-128, is not correct and caused corresponding errors in subsequent publications. It should be definitely suppressed.

[269] *J. Am. Ceram. Soc.*, 36, 1953, 35-43.

rectly compatible with the densities calculated after the methods developed by M. L. Huggings[270] and J. M. Stevels[271] based on additivity rules (cf. B. ¶ 120 f.). For silicate systems there is also no agreement concerning the data for lithium sil-

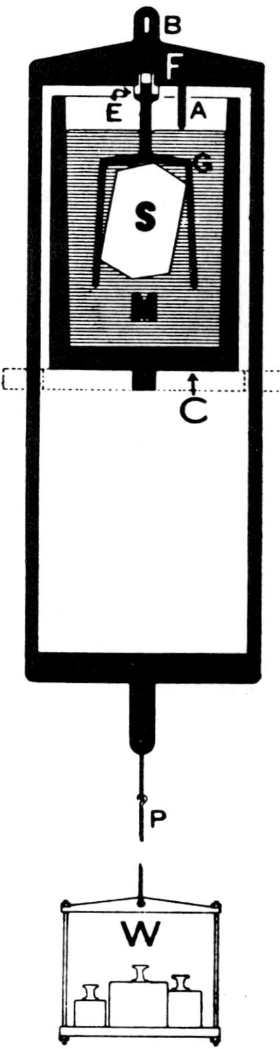

Fig. A. 132. Density measurement of crystalline or fused silicates at high temperatures by buoyancy method. (Day, Sosman, and Hostetter).

[270] *J. Opt. Soc. Am.*, **30**, 1940, 420-430.

[271] "Progress in Theory of Physical Properties of Glass," Elsevier Publishing Co., Inc., New York 1948, 104 pp.

icates. Stevels' formula requires that their densities should decrease with increasing ratio O/Si, whereas the observed densities increase.

184. J. O'M. Bockris, J. W. Tomlinson, and J. L. White[272] applied the Archimedes buoyancy principle with a platinum float crucible for silicate melts. The adhesion of gas bubbles on the bob walls was extremely troublesome in such measurements, as was also observed by Day, Sosman, and Hostetter in their method, but it can be

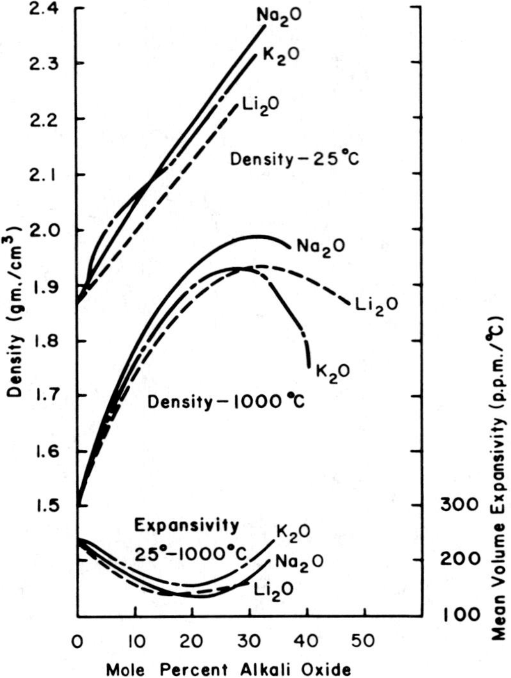

FIG. A. 133. Comparison of densities and mean volume expansivities of alkali borate melts. (Shartsis, Capps, and Spinner).

detected by the appearance of irregular deviations from the linear graphic projection of the buoyancy as a function of temperature. Also the volatilization of alkali is an important source of irregularities. The temperature range examined was between 1000° and 1500°C. The molecular volume of the alkali silicate compositions studied showed negative trends, but partial molecular volumina of the alkalies and the silica were constant between 12 to 33, 33 to 50, and 50 to 60 per cent R_2O. Up to 12 per cent R_2O the molecular expansivities were low, but rapidly increased with the con-

[272] *Trans. Faraday Soc.*, **52**, 1956, 299-310.

centrations in alkali, the rate of the increase decreasing with increasing concentration. J. W. Tomlinson, M. S. R. Heynes, and J. O'M. Bockris[273] further measured the density of melts in the systems $CaO–SiO_2$, $SrO–SiO_2$, and $BaO–SiO_2$, in the range from 1600° to 1950°C., applying a corresponding buoyancy method in a graphite resistor furnace. As in the alkali-silicate melts, the expansivity was zero at low BaO contents. The observed changes in the molecular volume, as a function of the cation

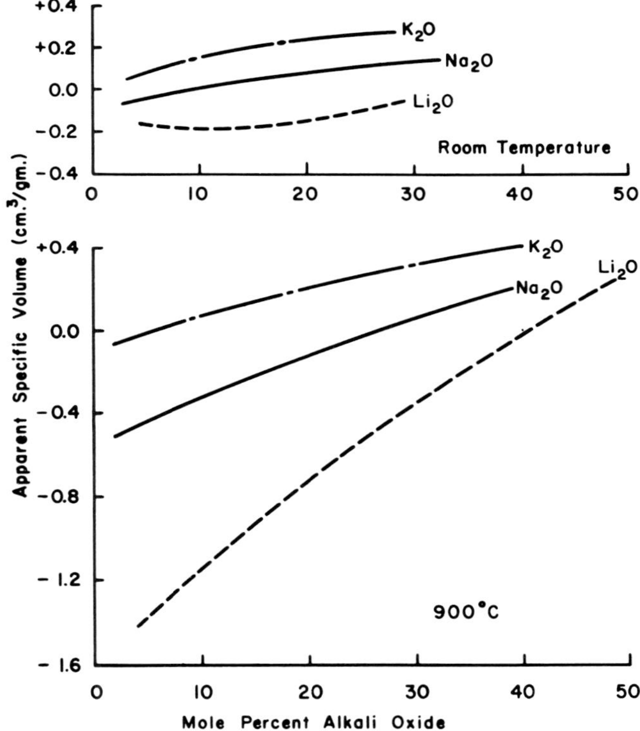

FIG. A. 134. Apparent specific volume of alkali oxides in binary borate melts at 900°C., and at room temperature. (Shartsis, Capps, and Spinner).

sizes, are interpreted by steric effects caused by the volume occupied by these cations. The expansivities were attributed to the attractions between the cations and oxygen anions. Bockris *et al.* developed several models for these steric conditions with the hypothesis that complex anions, such as $(Si_3O_9)^{6-}$ may be stable for the composition $CaO.SiO_2$, but that polymeric anions with four- or five-membered rings are probable at higher amounts in silica (cf. A. ¶ 22).

[273] *Trans. Faraday Soc.*, **54**, 1958, 1822-1833.

185. A method based on measurement of the *maximum pressure* during the formation of gas bubbles at the end of a refractory tube immersed in molten silicates was developed by S. I. Popel' and O. A. Esin,[274] by a systematic variation of this gas pressure with the depth of immersion. In the system $FeO–SiO_2$ the specific volume is a linear function of the composition, whereas in the system $CaO–SiO_2$ the relation is represented by a curve indicating the much stronger attraction of oxygen

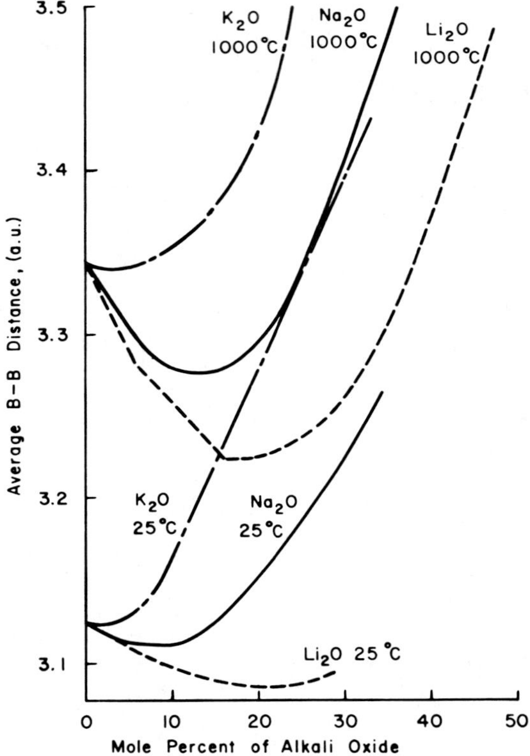

FIG. A. 135. Average distance B^{3+}—B^{3+} in alkali borate melts and glasses, as a function of R_2O concentration. (Shartsis, Capps, and Spinner).

anions exerted by the Si^{4+} if Ca^{2+} are present in the place of Fe^{2+}. The maximum pressure method, which was already discussed in A. ¶ 96 as suitable for the measurement of the surface tension of silicate melts, has a lower temperature limit of its application at 1200° to 1300°C. for common glass melts by the rapidly increasing viscosity. L. Merker[275] nevertheless used it successfully for sodium-calcium silicate glasses

[274] *Zhur. Priklad. Khim.*, **29**, 1956, 651-655.
[275] *Glastech. Ber.*, **32**, 1959, 501-503.

of industrial glass composition at 1400°C. A particular advantage is that secondary gas bubbles on the walls do not disturb the measurements. The calibration is easily made with water and glycerin. As an example of the results, the temperature functions for the specific volume of pure B_2O_3, Pb_2SiO_4, and sodium-calcium silicate glass melts are given in Fig. A. 136. L. R. Barrett and A. G. Thomas[276] used the same method for simultaneous determinations of the surface tension, and of densities in melts of the system $CaO–Al_2O_3–SiO_2$, with a linear function of temperature for densities in the range from 1350° to 1650°C.

186. Because of the great petrological and geophysical importance of the effects of *high pressure* in fundamental silicate systems, H. S. Yoder[277] investigated the

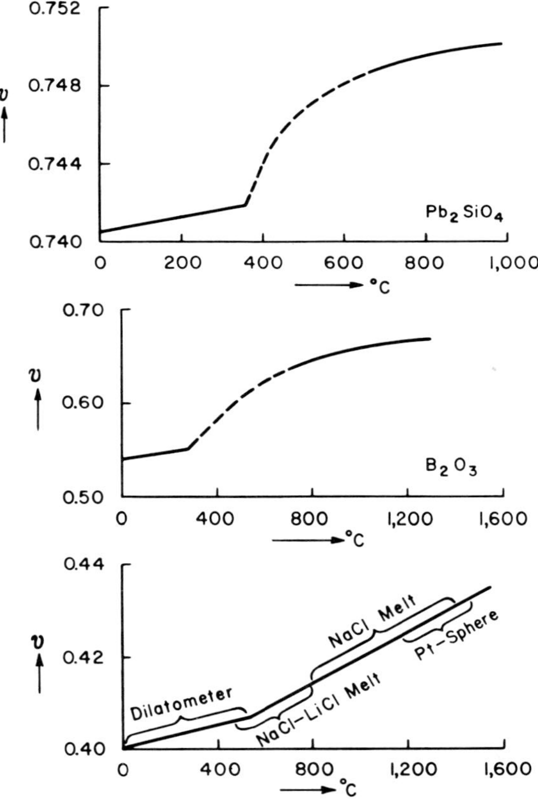

FIG. A. 136. Specific volume of melts of B_2O_3, Pb_2SiO_4, and sodium-calcium silicate glass (74 SiO_2; 16 Na_2O; 10 CaO), as a function of temperature. (Merker).

[276] *J. Soc. Glass Technol.*, 43, 1959, 179-190 T.
[277] *J. Geol.*, 60, 1952, 364-374.

changes in specific volume by pressure up to 5,000 bars experimentally (cf. vol. III, B. ¶ 88) for diopside. The fusion enthalpy is $\Delta H = 102$ cal./gm., the fusion temperuture 1664.5°C., and the volume change $\Delta V = 0.033$ cm.³/g., or $= 10.8$ per cent of the volume at atmospheric pressure. Comparing these results with the measarement data of E. B. Dane (1941) on the density of diopside melts one is surprised that from these a volume change would follow at the fusion point of $\Delta V = 0.057$ cm.³/gm. or of 18 per cent. Since there are evident uncertainties in the computed specific volumina of the crystalline phase at this temperature which were, however, the basis of Dane's calculations, Yoder's results should be preferred as the thermo-

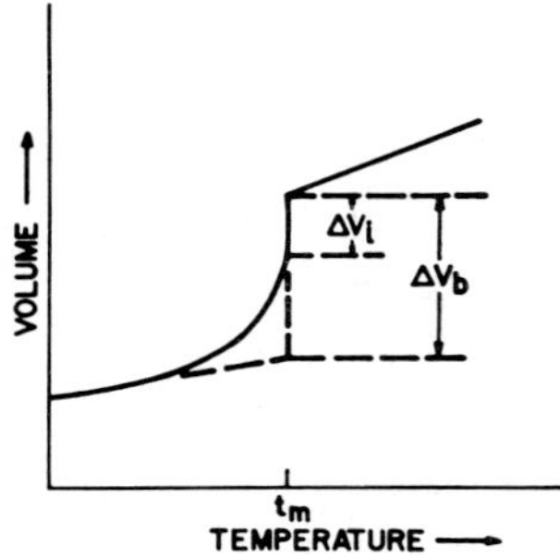

FIG. A. 137. Theoretical change of volume with temperature near the fusion point equilibrium temperature, t_m. (Yoder).

dynamic-instantaneous changes in volume, ΔV_i in Fig. A. 137, whereas Dane had calculated the bulk change in volume, ΔV_b, caused by an unavoidable melting interval of the volume change during the transition of the crystalline to the liquid phase. The difference $\Delta V_0 - \Delta V_i$ corresponds to an anomalous heat capacity immediately below the fusion equilibrium point.

CONSTITUTION OF METALS AND GLASSES; THEORY OF THE STRUCTURE OF LIQUIDS AND GLASSES[278]

187. When V. M. Goldschmidt (1926) introduced the crystallochemical principles into the theoretical discussion of the constitution of glasses, he predicted the existence of glasses which would be much more refractory and harder than silica glass. Starting

[278] As an excellent introduction to modern aspects of liquids and glass structures we recommend the article of W. A. Weyl and E. C. Marboe, *Glass Ind.*, **42**, 1961, 123-128, 168. Fundamental crystallochemical details for glass structure discussions were given by O. W. Flörke, L. Lehnert, and H. Scholze, *Glastech. Ber.*, **29**, 1956, 169-174. On an atomistic interpretation of phase relations in glass-forming alkali silicate systems, see W. A. Weyl, *Ibid.*, **34**, 1961, 301-311.

from the principle of "strengthened" glasses (in contrast to the "weakened" glasses of the beryllium fluoride group) he expected that carbides, nitrides, and borides would be able to form such glasses (A. ¶ 391). This expectation has not been fulfilled. By a rich knowledge of actual measurements of the structure of those compounds, we are certain that they are all excellent in forming crystals, but do not show any tendency to form glasses. It is the lack of information which was available at the time of Goldschmidt's investigations that one important principle of fundamental importance for the glassy state was neglected. Recently W. A. Weyl has called it the "absence of flaws" in simply composed glass formers which impedes the nucleation of a given melt and makes undercooling to a glass impossible. We do not wish to anticipate details of the discussion of this principle in ¶ A. 391, but may here emphasize how important consideration of the question is of how it is possible that such different chemical substances as are the organic "model glasses" used in 1903 by G. Tammann, organic polymers, selenium, glycerin, and silicates, all behave quite analogous in fundamental physicochemical properties as in viscosity, in changes in specific volume, and others, as a function of temperature, whereas in the crystalline state these same materials have absolutely nothing in common.

188. That model which actually best illustrates the constitution of the most typical of all glasses, namely silica glass in its resemblance to the crystal structure of that phase from which it is formed in the thermodynamic fusion equilibrium at 1713°C. namely from cristobalite, is the model proposed by J. D. Bernal.[279] From purely geometric approaches of the three-dimensional framework of "molecules," a homogeneous, coherent, and irregular assemblage is taken as the model of a liquid in which the irregularity of the whole is the result of a very small portion of atoms, which are in a coordination to their surroundings slightly deviating from the normal. For example, the whole array is built up from regular hexagonal planar aggregations with the coordination number 6. Slight deviations in one place to 5 or 7 disturb the whole in such a way that a return to the regular array cannot occur spontaneously within a domain of less than say 100 atoms. In spite of that, the similarity of the irregular arrangement in the liquid to that of the crystal is still convincing. In other words, the liquid silica above 1713°C. still has a great similarity to cristobalite and does not contain any (or at least very few) flaws. This liquid is, at the same time, highly viscous and does not spontaneously nucleate to form the thermodynamically stable crystal phase by undercooling. It, therefore, forms the frozen-in glass state. Bernal's model can also be extended to ionic types of liquids which in the flawless constitution should show separate mobilities of the anions and cations, in the corresponding anionic and cationic partial "lattices" of the liquid and glass structures.

[279] *Trans. Faraday Soc.*, **33**, 1937, 27-45; *Nature*, **183**, 1959, 141-147.

189. The great importance of the *order-disorder* aspects for the physical characterization of liquid and glassy states is most evident[280] but we are not able here to discuss more in details the wide perspectives for a classification of irregular states. Bernal did this quite recently for all types of irregularities, starting from the gaseous state, through the liquid (with a short-range array) to liquid crystals (anisotropic liquids) and fiber crystals with a one- or two-dimensional long-range order, to more or less incomplete crystals (with a three-dimensional long-range order). To these M. von Laue's principles of X-ray and electron diffraction give the key for the experimental verification. We, therefore, also only briefly mention the previous deductions of J.-E. Hiller[281] on the same principal problem, with many aspects of the occurrence of "amorphous" modifications of one and the same substance.

190. The physical significance of the ideas of Bernal and their consequences for glass formation from viscous melts are in strict contrast to fusion and crystallization phenomena of typical ionic salts, such as NaCl. To speak in Weyl's expressions such a salt breaks down at the fusion temperature to a highly fluid liquid phase. This latter is, in Ya. I. Frenkel's theoretical aspects,[282] characterized as a "*fissured*" dispersion state. His model of the liquid definitely breaks with the common opinion of structural homogeneity and sees the liquid as a state with "heterogeneous fluctuations," in distinction from a uniform distribution. In this understanding, the liquid has no more resemblance to the crystal from which it was derived. The crystalline phase can not be overheated, the fluid liquid can not be undercooled to a glass. Frenkel's theory, which is intimately related to his derivations for the viscosity of liquids, is also the basis of *statistical* theories for the glassy state in general, such as that of O. L. Anderson and D. A. Stuart[283] applying to nonequilibria of all kinds the principles of statistical mechanics to explain the macroscopic phenomena in terms of structure. All these ideas are based on the presence of "holes" (so-called Schottky holes), or *flaws* in the meaning of Weyl's deductions.

191. "*Swarms of molecules*," with temporary order, are assumed by G. W. Stewart[284] in another model of liquid structures, with a distinction of residual crystal regions, of high degree of order, and non-crystalline groups of molecules with a lower degree of order. These latter groups can become oriented if exterior forces act

[280] *Z. Krist.*, **112**, 1959, 4-21.

[281] *Naturw. Rundschau*, **6**, 1953, 49-53; we further mention here the most fascinating reports of R. Hosemann, *Naturwissenschaften*, **41**, 1954, 440-446; **46**, 1959, 540-552, on paracrystalline structures and the X-ray and optical diffraction phenomena observed with these (cf. Vol. I, C. ¶ 101*) in so-called diffraction microscopy; see also R. Bonart, *Z. Krist.*, **109**, 1957, 309-337.

[282] In the book "Kinetic Theory of Liquids," Oxford University Press, 1946, 544 pp., especially pp. 190-195.

[283] *Ind. Eng. Chem.*, **46**, 1954, 154-160.

[284] *Kolloid-Z.*, **67**, 1934, 130-135; *Am. J. Phys.*, **12**, 1944, 321-323; *Trans. Faraday Soc.*, **33**, 1937, 238-247. Stewart called those swarms "cybotactic regions."

on the liquid and its structure is stretched. The Stewart model is, therefore, "orientable" and similar to anisotropic (paracrystalline) liquids; further, it fits well to the state of "amorphous sulfur"[285] which has a pronounced tendency to develop chain-type oriented molecules (spirals of sulfur atoms and sulfur groups S_8) when stretched. The orientation effect, however, has nothing to do with a crystal nucleation because there is no sulfur modification with chain-type atom arrangements, whereas crystalline selenium has such chains. The formation of ordered regions in amorphous materials is *not* necessarily identical with the formation of crystals.

192. The comparison of the models given by Stewart and Bernal, as discussed by Weyl and Marboe (see above), is important because it grants in the different reactions to stretching strains a possibility of distinguishing between the "flawless liquids" of the silica glass type and those liquids which are undercooled complex heterogeneous, orientable liquids.[286] Flawless liquids of silicates, borosilicates and the like, also do not show orientation effects of chain-like macromolecules in fibers and are, therefore, following the Bernal concepts, whereas fibers of B_2O_3 and of sodium metaphosphate show orientation effects (on the chromatographic structure determination of metaphosphates cf. A. ¶ 380, on the diamagnetic anisotropy of B_2O_3 glass fibers under strain cf. B. K. Banerjee, A. ¶ 295 ff.), because of their evidently complex structures.

193. Transitions between the Bernal, Frenkel, and Stewart models of possible liquid structures are observed say when alkalies are added to B_2O_3; by depolymerization the Stewart type is changed more to the flawless type. One may illustrate such changes in a triangular diagram as shown in Fig. A. 138 which also shows the relative distinction of alkali nitrates from the purely ionic salt character of the alkali halides, coming somewhat nearer to the Bernal type of a liquid with a more pronounced ability to form complex nitrate glasses (as in the system KNO_3-$Ca(NO_3)_2$). The intermediate position of CaF_2 in its melt illustrates the phenomena observed by Tr. Bååk (cf. A. ¶ 129). The interpretation of glass formation phenomena in salt systems even fits to the complex effects observed by Weyl as "*chemically enforced glass formation,*" exemplified by the effects of small amounts of silica and alumina in melts of the easily crystallizing TeO_2,PbF_2, and lead orthogermanate, by a shifting from the "fluctuating fissures" towards the "flawless liquid" type.[287] The curves starting from the B_2O_3 point in Fig. A. 138 further indicate the facts that addition

[285] Cf. J. A. Prins, in "Non-Crystalline Solids," Conference Reports edited by V. D. Fréchette, Alfred University, 1958, John Wiley & Sons, Inc., New York, 1960, pp. 318-327.

[286] We recall here in the latter respect the older observations of J. J. Trillat and H. Forestier, *Bull. soc. chim. France*, **51**, 1932, 248-253, and K. H. Meyer and Y. Go, *Helv. Chim. Acta*, **17**, 1934, 1081-1093; *Kolloid-Z.*, **70**, 1935, 19-20.

[287] See the extensive discussion of W. A. Weyl and E. C. Marboe, *Glass Ind.*, **42**, 1961, especially p. 79. See also W. Weyl, *Glastech. Ber.* **34**, 1961, 301-311.

of Na_2O to B_2O_3 changes this glass above 1000°C. from the orientable liquid type towards that of a fissured liquid, whereas at lower temperatures (say at 600°C.) the same addition changes the B_2O_3 glass structure first to the flawless liquid type with a considerable increase in viscosity (cf. L. Shartsis, W. Capps, and S. Spinner,

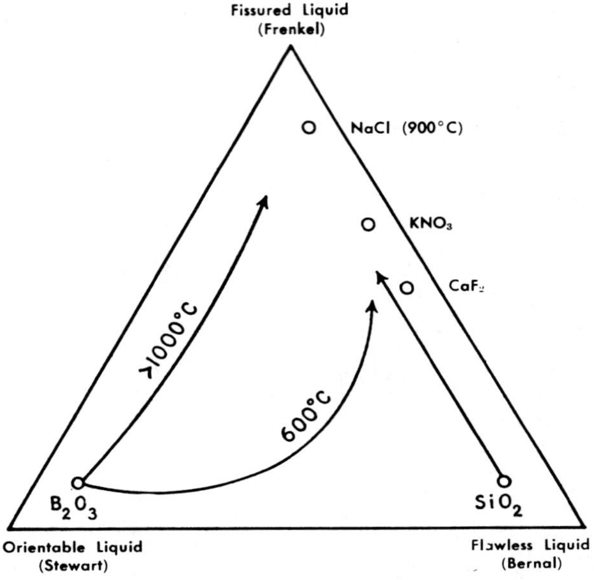

FIG. A. 138. Diagram showing the transitions of the fundamental structural concepts of liquids. (Weyl and Marboe).

A. ¶ 141), and after the addition of much more Na_2O toward the fissured type with the tendency to nucleate the crystallization. Additions of Na_2O to silica immediately change the flawless type of silica glass in direction toward the fissured liquid type.

194. Weyl's and Marboe's ideas are of basic value for a clear classification of glass forming and nonglass forming materials by the structural character of their liquids. They also present in a higher level of interpretation a plausible classification of what is generally defined as "*amorphous*" materials.[288] The highest degree of "amorphy" called very impressively a "*chaotic*" pile aggregation of molecules frozen on a supporting wall are the condensation products of an extremely rapid chilling from vapors which go back to the experiments of G. Tammann and J. D. Starinkewitsch (1913), St. von Bogdandy, J. Boehm, and M. Polanyi (1926/1927), and recently of W. Rühl[289] for different halides, such as TlCl. Even water can be chilled

[288] *Cf. J. Soc. Glass Technol.*, **43**, 1959, 191-210 T.

[289] *Z. Physik*, **143**, 1956, 591-604; 605-622. Strong polarizability of the molecules is essential

by rapid condensation to a "glass," or better said to a chaotic layer of piled-up molecules. By annealing such a product it is changed to a more or less defective *crystalline* state as expressed in a schematic "energy profile" (Fig. A. 139). "Explosive" antimony is such a chaotic phenomenon, as is also the case with metamict

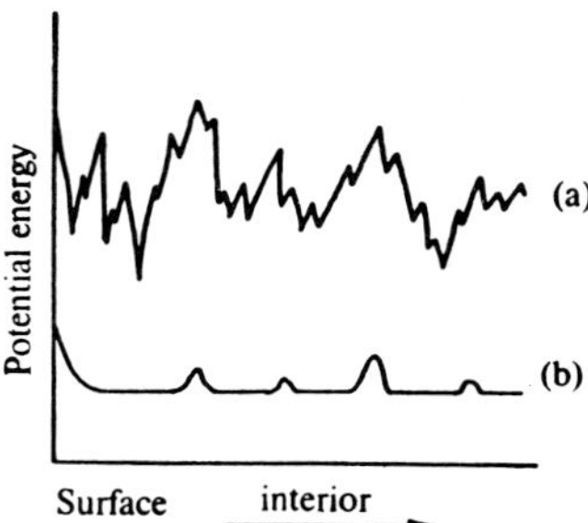

FIG. A. 139. Potential energy profile of an amorphous "chaotic" phase structure, and its change after annealing to a defective crystal phase. (Weyl and Marboe).

minerals (cf. Vol. III, A. ¶ 51 ff.) which are not glasses proper, because they are irreversibly changed by annealing to the crystalline phase and do not undergo any softening. *Isotropization* by radioactive bombardment or by simple fine-grinding is therefore not a process of glass formation, nor the development of the well-known Beilbey layers of metals in polishing processes. The distinction of chaotic and really glassy states, as schematically illustrated by the Figs. A. 139 and 140 in energy

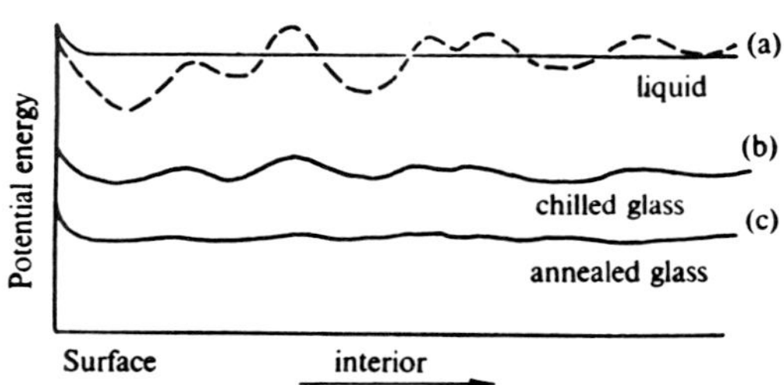

FIG. A. 140. Potential energy profile of a glass in the fused state, after annealing, and after chilling. (Weyl and Marboe).

for the formation of such depositions. To a similar group of phenomena belong the flame-fused and sprayed finest droplets of alumina described by H. von Wartenberg, *Z. anorg. u. allgem. Chem.*, **269**, 1952, 76-85; **270**, 1952, 328. They are *not* amorphous but consist of δ-Al_2O_3. The same identification was confirmed quite recently by A. Dietzel and H. Meyer, *Ber. deut. keram. Ges.*, **37**, 1960, 136-141; *Werkstoffe u. Korrosion*, **11**, 1960, 601-616, for flame-sprayed ceramic covers of pure alumina (cf. Vol. III, Par. A. 22).

profiles, makes evident that the glassy state is *exclusively* formed by way of undercooling of a suitable liquid phase and its solidification without any nucleation of crystals.

195. Among recent investigations on the application of *X-ray diffraction of liquid* materials we mention here the studies of F. Sauerwald and E. Osswald[290] on mercury, molten lead, tin, thallium, and the amalgams with lead and thallium, in continuation of the previous studies of P. Debye and H. Menke (1930) and of H. Hendus.[291] Amorphous antimony was also investigated by electron diffraction methods combined with a complete Fourier Analysis of the intensity curves, by A. I. Tatarinova and Z. G. Pinsker.[292] In this case an indication of the existence of groups [$SbSb_4$] in the "chaotic" state of the amorphous antimony structure is particularly interesting. The same tetrahedral coordination occurs in the intermetallic compounds SnSb and GaSb, both with diamond structure type. This structural principle is characteristically different from that in crystalline antimony. Important contributions on the structure of fused gold, of the alkali metals, and the "half-metals" tin, indium, gallium, germanium, bismuth and antimony, were recently made by H. Richter, G. Breitling, and F. Herre,[293] and further by H. Richter and S. Steeb[294] on fused tin and solid amorphous bismuth, also with electron diffraction methods. In the case of amorphous bismuth, a frozen-in liquid state is observed, with the same interatomic distances and the same coordination of the atom distributions.

196. Among the many investigations on the structure of *fused salts* we mention those by R. L. Harris, R. E. Wood, and H. L. Ritter[295] on aluminum chloride, or of R. E. Wood and H. L. Ritter[296] on indium and tin iodide, as typical studies based on the Fourier-analytical interpretation of X-ray diffraction diagrams at high temperatures in a vacuum furnace camera, using the classical methods of B. E. Warren, H. Krutter, and O. Morningstar (1936). Most coherent scattering is due to associated molecules Al_2Cl_6 similar to those observed by electron diffraction in the vapor state. It is, however, not possible to establish definitely from the radial distribution curves whether indium iodide is monomeric or dimeric, whereas in tin iodide independent molecules SnI_4 are established. The structures of fused *alkali fluorides* were

[290] *Z. anorg. u. allgem. Chem.*, **257**, 1948, 195-198. On the structure of liquid mercury see recently G. Voigtlaender-Tetzner, *Naturwissenschaften*, **48**, 1961, 520-521.

[291] *Z. Naturforsch.*, **2a**, 1947, 505-521.

[292] *Doklady Akad. Nauk S.S.S.R.*, **95**, 1954, 265-268; *Trudy Inst. Krist. Akad. Nauk S.S.S.R.*, 1955 (11) 101-103 (on crystalline antimony); 104-124.

[293] *Naturwissenschaften*, **44**, 1957, 109. On the calculation method for Fourier analysis for atomic distribution curves cf. the same authors, *Ibid.*, **42**, 1955, 151-152, demonstrated by active coal and graphite.

[294] *Naturwissenschaften*, **45**, 1958, 512-513.

[295] *J. Am. Chem. Soc.*, **73**, 1951, 3151-3155.

[296] *J. Am. Chem. Soc.*, **74**, 1952, 1760-1764.

studied by J. Zarzycki[297] who used a special high-temperature diffraction X-ray camera (cf. Vol. III, A. ¶ 99). The fundamental result of these investigations is the direct evidence given for the "hole" structure of these typical fissure-type liquids. The coordination $[RF_n]$ is here of a lower number n than in the crystalline state (= 6), e.g. $n = 3.7$ for LiF (at 860°C.), = 4.1 for NaF (at 1000°C.), and = 4.9 for KF (at 870°C.). The sequence of the short-range distances $(Na–F)_I$, Na–Na, F–F, $(Na–F)_{II}$ is the same as in the crystal, but the number of neighboring ions surrounding one central ion is lower and holes are left. The liquid lattice has a considerably larger specific volume (density 1.79) than the crystalline phase (density

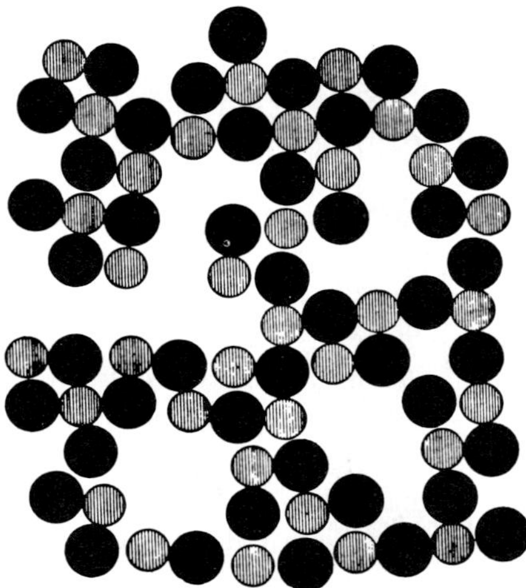

FIG. A. 141. Hole structure of fused fluorides. (Zarzycki).

2.27). The holes in the structure may be understood from the schematic Fig. A. 141, as for a typical Frenkel liquid, the volume of the holes occupying 21 per cent of the liquid volume at 860°C. Further investigations of H. A. Levy, P. A. Agron, M. A. Bredig, and M. D. Danford[298] on the structure of several alkali halide melts, not only by X-ray but also by neutron diffraction, confirm the conclusions of Zarzycki, especially concerning the reduced coordination numbers of the ions in the first and second sphere surrounding a central ion, and the hole or "open" structure of such liquids, with rather widely varying ion coordination characteristics. For alkali *ni-*

[297] *Compt. rend.*, **241**, 1955, 480-481; **244**, 1957, 758-760; *J. phys. radium*, **17**, Supplement to (3) 1956, 44A-51A; **18**, Supplement to (7) 1957, 65A-69A. The density data given are derived from previous measurements of A. Eucken and W. Danmöhl, (1934), and F. M. Jaeger (1917). See also recently J. Zarzycki, in V. D. Fréchette, "Noncrystalline Solids," Conference Alfred University, N. Y. 1958, John Wiley & Sons, Inc. New York 1960, 117-143.

[298] *Ann. N. Y. Acad. Sci.*, **79**, 1960 (11) 762-780.

trates V. I. Danilov and S. Ya. Krasnitskiĭ[299] determined the electron density curves of the melts after Fourier analysis of intensity distribution and found the presence of the NO_3^- anions exactly the same as in the crystalline salts, also with the same interatomic distances.

197. Concerning the orientable liquid structures of the "selfvitrifying" elements, S, Se, and Te, we mention here the studies of H. Grimminger, H. Grüninger, and H. Richter[300] on amorphous *selenium*, either condensed from the vapor and measured by X-ray diffraction at the temperature of liquid air, or the same at room temperature, with remarkably different curves of the atomic distribution, although no crystal interferences were observed at low temperatures. At room temperature there is a 1 : 1 mixture of amorphous selenium (particle size about 7 Å.) and crystalline selenium (about 10 Å. in size). Evidently, the short-range order is in this case the same and also for the layer structure of the amorphous selenium built up from characteristic chains and Se- rings. It was concluded, therefore, that amorphous selenium consists of two forms, namely of a "parcel structure" or of layers built up from folded Se_6 rings (prevailing at low temperatures) and another layer structure with chains which is a widened modification of the crystalline selenium lattice (at room temperature). In her general studies on glass formation Aniuta Winter-Klein[301] also discussed the chain structure in selenium, in contrast with the common interpretation of the Zachariasen theory of glass that a continuous framework would be the indispensable premise of glass formation. Self-vitrification is a general property of the elements of the VIth group of the Periodic System in Winter-Klein's opinion. Sublimated layers of selenium and tellurium, on a glass slide or on crystal faces of NaCl and mica, were examined by S. A. Semiletov[302] in their structural character, in films of 10^{-5} to 10^{-6} mm. thickness. For selenium the gradual reduction into the crystalline (hexagonal or monoclinic) phase was observed at elevated temperatures. Tellurium shows spiral chains in the direction of the *c*-axis of hexagonal crystals, in excellent epitaxis with the NaCl or mica support, and no amorphous behavior.

198. The suitability of *water* to be undercooled was repeatedly mentioned before. The structural analogy existing between tridymite-silica and ice, the analogy in the anomalous density and pressure behavior as well as the negative thermal expansion coefficient over certain temperature ranges are reason enough to study also the nature of water as a liquid, especially in view of its important surface properties in comparison to those of silica to be discussed later in A. ¶ 403 ff. Based on the first concepts of J. D. Bernal and R. H. Fowler (1938) which established the tetrahedral

[299] *Doklady Akad. Nauk S.S.S.R.*, **101**, 1955, 661-664.

[300] *Naturwissenschaften*, **42**, 1955, 256; see also H. Richter, W. Kulcke, and H. Specht, *Z. Naturforsch.*, **7a**, 1952, 511-532; *J. Colloid Sci.*, **6**, 1951, 389-405.

[301] *J. Am. Ceram. Soc.*, **40**, 1957, 54-58.

[302] *Trudy Inst. Krist. Akad. Nauk S.S.S.R.*, **11**, 1955, 115-120.

type of coordination, and on the assumption of a purely electrostatic interaction[303] between bent *dipoles* of individual water molecules, the model of the water structure is complex because it is built up not only from a tridymite-like ice structure type but also from a quartz-like structure, and a close-packed one of the so-called normal liquids. E. Forslind[304] gave a wave-mechanical aspect of association complexes in liquid water, with hybrid atomic orbitals for calculation in the valence bond theory, and added a theoretical discussion of the dipole nature of the H_2O and D_2O molecules. The ideal model of the water structure (Fig. A. 142) is able to undergo fluctuations in distribution of the thermal amplitudes producing peaks of amplitude which eventually lead to disruptures of the weakest bonds. The lattice of water is therefore of the Frenkel type with *defects* which determine behavior of ice as well as water, with holes as the vacant lattice portions. These are able to diffuse through the lattice just below and at the fusion point. The annihilation of the holes when they reach the phase boundaries of the system ice-water characterizes the fusion process, the latent heat of fusion, ΔH_f, being determined by the entropy of formation of lone interstitial molecules. The behavior in volume changes, the dielectric constant, viscosity, surface tension, and the Raman spectrum can be illustrated on this basis, and the energy of formation of Frenkel defects can be computed, based on the specific volume data given by N. E. Dorsey,[305] as seen in Fig. A. 143. The energy of

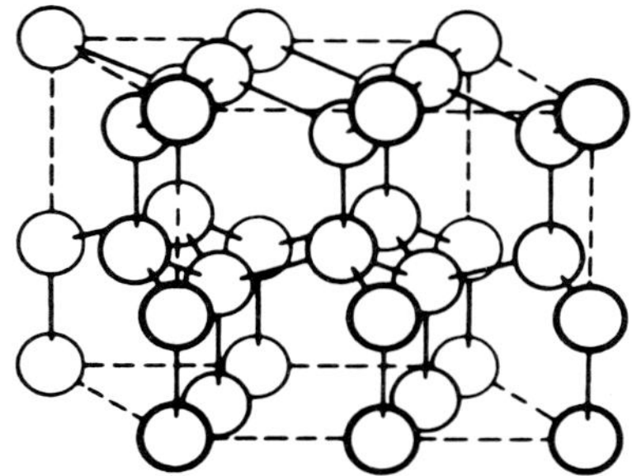

FIG. A. 142. Ideal water structure lattice. (Forslind).

[303] Cf. W. A. Weyl, *J. Colloid Sci.*, **6**, 1951, 389-405.

[304] We follow here especially the discussions given by E. Forslind in *Svenska Forskningsinst. Cement Betong vid Kgl. Tek. Högskol. Stockholm, Handl.*, No. 16, 1952, 43 pp. The same author gave in *Ibid.*, No. 21, 1954, 42 pp. a thorough theory of the lattice dynamics of ice and a discussion of diffuse thermal X-ray scattering. See recently G. G. Malenkov, *Doklady Akad. Nauk S.S.S.R.*, **137**, 1961, 1354-1355, who gave a pentagon-dodecahedral arrangement of double-tetrahedra in liquid water, of the *cis*-type (with symmetry center), and of *trans*-type (with mirror plane). Special assumptions are made on the density of such a structure.

[305] "Properties of ordinary Water Substance." Reinhold Publishing Company, New York 1940, 673 pp.

association of the hydrogen bond is 4400 cal./mole, that of the activation energy for viscous flow at 0°C. = 5260 cal./mole. The order-disorder problems in ice were further discussed by K. S. Pitzer and J. Polissar[306] in agreement with the proposal of L. Pauling (1935) for ice structure, with hydrogen bonds throughout the crystal, but with a complete disorder of the protons, and an entropy of disorder $R \cdot \ln (3/2) =$ 0.81 cal./mole/°C., for H_2O, and = 0.77 for D_2O.

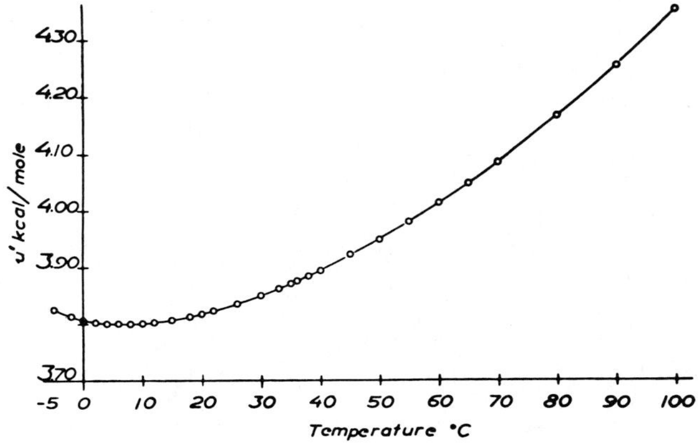

FIG. A. 143. Energy of formation of Frenkel defects in water, based on Dorsey's data. (Forslind).

199. For *ethyl alcohol* in its liquid and highly viscous-undercooled ("glassy") state a comparison study was made by G. Voigtländer-Tetzner[307] with the X-ray diffraction method and calculations on the basis of R. Hosemann's theories. The highly viscous ethyl alcohol " glass" contains chains of C_2H_5OH molecules in parallel position. Over a range of more than 5 Å. there is a short-range microcrystalline order with well-defined distances between the atoms of parallel chains. Liquid ethyl alcohol also has parallel chains of molecules but with variable atomic distances above 4 Å. Voigtländer-Tetzner concludes that these structures are characterized by the assumption of smallest crystalline units in the liquid lattice of a so-called amorphous substance. Ethyl alcohol in these states, therefore, behaves as a typical "*paracrystalline*" substance as defined by R. Hosemann.[308] For long-chained alcohols from 8 to 18 even-numbered carbon atoms the determination of molecular association is possible by an examination of the dielectric relaxation with alternating fields of 3 to 24 × 10^9 cycles/second, as G. B. Rathmann[309] demonstrated, with characteristic

[306] *J. Phys. Chem.*, **60**, 1956, 1140-1142.
[307] *Naturwissenschaften*, **42**, 1955, 95-96.
[308] *Naturwissenschaften*, **41**, 1954, 440-446.
[309] *J. Am. Chem. Soc.*, **78**, 1956, 2035-2038 (with A. G. Curtis, D. L. McGeer, and C. P. Smyth).

differences of the relaxation in pure alcohol and in dilute solutions in long-chained hydrocarbons.

200. Concerning the constitution of *silicate melts*, we repeatedly mentioned the ideas of J. O'M. Bockris[310] (A. ¶ 184) who also emphasized the failure of the chemical random-statistical disorder theory of liquids in explaining sharp inflection points observed at definite compositions, such as 10 mol. per cent R_2O in alkali silicates, or activation energy of viscous flow, compressibility, or expansivity. He suggested that for these critical compositions the three-dimensional framework of $[SiO_4]$ tetrahedra breaks down and discrete silicate anions, say the rings $(Si_3O_9)^{6-}$ appear to determine the properties of the melts. Liquids of the system $Na_2O.SiO_2$–$Na_2O.2SiO_2$ were studied by W. J. Knapp and W. D. Van Vorst[311] showing how from different models the activities would be calculated by a formula

$$\ln a_1 = -\frac{\Delta H_f}{R}\left(\frac{1}{T_0} - \frac{1}{T_f}\right)$$

or from the calorimetrically determined heats of fusion, ΔH_f. The best agreement was found in fact for a model based on $(SiO_3)_x^{2x-}$ rings or chains as anionic silicate groups with a random distribution of pairs $(SiO_{2.5})_2$ bridging between the rings

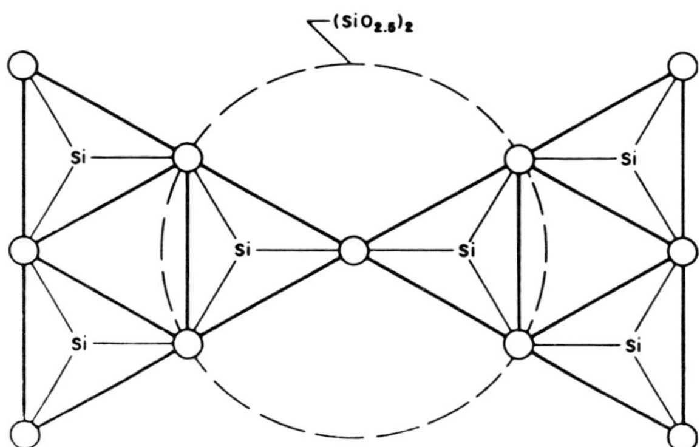

Fig. A. 144. Proposed mixed silicate anion of type $[(SiO_3)_2(SiO_{2\,1/2}]_2$ in sodium silicate melts. (Knapp and Van Vorst).

or chains. Groupings like SiO_3^{2-}, $SiO_{2.5}^{-}$, or $(SiO_3)_4^{8-} + (Si_2O_5)_y^{2y-}$, or $(SiO_3)_3^{6-} + (Si_2O_5)_3^{6-}$ gave inconsistent results for diagrams of $-\log(Na_2SiO_3)_x$, with $x = 1,3,4$, as a function of temperature. A proposed mixed silicate anion $[(SiO_3)_2SiO_{2.5}]$ is shown in Fig. A. 144. Each $(SiO_{2.5})_2$ pair bridges between the rings or chains as indicated in Fig. A. 145.

[310] Cf. J. O'M. Bockris and D. C. Lowe, *Proc. Roy. Soc. (London)*, *A*, **226**, 1954, 433-435.

[311] *J. Am. Ceram. Soc.*, **42**, 1959, 559-562.

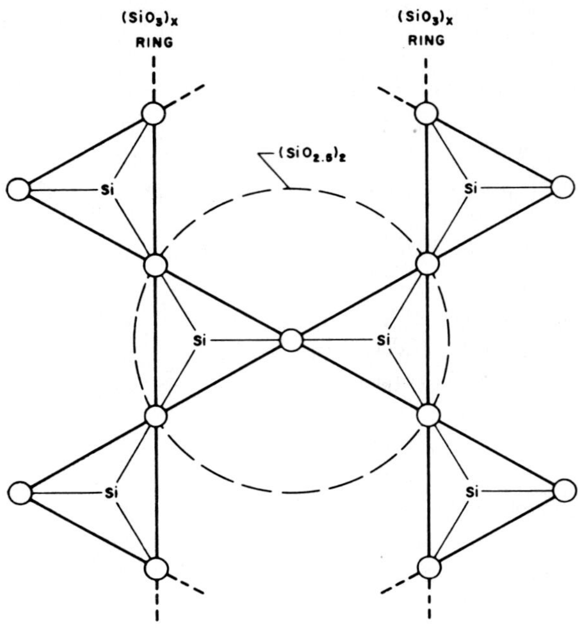

FIG. A. 145. Bridge formed between $(SiO_3)_x$ rings or chains, by a $(SiO_{2\,1/2})_2$ pair in sodium silicate melts. (Knapp and Van Vorst).

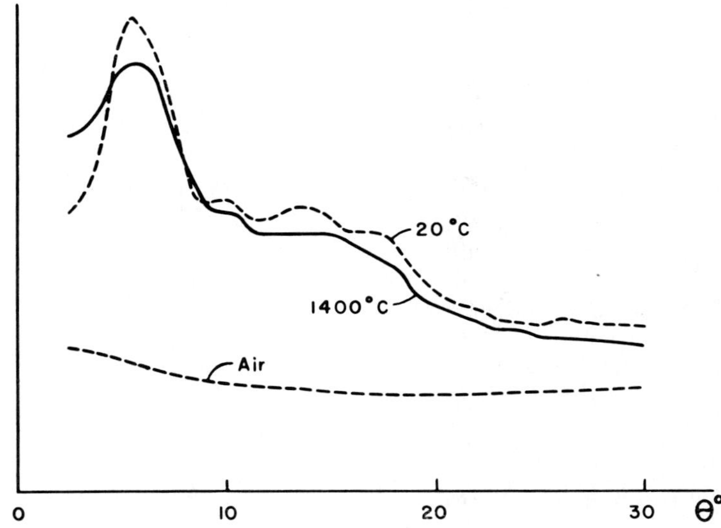

FIG. A. 146. Diffraction spectra (for Cu Kα radiation) of a sodium-calcium silicate glass (70 SiO_2; 10 CaO; 20 Na_2O), (*a*) at 1400°C.; (*b*) at 20°C. (Zarzycki).

201. J. Zarzycki[312] applied the high-temperature diffraction camera mentioned in A. ¶ 196 and described in Vol. III, Section A, for the study of liquid GeO_2 at 1150°C. The comparison of the photometric intensity curves at 1150° and 20°C. shows a considerable flattening of the characteristic maxima at $\Theta = 10°$ and 20° (for CuKα radiation). The same phenomenon is also characteristic say for window glass melts at 1400°C., with a flattening of interference maxima at $\Theta = 5°$, 14° and 18°, the latter two smoothed out to one very flat maximum at 16°. Most characteristic also is the much higher intensity of the low-angle interference maximum which indicates for silicate melts at high temperatures a tendency to approach the diffraction behavior of gases, or a strong hole formation in the liquid lattice (Fig. A. 146).

202. A series of contributions to a symposium on the constitution of glass held in Leningrad 1953[313] especially discussed problems of the so-called *crystallite hypothesis*.

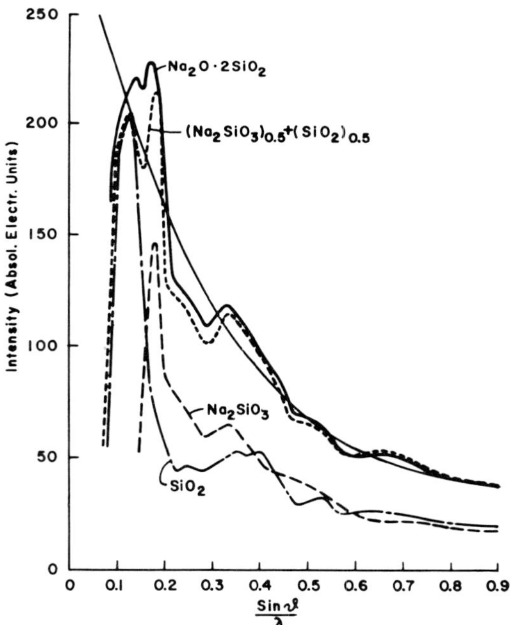

Fig. A. 147. Experimental curve of diffraction intensities of $Na_2O.2SiO_2$ glass (full curve), and of the sum of the intensities of SiO_2 glass and Na_2SiO_3 glass. (Poraĭ-Koshits).

[312] *Zhur. Tekh. Fiz.*, 17, Supplement to No. (3) 1956, especially pp. 49A to 50A.

[313] *Stroenie Stekla, Inst. Khim. Silikatov Akad. Nauk S.S.S.R., Trudy Soveshchaniya, Leningrad*, 1953 (Pub. 1955), 368 pp. A direct attempt of A. G. Alekseev, *Sructure of Glass. Proc. All-Union Conf. Glassy State, 3rd, Leningrad* (English Translation), 1959/60, 169-172, to determine the crystallites in sodium silicate glasses by X-ray diffraction methods, is only conclusive for the formation of α-cristobalite, but the existence of other crystalline phases is presumed. See also recently P. Tarte, *Ind. Chim. Belge*, 1961 (12); 1962 (1), 24 pp. (reprint).

E. A. Poraĭ-Koshits[314] outlined possibilities and results of X-ray investigations of structure derivation on the classical method of Fourier synthesis and typical electron density curves versus atomic distances r in Å. units, the density parameter being defined by the Debye factor $4\pi r^2 \sum \sum n_a k_a k_b \varrho_{ab}(r)$, for glasses of the systems SiO_2–Na_2O and SiO_2–B_2O_3. For $Na_2O,2SiO_2$ glasses the intensities as a function of $\sin \Theta/\lambda$ curve is not simply the sum of those curves for the "ingredients" $SiO_2 + Na_2SiO_3$ (cf. Fig. A. 147), whereas in the series SiO_2–B_2O_3 there is complete additivity from the components. These facts are not compatible with the assumption of a totally random distribution of glass structure (Zachariasen's hypothesis). In every case, Poraĭ-Koshits postulates the "microheterogeneity" of glasses in contrast with the common opinion of a complete uniformity of statistical distributions.

203. This is, in other words, the same opinion which O. A. Esin and P. V. Gel'd[315] called the presence of complex silicate anions in silicate melts (cf. A. ¶ 166), and the formation of "crystallites" or short-range domains in which there is a certain degree of order. Glasses therefore are closer related to the crystalline state in their structure than they are to totally random gas phases. This is the interpretation which A. A. Lebedev (1921) called a crystallite hypothesis. It was decidedly a misinterpretation if one hoped to find by X-ray diffraction methods the presence of the microcrystals directly, with one or the other stoichiometric composition, such as a definite compound, and to identify anomalies in the physical properties of a glass as a function of temperature with polymorphisms of such crystal phases, for example polymorphic inversions of tridymite, cristobalite or quartz, as N. A. Tudorovskaya (1949), N. A. Shishakov (1947, 1949) had done in analogy with metal and salt systems. Esin and Gel'd see heterophase fluctuations and cybotactic groups (see A. ¶ 191*) acting in the X-ray diffraction response of glasses. They call the "crystallites" "embryos of crystallization." The ionic nature of the binding and constitution of silicate glasses is beyond any doubt for Esin and Gel'd. These mostly show a unipolar conductance character and all the normal features of electrolytic dissociation, ion transference,

[314] *Stroenie Stekla, Inst. Khim. Silikatov Akad. Nauk S.S.S.R., Trudy Soveshchaniya, Leningrad,* 1953 (Pub. 1955), pp. 30-34; a short discussion see also by I. I. Kitaĭgorodskiĭ, "Technologie des Glases," 2nd edition, German translation, VEB Verlag Technik Berlin 1959, especially pp. 35-44; see also the previous article of N. N. Valenkov and E. A. Poraĭ-Koshits, in "Physical-Chemical Properties of the Ternary System Na_2O–PbO–SiO_2," edited by I. V. Grebenshchikov, Moscow 1949, 147-157, and an introduction given by K. S. Evstrop'ev, *Stroenie Stekla, Inst. Khim. Silikatov Akad. Nauk S.S.S.R., Trudy Soveshchaniya, Leningrad,* 1953 (Pub. 1955), 9-18.

[315] *Stroenie Stekla, Inst. Khim. Silikatov Akad. Nauk S.S.S.R., Trudy Soveshchaniya, Leningrad,* 1953 (Pub. 1955), pp. 44-55. The presence of quartz as minute relics in industrial silica glass which is quite evident in the electron-microscopic aspect of such products (cf. Fr. Oberlies, *Naturwissenschaften,* 44, 1957, 488 f., Volume V, Section B) was interpreted by V. P. Pryanishnikov, *Stroenie Stekla, Inst. Khim. Silikatov Akad. Nauk S.S.S.R., Trudy Soveshchaniya, Leningrad,* 1953 (Pub. 1955), 270-272, as an evidence in favor of the crystallite hypothesis, in the strict meaning of Lebedev's definition of crystallites.

and so on, although in nonohmic response in high potential fields, as made evident by H. H. Poole (1916).

204. It is highly interesting to follow in the Symposium mentioned the evolution of the debate between the defenders of the framework theory of Zachariasen, and those of the (modified) crystallite theory. We mention here contributions of Yu. N. Andreev[316] who ascribed to the inner portion of a "crystallite" domain a higher order than in the marginal portions, with transitions to the amorphous state in the margins according to ideas of Ya. I. Frenkel, but in contradiction with those of "amorphism" of P. P. Kobeko.[317] The principal failure of the Zachariasen theory is seen by Andreev to arise from ignoring the chemical nature of the alkali ions (modifying elements). The overemphasis given to tetrahedral frameworks as the geometric principle making glass formation possible is also the chief reproach against the Zachariasen theory made by W. A. Weyl in his recent deductions[318] not questioning its heuristic value as a method of "first approximation," but not acknowledging its fundamental and general validity. There is no question that the present state of X-ray diffraction methods cannot definitely decide the validity or nonvalidity of one or the other theory. Even B. E. Warren (1941) did not claim such an absolutely convincing rank of his X-ray results. Other methods, however, speak more in favor of the crystallite hypothesis in a somewhat modified form. The *purely* geometric and crystallochemical background of glass structures was also extensively discussed by N. V. Belov.[319] The Zachariasen model is in this respect ideally fulfilled only in the pure lattice oxide of silica and germanium dioxide systems. On borosilicate glasses see in A. ¶ 214 ff.

[316] *Stroenie Stekla, Inst. Khim. Silikatov Akad. Nauk S.S.S.R., Trudy Soveshchaniya, Leningrad*, 1953 (Pub. 1955), pp. 283-289.

[317] "Amorphous Substances, The Properties of Simple and High-Molecular Amorphous Material," Academy of Science, U.S.S.R., 1952, 431 pp.

[318] *Glass Ind.*, **41**, 1960, 429-433, 463-464, 487-491, 526; 549-553, 620-627, 658; 687-695; *Ibid.*, **42**, 1961, 23-28, 76-81; 123-128, 194-200, 221; *Sprechsaal*, **93**, 1960, 128-136, 518-521, 561-563, 588-589, with extensive criticisms of framework theory, and references. A similar valuable critical discussion of this theory see also by R. L. Thakur and S. Thiagarajan, *Central Glass & Ceram. Research Inst. Bull. (India)*, **8**, 1961, 37-50. In *Stroenie Stekla, Inst. Khim. Silikatov Akad. Nauk S.S.S.R., Trudy Soveshchaniya, Leningrad*, 1953 (Pub. 1955), see discussion remarks of G. M. Bartenev (pp. 293-296) on liquids and glasses, P. P. Kobeko (pp. 296-298) on glass as an entirely uniform material, able to establish equilibrium states), and on "organic glasses" (pp. 19-25); K. S. Evstrop'ev (pp. 301-302).

[319] *Stroenie Stekla, Inst. Khim. Silikatov Akad. Nauk S.S.S.R., Trudy Soveshchaniya, Leningrad*, 1953 (Pub. 1955), pp. 344-350; see also recently *Structure of Glass. Proc. All-Union Conf. Glassy State, 3rd, Leningrad* (English Translation), 1959/60, 74-79, in which the presence of "columns" of $[CaO_6]$ and $[NaO_6]$ octahedral groups are emphasized as characteristic structural units of alkali-lime silicate glasses. Lithium is able to enter either octahedral, or tetrahedral coordination units.

205. We owe to I. Peychès[320] a very complete and critical confrontation of modern theories on glass constitution especially seen under the viewpoints of molecular association. In this respect the Zachariasen and the crystallite hypotheses are reconcilable if only the short-distance structures are discussed which correspond to the X-ray diffraction conditions and are also in agreement with crystalline domains. A deformation of the molecular arrangements is simply possible for the cristobalite-like structure with statistical changes in valence angles. The statistical character of such distortion effects explains the absence of a sharp fusion point and the continuous viscosity-temperature function. The deformation of silica glass under very high pressures and the compaction of the glass was made evident (cf. A. ¶ 60) by P. W. Bridgman and I. Šimon, combined with *permanent* contractions. Alkali ions added reduce this contraction effect; the larger the ion radii the smaller is the contraction effect, bond angles are changed, and holes in the structure reduced. The glass of B_2O_3 behaves totally different (on infrared effects cf. A. ¶ 246 ff.).

206. In a reexamination of experiments of A. A. Lebedev (1921) on the "polymorphic inversions" of silica in silicate glasses, V. I. Garman and L. M. Krasil'nikova[321] determined the H. H. Poole coefficient

$$\sigma = A \cdot e^{\alpha \mathrm{E}} \left(\alpha = \frac{3}{2} \quad \frac{\varepsilon\, \delta}{kT} \right)$$

with δ the average ionic displacement of conductance in glass. This coefficient is pronouncedly structure-sensitive, and as a function of temperature it shows singular points by inversion effects, say of tridymite and cristobalite in a borosilicate glass (with 4 per cent Na_2O, and 26 per cent B_2O_3). More complex glasses do not show these effects, however. This phenomenon may be compared with the suppression of the differential-thermal inversion effects of cristobalite by alumina. In sodium silicate glass, evolution of the cristobalite inversion effect is a very characteristic function of temperature and time of thermal treatment. If the glass was exposed at 700°C. for 150 hours the effect is evident in sensitive differential-thermal analysis, but it is practically absent in nonannealed glass. The cristobalite formed by thermal treatment must have a very low degree of order.

207. A direct evidence of heterogeneous composition and constitution of sodium borosilicate glasses was given by N. S. Andreev and E. A. Poraĭ-Koshits[322] by *low-angle* X-ray diffraction methods (angle of incidence 30″) in a vacuum camera. The

[320] *Verres et réfractaires*, **8**, 1954, 181-193; *Bull. soc. franç. minéral. et crist.*, **77**, 1954, 362-394; *Silicates inds.*, **21**, 1956, 209-218.

[321] *Doklady Akad. Nauk S.S.S.R.*, **116**, 1957, 838-840.

[322] *Doklady Akad. Nauk S.S.S.R.*, **118**, 1958, 735-737. On the low-angle method cf. H. Brumberger and P. Debye, *J. Phys. Chem.*, **61**, 1957, 1623-1624, for porous glasses of the Pyrex and Vycor group.

glass samples had the composition 7 per cent Na_2O, 23 B_2O_3, and 70 SiO_2, as a well-polished plate of 2 cm. thickness, to avoid surface diffraction from the particles of a glass powder which would disturb the diffraction results (cf. Vol. I. C. ¶ 103). The rather uniform size of the primary structural heterogeneities and of the pores are evident, as well as the growth of "crystallites" of the components of the heterogeneities which are high in B_2O_3. The pore strutcure is positively established. Thus, low-angle studies entirely confirm the previous studies of S. P. Zhdanov[323] on the microporosity of borosilicate glasses, and their leaching properties (cf. Vol. I, C. ¶ 59), and further reports of S. P. Zhdanov, E. A. Poraĭ-Koshits and D. I. Levin,[324] and of the latter authors with N. S. Andreev.[325]

208. An excellent review of the whole problem complex was given by E. A. Poraĭ-Koshitz[326] from which we take the instructive Fig. A. 148. It is justifiable that this author criticizes the widespread and misleading schematic representations of framework structures of complex glasses and the unsatisfactory application of the Zachariasen postulates to such systems, because they are not able to explain the nature of a great number of glasses which are not considered by these postulates, as was especially demonstrated by K. Grjotheim and J. Krogh-Moe[327]. At least the possibility of "local chemical order" in glasses must be admitted, say of ordered domains, and this means a "polymer-crystallite" hypothesis reconciling the apparent contradictions between the Zachariasen and the crystallite theory as such. A one-component glass is in this meaning a polymer with a continuous framework and variable interatomic distances. If it is of a high order one may really call it a crystallite complex. Ranges of maximum order may reach 20 to 30 Å. and existence of definite chemical compounds in glass cannot be principally denied. The degree of porosity can be

[323] *Stroenie Stekla, Inst. Khim. Silikatov Akad. Nauk S.S.S.R., Trudy Soveshchaniya, Leningrad*, 1953 (Pub. 1955), 162-175 (cf. Vol. IV, Section A).

[324] *Izvest. Akad. Nauk S.S.S.R.*, 1955, Otdelenie Khim. Nauk, 31-39; 197-207; 395-402; *Bull. Acad. Sei. U.S.S.R. Div. Chem. Sci.*, 1955, 27-33; 353-358 (in English).

[325] *Izvest. Akad. Nauk S.S.S.R.*, ibidem, 1956, 287-293.

[326] *Glastech. Ber.*, **32**, 1959, 450-459; Poraĭ-Koshits used a low-angle camera constructed by O. Kratky, described on pp. 453 ff. In a recent review by E. P. Markin, V. V. Obukhov-Denisov, T. A. Sidorov, N. N. Sobolev, and V. P. Cheremisinov, *Structure of Glass. Proc. All-Union Conf. Glassy State, 3rd, Leningrad* (English Translation), 1959/60, 180-185, is emphasized that the Zachariasen theory completely fails to explain the existence of mere molecule glasses of As_2O_3, Sb_2O_3, P_2O_5.

[327] *Glass*, **33**, 1954, 465-470; *Glastek. Tidskr.*, **11**, 1956 (2) 47-55, especially on oxide glasses; *Glass Ind.*, **39**, 1958, 201-209; on the criticism of the X-ray method cf. previously Chr. Finbak, *Avhandl. Norske Videnskaps-Akad. Oslo. I. Mat. Naturv. Kl.*, 1944, No. 6, 27 pp.; *Acta Chem. Scand.*, **3**, 1949, 1279-1308. In this connection we also mention the aggregation processes studied by O. K. Botvinkin, *Structure of Glass. Proc. All-Union Conf. Glassy State, 3rd, Leningrad* (English Translation), 1959/50, 99-101, in systems SiO_2–TiO_2–Na_2O and SiO_2–ZrO_2–Na_2O, with particular birefringence phenomena.

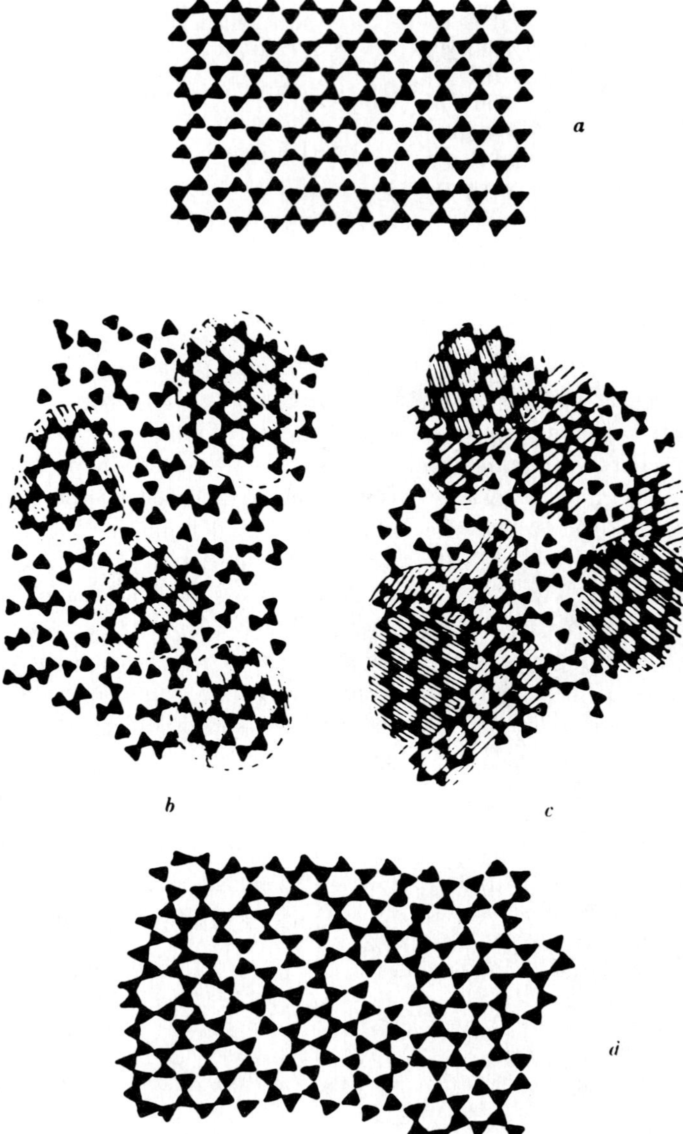

FIG. A. 148. Comparison of the "crystallites" assumed in glass structures, (*a*) ideally crystalline; (*b*) Randall crystallite; (*c*) Poraĭ-Koshits crystallite; (*d*) uniform framework (Zachariasen). (Peychès).

fully deciphered by low-angle X-ray diffraction methods which are the key to such finer distributions in glass structures (cf. Vol. I, C. ¶ 103 ff.), in parallel with opalescence phenomena as a function of the heat treatment of the samples, note the considerable increase in the heterogeneity ranges observed during annealing. A clear distinction can be made in such borosilicate glasses between the silica skeleton and the B_2O_3, and Na_2O containing "filler" material.

209. Another modern attempt to reconcile the statistical framework theory with crystallite-micellar concepts is recommended by L. Prod'Homme,[328] and in the so-called "Vitron" (structon) theory of L. W. Tilton[329] in which a remarkable suggestion is made to give a convincing explanation of the thermal expansion anomalies of silica glass, its permeability for helium gas and other striking physical properties. Tilton follows the geometrical idea that *pentagonal* structural elements if present in a liquid or glass constitution are incompatible with the elementary symmetry postulates for building up crystals, because in these a pentagonal symmetry is excluded. Pentagonal rings of symmetrical tetrahedra with Si–O–Si angles of 180° build up a three-dimensional framework, each ring forming a common interface between dodecahedral "cages," and in their turn forming higher structural units (clusters of such dodecahedral cages) called "Vitrons." This typical framework aspect is a feature reconciling the Zachariasen theory with the short-range order postulate in the structural microregularity principle assumed by the (modified) crystallite theory. In this respect the Vitron theory is also not exclusively a structural principle restricted to pure oxide glasses; it even agrees with the physical properties of binary alkali silicate glasses as well.

210. The Vitron theory postulates steric conditions for the accomodation of cations of different sizes, say of Na^+ and K^+, in relation to the bridging O^{2-} anions inside the "cages," of one or two $O^{2-} + 6\ Na^+$, or of not more than one O^{2-} and $4\ K^+$ included. This condition explains the occurrence of integral numbers of cations per dodecahedral cavity or cage in the framework to cause a change in the slope of property versus composition curves of the alkali silicate glass concerned. Typical properties characterized by such changes are discussed including the chemical durability (cf. A. Dietzel and H. A. Sheybany, 1948) and the solubility losses in water, with a maximum inclosure of O^{2-} inside the cavities for 16.7 molecular per cent K_2O, 23.1 per cent Na_2O, and 28.6 per cent Li_2O (corresponding to 1, 1½, and 2 O^{2-} per cage) as observed for the beginning deterioration of the framework by nonbridging O^{2-}. Changes in the rate of volatilization are equally expected at 28.6, 37.5, and 50 per cent R_2O in potassium, sodium, and lithium silicate glasses,

[328] *Verres et réfractaires*, **12**, 1958, 69-75; *Glass. Ind.*, **39**, 1958, 587-589, 604, 606.

[329] *J. Research Natl. Bur. Standards*, **59**, 1957, 139-154; **60**, 1958, 351-364; *J. Am. Ceram. Soc.*, **43**, 1960, 9-17.

respectively. In the "critical" composition there is a saturation of the cavities with the "modifier" oxides R_2O, and additional cations R^+. Tilton further observed for the molar volumina as a function of composition, for specific resistance and other factors the presence of "critical" concentrations at 16.7 per cent for high temperatures, and at 37.5 per cent for low temperatures in sodium silicate glasses. The

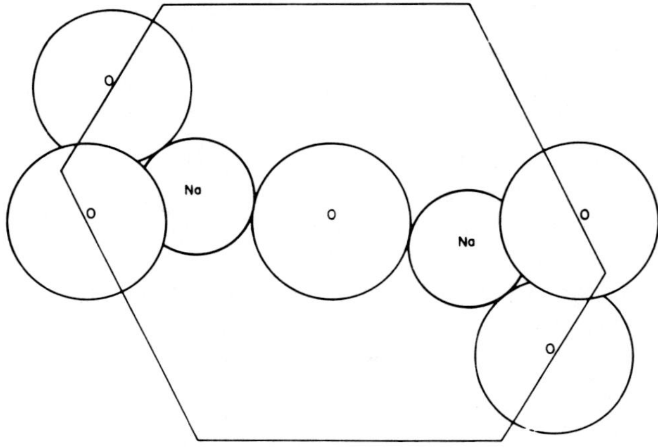

FIG. A. 149. Metastable arrangement for a glass with 16.7 molecular per cent Na_2O of the series Na_2O—SiO_2, in closest contact of the Na^+ ions with four O^{2-} (distance 2.25Å.), and six other O^{2-} in the next sphere (distance 3.4 Å.), in a dodecahedral Vitron unit. (Tilton).

16.5 per cent point is the separation limit between very readily devitrifiable sodium silicate glasses and more stable ones. The steric conditions in the uniquely metastable "deep potential well" composition at 16.7 molecular per cent R_2O are illustrated in Fig. A. 149. This point is particularly prominent on curves of equal fluidity near the annealing temperatures as seen from previous data of E. Jenckel and A. Schwittmann (1938).

211. In contrast with the usual static models of "glass constitution," A. Dietzel and E. Deeg[330] made an interesting attempt to come to a *dynamic* representation of the structure in glass by imitating the electrodynamic actions between O^{2-} anions and cations by the introduction of "elementary magnets" as structural elements, to imitate the polarized O^{2-} anions. In this way (cf. Fig. A. 150 showing the two magnetic model units) one may also demonstrate the phenomena of fusion as a breakdown of the coherence of such a dynamic lattice system over a range of slow softening during heating, effects of chilling, and mechanical behavior of glass under compression and extension. The three-dimensional framework is, of course, only

[330] *Glastech. Ber.*, **30**, 1957, 282-287; *Trabajos reunión intern. reactividad Sólidos, 3 Madrid,* 1956, Sección II, 127-160; *Veröffentl. Max-Planck-Inst. Silikatforsch.*, **17**, 1957, 1-34.

incompletely imitated by the two-dimensional plane network of a system of small rod magnets, but mobility of the structure is very well represented by floating the "elementary magnets" in water on wood holders keeping hemispherical ends above the water surface, analogous to a 100 per cent ionic bonding in the model. Relaxation times which in real glass are of the order of magnitude of 10^{-6} seconds, are in the model of 10^1 to 10^2 seconds analogous to the dimensions magnified by $10^8:1$ between model and real structural units. In a silica glass-like model the tendency to

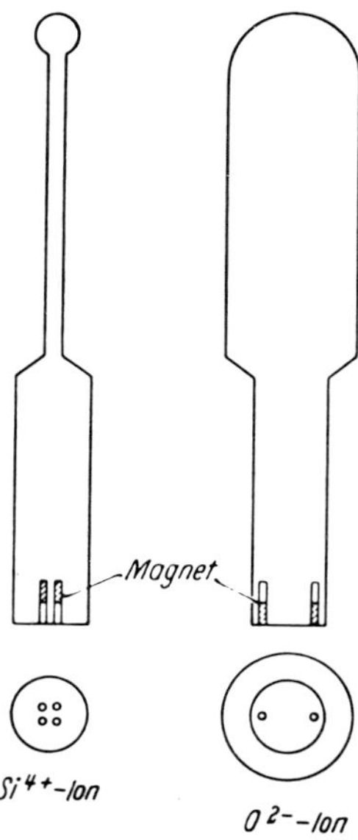

FIG. A. 150. Model structural units for a silicon cation and a (polarized) oxygen anion, for a schematic dynamic illustration of glass formation and constitution. (Dietzel and Deeg).

form six-membered rings is very striking, eliminating all less regular arrangements. This may be understood as an illustration for the devitrification tendency of silica glass under the action of thermal motions. Structures with modifying foreign ions are illustrated by corresponding models which show the latent tendency to unmix to portions (domains) enriched in silica, and others enriched in the foreign oxide component. Dietzel's and Deeg's models are in addition remarkable for an application of the principle of polarization in ionic binding which is so impressively defended by K. Fajans and W. A. Weyl,[331] and abandoning radically common concepts such as those of "breaking a bond" and "formation of a new bond" which have no

[331] *Glastech. Ber.*, **30**, 1957, 269-282; *Central Glass & Ceram. Research Inst. Bull. (India)*, **4**, 1957, 121-139.

real physical significance in solid state chemistry. "The lack of appreciation of polarizability of anions is perhaps the most serious short-coming in present theories on glass structure," says Weyl, and we fully agree with him.

212. We recall other ideas of A. Winter-Klein[332] who emphasized consideration of the elementary process of bonding in glass-forming systems, based on quantum-mechanical principles. Her endeavors to classify the glass-forming systems on such a general and exact basis in the Periodic System of the elements gives new aspects for future progress in this direction, even for more complex glass compositions. More recently[333] the same author discussed the discontinuities mentioned above in connection with Tilton's Vitron theory, in relations of structural "inheritance" which rule the state of silicate melt phase at temperatures above the Liquidus curve of the phase equilibrium of glass-forming oxides. Glass is much more ordered than Zachariasen's framework theory could foresee, and even polymorphous phenomena are included in the considerations of Winter-Klein. The structure of the undercooled liquid changing to a glass is principally determined by the rate of transgressing the temperature zone in which crystallization can take place. It is therefore a *kinetic* aspect which is in the foreground of the discussion of what "thermal history" means for the inheritance effects on physical properties of glass at room temperature. The range of transformation, and the temperature zone in which devitrification may occur, are, therefore, essential for the constitution of the glass as it is used at room temperature. Winter-Klein uses the optical properties as sensitive indicators of the factors of thermal history, as for sodium silicate glasses.

213. A *thermodynamical* basis for conditions of glass formation from the refractory oxides of elements in the IInd, IIIrd, IVth, Vth, and partly the VIth groups of the Periodic System was developed in the Russian school by R. L. Myuller[334] by the thermal vibration energy derived from the average specific heat, C_m, data, independent from the geometric conditions of the given material. The number of frozen-in degrees of freedom is $P^* = (6.4\,N - C_m)/2.13$. For silica glass it is $= 4.0$, and for B_2O_3 glass $= 8.1$, which are fully efficient for the valences; but in SiO_2 for the deformation it is $= 0$, in $B_2O_3 = 0.22$. G. B. Bokiĭ and S. S. Batsanov[335] used the electronegativities, calculated not only for the elements building up glass-forming sys-

[332] *Verres et réfractaires*, **9**, 1955, 147-156; *Compt. rend.*, **240**, 1955, 73-75.

[333] *Verres et réfractaires*, **13**, 1959, 125-130; *Compt. rend*, **247**, 1958, 1587-1590.

[334] *Zhur. Fiz. Khim.*, **28**, 1954, 1831-1836. The importance of stray covalent bonding (cf. A. ¶ 76) in highly polymerized compounds of elements of the Groups III, IV, and V in the Periodic System is emphasized by Myuller in a recent discussion, *Structure of Glass. Proc. All-Union Conf. Glassy State, 3rd, Leningrad* (English Translation), 1959/60, 50-57.

[335] *Zapiski Vsesoyuz. Mineral. Obshchestva*, **85**, 1956, 137-146; cf. previous theories by R. T. Sanderson, *J. Am. Chem. Soc.*, **74**, 1952, 272-274, 4792-4794; *J. Chem. Educ.*, **29**, 1952, 539-544; **31**, 1954, 2-7, 238-245.

tems, but also for the neutral groups $SiO_2 = 4.55$, and $Mg_2SiO_4 = 3.32$, or the ionized groups $(SiO_4)^{4-} = 1.45$, $(Si_2O_7)^{2-} = 2.02$, $(SiO_3)^{2-} = 2.58$, $(Si_6O_{11})^{6-} = 2.90$, $(Si_2O_5)^{2-} = 3.25$, and $(SiO_2)^0 = 4.55$ (cf. Vol. I, A. ¶ 18). Bokiĭ[336] strictly refutes the Zachariasen theory, and equally Kobeko's theory of amorphism (see A. ¶ 204) in the same understanding as K. G. Kumanin[337] and L. I. Demkina[338] refute neglect of the chemical individuality of glass-forming compounds like $CaO.SiO_2$, $Na_2O.SiO_2$, $Na_2O.2SiO_2$, $K_2O.4SiO_2$, and others. Every one of these with quite particular ratios $RO : SiO_2$ so that one cannot speak of *one* uniform $[SiO_4]$ tetrahedral skeleton equally fundamental of their structures.

214. The Russian authors expected many new contributions to the general validity of the crystallite hypothesis by electron diffraction methods. We mention in this respect the studies of L. A. Afanas'ev[339] on electron diffraction diagrams of industrial glasses. Cristobalite and tridymite were identified as trivial constituents in window glasses, $Na_2O.B_2O_3$ in borosilicate glass, whereas in a copper-stained (SnO_2 containing) ruby glass the diffraction lines of CuSn and $Cu_{20}Sn_6$ could be observed (cf. Vol. I, C. ¶ 88). Much more conclusive may be the investigations of A. I. Andrievskiĭ, I. D. Nabitovich, and P. A. Kotsyumakha[340] on the electron diffraction diagrams of thin films blown to a thickness of 10^2 to 10^3 Å. from chemical apparatus glass (Jena and Pyrex glass, other industrial glasses). The presence of finest dendrites and crystallites was established by these experiments which were not changed by an aging for 20 days at room temperature. Identified in silica glass were: quartz, tridymite, cristobalite; in white container glass also $CaO.SiO_2$, and calcium aluminates (?), in borosilicate glasses $Na_2O.B_2O_3$.

215. Starting from the position of the glass forming elements in the Periodic System, Kl. Kühne[341] attempted to modify the Zachariasen theory in its application to ultra-porous borosilicate glasses after acid-leaching (Vycor type). The size of the micropores of 20 to 60 Å. can be used to modify the silica glass skeleton by adsorption of heavy metal cations which, after calcination and sintering, are incorporated

[336] *Stroenie Stekla, Inst. Khim. Silikatov Akad. Nauk S.S.S.R., Trudy Soveshchaniya, Leningrad*, 1953 (Pub. 1955), 350-352.

[337] *Stroenie Stekla, Inst. Khim. Silikatov Akad. Nauk S.S.S.R., Trudy Soveshchaniya, Leningrad*, 1953 (Pub. 1955), 313-315.

[338] *Stroenie Stekla, Inst. Khim. Silikatov Akad. Nauk S.S.S.R., Trudy Soveshchaniya, Leningrad*, 1953 (Pub. 1955), 298-300.

[339] *Stroenie Stekla, Inst. Khim. Silikatov Akad. Nauk S.S.S.R., Trudy Soveshchaniya, Leningrad*, 1953 (Pub. 1955), 224-226.

[340] *Doklady Akad. Nauk S.S.S.R.*, **116**, 1957, 994-995. See also recently G. O. Bagdyk'yants and A. G. Alekseev, *Structure of Glass. Proc. All-Union Conf. Glassy State, 3rd, Leningrad* (English Translation), 1959/60, 198-201, on electron diffraction studies of silica, and lead silicate glasses.

[341] *Silikat Tech.*, **6**, 1955, 191-200; **7**, 1956, 451-465.

into the highly siliceous glass body in a characteristic function of their ionic radii and valencies. Only the cations of the VIth group (Cr, Mo, Wo, U) cause a regular increase say in electrical resistance with increasing atomic numbers (cf. A. ¶ 131); for trivalent cations there is a minimum of specific resistance. Calculating the loosening energy, Φ, as a function of the attraction forces in the glass structure, $e_1 e_2/r^2$, or for O^{2-} anions z/a^2 (Dietzel, 1940), there is a linear function of Φ versus z/a^2, with a minimum $= 2.0$ for z/a^2. Framework-forming and-modifying cations follow two different laws, with variable effects on the specific volume as a function of the polarizability, or in other words, a function of the electronic configuration of the cations relative to the O^{2-} anions. The specific resistance was chosen as a property which is highly sensitive to these effects. Corresponding ionic replacement studies were also made from simple sodium-calcium silicate and borosilicate glasses to elucidate the influence of introduced ions by their field potentials and strengths on the dielectric properties of the host glass structure.

216. The principal aptitude of Dietzel's conclusions of field potentials z/r and strengths (z/a^2) or $z_1 \cdot z_2/(r_1 + r_2)^2$ was made evident already as early as 1942 by the prediction of *immiscibility* relations in glass-forming systems. In this direction, R. C. De Vries, R. Roy, and E. F. Osborn[342] discussed these parameters in their investigation of the system SiO_2–TiO_2. In a more general and strictly systematic manner, E. M. Levin and St. Block[343] applied crystallochemical principles to the problems of immiscibility in 19 binary and 8 ternary glass-forming systems. S-shaped Liquidus curves in the binary phase equilibrium diagrams, which are characteristic of the systems with alkaline earths and indicate an approach to immiscibility, can be analysed in the same way as are complete immiscibility types. Most interesting is the fact that for silicate as well as for borate glasses the same coordination types can be used. Either two modifier cations are bonded to one oxygen anion of a tetrahedron (type *A* in Fig. A. 151), or an excess negative charge is distributed over the tetrahedra and neutralized by two cations, bonded to opposite pairs of oxygen anions (type *B* in Fig. A. 152). This type *B* is superior to *A* in packing efficiency of the glass lattice. In silicate systems the calcium cation is of the type *A*, but in borate systems it has mostly the *B* type. A full application is made of the Pauling rules (1940) originally established for crystal structures as a leading principle (cf. Vol. I, A. ¶ 60). A change from type *B* to *A* occurs in ternary borosilicate glasses, and in

[342] *Trans. Brit. Ceram. Soc.*, **53**, 1954, 525-540.

[343] *J. Am. Ceram. Soc.*, **40**, 1957, 95-106; 113-118. On analogous phenomena and calculations for glasses of the system La_2O_3–B_2O_3 see E. M. Levin, C. R. Robbins, and J. L. Waring, *Ibid.* **44**, 1961, 87-91. W. Vogel, *Structure of Glass. Proc. All-Union Conf. Glassy State, 3rd, Leningrad* (English Translation), 1959/60, 17-25, discussed recently the immiscibility phenomena and drop-shaped microheterogeneities of borosilicate glasses as paradigma of a general "cellular" or "micellar" structure of glass. These are the same phenomena as described in Vol. I, C. ¶ 83, and Vol. V, Section B, as the basis of Vycor and Pyroceram products.

zinc containing glasses a change of the coordination number from six to four. The electrostatic bond strength of the single cation as defined by the valence z divided by the coordination number, $C.N.$, and for the pairwise bond strength of two cations by

$$E_B = \frac{-z_1}{(C.N.)_1} \cdot \frac{z_2}{(C.N.)_2} \cdot e^2/a^2,$$

is decisive for the coordination types to be assumed. All cations with a bond strength below or equal to 0.25 assume the B type coordination in silicate and borate systems. If the bond strength is $= 1/3$, the type of coordination is A; therefore Ca^{2+} which is intermediate between A and B, will be able to occur in both forms. If there are linear Liquidus curves in binary systems (as in the systems with Cs^+ and Rb^+) the

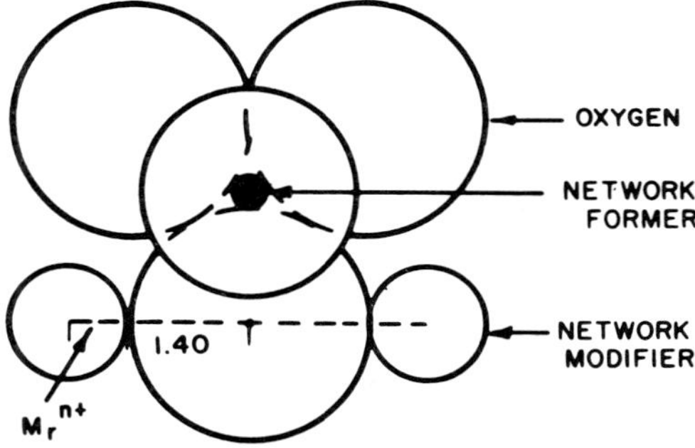

FIG. A. 151. Type A coordination of a modifier cation in glass structure. (Levin and Block).

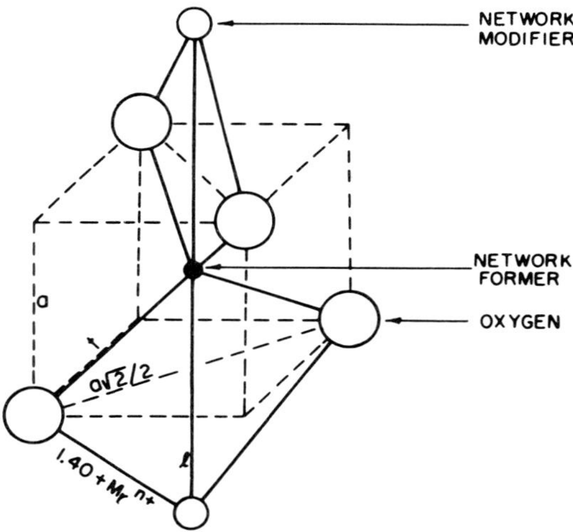

FIG. A. 152. Type B coordination of a modifier cation in glass structure. (Levin and Block).

bond strength is very low, namely = 1/12. On the other hand, it is = 1/8 or 1/6 in the systems with S-shaped Liquidus curves (with K^+, Na^+, and Li^+ as the cations). Unmixing occurs if the bond strength is = 1/4 and all cations with bond strengths higher than 1/4 and 1/3 occur in binary systems with typical miscibility gaps forming two liquid phases with SiO_2 and B_2O_3.

217. The suitability of Levin's and Block's deductions for alkali and alumina containing ternary systems[344] with those oxides as "homogenizers" as seen from corresponding three-phase equilibrium diagrams is further convincing by definite coordi-

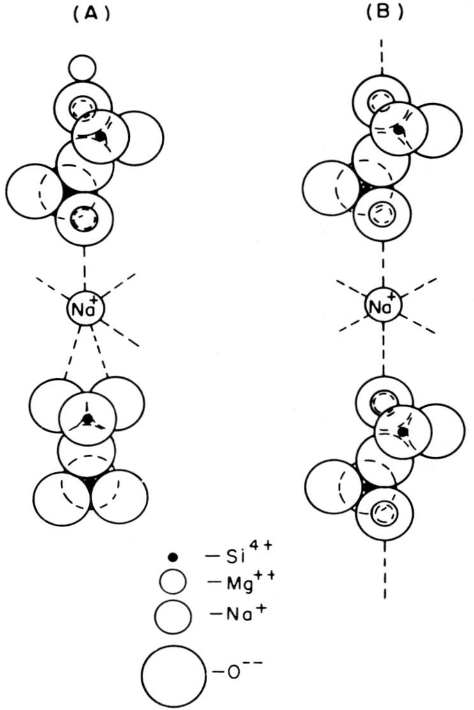

Fig. A. 153. Coordination configuration for alkalies, acting as homogenizers in the system Na_2O—MgO—SiO_2. (*A*) Sodium cation coordinating incompatible structural units of the binary system RO—SiO_2; (*B*) Configuration at the "critical" composition. (Levin and Block).

nation configurations as seen in Figs. A. 153 and 154, in the special cases of the systems Na_2O–MgO–SiO_2, and MgO–Al_2O_3–SiO_2, with characteristic cation distribution between the two liquid phases. Alkalies are homogenizers because of their

[344] *J. Am. Ceram. Soc.*, 41, 1958, 49-54.

weak electrostatic bond strengths, precluding the separation of conjugate liquid phases. Alumina is a homogenizer agent, in spite of its high bond strength, because the aluminum cation assumes both fourfold and sixfold coordinations. On the other hand, boron is *not* a homogenizer because it acts only in the three- and four-

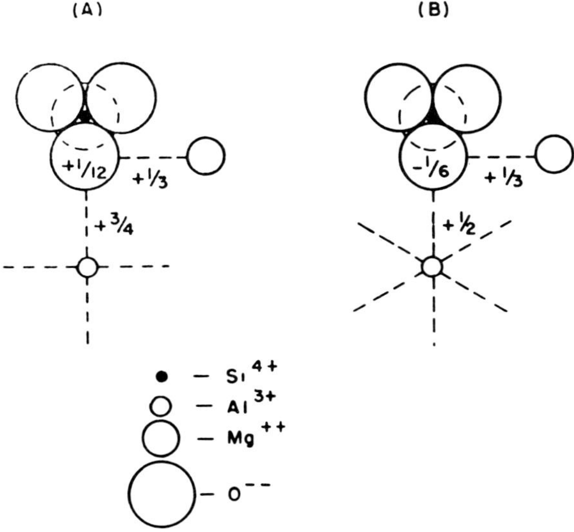

FIG. A. 154. Coordination configuration for tetrahedrally coordinated Al^{3+}(*A*), and for octahedrally coordinated Al^{3+}(*B*), in glasses of the system $MgO—Al_2O_3—SiO_2$. (Levin and Block).

fold coordination. As a framework former cation it cannot coordinate incompatible units merely by substituting for silicon. The amount of the homogenizer required for a complete miscibility in ternary silicate and borate melts can be estimated quantitatively and is in good agreement with the experimentally observed concentrations.

218. F. P. Glasser, I. Warshaw, and R. Roy[345] are of the opinion that the best suitable parameter for correlating ionic properties to immiscibility phenomena in silicate glasses is the ionic potential, z/r, of the second cation of binary systems, combined with its coordination number, the fusion temperature of its oxide, and other factors. The width of the miscibility gap and its approximate minimum temperature can thus be rationally interpreted and derived. The ionic potential is suitable to predict high-temperature immiscibility, exemplified for not easily accessible highly

[345] *Phys. Chem. Glass*, **1**, 1960, 39-45. On the system La_2O_3–SiO_2 and other systems studied in their unmixing characterics see also F. A. Hummel, *Pittsburgh Ceramist*, **9**, August 1957 (1), 4 pp.

refractory systems like ZrO_2–SiO_2, HfO_2–SiO_2, ThO_2–SiO_2, and UO_2–SiO_2, with the primary crystal phases of the dioxides in equilibrium with two liquids. Even the systems Fe_2O_3–SiO_2 and Cr_2O_3–SiO_2 may be discussed on the basis of the correlation between z/r and the molecular per cent concentration differences between the two liquid phases for variable metal oxides in silicate systems as seen in Fig. A. 155.

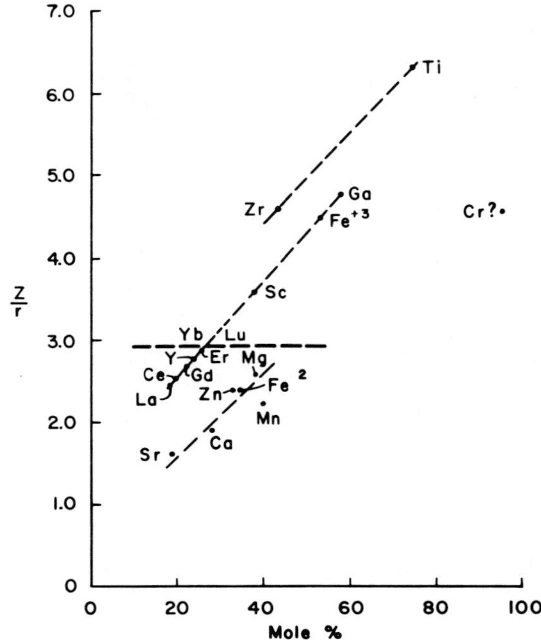

Fig. A. 155. Correlation between the ionic potential, z/r, and the extent of liquid immiscibility in silicate systems with variable metal oxides. (Glasser, Warshaw, and Roy).

In the many cases in which unfortunately the calculation methods of Levin and Block cannot be used, because sufficient data are lacking, this diagram may be an important help for a first orientation. Rules are given for use of z/r to distinguish systems with either SiO_2, or with the second oxide as the primary crystal phases. The lanthanum-rare earth elements silicate systems show transitions with both types of the fusion equilibria. The apparent anomaly of the system alumina-silica which does not show unmixing phenomena is explained by the fact that the fourfold coordination $[AlO_4]$ excludes the formation of a two-phase separation. The fact that the given predicted diagram postulates an immiscibility of ZrO_2 and SiO_2 melts near 2600°C. is remarkable, because this is in contradiction with the phase equilibrium diagram drafted in this temperature range by R. F. Geller and P. J. Yavorski (1945).

219. The close relations of the electrostatic field strengths z/a^2 to immiscibility phenomena in silicate systems as defined by A. Dietzel (1943) is also extended to the formation of glasses in *phosphate* systems and to the problems of the formation of ternary compounds in phosphate-silicate systems. As an excellent example of application of these rules we mention here the investigations of H. H. Paetsch and A. Dietzel[346] (cf. Vol. III, B. ¶ 444) on the system PbO–SiO_2–P_2O_5, with a wide ternary miscibility gap in the melts. Glass formation is particularly characteristic for melts with more than 30 molecular per cent SiO_2, if P_2O_5 is low. With more than 40 per cent P_2O_5 and low concentrations in SiO_2, all of the other ternary mixes crystallize easily. The immiscibility in this phosphate-silicate system and the corresponding system PbO–B_2O_3–SiO_2 is in contrast with the complete miscibility in the system PbO–Al_2O_3–SiO_2, illustrating the importance of different binding mechanisms of the groups R—O (R = Al,B,P,Si). Theoretically, immiscibility should be observed in the not yet known systems PbO–Al_2O_3–B_2O_3 and RO–Al_2O_3–P_2O_5.

220. Concerning the important theoretical deductions of E. Kordes (1941) on the constitution of zinc phosphate glasses, we mention more recent studies of the same author with H. Becker[347] on the systems of P_2O_5 with Na_2O, Li_2O, and CdO, and the molecular refraction of their glasses (cf. B. ¶ 133). Further investigations were made on their structural properties (with W. Vogel and R. Feterowsky[348]). The interpretations given by Kordes *et al.* are adapted to the postulates of the Zachariasen theory in silicate-like frameworks. In this respect, Kordes distinguishes between normal phosphate glasses in the systems of P_2O_5 with CaO,BaO,PbO,CdO, and anomalous glasses of the systems P_2O_5–BeO, MgO, and ZnO, with characteristically different physicochemical properties. In beryllium phosphate glasses a coordination $[OR_3]$ of oxygen which is analogous to that occurring in the phenacite structure is especially characteristic, but not in the silica modifications or highly siliceous silicates. The coordination $[OR_3]$ is more precisely that with one P^{5+} and two Be^{2+}

[346] *Glastech. Ber.*, **29**, 1956, 345-355. On *oxypyromorphite*, $Pb_{10}(PO_4)_6O$, which was not observed by Paetsch and Dietzel, cf. L. Merker and H. Wondratschek, *Z. anorg. u. allgem. Chem.*, **306**, 1960, 25-29. We emphasize once more the high principal importance of the ring and chain configurations in condensed phosphates, extensively studied by J. A. Van Wazer et al. (cf. *J. Am. Chem. Soc.* **75**, 1953, 1563-1567; **77**, 1955, 1468-1470, 1471-1473, concerning flow birefringence, density, surface tension and viscosity of molten sodium phosphates. The ring or chain patterns are also characteristic of crystalline sodium phosphates, cf. G. Donnay, *Carnegie Inst. Wash. Year Books*, **56**, 1956/57, 242-244, with a stereoisomerism of (P_4O_{12}) in ring or "boat" forms and order-disorder relations.

[347] *Z. anorg. u. allgem. Chem.*, **260**, 1949, 185-207.

[348] *Z. Elektrochem.*, **57**, 1953, 228-288; see also *Fortschr. Mineral.*, **33**, 1955, 152-153. On $Al(PO_3)_3$ as a crystallization product from optical phosphate glasses see W. Vogel, *Silikat Tech.*, **6**, 1955, 510-517, with a discussion of Kordes' theory and interpretation of the constitution of phosphate glasses.

cations. Only by this specific coordination is it possible for the beryllium phosphate glasses to be built up from normal $[PO_4]$ tetrahedra.

221. The interesting question of how silica glass may be correlated to SiS_2 or $SiSe_2$ was examined by Al. and Arm. Weiss,[349] by heating the fibrous crystalline phase $SiSe_2$ in an evacuated silica glass tube above 1060°C. and quenching the melt. The resulting slightly yellow, transparent glass is much more moisture-resistant than the crystalline fibrous form into which, however, the glass is easily converted by devitrification. Among the "anomalous" glasses that of As_2O_3 was studied multiply by I. N. Stranski *et al.*,[350] especially with X-ray diffraction methods and a Fourier analysis of intensity and electron-density curves, based on the F. Zernike-J. A. Prins (1927) relations for amorphous materials. As_2O_3 glass is evidently more closely related to claudetite than to arsenolite in its framework structure and short- range atomic distances, with rings of 6 $[AsO_3]$ groups of a distorted crystalline arrangement of claudetite type. This interpretation coincides with what was to be expected from the Zachariasen postulates.

222. Because of their great practical interest based on infrared transmissivity up to 6.2 μ, Sb_2O_3 and Sb_2O_5 containing glasses were systematically studied by W. A. Hedden and B. W. King[351] with Na_2O, K_2O, and Al_2O_3 as stabilizers. In ternary glass compositions of this kind up to 78 weight per cent Sb_2O_3 could be taken up; the resulting glasses were slightly yellow in this state if molten under oxidizing conditions. By additions of the oxides of Cr, Ni, Co, Mn, and Cu the same colors in general were observed as they are typical of silicate glasses; only with Fe_2O_3 (2 to 3 per cent) an anomalous deep-red color appeared. The stable antimony ion in the glasses is Sb^{5+}, and it is maintained by melting the glasses in an oxidizing atmosphere. One must therefore speak of *antimonate* glasses in all those cases in which

[349] *Z. Naturforsch.*, **8b**, 1953, 104-105.

[350] *Z. anorg. u. allgem. Chem.*, **266**, 1951, 302-312 (with H. Bötticher, K. Plieth, and E. Reuber-Kürbs); see also K. Plieth, *Proc. Intern. Symposium on Reactivity of Solids. Gothenburg*, 1952/54 II, 761-770. The X-ray diffraction camera used was furnished with a curved NaCl monochromator crystal; cf. K. Plieth, and E. Reuber, *Z. anorg. u. allgem. Chem.*, **280**, 1955, 197-204. The $[AsO_3]$ groups correspond to $[BO_3]$ groups in borosilicate glasses with configurations of distorted trigonal pyramids in the framework. The full Fourier analysis of As_2O_3 glass is given in *Ibid.*, **280**, 1955, 205-210. The same authors (with I. Zoll, *Ibid.*, **292**, 1957, 343-346, made the surprising discovery that by a reaction of claudetite with an alkaline Thuringian glass (and only with this) an unknown arsenosilicate is formed below 170°C. It is presumed that in this product $[AsO_3]$ groups replace the tetrahedral $[SiO_4]$ groups of the irregular framework of the Thuringian glass. The $[AsO_3]$ groups would then correspond to $[BO_3]$ groups in borosilicate glasses with configurations of distorted trigonal pyramids in the framework. See also conclusions from infrared absorption spectra of As_2O_3 and arsenite glasses, by R. V. Adams, *Phys. & Chem. Glasses*, **2**, 1961, 101-110.

[351] *J. Am. Ceram. Soc.*, **39**, 1956, 218-222.

$NaNO_3$ was used in the batch. In a reducing atmosphere deep amber-yellow or brown colors appear.

223. For similar purposes in the development of infrared-transmittent glasses (up to 6 μ wavelengths) H. C. Hafner, N. J. Kreidl, and R. A. Weidel[352] investigated a series of *calcium aluminate* compositions which are even easily produced in quantities of 10 to 15 kgs. in spite of their pronounced devitrification tendency. Weidel[353] stabilized such glasses by the introduction of some alkali and iron oxide although the latter is detrimental to the transmissivity effects. On the contrary, MnO_2 and particularly Cu_2O significantly increase the infrared transmission of such stabilized aluminate glasses, with a distinct optimum of the copper addition. Also P. L. Baynton, R. Rawson, and J. E. Stanworth[354] prepared calcium aluminate glasses with 38 to 65 per cent alumina, ternary glasses with 6 per cent silica, and further *calcium gallate* glasses of about 60 per cent Ga_2O_3, but with a very narrow interval of glass formation in the binary concentration diagram, all of the other compositions showing a rapid devitrification. Also in this case, silica (7 per cent) acts as a stabilizer addition, if molten at 1400°C. Gallate glasses have a higher infrared transmittance up to 5 μ than aluminate glasses. Thiogallate and thioaluminate glasses show particularly high softening temperatures combined with the useful transmission characteristics.

224. The structure of *tellurium dioxide* glass was determined by G. W. Brady[355] with common X-ray diffraction methods. The structural units in this framework are distorted *octahedral* configurations of 4 O^{2-} anions neighboring a tellurium central atom in a distance of 1.95 Å., and two others in a distance of 2.75 Å. In this case there is a nearly complete agreement of the short-range order between the glass and the crystalline phase of TeO_2, with domains of regularities in the glass. The existence of TeO_2 glass with octahedral units is in contradiction with the Zachariasen postulates also for a simple oxide glass which were derived from the experience that (at V. M. Goldschmidt's time) neither TiO_2 nor Al_2O_3 glasses could be prepared, as one imagined *because* of their octahedral constitution. The covalent bonding in TeO_2 with distorted octahedra also distinguishes this interesting glass from the constitution of more ionic-bonded compounds like TiO_2 and Al_2O_3 with regular octahedra. By the introduction of strongly polarizing cations like Li^+, on the other hand, the covalent character of TeO_2 can be modified toward a more ionic type, to form a crystalline structure with "opened-up" edges of the octahedra. In the glass structure of TeO_2 the existence of small domains of short-range order or of "crystallites" is

[352] *J. Am. Ceram. Soc.*, **41**, 1958, 315-323.

[353] *J. Am. Ceram. Soc.*, **42**, 1959, 408-412.

[354] *Nature*, **179**, 1957, 434-435; see also N. J. Kreidl, *Glastech. Ber.*, **32**.K, 1959, VII, 40.

[355] *J. Chem. Phys.*, **27**, 1957, 300-303.

highly probable with an order of magnitude of a few unit cells, as indicated by broadened X-ray interference lines in the Debye-Scherrer diagram.

225. We mentioned already in A. ¶ 7 that interesting *tellurite* glasses are known which are black to common light but are characteristically transmittent for infrared radiation. J. E. Stanworth[356] described a series of such glasses with Li_2O, BaO, PbO, MnO, CuO, and even with additions of V_2O_5 and WoO_3. Also the dielectric properties of tellurite glasses were measured. It is theoretically important that Stanworth came to the conclusion of how TeO_2 could be a glass forming oxide namely because its electronegativity (2.1) conforms with the series of typical glass formers like boron (with the electronegativity = 2.0), silicon (1.8), phosphorus (2.1), germanium (1.7), arsenic (2.0), and antimony (1.8). The best infrared transmission is shown by zinc tellurite glass, up to 5.5 μ.

226. Among the many types of *nonsilicate glasses* discussed by W. A. Weyl[357] those of alkali and alkaline earth *nitrates* are particularly remarkable. The system KNO_3–$Ca(NO_3)_2$ discussed by A. Dietzel and H. J. Poegel[358] shows in its phase equilibrium diagram (Fig. A. 156), only in the range between 40 and 60 weight per cent $Ca(NO_3)_2$, near the eutectic between this nitrate and KNO_3, the formation of a typical glass with all the characteristic properties of crystallization from an undercooled melt as given in G. Tammann's classical rules. The glasses can be colored by additions of copper, nickel, or cobalt silicates, in the colors corresponding to the $[RO_6]$ coordination of the coloring cations, as in aqueous solutions. The interesting problem of the constitution of such nitrate glasses was discussed by S. Urnes[359] in respect to the role which water may play in a deformation effect on the NO_3^- anion. That water, however, is *not* absolutely necessary for the glass formation was demonstrated by a totally anhydrous fusion experiment. Water, on the other hand, extends the range of glass formation in the binary system. The infrared absorption bands characteristic of NO_3^- are equally developed in melts, glassy, and crystalline phases of the nitrates. Urnes therefore abandoned the hypothesis of anion deformation as the reason for glass formation, but used a principle of M. Temkin[360] based

[356] *Nature*, **169**, 1952, 581-582; *J. Soc. Glass Technol.*, **36**, 1952, 217-241; **38**, 1954, 421-424; 425-435 (with J. A. James).

[357] *J. Chem. Educ.*, **27**, 1950, 520-524.

[358] *Atti congr. intern. vetro, 3rd Congr. Venice* 1953, 219-243; *Veröffentl. Max-Planck-Inst. Silikatforsch.*, **13**, 1953, 96-120. On the phase equilibrium diagram cf. A. G. Bergman, *Compt. rend. acad. sci. U.R.S.S.*, **38**, 1939, 304-306; *Doklady Akad. Nauk S.S.S.R.*, **38**, 1943, 320-322. Thermochemical measurements in the series of potassium-calcium nitrate glasses see by C. Kröger and W. Janetzko, *Z. anorg. u. allgem. Chem.*, **287**, 1956, 28-32, demonstrating coincidence of the range of "glassicity" (Dietzel), and maximum of the heat of crystallization with the lowest degree of dissociation of the compound $4KNO_3.Ca(NO_3)_2$.

[359] *Glastech. Ber.*, **31**, 1958, 337-341.

[360] *Acta Physicochim. U.R.S.S.*, **20**, 1945, 411-420.

on the fusion entropies, and partial molar entropies, of the components in mixed nitrate glasses. The results of the calculation are most remarkable in that an entirely statistical distribution of the particles in the nitrate glass constitution cannot be true, and a framework theory of the type of Zachariasen's is to be excluded. Urnes considers the energy distribution around the single cations Ca^{2+} and K^+ and the probability of their overcoming the threshold potentials postulated. The glass for-

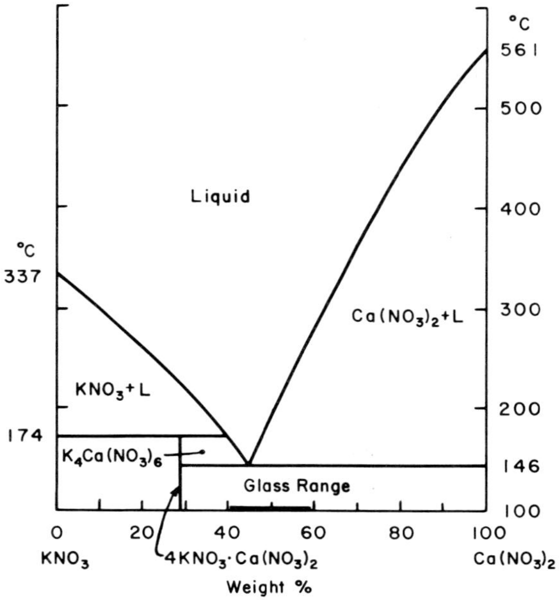

FIG. A. 156. Phase equilibrium diagram of the system KNO_3—$Ca(NO_3)_2$, according to A. G. Berman. (Dietzel and Poegel).

mation is only a consequence of the very low energy of the cations and it also impedes the crystallization process; the glassy state is "preserved." O. Borgen, K. Grjotheim, and S. Urnes,[361] in completion of these ideas, gave a thorough analysis of the infrared spectrum of NO_3^- ions from melts prepared in completely anhydrous conditions. In potassium-calcium nitrate glasses only a superposition of the infrared diagrams of the crystalline components takes place, in strict contradiction to the previous hypothesis of ionic deformation, and explains the NO_3^- oscillations of the nitrate which appear in all crystalline and glassy phases, and also besides the K^+ the K^+ and Ca^{2+} cations.

[361] *Glastech. Ber.*, **33**, 1960, 52-55.

227. Among the so-called anomalous glasses we further mention the mixed lead *silicate-sulfate* glass described by L. Merker and H. Wondratschek[362] in the systems $PbSiO_3$–and Pb_2SiO_4–$PbSO_4$, up to 33.3 mol. per cent of the sulfate. Particularly interesting is the relatively low effect exerted by $PbSO_4$ on the surface tension of the glass melts mentioned in A. ¶ 113. A latent tendency to unmixing may be the reason for this and other anomalies of the physicochemical properties of mixed silicate-sulfate systems. *Zinc chloride* forms from its highly viscous melt a glass after quenching in a thin layer, as Ingeborg Schulz[363] observed. By the addition of KCl much more stable chloride glasses are formed, the best, however, by an addition of potassium *iodide* in the molecular ratio 1 KI : 1 $ZnCl_2$. Also with 10 to 70 per cent KI, quenching yields durable glasses with a transformation range even below 100°C. $ZnCl_2$–Na_2SO_4 or –K_2SO_4 mixes in the molecular ratios 1 : 0.5 to 1.0 also have very good glass-forming properties, further mixes of $ZnCl_2$ with $Na_4P_2O_7$ or $NaPO_3$. To understand such unexpected anomalous glasses one may refer to Weyl's opinions mentioned above, generally based on the theory of *screening* of cations by surroundings either of polarizable anions (like Cl^-, I^- in the halide glasses mentioned above), to form a kind of an electrostatic Faraday cage as a coordination unit, or with free electrons. The screening is a consequence of the tendency of the central cations to neutralize their exterior fields in the smallest possible volume element.[364].

228. We mentioned already in A. ¶ 60 the effects of *very high pressures* on the volume properties of glasses. Such effects studied by P. W. Bridgman and I. Šimon[365] demonstrate the possibility of influencing the polarization of the oxygen anions in the glass framework by action of exterior forces to collapse the structure to a "contracted glass' state, with a typical threshold pressure in silica and silicate glasses, but a gradual collapse in B_2O_3 glass, as observed in the changes of the X-ray diffraction behavior. A "folding-up" of the framework structures is indicated, involving *bending* rather than shortening of the Si—O—Si or B—O—B distances. Alkalies added influence strongly the capacity for a permanent deformation. E. Plumat[366] assumes in the contracted state of high-pressure-treated glass a change of $[SiO_4]$ coordination to $[SiO_6]$. For binary alkali borate glasses C. E. Weir and L. Shartsis[367] made measurements of the compressibility at room temperature up to

[362] *Glastech. Ber.*, **30**, 1957, 473-475; **32**, 1959, 54-58.

[363] *Naturwissenschaften*, **44**, 1957, 536.

[364] A very good introduction in theory of screening in glass structures was given by E. Plumat, *Silicates inds.*, **19**, 1955, 97-107.

[365] *J. Appl. Phys.*, **24**, 1953, 405-413. See also W. A. Weyl, *Central Glass & Ceram. Research Inst. Bull. (India)*, **6**, 1959, 147-174, who discussed the high-pressure effects on glass from the viewpoints of "screening." See also J. Paymal and Mlle. M. Bonnand, *Silicates inds.*, **27**, 1961, 17-31.

[366] *Silicates inds.*, **21**, 1956, 447-457.

[367] *J. Am. Ceram. Soc.*, **38**, 1955, 299-306; **39**, 1956, 319-322; **45**, 1962, 196. See also R. Roy

10,000 atm. With increasing alkali contents the compressibility coefficient generally decreases as well in borates as in silicates. Potassium silicates show a maximum compressibility at low concentrations in K_2O. The order of compressibility is K > Na > Li, for silicates, but for low concentrations in borates this order is reversed. For alkaline earth borate glasses the order of compressibility is Ba > Sr > Ca, in good agreement with the polarizability properties of the cations of these bases as well as for the alkalies. At corresponding molar concentrations the compressibility of alkali borate glasses is higher than that of the alkaline earth borate glasses, indicating the influence of electrostatic charges of the cations on volume changes under pressure.

EXPERIMENTAL RESULTS OF STRUCTURE INVESTIGATIONS OF SILICA AND SILICATE GLASSES

229. A reexamination of B. E. Warren's classical studies on the structure of *silica glass* by *neutron diffraction* was made by W. O. Milligan, H. A. Levy, and S. W. Peterson,[368] with wavelengths of 1.21 Å. (from NaCl) and 0.764 Å. (from Cu crystal monochromators). Six well-developed interference maxima were observed in contrast with Warren's single broad peak and 3 ill-defined maxima of intensity. The radial electron distribution curve shows 5 well-resolved peaks with interatomic distances in good agreement with those given by Warren. The better detection of additional peaks in spheres of distances up to near 10 Å. is a great advantage for neutron over X-ray diffraction methods for the study of glassy and colloid systems (cf. Vol. I, C. ¶ 105).

230. On the structure of *alkali silicate glasses* and related systems we recall the general remarks of N. V. Belov[369] from viewpoints of the framework theory, whereas Cl. L. Babcock, St. W. Barber, and K. Fajans[370] basically criticize this viewpoint. They demonstrate the necessity of the assumption of several coexisting structural elements in the constitution of simple silica glass to explain its multiple anomalous physico-chemical behavior, based on the idea of homogeneous equilibria built up between *different* atomic arrangements coexisting and varying with pressure and temperature, as the leading new principles in the deductions. For the well-known very

and H. M. Cohen, *Nature*, **190**, 1961, 798 f., and *J. Am. Ceram. Soc.* **44**, 1961, 523 f.; **45**, 1962, 398 f.; further R. A. Eppler, A. A. Giardini, and J. E. Tyding, *Ibid.*, 218-220 (for lithium aluminosilicate glasses), and E. B. Christiansen, S. S. Kistler, and W. B. Gogarty, *Ibid.*, 172-177 (for application of compressibility of silica glass in high-pressure cells).

[368] *Phys. Rev.*, **83**, 1951, 226-227.

[369] *Stroenie Stekla, Inst. Khim. Silikatov Akad. Nauk S.S.S.R., Trudy Soveshchaniya, Leningrad*, 1953 (Pub. 1955), 344-350.

[370] *Ind. Eng. Chem.*, **46**, 1954, 161-166.

low thermal expansion coefficient of silica glass Babcock, Barber, and Fajans postulate a structural element specifically dominant at low temperatures which has a larger volume than the other coexisting elements. Equal assumptions are plausible in the discussion of the elastic and compressibility behavior of silica glass. This also means the coexistence of strong and weak bonding forces, of larger and smaller vibrating masses in the constitution of this glass to explain the low-temperature heat capacity and the infrared and Raman characteristics. The mutual conversions of the coexisting structural elements must always be reversible and relatively rapid to explain the fact that even at moderate temperatures physical anomalies can be measured so well experimentally. One must not fail to consider that the coexisting structural elements are *not* identical with those states which are assumed for glass in the transformation range as a function of its thermal history (cf. A. ¶ 157, ¶ 160). The latter phenomena evidently concern much more complex states than those of the elements mentioned above.

231. Another criticism of the interpretation of the structure of silica glass as given by Warren on the basis of the Zachariasen framework theory was made by Fr. Oberlies and A. Dietzel[371] by comparison with the structure of α-cristobalite. If one calculates the diffraction and electron distribution curves, it is established that the short-range order in silica glass must be *higher* than previously assumed, and similar to that in α-cristobalite. Six-membered rings of $[SiO_4]$ tetrahedra in this framework are not all complete; many of them may be "open" or incomplete, as shown by the behavior in the transformation range. If the polymerization process in a cooling glass is interrupted in different stages, the numbers of open-standing ring-bondings will be different. The photometric measurements from X-ray diffraction curves of silica glass are in good agreement with the calculated interplanar distances in α-cristobalite at 290° and 1300°C. A "forbidden" interference (for 100) of α-cristobalite appears in addition in the glass, indicating a loosening and slight distortion of the α-cristobalite structure. Such forbidden interferences were already observed by G. Hartleif (1938) near 7.2 to 7.5, and 16 Å. A structure of silica glass revised after these conclusions shows in *prevalence* six-$[SiO_4]$ tetrahedral units in annular complexes of much higher order than in the Zachariasen model structure. The tetrahedra cannot be strictly regular but are more or less distorted in the glass.

232. If one wishes to give an *approximate* representation of the silica glass structure one may do so in the form of Fig. A. 157, indicating the prevalence of the six-membered ring elements and the local occurrence of open rings. In this respect it is also important to discuss the "amorphousized" quartz destroyed by heavy neutron irradiation (by 10^{20} neutrons/cm.^{2}) in an atomic pile as described by I. Šimon,[372]

[371] *Glastech. Ber.*, **30**, 1957, 37-42. See also A. Dietzel, *Verres et réfractaires*, **16**, 1962, 215-221, with a critical remark of H. Jagodzinski on p. 220.

[372] *J. Am. Ceram. Soc.*, **40**, 1957, 150-153; **41**, 1958, 116.

from measurements by X-ray diffraction and infrared spectroscopy. The radial electron distribution curves (Fig. A. 158) for the amorphous irradiation product are very similar to that of silica glass, with the same fundamental distance Si—O =

Fig. A. 157. Revised schematic representation of the structure in silica glass. (Oberlies and Dietzel).

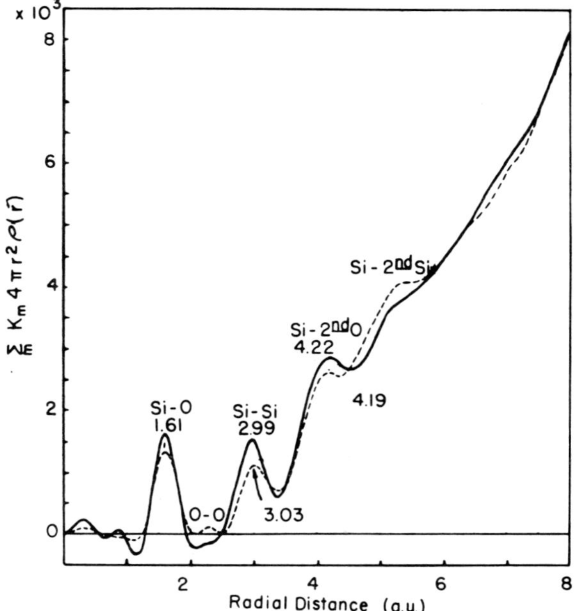

Fig. A. 158. Radial electron density distribution curves for silica glass (dashed curves) and quartz, irradiated by neutrons (solid curve). (Šimon).

1.61 Å., but a decrease in the bond angle Si—O—Si from 142° to 138° (with a slight contraction of the distance Si—Si), and a decided widening of the distances in spheres more remote from the central Si^{4+} cation. This structural interpretation illustrates the impact action of the neutrons to knocked-on atoms of the crystalline quartz by a kind of "thermal spike" action (temperature estimated to be 10^4 °K., to form the "amorphous highly disordered modification. The infrared reflection spectrum (Fig. A. 159) also shows the great similarity (but not identity) of the irradiation products with silica glass, and a shift towards lower frequencies in the irradiated silica glass, analogous to a corresponding shift under the action of very high pressure compaction of silica glass. The density of the irradiated amorphous quartz products is higher (2.30) than that of silica glass 2.21); the thermal expansion coefficients $\alpha_{25°}^{200°} = 5.4 \times 10^{-7}$/°C. are practically the same for both products.

233. Once again, the Warren type structure of silica glass, in comparison with that of germanium dioxide glass, was discussed by J. Zarzycki[373] on the basis of extensive X-ray diffraction determinations at temperatures up to 1600°C. (Figs. A. 160 and 161), combined with analysis of the interatomic distances derived from the electronic distribution function

$$\sigma(r) = Kr \cdot \int_0^\infty s\, I(s) \cdot \sin rs \cdot ds$$

which is much more precise than that from the radial atomic distribution curves. The characteristic distances are:

Si—O = 1.60 ± 0.05 Å.	Ge—O = 1.70 ± 0.05 Å.	angle Si—O—Si = 143° ± 17°, rather the same up to 1600°C.
Si—Si = 3.00 Å.	Ge—Ge = 3.15 Å. at 20°C.	= 150° if entirely disordered
	= 3.30 Å. at 1200°C.	angle Ge—O—Ge = 125° to 152° at 20°C.
		= 137° to 180° at 1200°C.

One $[SiO_4]$ unit is shown in Fig. A. 162, the characteristic angles in Fig. A. 163. Zarzycki did not find α-cristobalite crystallite elements in silica glass, but considered a residual α-quartz crystallite structure which was also indicated by infrared absorption measurements (cf. Florinskaya and Pechenkina, A. ¶ 235). For germanium dioxide glass the quartz-like residual structure is highly characteristic. In liquid GeO_2, however, the angle Ge—O—Ge is wider and reaches nearly 180°. Zarzycki also compared the diffraction behavior of B_2O_3 with that of GeO_2 on the basis of the hole and micellae theory, as districts in a microheterogeneous glass structure with micellae of the order of magnitude of 3 Å. As a function of temperature the decay

[373] *Trav. Congr. Intern. du Verre* IV, Paris 1956, VI, 4, 8 pages; *Verres et réfractaires*, **11**, 1957, 3-8.

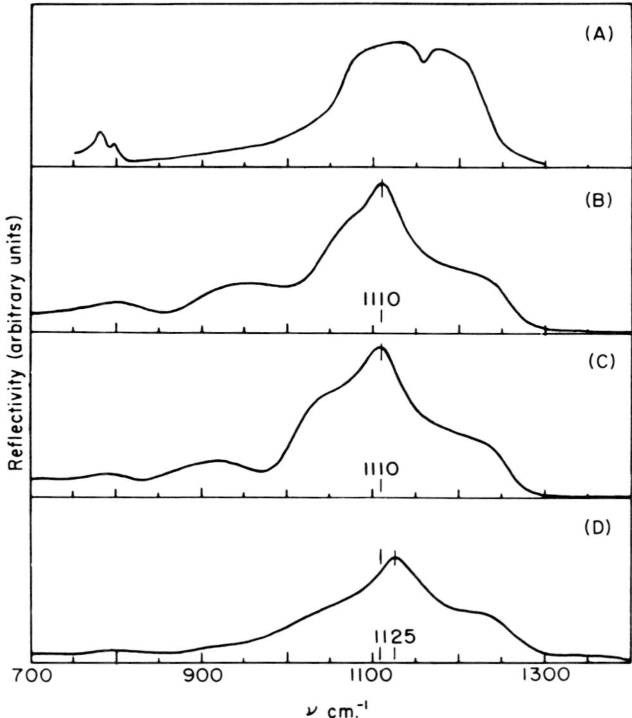

FIG. A. 159. Infrared reflection spectra (*A*) of quartz parallel to the *c*-axis; (*B*) the same, irradiated by neutrons; (*C*) irradiated silica glass; (*D*) silica glass, not irradiated. (Šimon).

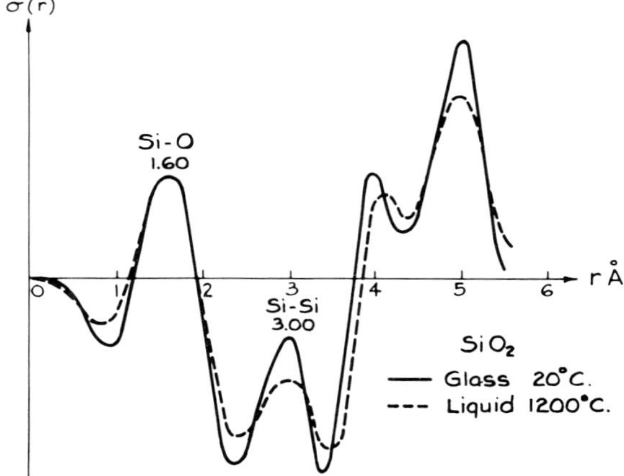

FIG. A. 160. Radial electronic density distribution curve for silica glass at 20° and 1600°C. (Zarzycki).

of the structure to smaller and smaller elementary units with increasing temperature and fluidity is most characteristic in the calculation analysis of distribution curves. For B_2O_3 the progressive destruction can finally cause the presence of free oxygen anions by thermal intensification of the breaking of B—O bonds.

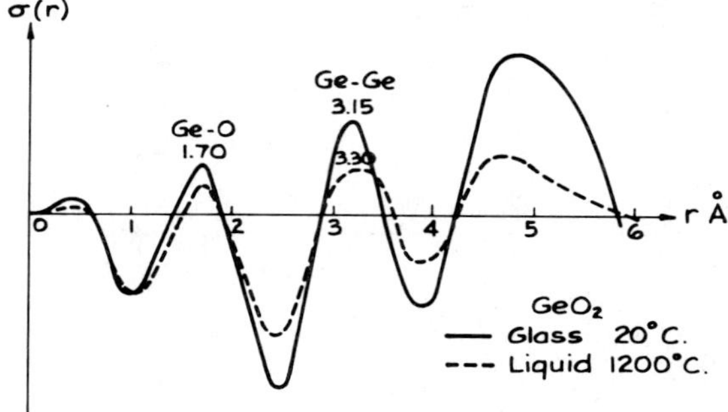

FIG. A. 161. Radial electronic density distribution curve for germanium dioxide glass at 20° and 1200°C. (Zarzycki).

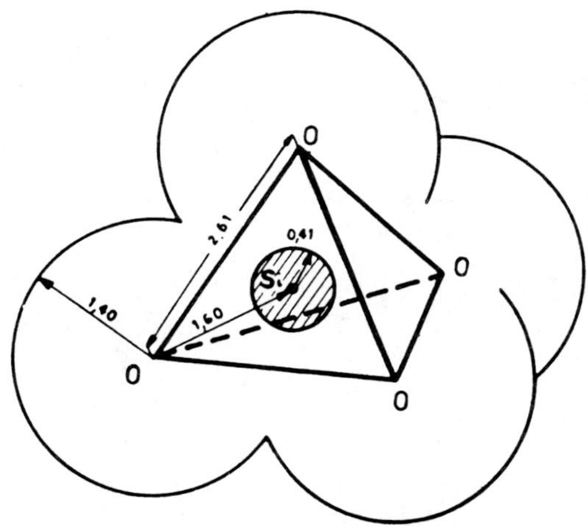

FIG. A. 162. Dimensions of one $[SiO_4]$ unit in silica glass. (Zarzycki).

234. Concerning the angle Si—O—Si in silica glass J. D. Mackenzie and J. C. White[374] rediscussed the problem of whether this glass is more closely related to the structure of α-cristobalite (allegedly 180°) or to α-quartz (145°). The value of 180°

[374] *J. Am. Ceram. Soc.*, 43, 1960, 170-171. Generals on the interpretation of infrared absorption spectra of silicate glasses, see by G.-J. Su, N. F. Borrelli, and A. R. Miller, *Phys. & Chem. Glasses*, 3, 1962, 167-176.

for cristobalite goes back to R. W. G. Wyckoff (1925). Those authors demonstrated that Wyckoff's conclusions must be in error, and that the Si—O—Si angle in cristobalite is between 142° and 153°. Zarzycki's determination of the angle is 143°; Milligan, Levy, and Peterson (see A. ¶ 229) determined it = 146° from neutron diffraction, in good agreement with cristobalite. Zarzycki's conclusion, therefore, that silica glass has a quartz-like structure must be revised. It is more nearly related to that of α-cristobalite, and the molar volumes of cristobalite and silica glass are nearly identical even at high temperatures. The micellar units in silica glass seen from the viewpoints of Cl. L. Babcock, St. W. Barber, and K. Fajans (see A. ¶ 230) were also discussed by Barber and M. Dank[375] by a calculation of low-temperature specific heats of silica glass in comparison with that of cristobalite. For both materials a highly *anisotropic* structure is demonstrated. This anisotropy implies that a really more extensive order is in accord with light scattering and electron microscopy than can only be concluded from the continuous framework theory. Units like $[Si_3O_6]$ as mass units, which also were used by B. D. Saksena (1950) in his discussion of the specific heat of quartz, correspond to data of infrared frequencies which are in good agreement with infrared absorption bands measured at room temperature.

235. Similar critical attitudes as are seen against the first determination of the structure of silica glass by B. E. Warren are also seen against his determination for boric acid glass. H. Richter, G. Breitling, and F. Herre[376] compared the atomic distribution curve of B_2O_3 glass with those of the elements Si,Ge,Sb,Se,As, in their "amorphous" modifications, which show typical chain and layer structures. Richter *et al.* found a strikingly high order for B_2O_3 glass, with an unexpected peak at 1.85 Å., interpreted as the shortest distance of *layers* of $[BO_3]$ groups in a paracrystalline arrangement which to a certain degree is similar to that of graphite, although with a statistical diffraction from the thermal vibration of the amorphous material. The $[BO_3]$ coordinations in B_2O_3 glass have the type of distorted tetrahedra and trigonal pyramids with different distances B—O, arrenged in extensive ordered layers of 1.85 Å. distance B—B (Fig. A. 164). The latent metastability of B_2O_3 glass relative to crystalline B_2O_3 is seen in the slightly shorter distance B—B for the crystal phase. If these conditions are correct there would be a strikingly high order in B_2O_3 and silica glasses, for both of them chain or layer types as in the solid amorphous elements B and Si. For silica glass the peak corresponding to that of 1.85 Å. in B_2O_3 glass was not observed by Warren, and is also not compatible with the Zachariasen model of silica glass. In spite of that, the reality of a peak at 2.15 Å. indicated by Richter *et al.* corresponds to a shortest distance of an assumed chain structure of relatively high order in silica glass.

[375] *J. Chem. Phys.*, **23**, 1955, 597-598; cf. B. D. Saksena, *Proc. Indian Acad. Sci.*, **12**.A, 1950, 93-139.

[376] *Naturwissenschaften*, **40**, 1953, 482-483, 621; *Z. Naturforsch.*, **9**.a, 1954, 390-402.

236. The derivations of Richter *et al.* were sharply criticized by K. Grjotheim[377] in a discussion of errors possible in the interpretation of electronic distribution curves which Richter extends to a value of $s = 4\pi \cdot \sin \vartheta/\lambda$, originating from "*error ripples*" of the wavelength $2\pi/s$. Omitting here the physical significance of these deductions, we note with Grjotheim that false peaks are easily introduced, and the error ripples bring about incorrect values for the peak areas in the radial distribution curves in the calculation of atomic distances. The error ripples in Richter's curve for B_2O_3 glass are very strong, with a period of 0.45 Å. corresponding to the small peak at $s = 14.0$ in their intensity curve. The important peak at 1.85 Å. is therefore only a part of the periodic ripples plainly due to an error in the intensity curve at $s = 14.0$. Grjotheim definitely refutes all conclusions made by Richter *et al.* and thinks that the proposed layer structure is "chemically not very likely to occur" (see Fig. A. 164). Grjotheim's criticisms are in every case a serious warning directed to every conclusion made for X-ray studies on the structure of more than monoatomic glasses, if a careful complete elimination of many systematic errors has not been possible, among which that of the "error ripples" was theoretically discussed by this author on the basis of Chr. Finbak's theories (cf. A. ¶ 208*). To avoid multiple errors the present X-ray techniques must be considerably improved for investigations on glass constitution. As a practical rule O. Borgen, K. Grjotheim, and J. Krogh-Moe[378] recommend particular reservation to be applied on conclusions made from determinations of the coordination number, too. This is also the key to problems

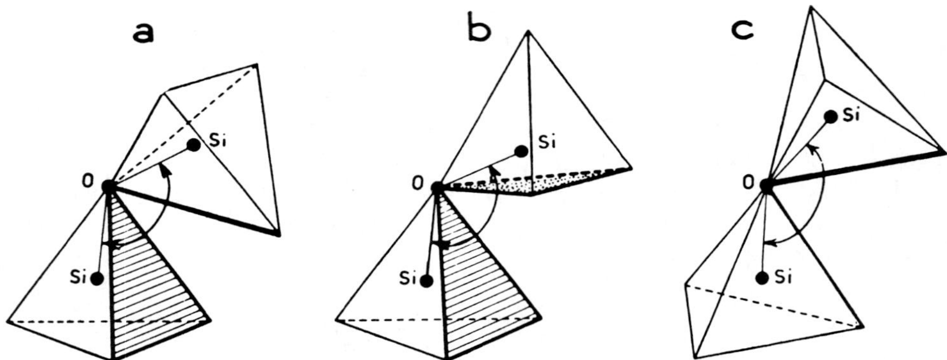

Fig. A. 163. Contact position of two $[SiO_4]$ tetrahedral units, (*a*) face to edge (angle 148° in average); (*b*) face to face (angle 148° in average); (*c*) edge to edge (angle 158° in average). (Zarzycki).

[377] *Glass Ind.*, **39**, 1958, 201-209; see also E. A. Poraĭ-Koshits, *Stroenie Stekla, Inst. Khim. Silikatov Akad. Nauk S.S.S.R., Trudy Soveshchaniya, Leningrad*, 1953 (Pub. 1955), 30-43.

[378] *Glastek. Tidskr.*, **11**, 1956, 47-55; *Kgl. Norske Videnskab. Selskabs. Forh.*, **27**, 1954 (17), 86-93.

in B_2O_3 containing silicate glasses which are commonly called the *boric acid anomaly* phenomena, as discussed previously by J. Biscoe and B. E. Warren (1938) for sodium borosilicate glasses (cf. A. ¶ 239).

237. Commonly this phenomenon is explained by the change from planar or flat-pyramidal [BO_3] configuration towards [BO_4] tetrahedra, with increasing alkali contents in borosilicate glasses and a *reversal* to [BO_3] again with higher alkali compositions. Grjotheim and Krogh-Moe[379] consider a configuration change around a central B^{3+} cation from a larger number of coordination to a smaller one in alkali borate glasses, that is in the direction of an easier polarization of oxygen anions in the glass structure. In addition, the structural units in crystalline B_2O_3 are real tetrahedra, and not trigonal configurations, as established by S. V. Berger.[380] The transition [BO_4] $\rightarrow$ [BO_3] means a reinforcement of the structural bonding in alkali containing glasses, which must importantly affect the physical properties of borate and borosilicate glasses as St. W. Barber and K. Fajans postulated it in 1952. The reinforcement effect, however, reaches a maximum followed by a reduction of bond strengths by the secondary introduction of more and more bridging oxygen anions, if the alkali contents reach concentrations above the maximum amount in the boric acid anomaly observed, such as for Na_2O at 16 to 17 molecular per cent. We agree with W. A. Weyl[381] that these ideas are more plausible than the somewhat artificial interpretations given before by Biscoe and Warren.

238. Sc. Anderson, R. L. Bohon, and D. D. Kimpton[382] attacked the problems of the constitution of B_2O_3 glass and similarly that of alkali borate glasses, by in-

[379] *Kgl. Norske Videnskab. Selskabs, Forh.*, **27**, 1954, 94-99; *Naturwissenschaften* **41**, 1954, 526-527. Changes in coordination explain on the same general basis the boric acid anomaly, and corresponding phenomena in alumino-boro-silicate glasses, studied by E. I. Galant, *Structure of Glass. Proc. All-Union Conf. Glassy State, 3rd. Leningrad* (English Translation), 1959/60, 451-453. See also S. P. Zhdanov, *Ibid.*, 1959/60, 454-458, on variations in the molecular volumina by the anomalies. Galant examined the refractive index as the "structure-sensitive" parameter.

[380] *Acta Chem. Scand.*, **7**, 1953, 611-622; *Proc. Intern. Symposium on Reactivity of Solids. Gothenburg*, 1952/54, II, 421-426. Of recent literature on boric acid and alkali borate glass structure investigations we mention here: G. Becherer, O. Brümmer, and G. Herms, *Silikattechnik*, **13**, 1962, 339-345; J. Krogh-Moe, *Phys. & Chem. Glasses*, **1**, 1960, 26-31; **3**, 1962, 1-6, 101-110, 208-212 (on barium borate glasses); B. Ottar and W. L. Ruigh, *Ibid.*, **3**, 1962, 95-98.

[381] *Glastech. Ber.*, **30**, 1957, 269-282, especially p. 271.

[382] *J. Am. Ceram. Soc.*, **38**, 1955, 370-377; previous observations of Sc. Anderson, *Ibid.*, **33**, 1950, 45-51, could not reconcile the infrared reflection and absorption spectra of B_2O_3 glass with the triangular framework ideas. The difficult problems of infrared spectra and constitution of "water-containing" boric acid and borate glasses were tackled recently by J. L. Parsons and M. E. Milberg, *Ibid.*, **43**, 1960, 326-330; F. Meller and M. E. Milberg, *Ibid.*, 353-356; M. E. Milberg, R. K. Belitz, and A. H. Silver, *Phys. & Chem. Glasses*, **1**, 1960, 155-158. The powerful method of nuclear magnetic resonance was applied by A. K. Silver and P. J. Bray, *J. Chem. Phys.*, **29**, 1958, 984-990. See also the article of these authors in J. D. Mackenzie's book "Modern Aspects

frared absorption and reflection spectrum investigations. Vacuum-pressed pellets of fused B_2O_3 and KBr definitely show a spectrum which is incompatible with the idea of a completely continuous three-dimensional framework of planar [BO_3] coordination units. Hydrogen bonds play an important role in the arrangement of "pure" B_2O_3 or of low-in-alkali borate glasses with a low moisture content. Anderson *et al.* are inclined to assume in B_2O_3 glass a complex of the type $[B_9O_{14}]^-$, with H-bonds holding it together to form $B_{18}O_{27}.H_2O$. Every one of 9 B^{3+} ions would then be in tetrahedral coordination. The infrared spectrum of sodium borate glasses with up to 10 per cent is interpreted from a structure which becomes increasingly approximate to a continuous framework with [BO_4] groups, as is the case say in alkali pentaborates (e.g. $K_2O.5B_2O_3.4H_2O$ with $[B_5O_{10}]^{5-}$ groups as seen in

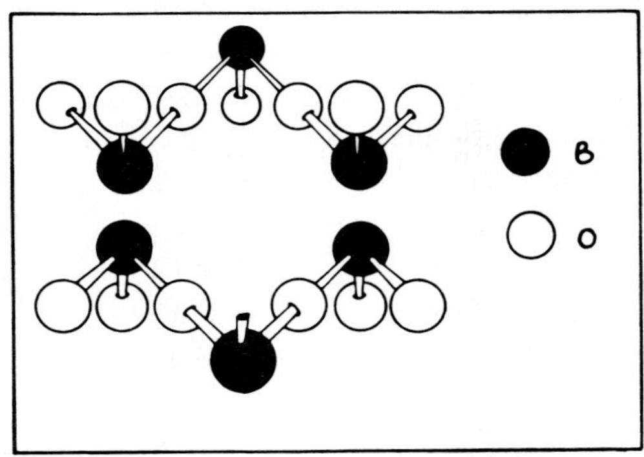

FIG. A. 164. Structural model of an ordered domain in B_2O_3 glass as given by Richter, Breitling, and Herre.

Fig. A. 165). Above 15 per cent alkali the structure becomes basically different. No definite conclusion concerning the special arrangement can be made although it may be possible that a high order with $[B_3O_6]^{3-}$ anionic groups as the dominant units exists because it is present in the sodium metaborate crystal structure. Each [BO_4] tetrahedron in the framework arrangement below 10 per cent R_2O would share a corner with [BO_3] triangles pairwise which in their turn share additional corners with another pair of triangles.

239. The connection between the boric acid anomaly with the physical properties of borosilicate glasses was especially studied by T. Abe[383] who assumed, in a similar way as Anderson did, a quite distinct grouping of one [BO_4] and four [BO_3] planar groups bonded together to one unit called a (xy_4) group (Figs. A. 166 and 167)

of the Vitreous State," Butterworths, London, 1960, 92-119, and specifically on nuclear quadrupole effects, P. J. Bray, *Acad. Roy. Belg., Cl. Sc., Mém.*, **33**, 1961, 289-320.

[383] *J. Am. Ceram. Soc.*, **35**, 1952, 284-299.

owing to the conditions of neutralization of one negative electronic surplus charge of every $[BO_4]$ group. The tetrahedra in this type of bonding are unable to approach one another for mutual contact. Abe illustrated from this basic assumption the particular behavior of systematically varied borosilicate glasses in viscosity, surface tension, density, molecular and atomic refraction, beyond that the phenomenon of opalescence by unmixing, and the often discussed microheterogeneity of borosilicate glasses. From the practical viewpoint especially Y. Tamura[384] studied the conditions of these immiscibilities under aspects of the Vycor process (cf. Vol. V, Section B) and of the systematic change of physical properties and chemical durability with composition (cf. B. ¶ 171 ff.).

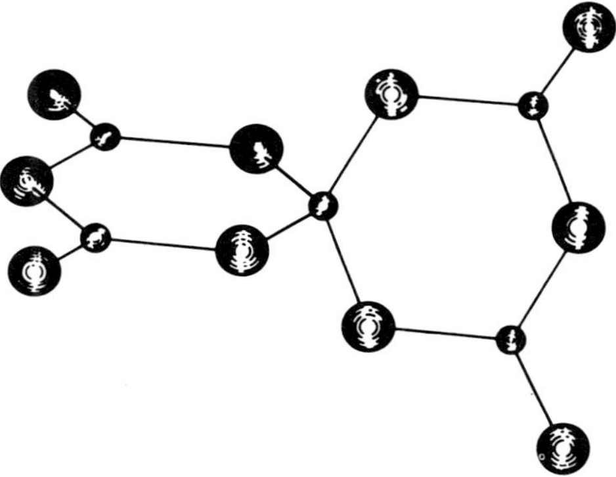

Fig. A. 165. $[B_5O_{10}]^{5-}$ group existing in potassium pentaborate tetrahydrate. (A. F. Wells). The central boron atom forms a $[BO_4]$ group, interlinked with four $[BO_3]$ groups, to two six-membered rings, in positions perpendicular in their planes.

240. A very interesting systematic study of the physical properties of lithium-beryllium borate glasses in which lithium was replaced by sodium and potassium, beryllium by magnesium, calcium, strontium, or barium was made by B. Laurent,[385] with variations of the structurally important ratio O : B between 1.71 and 1.93. The expansivity of the glasses is increased not only with increasing radii of the alkali and alkaline earth cations (and corresponding changes of the electrostatic field strengths z/a^2), but also with the ratio O : B. At high temperatures, and the more the ratio O : B approaches the limit value 2 : 1, the structure was assumed to become, in the first place, determined by $[BO_3]$ groups in flexible chains which are from case

[384] *Osaka Kogyo Gijutsu Shikenjo Kokoku*, No. 299, June, 1953, 70 pp. Direct electron microscopy of such glasses was used by F. K. Aleĭnikov, P. V. Dundzis, P. V. Paulavichyus, and V. A. Slizhis, *Lietuvos TSR Mokslu Akademijos Darbai, Ser. B*, **2**, (29), 1962, 95-108; *Doklady Akad. Nauk S.S.S.R.* **141**, 1961, 674-676 (cf. Vol. I, C. ¶ 126*).

[385] *Chim. & ind. (Paris)*, **68**, 1952, 710-716; *Verres et réfractaires*, **7**, 1953, 167-174.

to case cross-linked, as the framework theory of Zachariasen and Warren-Biscoe postulates. The contraction effects of the small ion Li^+, and the participation of Be^{2+} in the framework of the glass structure is particularly evident in the expansion behavior, the "dissociation" of the framework in the isothermic curves for the vis-

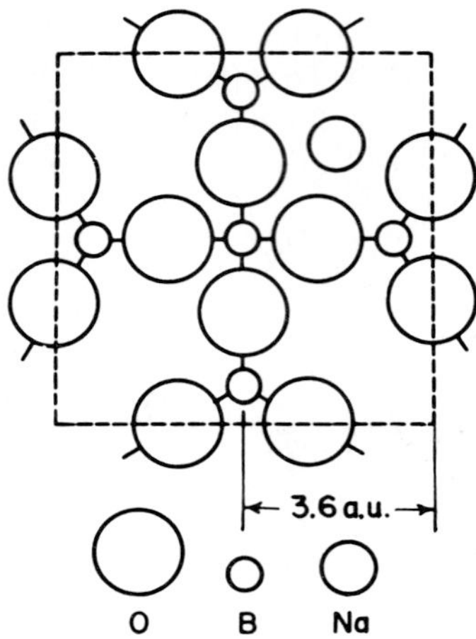

FIG. A. 166. Two-dimensional representation of a $[xy_4]$ group in borosilicate glasses. (Abe).

cosity in $\log \eta/r$ and $\log \eta/(O:B)$ parameters as functions of temperature, to form independent $[BO_3]$ groups, and to suppress the specific effects of the cations. Laurent's results are also similar to those of H. Moore and P. W. McMillan[386] who included alumina and silica in the compositions for their experimental glasses restricting the compositions to elements of low atomic weight, but also including ultraviolet

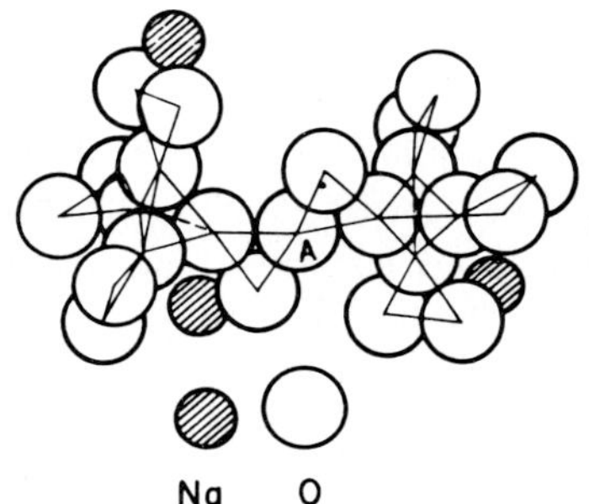

FIG. A. 167. Two $[xy_4]$ groups linking to each other through one oxygen atom (*A*). In the upper part a sodium atom breaks the bond between the groups. (Abe).

[386] *J. Soc. Glass Technol.*, **40**, 1956, 66-161 T.

and infrared transmission measurements. For problems of the constitution of borate glasses the conclusion is important that in alkali borates and in glasses high in B_2O_3 the proportion of $[BO_4]$ groups present cannot be higher than about 1/5 of the total $[BO_n]$ groups. Every $[BO_4]$ group with one excess negative charge must be separated from the next similar group by at least two planar $[BO_3]$ groups in borate glasses. $[BeO_4]$ groups with two negative excess charges must be separated from the next similar groups by at least 2 $[SiO_4]$ or 2 $[BO_3]$ groups. Like the $[BO_4]$ groups the $[AlO_4]$ groups also have one negative excess charge but the separation from neighboring $[AlO_4]$ groups need not be greater than would be afforded by *one* $[SiO_4]$ or *one* $[BO_3]$ group.

241. Concerning the constitution of *ternary silicate glasses*, especially those of the common window glass type (in the system Na_2O–CaO–SiO_2) A. H. Weber[387] made interesting observations on the *neutron* diffraction behavior. The relative neutron intensity curve as a function of sin ϑ/λ ($\lambda = 0.948$ Å.) shows seven sharp peaks but no complete Fourier synthesis was performed for an evaluation of the atomic distances. Especially important for knowledge of the surface state of such a glass after rinsing with water are the electron diffraction studies by J. J. Antal and A. H. Weber.[388] On the surface of the glass a film of bismuth (of a thickness less than 10 Å.) was deposited to reduce disturbing electrostatic charges. The preferentially oriented aggregates of bismuth in this film contribute essentially to the intensity of the diffracted interferences. The application of the usual radial distribution curves was possible by photometric measurements of coherent electron scattering. The calculations give the important result that the composition of the surface layer of the glass is that of the bulk glass in CaO and SiO_2, but with a deficiency in alkali in the exterior zones of the layer.

242. In its high significance for the calculation of *additive properties* of silicates (cf. B. ¶ 114 ff.) from structural ideas, the "*Structon*" *theory* of M. L. Huggins[389] is to be emphasized as a very important contribution although it does not directly create new ways for the investigation of constitution problems of silicate glasses. A Structon is, according to Huggins's definition, an atom in an "amorphous" medium which is surrounded by close neighbors in a specific way. The minimum number of Structons giving the correct over-all composition can change discontinually as the composition changes, as seen from break-points in linear functions of composition for definite physical property parameters. Huggins examined his theory in sodium silicate glasses of a carefully annealed state using the volume-density parameters as indicators. The observed break-points (see more in details in A. ¶ 375) bring about

[387] *Nucleonics*, **7**, 1950, December, 31-39, especially pp. 38-39.

[388] *Acta Cryst.*, **7**, 1954, 122-126.

[389] *J. Am. Ceram. Soc.*, **38**, 1955, 172-175; *J. Phys. Chem.*, **58**, 1954, 1141-1146.

reasonable *sets* of Structons accounting for the quantitative conditions of the composition dependency of the parameter. Structons, therefore, are not chemical compounds which have so often been sought with discouraging results for bringing about the break-points here in question. They are groups of which we see in sodium silicate glasses the typical Structons of the set

Si(4O) Na(6O) O(2Si) O(2Si,Na) O(Si,3Na)

the number of which (N) and the volume of which (V) can be derived from the curves in different sets for different ranges of composition. As a "*Structon Number Rule*" (roughly analogous to Gibbs' Phase Rule) one may consider the equation $S = C + F + 2$, with $S =$ the number of Structon types possible for a given system of C components, with F, the number of freedoms, considering those in composition only. For the system Na_2O–SiO_2 ($C = 2$) the freedom F being $= 1$ for every composition range, gives $S = 5$ as the number of Structons in a set, as noted above.

243. For borate glasses M. L. Huggins ad T. Abe[390] applied the Structon theory on glasses of the system Na_2O–B_2O_3 with a new interpretation of what Abe had concluded on the structure of borate glasses (see above). As a matter of fact, the Structon theory is in agreement with observed break-points in the physical parameter (especially thermal expansion) curves as a function of composition. Coordination numbers obtained in this way are somewhat larger for glasses than are usually observed in crystals of salts of the same elements. Structons may exist in glasses which are impossible in crystals because of the restricted geometric and stoichiometric variabilities in crystals.

244. That the distribution of "foreign" isomorphous elements replacing a typical component of a silicate glass is *not* at random as would usually be expected, but in characteristic clusters of probably some order was demonstrated by C. Brosset[391] by replacing Ba^{2+} for Ca^{2+} in sodium calcium silicate glass. This cluster formation was predicted by K. Grjotheim and J. Krogh-Moe,[392] and then established by Brosset in X-ray diffraction measurements with derived distribution curves indicating the Ba^{2+} ions clustering in distances Ba—Ba = about 4 Å., and not at random. No detailed information on the size and configuration of these cluster groups, however, could be obtained.

[390] *J. Am. Ceram. Soc.*, **40**, 1957, 287-292.

[391] *J. Soc. Glass Technol.*, **42**, 1958, 125-129 T.

[392] *Glastek. Tidskr.*, **11**, 1956, 47-55; see also T. Förland, Office of Naval Research Technical Report No. 63, 1955. The Pennsylvania State University, Contract No. 269, Task Order 8 NR 032-264, April 1955 (thermodynamic calculations). We may here add a hint to the fascinating studies of S. Urnes on the structure of fused alkali silicates, *Trans. Brit. Ceram. Soc.*, **60**, 1961, 85-95.

245. The *unmixing* phenomena in borate and borosilicate glasses, already mentioned in the extensive studies of E. M. Levin and St. Block (A. ¶ 216 ff.), have been rediscussed by L. Shartsis, H. F. Shermer, and A. H. Bestul[393] systematically for binary borate systems in connection with measurements of surface tension at 900° to 1300°C. (cf. A. ¶ 97), including the homogenization by additions of alumina. R. Roy[394] also discussed problems of latent (metastable) liquid immiscibility and subsolidus nucleation in connection with crystallization phenomena of the Pyroceram type. This cycle of problems also contacts once again the structural conditions of the microheterogeneity of sodium borosilicate glasses so intensely studied chiefly by Russian authors. Among these we mention here only the studies and calculations of V. V. Tarasov and E. F. Stroganov[395] who performed from determinations of the heat capacity of borosilicate glasses in the interval from 65.8 to 167.6°K. quantitative calculations to establish the principle of additivity for the constituent oxides R_2O, SiO_2, and B_2O_3, in agreement with the aspects of microheterogeneity, and also the chain structure of B_2O_3 glass in analogy with that observed in the orthorhombic modification of Sb_2O_3 (as the mineral valentinite).

246. The *infrared* absorption and reflection *spectrometry* as a powerful tool for the investigation of glass constitution found a rich application in the last decade. We mention in this direction first the excellent measurements of N. Neuroth[396] who studied especially the influence of elevated temperature (up to 1300°C.) on the spectral absorption (in the range from $\lambda = 1$ to 5 μ) of well-defined industrial glasses (one of window glass type, a green bottle glass, a container glass, and a special X-ray protection glass with 61 per cent PbO, 2 Na_2O, 3 BaO, and 34 SiO_2). Neuroth's results are basic for our actual knowledge of the transmissivity and emissivity of glass melts in industrial tank furnaces (cf. B. ¶ 107 ff.). His curves show a fundamental difference in the temperature function of the extinction below and

[393] *J. Am. Ceram. Soc.*, **41**, 1958, 507-516.

[394] *J. Am. Ceram. Soc.*, **43**, 1960, 670-671. Fractographic-electron microscopic studies of liquid-in-liquid immiscibility by St. M. Ohlberg, H. R. Golob, and Ch. M. Hollabaugh, *Ibid.* **45**, 1962, 1-4, confirm these expectations.

[395] *Trudy Moskov. Khim. Tekhnol. Inst. im. D.I. Mendeleeva*, 1956 (21) 26-33.

[396] *Glastech. Ber.*, **25**, 1952, 242-249. On apparatus used see L. Genzel and N. Neuroth, *Z. Physik*, **134**, 1953, 127-153. Neuroth extended his measurements in *Glastech. Ber.*, **26**, 1953, 66-69, to a Cr_2O_3 containing, and a brown ("coal-yellow" amber) glass, with irreversible changes in the infrared characteristics by chemical reactions. On temperature influence of the refractive index and of reflectivity of glass see Neuroth, *Ibid.*, **28**, 1955, 411-422, with measurements above 6 μ in the range of oscillations of the SiO_2 molecule, determinations of the Brewster angles and of the polarization degree in reflection under definite azimuths. As a most complete recent report on the absorption of infrared radiation in correlation to the structure of glass we recommend the excellent contributions and data given by R. V. Adams and R. W. Douglas, *Ibid.*, **32**, K, 1959, VII, 24, and the article of Douglas, Chapter 5 in the book of J. E. Burke, "Progress in Ceramic Science," Vol. I, 1961 (Pergamon Press), pp. 200-223.

above the transformation and softening range. The general decrease in absorption over the whole spectral range could reach two-thirds of the value measured at room temperature whereas in the liquid state the absorption increases with increasing temperature. The lead-rich X-ray protection glass was particular in showing a sudden increase in the absorption coefficient at 580°C. for $\lambda = 1.0$ to $2.0\ \mu$ by about one order of magnitude.

247. At higher temperatures the absorption coefficient is increased by 200 to 300 per cent for $\lambda = 2.0$ to $2.7\ \mu$ (cf. Figs. A. 168 and 169). These changes are irreversible and show the type of inner molecular equilibrium reactions released by thermal

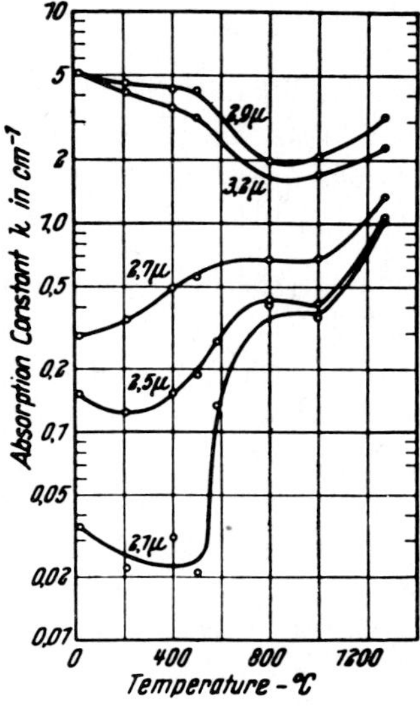

Fig. A. 168. Infrared absorption of lead glass (for X-ray protection), as a function of temperature. (Neuroth).

treatment. The glass quenched from 580°C. shows, however, in a period of 2 months a gradual restitution of the original absorption character, continuously releasing from the frozen-in constitution. The other glasses show corresponding changes of the absorption characteristics, but these combined with chemical changes of divalent to trivalent iron ions, in repeated heating and cooling cycles. For *silica* glass the optical-thermal effects are much simpler than for the complex glasses studied before. The fundamental infrared band at 9 μ was shifted and reduced in intensity according to a linear function of temperature up to 400°C. Above 900°C. the structure of the band is changed and at 1270°C. the reflectivity is only slightly affected. Evidently, the silica framework is more and more broken up at higher temperatures, the Si—O vibration is out of tune and damped. In ternary and more complex glasses the molecular structure is identical for fused and cooled glasses. The introduction of metal cations into the framework brings about a specific reduction of the intensity and sharpness of the 9 μ band. The degree of aggregation in complex glasses is therefore lower than in silica glass, and the temperature effect on the framework in pure silica glass is larger than in complex glasses (cf. Figs. A. 170 and 171).

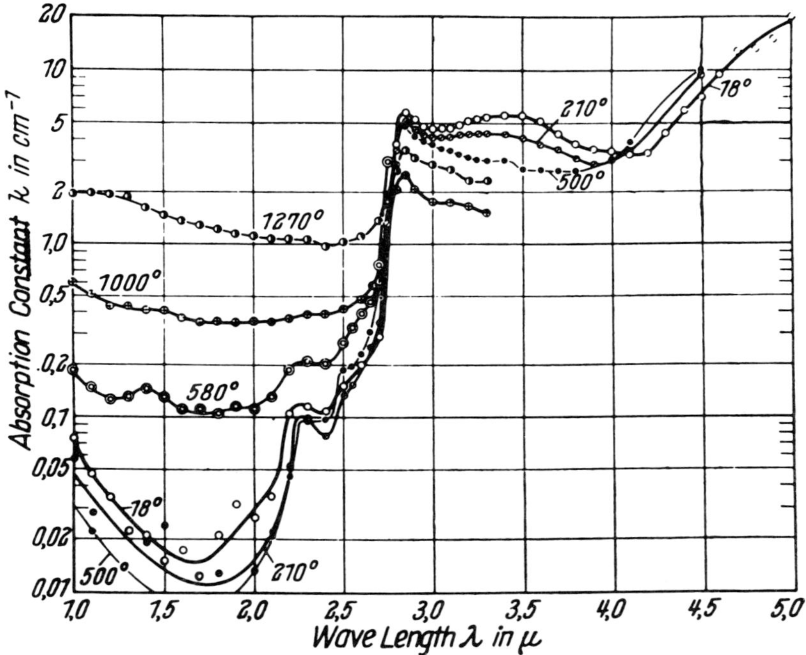

FIG. A. 169. Infrared absorption of lead glass (same as before), at variable temperatures, as a function of wavelengths. (Neuroth).

248. With some experimental and methodical improvements, but for the same purpose of determining the basic data for the calculation of the radiation heat flow in glass tank furnaces (cf. B. ¶ 108 ff.), F. J. Grove and P. E. Jellyman[397] measured the transmission of different glasses up to 1400°C., especially iron and nickel-colored glasses, with the details of the absorption peaks in the range from the visible spectrum to 2.6 μ, and 4.0 μ, respectively, and for industrial "Calorex" glass (a bluish-green heat protection glass containing 0.86 per cent Fe_2O_3). The high complexity of the absorption phenomena may be impressive from Fig. A. 172. More extensive and systematic measurements were made by P. E. Jellyman and J. P. Procter[398] with

[397] *J. Soc. Glass Technol.*, **39**, 1955, 3-15 T.

[398] *J. Soc. Glass Technol.*, **39**, 1955, 173-192 T. Recent studies of the infrared absorption of

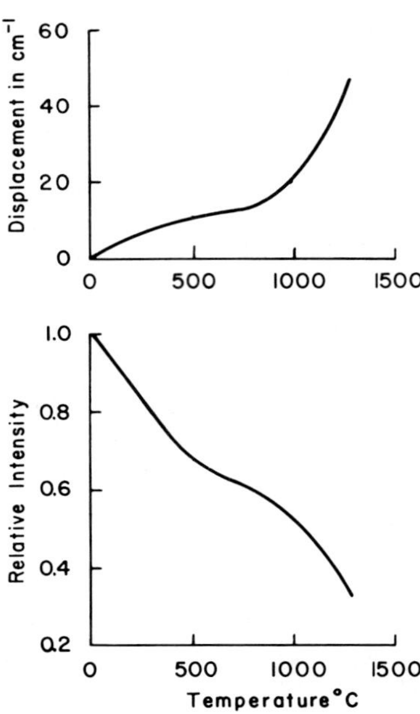

FIG. A. 170. Displacement of the fundamental absorption band of silica glass, and decrease in its intensity, as a function of temperature. (Neuroth).

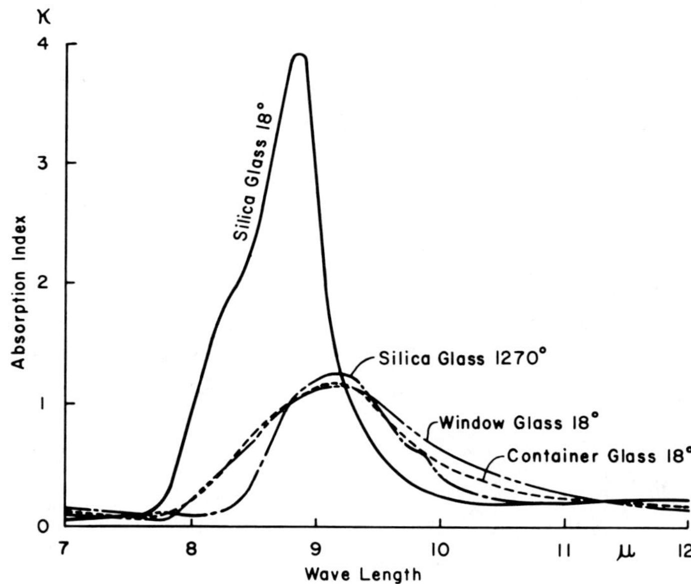

FIG. A. 171. Comparison of the absorption spectra of silica glass at 18° and 1270°C., and of different industrial glasses at 18°C. (Neuroth).

alkali borate, silicate, and borosilicate glasses using the reflection-spectrometric method. In addition to the very instructive and rich graphic materials of the reflection curves observed, the evaluation of these results for the knowledge of specific oscillations of the constitution units in such glasses is of great importance. We, therefore, offer here a tabulation as derived by Jellyman and Procter from their results:

TABLE I

ASSIGNMENT OF REFLECTION PEAKS TO STRUCTURAL LINKAGES

Structural linkage	Approximate wavelength of the principal peak in μ	Approximate wavelength of subsidiary peak in μ
—Si—O—Si— (each Si with two further bonds)	9 to 9.5 (8.9 in fused silica)	12.5 to 13.5
—Si—O (Si with two further bonds)	10 to 11	
>B—O—B<	8 (7.8 in fused B_2O_3)	14.0 to 14.5
>B—O—$B^{\ominus}$— (second B with two further bonds)	7 to 7.5	9.5
>B—O^-	Within 10 to 12 region	
>B—O—B< (second B bearing O^-)	Within 10 to 12 region	
>B—O—Si— (Si with two further bonds)	7 to 7.5	11
—$B^{\ominus}$—O—Si— (each B and Si with two further bonds)	9 to 9.5	

lead and bismuth borate glasses see by S. M. Brekhovskikh and V. P. Cheremisinov, *Structure of Glass. Proc. All-Union Conf. Glassy State, 3rd, Leningrad* (English Translation), 1959/60, 191-194.

249. How the character of different reflection peaks changes in sodium borate and especially in borosilicate glasses as a function of the presence of triangular $[BO_3]$ and/or tetrahedral $[BO_4]$ coordination groups is seen from the curves in Figs. A. 173 and

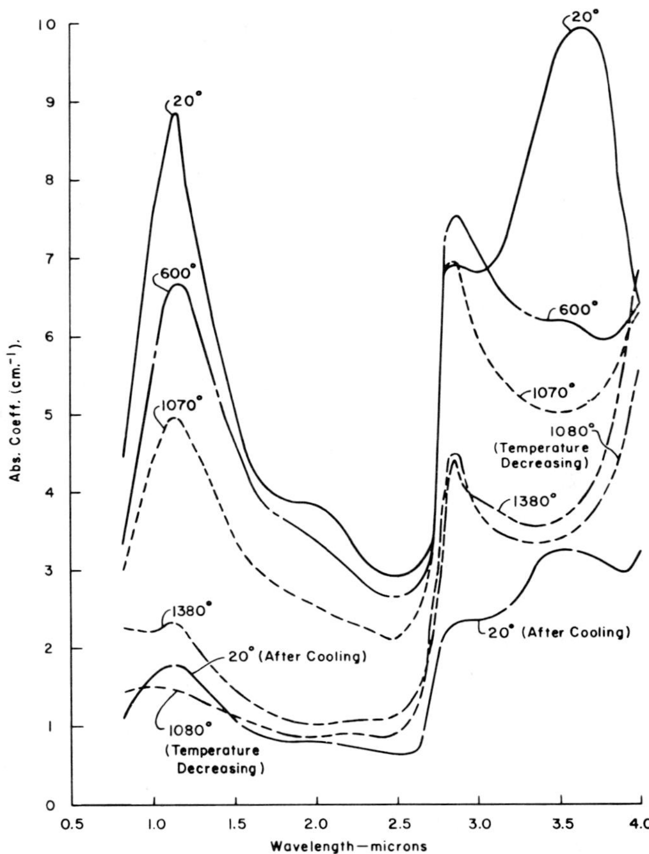

FIG. A. 172. Infrared absorption curves of "Calorex" glass (iron containing heat protection glass), in the temperature range from 20° to 1380°C. (Grove and Jellyman).

174. The dominant peak in the borosilicates is the 9 to 9.5 μ band corresponding to

$$-\overset{|}{\underset{|}{Si}}-O-\overset{|}{\underset{|}{Si}}-$$

whereas that at 7 to 7.5 μ in sodium borates is typical of

$$\rangle B-O-B\langle \quad \text{and} \quad \rangle B-O-\overset{|\ominus}{\underset{|}{B}}-$$

with $[BO_3]$ and $[BO_4]$ together but also for

$$\begin{array}{c} \diagdown \qquad | \\ B—O—Si— \\ \diagup \qquad | \end{array}$$

The absence of the peaks at 8 μ and 14.5 μ indicates in glasses with excess SiO_2 over B_2O_3 the absence of

$$\begin{array}{c} \diagdown \qquad \diagup \\ B—O—B \\ \diagup \qquad \diagdown \end{array}$$

linkages, both B atoms in $[BO_3]$ groups. The thorough discussion of the curves and intensity of their peaks gives full information on the influence of alkali contents on the infrared spectrum characteristics and a good agreement with theories previously developed by T. Abe (cf. A. ¶ 239).

250. Among the numerous investigations determined for studying the effects of varying chemical composition on the infrared absorption characteristic, we first

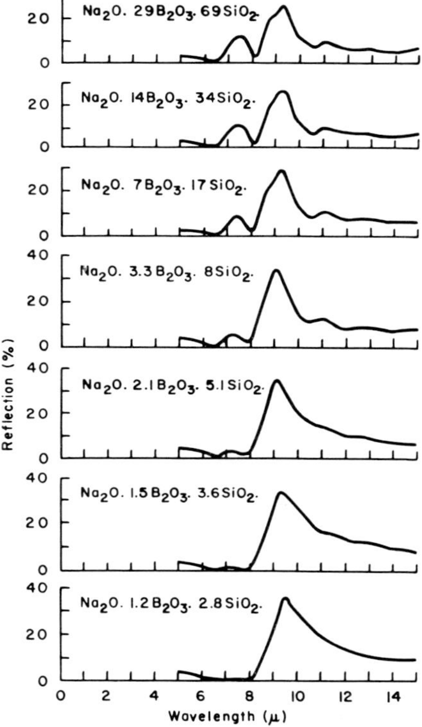

Fig. A. 173. Infrared reflection curves for sodium borosilicate glasses, with a constant molar ratio $B_2O_3 : SiO_2 = 1 : 2.4$. (Jellyman and Procter).

emphasize the studies of the transmission in two- and three-component glasses by J. M. Florence, Fr. W. Glaze, and M. H. Black[399] in which the peak at 2.72 μ corresponding to "water" contents was early recognized, as well as an alleged effect of the carbonate anion at 4.0 μ, and of CO_2 at 4.25 μ (see below). For the longer wave range (9 to 11 μ) the curves, offered by W. Pepperhoff[400] of lead silicate glasses are equally significant, especially for the shifting of the 9 μ silica band to higher

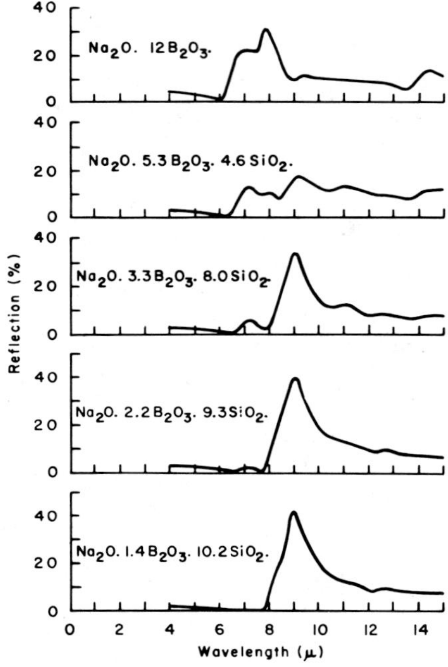

Fig. A. 174. Infrared reflection curves for sodium borosilicate glasses, with an approximately constant molecular fraction of Na_2O of about 1 : 12. (Jellyman and Procter).

wavelengths in reflection spectra. The composition $PbO.SiO_2$ brings about a distinct anomaly, interpreted by a certain order state in glasses of this kind which is similar to that in the crystalline compound. This is a viewpoint which also plays an important role in the later discussed results of V. A. Florinskaya and R. S. Pechenkina.

251. The high infrared transmittance of some *nonsilicate glasses*, namely of As_2O_3 type, was mentioned already in A. ¶ 3 as remarkable up to 12 μ whereas the common optical glasses on the basis of silicates, borosilicates or germanates are only

[399] *J. Research Natl. Bur. Standards*, **50**, 1953, 187-196, with 27 characteristic transmission curves for $\lambda = 1$ to 5 μ of lead containing glasses.

[400] *J. Phys. Chem.*, **58**, 1954, 520-522.

transparent up to 6 μ in the best case, as Fr. W. Glaze[401] demonstrated. The absorption and reflection spectra of B_2O_3 glass and of typical sodium borate glasses was given by Sc. Anderson, R. L. Bohon, and D. D. Kimpton[402] who emphasized the great importance of the hydrogen bonds in the constitution of the glasses (A. ¶ 238), in comparison with the infrared spectra of H_3BO_3, $K(B_5O_8).4H_2O$, $CaB_2Si_2O_8$, (danburite) and others, with $[BO_3]$ groups alone in the first, $[BO_4]$ in the last compound. The complex spectra, say of the simple series Na_2O–B_2O_3, show at a critical composition of 15 per cent Na_2O a change of the constitution in the structural arrangements.

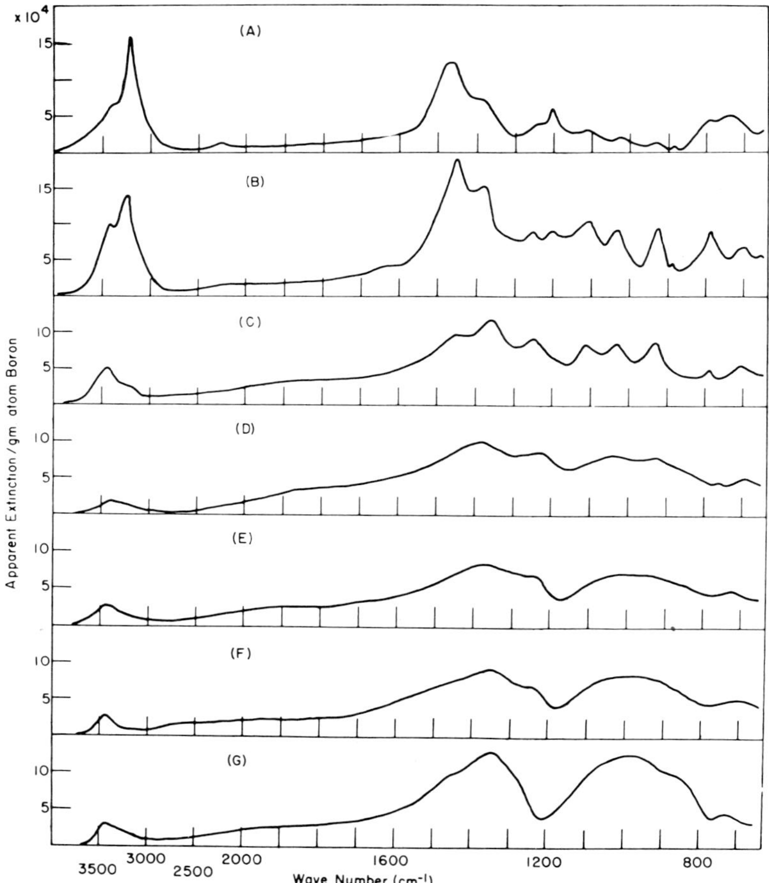

FIG. A. 175. Infrared absorption spectra of sodium borate glasses, starting from pure B_2O_3 glass, with increasing contents in Na_2O, from 5 to 5 per cent. (Anderson, Bohon, and Kimpton).

[401] *Am. Ceram. Soc. Bull.*, 34, 1955, 291-294.

[402] *J. Am. Ceram. Soc.*, 38, 1955, 370-377.

For 30 per cent Na_2O the ring complex $[B_3O_6]^{3-}$ may become dominant as a structural unit, because of the similarity of the glass spectrum with that of crystalline sodium metaborate, $NaBO_2$ (Fig. A. 175/176). We also emphasize corresponding systematic studies on the infrared absorption spectra of B_2O_3 and of derived borate glasses as given by H. Moore and P. W. McMillan[403] which are extended up to 14.5 μ in suitable thickness of the glass samples. All of the oxide glasses contained water in significant amounts.

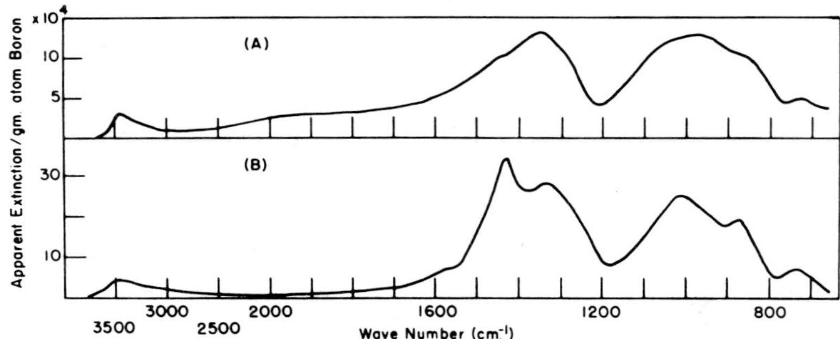

FIG. A. 176. Infrared absorption spectra, (*A*) of sodium borate glass with 30 per cent Na_2O, and (*B*) of sodium metaborate (crystalline). (Anderson, Bohon, and Kimpton).

252. The infrared reflection method was used by I. Šimon and H. O. McMahon[404] for a study of the 10 μ region in which the fundamental vibrations of the silica lattice are observed. This region is particularly sensitive to influences of alkali and alkaline earth cations as framework modifiers in the structure, as for sodium silicate glasses in the range up to 40 molecular per cent Na_2O. The disruption of the $[SiO_4]$ framework by excess oxygen anions introduced with the alkali cation is shown by decreases in intensity and shifting of the characteristic peak at 8.9 μ for Si—O—Si to larger wavelengths, that is, the binding energy is dispersed. A new peak appears at 10.5 to 11.0 μ for Si—O for one-sided binding. L. Hoziaux[405] emphasizes the differences in characteristics of infrared absorption and reflection spectra, and the non-identity of the peaks indicated in the corresponding curves. Thus, the characteristic ring frequence at $\lambda = 12.5\ \mu$ in quartz, cordierite, beryl, and others, appears only in absorption, but not in reflection spectra. For silica glasses the latter show the active

[403] *J. Soc. Glass Technol.*, **40**, 1956, 97-138 T.

[404] *J. Am. Ceram. Soc.*, **56**, 1953, 160-164. On the theoretical background and the Si—O—Si vibrational modes cf. J. Barriol, *J. phys. radium*, **7**, 1946, 209-216, and I. Šimon and H. O. Mc Mahon, *J. Chem. Phys.*, **21**, 1953, 23-30, further a comparison with the infrared spectra of alkyl silanes and siloxanes *Ibid.*, **20**, 1952, 905-907. See also I. Peychès, *Silicates inds.*, **21**. 1956, 209-218; *Verres et réfractaires*, **8**, 1954, 181-193.

[405] *Silicates inds.*, **21**, 1956, 491-500.

oscillation of the $[SiO_4]$ groups displaced to longer wavelengths, in comparison with crystalline silicates, and this displacement is a function of the ionic radius and concentration of alkali added as Šimon and McMahon had observed before. Hoziaux gives the spectra of ten industrial glasses of the systems Na_2O–B_2O_3–SiO_2 and K_2O–PbO–SiO_2. The frequency characteristics of the bonds B—O—B and B—O—Si are calculated at 1410 cm.$^{-1}$ (7.05 μ) and 675 cm.$^{-1}$ (1.49 μ), in good agreement with the measured data. The annular band at 12.5 μ is shifted to longer wavelengths; at the same time it is weakened in proportion to the introduction of framework modifiers.

253. Most elaborate investigations of infrared absorption spectra of silica glass, alkali, and lead silicate, and borosilicate glasses are owed to N. A. Sevchenko and V. A. Florinskaya,[406] and especially to the latter author with R. S. Pechenkina,[407] over the range from 7 to 24 μ. As a summary from these precision data we may only emphasize a few general results. The thermal history of a glass from its melt to the solid state and subsequent crystallization may be traced in all the transitions by using infrared transmission methods. The complex microstructure of the glass is interpreted as a heterogeneous system. Silica in its different modifications, and definite silicates participate in the glass formation as distinct chemical compounds and crystallites, or even as crystalline solutions. In every case, the infrared spectra are interpreted by Florinskaya and Pechenkina as a strong support for the crystallite theory of glass constitution with regular structural domains. The atomic positions in the crystallites must be similar to those in the crystalline phases of the silica modifications and of definite silicates, say of sodium and lead. There is a continuous transition assumed from the most ordered zones to irregularly disordered portions of the glass structure, and vice-versa in neighboring crystallites. The formation of groups from which crystallites are formed in the cooling glass begins very early, even in a glass melt above the Liquidus temperature. The diameters of the crystallites are evidently somewhat larger than those indicated by X-ray diffraction

[406] *Doklady Akad. Nauk S.S.S.R.*, **109**, 1956, 1115-1118.

[407] *Stroenie Stekla, Inst. Khim. Silikatov Akad. Nauk S.S.S.R., Trudy Soveshchaniya, Leningrad*, 1953 (Pub. 1955), 70-95; discussions 322-335; *Optika i Spektroskopiya*, **1**, 1956, 690-709. See also *Glass Ind.*, **39**, 1958, 27-31, 93-96, 151-154, 168. Quite recently, Florinskaya gave a special discussion of the glasses in the system Na_2O–SiO_2, in *Silikat Tech.*, **11**, 1960, 364-367, dealing with problems of the existence of $Na_2O.2SiO_2$ or $Na_2O.3SiO_2$ in these, and of the heterogeneity of the structure and constitution (with an excellent bibliography). Florinskaya and Pechenkina further reported additional measurements on sodium silicate glasses in the range between 7.5 and 15 μ, from infrared reflection and transmission, in *Zhur. Strukt. Khim.*, **1**, 1960, 86-98, and *Structure of Glass. Proc. All-Union Conf. Glassy State, 3rd, Leningrad* (English Translation) 1959/60, 135-153, 154-168. The recent results confirm microheterogeneities in the structure of sodium silicate glasses with zones of ordered arrangements not compatible with the Zachariasen theory of a statistical distribution of the alkali ions in the $[SiO_4]$ framework. See also the review article of P. Tarte, *Ind. Chim. Belg.*, 1961 (12), 1962 (1), 24 pp. (from reprint).

methods. The Si—O distances in the $[SiO_4]$ tetrahedra in glass are variable, not fixed. The average distances are smaller in silica glass, largest in glasses which contain isle structures.

254. The problem of crystallization in different degrees of order is similar to the phenomena observed in organic polymers, as in polyethylene which were successfully studied by infrared methods by V. N. Nikitin and E. I. Pokrovskiĭ[408] in the appearance of specific absorption lines, such as at 730 cm.$^{-1}$ (13.3 μ) and 1308 cm.$^{-1}$ (7.65 μ).The determination of the "degree of crystallinity" in per cent can be based on the Lambert-Beer law of absorption intensities.

255. We repeatedly mentioned the observation that infrared absorption bands in the vicinity of lines at 2.75 to 2.95 μ have been discussed and interpreted as belonging to "*water*" and more precisely said, to definite states of *hydroxyl* groups, either free, or bound in the silicate glass structure. Also free silica glass contains these bands, as V. Garino-Canina[409] emphasized. Only if strictly anhydrous fusion is used, the silica glass shows the band either weak, or really absent. The infrared test is in every case highly sensitive for the presence of hydroxyl groups, free, or bonded to Si atoms. Neither long heating nor fusion in vacuo are able to remove these effects. More recently, R. V. Adams and R. W. Douglas[410] confirmed in essentials those conclusions for silica glass, but they emphasize that in normal silicate glasses the bridging *hydrogen bond* type of the type O—H···O prevails. Four different peaks in flame-fused silica glass are assigned to "water," namely at 2.73 μ, 2.6 μ, 2.22 μ, and 1.38 μ, whereas in common silicate glasses the peaks are variable from 2.85 μ to about 3.2 μ when association takes place.

256. Because of the great importance of infrared absorption phenomena ascribed to "water" in crystalline and glassy silicate (cf. also Vol. IV, Section B and B. ¶ 89 ff.) H. Scholze[411] made extensive investigations on the influence of "moisture" or of deliberately added hydrates in glass batches, in comparison with the natural "water" contents such as in clay minerals, and their response in infrared absorption. Scholze emphasizes that the band near 2.75 μ alone cannot be used for a decision of the problem of how the hydroxyl groups are bonded. It is equally present in the H_2O molecule as in "free" hydroxyl groups. A decision could perhaps be possible if the line at 6.25 μ would not unfortunately be usually completely submerged by the strong

[408] *Doklady Akad. Nauk S.S.S.R.*, **95**, 1954, 109-110.

[409] *Compt. rend.*, **239**, 1954, 705-706.

[410] *J. Soc. Glass Technol.*, **43**, 1959, 147-158 T. On p. 157 a complete table with assignments of absorption bands in silica glass with wavelengths, frequencies (observed and calculated), the normal intensities, and vibrational characteristics. See also Adams and Douglas, *Glastech. Ber.*, **32**, K, 1959, especially VII, 15, an assignment of OH frequencies in silica glass.

[411] *Naturwissenschaften*, **42**, 1955, 342-343 (with A. Dietzel); *Trav. Congr. Intern. du Verre*, IV, Paris 1956, VIII, 3, 6 pages; *Veröffentl. Max-Planck-Inst. Silikatforsch.*, **16**, 1956, 135-140.

absorption of the $[SiO_4]$ in silicates. Scholze refutes the opinion of Florence *et al.* (see A. ¶ 250) that CO_3^{2-} bands occur in the infrared spectrum of glass at 3.6 to 3.7 μ. This band also is ascribed by him to hydroxyl groups, namely by a hydrogen atom connected in a distance of 2.66 Å. to a singly bonded oxygen anion of the glass structure. The 3.6 μ band is rather inert if the chemical composition of the glass is changed, but its intensity is increased with increasing alkali content and basicity of the composition.

257. We mention briefly how different R. Brückner and H. Scholze[412] found the infrared absorption behavior of B_2O_3 glass in the freshly prepared state, and after some "weathering." Particular care must be given to the necessity of avoiding a confusion in this case of the B_2O_3 band at 2.8 μ, and that at 3.1 μ (and 8.4 μ) for H_3BO_3. Completely anhydrous B_2O_3 glass, such as that fused in an atmosphere of BCl_3 at 1350°C., does not show the 2.8 μ band. In weathered glass that of H_3BO_3 occurs immediately which even develops a crystalline texture effect by the scaly crystals of that compound, in the line between 12 and 13.5 μ. Scholze and Brückner refute the presence of water as a premise for the fusion of B_2O_3 glass, and the presence of $[B_9O_{14}]$ groups as assumed by Sc. Anderson *et al.* (cf. A. ¶ 238).

258. Systematic experiments on the introduction of water into glasses[413] showed that the measurement of intensity and of the extinction coefficient, $\varepsilon = (1/cd) \cdot D$ (D the transmissivity, c the concentration) for the 2.75 μ peak is an excellent tool in the quantitative determination of the water content in glass in general. The parameters ε and the integral intensity $I = \int \varepsilon \cdot d\nu^*$, should be combined for the bands 2.75, 3.85, and 42.5 μ, and be included in the calculation, but conventional simplifications using chiefly the 2.75 and 3.85 μ band can be applied. The greater the wavelength of the bands the lower the ν^*, the stronger the hydrogen bridging to neighboring oxygen anions, and the lower the distance O—O. The influence of the chemical composition of the host glass on water contents was studied in a systematic series of binary, ternary, and quaternary glasses, with considerable changes qualitatively in the intensities of the 2.8 and 3.8 μ bands as a function of composition. The hydroxyl groups must be placed in the holes of the framework structure. The formation of strong hydrogen bonding indicated by the 3.8 μ band is favored by

[412] *Glastech. Ber.*, **31**, 1958, 417-422. See also E. P. Markin and N. N. Sobolev, *Izvest. Akad. Nauk S.S.S.R.*, Ser. Fiz. **22**, 1958, 1097-1099, on the reflection spectrum of molten B_2O_3 up to 1000°C.

[413] *Glastech. Ber.*, **32**, 1959, 81-88; 142-152; 278-281; 314-320; 381-386 (with H. O. Mulfinger), 421-426 (with H. Franz and L. Merker). See also the review in *Ibid.*, **32**, K, 1959, VII, 1-VII, 5; further A. Kats and Y. Haven, *Phys. & Chem. Glasses*, **1**, 1960, 99-102; R. V. Adams, *Ibid.* **2**, 1961, 39-49; A. Kats, Y. Haven, and J. M. Stevels, *Ibid.* **3**, 1962, 69-75; specifically for silica glass: A. J. Moulsen and J. P. Roberts, *Trans. Brit. Ceram. Soc.*, **59**, 1960, 388-399; H. Scholze and H. Franz, *Glastech. Ber.*, **35**, 1962, 278-281.

groups like ONa; the more alkali present in the glass the more intense the bridge bonding. The wavelength increases as the field strength z/a^2 of the alkali cation decreases. Increasing contents in alumina, however, reduce the intensity of the 3.8 μ band, and a glass with 20 molecular per cent Al_2O_3 and 20 per cent Na_2O does not show any more hydroxyl with hydrogen bonding. The strongest hydrogen bridge bonding is indicated by the 4.25 μ band, with a distance O–O=2.55 Å., as in alkali-rich glasses. The introduction of Ca^{2+}, Mg^{2+} or Zn^{2+} increases the number of free hydroxyl groups. In B_2O_3–containing glasses the bands at 3.68 μ are particularly sharp corresponding to B—O oscillations. If TiO_2 is introduced into sodium or barium silicate glasses the 3.6 μ band is reduced in favor of that at 2.9 μ; $[TiO_6]$ groups withdraw oxygen anions from the OR bands.

259. Nonsilicate glasses also contain mostly water, such as a calcium-beryllium aluminate glass, further fluorine containing silicate, and borate glasses. Increasing temperature changes the infrared spectra of water containing glasses (cf. Fig. A. 177); the firmly bound hydroxyl groups are gradually changed into free groups as

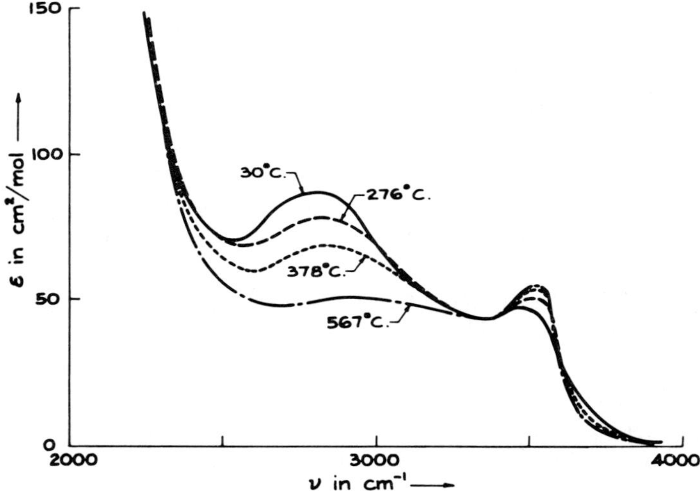

Fig. A. 177. Infrared absorption spectrum of a sodium-calcium silicate glass (16 Na_2O; 10 CaO; 74 SiO_2, in weight per cent), at different temperatures. (Scholze).

indicated by the increasing intensity of the 2.85 μ band, whereas the 3.8 μ band is weakened and shifted to shorter wavelengths. The conversion of the hydroxyl bands from the freezing-in range (450° to 650°C. for silicate glasses) upwards is a true reversible chemical equilibrium reaction with the definite reaction constant $K = [OH_{free}]/[OH_{bridging}]$, and an enthalpy of 5000 cal./mole. Alkali-rich glasses evidently contain bound hydroxyl groups even at high temperatures. The concentration in free hydroxyl groups at 1300°C. is highest for lithium aluminosilicate glasses (cf. Fig. A. 178), but not considerable in potassium silicate glasses. The diffusion coefficient of water in glasses at 1000° to 1400°C. could be determined by using the quantitative intensity of the 2.85 and 3.8 μ hydroxyl bands. The order of magnitude for

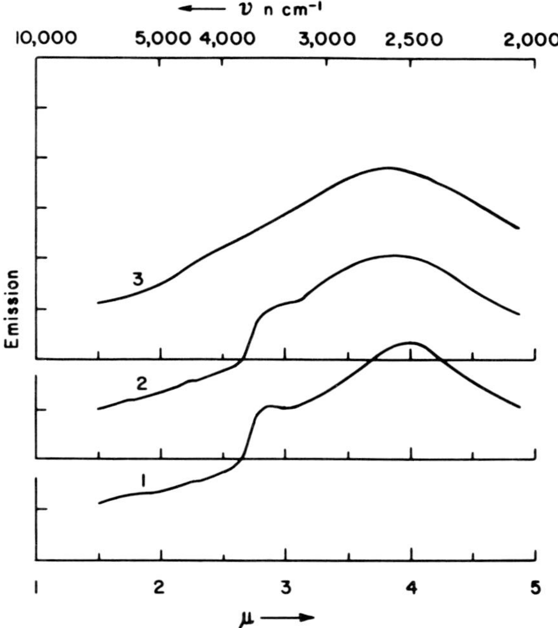

Fig. A. 178. Infrared emission curves of glasses at 1300°C. (Scholze). Curve (*1*) for a glass 20 Li_2O; 10 Al_2O_3; 70 SiO_2; Curve (*2*) for glass 16 Na_2O; 10 CaO; 74 SiO_2; Curve (*3*) for glass 30 K_2O; 70 SiO_2 (molecular per cent).

this coefficient is at 1400°C. 10^{-6} cm.^{2}/sec, increasing with temperature, alkali content and field strength z/a^2 of the alkali ions. On the effects of the water contents on the physical properties of the glasses, based on the use of infrared absorption peaks for the determination of the water concentration, consult B. ¶ 94.

260. The problem of distinction in the infrared absorption spectra of the water molecules from hydroxyl groups in glasses and silicate minerals was again discussed by H. Scholze.[414] The combination oscillations $\nu_{OH} + \delta_{H_2O}$ at 5100 cm.$^{-1}$ are conclusive for water molecules. Typical Si–OH groups as in silanols, silica hydrogel, and

[414] *Naturwissenschaften*, **47**, 1960, 226-227; *Fortschr. Mineral.*, **38**, 1960, 122-123, with valuable tabulations of wavelengths, integral intensities and extinction coefficients of the infrared bands of H_2O or OH in different silicate minerals.

others are assumed near 900 cm.$^{-1}$ Scholze studied for this purpose the infrared absorption in triphenylsilanol, $(C_6H_5)_3Si–OH$, and its deuterium analog (with 70 per cent D replacing H). The spectra give a clear possibility of distinguishing H_2O molecules from hydroxyl groups in appearance either of the band at 5100 cm.$^{-1}$ ($\lambda = 1.95\ \mu$), or at 4500 cm.$^{-1}$ ($\lambda = 2.2\ \mu$), both of the same intensity, but of lower intensity than the valence oscillation bands have. A silica hydrogel shows both bands

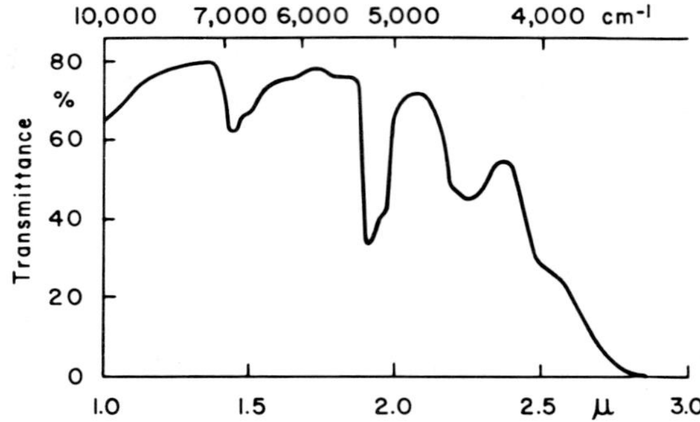

Fig. A. 179. Infrared absorption spectrum of a silica hydrogel, with about 8 per cent water. (Scholze).

well developed (Fig. A. 179). Similar conditions were observed in chalcedony and opal (cf. Vol. IV, Section B). In this connection we also mention the recent results of Ya. I. Ryskin, G. P. Stavitskaya, and N. A. Toropov[415] on trimethylsilanol and synthetic calcium silicate hydrates.

261. Concerning the theory of the molecular vibrations indicated in the infrared spectra we mention the discussions of B. I. Stepanov and A. M. Prima[416] based on calculations of M. V. Volkenshteĭn *et al.* The previous ideas of Cl. Schaefer, F. Matossi, and K. Wirtz (1934) which started from the frequencies derived from isolated $[SiO_4]$ tetrahedra in silicate structures to interpret the observed infrared frequencies, are refuted by Stepanov and Prima as incompatible with the more or less strong forces binding the tetrahedra to structural units of larger size oscillating together, especially in silicates with pronounced frameworks, as in quartz, tridymite, cristobalite, and also in silica and silicate glasses. The fact that infrared absorption characteristics of silica glass prepared from quartz still show the type of the original crystal phase (as also observed by Florinskaya and Pechenkina) evidently speaks for an extensive preservation of the bonds of the tetrahedra in such frameworks, also

[415] *Zhur. Neorg. Khim.*, **5**, 1960, 2727-2734.

[416] *Optika i Spektroskopiya*, **4**, 1958 734-749; cf. the volume I of "Vibration of Molecules," Moscow 1949, by M. V. Volkenshteĭn.

in glass as J. Zarzicky[417] emphasized, and the same is valid for germanium dioxide glass although the angles Si—O—Si and Ge—O—Ge are somewhat different, in GeO_2 nearer to 180° (cf. A. ¶ 233). The same author, with F. Naudin[418] rediscussed the premises of the theories of Stepanov and Prima, as especially suitable for interpretation of SiO_2, GeO_2, and BeF_2 glasses. A refined analysis of the infrared spectra in addition makes evident an asymmetric deformation of the tetrahedron in binary glasses, which cannot be detected by elementary X-ray diffraction methods. The distance of Si—O with unilateral O^{2-} must be smaller than that from a bridging O^{2-}, in agreement with W. A. Weyl's theory of polarization effects in such deformations.[419]

262. The usefulness of *Raman spectra* according to the principle of combination dispersion effects was demonstrated in early experiments by M. F. Vuks and V. A. Ioffe[420] both for the systems Na_2O–SiO_2, and PbO–SiO_2, with four and two characteristic bands, respectively. With increasing silica contents the bands are systematically shifted. Some of the bands show a distinct similarity with those observed in the quartz spectrum, and in that of crystalline $Na_2O.2SiO_2$, if the composition of the glass is near to the molecular ratio $Na_2O : SiO_2 = 1 : 2$. On the section $Na_2O.2SiO_2$–$PbO.SiO_2$ five bands appear. A particularly extensive material of Raman data was collected by E. F. Gross and V. A. Kolesova[421] including intensities of the observed bands, relative to silica glass or to the 944 cm.$^{-1}$ band of a glass with 20 mol. per cent Na_2O. The systems tabulated are K_2O–SiO_2, Na_2O–SiO_2, and some compositions in Li_2–SiO_2. The dependence of the Raman bands in their frequencies as a function of composition is the same for Na and K silicate glasses, all of the vibrations being assigned to oscillations of Si relative to O atoms whereas vibrations of

[417] *Verres et réfractaires*, **11**, 1957, 3-8; cf. J. D. Mackenzie and J. L. White, *J. Am. Ceram. Soc.*, **43**, 1960, 170-171.

[418] *Verres et réfractaires*, **14**, 1960, 113-123.

[419] *Silicates inds.*, **23**, 1958, 309-315; **24**, 1959, 321-327.

[420] "Physical-Chemical Properties of the Ternary System Na_2O–PbO–SiO_2," Edit. I. V. Grebenshchikov, Moscow-Leningrad 1949, pp. 164-170, 171-173. Extensive studies on Raman spectra of different silicate, borosilicate, germanate glasses see by Ya. S. Bobovich and T. P. Tulub, *Structure of Glass. Proc. All-Union Conf. Glassy State, 3rd, Leningrad* (English Translation), 1959/60, 173-176, with particular consideration of the covalent character of the chemical bonding B—O, Pb—O, MgO—O, Cd—O, Zn—O, Bi—O.

[421] *Zhur. Fiz. Khim.*, **26**, 1952, 1673-1680; *Stroenie Stekla, Inst. Khim. Silikatov Akad. Nauk S.S.S.R., Trudy Soveshchaniya, Leningrad*, 1953 (Publ. 1955), 56-61. Kolesova, *Structure of Glass. Proc. All-Union Conf. Glassy State, 3rd, Leningrad* (English Translation), 1959/60, 177-179, observed in alkali aluminosilicate glasses a particular band at 760 cm.$^{-1}$ (13.1 μ) indicating Al^{3+} ions in the framework, and a shifting of the Si—O—Si vibrations at 1000 cm.$^{-1}$ (10 μ) to lower frequencies if Al contents are increased. In aluminosilicate glasses the Na^+ ions act as depolymerizers. Modern X-ray fluorescence methods used by D. E. Day and G. E. Rindone, *J. Am. Ceram. Soc.*, **45**, 1962, 579-581, made considerable contributions to the interpretation of Al—O coordination in sodium aluminosilicate glasses (on the method see Vol. V. B. ¶ 48*).

the alkali metal atom to O are absent (they are assumed to be in the range of 250 cm.$^{-1}$ to 300 cm.$^{-1}$).

263. Anomalies occur for higher alkali concentrations (above 40 per cent) near the composition of the metasilicates $R_2O.SiO_2$. Nevertheless, Gross and Kolesova found characteristic differences in the Raman spectra of quartz and silica glass which demonstrate that it is not alone a distortion of the quartz structure causing the vibrational schemes for silica glass which approaches more to the cristobalite structure as seen in the neutron diffraction analysis by Milligan, Levy, and Peterson (A. ¶ 229). The changes observed in binary glasses are interpreted as "depolymerisation"

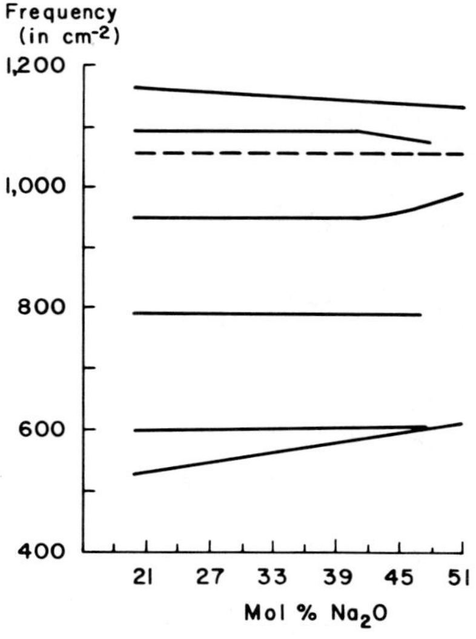

FIG. A. 180. Changes of the Raman combination oscillation frequencies for sodium silicate glass, as a function of the contents in Na_2O. (Gross and Kolesova).

effects of the glass framework, as in the formation of chains as structural elements, observed in the range of frequencies between 1200 cm.$^{-1}$ and 500 cm.$^{-1}$. These frequencies as a function of the alkali contents are illustrated in Fig. A. 180 for the system Na_2O–SiO_2 above 20 mol. per cent alkali oxide with a characteristic breakpoint at about 40 per cent Na_2O for ν = 1100 cm.$^{-1}$ and 940 cm.$^{-1}$. In every case, the Raman spectra of the investigated sodium silicate glasses cannot be interpreted by a merely additive superposition of silica and say $Na_2O.SiO_2$ spectra. They give the impression of a structural heterogeneity existing in the glass, but no decision is made by Gross and Kolesova in favor of the presence of crystallites in the frequently mentioned meaning.

264. A particularly valuable analysis of Raman spectra of different optical glasses was made by L. Prod'homme,[422] in comparison with those of quartz and silica glass to study the specific effects of the single framework-modifying oxides and of thermal treatments by quenching and annealing. The random structure of the glasses is indicated by Raman bands in the place of the sharp lines of the crystalline media (see Fig. A. 181). The vibration at 500 cm.$^{-1}$ (in quartz 466 cm.$^{-1}$) is ascribed

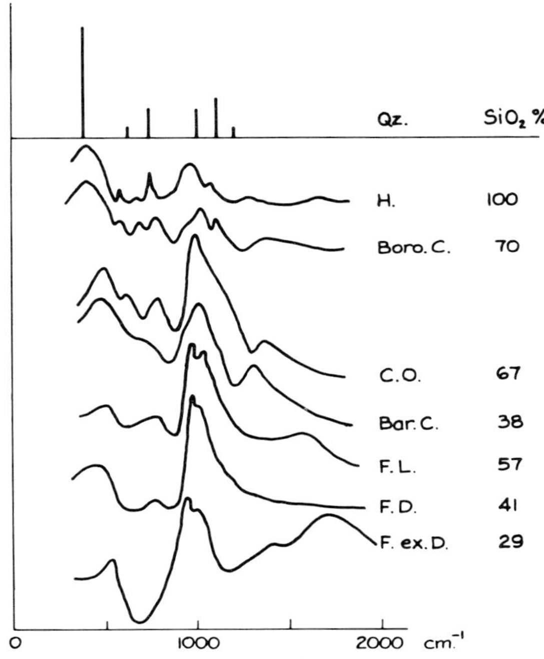

Fig. A. 181. Comparison of the Raman spectra of different glasses: *H* = "Homosil" silica glass; boro C. = borosilicate crown; C.O. = common crown; Ba C. = barium crown; F. L. = light flint; F. D. = heavy flint; F.ex.D. = extra heavy flint glass. (Prod'homme).

to Si–Si oscillations in the $[SiO_4]$ groups. The 800 cm.$^{-1}$ band in quartz appears also in silica glass and silicates, its intensity being proportional to the silica contents. The lines 1062 cm.$^{-1}$ and 1165 cm.$^{-1}$ of quartz are in silicate glasses changed to one strong band near 1000 cm.$^{-1}$, particularly strong in Flint and Crown glasses. It is ascribed to O—O oscillations. With increasing ratios O : Si (in lead-rich glasses) it is in parallel to the increase in intensity and wavelength. The B_2O_3 effect is identified by a diffuse band at about 1300 cm.$^{-1}$ but in lead Flint a strong band at 1700 cm.$^{-1}$ could not be assigned. Quenched and annealed glasses have somewhat different

[422] *Verres et réfractaires*, 8, 1954, 305-315.

Raman spectra; that of the quenched glass is more diffuse, evidently the atomic distances in both types being different. For a borosilicate glass the corresponding curves intersect at about 2200 cm.$^{-1}$. The annealed glass shows sharper maxima at 635 cm.$^{-1}$ and 800 cm.$^{-1}$.

265. Of general interest are the facts observed by W. Bues[423] in the Raman spectra of binary salt melts, such as zinc and cadmium chloride with KCl, in comparison to those of their crystalline components. It is remarkable that the layer structure of the heavy metal chlorides remains preserved to a certain degree in the constitution of the liquid melts as indicated by characteristic bands. The transfer of these observations to the investigation of crystalline, molten, and glassy high-polymeric *sodium phosphates* made by W. Bues and H. W. Gehrcke[424] is highly instructive because of the presence of well-defined *polyanions* in the glasses (up to 97 per cent). The Raman spectra of such polyphosphate melts and glasses show remarkably sharp lines. Starting from the coupling principles for complex phosphate formation as given by E. Thilo, G. Schulz, and E.-M. Wichmann[425] it is possible to assign in the P—O—P chains, and in single (PO_2) or (PO_3) groups, as structural units, the different oscillation mechanisms indicated in the Raman spectra, and also those of the di-, tri- and tetra-phosphate anions. Bues and Gehrcke come to the conclusion that there is no principal structural difference between the constitution of the melts and the glasses quenched from the liquid with a typical "freezing-in" effect. For $NaPO_3$ and KPO_3 in comparison to the Kurrol salts the Raman spectra show typical chain arrangements in the structure as expected from the fibrous character of these compounds, with different periodicities in the Na and K salts. In the glasses the chains are preserved to a definite degree of order.

PHYSICOCHEMICAL THEORY OF THE GLASSY STATE

266. The classical problems of the presence of *mechanical inner tensions* (strains) in glass bodies in the quenched state are well known by analysis of the optical phenomena of anomalous birefringence. By the modern evolution of short- and ultra-short-electrical radiation waves one has the possibility of not being restricted in such studies to the normal optical (including the invisible infrared) radiation methods

[423] *Z. anorg. u. allgem. Chem.*, **279**, 1955, 105-114, with description of an ingenious optical apparatus for the observation of the Raman spectra using reflections from monocrystals of silicon carbide submerged in the melts.

[424] *Z. anorg. u. allgem. Chem.*, **288**, 1956, 291-306, 307-323.

[425] *Z. anorg. u. allgem. Chem.*, **272**, 1953, 182-200. See also E. Steger, *Ibid.*, **296**, 1958, 305-312, on the trimetaphosphate anion in crystalline and glassy phases as well as in solutions in H_2O and D_2O.

for determination of such anomalies. A. Dietzel and E. Deeg[426] studied such methods with polarized electric waves, namely of 3.2 cm. wevelength, as suitable for determination of the distriburion of strains in opaque media of ceramic bodies, tank blocks, and "black glass" using, as receiver of the radiation, germanium diodes with amplifying systems connected with a common relatively little sensitive galvanometer. Anisotropies caused by external pressure or thermal contraction in the order of magnitude of a birefringence of 0.10 to 0.015 could thus be measured and oriented in their distribution in the opaque samples. Especially for glass bodies transparent in the short wave infrared, E. W. Deeg and T. D. Riney[427] described very characteristic anisotropy diagrams in polarized radiation for As_2S_3 and opaque structural glass using infrared image-forming electron tubes, the image converter ("snooperscope") with a photocathode sensitive up to 1.2 μ and a fluorescent screen for direct visual inspection.

267. It is the technological art of the glass manufacturer to remove mechanical strains by the well-known measures of *annealing*, specifically by a thermal treatment with the purpose of bringing the molecular *inner nonequilibria* to a stable end state of the glass constitution. The restitution of such physical stabilization effects can be observed by refined *differential-thermal analysis* methods as they were first used by K. Quasebart in 1916, and later by A. Q. Tool, C. G. Eichlin, and W. B. Pietenpol (1920). Of more recent differential-thermal determinations of this kind we mention here that for glasses of the system Na_2O–PbO–SiO_2 by N. A. Vakhrameev,[428] with systematic observations of the exothermic peaks between 408° and 720°C., variable with the chemical composition of the ternary glasses, or the heating curves described by I. V. Borovikov,[429] with an anomalous temperature interval of spontaneous stepwise heat evolution which is important during shaping and cooling of the glass. With modern tools of highly sensitive differential-thermal analysis, Kl. Kühne[430] developed a precision method for the study of exo- and endothermic effects on heating and cooling of glass (Fig. A. 182). Thin-walled alumina tubing protects the exactly centered thermocouples from direct contact with the glass, the effects being received over a mirror-galvanometer. The classical T_g as defined by G. Tammann appears

[426] *Ber. deut. keram. Ges.*, **31**, 1954, 396-404; **32**, 1955, 211-213; *Naturwissenschaften*, **42**, 1955, 11.

[427] *Am. Ceram. Soc. Bull.*, **39**, 1960, 674-676. On stress-optical coefficients of glass determined by its specific components see W. Schwiecker, *Glastech. Ber.*, **30**, 1957, 84-88, measured for a given wavelength in Brewster units (10^{-13} cm.^{2}/dyne), with anomalies (reversed signs of stress effects) for strongly polarizable cations like Ba^{2+} or Pb^{2+}. On image converters cf. H.-I. Friel, and H. Weise, *Jenaer Rundschau*, **4**, 1959, 93-97.

[428] "Physical-Chemical Properties of the Ternary System Na_2O–PbO–SiO_2," edited by I. V. Grebenshchikov, Acad. Sci. U.S.S.R., Moscow-Leningrad 1949, 139-146.

[429] *Doklady Akad. Nauk S.S.S.R.*, **80**, 1951, 229-232; *Steklo i Keram.*, **10**, 1953 (1) 10-14.

[430] *Silikat Tech.*, **11**, 1960, 106-108.

at 600° to 620°C. with strong peaks for ternary alkali-calcium silicate, 16 III, and Suprex Jena glasses, with distinct effects of the thermal history by quenching and annealing (Fig. A. 183). By a combination of the differential-thermal analysis method with interferometric dilatometry, the low-temperature "splitting points"

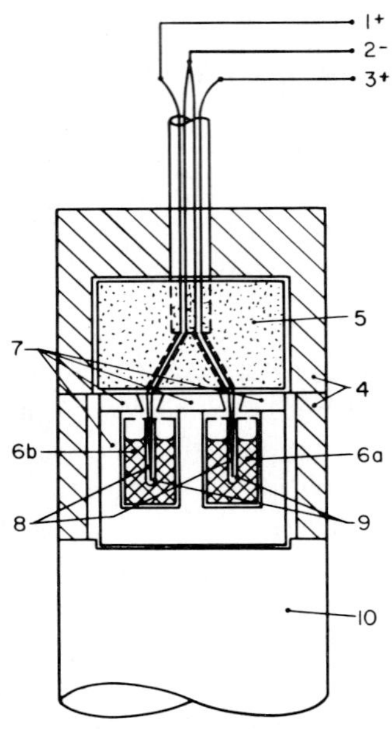

FIG. A. 182. Sensitive differential-thermal device for thermal effects in glass (Kühne). (*1*) (*2*) (*3*) thermocouples; (*4*) Cr-Ni-Steel, (*7*) massive platinum body of sample container; (*5*) filler body of sintered alumina; (*6a*) and (*6b*) glass sample and inert substance (highly calcined alumina); (*8*) thermocouple joints, with (*9*) alumina protection tubes; (*10*) support of silica glass.

(cf. A. Klemm and E. Berger, 1922) were identified at 200° to 300°C. and interpreted as $\alpha \rightleftharpoons \beta$ inversion effects of cristobalite crystallites (cf. H. Schönborn, 1950). Other irregularities indicate microphases in borosilicate glasses by unmixing tendencies (cf. A. ¶ 216).

268. Differential-thermal methods as a special modification of the calorimetric determination of enthalpy effects as a function of temperature[431] and, therefore, of changes of the specific heat, were also extensively discussed by M. Prod'homme[432]

[431] See on this general aspect of methods for determination of caloric effects the book of W. Eitel, "Thermochemical Methods in Silicate Investigation," Rutgers University Press, New Brunswick, 1952, 132 pp., especially paragraphs 14 to 65.

[432] *Compt. rend.*, **240**, 1955, 180-181; *Verres et réfractaires*, **11**, 1957, 287-297, on historical evolution; *Ibid.*, **13**, 1959, 3-16, on the differential-thermal method. The specific heats calculated

for the transformation range temperatures, based on the measurement of the true specific heat of seven optical glasses. Concerning the differential-thermal analysis (especially with silica glass as a refractory material and a heating rate of 2.8° to 3.0 per minute only) the calculation methods of S. L. Boersma[433] were applied, the peaks indicating a discontinuous change in the enthalpy of the glass below and above the indicated temperature. The intensity of the peaks is highly instructive for forming conclusions of the thermal history of the glass; it is high when the glass is well annealed, but "smeared out" if it was chilled. The differential-thermal analysis curve,

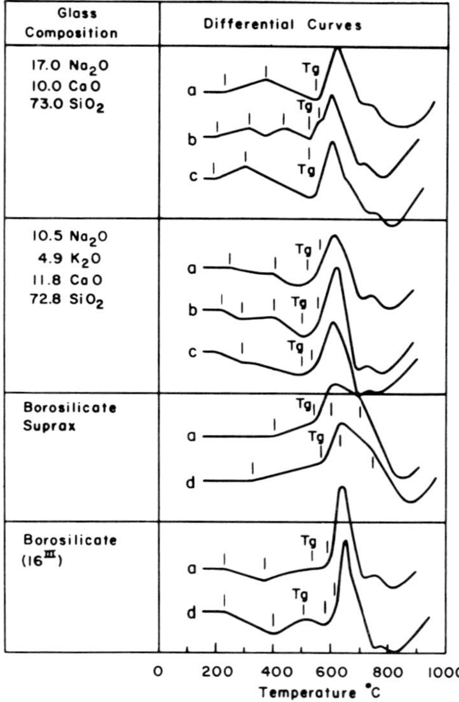

FIG. A. 183. Typical differential-thermal curves of glasses, with varying thermal history. (Kühne). (*a*) Glass quenched from melt to room temperature; (*b*) annealed at 480°C. for 120 hours; (*c*) annealed at 850°C. for 120 hours; (*d*) annealed at 700°C. for 100 hours.

by this method show a tolerably good agreement with those calculated from additivity relations of D. E. Sharp and L. B. Ginther (1951) but there are considerable deviations for borosilicate and heavy Ba Crown glasses.

[433] *J. Am. Ceram. Soc.*, **38**, 1955, 281-284. On the analogy with dilute H_2SO_4 and H_2SeO_4 cf. G. Vuilliard, Thèse, Université de Paris, 1956; *Compt. rend*, **241**, 1955, 1126-1128; **242**, 1956, 1326-1329. As an excellent model substance for study of the influence of thermal history effects on the specific heat, E. Kanda, A. Otsubo and T. Haseda, *Sci. Repts. Tohoku Univ. First Ser.*, A, **2**, 1950 (1) 9-15, used *i*-butyl alcohol from − 170°C. to 0°C.

as a function of the rate of heating, for low rates has completely the characteristics of corresponding curves with a simple slope flexure as is observed in aqueous solutions of H_2SO_4 or H_2SeO_4 with a glasslike behavior at very low temperatures, studied by G. Vuilliard by differential-thermal analysis.

269. The *dilatometric* method first introduced by C. G. Peters and C. H. Cragoe (1920) in investigation of the thermal history effects of glass, is in its turn actually highly accomplished, and therefore especially indispensable as a tool for the study of *transformation phenomena.* A. I. Stozharov[434] studied from the viewpoint of A. A. Lebedev's original crystallite theory the *low-temperature effects* between 200° and 250°C. which were so often interpreted as indications for the presence of cristobalite in the glass constitution by its classical polymorphic inversion. Stozharov especially used previous interferometric measurement data of these effects in well-defined optical glasses by A. M. Selezneva, not only demonstrating the common expansion curve, but also the derived temperature function of the expansion coefficient as seen in Fig. A. 184, with a sensitivity of 5×10^{-7}. If there would be a true crystal-

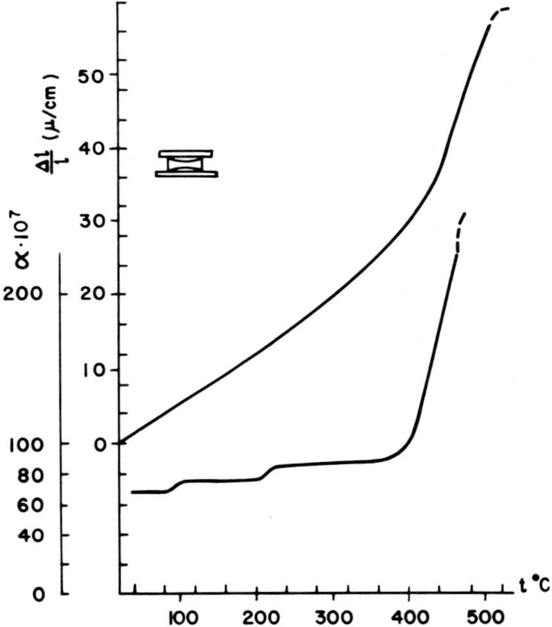

Fig. A. 184. Expansion curves of glass F - 6 (upper curve), and thermal expansion coefficient as a function of temperature. (Stozharov).

[434] *Stroenie Stekla, Inst. Khim. Silikatov Akad. Nauk S.S.S.R., Trudy Soveshchaniya, Leningrad,* 1953 (Pub. 1955), 120-125.

lization of cristobalite and tridymite as the dominant constitution factor for building up those thermal effects, in the measure of their increasing amounts, the intensity would by systematically increased to sharp peaks. The expansion curves of a really recrystallizing glass containing those crystal phases and those of glasses with typical splitting points, however, are entirely different. The same conclusions must be made from measurements of the refractive indices of glasses as a function of the thermal history, as previously made by N. A. Tudorovskaya.[435] Her interpretation of low-temperature effects as indications for the actual presence of crystallites of cristobalite and tridymite are incorrect and refuted by Stozharov. On the other hand, those effects have an important relation to practically observed phenomena of the zero-point depression of thermometer glass (cf. A. ¶ 55).

270. The introduction of the "*fictive temperature*" by A. Q. Tool[436] is determined to characterize the physicochemical condition or the state of a given glass, when both the actual temperature and that temperature at which the glass would be in molecular equilibrium if heated or cooled very rapidly, are known. The fictive temperature, τ, as this equilibrium temperature can be reached through heating or cooling of the undercooled or superheated glass, respectively. The difference between actual temperature, T, and fictive temperature, τ, is a measure of the departure of the glass in its given state from equilibrium. For the approach to equilibrium Tool gives differential equations of the type $d\tau/dt = K_T(T - \tau)$, and $K_T = K \cdot e^{T/k}$, where k is Twyman's constant (1917), and $K = K_T$ for $T = 0$. Both equations combined give

$$d\tau/dt = K(T - \tau) \cdot e^{T/k}$$

The glass contracts and evolves heat when equilibrium is reached from the undercooled condition. It expands and shows an endothermic heat effect when equilibrium is approached from the superheated state.

271. If the measurements of expansion are made slowly enough, A. M. Kruithof[437] showed how the fictive temperature will decrease at the beginning of the transformation phenomenon. The dilatometric curve is flattened in the region where the actual temperatures are above τ. This effect is more distinct the better the initial glass structure corresponds with the higher τ. If the changes are rapidly made, τ is retarded relative to T, and the dilatometric curve is now on the right side of the line for $T = \tau$ (Fig. A. 185). The point P marks the change of modifications in the glass structure; below P is T below τ, above P it is above τ. The effect is more distinct in

[435] "Physical and Chemical Properties of the Ternary System Na_2O–PbO–SiO_2," Edit. I. V. Grebenshchikov, Acad. Sci. U.S.S.R., Moscow-Leningrad, 1949, 186-200, especially on the partial system Na_2O–SiO_2, and pp. 201-210 for ternary sodium lead silicate glasses.

[436] *J. Am. Ceram. Soc.*, **29**, 1946, 240-253.

[437] *Verres et réfractaires*, **9**, 1955, 311-319.

the moment in which $T = \tau$. For a glass with 9.5 per cent Na_2O, 4 K_2O, 1 Al_2O_3, 10.5 PbO, and 75 SiO_2, the stabilization is demonstrated by Kruithof in the corresponding changes of the refractive index, the electrical resistance, and the differential-thermal effects. A plate of 5 mm. thickness shows changes in τ from 450° to 370°C.,

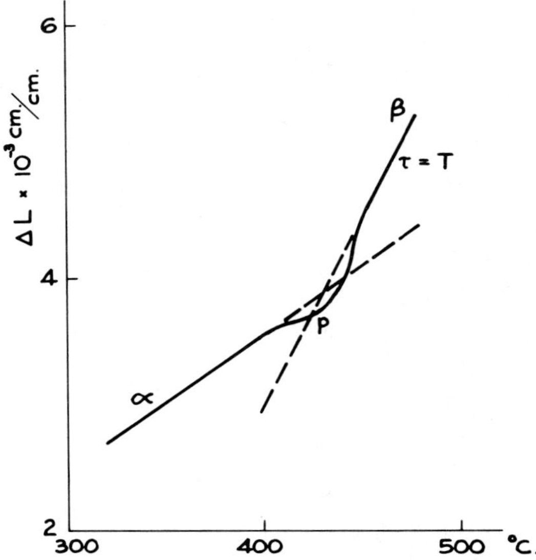

FIG. A. 185. Influence of structural modification in the expansion curve on heating. (Kruithof).

after three months of thermal stabilization. The interrelation between specific volume, V, refractive index, density, ϱ, and electric resistance is also discussed by E. U. Condon,[438] by the equations

$$V(T, \tau) = V_0 [1 + 3\alpha' (\tau - \tau_0) + 3\alpha T]$$

and

$$\varrho(T, \tau) = \varrho_0 [1 - 3\alpha' (\tau - \tau_0) - 3\alpha T]$$

in which V_0 is the specific volume for $T = 0$, and τ_0 the lowest practicable fictive temperature; α' is the fractional change in length per unit change in fictive, about $= 3 \times \alpha$. Condon demonstrates how well the modern aspects of chemical rate processes are compatible with the fictive temperature concept in relation to the kinetics of the approach of the glass nonequilibria, as also in the high-pressure compaction of glass studied by P. W. Bridgman and I. Šimon.[439]

[438] *Am. J. Phys.*, **22**, 1954, 132-142.
[439] *J. Appl. Phys.*, **24**, 1953, 405-413.

272. An important application of such ideas is also seen in the studies of H. E. Nagy and H. N. Ritland[440] on the expansion characteristics of two glasses, one of them a mixed alkali type, both for cooling and heating at constant rates, with widely varying thermal histories. Both methods, of constant rate cooling, and of constant temperature stabilization before the expansion thermal tests, were also used. The contraction curve on cooling at a constant rate is subdivided in three regions: a high-temperature range with an essentially linear course, a low-temperature range which is also approximately linear (with a small quadratic term), and between these the transformation range in which the structural expansivity decreases exponentially, the slope of the curve decreasing accordingly. This part of the whole contraction curve shifts upward along the temperature axis by an amount proportional to the logarithm of the cooling rate. For the mixed alkali glass an ionic process which contributes in a certain way to the total expansion is characteristic. The effect freezes out during cooling near 100° to 150°C. and causes a distinct deviation from the quadratic curve below these temperatures.

273. Concerning the thermal history variables in their influence on the expansion behavior on heating Nagy and Ritland come to the opinion that the fictive temperature concept is *not* adequate for an explanation of such effects. It is limited according to H. N. Ritland[441] to qualitative descriptions of effects of thermal history in the transformation range, but as a one-dimensional single-parameter principle its quantitative significance is not sufficient to give insight into structural differences between the behavior of "rate" and "soak" samples (as shown in Fig. A. 186), and the complex description of such phenomena in toto. For a soak sample the fictive temperatures are equal, whereas for a rate sample they are spread over quite a temperature range.

274. In relation to the apparently *negative* expansion coefficient shown in the classical curves given by W. E. S. Turner and F. Winks (1930) for quenched glass samples, M. Prod'homme[442] gives the plausible explanation of a density effect of the nonstabilized modification. This effect was also observed by Ritland for a borosilicate glass as a contraction with an increase in density by about + 0.01 (cf. A. ¶ 286). For the low-temperature *brittleness* of glass an increasing understanding of this important property is actually possible from the viewpoints of modern solid-state physics. We mention here only the theories of hardness and atomic structure, and the theory of fracture in metals as "ductile crystalline materials" developed by E. Oro-

[440] *J. Am. Ceram. Soc.*, **40**, 1957, 436-442.

[441] *J. Am. Ceram. Soc.*, **39**, 1956, 403-406. M. Hara and Sh. Suetoshi, *Asahi Garasu Kenkyu Hokoku*, **5**, 1955, 125-135, come to corresponding conclusions on suitability of the fictive temperature, and its limited value in describing mechanisms of transformation and stabilization.

[442] Thèse, Université de Paris 1960, 41 pp., especially pp. 37 f.; *Verres et réfractaires*, **14**, 1960, 261-273, especially p. 272.

van.[443] It is particularly remarkable which analogies exist between the softening range of glasses and the dislocation softening range for metals caused by a considerable dependence of mechanical yield stress on temperature and velocity. In the temperature range required for softening glasses, the thermal energy becomes sufficient to overcome the interatomic forces opposing the movement of dislocations, and the low-temperature hardening yields to insignificant hardness degrees relative to that acquired by plastic deformation.

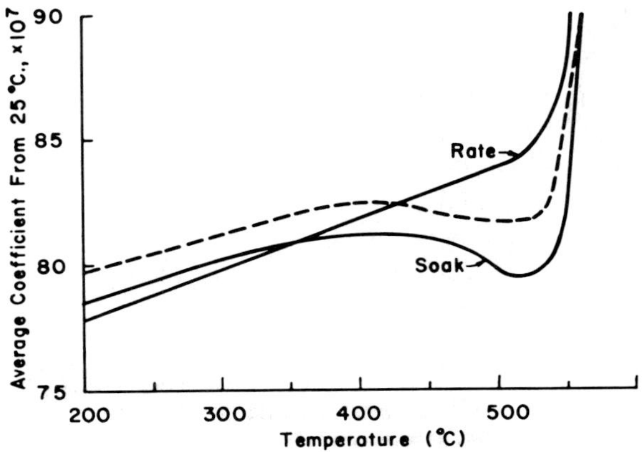

FIG. A. 186. Average expansion coefficient as a function of temperature, for a sample with fictive temperature = 539°C. For the dashed curve the rate sample was given a 30 minutes heat treatment at 543°C. before the expansion test. (Nagy and Ritland).

275. The wide field of theoretical background for *technological annealing* measures can here be only briefly characterized, especially by mentioning some of the methods used for optical and container glass manufacturing. F. W. Preston[444] discussed the fascinating mathematical analogies existing between the atomistic process of annealing and the law of extinction in biogenetics, comparing this law with the random loss of strain (as an extinction of "polarons"), to have a full understanding of the Adams-Williamson equations for the relaxation in glass (cf. A. ¶ 45 ff.). Particularly interesting schedules of annealing based on structural changes during glass annealing were developed by B. Daragan.[445] Complete stabilization is achieved in a definitely

[443] "Fatigue and Fracture of Metals," Symposium Massachusetts Institute of Technology, June 1950, Technology Press and John Wiley & Sons, Inc., New York 1952, 313 pp., especially pp. 139-167; *Science*, **118**, 1953, 573-574; see also E. Schmid, *Radex Rundschau*, 1953, 202-204, with a higly interesting discussion.

[444] *J. Am. Ceram. Soc.*, **36**, 1953, 232-238; cf. L. G. Gehring and F. W. Preston, *Ibid.*, **33**, 1950, 321-322.

[445] *Atti congr. intern. vetro, 3rd Congr. Venice* 1953, 411-425; *Verres et réfractaires*, **7**, 1953, 157-166; see also *Ibid.*, **6**, 1952, 10-21, and a translation by F. W. Preston, *Glass Ind.*, **33**, 1952,

shorter time if one works over small ranges of temperature than when the stabilization is made at one constant temperature in the lower part of the transformation range, or by a stepwise (5° per step) processing as recommended by A. Winter-Klein (1938). Very dissimilar glasses varying even in thickness from 20 mm to 20 cm gave the same final result. The structural heterogeneities could be reduced to such a high degree that differences in the refractive index did not exceed $\pm$ 0.00001. The residual birefringence by a "classical" annealing process is decidedly higher if, in the time period, the glass undergoes a partial heat treatment with a minimum final birefringence. The experimental temperature range used by Daragan was from + 100°C. above, and — 50°C. below, the conventional (dilatometric) transformation point (and the point T_i) followed by a rapid cooling, say in 15 minutes to room temperature. The complete curve proposed for practical use is shorter than that of the annealing process according to Winter-Klein and makes possible predicting the types of thermal treatment to be applied for the glass to reach a desired degree of stabilization expressed in per cent of maximum annealing, $= 100\ \delta_1/\delta_2$, as the difference, δ_1, of the actual density of the glass and its density after quenching, and δ_2 the difference of density of quenched and of completely annealed glass.

276. Fr. Naudin[446] discussed in a similarly extensive way the relation between the inner tensions in glass and the transformation range for the refractive index of different optical glasses, and the birefringence of glasses with different thermal histories, with the aim of determining the optimum for industrial annealing schedules. The samples used had the shape of truncated prisms; the amplitude of the variation of

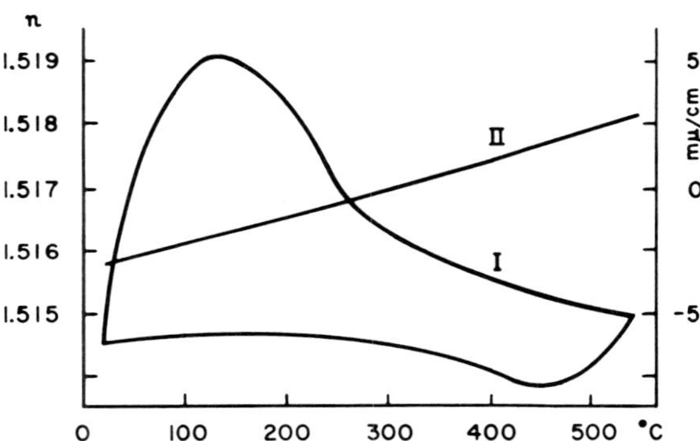

FIG. A. 187. Changes of the refractive index for $\lambda = 546.1$ mμ (curve II), and of birefringence (curve I), for a borosilicate glass. (Naudin).

69-74, 98-99, of a previous article, *Verres et réfractaires*, **5**, 1951, 135-143, for plate glass using a sink-float test for the determination of the density (cf. Par. A. 281 f.).

[446] *Verres et réfractaires*, **6**, 1952, 144-153, 209-218, 282-289; translations in *Glass Ind.*, **35**, 1954, 542-548, 603-608, 628, 666-671, 688.

the birefringence depends on the shape of the sample. If the temperatures of exposure are in the transformation range, the curve for the birefringence as a function of temperature shows a maximum (in some exceptional cases a minimum) (Figs. A. 187 and 188) the distinctness of which depends on the rate of heating. The time factor of the changes in refractive index and birefringence was studied for a borosilicate and a heavy Flint glass for constant exposure temperatures in the upper limit of the elastic range or in the transformation range. In the first case the difference of the refractive index and of birefringence before and after heating is very low only if the glass was initially well annealed. In quenched or insufficiently annealed glasses a rather marked increase in refractive index and a distinct decrease in birefringence was measured. In the second case the refractive index of equilibrium, which characterizes the molecular state of the glass at a given temperature in the transformation range, is well-defined. After cooling the refractive index and the birefringence resulting from inner tensions depend chiefly on the transformation temperature to which the glass was exposed.

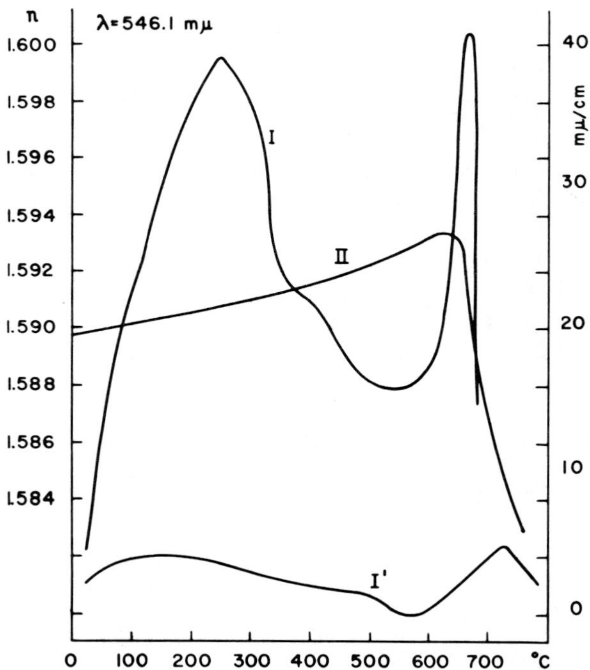

Fig. A. 188. Influence of the shape of specimens of heavy barium crown glass on the temperature function of birefringence (*I*) in the shape of a parallelepipedon, (*I'*) in the shape of a prism. Curve (*II*) gives the change of the refractive index for $\lambda = 546.1$ mμ, as a function of temperature. (Naudin).

277. Aniuta Winter-Klein[447] gave an excellent review of our actual knowledge of the transformation phenomenon as a true structure conversion of thermodynamic character, not only as an apparent effect of viscosity and relaxation parameters. The influence of the time factor in these complex conversions alone is not sufficient to explain the transformation reactions in glass-forming systems, and it is not easy to separate its effects from the thermodynamic basis of the phenomenon. One may compare ancient glasses (2,000 to 2,500 years old) with modern glasses of the same composition. Exactly the same curves are plotted as a function of temperature for the refractive index with an order of relaxation time at room temperature of about 10^2 years. Similar determinations of thermal expansion coefficients confirm the fact that for the refractive index and the expansion coefficient the time factor has no influence. One may further apply the classical methods of R. H. Lillie (1933) to the transformation range and to determine the final equilibrium state of the index (cf. Fig. A. 189). A time factor in this case is indicated by the controversial course

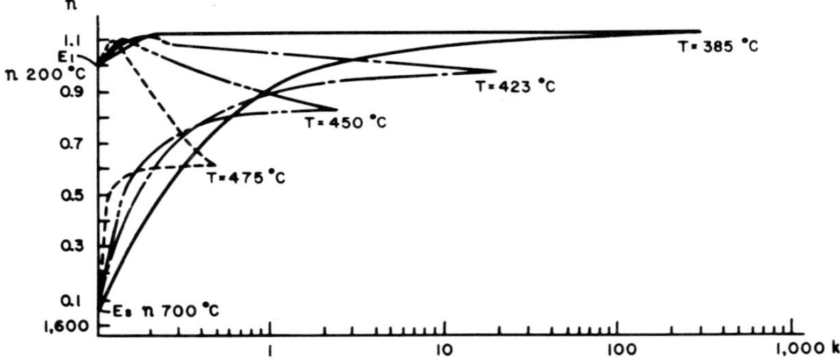

Fig. A. 189. Changes in the refractive index of a heavy flint glass, starting from the initial temperatures $E_1 = 200°C.$, and $E_2 = 700°C.$, and different experimental annealing temperatures, as a function of time. (Winter-Klein).

of these curves. Besides the difficulties in analyzing exactly the enthalpy effects of the transformation reactions, the ratio of the thermal conductivity/specific heat which is normally a constant[448] shows for glucose glass in the transformation range a distinct anomaly, conclusively a change in the structure. The activation energy shows in the transformation range a distinct maximum of 130,000 cal./mole (normal 10,000 to 30,000 cal./mole) as a typical order $\rightleftharpoons$ disorder phenomenon in the definition of Ya. I. Frenkel (cf. A. ¶ 198). This latter phenomenon may be compared with changes in the magnetic resonance observed in the transformation range of

[447] *Verres et réfractaires*, **11**, 1957, 71-88.
[448] Cf. Ch. Kittel, *Phys. Rev.*, **75**, 1949, 972-974.

polystyrene or poly-methyl acrylate by L. V. Holroyd, R. S. Codrington, R. A. Mrowca, and E. Guth[449] (cf. Fig. A. 190).

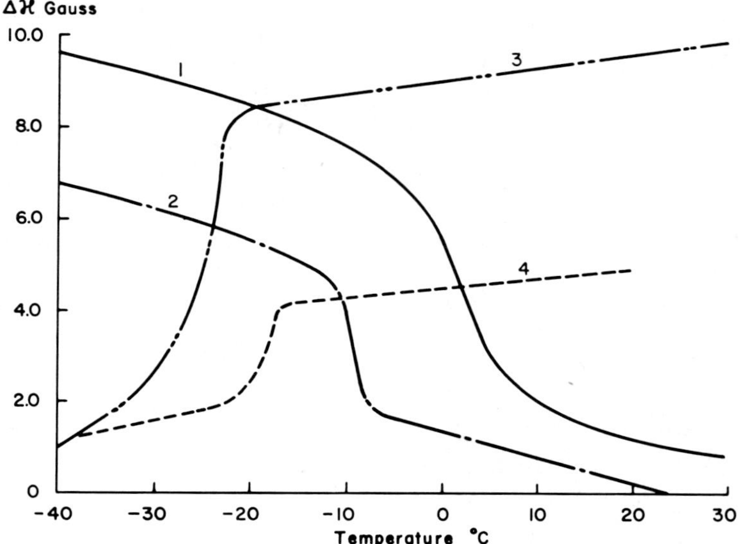

Fig. A. 190. Magnetic resonance curves, as a function of temperature, for poly-methyl acrylate [upper full curve (*1*)], and polystyrene [lower curve (*2*)], in comparison with the specific heat [upper curve (*3*)], and thermal expansion [lower curve (*4*)]. (Holroyd, Codrington, Mrowca, and Guth).

278. In a similar way L. Prod'homme[450] extensively analyzed the conditions of glass stabilization as indicated in changes of the refractive index and of the specific refractivity, in combination with the volume and density changes (cf. B. ¶ 116, ¶ 120), by deviations from the Lorenz-Lorentz law. Particularly instructive are the curves for the functional relations between the changes, Δn, of the refractive index relative to that at 20°C. (in units of the fifth decimal) with temperature for different optical glasses and a few fundamental oxides. These changes can have a positive or a negative sign (see Fig. A. 191). For the specific refractivity

$$R = \frac{1}{\varrho}\left(\frac{n^2 - 1}{n^2 + 2}\right)$$

[449] *J. Appl. Phys.*, **22**, 1951, 696-705.

[450] Thèse, Université de Paris 1957, 74 pp.; *Verres et réfractaires*, **10**, 1956, 267-276; **11**, 1957, 351-360; **12**, 1958, 3-22; *Compt. rend.*, **244**, 1957, 584-586; *J. phys. radium*, **18**, 1957, 7S-10S. More recently: *Phys. Chem. Glasses*, **1**, 1960, 119-122. See also extensive studies by F. C. Eversteijn, J. M. Stevels, and H. I. Waterman, *Ibid.*, 123-133, on density and optical properties of B_2O_3 and sodium borate glasses, as a function of composition, mode of preparation, and rate of cooling.

as a function of temperature is characteristic the function

$$\frac{\Delta R}{R} \cdot \frac{1}{\Delta T} = \varphi$$

in $10^{-6}/^{\circ}C.$; the higher it is the softer the glass composition, especially for B_2O_3 and GeO_2 glasses. A surprising analogy exists between the temperature function of

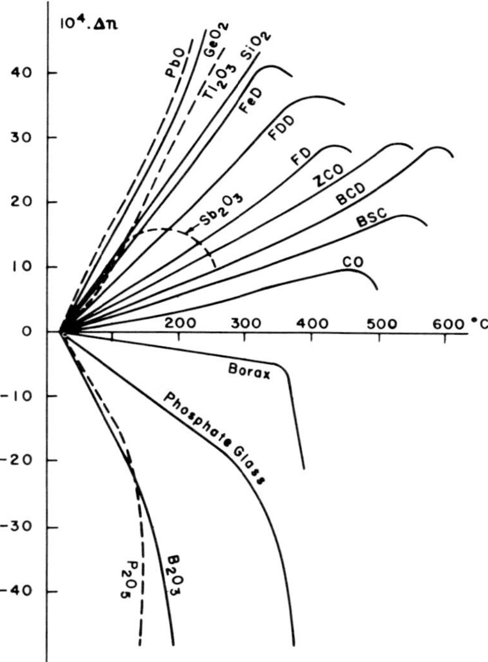

Fig. A. 191. Changes of the refractive indices (for $\lambda = 546.1$ mμ), and the difference, Δn, relative to that at 20°C., as a function of temperature. (L. Prod'homme).

φ and the activation energy, E, of viscous flow as seen from Fig. A. 192, with steep maxima for glycerin, B_2O_3, and a borosilicate glass. In the relation $dn/dT = (n^2 - 1) \cdot (n^2 + 2) \cdot (\varphi - \beta)/6n$ the coefficient β controls the atomic distances and the aggregate dissociation in the glass, φ the increasing deformation of the electron clouds by electromagnetic waves, with increasing loosening of the structure. The refractive index as a function of temperature is a composite parameter of both actions. Analogous considerations are valid for the absorption phenomena and the Rayleigh-Raman dispersion. Three characteristic types are distinguished for the variation of refractive constants with temperature as seen in Fig. A. 193.

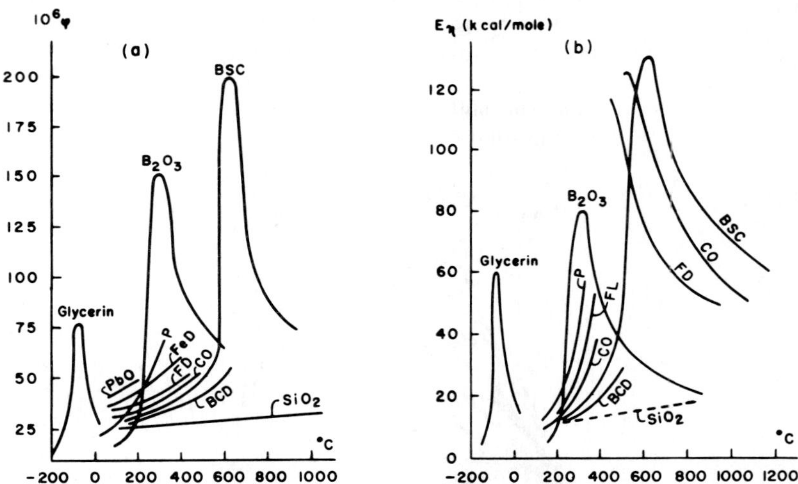

FIG. A. 192. (*a*) Variation of the temperature coefficient, φ, of the refractivity, (*b*) of the activation energy of viscous flow, for different glasses, with temperature. (L. Prod'homme).

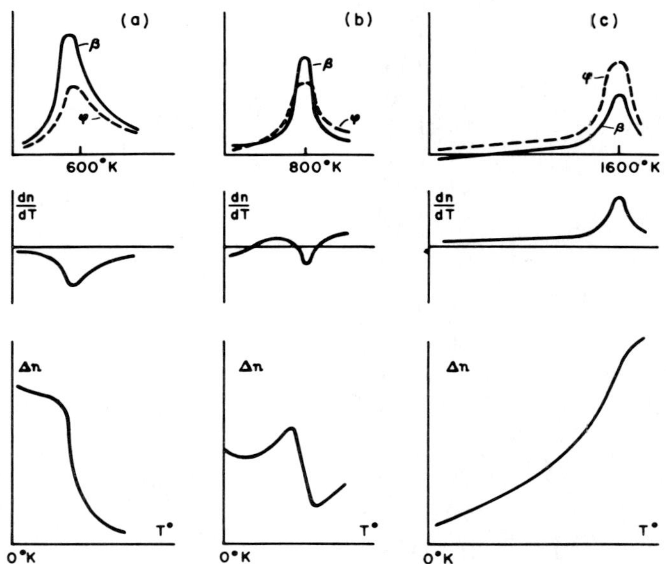

FIG. A. 193. Basic types of the variation in the refractive index with temperature, (*a*) type of B_2O_3 glass, (*b*) type of crown glass, (*c*) type of silica glass. (L. Prod'homme).

279. For the theory of annealing, the observations of L. Prod'homme are particularly important due to the clear interrelation they show of optical and viscosity properties. If one compares the real stabilization times with those relaxation periods which are derived from viscosity values at different temperatures according to the Debye-Einstein relations, one observes that the results of the first group are always lower than those of the second, the more the higher the temperature concerned. There is a "*secondary viscosity*" effect which influences the probability of the dislocation of a given atom from an unstable position to another position which has a higher degree of stability. This "secondary viscosity" superimposed on the primary (common or apparent) viscosity is able to explain the stabilization curves given by Daragan, and in its mathematically formulated expression it is also in agreement with previous postulates of E. U. Condon.[451].

280. The structural rearrangements during annealing in silicate glasses were studied as a function of the chemical composition by L. W. Tilton[452]; the "optical sensitivities," Δn per ^{0}C. = $\Delta n/\Delta ^{0}$C., to prolonged annealing are poroptional to the molecular fraction of the nonsilica components, independent of the size of their particles and their chemical nature. As Tilton interprets the annealing process there is a continuous gradual and sluggish collapse or "folding" by orientation of the Si—O bonds (on the angles Si—O—Si deviating from 180^{0} cf. A. ¶ 234). The rate of this collapse would be chiefly controlled by thermal vibration of the non-framework particles in interstices of the tenacious tetrahedral framework. Potentially all these structural rearrangements are homogeneous reactions. It is not to be expected that annealing at low temperatures and over longer periods would produce a better homogeneous glass than if the process is made at higher temperatures and in shorter times.

281. Practical considerations of the refractive index and density changes in different optical glasses during annealing, and especially in connection with the "fine-annealing" of optical glass, were described by H. N. Ritland,[453] with empirical relations to represent satisfactorily the approach to equilibrium values at different temperatures. If the initial temperature is sufficiently high the density-temperature relations for variable rates of cooling are identical. They are only displaced along the equilibrium density-temperature lines by an amount proportional to the logarithm of the cooling rate. The final density reached after cooling through the whole transformation range depends linearly on the logarithm of the rate. The postulated uniformity in the fictive temperature all over the completely stabilized glass, which is the practical aim to optical fine-annealing, is best reached by soaking for a short time at a temperature near the annealing point, and cooling at a constant rate below the annealing range, followed by an accelerated cooling down to room temperature.

[451] *Glass Ind.*, **33**, 1952, 307, 322-323.

[452] *J. Soc. Glass Technol.*, **40**, 1956, 338-352. On "optical sensitivity" see A. Q. Tool and L. W. Tilton, *J. Research Natl. Bur. Standards*, **38**, 1947, 513-526.

[453] *J. Am. Ceram. Soc.*, **37**, 1954, 370-378, 466-473 (with H. R. Lillie); *Ibid.*, **38**, 1955, 86-88.

A complete annealing schedule in terms of the final stress and refractive index conditions is thus possible. The total time required is much more feasible for such methods with constant rate annealing, the uniformity of the refractive index better guaranteed, the stress tolerably low. The interrelation of refractive index and density was studied by Ritland in a borosilicate Crown glass. It is smaller than the Lorenz-Lorentz law leads one to expect, which is apparently not strictly valid for solids like glass. Induced polarization between adjacent ions makes only a very low contriburion to the local electrostatic fields due to a reduction of ionic polarizability with increasing density. Concerning relations of the molecular volume, the refractive index and the molecular refractivity we may also recall in this connection the interesting observations by E. Kordes and H. Becker[454] on cadmium and sodium-lithium phosphate glasses with their apparent anomalies and deviations from elementary linear additivity functions of composition, and on the anomalies in borosilicate glasses discussed by T. Abe (cf. B. ¶ 138).[455]

282. Very interesting deductions on the influence of high *pressure* on glass properties in the transformation range were made by R. Maurer,[456] especially on the rates of relaxation. To describe the behavior of glass cooled with a constant rate over the transformation rate an empirical equation was developed from Tool's (1946) formula, of the form for the fictive temperature, τ_p, under pressure:

$$\frac{d\tau_p}{dT} = \frac{1}{R} \cdot e^{K_1 T} \cdot e^{K_2 \tau_p} f(\tau_p - T)$$

It is possible to calculate the increase in the refractive index due to pressure measured after cooling at a constant rate $(\Delta n)_R$, in comparison with the increase measured after quenching from a given temperature in the transformation range, $(\Delta n)_S$. The derived volume changes $(\Delta V)_R$, and especially $(\Delta V)_S$ can be used to determine approximately the *liquid compressibility* of glass for temperatures in the lower part of the transformation range as previously shown in the schematic diagram Fig. A. 55, (A. ¶ 61), in the order of magnitude 40×10^{-7} cm.^{2}/kg. for the frozen changes, derived from sink-float density measurements. The compressibility in the liquid state is made up of a "solidlike" and a "liquidlike" portion, of which the first is instantaneously restored when the pressure is released, whereas the liquidlike compression share is slowly restored by relaxation effects. This latter share is frozen-in, with an order of magnitude of 68×10^{-7} cm.^{2}/kg. for a given glass.

283. The many repeated attempts to bring the optical properties of glasses in harmony with their structural theories, be it in the form of the framework or of the

[454] *Z. anorg. u. allgem. Chem.*, **260**, 1949, 185-207.
[455] *J. Am. Ceram. Soc.*, **35**, 1952, 284-299.
[456] *J. Am. Ceram. Soc.*, **40**, 1957, 211-214.

crystallite hypothesis is most evident in the contributions made by A. A. Appen,[457] based on the coordination principle, and by E. U. Condon,[458] on the polarization conditions of framework-modifying ions. In the latter respect the fundamental concepts of K. Fajans and N. Kreidl (1948), and of W. A. Weyl (1951) concerning the full understanding of the behavior of strongly polarizable cations like Ba^{2+} or Pb^{2+} on one hand, and of strongly polarizing cations like Li^{+} on the other hand, may be particularly recalled.

284. Among the thermodynamic parameters playing a fundamental role in the mechanism of the transformation phenomenon, the *specific heat* as a function of temperature is particularly well studied. We dispose of a series of excellent modern data collected from mixing-calorimetric measurements by H. E. Schwiete and G. Ziegler,[459] from systematically varied experimental and industrial glasses. These results are well compared to those data calculated from the additivity rule (cf. B. ¶ 114 ff., ¶ 126) in the equation given by D. E. Sharp and L. B. Ginther (1951). We further mention measurements by H. Hartmann,[460] up to 1300°C. M. Tashiro[461] modified the common mixing calorimeter by a construction first developed by W. Oelsen *et al.* which has the great advantage of rapid working-operation. An excellent discussion of the literature on the specific heat of glasses, especially in the transformation range, we owe to M. Prod'homme.[462]

285. Concerning the establishment of equilibria in glass structures at lower temperatures, especially in the range up to 100°C. the practical needs of controlling the so-called *ice-point depression* in thermometry were investigated early by H. Schönborn (1929, etc.) (cf. Fig. A. 194). More recently, the same author[463] discussed this phenomenon as a function of temperature and time in corresponding volume changes. The secular increase, Δ, in the ice point is a function of time according to the law

[457] *Stroenie Stekla, Inst. Khim. Silikatov Akad. Nauk S.S.S.R., Trudy Soveshchaniya, Leningrad,* 1953 (Pub. 1955), 96-106, 306-310.

[458] *Am. J. Phys.*, **22**, 1954, 43-53.

[459] *Glastech. Ber.*, **28**, 1955, 137-146. We further recommend the excellent thermochemical measurements and thermodynamic data (ΔH, ΔF, ΔS) determined for alkali borates, by G. Sl. Smith and G. E. Rindone, *J. Am. Ceram. Soc.*, **44**, 1961, 72-78, including high-temperature data on heat and entropy of fusion and crystallization.

[460] *Glastech. Ber.*, **26**, 1953, 29-33; **27**, 1954, 12-15 (with H. Brand); *Ibid.*, **30**, 1957, 186-188 (with K.-H. Kiessling). J. Moore and D. E. Sharp, *J. Am. Ceram. Soc.*, **41**, 1958, 461-463, recalculated some of the increments to a better fit with the measurements of the authors before mentioned.

[461] *Glass Ind.*, **37**, 1956, 549-552. The calorimeter used was determined originally for continuous measurements of the specific heat of metals; cf. W. Oelsen, K. H. Rieskamp, and O. Oelsen, *Arch. Eisenhüttenw.*, **26**, 1955, 253-266.

[462] *Verres et réfractaires*, **11**, 1957, 287-297.

[463] *Silikat Tech.*, **6**, 1955, 367-371.

$\sqrt{\Delta} = A \cdot \log t + B$, and the ice point depression $\Delta^2 = A' \cdot \log t + B'$. For the classical Jena thermometer glass 16 III, and a series of other glasses, expansion curves in the well-annealed and in the quenched states were accurately analyzed. For the (rather unsatisfactory) glass VR 63 the Δ^2 law shows changes in the ice point which are by 60 times larger than those for another glass VR 70. The even considerably worse glass 17 III has a 3 times larger Δ^2 than VR 63. It was recognized al-

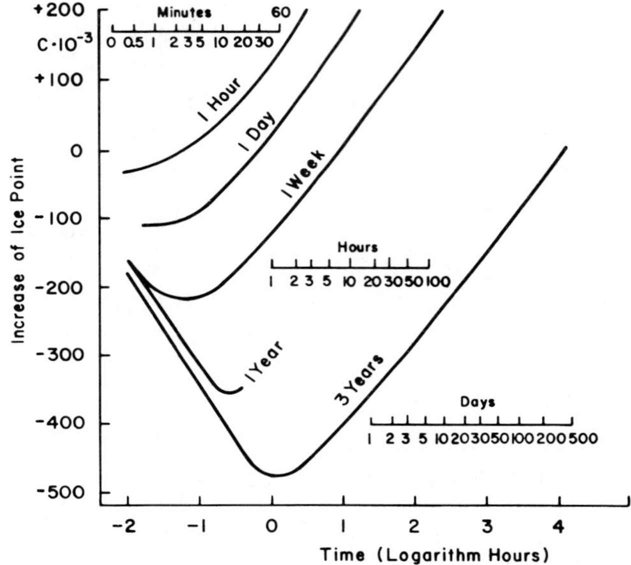

FIG. A. 194. Ice-point changes of thermometers made of the unsatisfactory glass VR 63, at 100°C., after aging over different time periods. (Schönborn).

ready by O. Schott and H. F. Wiebe (1884) that mixed alkali glasses have particularly large thermal-volumetric after-effects. Anomalies in such glasses of inferior quality can be mathematically analyzed by the discussion of elastic after-effects in the glass as H. Roetger[464] demonstrated. After a previous "aging" such as a thermal treatment at 100°C. (or simply at room temperature), the rate of the ice-point depression ($\sqrt{\Delta} = f(t)$) is analogous to that of the directly measured change in length of unannealed glass rods at temperatures below the transformation range. In relation to the relaxation behavior of glass and its response to changes in constitution at temperatures below the transformation range the opinions of H. Roetger discussed in A. ¶ 55 are of special importance. They are, therefore, emphasized again[465] as well as the previous observations of L. G. Liberatore and H. J. Whitcomb (1952).

[464] *Silikat Tech.*, **6**, 1955, 326-331.

286. Although oriented in a different field of technological practice, the calculations of T. Kishii[466] are mentioned here as a thorough development of a theory for volume and density changes on thermal treatment of lead glasses used for electronic tube manufacturing. Samples of annular shape were air-quenched from measured temperatures (in the range from 400° down to 100°C.) and their volume changes determined by interferometric dilatometry and liquid suspension methods. Introducing Tool's "fictive temperature," the contractions measured were plotted as a function of time in the difference $\tau - \tau_1$, versus $d\tau/dt(T - \tau)$. Kishii did not find an agreement with the theory of Tool or of Ritland; the samples showed a more rapid contraction in the first period, a slower contraction in end states than calculated from those theories. Curves for glass samples with variable thermal treatments after air-quenching from 475°C. are also not compatible with the theories of Tool and Ritland based on fictive temperature. A new theory developed by Kishii starts from the probability of transition from an energy level (i) to another level (j) in an equation of the type $A_{ij} = a_{ij} \exp(-E_{ij}/kT)$, with E_{ij} the activation energy of the transition, k the Boltzmann constant, and a_{ij} a specific constant. For pairwise levels ij with similar energy potentials and high probabilities, A_{ij} is developed as a linear function of $1/T$. The derived curves of ΔV as a function of time are in good agreement with the experimental results. The theory beyond that yields a satisfactory illustration of Na^+ migration in borosilicate glasses, rotation of chain-shaped structural units, or shiftings in coordination in mixed alkali glasses. The probability of the transitions between neighboring levels, with a low activation energy, can become so high that a rapid contraction of quenched glass is the consequence as observed in the ice point displacement as a thermal after-effect in thermometer glass (see above).

287. *Silica glass* in its particular mechanical-physical properties was investigated theoretically and experimentally by O. L. Anderson and H. E. Bömmel[467] by the method of absorption of ultrasonic waves ($\nu = 6 \times 10^4$ to 2×10^7 cycles/sec.). The large peak observed at 30° to 50°K. was already discussed in A. ¶ 64 ff. in connection with problems of the anelastic response of glass as a function of the chemical composition. The peak at low temperatures is the more remarkable in silica glass since a corresponding effect in quartz is absent. Anderson and Bömmel are of the opinion that lateral motion of the oxygen atoms perpendicular to the bond Si—O in the $[SiO_4]$ group is primarily responsible for the inner friction effect in silica glass (Fig. A. 195). The "structural relaxation" ensuing is a consequence of the temperature and experimentally observed frequency response behavior, especially associated with the distribution of activation energies of relaxation. This activation energy is

[465] *Silikat Tech.*, **10**, 1959, 57-62.

[466] *Glastech. Ber.*, **33**, 1960, 47-52.

[467] *J. Am. Ceram. Soc.*, **38**, 1955, 125-131. On the temperature peak of inner friction in silica glass see also M. E. Fine, H. Van Duyne, and N. T. Kenney, *J. Appl. Phys.*, **25**, 1954, 402-405.

very low, only 1030 cal./mole. An atomic diffusion is, therefore, excluded, as well as a molecular rotation, and only a small change in the bond angle is the most probable mechanism, which also explains why a corresponding effect in the quartz

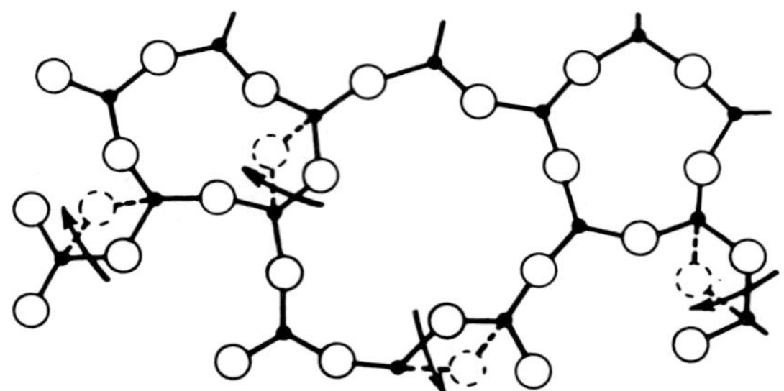

FIG. A. 195. Random network scheme illustrating the possibility of the motion of oxygen atoms in alternate positions of equal energy, in silica glass. (Anderson and Bömmel).

structure is absent. The order of magnitude of the activation energy corresponds to a relaxation time equal to that of ultrasound waves, as seen from the discussion of the equation

$$f_0 = \frac{kT}{h} \cdot e^{-Q/RT}$$

($Q = 1030$ cal./mole). The shear modulus of silica glass from 4°K. to the minimum value is only slightly changed (by about 3 per cent). This indicates that only a small fraction of silica molecules are capable of contributing to the losses.

288. The low thermal expansion of silica glass was equally discussed by H. T. Smyth, H. S. Skogen, and W. B. Harsell[468] with the model concept that internal friction is due to transverse vibrations of the oxygen atoms which cannot contribute much to the expansion effect by increased thermal energy, on the contrary, by pulling the atoms in the network closer together. Similarly, heat capacity of silica glass in comparison with that of quartz could be calculated by Smyth *et al.* by a simple vibration model, based on a system of additive Einstein functions, with a good agreement with experimental data up to a temperature of 1300°K. above which temperature the devitrification into cristobalite interferes and brings about an inflection point and increased heat capacities. The unusual behavior of silica glass was also extensively studied by Cl. L. Babcock, St. W. Barber, and K. Fajans[469]

[468] *J. Am. Ceram. Soc.*, **36**, 1953, 327-328. On thermal expansion see particularly *Ibid.*, **38**, 1955, 140-141; on delayed elastic effects *Glastech. Ber.*, **32**, K, 1959, III/28-III/34. A detailed study of the thermal expansion of silica glass as a function of its thermal history was recently presented by R. Brückner, *Naturwissenschaften*, **49**, 1962, 150-152.

[469] *Ind. Eng. Chem.*, **46**, 1954, 161-166.

from the viewpoint that there would be several different atomic arrangements coexisting in a homogeneous equilibrium, and in quantitative ratios variable as a function of temperature and pressure (cf. A. ¶ 230). This concept is compatible with the phenomenon of anomalous expansion coefficients, too, if one or the other of the "dominant" structural components would have a larger volume than the other arrangements, consistent with other anomalies of compressibility and elastic properties.

289. A special investigation of *chemical durability* (water leachability) (cf. B. ¶ 165 ff.) of simple sodium-calcium silicate glasses as a function of the thermal treatment was made by R. Guy[470] by exposing the glass samples first to different temperatures, then by chilling them in water, or after annealing for two hours at 525°C. (for a glass with 16 per cent Na_2O, 11 per cent CaO, and 73 per cent SiO_2). The glasses were more intensely leached by water (tested after the conventional grit standard method of the German Society of Glass Technology, DGG) the lower the fusion temperature, the shorter the heat exposure at the fusion temperature, or the more rapid the cooling from the fusion temperature. Optimum durability is expected for the glass exposed for a long time at not too high a fusion temperature and with slow cooling. The results of these tests can be only averages over different fractions of variable densities in the sample. If a denser fraction is selected for the leachability test the results are lower than those for a lighter fraction. Density differences between chilled and annealed glasses were observed for the glass of the composition given above between 2.491 (extremely chilled), and 2.4936 (fine-annealed). For a basic study of durability of glass in water, as a function of temperature and of thermal history, the distinction of different density fractions must not be neglected.

290. New aspects for understanding thermal effects on the chemical properties of sodium borosilicate glasses were brought about by O. S. Molchanova and M. V. Serebryakova,[471] in studying the rate of interaction of such glasses with HCl, HF, or KOH solutions, and in comparison with the temperature response of refractive index and density in the annealing range. The maximum chemical reactivity coincides with the lowest density, the minimum reactivity with the maximum density, if the glass does not contain more than 12 per cent Na_2O. Molchanova and Serebryakova interpret this behavior as an indication that sodium borate-rich domains are present in the constitution of such borosilicate glasses. The formation and decay of borate domains as a chemical reaction is chiefly dependent on temperatures in the range from 500° to 750°C., whereas purely physical effects of thermal treatment are observed below 500°C. Only by the combined aspects of physical *and* chemical processes in the constitution of borosilicate glasses is it possible to describe with sufficient accuracy their anomalies.

[470] *Compt. rend. congr. intern. chim. ind.*, **27**, Bruxelles 1954; *Ind. chim. belge*, **20**, 1955, Special Issue, 88-91.

[471] *Trudy Gosudarst. Opt. Inst. Leningrad*, **23**, 1953, No. 141, 3-12.

291. Very rich data on the acid and alkaline leachability of borosilicate glasses were further contributed in the extensive investigations of Y. Tamura,[472] for systematically varied compositions of the system Na_2O–B_2O_3—SiO_2. A particularly remarkable result is that slowly cooled glasses of this type could show a *higher* alkalinity than quenched specimens of the same composition. Chilled rods of a glass

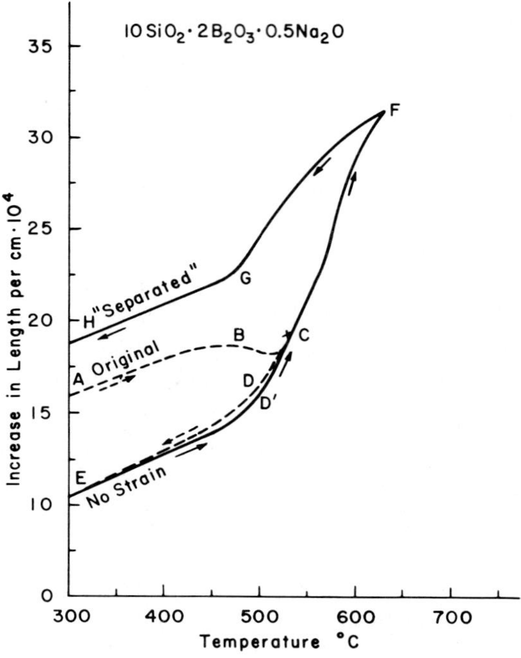

FIG. A. 196. Anomalous thermal expansion of glass 0.5 Na_2O, 2.0 B_2O_3, 10 SiO_2, with raising temperature. Curve *ABC* for chilled glass, curve *CDE* after cooling. Strain-released glass of the same composition (curve *D'F'*) shows anomalous expansion due to separation of glass components, curve *FGB* the cooling curve, *above* the preceding curves of the original sample. (Tamura).

with 0.5 Na_2O, 2.0 B_2O_3, and 10 SiO_2 showed an abnormally large expansion after passing the transformation point, probably due to rearrangements in the glass structure (Figs. A. 196, 197). Alumina containing barium borosilicate glasses were studied by I. I. Kitaĭgorodskiĭ, T. N. Keshishyan and E. A. Faĭnberg.[473] Under the influence of the barium content there is a characteristic change in the coordination of boron from [BO_3] to [BO_4] at a specific ratio BaO:B_2O_3. The boric acid

[472] *Osaka Kogyo Gijutsu Shikenjo Hokoku*, 1953, No. 299, 70 pp. On specific effects of alkaline earth oxides and alumina on thermal properties, viscosity, etc. of Par. B. 99, 172.
[473] *Steklo i Keram.*, **15**, 1958 (3) 1-5.

anomaly is evident here also in a typical alkaline earth borosilicate glass. It extends as well on thermal expansion as on viscosity, electric resistivity, and dielectric losses.

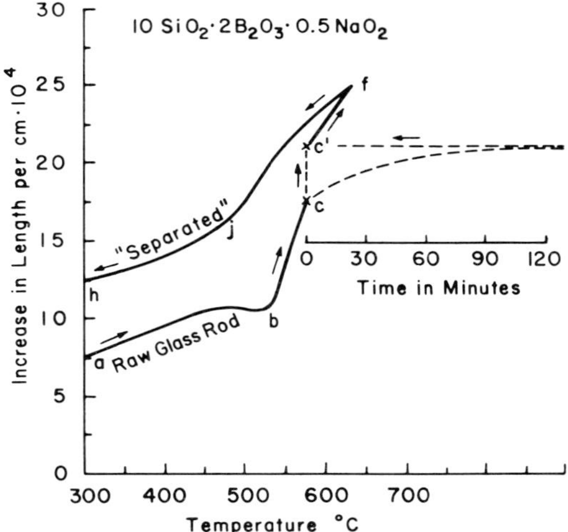

FIG. A. 197. Anomalous thermal expansion of glass 0.5 Na_2O, 2.0 B_2O_3, 10 SiO_2. Same chilled sample as in fig. A. 196, curve *abc*, followed by thermal exposure for 4 hours at 575°C. (above transformation point), temperature raised to point *f*, causing "separation" in glass constitution, and cooling curve *fgh*. (Tamura).

PROPERTIES OF GLASS FIBERS

292. Even though R. B. Sosman's theory (1922) of the existence of chains of the type

```
     O       O       O
 \  / \     / \     / \  /
  Si    Si      Si      Si
 /  \ /     \ /     \ /  \
     O       O       O
```

as the decisive structural units in silicate glasses has been generally abandoned in favor of the three-dimensional framework theory, the reality of such chains in a fibrous form of silica was recently confirmed by Al. and Ar. Weiss.[474] At the same time

[474] *Naturwissenschaften*, **41**, 1954, 12; *Z. anorg. u. allgem. Chem.*, **276**, 1954, 95-112, Cf. Vol. III, Par. B. 47. Possibilities of chain-shaped complex anions in silicate must be reconsidered in view of recent observations of a distinct anisotropy of electric conduction in flowing melts of Na silicate and Ca-borates as described recently by B. M. Lepinskikh, O. A. Esin, and V. I. Musikhin, *Structure of Glass. Proc. All-Union Conf. Glassy State, 3rd, Leningrad* (English Translation), 1959/60, 104-106.

the analogy of this highly unstable fibrous silica built up of chain molecules, with SiS_2 and especially with $SiSe_2$ in the form of a glass, was established.[475] The surprisingly high tensile strength of single (crystalline) fibers of the unstable SiO_2 is particularly interesting in comparison to metal "whiskers" discussed below. The fact is striking that such silica fibers are longer the more alumina is present in the silica. A replacement of Si^{4+} by Al^{3+} in the chains evidently acts stabilizing. In this way it *is* possible to raise the durability, although this is in contradiction to the structural rule that $[SiO_4]$ tetrahedra should only be interlacing over corners, whereas in fibrous silica, SiS_2, and $SiSe_2$, they have contact over edges. Glassy $SiSe_2$ (formed from crystalline $SiSe_2$ above 1060°C.) is chemically more resistant than the fibrous crystalline modification as A. and A. Weiss emphasize. Not accounting for the high chemical instability of such fibrous forms of silicon dichalcogenides SiR_2, there is a surprising morphological similarity and a genetic analogy with the so-called metal "*whiskers*," namely in their formation by heterogeneous gas reactions. These have relatively highly increased mechanical strengths in comparison with that of the large monocrystals and polycrystalline aggregates of fused metals of the same composition, as was demonstrated by S. S. Brenner,[476] starting from natural hair silver and copper, and similar metallurgical products. All these whiskers, however, are *not* analogous to glass structures. They are only metal monocrystals grown with an unusually low concentration in structural dislocations and defects, thus making understood their striking mechanical strengths.

293. The presence of fiber-shaped molecules of infinite one-dimensional extension in glass was an important structural principle from which V. V. Tarasov[477] developed a deep-going quantum-mechanical theory of the specific heat properties of glasses. Omitting here the purely mathematical-physical portion of the problems involved, Tarasov gave a plausible model of the quantum-mechanical behavior of glasses of the simple metasilicate type, but also for condensed $[Si_2O_5]_\infty$ networks as in $Na_2O.2SiO_2$. The calculation can be extended from the common hexagonal network type for anionic configuration to antiparallel arrangements of chains of four-membered

[475] *Z. Naturforsch.*, **8**b, 1953, 104-105; *Intern. Union Pure and Appl. Chem. Sect. for Inorg. Chem. Colloq.*, September 1954, 41-50.

[476] The rich literature on whiskers is discussed by S. S. Brenner in *Science*, **128**, 1958, 569-575; see also F. Blaha, *Radex Rundschau*, 1957, 882-890.

[477] *Steklo i Keram.*, **11**, 1954, (2) 6-9; *Silikat Tech.*, **6**, 1955, 54-55; *Trudy Inst. Krist. Akad. Nauk S.S.S.R.*, **10**, 1955, 136-148; *Stroenie Stekla, Inst. Khim. Silikatov Akad. Nauk S.S.S.R., Trudy Soveshchaniya, Leningrad*, 1953 (Pub. 1955), 62-69; cf. also N. A. Chernoplekov, *Zhur. Fiz. Khim.*, **27**, 1953, 1090-1102. See recently V. A. Ratobyl'skaya and V. V. Tarasov, *Structure of Glass. Proc. All-Union Conf. Glassy State, 3rd, Leningrad* (English Translation), 1959/50, 352-354, with extensive measurements of mechanical resonance curves in silica glass, B_2O_3, and As_2S_3 glass. On optical activity and vitrification see I. A. Bolotina, ibid. 121-124, for organic model glasses.

rings like those observed in apophyllite (cf. Vol. I, ¶ 118), representing shorter branched chains which must affect the properties of corresponding glasses. We come back to these structural principles of *polymerization* in A. ¶ 314.

294. It was often considered whether or not chainlike arrangements which are in many respects similar to organic high-polymeric textiles, should not show expressive *fiber texture* diagrams in X-ray or electron diffraction patterns, especially if they are in an intensely stretched state, as the mechanical theory of A. A. Griffith (1926) had suggested. G. O. Bagdyk'yants[478] made experiments in this direction with electron diffraction diagrams which show for stretched caoutchouc samples excellent orientation texture patterns, but not for $Na_2O.SiO_2$ glass or another industrial glass, in both cases without any indication of such effects as provided for oriented fiber arrangements, in spite of a strong birefringence phenomenon under stress. Further negative evidence against the orientation of structural chains in highly stretched thin glass fibers was given by R. T. Brannan[479] who also found for such fibers about the normal theoretical Poisson ratio (0.18) as observed in massive glass. It is also not affected by any thermal treatment, in agreement with the previous observations of W. H. Otto and F. W. Preston (1950) that strength in glass fibers is equal in all directions. The openness of the structure is also equal in all directions, within limits of experimental errors. There are no measurable effects of damping in the torsion modulus test. Neither is support given to the view that strong bonds in the glass structure would be preferentially oriented in the application of stress.

295. The evident anisotropy of phosphate and borate glass fibers in contrast to the behavior of silicate glasses was emphasized by B. K. Banerjee[480] and by M. Goldstein and T. H. Davies,[481] in both cases explained by the real presence of chain-shaped structural units in the liquid structure. In borosilicate glass of the Jena 16 III type, L. Merker[482] recently not only made evident an optical birefringence in the order of magnitude of 1800 mμ/cm. in maximum, but also the shrinkage phenomenon corresponding to a frozen-in elastic deformation shown in the glass. There is a definite after-compaction in the manner J. F. Stirling[483] demonstrated (see below). Merker observed high degrees of anisotropy at low temperatures of fiber spinning and at a high viscosity. No thermal tensions between the core and the peripheral

[478] *Stroenie Stekla, Inst. Khim. Silikatov Akad. Nauk S.S.S.R., Trudy Soveshchaniya, Leningrad*, 1953 (Pub. 1955), 216-218.

[479] *J. Am. Ceram. Soc.*, **36**, 1953, 230-231.

[480] *J. Am. Ceram. Soc.*, **36**, 1953, 294-298.

[481] *J. Am. Ceram. Soc.*, **38**, 1955, 223-226.

[482] *Naturwissenschaften*, **47**, 1960, 105-106; see also *Union Scient. Continent. du Verre, Charleroi, Belg.*, 1962, 567-587; *Veröffentl. Max Planck-Inst. f. Silikatforschg.* **22**, 1962, 149-169.

[483] *J. Soc. Glass Technol.*, **39**, 1955, 134-144 T.

portions of the fibers, however, could be observed. The optical anisotropy can be released by thermal treatment and presence of mobile cations, especially of Na^+, essentially enhances the release.

296. The anisotropy of glass fibers spun from melts of *sodium metaphosphate* was made evident by B. K. Banerjee[484] by measurements of the magnetic susceptibility, determined in a field of 2450 Gauss by measuring the torsion angle. The anisotropy is here definitely caused by the infinite phosphate anion chains oriented parallel to the axis. The mass susceptibility of the metaphosphate of 0.390×10^{-6} corresponds to its random orientation in the compact glass, whereas for fibers parallel to the axis the relative anisotropy coefficient is $\Delta\varkappa/M = (2C/m\mathfrak{H}^2)(\alpha_c - \pi/4)$ (α_c the critical torsion angle, C the torsion constant of the device, m the mass, and M the molecular weight) $= 0.0187$ to 0.02788×10^{-6}. For B_2O_3 glass the corresponding value is $\Delta\varkappa/M = 0.330$ to 0.369×10^{-6}. In sodium metaphosphate glass on an average, two of four O^{2-} anions are bonded in a one-sided manner, according to the chain structure in the high-polymeric glass. Even side-branched chains may occur. Spinning of the melts orients the chain elements to make the glass magnetic-anisotropic, but annealing favors random orientation (with $\Delta\varkappa/M = 0$) reached for the metaphosphate glass at 375°C. for 24 hours.

297. Banerjee[485] also studied the X-ray diffraction from pure B_2O_3 and sodium borate glasses after aging for 1 month. The diagrams did not show any orientation effects, only the common diffuse glass bands. Alkali borate glasses with up to 5 per cent alkali oxide show a strong interference at 3.13 Å., and a weaker one at 3.238 Å. which indicates crystalline domains with a preferred fiber texture by intense maxima on the equator horizontal. It depends on the conditions of spinning the fibers from the melt whether the radii of these interference rings are larger or smaller, the larger ones indicating a more stabilized distribution of crystallites (presumably of B_2O_3). Additional lines may appear in case of an extreme mechanical treatment in spinning the fibers, with slightly indicated azimuthal interferences. The crystalline domains are distinctly different from the compact initial glass, with an oriented crystal growth which is absent in the glass cores. Even at relatively low temperature B_2O_3 has a high molecular mobility and on the surface there is a rapid reaction with moisture. If the surface layer is mechanically removed there are no longer distinct interferences of the type here described. This crystallization explains the typical aging effects already observed by Banerjee in 1953.

298. The apparent contradictions in the behavior of glasses like those of sodium metaphosphate which *do* show texture effects, and common silicate glasses which

[484] *Glastech. Ber.*, **33**, 1960, 8-9.

[485] *Glastech. Ber.*, **33**, 1960, 120-124. The studies include also glass fibers of more complex systems, like B_2O_3–ZrO_2; Na_2O–B_2O_3–ZrO_2; B_2O_3–Fe_2O_3; Na_2O–B_2O_3–Fe_2O_3, etc.

do not, is understood to a high degree if one takes into consideration the general theoretical viewpoints of J. D. Bernal, Ya. I. Frenkel, and G. W. Stewart estensively discussed in A. ¶ 190 ff. A distinction was made between "flawless liquid" systems and "orientable" or paracyrstalline liquid systems and glasses. We also come back to the discussions of J. F. Stirling (see above) on "frozen-in" strains in glass of Pyrex (borosilicate) and sodium-calcium silicate type. The application of the theory of elasticity shows that each glass exhibits a modulus analogous to Young's modulus, namely the ratio of stress to frozen longitudinal strain, and a ratio of frozen lateral to longitudinal strain, analogous to Poisson's ratio. The birefringence phenomena present in spun fibers are, therefore, physically anisotropic with frozen-in elastic deformations.

299. An attempt to apply modern methods of low-angle X-ray diffraction to glass fibers was made by W. O. Statton and L. C. Hoffman,[486] especially to the glass Na_2O, MgO, and $3SiO_2$, with replacement of Na^+ by K^+, and of Mg^{2+} by Cu^{2+} and Pb^{2+}. The diffraction diagrams in most cases show very intense shaped diffuse scattering. If special thermal treatments are applied there are intense unshaped maxima. The interpretation of this particular phenomenon can be given by the formation of elongated ellipsoidal cavities in the glass fiber structure originating from stretching during the spinning process, with discrete Bragg spacings of the order of magnitude of 30 Å. It is remarkable that those cavities are evidently larger in silicate glass fibers than they are in viscous organic polymers like Nylon, polyethylene tetraphtalate, and others. The formation of such cavities must not be related anyhow to short-range order in the glass constitution, of course. It is locally concentrated in the peripheral portions where the glass is open to liquid penetration from the surface. Either one may understand this phenomenon as a freezing-in during quenching, or by a chemical reaction of the glass with agents in the atmosphere of the quenching furnace combined with quenching effects. A. F. Prebus and J. W. Michener[487] made a very interesting approach to study glass fibers in their textures by electron microscopy. The filaments were prepared from special glass compositions ("E" glass) used in industrial glass fiber manufacturing. An interpretation of the given electron micrographs is difficult, but it may not be denied that there are indications of chain-type structural elements; however, in one glass ("C") a more uniform structure prevails. After leaching, the E glass shows chains to be traced over a length equivalent to several fiber diameters. It may be presumed that these "chains" are the same sequences of cavities concluded from the low-angle diffraction behavior.

[486] *J. Appl. Phys.*, **31**, 1960, 404-409.

[487] *Ind. Eng. Chem.*, **46**, 1954, 147-153. On these problems see also I. Warshaw, *J. Am. Ceram. Soc.* **43**, 1960, 4-9; J. Zarzycki and R. Mezard, *Phys. Chem. Glasses*, **3**, 1962, 163-166; S. Asunmaa and Tr. Baak, *Ibid.* 46-49, on the microstructure of thin films of a $BaO–Al_2O_3–TiO_2–SiO_2$ glass under the action of 100kv. electron irradiation.

300. In general the granular structure shown gives the impression of a certain degree of order over distances ranging from 20 Å. up to at least 200 Å. P. Kuhn and G. Schimmel[488] with refined methods of transmittance electron microscopy (with a resolution power of about 15 Å.) came to the conclusion that it is not yet possible to have a clear judgement on the applicability of A. Smekal's theory (1936, 1938) concerning surface defects ("Kerbstellen") on glass fibers. The examination by electron microscopy is particularly influenced by evident structural changes occurring during the irradiation with the intense electron beam which may on the other hand be used to "spin," say from silica glass filaments of 2 μ diameter, finest fibers of 15 Å. in diameter, with an extremely high surface energy (a few single [SiO_4] groups at the endings). Other structural details cannot yet be deciphered, but for a "boiling" by alkali evaporation from the core of the fiber. In selenium ruby glass the boiling by the Se vapor is particularly evident. Lead silicate glass shows even unmixing phenomena, with a high material mobility reminding one of "collective crystallization." Fresh-spun fibers show this mobility very distinctly, whereas aged filaments are more inert and "stabilized." Regularly chain-like components in the structure may be admitted from the electron micrographs, such as in a refractory alkali-free calcium aluminate glass.[489]

301. Griffith's theory of structural flaws as the reason for brittleness of compact glass and especially the cause of the mechanical strength phenomena of glass fibers was recently advocated again by D. Dollimore and S. J. Gregg,[490] further by E. Deeg and A. Dietzel.[491] The accelerated increase in *tensile strength* with decreasing fiber diameter corresponds to similar increases in the elastic coefficient 1/E and 1/G (E = Young's modulus, G = the modulus of rigidity). All three parameters decrease with aging the fibers by tempering. The anomalies described by previous authors are in general a necessary consequence of quenching in preparing the fibers in a spinning process. The structural conversions which accompany the anomalies were studied by dilatometric diagrams showing the typical distinction of the mechanisms of freezing-in of normally cooled glass and that in fibers with a shrinkage (e.g., by 0.36

[488] *Glastech. Ber.*, **30**, 1957, 463-470. On the Smekal theory of flaws see recently W. A. Weyl and E. C. Marboe, *Ibid.*, **32**, K, 1959, VI, 1-VI, 18, especially pp. 14-15. A modern treatment of problems of strength of glass see by E. U. Condon, *Am. J. Phys.*, **22**, 1954, 224-232.

[489] On such high-melting glasses for fiber production in the systems $CaO–Al_2O_3$, $CaO–ZrO_2$, $Al_2O_3–SiO_2$, and others, see A. Silverman, *J. Soc. Glass Technol.*, **40**, 1956, 413-428, and Patent literature cited.

[490] *Research* (*London*), **11**, 1958, 180-184. On the detection of Griffith cracks by treament with Na vapors, see F. M. Ernsberger, *Phys. & Chem. Glasses*, **1**, 1960, 37-38; on the structure of fibers of glasses with high Young's moduli, K. L. Loewenstein, *Ibid.* **2**, 1961, 69-82, 119-125. Tensile strengths and extension effects of fibers of lithium-barium silicate glass with unmixing tendencies were studied by Fr. Oberlies and G. Pohlmann, *Union Scient. Continent. du Verre, Charleroi, Belg.* 1962, 429-445; *Veröffentl. Max Planck-Inst. f. Silikatforschg.* **22**, 1962, 209-219.

[491] *Glastech. Ber.*, **28**, 1955, 221-232.

per cent between 130° and 520°C. for a given glass and freezing temperatures of 740° and 830°C.) (see Fig. A. 198). The more ionic, less covalent binding mechanism in quenched glass shows that the lower density of the fibers corresponds to a more

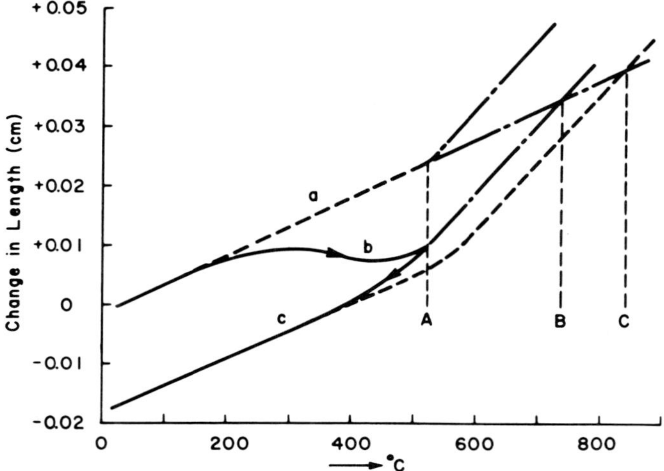

FIG. A. 198. Dilatometer diagram of glass 16 per cent Na_2O, 10 CaO, 74 SiO_2, (*a*) normally cooled rod, (*b*) bundle of freshly prepared fibers, extremely chilled, (*c*) shrinkage on cooling. (*A*) is the freezing-in range of normally cooled glass, (*B*) that without shrinkage, and (*C*) that with the shrinkage amount considered. (Deeg and Dietzel).

"angular" structure of lower degree of framework formation with increased distances Si—O. The higher tensile strength, better elastic behavior indicates a better isotropic structure with reduced strain peaks, and a higher ionic share in the binding between Si^{4+} and O^{2-}.

302. We further emphasize the measurements of W. H. Otto[492] who demonstrated that, contrary to common belief, the strength of a glass fiber is *not* a simple function of the diameter. When fibers of different diameters are formed under accurately controlled more or less identical conditions, the rupture strengths are identical within experimental limits, and no diameter effect is observed at all. Pulling speed in the spinning process and temperature of the glass melt from which the fibers are spun are thermal history factors influencing on mechanical (tensile) strength to fracturing. This may also be one of the principal reasons why *reheating* has a very important effect on the tensile strength of glass fibers between room temperature and the

[492] *J. Am. Ceram. Soc.*, **38**, 1955, 122-124. See also G. M. Bartenev and A. N. Bovkunenko, *Zhur. Fiz. Khim.*, **29**, 1955, 508-512. See also W. F. Thomas, *Phys. & Chem. Glasses*, **1**, 1960, 4-18; W. M. Otto, *J. Am. Ceram. Soc.* **44**, 1961, 68-72.

softening point as S. Sakka[493] observed. The glass which he studied was a rather complex sodium-barium-borosilicate composition; etching with HF after reheating was applied. The reheated fibers show the well-known decrease in tensile strength but they regain their original strength by an adequate removal of the chemically changed surface skin by etching with HF. The reheating is only able to exert these changes in the surface portions, but not in the core of the fibers, the strength of which remains unchanged. It is assumed by Sakka that this part of the structure of the glass fibers which governs the mechanical strength properties is different from that part which governs say its density. Ch. L. McKinnis and J. W. Sutton[494] speak of the "melting history" of glass in its specific effects on the properties of glass, and especially the tensile strength of fibers, increasing this parameter with increasing maximum temperature in the melting history process. It is, therefore, a *kinetic* aspect of the glass fusion which plays an integral role in the evolution of the industrially valuable properties of glass fibers. The *cooling rate* for fine-spun glass fibers was calculated by S. Bateson[495]; strength as a function of the diameter is compared with the ratio of quenching time, t_e, to Maxwell relaxation, τ. If this ratio is above 1.0 the increase in mechanical strength occurs which was predicted by A. A. Griffith and E. J. Gooding, within the order of magnitude of 10^{-4} sec. at 1100°C. The refractive index is only little changed in the critical region and a rise in fictive temperature with increased quenching is insignificant. Bateson believes, therefore, that structural changes in the glass are secondary in importance in contributing to high mechanical strengths. The flaw theory in a modified form is reestablished by Bateson.

303. The *chemical durability* of glass fibers and especially of silica glass wool ("quartz wool") was discussed by G. Satlow,[496] by leachability tests in H_2SO_4, acetic acid, and KOH solutions, as well as in reaction with soda waterglass using different concentrations and over different time periods. The decrease in tensile strength after the chemical treatment is insignificant or tolerable if the concentration of the agents is not above 20 per cent. Only silica glass is more intensely leached and corroded by KOH solution. A remarkable increase in tensile and rupture strength is observed at temperatures up to 90°C. It is technologically interesting that a practically sat-

[493] *Bull. Inst. Chem. Research, Kyoto Univ.*, **34**, 1956, 316-320.

[494] *J. Am. Ceram. Soc.*, **42**, 1959, 250-253. W. A. Weyl, *Sprechsaal*, **93**, 1960, 128-136, 518-521, 561-563, 588-590, emphasizes the importance of the concept of "melting history" for the molecular aspects of nucleation in glass. It is, on the other hand, most significant that recently R. Brückner *Glastech. Ber.* **35**, 1962, cf. especially, pp. 50-52, could not observe melting history effects in his careful measurements of viscosity and homogeneization in glass melts. On thermal history and devitrification rate of glass, cf. B. Locardi, *Vetro e Silicati*, **9**, 1962, 5-10.

[495] *J. Appl. Phys.*, **29**, 1958, 13-21, and *J. Soc. Glass Technol.*, **37**, 1953, 302-305 T. Cf. E. J. Gooding *Ibid.*, **16**, 1932, 145-170.

[496] *Glas-Email-Keramo-Tech.*, **4**, 1953, 169-172.

isfactory replacement of borosilicate glass compositions for fiber glass manufacturing is possible if B_2O_3 must be avoided for economical reasons, namely by the use of synthetic anorthite ($CaO.Al_2O_3.2SiO_2$), diopside ($CaO.MgO.2SiO_2$), and natural rhodonite ($MnO.SiO_2$) as raw materials, as B. Yoshiki[497] demonstrated for glass fiber plants in Japan. Such a glass fiber is even superior to borosilicate compositions in chemical durability, both to acids and alkalies in leachability.

304. We also emphasize the important effects of *moisture* on the surface of glass fibers influencing their *elastic fatigue* and *creeping* properties which were studied by M. S. Aslanova and P. A. Rebinder.[498] Under the influence of strain an instantaneous elongation occurs, after its removal a fatigue, observed in mixed alkali silicate fibers or industrial aluminum borosilicate material. The elongation increases for a given load with increasing relative humidity of the surroundings. Surface-active agents may raise the elongation to 10 per cent, compared wity the maximum effect of saturated water vapor of 7 per cent. In addition to these elongation-fatigue phenomena long-time creeping is superimposed. By a treatment with HF the firmness of the fibers is improved, creeping reduced. Alkali- and silica-free phosphate glasses also show typical fatigue. On the action of organic media on these phenomena see A. ¶ 410. We further mention the "texture-orientation" effects first described by H. Elsner von Gronow (1931) by the oriented crystallization of methylene blue on glass fibers under strain, and classified in distinction from true epitaxis phenomena (cf. Vol. I, A. ¶ 251) by A. Neuhaus.[499]

INTERPRETATION OF THE GLASSY STATE AS A MOLECULAR STRUCTURE

305. In previous paragraphs of the present book we have given preference to the crystallochemical, specifically to the atomistic-ionic interpretation of the constitution of glasses, in general, and of silicate glasses in particular. Through many difficulties observed in the last two decades in applying the framework theory of Zachariasen, as it is based on V. M. Goldschmidt's fundamental principles of crystallochemistry, a definite reaction against this preferred aspect of glass constitution has developed. Such tendencies were mentioned several times in our previous text as critical viewpoints against the framework interpretation of glass.

306. Among typical representatives of a *micellar*-molecular theory of glass con-

[497] *Glass Ind.*, **33**, 1952, 353-355, 376.

[498] *Doklady Akad. Nauk S.S.S.R.*, **95**, 1954, 1215-1218; **96**, 1954, 299-302; recently *Silikat Tech.*, **12**, 1961, 528-531.

[499] *Fortschr. Mineral.*, **29**/**30**, 1952, 136-296, especially pp. 140 f.

stitution we mention, in the first place, that of T. Moriya[500] which goes back to the ideas of W. Biltz and F. Weibke (1932) assuming singular atomic arrangements in the micellae of definite chemical compositions, such as $Na_2O.SiO_2$, $Na_2O.2SiO_2$, $Na_2O.4SiO_2$, SiO_2, in the system $Na_2O–SiO_2$, or $Na_2O.SiO_2$, $Na_2O.2SiO_2$, $Na_2O.Al_2O_3.4SiO_2$, $Na_2O.Al_2O_3.6SiO_2$, in the system $Na_2O–Al_2O_3–SiO_2$. These micellae have a nonsymmetric "amorphous" constitution by themselves and are *polymers* of the simple compounds as written above in stoichiometric formulation. The degree of polymerization and numbers of the micellae in a glass of a defined physicochemical state depends on the thermal history in the previously discussed understanding. They are agglomerated and surrounded by "dissociated molecules" even in larger amounts than the polymeric units. These ideas evidently refute ionic bonding factors and bring to the foreground the atomic or covalent bonding by electron pairs making silicate glasses highly analogous to organic polymers and organoplasts (see below). A breakpoint in this concept is the abrupt change of such aggregations, as the amorphous silicate micellae are in Moriya's opinion, to the constitution of the well-crystallizing orthosilicates, for which the ionic concept of isolated $[SiO_4]$ groups (isles) cannot be doubted. The shapes of the micellae may be spheres, sheets, or linear chains. All the phenomena, like transformation and annealing, are understood by Moriya as changes in the size of the micellae and of their distribution.

307. In a much more comprehensive manner, H. Salmang[501] gave a full revision of our present knowledge in both directions, the atomistic and the molecular interpretation of the glassy state, not only as a degenerated, strongly deformed crystalline structure, but as a molecular complex following the classical aspects of G. Tammann, of undercooled liquids, such as the organic model glasses which he preferably used in his experimental studies. Covalent binding in silicate glasses causes the formation of polymerization aggregations with permanent dipole moments (cf. A. Silverman, 1917) and with solvation complexes (cf. W. Weyl, 1933). Only molecular theories can fully interpret the high viscosity, the specific volume anomalies, or the transformation phenomena from the liquid to the plastic and finally to the elastic state of undercooled systems. Salmang tries, of course, to combine the undeniable shares of ionic binding and the valuable knowledge acquired through the crystallochemical theories, and to reconcile these with the equally indispensable advantage of the molecular interpretation, especially in glasses with polarizable cations of non-rare gas type elements, the many analogies of glass fiber spinning

[500] *Tokyo Kogyo Daigaku Gakuho*, Ser. B, 1951 (1) 1-10; in many details this theory recalls previous ideas of E. Berger (1927, 1931).

[501] *Naturwissenschaften*, **45**, 1958, 4-6; *Glastech. Ber.*, **32**, 1959, 137-142; "Die Physikalischen und Chemischen Grundlagen der Glasfabrikation," Springer-Verlag, Berlin, Göttingen, Heidelberg 1957, 354 pp., especially pp. 22 ff. Recently: *Chem. Weekblad*, **57**, 1961, 469-474, and *Naturwissenschaften*, **35**, 1962, 6-8.

with organic plastic polymers, or the often surprising analogy of glasses and aqueous solutions of colored metal ions, dipole actions, and so on. The characteristics of the glassy state is evidently that emphasized already in 1942 by A. Smekal that glass-forming systems contain covalent as well as ionic binding mechanisms.

308. The evident parallelisms between the glassy state and that of modern organo-plastics is nearly trivial but it is profitable to consider them under these aspects given especially by Salmang, and in discussions of H. Schulz[502] on the fact that there are in both classes of materials transitions and overlapping phenomena concerning their mutual approach, either more to liquids, or to crystalline coordination structures. Organoplastics have no framework structures in the meaning of the $[SiO_4]$ three-dimensional groupings of silicates, but interlinking molecular chains. They do not show the high degree of brittleness which is characteristic of true silicate glasses. The polymerization qualities of glasses were discussed by J. M. Stevels,[503] demonstrating the high instability of silicate chain molecules of infinite length extension and the dielectric loss response to polymerization effects. An extreme comparison of the rules governing crystal or glass phases of organic "model" polymers with those ruling in silicates and phosphates shows how important symmetrical atom groupings are in the constitution for physical-chemical character. A slight asymmetry determines a more glassy type, as is seen in mixed alkali silicate glasses.

309. The real existence of *polyphosphate chains*, rings, and other configurations as made evident by Van Wazer, Westman, and Crowther (see below) is a strong support in E. Thilo's[504] discussions of the glassy state as that of a high-polymer, namely by action of steric hindrance effects in crystallization caused by the complex structural units themselves. The chief advantage of such a general assumption over the Zachariasen theory is its equal applicability for inorganic and organic molecular complexes, namely for sulfur, selenium, polysaccharides, plastomers, albumines, and others. Thilo also emphasizes the importance of the presence of catalysts for a conversion of polymeric units into smaller, low-molecular units, and a strong reduction of the activation energy of crystallization. Water especially plays a large role in this respect, but so does also, for example, the Li^+ ion as a strongly polarizing small center.

310. I. Schulz and W. Hinz[505] go back to the previous ideas of G. Hägg (1935)

[502] *Glas-Email-Keramo-Tech.*, **4**, 1953, 160-162.

[503] *Verres et réfractaires*, **7**, 1953, 91-104; *Glass Ind.*, **35**, 1954, 69-72, 100; 135-138, 160; 657-662. See also L. V. Sergeev, *Stroenie Stekla, Inst. Khim. Silikatov Akad. Nauk S.S.S.R., Trudy Soveshchaniya, Leningrad,* 1953 (Pub. 1955), 280-282, who refers to the excellently glassy character of polyacrylate, polystyrene, and polyvinyl plastomers in contrast with the easily crystallizing polyethylenes.

[504] *Silikat Tech.*, **6**, 1955, 278-280.

[505] *Silikat Tech.*, **6**, 1955, 235-241; *Glastech. Ber.*, **29**, 1956, 319-323.

concerning the configuration of primary molecules of silica and of silicates in one-, two-, and three-dimensional arrangements. As particularly suitable models, they use the glasses of binary phosphates. In metaphosphates, chain or ring configurations are common. Binary phosphates of alkalies and divalent metals do not show any indication of three-dimensional frameworks of $[PO_4]$ tetrahedra, metaphosphates always showing as structural units the (PO_3) type, even for zinc phosphates which have a higher tendency to form frameworks. Also Schulz and Hinz try to reconcile the Zachariasen framework theory with the molecular concepts because of analogies to structures of congruently melting chemical compounds, although it is not possible to indicate, from phase-equilibria diagrams of given phosphate systems, which structural elements characterize a glass of a given composition. In another way, W. A. Weyl[506] seeks the reconciliation of the atomistic and the molecular concepts in the application of K. Fajans' and Weyl's polarization and screening theories of binding with deformation effects of electron orbitals and quantum-mechanical and thermodynamic parameters like the entropies of mixing in given glass-forming systems. This entropy is directly related to the degree of order in the atomic arrangements thus giving information on the stability or instability of their groupings, as of $[SiO_4]$ and $[AlO_4]$ in frameworks, but also on questions like chemical durability by the formation of larger atomic complexes, such as when lime enters a sodium silicate glass.

311. The most simple really molecular glasses are those of the elements of the VIth group of the Periodic System, namely oxygen, sulfur, selenium, and tellurium which were well studied in their physical properties by A. Winter-Klein,[507] but those best investigated non-framework glasses are the *polymetaphosphates* based on extensive studies of J. R. Van Wazer (1950) and later by A. E. R. Westman with J. Crowther and P. A. Gartaganis.[508] These authors developed Van Wazer's linear-polymer theory of condensed phosphates and found that from a certain ratio Na : P $(Na_2O : P_2O_5) = 1 : 1$ not the framework of $[PO_4]$ tetrahedra is the structural principle but a linear chain polymerization or formation of stable rings occurs. The paper-chromatographic method proved to be a reliable and sensitive indicator of the existence of such higher anionic units, with a characteristic distinction of the numbers of chain lengths from one to 6 or 7. Not only in qualitative respects but also quantitatively the theory of polymerization, as it was developed by Van Wa-

[506] *Glastech. Ber.*, **30**, 1957, 269-282; *Central Glass & Ceram. Research Inst. Bull.* (*India*), **4**, 1957, 121-139.

[507] *Compt. rend.*, **242**, 1956, 2931-2933.

[508] *J. Am. Ceram. Soc.*, **37**, 1954, 420-427; **40**, 1957, 293-299. Also the so-called Poisson distribution is equally absent as the "random reorganization" discussed by A. J. Flory, *J. Am. Chem. Soc.*, **64**, 1942, 2205-2212. On the constitution of *mixed alkali* phosphate glasses see recently M. Krishna Murthy, M. J. Smith, and A. E. R. Westman, *J. Am. Ceram. Soc.* **44**, 1961, 97-105, 475-480; **45**, 1962, 401-407.

zer and Westman, could be confirmed, and even objections made by E. Ingerson and G. W. Morey (1943) against "molecular groupings" in metaphosphate solutions could be explained in favor of the polymer theory. The trimetaphosphate rings are especially persistent in the crystalline, glassy, and solution states. Later investigations using the through-flame fusion techniques of sodium phosphate glasses, with the ratio Na : P above 1 : 1,[509] confirmed this high stability of cyclic phosphates in solutions or glasses despite prolonged heating in the molten condition, and no elimination of the ring groups occurs as was previously concluded. No indications of the existence of any random distribution could be found by X-ray diffraction methods. For sodium and potassium, later also for lithium phosphate glasses a rapid change of the chemical properties with variable chain lengths in the short polymers, however, is observed, specifically a regular cation effect takes place. Typical distribution curves from paper chromatograms are given for sodium metaphosphates in Fig. A. 199.

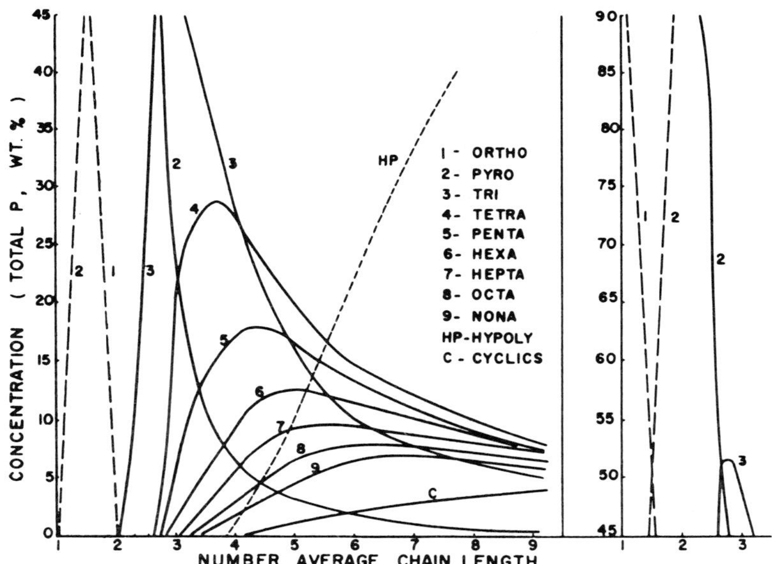

Fig. A. 199. Typical distribution curves for the polymer sizes of sodium metaphosphate glasses. (Westman and Gartaganis).

312. The same paper-chromatographic method was later used by H. Grunze[510] for the study of sodium polyphosphate glasses between $Na_2O.P_2O_5$ and $5Na_2O.3P_2O_5$,

[509] On the flame fusion method used cf. W. K. Haller and B. M. Landis, see unpublished Master Thesis of the latter author, University of Toledo, Ohio, 1956, 34 pp.

[510] *Silikat Tech.*, 7, 1956, 134-138.

with linear, nonbranched chains, and with the radioactive tracer P^{32}, after addition to CaO-containing polyphosphates (oligophosphates), with the determination of the average degree of condensation, as $n = 2.93$ in triphosphate, $n = 3.87$ in tetraphosphate glasses.

313. More from the glass-technological viewpoint of fiber-spinning and of the orientation of the chain molecules in sodium polyphosphate glasses, M. Goldstein and T. H. Davies[511] could give full evidence of such orientation effects by X-ray diffraction diagrams of optically anisotropic fibers, with the typical equatorial texture intensities of interferences for oriented chain molecule aggregates. Further studies on stability of the chain and ring phosphate structures were made by I. Schulz and W. Hinz[512] who extended their investigations to glasses with heavy cations of high polarizability (Pb^{2+}, Cd^{2+}, Zn^{2+}, Bi^{3+}). The tendency to form glasses is not only determined by the disproportionation to a series of different anions but also by the difference in energy contents of crystalline and glassy modifications. Thus, cadmium or zinc pyrophosphates are, in spite of their higher disproportion, not better glass formers than lead pyrophosphate, the enthalpy of devitrification for $2ZnO.P_2O_5$ being much larger than for $2PbO.P_2O_5$. H. W. Bues and H.-W. Gehrcke[513] described infrared absorption and Raman spectra of typical sodium phosphates in the glassy form, with assignment of specific frequencies to PO_2, PO_3, and P–O–P groups in chains and rings of their structure, the changes by dismutation, quenching, and other effects. The devitrification of chain-structured sodium polyphosphates and annular ($Na_3P_3O_9$) glasses was studied in its kinetics by I. Gutzow[514] (cf. Vol. III, A. ¶ 37).

314. A transfer of these experimental results obviously so successfully applied for the problems of polyphosphate glasses to silicate glasses is still a difficult question of theory. We mentioned already in A. ¶ 293 attempts made by V. V. Tarasov[515] to solve the constitution problems of simple silicates by a calculation of the specific heats from chain and network configurations of different kinds which are partially similar to those of the phosphate polymers. This mathematically elegant quantum-mechanical approach for chain polymers is in entire analogy with corresponding

[511] *J. Am. Ceram. Soc.*, **38**, 1955, 223-226, with valuable bibliography.

[512] *Glastech. Ber.*, **29**, 1956, 319-323. E. Thilo, *Naturwissenschaften*, **46**, 1959, 367-373, gave a complete report on the investigations on condensed phosphates in general, of the constitution of the metaphosphate glasses in particular, with their mesomeric phenomena to which we refer expressively for further detailed information.

[513] *Z. anorg. u. allgem. Chem.*, **288**, 1956, 291-306, 307-323.

[514] *Z. anorg. u. allgem. Chem.*, **302**, 1959, 258-274.

[515] *Steklo i Keram.*, **11**, 1954 (2) 6-9; *Silikat Tech.*, **6**, 1955, 54-55; *Trudy Inst. Krist. Akad. Nauk S.S.S.R.*, **10**, 1955, 136-148. A. V. Gladkov and V. V. Tarasov, *Structure of Glass. Proc. All-Union Conf. Glassy State, 3rd, Leningrad* (English Translation), 1959/60, 277-280, come to basic conclusions on the polymeric constitution of inorganic (silicate, borate, phosphate) glasses.

calculations for organic polymers as shown by C. E. Hecht and W. H. Stockmayer,[516] as for polytetrafluoroethylene (Teflon), or HF, both with covalent binding mechanisms.

315. The most fascinating analogy existing between *silicoorganic compounds* in their reaction behavior with silicates is often accompanied by surprising exterior similarities as when E. Wiberg and W. Simmler[517] described mica-like flakes of the hydrolysis product of trichlorosilane, $SiHCl_3$, in dilute ether solution, as a high-molecular "silicon oxyhydride" or dioxodisiloxane, or a silico-formic acid anhydride, and so on, either with ribbon or foliated configurations. By calcination in vacuo this interesting compound is changed to a gold-brown "crystalline" product of the brutto formula Si_2O_3. The deep color is assumed to be caused by bonds Si—Si between the layer units of the type $(SiO_{1\frac{1}{2}})_\infty$. The condensation during the hydrolysis goes over lower linear polymers with two or three Si—O—Si groups to the annular core indicated of which the silicon oxyhydride is the hydride of plane configuration.

316. The classical investigations of R. Schwarz *et al.* of the higher-molecular *silicon chlorides* and their typical topochemical reactions were reexamined by H. Schäfer[518] under the aspects of *radical* reactions with the formation, as of $SiCl_2$ in the quenching tube (Fig. A. 200) which is the characteristic experimental tool for such condensa-

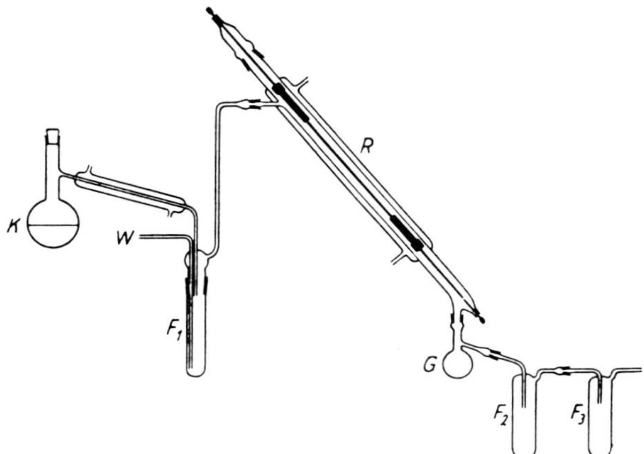

FIG. A. 200. Quenching tube (of silica glass), with internal globar rod (heated with 20 amp., 90 volts, to 1150°C.). Distillation and condensation system. (Schwarz).

[516] *J. Chem. Phys.*, **21**, 1953, 1454-1458. On polytetrafluoroethylene cf. G. T. Furukawa, R. E. McCoskey, and G. J. King, *J. Research Natl. Bur. Standards*, **49**, 1952, 273-278; on HF: V. V. Tarasov, *Zhur. Fiz. Khim.*, **26**, 1952, 1347-1382, and recently *Structure of Glass. Proc. All-Union Conf. Glassy State, 3rd, Leningrad* (English Translation), 1959/60, 64-73, with discussion of the Heitler-London (covalent) bonding type in glass polymers, or one of the donor-acceptor type, in the case of framework builders by copolymerization, making the glass more brittle.

[517] *Z. anorg. u. allgem. Chem.*, **283**, 1956, 401-413.

[518] *Z. anorg. u. allgem. Chem.*, **274**, 1953, 265-270, with bibliography.

tion processes by heterogeneous gas reactions, to form on the hot walls of the Globar heating element the thermodynamically more stable higher chain structures. The heat of formation of gaseous $SiCl_2$ would be 29,900 cal./mole (at 298°K.), its normal entropy 71.1 cal./°C. By hydrolysis of the condensed silicon chlorides or chlorohydrides, e.g., with a relatively stable Si_{10} chain, R. Schwarz and Chr. Danders[519] prepared a silico-sebacic acid, the highly unstable $Si_{10}(OH)_{20}H_2$, and its more stable derivatives such as

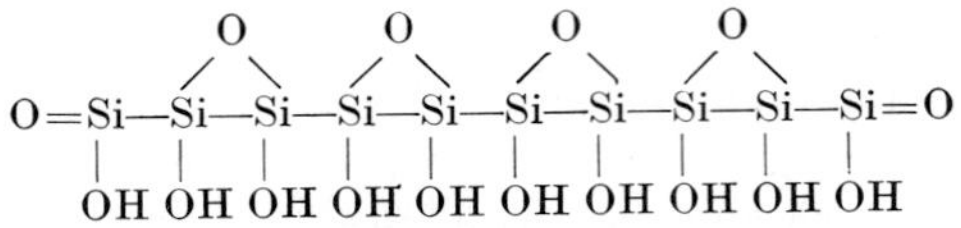

```
        O       O       O       O
       / \     / \     / \     / \
O=Si—Si—Si—Si—Si—Si—Si—Si—Si—Si=O     = Si10(OH)10O6
  |   |   |   |   |   |   |   |   |   |
  OH  OH  OH  OH  OH  OH  OH  OH  OH  OH
```

(octoxy-tetroxo-silico-sebacic acid)

which is oxidized on calcination to the siloxane molecule

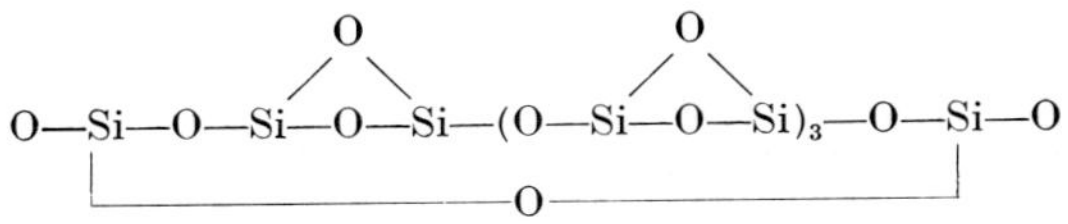

```
              O                     O
             / \                   / \
O—Si—O—Si—O—Si—(O—Si—O—Si)3—O—Si—O
  |                                  |
  └─────────────O────────────────────┘
```

with apparently trivalent silicon atoms. The siloxene ring thus formed is a chromophoric unit in the meaning previously defined by H. Kautsky (1943). An analogous octoxy-silico-adipic acid, $Si_6(OH)_{10}O_2$, was prepared from Si_6Cl_{14} by hydrolysis. The interesting *silicon monochloride*, $(SiCl)_\infty$, could be prepared by R. Schwarz and A. Köster[520] from $Si_{10}Cl_{20}H_2$, with $Si_{10}Cl_{18}$ as an intermediate product of decaline-like constitution:

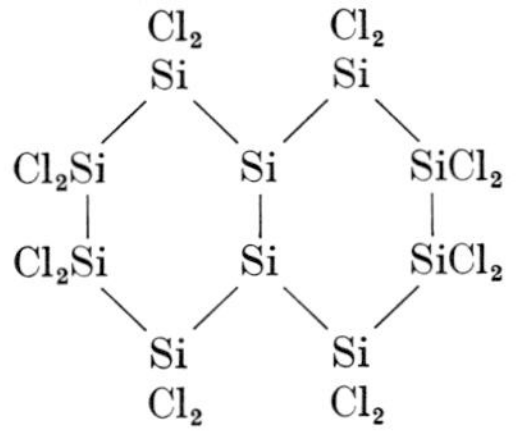

```
          Cl2       Cl2
          Si        Si
         /  \      /  \
   Cl2Si      Si      SiCl2
     |        |         |
   Cl2Si      Si      SiCl2
         \  /      \  /
          Si        Si
          Cl2       Cl2
```

and quite a series of annular silicon chlorides with increasing molecular weights, of the type of anthracene, tetracene (naphtacene), octacene, coronene, and such. The end member is the infinite network of $(SiCl)_\infty$ which at 800°C. is decomposed forming an X-ray-amorphous silicon. All these compounds are amorphous. The silicon monochloride is also different from the silicon monofluoride, $(SiF)_\infty$, which is a graphite-like structure, somewhat widened with larger distances of the sheets.

[519] *Z. anorg. u. allgem. Chem.*, **253**, 1947, 273-280, with valuable proposals for correct nomenclature of complex silicon-organic compounds. The problem of the role of chromophoric groupings and the hypothesis of double bonds are left open in the discussion given.

[520] *Z. anorg. u. allgem. Chem.*, **270**, 1952, 2-15.

317. *Silicon-amines* with organic substitution groups were made by R. Schwarz and Fr. Weigel[521] by reaction of organosilanes with ammonia from which further disilazanes could be derived, partially with cyclic groups. By reaction with hexamethylene-diamine a reaction to form triphenoxy-cyclohexyl silazine was observed, by reaction of trialkyl silicon amide with adipic acid chloride a triisopropoxycyclotetramethylene diketosilazine. Without further discussing here the fascinating constitution of these new compounds, important is the general observation of how promptly the reactions of the initial simple silicon compounds follow the rules of organic chemistry. The chief condition for the formation of annular configurations is the high mobility of the hydrogen in the imino groups of —Si—NH—C— and —Si—NH—C—O. Nevertheless, there is a difference in the fact that the bifunctional character of hexamethylene diamine and of adipic acid is not essential for the synthesis of linear products as it is in the purely organic higher-polymer chemistry, as in the important condensation type of aniline with formaldehyde, or of polyamides.

318. *Polymeric* propyl and isobutyl *esters* of silicic acid were described by R. Schwarz and A. Kessler.[522] They are formed if the simple orthosilicic acid esters are partially hydrolysed. The di- and tri-esters are linear, the tetraester is cyclic in configuration. The condensation can go on from the tetra-propyl ester to an octa- and ditria-ester (molecular weight 3550, still soluble in benzene) which is changed by calcination to a half-ester $(HRSi_2O_5)_x$ and propylene. The hydrolysis reaction of the polyesters has a monomolecular kinetic type.

319. With a short hint we may also mention the interesting synthesis of alkoxysi-

[521] *Z. anorg. u. allgem. Chem.*, **268**, 1952, 290-300.

[522] *Z. anorg. u. allgem. Chem.*, **263**, 1950, 15-30. In place of simple —O— bridges also groups —O—C_6H_4—O— can be built into the constitution of silico-organic compounds, if $SiCl_4$ reacts with *o*-, *p*-, or *m*-dixoybenzene; cf. R. Schwarz and W. Kuchen, *Ibid.*, **266**, 1951, 185-192. One of the typical ring-formations is that in the series

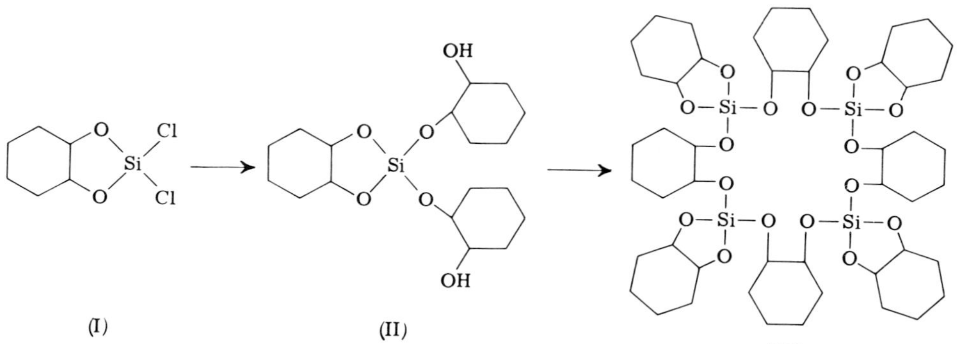

(I) (II) (III)

with rubberlike or brittle glasslike behavior.

lanes and "oligo-silicic acids" (cf. Vol. I, C. ¶ 35) by R. Schwarz and K. G. Knauff,[523] and the corresponding synthesis of alkoxy-*germananes*,[524] especially of tetramethoxy- and tetrapropoxy-germanates of the interesting configuration of

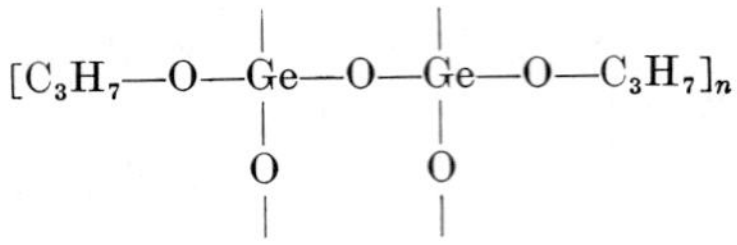

$$[C_3H_7\text{—O—}\overset{|}{\underset{\underset{|}{\overset{|}{O}}}{Ge}}\text{—O—}\overset{|}{\underset{\underset{|}{\overset{|}{O}}}{Ge}}\text{—O—}C_3H_7]_n$$

in surprising exterior analogy to phyllosilicates of the type $(Me_2Si_2O_5)_n$ and with talc. Such compounds should not be called esters, in view of the particularly stronger metallic character of germanium relative to silicon, and a fortiori to carbon. The *o-o'*-dioxydiphenyl reacts promptly with $SiCl_4$ to form the *o-o'*-diphendioxydichlorosilane,

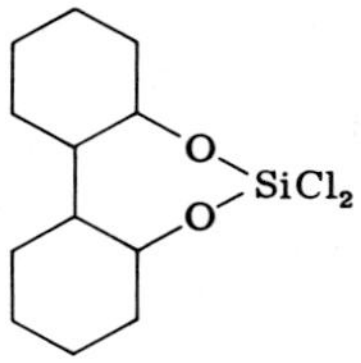

$SiCl_2$

(IV)

and this with metal oxides to form spirodisiloxane,

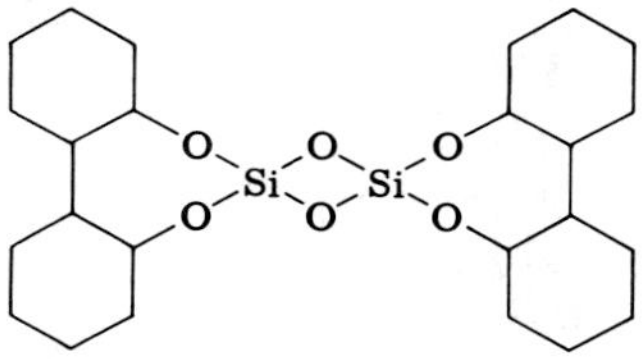

(V)

and annular polysiloxanes, polymerizing to multiple products (spiro-polysiloxanes with heterocyclic rings of eight Si atoms bonded over oxygen bridges to a higher formula unit, and molecular weights of the whole up to 5000), as R. Schwarz and W. Kuchen[525]

[523] *Z. anorg. u. allgem. Chem.*, **275**, 1954, 176-192.

[524] *Z. anorg. u. allgem. Chem.*, **275**, 1954, 193-197.

[525] *Z. anorg. u. allgem. Chem.*, **279**, 1955, 84-93. We may also emphasize the interesting steric type of octamethyl spiro-pentasiloxane:

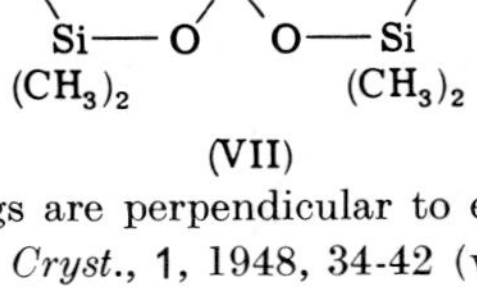

$$\begin{array}{ccccc} (CH_3)_2 & & & & (CH_3)_2 \\ Si\text{—}O & & & & O\text{—}Si \\ O & & Si & & O \\ Si\text{—}O & & & & O\text{—}Si \\ (CH_3)_2 & & & & (CH_3)_2 \end{array}$$

(VII)

in which the planes of the two rings are perpendicular to each other; cf. W. L. Roth, *J. Am. Chem. Soc.*, **69**, 1947, 474-475; *Acta Cryst.*, **1**, 1948, 34-42 (with D. Harker).

described them. Also interesting is the *non*-polymeric silane

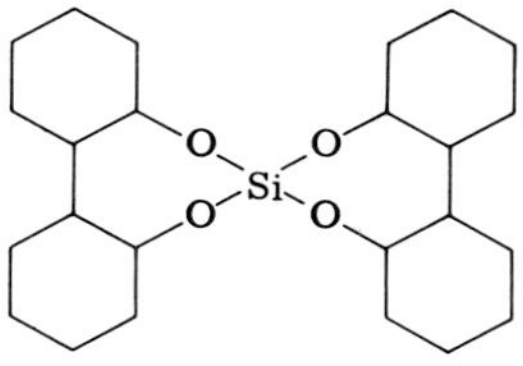

(VI)

formed from *o,o'*-dioxydiphenyl directly.

320. In parallelism with the work of R. Schwarz and his school, we wish to report briefly on analogous investigations made by J. Goubeau, first on the derivatives of the silicon tetrahalides. With R. Warncke,[526] Goubeau studied the partial hydrolysis of $SiCl_4$ in ether solution by ice at — 75°C., and at room temperature with hydrates of different salts, hydroxides, and others. Among the intermediate products of hydrolysis we mention the isolated compounds $SiCl_3OH$, $SiCl_3$–O–$SiCl_3$, $SiCl_3$–O–$SiCl_2OH$, $SiCl_3$–O–$SiCl_2$–O–$SiCl_3$, $Si_4O_3Cl_{10}$, and so on, as higher polymers of the homologeous series $Si_nO_{n-1}Cl_{2n+2}$. The Raman spectra[527] of these compounds are very similar with the constant frequency $\nu = 421$ cm.$^{-1}$ of the pulsation vibration of the $SiCl_3$ groups, and further $\nu = 140$ cm.$^{-1}$ and 220 cm.$^{-1}$, but with a low intensity of the Si—O frequencies as a consequence of the strong polarity of the Si—O bond. The angle Si—O—Si is rather constantly = about 160°. Goubeau and H. Grosse-Ruyken[528] were not able to prepare fluorosiloxanes directly by a partial hydrolysis of SiF_4 but by a double conversion of SiF_4 with salts of oxygen acids, or with metal oxides at temperatures between 400° and 900°C. The fluorosiloxanes could be synthesized with a good output by a reaction of anhydrous Na_2SiO_3 at 720° to 750°C., or with glass wool at 420°C., or with CuO at 480°C. in a circulation apparatus. Isolated molecules were SiF_3–O–SiF_3, SiF_3–O–SiF_2–O–SiF_3, and a viscous mixture of higher polymers $(SiOF_2)_n$. These compounds are in the lower-molecular types highly sensitive to water and HF. With acetic acid, $SiCl_4$ reacts to form an interesting trichloro-silicon acetate, $SiCl_3.OOC.CH_3$, which is able to react further to higher polymers with long Si—O—Si chains, of relatively high temperature stability, for example

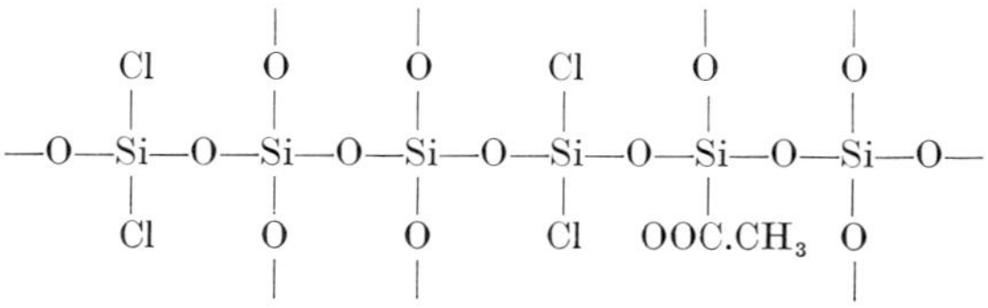

```
            |      |          |     |
    Cl      O      O     Cl   O     O
    |       |      |     |    |     |
—O—Si—O—Si—O—Si—O—Si—O—Si—O—Si—O—
    |       |      |     |    |     |
    Cl      O      O     Cl OOC.CH3 O
            |      |                |
```

[526] *Z. anorg. u. allgem. Chem.*, **259**, 1949, 109-120.

[527] *Z. anorg. u. allgem. Chem.*, **259**, 1949, 233-239; 240-248 (with H. Siebert and M. Winterwerb) for the methylchlorosilanes.

[528] *Z. anorg. u. allgem. Chem.*, **264**, 1951, 230-248. On trichoro-silicon acetate see J. Goubeau and R. Mundiel, *Ibid.*, **272**, 1953, 313-325.

$SiCl_3.OOC.CH_3$ (molecular weight 194) does not melt as a crystalline material, but forms a glass on cooling. Its Raman spectrum and that of similar compounds with the characteristic group $SiCl_3$, shows the typical frequencies at 230 cm.$^{-1}$, 290 cm.$^{-1}$, and 360 cm.$^{-1}$.

321. The deposition of hydrogels of silicic acid by hydrolysis of $SiCl_4$ by water vapor can be controlled on materials of a high exterior surface, as on the cellulose fibers of cotton wool as H. W. Kohlschütter and H. Mattner[529] observed. The deposits after calcination and combustion of the organic support form a very exact pseudomorph or replica of the cellulose filaments, but they are more filmlike and, therefore, have adsorption and capillary condensation properties inferior to those of typical silica gels.

322. Concerning the wide field of the modern *silicones* in which the principles of the classical synthetic organic chemistry are most efficiently combined with those of the silicates, the evolution of research has become so tremendously specific that we can only present here a very limited hint on a few points of some importance. We refer in detail to the special comprehensive literature.[530] As a basis for production of temperature-stable silicone oils or fluids we mention the example of tetracosamethyl endeca siloxane,

$$(CH_3)_3Si{-}O{-}(\overset{\displaystyle CH_3}{\underset{\displaystyle CH_3}{\overset{|}{\underset{|}{Si}}}}{-}O{-})_9{-}O{-}Si(CH_3)_3$$

synthesized by W. I. Patnode and D. F. Wilcock[531] from hexamethyl disiloxane with octamethyl cyclotetrasiloxane. By varying the ratios between these compounds other methyl polysiloxanes can be prepared with desirable molecular weights and lengths of linear chains. The multiple uses of the silicones, as in the glass industry

[529] *Z. anorg. u. allgem. Chem.*, **262**, 1950, 122-128.

[530] See the monograph of E. G. Rochow, "Introduction to the Chemistry of the Silicones," John Wiley & Sons, New York, 2nd edition, 1951, 213 pp.; R. R. McGregor, "Silicones and Their Uses." McGraw-Hill Publ. Co., New York, 1954, 302 pp.; review articles in "Encyclopedia of Chemical Technology," editors R. Kirk and D. F. Othmer, Interscience Encyclopedia, Inc., New York, Vol. **12**, 1954, 393-413; Supplement-Vol. I, 1957, 755-770; Walter Noll, "Silicone (Alkyl- und Aryl-Polysiloxane)," in "Chemische Technologie," Editors K. Winnacker and E. Weingaertner, Carl Hauser Verlag, München, Band **4**, 1954, 735-763; P. D. George, M. Prober, and D. R. Elliott, *Chem. Rev.*, 1956, 1065-1219. On the important correlations between the constitution of silicones and that of silicate minerals cf. W. Noll, *Fortschr. Mineral.*, **34**, 1956, 63-84. Strictly spoken, the nomenclature of "silicones" is not entirely correct; they have nothing to do with organic ketones. One may characterize them e.g. as "softened glasses." See also G. Paoletti, *Vetro e Silicati*, **2**, 1958 (11) 29-35.

[531] *J. Am. Chem. Soc.*, **68**, 1946, 358-363.

were demonstrated by A. Richard[532] especially for hydrophobic coatings, as corrosion-protecting films on glass container surfaces, as dressings for glass fiber textiles, and similar applications, in which the silicones are much superior to most common organic agents. The structural analogies in the constitution of glass and silicones in their inorganic portion (the Si—O—Si bridges) are the basis for successful combination of both kinds of materials. The specific chemical inertness of the organic share in the alkyl and aryl groups offers the second important contribution to an improvement of glass surfaces to make them water-repellent and chemically inert, as H. Wessel[533] discussed in details. By "burning" of silicone oils at 300°C. or higher on the glass surface the valuable repellent properties are even increased, and the electric conductivity of the surface is drastically reduced, as it is important for ceramic insulator materials (e.g., Steatite). The length of the alkyl chains of silicones applied to glass and ceramic surfaces determines the degree of hydrophobicity and the coefficient of exterior friction, the small CH_3- group causing on the other hand an improved heat resistivity. Wessel not only discussed the linear structures in silicones but also those with a wollastonite-like constitution of the type

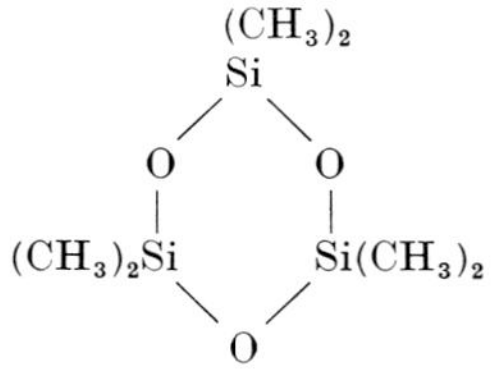

```
              (CH3)2
                Si
              /    \
             O      O
             |      |
       (CH3)2Si      Si(CH3)2
              \    /
                O
```

and the formation of tetrafunctional group reaction to form three-dimensional frameworks.

323. An interesting different type of silicoorganic synthesis is shown by U. Wannagat and Fr. Brandmair[534] in the reaction of $SiCl_4$ with the lithium compound of *p*-dimethylaminophenyl, $(CH_3)_2N$—⬡—) which brings about the formation of the "*Silicon Crystal Violet,*" $[(CH_3)_2N\text{-}C_6H_4]_3SiCl$, and of other derivatives in which

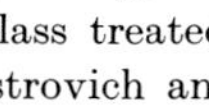

[532] *Verres et réfractaires*, **11**, 1957, 218-222.

[533] *Silikat Tech.*, **6**, 1955, 3-9. See also W. Noll and H. Weissbach, *Zement-Kalk-Gips*, **9**, 1956, 476-486, on silicone application for construction materials, and on hydrophobicity of silicone-treated glass J. Rathouský and V. Chvalovský, *Chem. průmysl*, **9**, 1959, 657-661. An improvement of the mechanical properties of glass treated in this way by coating with silicoorganic compounds is described by S. I. Sil'vestrovich and I. A. Boguslavskiĭ, *Doklady Akad. Nauk S.S.S.R.*, **124**, 1959, 1362-1365; *Steklo i Keram.*, **17**, 1960 (1) 7-12. Particularly efficient as a glass reinforcing measure is the chilling in silicoorganic liquids, superior to common quenching in air or in oil which often causes an irregular heat distribution ("mottling"). On the action of silicones on glass surface see recently H. E. Schwiete, Th. Langauer, and G. Auchter, *Glastech. Ber.* **35**, 1962, 65-71.

[534] *Z. anorg. u. allgem. Chem.*, **280**, 1955, 233-240; **288**, 1956, 91-100.

the Cl atom in Si-Crystal Violet is replaced by the methoxyl group, $—OCH_3$, or by the amino group, NH_2, in so-called silazanes, or the groups $—OSi[C_6H_4–N(CH_3)_2]_3$ and $—NHSi[–C_6H_4–N(CH_3)_2]_3$. The Silicon Crystal Violet is entirely *colorless*, insoluble in water, but well soluble in organic solvents, thus much different from ordinary carbon-Crystal Violet. The ultraviolet absorption is strong at 282 mμ corresponding to the band of carbon Crystal Violet at 590 mμ (cf. Fig. A. 201).

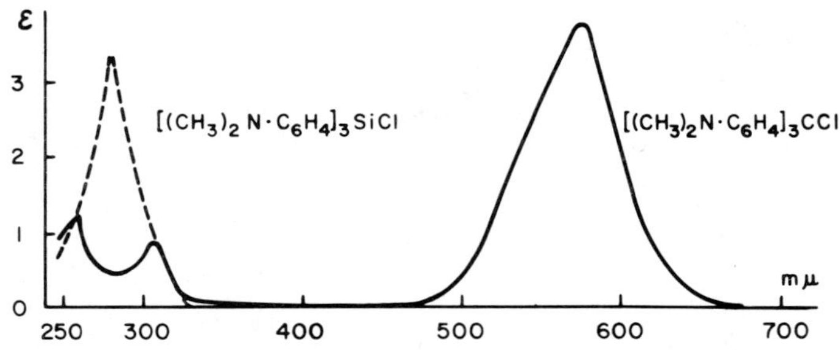

FIG. A. 201. Absorption bands of silicon Crystal Violet in the ultraviolet, and of carbon Crystal Violet i2 visible light, in 5×10^{-5} molar solution. (Wannagat and Brandmair).

The infrared absorption spectra of both compounds are much different; the silicon compound distinctly contains bands of dimethyl aniline which are absent in that of the carbon Crystal Violet. Also the X-ray diffraction diagrams show essential differences demonstrating that the polarity of the Si—Cl bond is lower than that of C—Cl. A characteristic chemical difference is also the prompt reaction of $[(CH_3)_3N–C_6H_4]_3COH$ as a base with HCl to the carbon Crystal Violet (the chloride) whereas the corresponding compound with —SiOH forms with HCl:

$$[Cl^-(NH_3)_2NH^+–C_6H_4]_3Si–OH.$$

324. Important experiments on the modification (hydrophobization) of glass surfaces by *p*-nitrobenzyl bromide, $Br–CH_2–C_6H_4.NO_2$, were made by W. K. Haller and H. C. Duecker[535] at 100°C. in a 30 per cent toluene solution. The measurement of the contact angle shows a final establishment of a constant value after 60 hours, whereas the simple treatment of glass surfaces with the quaternary cetyl pyridinium hydrochloride shows after a high initial hydrophobicity by cation exchange reaction a continuous decrease in the contact angle finally approaching the common wettability of untreated glass surfaces. The specific reaction of the glass with the $—CH_2Br$ group is doubtless interpreted by the observed formation of free HBr, strictly speaking the strongly hydrophobic complex $\equiv Si—O—CH_2C_6H_4.NO_2$ is anchored

[535] *Nature*, 178, 1956, 376-377. Thesis for Master of Science Degree, University of Toledo 1955, 83 pp.

on the glass surface. Haller and Duecker also attempted with some positive success the reduction of the nitro group in the reaction product of glass with *p*-nitrobenzyl bromide (as a "lachrymator") to an amino group, and to convert this to a diazo group, coupling it to another compound with an auxochromic group (e.g., $(CH_3)_2N.C_6H_4$) to produce a dyestuff derivative of high color intensity.

325. A highly fascinating possibility of the halogenation reaction of silicates with *thionyl chloride*, $SOCl_2$, was investigated by H. Deuel[536] as the starting point of many secondary reactions of general interest. The importance of these investigations is to be discussed in connection with problems of surface chemistry of clay minerals (cf. Vol. I, C. ¶ 304 f.). We will therefore preferably discuss the general principles followed in the work of Deuel *et al.* It was known before that the Si–OH groups of silanols, and therefore also of clay minerals as related complexes can be quantitatively determined by the method of Zerewitinoff, namely by a reaction with CH_3MgI evolving methan. The reaction of montmorillonite as a clay mineral type with diazomethane introduces the methoxyl group (cf. G. Berger, 1941), or with trimethyl monochlorosiloxan (cf. E. G. Rochow, 1946) hydrophobic products with the groups clay $\equiv$Si—O—$Si(CH_3)_3$, and very stable complexes also with mustard gas, $Cl(CH_2)_2S(CH_2)_2Cl$, of the type clay $\equiv$Si—O—$(CH_2)_2S(CH_2)_2OH$.

326. The basis reaction with thionyl chloride used by Deuel creates a direct bond of Si with Cl, thus to form a montmorillonite chloride which is changed with alcohols to alkoxy derivatives with clay $\equiv$Si—OH, or by Grignard's reagent one may develop direct bonds clay $\equiv$Si—O—alkyl, by Friedel-Crafts reaction complexes with clay $\equiv$Si—O—aryl, among which the simple phenyl montmorillonite was particularly studied. The X-ray diffraction diagrams of such multiple products do not show characteristic or essential changes in comparison with the initial montmorillonite, but in infrared spectrograms of phenyl montmorillonites one finds a pronounced decrease in intensity of bands at 2.8 to 3.1 μ (3200 to 3600 cm.$^{-1}$), or of the bands of hydroxyl at 1645 cm.$^{-1}$ (6 μ). The substitution of OH by phenyl groups appears evident. Particularly large changes are also seen in the differential-thermal analysis curves of phenyl montmorillonites of different degrees of substitution. Summarizing, Deuel compares the clay minerals (especially in their H-modification) with silanols. The organic derivatives are stable to high vacuum treatments and are not simply decomposed by acids or bases. The degeneration of phenyl montmorillonite was especially discussed by R. Gentili and H. Deuel in the durability of the Si–Phenyl bond to dilute NaOH solution, as studied by chromatographic analyses of different

[536] *Helv. Chim. Acta*, **33**, 1950, 1229-1232 (with G. Huber and R. Iberg); *Kolloid-Z.*, **124**, 1951, 164-169; *Clay Minerals Bull.*, **1**, 1952, 205-211; *Helv. Chim. Acta*, **34**, 1951, 1697-1701 (with G. Huber); *Ibid.*, **35**, 1952, 1799-1802 (with H. H. Günthard); *Ibid.*, **36**, 1953, 808-810 (with R. Iberg); *Ber. deut. keram. Ges.*, **31**, 1954, 1-7; *Helv. Chim. Acta*, **40**, 1957, 106-113 (with R. Gentili); *Agrochimica*, **1**, 1957, 248-267.

reaction products. A prolonged alternate treatment with HCl and NaOH finally breaks down the phenyl montmorillonite to decomposition products of low molecular weight but still containing the groups Si—C_6H_5.

327. Much criticism has arisen against the conclusiveness of Deuel's observations, comparable to that against Berger's diazomethane method. Especially G. Brown, R. Greene-Kelly, and K. Norrish[537] doubted whether the phenyl montmorillonite is really a compound of the alleged individuality, not only an adsorption product of organic "contaminants" on H-montmorillonite, say with thiophene. A serious objection may also be that of R. Schwarz and H. W. Hennicke.[538] If the crystalline disilicic acid $(H_2Si_2O_5)_n$ (cf. Vol. I, C. ¶ 268) is used as a model reagent for the conversion with diazomethane and thionyl chloride, no conversion at all was observed with $SOCl_2$. A reexamination of Wyoming bentonite in its behavior with $SOCl_2$ did not confirm Deuel's conclusions.

328. Silicones with an urotropine (hexamethylene diamine)-like structure were synthesized by E. Wiberg and W. Simmler[539] which differ from the common linear silicones by formation of honeycomb-like networks of tye type $(RSiO_{1.5})_4$ with interesting properties (soft crystalline solids with a sublimation tendency at room temperature or slightly elevated temperature). These compounds are formed if alkyl trichlorosilanes are hydrolyzed in ether solution. The constitution is of the type

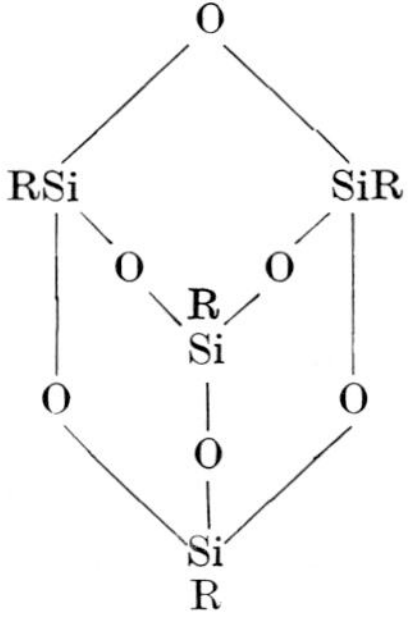

as with $R = C_4H_9 : [(C_4H_9)SiO_{1.5})_4$, as a symmetrical "silicon-adamantane" of good soulbility in organic solvents because the inorganic siloxane skeleton is surrounded in multiple by alkyl radicals. The X-ray diffraction diagram is rather similar to that of common urotropine.

329. A reaction of general great importance for determination of the hydroxyl groups in silico-organic compounds is the conversion with phenyl isocyanate,

[537] *Nature*, **169**, 1952, 756-757; *Clay Minerals Bull.*, **1**, 1952, 214-220.

[538] *Z. anorg. u. allgem. Chem.*, **283**, 1956, 347-350.

[539] *Z. anorg. u. allgem. Chem.*, **282**, 1956, 330-344.

C_6H_5.NCO, with a change of the silanol groups —Si—OH to siloxane, —Si—O—Si—. K. Damm and W. Noll[540] studied this reaction especially in comparison and combination with the determination of free molecular water formed in the phenyl isocyanate reaction, by the Karl Fischer reagent (methanol solution of pyridine and SO_2, with iodine as indicator), by simple titration, with the possibility of a sharp discrimination of reaction rates of the OH groups in different bonding states, as in "silicic acids" of different origin, but also in silica glass. The latter has a hydroxyl content of only 0.2 to 0.3 per cent OH, whereas the common poly-condensed silicic acids may contain 2 to 3 per cent in silanol-type bonding. As a control of the silanol groups in silicones the phenyl isocyanate and Karl Fischer reagent determinations are of particular importance for an accurate classification of their hardening characteristics and in this connection also for their dielectric properties (loss angle tan δ).

330. H. W. Kohlschütter and G. Jaekel[541] studied the step-wise condensation of $SiCl_4$ in reaction with 1,4 butane diol in a solution in dioxane, and the observation of the evolution of three-dimensional frameworks. In the same way, Kohlschütter[542] discussed the parallels between silicones and silico-organic resins, with silicate glasses and their frameworks. The "silico-organic gels" are intermediate between the silicate glasses and highly-polymerized organic resins, with the same principles of interlinkage in respect to the stereochemistry of high-polymerization. This parallel is important in view of the technological application in construction materials as was outlined before by E. Brandenberger,[543] and even their "pseudocrystallinity" in configurations for which this author gave many examples, adding some in connection with hydrodynamic flow and similar phenomena.

331. From the viewpoint of the theory of cracking catalysts on the basis of aluminosilicic acids (cf. Vol. I, C. ¶ 211 ff.). J. D. Danforth[544] discussed interesting condensation products of methylsiloxanes with $Al(OH)_3$ and aluminum isopropoxide, with definite ratios of Si : Al atoms as a function of available OH groups bonded to silicon. A typical annular representative of this kind would be derived say from dimethyl diethoxysilane, the "hydrate" in formula (A), the "anhydride" in (B) and (C), methyl-triethoxysilane in (D), and the final "aluminosilicic acid" in (E), the acidity of which depends on the OH groups.

[540] *Kolloid-Z.*, **158**, 1958, 97-108; *Fortschr. Mineral.*, **37**, 1959, 57 (with K. Damm and R. Fauss). On the Karl Fischer reaction see also H. Gilman and L. S. Miller, *J. Am. Chem. Soc.*, **73**, 1951, 2367-2368. On the condensation and thermal durability of silicone resins see W. Noll, K. Damm and W. Krauss, *Farbe u. Lack*, **65**, 1959, Jan. 17-24.

[541] *Z. anorg. u. allgem. Chem.*, **271**, 1953, 185-205.

[542] *Glastech. Ber.*, **24**, 1951, 29-35.

[543] "Grundlagen der Werkstoffchemie," Rascher-Verlag, Zürich 1947, 298 pp. Concerning the hydrodynamic effects (Darcy constants, etc.) cf. R. Signer and H. Egli, *Rec. trav. chim.*, **69**, 1950, 45-58.

[544] *J. Phys. Chem.*, **59**, 1955, 564-566.

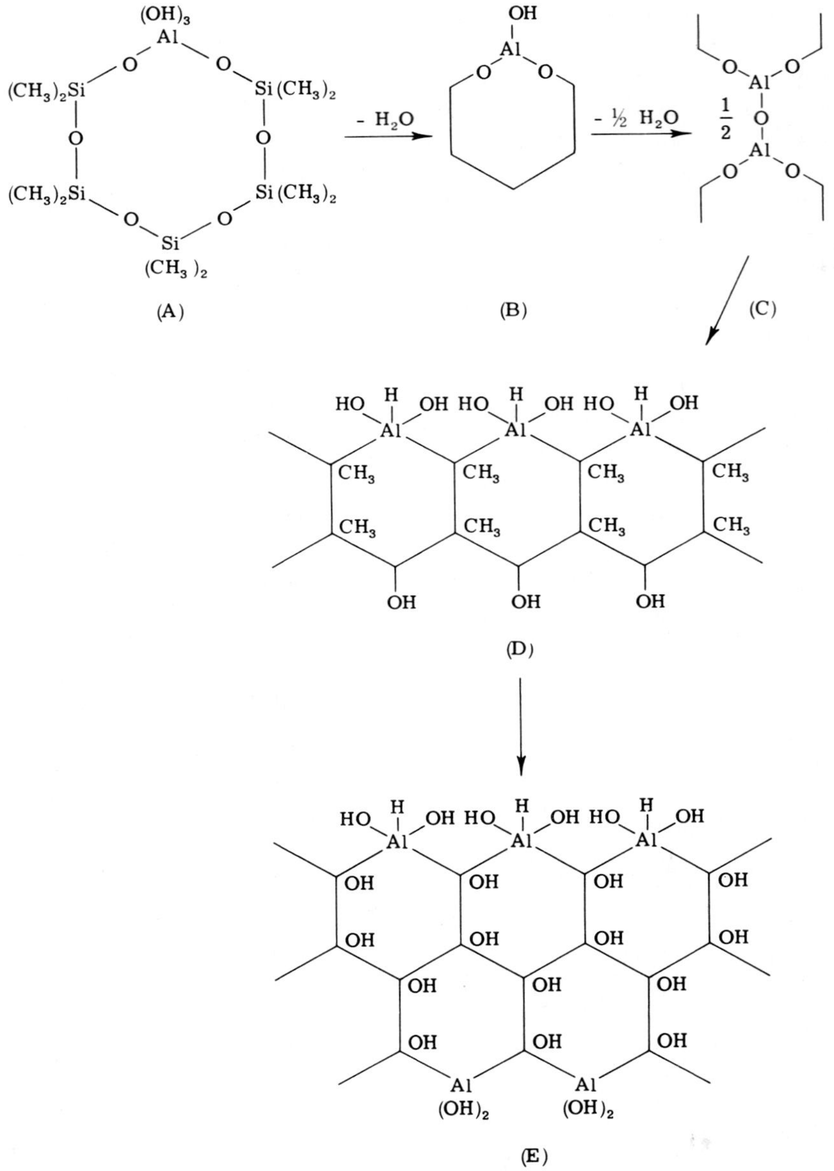

(A)

(B)

(C)

(D)

(E)

332. Another wide field in which the formation of silico-organic compounds was extensively discussed is that of the solubility of silica in organic media as a factor in *biochemical* and even *pathological* (silicosis, asbestosis) phenomena. Direct electron microscopy evidence of the reaction of quartz, feldspar, kaolinite, and clay minerals was especially given by Fr. Oberlies and G. Pohlmann,[545] not only by low organisms (autotrophous bacteriae, fungi, etc.) but also on glass surfaces by animalic tissues, sera, slimes, and the like. The great significance of such reactions with organic solutions, as in the fermentation process of ceramic clays was discussed in Vol. I, C. ¶ 233. Electron microscopy is thus able to confirm many of the conclusions made by L. Holzapfel[546] on the solubility of silica in biological solutions, culture media, and others, with Aspergillus niger, leading to a general solubility theory of silicosis lungs (see below).

333. When we discussed the reactions of silicic acid derivatives with dioxybenzene by R. Schwarz and W. Kuchen (see footnote 522) we have to refer to the early synthesis of tri-pyrocatechol silicate by A. Rosenheim, B. Raibmann, and S. Schendel[547] which demonstrated wide possibilities of the formation of soluble silico-organic compounds which we nowadays calculate for the large group of *chelates* with a complex

$$\left[\mathrm{Si}\left(\begin{matrix}-\mathrm{O}-\\-\mathrm{O}-\end{matrix}\mathrm{C_6H_4}\right)_3\right]^{2-}$$

Fr. Bacon and Fr. C. Raggon,[548] on the other hand, investigated the solubility of industrial glasses in neutral sodium citrate solution with a severity similar to that of substantially alkaline solutions (cf. B. ¶ 189), at pH from 5 up to 7.6, mostly in the neutral range. No doubt can exist that this specific effect with complex-forming silico-organic anions is of great practical significance, say if anticoagulants on the basis of sodium citrate evolve a high silica content, whereas acid citrate dextrose solutions dissolve only small amounts. The citrate solubility is not unique; also sodium gluconate, oxalate, tartrate, malate, and ascorbate solutions have (in decreasing measures) similar effects. In this series should also be included the solubility with versenate. This interesting material (a salt of ethylene diamine tetraacetic acid, abbreviated EDTA) forms extremely stable complexes with "sequestered" Al^{3+}, Ca^{2+}, or Mg^{2+} ions, analogous to the sequestering of Si^{4+} by pyrocatechol in a soluble chelate, as F. M. Ernsberger[549] described from etching experiments with

[545] *Naturwissenschaften*, **45**, 1958, 487; 513-514; *Ibid.*, **47**, 1960, 58; *Trans. Intern. Ceram. Congr.*, VI, Wiesbaden 1960, 149-168; *Veröffentl. Max-Planck-Inst. Silikatforsch.*, **20**, 1960, 223-225; *Glastech. Ber.* **35**, 1962, 60-64. In this connection we refer to general information in older literature collected by R. K. Iler in his book "The Colloid Chemistry of Silica and Silicates," Cornell Univ. Press, Ithaca, N.Y. 1955, 324 pp., especially the fascinating Chapter IX on silica in living organisms, pp. 276-299.

[546] *Naturwissenschaften*, **42**, 1955, 255.

[547] *Z. anorg. u. allgem. Chem.*, **196**, 1931, especially pp. 161-164. On analogous reactions of pyrocatechol with silicic acid on a cation exchange resin column at 20°C. in an aqueous medium see Arm. Weiss, G. Reiff, K. Holzapfel, and Al. Weiss, *Beitr. Silikose-Forsch.*, Sonderband "Grundfragen aus der Silikose-Forschung," Vol. II, 1956, 303-316.

[548] *J. Am. Ceram. Soc.*, **42**, 1959, 199-205.

[549] *J. Am. Ceram. Soc.*, **42**, 1959, 373-375.

soda or sodium acetate containing mixed solutions of these agents corroding glass surfaces.

334. If on the other hand, qualitative observations were available showing the strong reaction of plasma proteins, fibrinogen (strongest), and (in decreasing measure) β-, γ-, α-globuline, albumine, K. Karasawa, Y. Yasui and T. Mori[550] came to the conclusion, as a plausible consequence of these facts, that silicosis might be explained by a dissolution of quartz silica in the serum of the living human organism, especially in the lungs. With this consideration, we enter the field of strongly contrasting theories on the origin of this disease, and we must emphasize from this start that the dissolution theory is only one of a series of attempts to integrate the problems here involved. An enormous quantity of facts have been accumulated on the constitution of the so-called lung stones or nodules in typical silicosis or mixed dust pneumoconiosis tissues and "callosities." The fact is that quartz particles are the most frequent "insoluble" material in these formations which, however, may also bear some hydroxyl apatite as W. Ehrhardt and F. Leicher[551] described, as well as kaolinite and other clay minerals. We may recall that L. Holzapfel (since 1942) as one of the typical representatives of the solubiltiy theory, observed the formation of the cholesteryl ester of silicic acid[552] associated with free cholesterine, galactose, and so on.

335. In view of the tremendous complexity of silicosis problems it is most valuable that Arm. Weiss, G. Hofmann, and G. Reiff[553] made a thorough microscopic-mineralogical determination of the nature of the dust particles from silicotic lungs in order to find definite classification principles from which the pathogenetic questions may be challenged. It is an important finding that in addition to quartz, nearly in every case, a "mica-like" mineral occurs, along with an amorphous silica substance which may have been formed by decomposition of the organic material with formamide, namely as a secondary product of the preparation method. The mica-like mineral is *not* identical with sericite, but shows a distinctly increased basal distance $c_0' = 10.4$ Å. (a somewhat broadened interference line), and in the electron microscope very thin films with diffraction fringes. The reason for the enlarged spacing is seen by Weiss in the intercalation of organic cations (in base exchange with Na^+) among which those with triethylene tetramine, cadaverine, putrescine, and serum

[550] *Asahi Garasu Kenkyu Hokoku*, **4**, 1954, 141-151.

[551] *Arch. Gewerbepathol. Gewerbehyg.*, **15**, 1956, 1-18.

[552] We mention here more recent observations in *Naturwissenschaften*, **34**, 1947, 189-218; **35**, 1948, 314; **36**, 1949, 252-254 (with W. Engel); 375, especially on the role of galactose as adsorption component on asbestos.

[553] *Beitr. Silikose-Forsch.*, Sonderband "Grundfragen aus der Silikose-Forschung." Vol. III, 1958, 21-34, 45-68, especially on the "mica-like" mineral. Some new crystallographic aspects are recently discussed by H. Jung and W. Drees, *Silikattechnik*, **12**, 1961, 294-300.

albumine are particularly interesting. By calcination at 350° to 500°C. the spacing is reduced to 10.0 Å.

336. Model experiments with an exchange of the cations of dodecyl ammonium show X-ray diagrams of the products very similar to the mica-like material from silicosis lungs. It is suggested by the experiments of Weiss that catalytic splitting of proteins to smaller *peptide* chains

```
        O                   O                   O
        |                   |                   |
        C     NH      |     C     NH      |     C
      /   \  /   \  /   \  /   \  /   \  /   \  /   \
    NH     |      C      NH     |      C      NH     |
           |      |             |      |             |
          NH2     O                    O            NH2
```

may illustrate the mechanism of the intercalation of organic proteinase substance between the layers of this mineral according to the scheme of Fig. A. 202. The original silicate shows fixed negative excess charges and strongly acid hydroxonium ions.

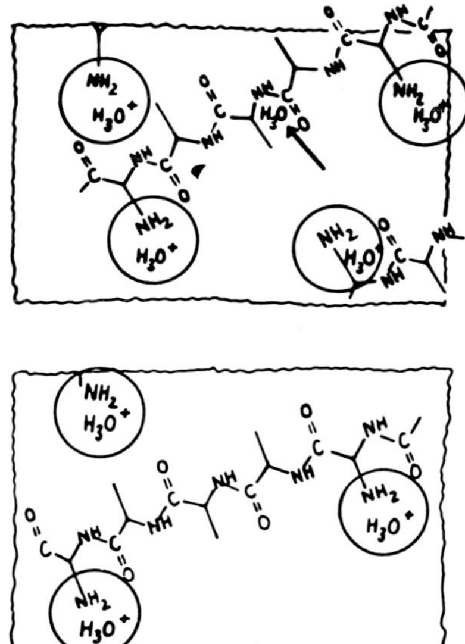

FIG. A. 202. Schematic interpretation of a catalytic protein hydrolysis (at pH = 4.0 to 6.0) by swelling layer structures of mica type. (Arm. Weiss).

The basic groups of the protein react with these to form ammonium ions (written as $(NH_2.H_3O)^+$). By acid hydrolysis the peptide bonds of the protein macromolecule are split and peptide chains adsorbed on the layer structure, either with a high, or a low excess in negative charges as shown in Fig. A. 202 (cf. Vol. I, C. ¶ 298 ff.).

337. Armin Weiss, G. Reiff, and G. Hofmann[554] further studied the amorphous silica in silicotic lungs which in its quantitative amounts shows a distinct relation to the *degree* of the silicosis, in contrast with quartz or other crystalline silicates for which such a relation was not observed. In spite of that, it is most probable that the amorphous silica is *not* primary in the pathologically changed tissue, but formed during the formamide decomposition process for the electron-microscopic inspection, chiefly derived from organic silicic acid esters, specifically from 1,3 diketones (e.g., acetylacetone), 1,2 dienoles (ascorbic acid), with hydroxyl groups in o-position (like in pyrocatechol, epinephrine = adrenaline, dioxyphenylalanine). Amorphous silica (Aerosil gel, and others) reacts quantitatively with pyrocatechol in a few days, diatomaceous earth initially also very promptly, later sluggishly, quartz and cristobalite very slowly. In reactions with gallic acid, or with tanning substances the formation of difficultly soluble depositions on quartz was observed, and it may be considered whether such compounds might become important as protection materials. In every case the fact that firm bonds Si—O—Si in silica (crystalline, or amorphous) can be quantitatively split by the formation of silicic acid esters of the o-diphenol derivatives in aqueous solutions below 50°C. and at pH between 6.0 and 8.0 is most remarkable. The paper-chromatographic separation makes it probable that besides low-molecular compounds also difficultly soluble, fibrous high-polymers occur, as in this example:

OH O O O O O O O O O O OH
Si Si Si Si Si
OH O O O O O O O O O O OH

We wish in this connection also to emphasize the discussion of several authors[555] on the raction between silica hydrosol and ribonucleinic acid, causing a callosity

[554] *Wiss. Forschungsber. Naturw. Reihe*, **66**, 1958, "Die Staublungenerkrankungen," Vol. III, 79-91; cf. also A. Weiss and G. Hofmann, *Beitr. Silikose-Forsch.*, Sonderband II, 1956, 281-302; Arm. and Al. Weiss, G. Reiff, and K. Holzapfel, *Ibid.*, pp. 303-316, especially on dioxybenzene reactions with silicic acid in an aqueous medium at 20°C. on a cation exchange resin column. On important animal tests for the toxicity of different silica and silicate minerals, see systematic investigations by K. G. Schmidt and H. Lüchtrath, *Ibid.*, **44**, 1956, 32 pp.; **55**, 1955, 43 pp.; **58**, 1959, 37 pp.; **61**, 1959, 60 pp. New observations of iron phosphate silicate and swellable silicate in silicosis lungs see Armin Weiss, *Ibid.*, Sonderband IV, 1960, 51-65, and *Hoppe-Seyler's Z. Physiol. Chem.*, **324**, 1961, 153-162. See also the important review by U. Hofmann, *Ber. deut. keram. Ges.*, **39**, 1962, 272-279, on general surface reactivity of silica.

[555] *Naturwissenschaften*, **43**, 1956, 68; **44**, 1957, 200 (a correction). The discussion was held

fibrosis, a "poisoning" of the cellular tissue by damaging the breathing process or by diffusing monomeric silicic acid, or by a splitting of the ribonucleinic acid to free phosphoric acid and a silicon-substituted acid. The influence of silicic acid on this hydrolysis is reduced by agents which also inhibit the formation of granuloms.

338. In contrast with the many theories based on silica solubility reactions in lung tissues as the origin of silicosis, H. Seifert[556] developed a highly interesting theory based on the idea that the specific *oriented deposition* of organic material on crystalline quartz (exclusively, or on some other crystalline silicates, such as asbestos particles) would conduct the pathogenic process. This theory is entirely based on the phenomenon of oriented intergrowths and, therefore, discussed extensively in Vol. I, A. ¶ 257 ff. The basic idea that organic compounds of suitable constitution (say aminoacids) can be adsorbed on crystal structural planes following crystallographic laws is expressed in the similitude of a "matrix" or key-and-lode relation to the supporting material, generally called therefore a *matrix-theory*. The experimental basis shows an increasing approximation of the organic compounds which show a positive oriented intergrowth on quartz, to real proteins, starting from glycine,

$$NH_2.CH_2.COOH,$$

to L-oxyproline,

(five-membered ring: N–H at top, COOH on the adjacent carbon, OH on the ring carbon opposite)

alanine,

$$CH_3.CH(NH_2).COOH$$

glutamic acid,

$$COOH.CH_2.CH_2.CH(NH_2).COOH$$

to peptides and mixes of different aminoacids, under the viewpoints of a pathogenic misdirection ("Fehllenkung", Butenandt), in which selectivity of the epitactic phenomenon is particularly striking. If on the other hand, say cystine, *l*-serine, tyrosine, hystidine, arginine, are substances which do not show any positive effect

between W. Kikuth and H. W. Schlipköter, R. Schwarz, and E. Baronetzky, W. Kersten and Hj. Staudinger, mutually.

[556] We select the following orienting publications: *Z. Aërosol-Forsch. u. Therap.*, **2**, 1953, 500-509; *Wiss. Forschungsber. Naturw. Reihe*, **63**, 1954, 199-207; **66**, 1958, 3-26; *Staub*, **18**, 1958, 201-206; *Beitr. Silikose-Forsch.*, Sonderband, "Grundlagen der Silikose-Forschung," Vol. III, 1958, 291-300, further *Naturw. Rundschau*, **14**, 1960, 7-11; 1961, 60-65; *Trabajos reunión intern. reactividad sólidos, 3, Madrid,* 1956, Sección IV, Vol. III, 1959, 93-113. Especially on oxyproline also K.-F. Seifert, *Z. Krist.*, **114**, 1960, 361-392.

of this kind, we must not feel surprised in view of the evidently quite specific conditions of crystal structures with different spacings which will be, perhaps, in the majority of cases inadequate to "fit" the quartz structure.

339. In a careful critical comparison and discussion of the different silicosis theories, W. Stöber and G. Bauer[557] classify the chemical solubility and the surface theories, the matrix theory being a special subgroup to the latter type which postulates long-distance structural-geometric relations between the surface of quartz particles and the sorbendum material whereas a simpler "*contact theory*" is more related to the common theories of heterogeneous catalysis. For the suitability of the matrix theory as its failure it would be decisive, if amorphous silica could be able to start silicosis. Not many indications of such an effect were yet observed, but F. J. Strecker[558] discussed them seriously after the results of injections of trimethyl silyl aërosol in animal experimentation in comparison with the toxic properties of untreated Aërosil. Stöber and Bauer are inclined to give preference to the contact (catalysis) theory as a plausible expression of *local* reactions of definite atom groups on the surface of silica dust particles, and of reversible sorption processes, without any further auxiliary hypotheses.

340. The great interest which we owe to modern evolution of the silico-organic compounds, a few of which are well identified crystalline phases, most of them, however, gels or even glasslike resins, justifies the rediscussion, to a certain degree, of the old problem of a "colloid-chemistry of glass constitution." We believe that it is not necessary to concentrate too much interest on such academic classification questions. Nevertheless, especially Russian investigators rediscussed these points in the Glass Symposium of 1953.[559] Previously in 1952, I. Peychès[560] described the viscous mobility of glass at lower temperatures in analogy with colloid-physical phenomena and chemical heterogeneity, as of different states of polymerization indicated in the chilling and annealing reactions (cf. A. ¶ 50). These discussions include the general usefulness of ideas and experimental methods of modern high-

[557] *Staub*, **19**, 1959, 1-5.

[558] *Beitr. Silikose-Forsch.*, Sonderband II, 1956, 343-350.

[559] *Stroenie Stekla, Inst. Khim. Silikatov Akad. Nauk S.S.S.R., Trudy Soveshchaniya, Leningrad*, 1953 (Pub. 1955), especially P. P. Kobeko, pp. 19-25 on "organic glasses," O. A. Esin and P. V. Gel'd, pp. 44-45 on structures of liquids and glasses, S. P. Zhdanov, pp. 162-175, on microheterogeneity and porosity of glasses (cf. A. ¶¶ 202, 207), L. G. Mel'nichenko, pp. 302-303, on colloid-physical problems of glass, and P. V. Gel'd, pp. 304-306. These reviews may be supplemented by a hint to the article of V. L. Indenbom, *Doklady Akad. Nauk S.S.S.R.*, **89**, 1953, 509-511, and S. P. Zhdanov, *Ibid.*, **92**, 1953, 597-600; *Izvest. Akad. Nauk S.S.S.R.*, Otdel. Khim. Nauk 1955, 197-207; *Bull. acad. sci. U.R.S.S.*, 1955, 353-358 (with E. A. Poraĭ-Koshits and D. I. Levin), on microheterogeneities, further O. A. Esin and P. V. Gel'd, *Steklo i Keram.*, **11**, 1954 (3) 4-6, on glasses and slags in the molten state.

[560] *J. Soc. Glass Technol.*, **36**, 1952, 164-180 T.

polymer investigation as they were originally developed for organic compounds but may also be transferred to silicate glasses. We mention in this respect especially the low-angle X-ray diffraction method which was used successfully by E. A. Poraĭ-Koshits.[561] The reader may use with much profit the introduction into this field as given from the viewpoints of organic macromolecular physical chemistry by H. Staudinger, K. Hammann, and O. Kratky[562] for the many parallels existing between organic, silico-organic, and silicate glasses. Also the electron microscope proved to be a most valuable tool for such studies, as we may see from the results given by W. Vogel *et al.* on borosilicate glasses of the Vycor type (Vol. V, Section B),[563] and on complex model glasses built up from alkali and alkaline earth fluoride systems with beryllium fluoride[564] (cf. A. ¶ 384 ff.). Analogous unmixing dispersoids, as droplets of different melt phases in the glassy state, are visible in the electron microscope also beyond the optical perceptibility of opalescence (Tyndall) phenomena. The highest degree of polymerization in organic resins of the polyethylene and polyamide copolymers was observed in the electron microscope as spherulites and crystallites by V. A. Kargin and T. A. Koretskaya,[565] but these formations are apparently highly sensitive to the electron beam and destroyed (amorphization effect). Spherulites of this kind are developed especially in oriented and stretched films of the polymers.

MOLECULAR EQUILIBRIA IN THE TRANSFORMATION RANGE

341. Starting from the structural similarity of glasses with corresponding crystalline phases, W. Weyl[566] developed extensive discussions on the physicochemical nature of the transitions occurring in the so-called transformation range. These reactions, however, should not be identified with the elementary enantiotropic (high $\rightleftharpoons$ low, or $\alpha \rightleftharpoons \beta$) inversions of crystalline modifications, as A. A. Lebedev (1921) and his school had postulated. The lack of an "over all symmetry" in the glass structure precludes the thermodynamic conditions of such inversions. They may, however, be compared with full justification to the sluggish reactions of crystalline phases of the polytype class, as inversions of a higher order which show in the same way glasses do, *gradual* changes of physical-chemical parameters over an "annealing" range. The states of different degrees of crystalline arrangements (order-disorder

[561] *Glastech. Ber.*, **32**, 1959, 450-459.

[562] *Naturwissenschaften*, **42**, 1955, 221-230; 230-237; 237-252.

[563] *Silikat Tech.*, **9**, 1958, 51-62 (with W. Skatulla and H. Wessel), 323. See recently V. I. Shelyubskiĭ and N. M. Vansfel'd, *Doklady Akad. Nauk S.S.S.R.*, **138**, 1961, 100-101.

[564] *Glastech. Ber.*, **31**, 1958, 15-28; *Silikat Tech.*, **9**, 1958, 353-358; 495-501.

[565] *Doklady Akad. Nauk S.S.S.R.*, **110**, 1956, 1015-1017.

[566] "Phase Transformations in Solids," a Symposium, John Wiley & Sons, New York 1951, 296-334.

effects) may be illustrated as in the "secular" changes of the refractive index, and in the ice-point depression (cf. A. ¶ 285) in mixed alkali glasses.

342. As a typical representative of modern understanding of the physicochemical changes in glass during the transformation we also have to discuss the ideas of Aniuta Winter-Klein[567] who distinguishes the transformation range as intermediate between liquid and solid state of glass, but with a transition also between the liquid and the transformation range in which the glass is an undercooled liquid system in G. Tammann's classical definition, with the special laws of crystallization (devitrification) to be applied in this transition range. As a sensitive indicator of the molecular changes (called α, β modifications although *not* identical with enantiotropic forms of crystal phases) the refractive index was chosen, combined with the idea of the fictive temperature (Tool). In the neighborhood of the absolute zero point the specific heat, c_p, and the entropy, ΔS, are practically identical as previously emphasized by F. Simon (1922). Studies of the refractive index were also previously made in sodium and lead silicate glasses by N. A. Tudorovskaya,[568] although interpreted as direct analogues to the enantiotropic inversions of cristobalite and tridymite. We mentioned already in A. ¶ 269 that A. I. Stozharov[569] refuted this explanation in a discussion of the volume effects of the low-temperature "splitting points" and related molecular phenomena, differing from inversions of the crystalline character. We also previously emphasized how important these problems are in connection with practical requirements for the optimum fine-annealing of optical glass and other glasses, as advanced by Fr. Naudin[570] and B. Daragan.[571].

343. For Jena thermometer glass 16 III, P. Rahlfs and K. Fr. Zobel[572] discussed an important effect of the rates of heating used in the conventional determination of the "trasformation point" by dilatometric measurements, especially for rates of 0.5 to 3.0°/min., (Fig. A. 203). The transformation point also depends on the aging

[567] *Verres et réfractaires*, **6**, 1952, 271-281; "Changement de phases," comptes rendus de la 2ème réunion annuelle tenue en commun avec la Commission de thermodynamique de l'Union internationale de physique, Paris, 2-7 juin 1952, pp. 407-414. See also *Verres et réfractaires*, **11**, 1957, 71-88 (A. ¶ 275).

[568] "Physicochemical Properties of the Ternary System Na_2O–PbO–SiO_2," Edited by I. V. Grebenshchikov, Acad. Sci. U.S.S.R., 1949, pp. 186-200, 201-220; *Stroenie Stekla, Inst. Khim. Silikatov Akad. Nauk S.S.S.R., Trudy Soveshchaniya, Leningrad*, 1953 (Pub. 1955), 190-197.

[569] *Stroenie Stekla, Inst. Khim. Silikatov Akad. Nauk S.S.S.R., Trudy Soveshchaniya, Leningrad*, 1953 (Pub. 1955), 120-125. We may further recall how V. T. Slavyanskiĭ, *Ibid.*, 251-255, attempted to refer certain break points on relative viscosity curves vs. temperature for salol etc. to a "polymorphism" in the liquid (cf. A. ¶ 74 f.).

[570] *Verres et réfractaires*, **6**, 1952, 209-218, 282-289; *Glass Ind.*, **35**, 1954, 542-548; 603-608, 628; 666-671, 688.

[571] *Verres et réfractaires*, **5**, 1951, 135-143; **6**, 1952, 10-21; **7**, 1953, 157-166; *Atti congr. intern. vetro, 3rd Congr. Venice* 1953, 411-425; *Glass Ind.*, **33**, 1952, 69-74, 98-99; **35**, 1954, 423-429, 456.

[572] *Glastech. Ber.*, **28**, 1955, 310-312.

temperature in a simply linear function. The highest transformation points were observed for a heating rate of 5°/min., whereas for 1°/min. the point is by 30°C. lower. For cooling from softening point a rate of 5°/min. is proposed for a standard test of trasformation points and a definite heating rate of 5°/min. too.

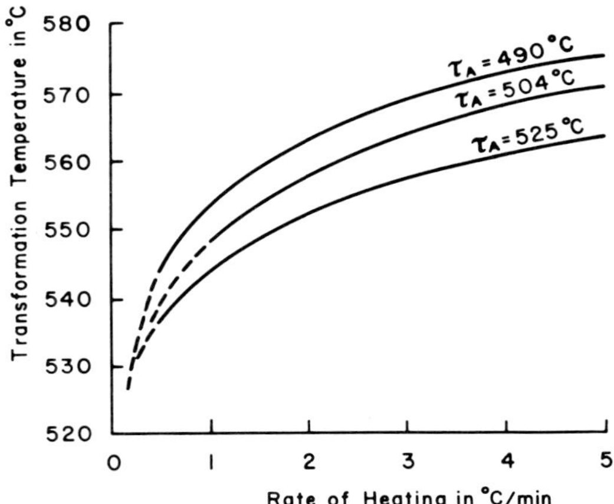

Fig. A. 203. Conventional determination of the transformation point for glass 16 III, as a function of the rate of heating, and for different temperatures of aging. (Rahlfs and Zobel).

344. The thermodynamic consequences of Winter-Klein's interpretation of transformation reactions are in many respects substantiated by measurements of enthalpies (specific heats) and of the heats of dissolution of glasses in different states of thermal history. In the latter connection we emphasize the data given by E. Jenckel and Kl. Gorke[573] on the transition of frozen-in, to equilibrium states as indicated in solution calorimetry, say for polystyrene glass. C. Kröger and G. Kreitlow[574] found the same effects for glasses of the system Na_2O—SiO_2, with a sensitive twin calorimeter method for the dissolution in HF (using "Trovidur"-plastomer containers). The common heats of devitrification measured in this way are, therefore, in reality complex as the sum of the real heat of crystallization, $\Delta H_{cr} = \Delta H_f$, and the heat of freezing-in effects, ΔH_{f_r} (cf. Fig. A. 204). We also refer to theoretical discussions of J. H. Gibbs and E. A. Di Marzio,[575] who derived for chain-shaped polymers with a "quasi-lattice" by statistic mechanics a model process characterizing the transition temperature of second order reactions, increasing in a function to the chain stiffness and length, and decreasing as a function of the free volume. This theory also gives a quantitative expression for specific heat as a function of temperature,

[573] *Z. Naturforsch.*, 7a, 1952, 360-362.

[574] *Glastech. Ber.*, **29**, 1956, 393-400.

[575] *J. Chem. Phys.*, **28**, 1958, 373-383; cf. W. Kauzmann (1948).

in good agreement with experimental data, as for polystyrene. The theory evolved by Gibbs and Di Marzio is important for determination of the lower limit of the transformation range observed for infinite times. If it really exists, the glass formation cannot be dependent alone on kinetic phenomena, and E. Berger's concepts (1931) and discussions on glass as the "fourth state of matter" may become substantial again.

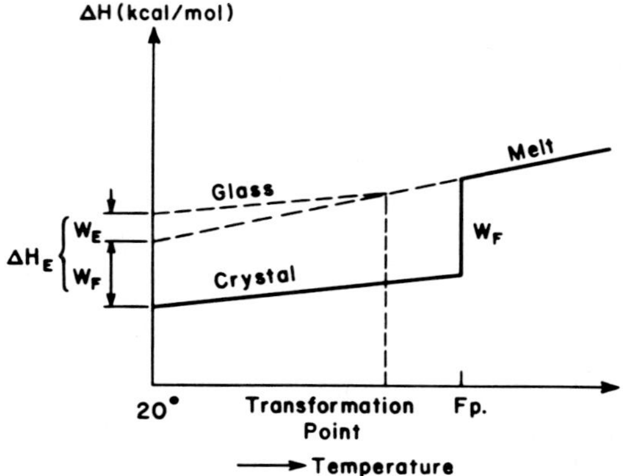

FIG. A. 204. Heat contents of glass, the corresponding crystalline phase, and the undercooled melt. (Kröger and Kreitlow). ΔH_f is the heat of fusion and crystallization; ΔH_{fr} that of freezing-in reactions.

345. We mentioned in A. ¶ 138 ff. that the temperature function of the electric conductance of glass over the transformation range as determined by O. V. Mazurin[576] and by M. Munakata and M. Iwamoto[577] during stabilization of glass shows remarkable analogies with those in oxide semiconductors. The width of the potential barrier for alkali ion diffusion in glass decreases in the measure of the gradual stabilization and with increasing entropy, ΔS^*. The constant A in the Rasch-Hinrichsen law (written in the form $\log \varrho = -A + \Delta H^*/T$) is changed. A calculation of free energy, ΔF^*, is also possible. When the glass is stabilized at constant temperature, free energy is little changed, whereas ΔH^* and ΔS^* increase. On cooling with a constant rate, ΔF^* slowly increases. The level of ΔF^* and therefore the resistance are *not* the same for a sample cooled at constant rate, and for another sample exposed at constant temperature, although both may have the same apparent density. In this connection we also mention the derivation of an equation for activation energy of ionic conduction in silicate glass as given in A. ¶ 137, by O. L. Anderson and D. A. Stuart,[578] involving the radius and valency of the framework-modifying cation

[576] *Trudy Leningrad. Tekhnol. Inst. im. Lensoveta,* 1954 (29) 72-89.
[577] *Yogyo Kyokai Shi,* **65**, 1957, 331-335.
[578] *J. Am. Ceram. Soc.,* **37**, 1954, 572-580.

in glass constitution, electronic charge, shear modulus, as parameters related to the geometry of the $[SiO_4]$ framework, and the dielectric constant = a "covalency parameter" of the structure.

346. The often discussed question whether physicochemical parameters may be used by their functional relation to chemical glass composition as indicators for the *presence of chemical compounds* in these is also involved in the measurement of calorimetric data, as on the system Na_2O—SiO_2 as demonstrated by Chr. Hummel and H. E. Schwiete,[579] especially in the heats of solution. Whereas, density as a function of the molecular percentage of SiO_2 (Fig. A. 205) does not show anything of a break-

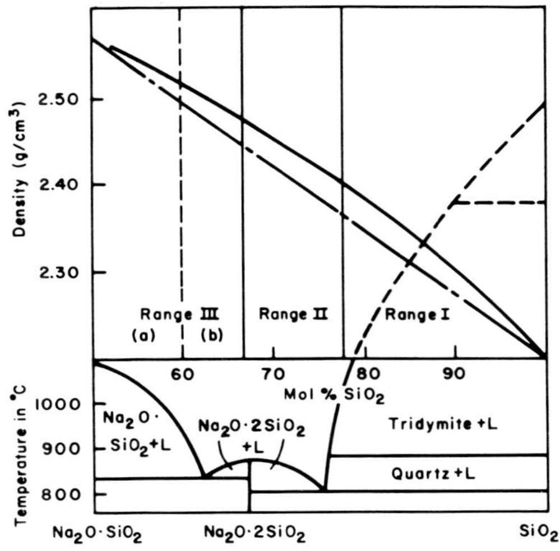

FIG. A. 205. Density of sodium silicate glasses as a function of chemical composition, and phase equilibrium diagram of the system SiO_2—Na_2O. (Huggins).

point, the heats of solution indicate in a corresponding diagram (Fig. A. 206) distinct discontinuities for 66.7 and 77 per cent silica. There are three distinct ranges with different types of short-distance order domains in the glass in agreement with previous structon theories of M. L. Huggins (1942) on the density as a function of composition. This result, on the other hand, differs much from that of L. Shartsis and W. Capps[580] in the system Na_2O—B_2O_3 in which no discontinuities were found on

[579] *Glastech. Ber.*, 32, 1959, 413-420; *Forschungsber. Wirtsch. u. Verkehrsministeriums Nordrhein-Westfalen*, No. 515, 1958, 110 pp., with extensive critical discussions on the influence of acid concentration in solvents. On the heat capacity of borosilicate glasses, as a function of their thermal treatment see recently Kh. B. Khokonov, *Structure of Glass. Proc. All-Union Conf. Glassy State, 3rd, Leningrad* (English Translation), 1959/60, 465-467.

[580] *J. Am. Ceram. Soc.*, 37, 1954, 27-32.

the curve of ΔH^* versus composition. For the measurements of Hummel and Schwiete it is a further characteristic that the $\Delta H^*/T$ function of sodium silicate glasses as a function of composition (Fig. A. 207) also shows characteristic break-points at 33 molecular per cent Na_2O. These conditions remind us very impressively of the ranges of "glassiness" or "degree of vitrification" as distinguished by A. Dietzel and H. Wickert[581] in the system $Na_2O—SiO_2$ in which very distinct minima of the recipro-

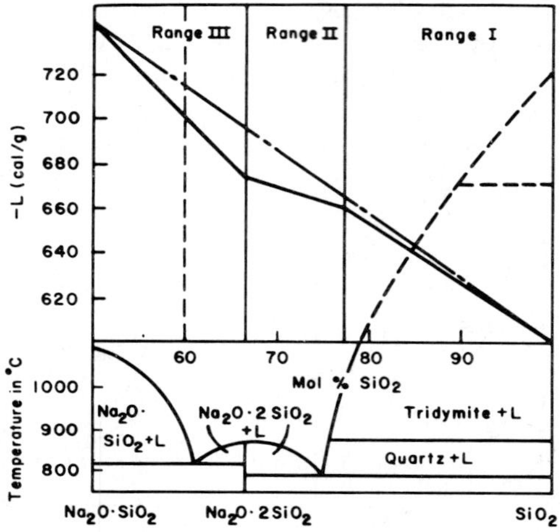

FIG. A. 206. Heats of solution of sodium silicate glasses, as a function of chemical composition, and phase equilibrium diagram of system $SiO_2—Na_2O$. (Hummel and Schwiete).

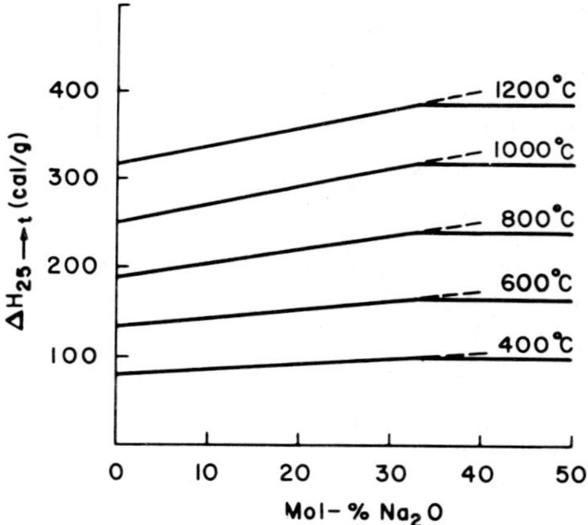

FIG. A. 207. Isotherms of the enthalpies of sodium silicate glasses, as a function of the contents in alkali for different temperatures. (Hummel and Schwiete).

[581] *Glastech. Ber.*, **29**, 1956, 1-4.

cal maximum rates of crystal growth in devitrification were marked for the constant ratios $Na_2O : SiO_2 = 1 : 2$ and $1 : 1$ (cf. B. ¶ 212; Vol. III, A. ¶ 42). Dietzel and Wickert emphasize the evident analogy existing between the role of soda in the silica framework of glass and softening agents in organic high-polymers as E. Jenckel (1950) observed. In both cases it is characteristic that these agents are strikingly efficient when only a small amount of the softener is added to a strongly developed framework of the polymerizate. Analogous is the particularly strong vitrification effect of alkali concentrations below 15 per cent, whereas, in higher-alkaline mixes, the vitrification effect (caused by a splitting-up of the framework) is only moderate.

COLOR AND FLUORESCENCE OF GLASS AS INDICATORS OF THE CONSTITUTION

347. It is a principal merit of W. Weyl to have discussed in wide extension effects of molecular electrostatic field actions on the observable color parameters and the related fluorescence phenomena, as sensitive indicators of changes occurring in the constitution of glass, or in silicate minerals for which the same rules are valid. It is, therefore, appropriate to discuss both groups of phenomena of ionic color in applied photometry in the same way.[582] The most characteristic problems of cation-coloring and constitution are their correlations to definite *coordination* types with oxygen anions. In this respect, the elements of the iron group, cobalt and nickel (to a lower degree) have been the classical subject of investigation which Weyl started in 1933 with E. Thümen. These problems are similar to those studied by O. Schmitz-Dumont and E. Raederscheidt[583] on the role which cobalt plays as a coloring agent in the regular crystals of the spinel group, with very characteristic differences of the coordination in normal and in inverse spinel structures (namely within the tetrahedral or octahedral cavities of the structure), and corresponding transitions from the common deep-blue to green with strongly changed extinction curves.

348. For alkali borate glasses, Kl. Breit and R. Juza[584] confirmed the previous results of Weyl *et al.* on the pink color of acidic, the blue color of alkali-richer glasses

[582] See W. A. Weyl's monograph "Coloured Glasses" (1943/44), also the contribution in "Phase Transformations in Solids," R. Smoluchowski, J. E. Mayer, and W. A. Weyl, editors, John Wiley & Sons, New York 1951, 296-334, especially pp. 312 ff.; on colors of minerals see *Penna. State Univ. Mineral Inds. Expt. Sta. Tech. Paper*, Final Report Jan. 1954, 31 pp. See further H. Moritz, *Silikat Tech.*, 3, 1952, 361-364, on general relations of "transparency" and glass constitution.

[583] *Naturwissenschaften*, 41, 1954, 478-479.

[584] *Glastech. Ber.*, 27, 1954, 117-126. On corresponding correlations of colors of iron-containing glasses and their magnetic susceptibilities cf. A. Bishay, *J. Am. Ceram. Soc.*, 42, 1959, 403-407; cf. ¶ B. 51. On effects of γ-radiation on spectral transmission of different glasses see further R. S. Barker and D. A. Richardson, *Ibid.*, 44, 1961, 552-560; A. Bishai, *Ibid.*, 289-206, 545-552;

containing the Co^{2+} ion, analogous to the blue complex anion $[Co(OH)_4]^{2-}$. Also the magnetic susceptibility, χ, was measured for such glasses with Gouy's cylinder-microbalance method in a field of 13.2×10^3 Gauss units. For a glass with 37 per cent CoO dissolved in borax glass is $\chi_{Co} = 152.0 \times 10^{-6}$. For increasing acidity χ_{Co} increases to 163.10^{-6}, analogous to the transition of Co^{2+} in partly covalent binding of complex anions (with $\mu = 4.31$ to 4.93 Bohr Magnetons) to free Co^{2+} ions (with $\mu = 5.04$ to 5.15 Magnetons). The transition range is ruled by the ratio $Na_2O : B_2O_3$ in the glass composition; side-by-side ionic and covalent mechanisms act here. If

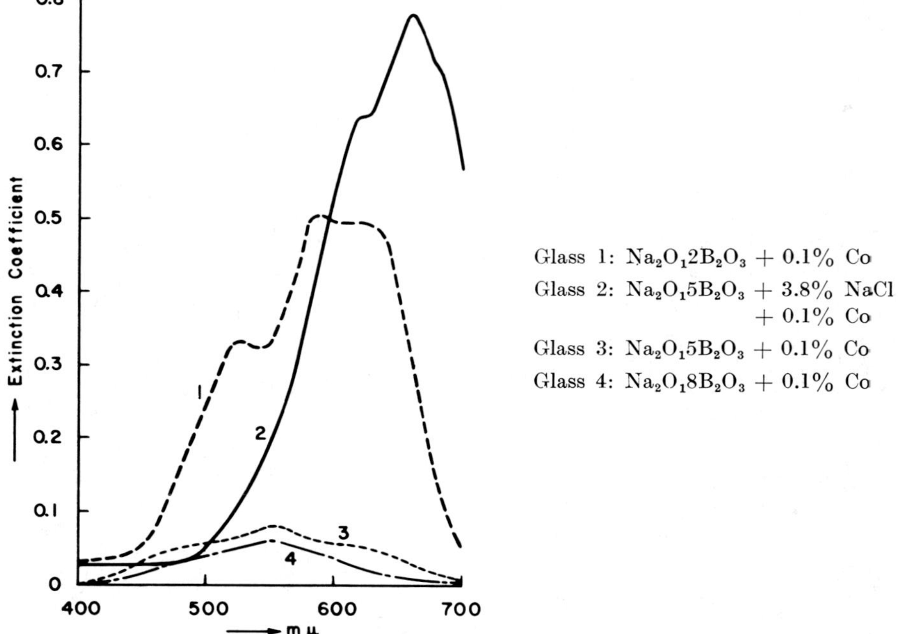

FIG. A. 208. Extinction curves of different cobalt-colored sodium borate glasses. (Breit and Juza).

R_2O varies in borate glasses with 2.4 per cent CoO, the transition fields are strongly affected. For lithium borate glasses they are between 33 and 43 molecular per cent R_2O, for sodium borate glasses between 22.5 and 25.0 per cent, and for potassium borate glasses between 19 and 22 per cent. It is remarkable that these are the same limits observed by A. Dietzel, W. Stegmaier and P. Csaki (1940) for the electromotoric forces of galvanic cells with those electrolytes. If Cl^- anions are introduced

45, 1962, 389-393; W. H. Chopper, *Ibid.*, 293-297; Ch. Hirayama, *Ibid.*, **44**, 1961, 602-606; **45**, 1962, 288-293 (on aluminoborate glasses); A. E. Owen, *Phys. & Chem. Glasses*, **2**, 1961, 87-98, 152-162; **3**, 1962, 134-138 (on calcium aluminoborate, so-called Cabal glasses).

into Co-containing alkali borate glasses, the color change from red to blue is analogous to that observed by H. Pfeilschifter (1936) in solutions of $CoCl_2$ to which HCl or NaCl are added. The magnetic susceptibility of Co-free glasses is immediately changed to that of the complex anion $[CoCl_4]^{2-}$ with its typical absorption curve. Nickel-borate glasses analogous to those studied by Weyl and Thümen show a char-

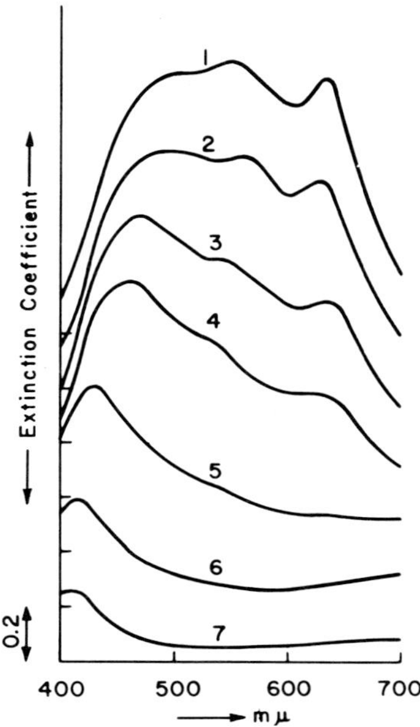

FIG. A. 209. Extinction curves of different nickel-colored potassium borate glasses. (Breit and Juza). Composition of the glasses, relative to 1 mol. K_2O: (*1*) 1.95 B_2O_3; (*2*) 2.22 B_2O_3; (*3*) 2.5 B_2O_3; (*4*) 2.75 B_2O_3; (*5*) 3.25 B_2O_3; (*6*) 4.26 B_2O_3; (*7*) 6.3 B_2O_3. The colors change from violet to amber to yellow-green.

acteristic yellow Ni^{2+} color center and violet-purple complex anions $[NiO_4]^{2-}$. Ni is much less soluble in borate glasses than Co; $\chi_{Ni} = 81.8 \times 10^{-6}$ ($\mu = 3.35$ Bohr Magnetons). No change of susceptibility is associated with the color changes. With increasing contents in K_2O, there is a strong change in the absorption curves (Fig. A. 208/9), and maxima of extinction appear at 550 and 640 mμ in the violet-colored glass. Qualitatively this phenomenon is analogous to changes in Co-alkali borate glasses indicating the interaction of the O^{2-} anions with Ni^{2+}.

349. Similar very detailed investigations on cobalt-colored alkali borate glasses by A. A. Aglan and H. Moore[585] also include the near-infrared range where for the

[585] *J. Soc. Glass Technol.*, 39, 1955, 351-384 T.

blue glasses new bands appear at 1.25 and 1.75 μ. In all of these glasses cobalt is present as the divalent cation, the coordination is $[CoO_6]$ in the pink-colored acidic glases, $[CoO_4]$ in the blue glasses. The halogen-containing glasses are characterized by the presence of complex anions of the type $[CoHl_4]^{2-}$. They are green or yellow for Br^- and I^-, but blue (although different from the color of $[CoO_4]$ groups) for Cl^-. The analogous studies for different nickel-containing glasses (also including borosilicates) by H. Moore and H. Winkelmann[586] show greater differences of the colors in glasses from those in nickel salt solutions, over the range of $\lambda = 300$ to 2300 mμ with six principal absorption bands in three characteristic groups. The borate and borosilicate glasses low in alkalies are green, sodium silicate glasses brown, potassium silicate glasses blue or blue-violet, although they have nearly corresponding absorption spectra types. There is no simple proportionality of absorption intensity with the nickel contents. Alkali-free and silica-free calcium boroaluminate glasses of the so-called "*Cabal*" type have their particular absorption regularities. The highly complex changes, which cannot be discussed here without the rich supply of detailed absorption curves given by Moore and Winkelmann, are in analogy also with Ni-salt solutions and Ni-organic compounds, of which we mention here nickel-salicyl aldehyde, $Ni(-O.C_6H_4.CHO)_2$; nickel-salicyl dioxime, $Ni(-O.C_6H_4.CH.NO)_2$; nickel-salicyl ethylenediamine, $Ni(-O.C_6H_4.N.CH_2-CH_2.N.C_6H_4.O-)$; nickel dimethyl glyoxime, $Ni(ON.CCH_3.CH_3.CH_3C.NOH)_2$, and the nickel-ammonia complex $Ni(NH_3)_6$, in dilute ammoniacal solution. The green type of the absorption curves corresponds to the $[NiO_6]$ coordination within the interstices of the glass structure. In the "undulatory" type there are coordinations of the type $[Ni(OR)_4]^{2-}$ (with R = Na,Li), thus in tetrahedral coordination, the brown type for isolated nickel ions between two oxygen anions, single-bonded to structure-building units, as in a bridging position (cf. B. ¶ 53 ff.).

350. S. Kumar[587] extended the analysis of absorption spectra beyond the cobalt-nickel group generally to glasses containing ions with partially filled 3d-orbitals, namely to those colored with titanium, vanadium, chromium, and manganese as trivalent, iron, nickel, cobalt, and copper as divalent elements. The absorption spectra are given for the visible as well as for the ultraviolet and near-infrared ranges

[586] *J. Soc. Glass Technol.*, **39**, 1955, 215-249 T; 250-286 T; 287-313 T.

[587] *Central Glass & Ceram. Research Inst. Bull.* (*India*), **6**, 1959, 99-126. See theory of *term systems* of ions here concerned H. Hartmann, *Z. physik. Chem.* (*Leipzig*), **197**, 1951, 234-246 (with F. E. Ilse); *Angew. Chem.*, **66**, 1954, 768-776 (with L. H. Schläfer); *Z. physik. Chem.* (*Frankfurt*), **4**, 1955, 376-379; **5**, 1955, 9-19 (with H. H. Kruse). This theory is of great importance for interpretation of absorption characteristics of chromium containing minerals and especially of gem stones (ruby, spinel, emerald, uvarovite, fuchsite, Cr-diopside); cf. A. Neuhaus, and W. Schilly, "Festschrift zum 70. Geburtstag von K. Schlossmacher," Deutsche Gesellschaft für Edelsteinkunde, 1957, 105-114; further *Fortschr. Mineral.*, **36**, 1958, 64-65; *Z. Krist.*, **113**, 1959, 195-233.

in alkali silicate and alumino-borophosphate glasses. The number of observed absorption bands, their position and intensity show distinct relationships with the spectroscopic energy levels of the ions and the glass compositions. Very weak absorption bands are characteristic for the *S*-state ions namely Mn^{2+} and Fe^{3+}, whereas the *D*-state cations Ti^{3+}, Mn^{3+}, Fe^{2+}, and Cu^{2+} show one broad absorption band, and those of the *F*-state (V^{3+}, Cr^{3+}, Co^{2+}, and Ni^{2+}) two or more bands. In general the spectra of all these elements have a certain similarity in their absorption characteristics.

351. The problems of the existence of *trivalent* cobalt ion in glasses of high alkalinity were investigated by A. Dietzel and M. Coenen.[588] If the Na_2O content of the melts is raised to 50 to 70 per cent, and strongly oxidizing conditions are maintained, the glass is changed in color from blue to green and even to a greenish-yellow. Measurements of absorption moduli in such Co^{3+} containing glasses with varying cooling rates show that Co^{3+} is more stable at lower temperatures, Co^{2+} at higher temperatures. The ion Co^{3+} is a framework-forming constituent forming $[Co^{3+}O_4]$ coordinations. In normal cobalt-blue glasses the concentration in Co^{3+} is always very low (below 10^{-5} per cent) as electrochemical measurements of the oxygen potential demonstrated. A calculated oxygen pressure of 10^5 atm. would be necessary to oxidize 20 per cent of the CoO in such a glass to Co_2O_3, if only the gas absorption mechanisms are concerned. The characteristic extinction curve of a glass with 1 molecular per cent Co_2O_3 (after the deduction of the absorption by Co^{2+} ions) is shown in Fig. A. 210.

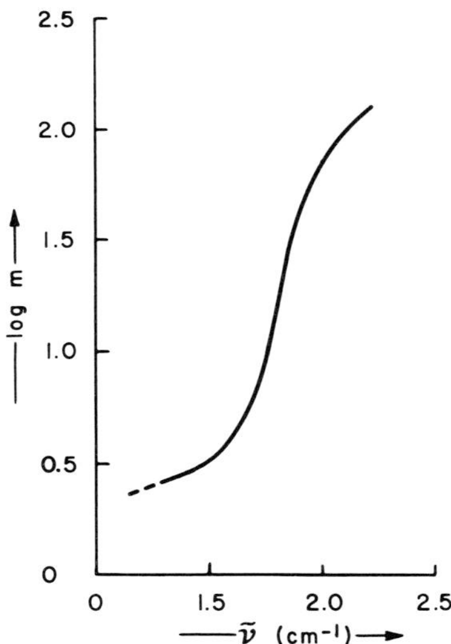

FIG. A. 210. Extinction curve of a trivalent cobalt containing glass (with 1 mol. per cent Co_2O_3, 1 cm. thick), after deduction of the Co^{2+} absorption. (Dietzel and Coenen).

[588] *Glastech. Ber.*, **34**, 1961, 49-55.

352. The previous observations of F. Weidert and K. Rosenhauer (1938) on the absorption spectra of glasses colored by *rare earth elements* as indicators of the constitution were not continued, but R. C. Vickery and R. Sedlacek[589] discussed the density changes in sodium-calcium silicate, lead silicate, and different phosphate glasses with rare earths. In this way the great contrast which exists between the framework-forming character of scandium and yttrium (in concentrations of 0.2 to 1.0 atomic per cent) and the framework-modifying type of the lanthanon (lanthanum to lutecium) elements becomes evident. The high electronegativity of Sc^{3+} (1.65) and Y^{3+} (1.60) in contrast with that of the lanthanons (below 1.20) is in this connection particularly remarkable. For Pr_2O_3 as a coloring agent in concentrations of 2, 6, and 12 per cent in lead and barium silicate, and barium phosphate glasses, I. I. Kitaĭgorodskiĭ and M. V. Artamonova[590] determined the spectrophotometric curves over the ranges from 400 to 700 mμ, and from 1.0 to 2.2 μ in the near-infrared. The green color is most intense in the phosphate glasses, say for filter glasses, with characteristic absorption maxima at 442, 468, 480, and 590 mμ, 1,55, and 1.95 μ. The lead glass shows these maxima more diffuse and peaks somewhat shifted. Particularly remarkable is the fact that the same maxima are found in the absorption spectrum of PrF_3, as given by Weyl (Fig. A. 211).[591]

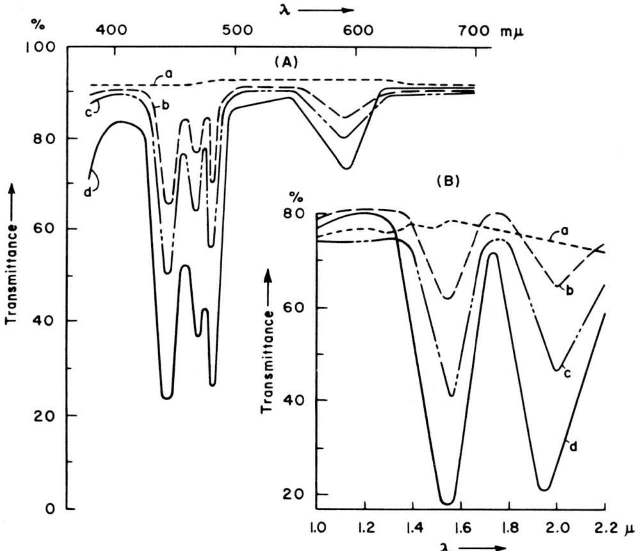

FIG. A. 211. Absorption spectrum of praseodym containing phosphate glasses with (*a*) 0, (*b*) 2, (*c*) 6, and (*d*) 12 per cent Pr_2O_3, respectively, in the visible and the near infrared range. (Kitaĭgorodskiĭ and Artamonova).

[589] *J. Am. Ceram. Soc.*, **41**, 1958, 422-426.

[590] *Doklady Akad. Nauk S.S.S.R.*, **130**, 1960, 830-833.

353. The ultraviolet transmissivity in different silicate, borate, and phosphate glasses was investigated by E. Kordes and E. Wörster.[592] J. M. Stevels[593] predicted that the limit (threshold) wavelength of ultraviolet transmittance must be shifted to larger wavelengths in the sequence from phosphate to silicate and then to borate glasses. The systematic experiments of Kordes and Wörster extended to lead glass compositions of all three types, supplemented by metaphosphate compositions with the cations Li^+, Na^+, Be^{2+}, Ca^{2+}, Ba^{2+}, and Zn^{2+}. For borate glasses the absorption threshold is, however, at *lower* wavelengths than in silicate glasses, in contradiction

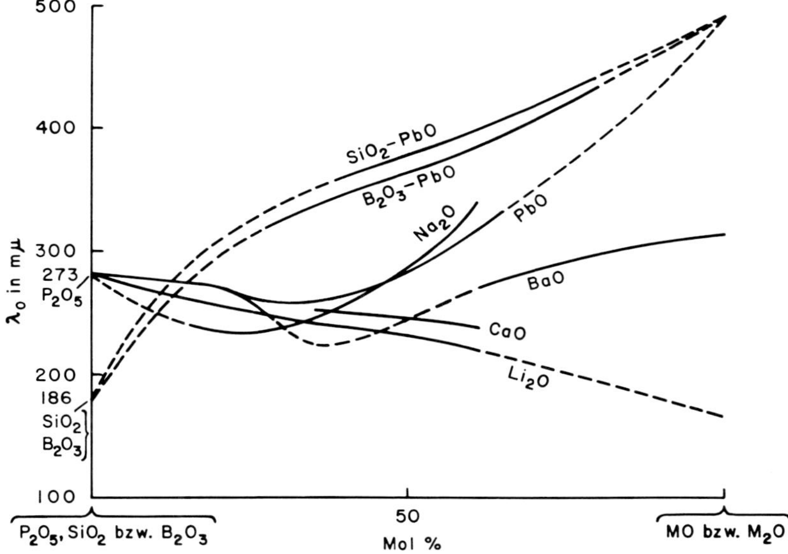

Fig. A. 212. Threshold limit, λ_0, in ultraviolet, as a function of the composition of binary glasses. (Kordes and Wörster).

with Stevels's theory. Glasses rich in P_2O_5, on the other hand, show another anomaly, namely a reversal of the shift of the absorption limit, and a minimum for a definite wavelength, λ_0, in sodium barium, and lead phosphate glasses (with 60 to 80 per cent P_2O_5). Pure P_2O_5 glasses (fused and cooled between two silica glass plates) gave the absorption limit at $\lambda_0 = 273$ mμ (predicted much below 200 mμ), and much higher than for pure silica and B_2O_3 glasses. Sodium phosphate glasses with 20 to 25 per cent Na_2O show the limit at 235 mμ, barium phosphate glasses at 230 mμ (with 35 per

[591] *J. Soc. Glass Technol.*, **28**, 1944, 250-266, based on measurements by F. Ephraim and R. Bloch, *Chem. Ber.*, **59**, 1926, 2692-2705; **61**, 1928, 65-80.

[592] *Glastech. Ber.*, **32**, 1959, 267-276.

[593] *Verres et réfractaires*, **2**, 1948, 4-14.

cent BaO), and lead phosphate glasses with 31 to 40 per cent PbO at 250 mμ. In lithium phosphate glasses a constant decrease in λ_0 with increasing contents in Li_2O is observed, whereas in calcium phosphate glasses λ_0 is rather constant up to 50 per cent CaO. If more than 25 per cent lead oxide is present the ultraviolet absorption in lead phosphate glasses is higher than for lead silicate and borate glasses. From the framework theory an explanation of the minimum for λ_0 as a function of the concentration in cations (cf. Fig. A. 212) is given for phosphate glasses, as a consequence of an antagonistic effect of bridging oxygen anions (100 per cent of all O^{2-} in SiO_2 and B_2O_3 glass, only 60 per cent in P_2O_5 glass), which shifts λ_0 to lower wavelengths, and of corner O^{2-} anions (0 per cent in SiO_2 and B_2O_3 glass, 40 per cent in P_2O_5 glass), which shifts λ_0 to longer wavelengths. A glass of the composition $Li_2Si_2O_5$ which also has 40 per cent of corner oxygen anions, has exactly the same λ_0 (275 mμ) as pure P_2O_5 glass (Fig. A. 213), and the reduced transmissivity curve versus λ is also practically the same for both glasses. The particular shape of the λ_0 versus concentration curves in lithium and calcium phosphate glasses also finds a structural interpretation by the electron-consolidating effects of the strongly polarizing ions Li^+, Be^{2+} and Zn^{2+} participating in the tetrahedral framework of the glass constitution.

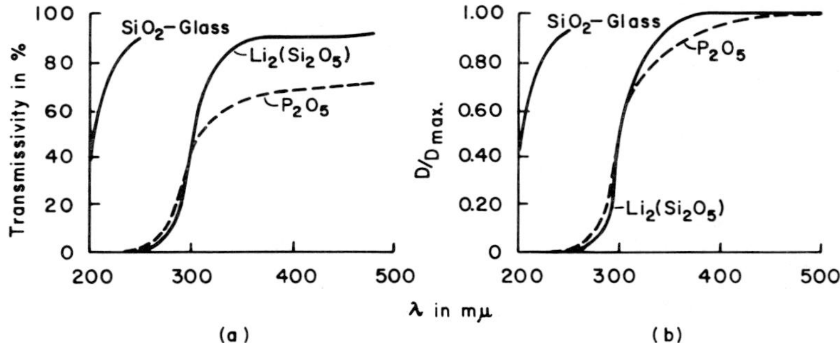

FIG. A. 213. Light transmissivity of the glass $Li_2O,2SiO_2$, and of pure P_2O_5 glass, (*a*) in per cent, (*b*) relative transmissivity $D/D_{max.}$. (Kordes and Wörster).

354. With problems of ultraviolet transmissivity and absorption we enter the fascinating fields of modern aspects on the principles of *high-energy absorption* from radiation. We owe especially to J. M. Stevels and his investigation group a total concept on the problems involved of which we have discussed already in A. ¶ 153 ff. some special items of the response of irradiated glasses in their dielectric loss behavior, as a highly sensitive indicator for such effects. We start from studies of Stevels *et al.* on theories of the power factor of glasses as a function of composition, and of the frequency ν, of energy irradiated,[594] especially in the wave range of $\nu = 10^{-2}$ to

[594] See introductory article of J. M. Stevels, *Verres et réfractaires*, 7, 1953, 91-104; *Glass Ind.*,

10^{12}/sec., with the maximum oscillation losses for infrared at $\nu = 10^{12}$ in silicate glasses and of 10^9 for the very loose framework of phosphate glasses with heavy structure-modifying cations. The important phenomenon of *color center* evolution by irradiation with X-rays (10^6 Röntgen units), or electrons is in close relation with the sharp peaks of tan δ as a function of temperature and frequency, as J. Volger, J. M. Stevels, and C. van Amerongen[595] demonstrated. The brown- and black-colored quartz crystals show a much increased tan δ value and a very important peak at 21°K. for a given sample. By heating to 300°C. irradiation effects are destroyed, and initial tan δ characteristics are reestablished. Natural smoky quartz crystals have a high initial tan δ which is reduced by thermal treatment, but repeated irradiation reestablishes the higher tan δ. In contrast to these effects with quartz, radioactive irradiation of different glasses does *not* change the low-temperature characteristics of the dielectric response although color centers are equally developed and the glass becomes fluorescent in ultraviolet light. The temperature function of tan δ of commercial silica glass shows a maximum at 30°K., if contaminated by aluminum at 38°K. Purest silica glass, however, does not show any peak of this kind. Alkali and alkaline earth silicate glasses have slight peaks at 20°K. if there is less than 1 molecular per cent present but these are obliterated for higher alkali contents. Pure B_2O_3 glass shows maxima at 30° and 210°K.; commercial lead silicate glass (30 per cent PbO) has a shift of the peak from 35° to 40°K., with heat exposures to 400° to 450°C. The fact that dipole losses caused by color centers are absent in glasses, whereas they play an important role in quartz, shows that there are different mechanisms in both systems. The crystalline framework requires an additional entropy effect of about = 12 k which is absent in the random framework of glass.

355. Especially for silica glass and quartz, J. M. Stevels[596] gave an excellent report on the nature of *framework defects* in silica structures in which the close relationship between infrared, visible light, and ultraviolet absorption with the measurement of dipole losses, and with the paramagnetic resonance, are outlined for a determination of the quantity and distribution of the defects. Also their specific characteristics can then be classified and distinguished. Before and after the irradiation of silica glass and quartz with ultraviolet, X-, or γ-rays, or by the bombardment with neutrons, the mechanism of different reactions to these irradiations

35, 1954, 69-72, 100; 135-138, 160; 657-662; on dielectric losses *Glastech. Ber.*, **26**, 1953, 227-231. See also J. Volger and J. M. Stevels, *Philips Research Repts*, **11**, 1956, 452-470; *Ind. chim. belge*, **20**, Special Issue, 1955, 128-132, on color center-dipole losses.

[595] *Philips Research Repts.*, **10**, 1955, 60-80; *Verres et réfractaires*, **10**, 1956, 3-14; **11**, 1957, 137-146.

[596] *Glastech. Ber.*, **32**, 1959, 307-313. On paramagnetic resonance measurements cf. J. S. van Wieringen and A. Kats, *Philips Research Repts.*, **12**, 1957, 432-456. On nomenclature used by Stevels for different types of structural defects cf. *Ibid.*, **11**, 1956, 103-114; *Verres et réfractaires*, **10**, 1956, 129-134.

can be demonstrated. A distinction is possible, say of those defects caused by empty positions of oxygen anions, from the defects caused by contaminating foreign elements. The ideas developed by Fr. Oberlies and A. Dietzel[597] on the structure of silica glass (cf. A. ¶ 231) with distorted [SiO_4] groups, open rings, and other configurations, are highly valuable to follow such reactions because the centers of discoloring effects by irradiation which previously were characterized by Stevels purely empirically, now receive a structural meaning, such as those caused by aluminum. The paramagnetic response reactions are now better understood as transitions of the complex Al-Na-caused P centers to Al(A) and Na(P) centers (Li and H may replace the Na in this case). A quartz crystal bombarded with rapid neutrons shows electron resonance and atomic displacements which locally quasi-"vitrify" the quartz and cause defects in oxygen anions. The Al-H caused P-centers are changed to Al(B) + H(P) centers and free oxygen atoms. All the centers mentioned correspond to definite optical absorption maxima which can be used for the identification in different reaction cycles of irradiation, thermal treatment, etc.

356. Many experimental facts on irradiation phenomena for quartz, silica, and silicate glasses are reported by A. Kats and J. M. Stevels.[598] Aluminum in concentrations down to 0.01 per cent is responsible for absorption bands at 5500 and 3000 Å., visible after X-ray irradiation, and caused by electron capture in oxygen defects in the neighborhood of Al(A) centers which are changed in the irradiated glass to $Al(B)^-$ centers. Likewise, the bands at 2400 Å., and a corresponding fluorescence band at 2800 and 3900 Å. are caused by aluminum. The absorption spectra of synthetic and natural quartz, amethyst, and smoky quartz irradiated with X-rays are similar to those of silicate glasses with low alkali contents. Synthetic quartz shows abundant centers Na(D) and $Na(P)^-$ which cause an absorption band at 4600 Å., corresponding to the band 4500 Å. in sodium silicate glasses. Smoky quartz differs in its coloring and discoloring cycle by thermal treatment at 300°C. and irradiation, from clear quartz by having been evidently irradiated with natural radioactivity. Amethyst (discolored at 420°C.) shows after irradiation by X-rays the same bands (5500 and 3500 Å.) as silica glass. The color centers in amethyst, however, are remarkably more stable than those in common quartz and in alkali silicate glasses. Probably, Fe(B)-centers (with a strong binding of the electrons by the Si^{4+} cation) cause the color effects at 5500 Å., 2Fe(B)2 + centers those at 3500 Å. A valuable tabulation is given by Kats and Stevels for wavelengths of the absorption bands and the nature of the color centers, for quartz, silica, and common silicate glasses.

[597] *Glastech. Ber.*, **30**, 1957, 37-42.

[598] *Philips Research Repts.*, **11**, 1956, 115-116; *Verres et réfractaires*, **10**, 1956, 135-150, 215-220, Irradiated aluminosilicate glasses were recently studied by S. Lee and P. J. Bray, *Phys. & Chem. Glasses*, **3**, 1962, 37-42, applying the method of electron-spin paramagnetic resonance.

357. A. Kats[599] further discussed the particular effects of *germanium* as a foreign constituent in quartz and silica glass on color center reactions by irradiation. We may here restrict ourselves to merely a hint of these excellent investigations because the corresponding phenomena are discussed with some details in Vol. V, Section B, in connection with the results given by A. J. Cohen.[600] A specific absorption band at 2450 Å. is developed in silica glass after neutron irradiation from an atomic pile reactor and is extinguished by heating to 700°C. Particularly interesting are the strong pleochroitic phenomena observed by Kats by an anisotropy of the distribution of color centers in irradiated quartz, for the electrical vectors parallel and perpendicular to the c-axis, as seen by the dichroitic factor, π, as a function of temperature in Fig. A. 214 (π is the ratio of the optical density, log I_0/I), for the two electric

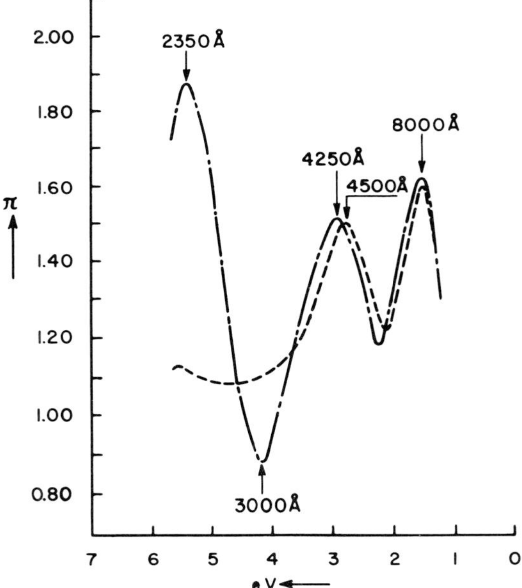

FIG. A. 214. Dichroitic ratio, π, as a function of the wavelengths (given in *eV*), at 293°K. for synthetic quartz containing 2×10^{-4} gm. Ge/gm., irradiated by X-Rays, and a natural quartz crystal (from Madagascar), irradiated in the same way (Kats).

Key: –·–·–, synthetic quartz, with 2×10^{-4} gm. Ge/gm., irradiated; – – –, natural quartz, irradiated, in both cases with 1.5×10^{6} R, at room temperature.

599 *Verres et réfractaires*, **12**, 1958, 191-205.

600 *Phys. Rev.*, **105**, 1957, 1151-1155; *Glastech. Ber.* **32**, K, 1959, VI.53-VI.58; see also V. Garino-Canina, *Verres et réfractaires*, **12**, 1958, 313-323, concerning experiments with pure GeO_2 glass. See further V. Garino-Canina, *Vetro e Silicati*, **4**, 1960 (22) 25-32; *J. Am. Ceram. Soc.* **43**, 1960, 415-416 (with S. Cohen). "Doping" of fused silica with "synthetic impurities" was also used by E. Lell, *Ibid.*, **43**, 1960, 422-426; *Phys. & Chem. Glasses*,, **3**, 1962, 84-94.

vectors. The absorption peaks are somewhat different if the irradiation is applied at room temperature or at 78°K. In addition, there are surprising changes in factor π as a function of temperature if quartz is irradiated at 78°K. and then slowly heated. Above 200°C. there is a rather sudden change which indicates a different orientation of the color centers for the electric vectors parallel and perpendicular to the *c*-axis. Ultraviolet irradiation with polarized light (at 78°K., with the electric vector parallel c) causes decoloring by a recombination of the electrons with the holes (for $\lambda = 2420$ and 2550 Å.) whereas the inverse effect is observed for centers causing the 2960 Å. band. Anisotropy effects also occur after irradiation with 10^{18} neutrons/cm.^{2}.

358. From the more practical viewpoints of glass technology, we emphasize the general discussions of E. U. Condon[601] on the nature of radiation sensitivity of commercial glasses including the *solarization* effects which are obvious by changes of the absorption spectrum after irradiation. They can be reversed by a thermal treatment or in certain cases also by red light or infrared radiation. On development of centers of nucleation around which colloid or crystalline precipitation or growth may ensue we speak in Vol. 1, C. ¶ 90 for the photosensitive glasses of S. D. Stookey. In every case, the electronic energy levels in colored glass are by about 1.8 to 3.4 e.v. above the ground level (Fig. A. 215). All the other phenomena like the development of color centers, photoconductivity, fluorescence, phosphorescence, and thermoluminescence are consequences of these electronic displacement phenomena.

359. K. H. Sun and N. J. Kreidl[602] gave an excellent introduction to the general characteristics of glass coloration by radiation on a wide experimental basis. In such measure as the use of atomic energy increases, there is a rapidly increasing need for the development of glasses which do not become so deeply colored by irradiation as many common glasses developing strong absorptions and which are detrimental in the use of common optical glass for instrumentation at nuclear piles. Kreidl[603] developed *cerium*-containing glasses which show a high inhibition to this discoloring effect. On the other hand, one may use color effects by absorbed radiation for dosimetric purposes (cf. B. ¶ 55, ¶ 62). Common glasses become colored by a radiation intensity of 10^4 Roentgen units, and are practically unsuitably dark at 10^6 Roentgen

[601] *Am. J. Phys.*, **22**, 1954, 310-317.

[602] *Glass Ind.*, **33**, 1952, 511-514, 546; 589-594, 614; 651-653, 674. Effects of electron bombardment on different glass properties are further described by T. M. Mike, B. L. Steiernan, and Ed. F. Degering, *J. Am. Ceram. Soc.* **43**, 1960, 405-407; S. W. Barber, K. E. Forry, and Ed. F. Degering, *Ibid.* 408-410.

[603] *Ind. Eng. Chem.*, **46**, 1954, 170-171; cf. W. A. Weyl, J. H. Schulman, R. J. Ginther, and L. W. Evans, *J. Electrochem. Soc.*, **95**, 1949, 70-79, especially on silver containing phosphate glasses for γ-ray dosimetry, by a combination of fluorescence with a photoelectric cell device. See further N. J. Kreidl, *Vetro*, **1**, 1956 (3) 3-14; *J. Opt. Soc. Am.*, **47**, 1957, 73-75 (with J. R. Hensler).

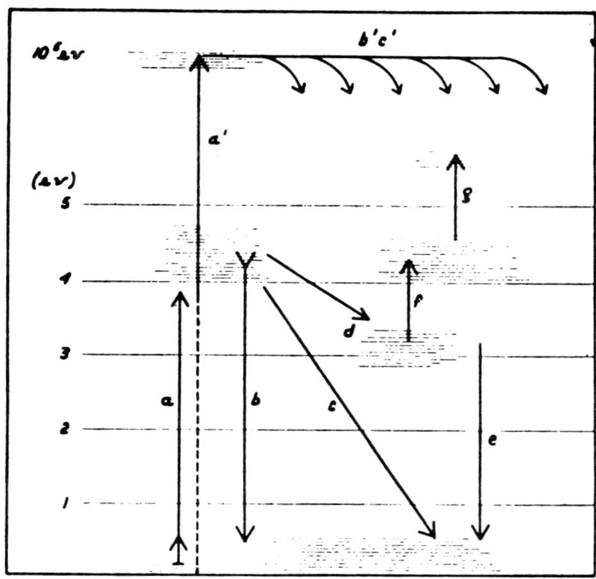

FIG. A. 215. Scheme of electronic transitions in glass, by energy levels. (Condon). (*a*) Absorption of a few el. v. of ultraviolet radiation energy. (*a'*) Absorption of several ten-thousands of e.v. of X-Rray radiation energy. (*b*) Spontaneous return to ground state (fluorescence). Stokes' law: $\nu_b < \nu_a$. (*c*) Nonradiative return to ground state, in form of heat given up to lattice. (*d*) Transition to a metastable state, emission of radiation $\nu_d \ll \nu_a$. (*e*) Radiative transition ν_e to ground state over a long period (phosphorescence). (*f*) and (*g*) Absorption in metastable state of a few e.v. as ultraviolet or visible radiation (color centers). (*b'*, *c'*, etc.). Additional free electrons liberated by impact with high energy electrons from (*a'*).

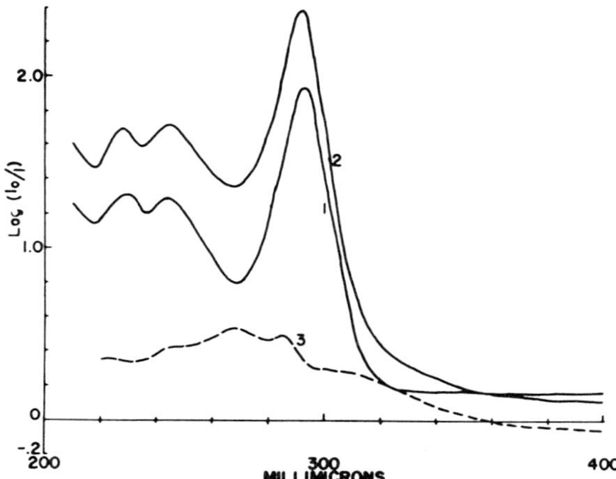

FIG. A. 216. Phosphate glass containing 1 mol. per cent cerium, 0.2 mm. thick, (*1*) before exposure, (*2*) after irradiation with 10^5 Roentgen units, (*3*) difference (Kreidl and Hensler).

units (from Co^{60} as γ-radiation source). The cerium containing glasses are still useful at 5×10^8 Roentgen units. More detailed information on such special glasses are given by N. J. Kreidl and J. R. Hensler,[604] especially on ultraviolet-transmitting phosphate glasses with cerium. In these, the X-radiation causes a strong absorption band in the ultraviolet which depends on the electronic structure of the cerium ion. Similar electronic structures of iron, manganese, cobalt, nickel, vanadium and copper show a possibility of reducing the visible color effects in a corresponding,

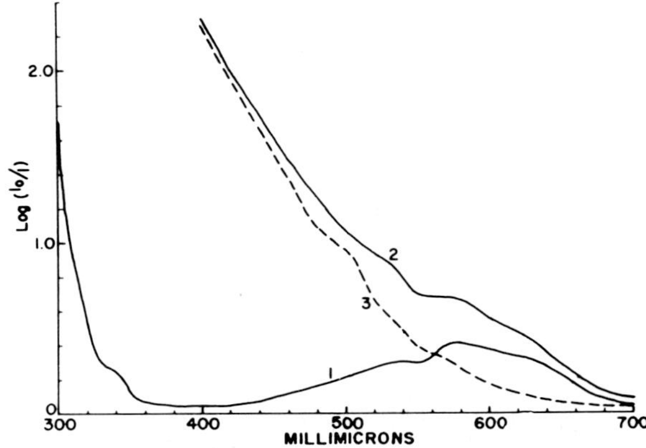

FIG. A. 217. Phosphate glass containing 0.3 mol. per cent cerium, 2.0 mm. thick, (*1*) before exposure, (*2*) after irradiation with 10^6 Roentgen units. (Kreidl and Hensler).

non-visible, ultraviolet absorption. A striking example is an iron phosphate glass which is so well "protected" that it remains colorless after irradiation. On the other hand, the radiation develops in cobalt containing glasses a new absorption band in the visible region which makes such a glass dark-reddish-brown and useful for dosimetry (Figs. A. 216 and 217).

360. The change in coefficients of reflection of common sodium-calcium silicate glass by bombardment with *positive ions* ($5 \times 10^6 Ar^+/cm.^2$ intensity) and an energy of 33.5 kev.) is most striking, namely a reduction to 36 per cent of its normal value at $\lambda = 600$ mμ, as R. L. Hines[605] observed. The surface of the glass is definitely changed and shows in a 0.095 μ thick layer a reduction of the average effective refractive index from 1.5246 to 1.343, as a function of the integrated flow of positive ions, and the type of the ion itself. The positive ion flux is by orders of magnitude larger than that of primary "knocked-on" atoms produced by fast neutron bombardment in a nuclear reactor, and is, therefore, a powerful tool for study of high-energy

[604] *J. Am. Ceram. Soc.*, **38**, 1955, 423-432.
[605] *J. Appl. Phys.*, **28**, 1957, 587-591.

effects on solids. It is, on the other hand, interesting to see that W. Primak, L. H. Fuchs and P. Day[606] found, for the effect of neutron irradiation on quartz a reduction of the refractive index from $\varepsilon = 1.5534$ to 1.4666, and for ω from 1.5444 also to 1.4666, whereas in irradiated silica glass the refractive index is *in*creased from 1.4589 to 1.4666. Silica glass contracts under irradiation, quartz expands. The rotary power is reduced from 21.5°/cm. to practically zero.

361. Systematic experiments on irradiation by direct *sunlight* were made by G. E. Rindone,[607] the effects of which are distinctly different from those of visible and ultraviolet light, by antagonistic behavior to longer and shorter wavelengths. The ultraviolet rays promote the initial electron displacements in the glass structure, whereas the longer waves enhance the returning of the electrons to their initial positions. The electron transition mechanism is strongly influenced by the nature of the basic glass to which specific agents like arsenic and cerium were added to make the glass highly photosensitive (cf. B. ¶ 55). These sensibilized glasses are suitable as the basis for a comparison with other specific agents and their solarization effects, demonstrating a wide range of sensitizing. Some of the glasses were decolorized after a few minutes, others did not show any change even after more than a thousand hours. Especially for radioactive coloring effects, but also in comparison with X-ray and neutron radiation effects, W. Jahn[608] discussed the specific behavior of phosphate, optical (Baryte Crown), and lead silicate glasses, indicating their spontaneous bleaching. In the visible range, the action of X- and γ-rays is strikingly analogous. This is important because of the systematic development of protection glasses on the basis of lead and cerium containing compositions, for dosimetry, and like uses. Highly efficient protection windows for nuclear reactor constructions were thus developed in connection with multiple liquid immersion windows for the "hot cells." The fundamentally different behavior of neutron radiation, dissipation and absorption by B^{10} and Cd^{113} (with the highest known cross section values) shows the way for development of neutron absorbing lithium-beryllium borate glasses, with additions of cerium to eliminate the "edge effects" caused by α-particles emitted during the capture of electrons by boron, combined with troublesome coloring effects on edges and surfaces.

362. The *luminescence* phenomena connected with the thermal treatment of ce-

[606] *J. Am. Ceram. Soc.*, **38**, 1955, 135-139. The great progress in the art of manufacturing highly purified and nearly ideally homogeneous silica glasses (type "Ultrasil," "Suprasil"), of great stability to γ- and neutron irradiation, is demonstrated recently by H. Mohn, *Verres et réfractaires*, **15**, 1961, 329-340.

[607] *Trav. Congr. Intern. du Verre* IV, Paris 1956, VII, 4, 373-389. On electron spin resonance effects connected with trapped electrons in irradiated silica, see C. M. Nelson and R. A. Weeks, *J. Am. Ceram. Soc.*, **43**, 1960, 399-404.

[608] *Glastech. Ber.*, **31**, 1958, 41-54.

rium containing glasses were investigated by V. V. Vargin and G. O. Karapetyan[609] using photoelectric methods, the quantum output being determined by the sphere method of Z. L. Morgenshtern,[610] with a heating rate of 0.34°/sec., the phosphorescence by photography. One of the most important results of these measurements is the observation that absorption characteristics of the cerium glasses depend on the *anions* present, in addition to the oxygen potential conditions during their fusion, say for silicate glasses with 2 per cent Ce with variable intensities of the bands at 313, 240, and 223 mμ. Borate glasses with Ce show a strong absorption with a maximum at 250-260 mμ, with strongly reducing conditions developing luminescence. A change from sodium to potassium or of potassium to barium borate glass compositions brings about a considerable increase in absorption intensity; a reduction effect, even by strong agents, is not observed. CaF_2, SrF_2, and BaF_2 crystals containing CeF_3 in crystalline solution show maxima of absorption at 290, 295, and 328 mμ and their absorption spectra are similar to those of the glasses, but much different from cerium salt solutions (with maxima at 240 and 253 mμ). Beer's law is obeyed only up to 0.5 per cent of Ce content. The maximum output of luminescence for Ce-silicate glasses is 55 per cent, for phosphate glasses, however, about 100 per cent. It is particularly strong if the glass is excited by X- and γ-rays as a function of the concentration in cerium, silicon and phosphorus. Silicate glasses show a long-period phosphorescence after γ- or electric spark irradiation.

363. The fact that glasses in general do not react by fluorescence and phosphorescence in the same way that crystals do (e.g., the alkali halides when "activators" like Cu^+, Ag^+, Mn^{2+}, are incorporated in their structure) is discussed by W. A. Weyl[611] (cf. Vol. I, C. ¶ 85, ¶ 88). Only when metals are introduced in the *atomic state*, with concentrations below 0.01 per cent, distinct fluorescence and luminescence phenomena are observed, e.g., in silver containing glasses, the atoms of which as a function of the thermal treatment, are aggregated to larger units. Metal vapors can be "frozen" in glasses in the same way as are sulfur or selenium in atomic dispersion and show intensely colored glass solutions which are sensitive to X-rays with a typical bleaching effect by capture of electrons and a change of S_2 molecules in the blue color centers into colorless anions. A very well-studied effect of this kind is observed with silver glasses, reduced by hydrogen. The neutral atom Ag^0 shows

[609] *Glastech. Ber.*, **32**, 1959, 443-450. In addition we emphasize the recent investigations of E. V. Anufrieva and M. V. Vol'kenshteĭn, *Structure of Glass. Proc. All-Union Conf. Glassy State, 3rd, Leningrad* (English Translation), 1959/60, 117-120, on the luminescence of organic plastomers in their vitrification behavior, with the addition of fluorescence-sensitive dyestuffs (Auramine, Crystal Violet, Michler's Ketone). Every curve of the luminescence intensity as a function of temperature shows a breakpoint at T_g.

[610] *Zhur. Eksptl. i Teoret. Fiz.*, **29**, 1955, 903-904.

[611] *J. Phys. Chem.*, **57**, 1953, 753-757; see also G. E. Rindone, *J. Soc. Glass Technol.*, **37**, 1953, 124-128 T.

a strong fluorescence when irradiated with the ultraviolet wave $\lambda = 3650$ Å. The fluorescence centers are formed by trapping electrons to silver ions. The aggregation of the Ag^0 atoms under influence of radiation destroys the centers. Silver borate glasses with up to 60 per cent Ag_2O were prepared by Weyl which undergo on their surface a decomposition in the dark (under the influence of moisture) and form a coherent silver metal film on the surface. Ultraviolet radiation accelerates this process to the rapid formation of the silver mirror.

364. Specifically for silica glass irradiated by ultraviolet light, V. Garino-Canina[612] made extensive experiments to study the evolution of absorption bands and fluorescence and phosphorescence behavior of different melts prepared either in graphite crucibles (therefore called C-glasses) or molten in the hydrogen or propane flame (modified Verneuil process, therefore called V-glasses). If the C-glass is electrolyzed at 1080°C. (1000 volts/cm.) characteristic metal clouds are evolved from the platinum electrodes with yellow, reddish or brown colors, migrating from the anode to the cathode. Also copper and silver clouds show the same phenomenon. The centers causing the absorption at 2420 Å. and a luminescence excited by Hg $\lambda = 2537$ Å. light are either silicon ions in interstitial positions in the framework of silica glass or oxygen defects with two trapped electrons. The absorption effects of discolored silica glass of C-type are much more intense than those for V-glasses with sharp maxima at 2200, 3000, and 5400 Å. (this latter is absent in V-glass). By exposure of V-glass in silicon vapor or by melting at 2000°C. it is changed to the C-type and made more sensitive to ultraviolet radiation. The color centers in irradiated and thermally treated glasses are evidently much different in stability; those of the heated samples are not destroyed by irradiation with Hg $\lambda = 2573$ Å., whereas X- and γ-ray-discolored glasses are rapidly decolored.

365. It is highly interesting to compare the phenomena of luminescence of *silicate minerals* with those of common glasses, say of chromium containing spodumene (called hiddenite) and of common colorless or pink-lilac spodumene (kunzite), as it was studied by E. W. Claffy,[613] with certain analogies to Mn, V, or Cr-colored glasses, studied by Weyl, and by S. Grum-Grzhimaĭlo (since 1915) in their absorption spectra, before and after X-ray irradiation. A specific phenomenon of the spodumenes and related minerals is the "*tenebrescence*," or the reversible darkening (induced by X- or cathode rays, or other high-energy sources) and bleaching (caused by heat treatment, or irradiation with light of the same wavelength as the absorption band evolved during darkening). Reversible tenebrescence is also described by D. B.

[612] *Verres et réfractaires*, **10**, 1956, 63-82; 151-158. A very complete discussion of the defects in silica in general, and their reactions to irradiation see by M. V. Garino-Canina and M. Priqueler, *Silicates inds.*, **26**, 1961, 565-580; *Phys. & Chem. Glasses*, **3**, 1962, 43-45, on proton diffusion in alumina-silica glass, under the action of an electrical field.

[613] *Am. Mineralogist*, **38**, 1953, 919-931.

Medved[614] for pink sodalite (hackmanite) in terms of the band theory of solids. In this case, the presence of sulfur or manganese is *not* a necessary condition for the luminescence phenomena. Also synthetic sodalite exhibits darkening under Hg $\lambda = 2537$ Å. and bleaching in visible light, or on heating. In addition, R. D. Kirk[615] studied the orangecolored luminescence and the tenebrescence of sodalite at 20° and — 196°C., in analogy with that of mixtures of Na_2SO_4 and sodium polysulfide.

366. R. Tomaschek and O. Deutschbein (1937), concerning the degree of order in the structure of glass from analysis of absorption spectra of *europium*-containing glasses in comparison with those of Eu salt solutions, concluded that the molecular electrostatic fields in glass would be less symmetric than in aqueous solutions. These ideas are not tenable in view of more recent studies and experimental results of A. N. Sevchenko,[616] on uranyl-colored borosilicate and phosphate glasses, their absorption and luminescence spectra in polarized ultraviolet exciting light and their bleaching behavior. In every case, the structure of such glasses (based on compositions $Na_2O,3SiO_2,K_2O,7B_2O_3,10SiO_2$, $BaO.P_2O_5$, and $Na_2O.3B_2O_3$) is in a *higher* degree of order than in the liquid solutions and not vice-versa. Sevchenko interpreted this conclusion entirely in the direction of the crystallite hypothesis of glass constitution.

PROBLEMS ON THE PRESENCE OF MOLECULAR COMPOUNDS IN GLASS CONSTITUTION

367. We have repeatedly come into contact with the question of whether definite chemical compounds may be assigned as structural units in glass. The colloid or micellar theory of T. Moriya[617] discussed in A. ¶ 306 in this respect goes back to previous schematic concepts of W. Biltz and F. Weibke (1932) who calculated the glass constitution as built up from constituents derived from stoichiometric ratios in phase equilibrium diagrams of glass-forming oxides with silica. We are inclined at present to estimate such attempts only as an auxiliary hypothesis in the calculation of physical property parameters from additivity rules (cf. B. ¶ 114 ff.).

368. Nevertheless, the existence of characteristic *break-points* in certain curves representing physical properties as a function of composition is again and again

[614] *Am. Mineralogist*, **39**, 1954, 615-629.

[615] *Am. Mineralogist*, **40**, 1955, 22-31; sulfur-containing scapolite shows accurately the same luminescence with yellow colors.

[616] *Stroenie Stekla, Inst. Khim. Silikatov Akad. Nauk S.S.S.R., Trudy Soveshchaniya, Leningrad*, 1953 (Pub. 1955), 207-215. Recently, G. O. Karapetyan, *Structure of Glass. Proc. All-Union Conf. Glassy State, 3rd, Leningrad* (English Translation), 1959/60, 319-323, compared the absorption spectra of cerium ions in crystalline solutions with alkaline earth fluorides, aqueous solutions of cerium salts, and in cerium containing glasses. The spectra are in all cases rather similar.

[617] *Tokyo Kogyo Daigaku Gakuho*, 1951, B (1) 1-10.

discussed and they *do* coincide with the stoichiometric ratios of definite compounds in the equilibrium diagrams. It is justifiable, as was demonstrated in alkali silicate systems, that Aniuta Winter-Klein[618] spoke of a certain "heredity" of the constitution of the liquid state from the crystalline phases from which the liquid was molten. The inverse phenomena of devitrification confirm that nucleation starts from hereditary residues, if the liquid is cooled and undercooled, over a distinct temperature zone. In these undeniable facts the postulates of the Zachariasen theory of the complete random state of the glass were probably too strict.

369. In the Russian school the problems of stoichiometric compounds in glass were numerously discussed, prevailingly in the affirmative direction. K. A. Krakau[619] discussed in detail the inflection points on the curves of viscosity as a function of composition in ternary $Na_2O—PbO—SiO_2$ glasses with a field of highest viscosity (at constant temperature) in the molecular ratios for the compound $Na_2O.3PbO.7SiO_2$. We recall how extensively V. A. Florinskaya and R. S. Pechenkina[620] derived these coefficients from their accurate infrared absorption and reflection measurements as a function of composition. The stability of $Na_2O.SiO_2$ and $PbO.SiO_2$ as compounds in the glassy systems was assumed and likewise the role which is played by $Na_2O.2SiO_2$ in its α- and β-modifications in siliceous alkali silicate glasses (cf. A. ¶ 262). Such methods of treating deeper-going questions of molecular configuration are on an evidently higher level than the mere discussion of more or less uncertain breakpoints on curves of the type mentioned above. The same higher aspects are also emphasized for the electrochemical approaches made by O. A. Esin and B. M. Lepinskikh[621] in studies on molten metallurgical slags of the system $CaO—Al_2O_3—SiO_2$ in which the presence of characteristic anions $(SiO_4)^{4-}$, $(SiO_3)^{2-}$, and $(Si_2O_5)^{2-}$ is very distinctly indicated by the electromotoric forces of adequate galvanic cells. These different anions show characteristic fields of their stability and instability, say as a function of the concentration of melts in FeO. Evidently, the cations Na^+, Ca^{2+}, and Fe^{2+} considerably affect the stability of complex anions with Si^{4+} and also with Ti^{4+} as centers. L. S. Kaĭnarskiĭ and E. V. Degtyareva[622] developed on the basis of Esin's theories a highly useful mechanism for the mineralizing actions in the tridymitization of silica (cf. Vol. V, Section B). This industrially important effect can only be promoted if structural principles of the constitution of the melt and the mineralizer are in favor of nucleating the $[SiO_4]$ frameworks of the special modification of silica in question.

[618] *Compt. rend.*, **247**, 1957, 1587-1590.

[619] "Physical-Chemical Properties of the Ternary System $Na_2O–PbO–SiO_2$," Edit. I. V. Grebenshchikov, Moscow and Leningrad 1949, 211-229.

[620] *Stroenie Stekla, Inst. Khim. Silikatov Akad. Nauk S.S.S.R., Trudy Soveshchaniya, Leningrad,* 1953 (Pub. 1955), 70-96, especially pp. 79-94 with important references.

[621] *Doklady Akad. Nauk S.S.S.R.*, **95**, 1954, 135-138.

[622] *Doklady Akad. Nauk S.S.S.R.*, **99**, 1954, 301-304.

370. Concerning the constitution of common glass, K. S. Evstrop'ev and N. A. Toropov[623] are defenders of the hypothesis of the existence of quite definite compounds in glass, whereas K. G. Kumanin and L. I. Demkina[624] believe that break-points for physical parameters (such as the specific volume) as a function of composition do not coincide with singular compounds but with the composition of *eutectics* as "phase limits," as in the systems $Na_2O—SiO_2$, $K_2O—SiO_2$, $PbO—SiO_2$, and $PbO—P_2O_5$. Similar ideas are discussed on a wider basis of silicate and borate glasses by K. A. Kostanyan[625] for specific volume and surface tension, in connection with structural changes in the coordination principles of compounds concerned as constituents of the glass, causing singularities (break-points). We evaluate all these discussions, on the same levels with previous discussions on break-points indicating stoichiometric compounds, as suggestions and plausible approaches. One may also add to these the recent discussions of E. Eipeltauer and A. More[626] on viscosity parameters of potassium silicate glasses in which the compounds $K_2O.2SiO_2$, $K_2O.3SiO_2$, and $K_2O.4SiO_2$ are well marked (cf. Fig. A. 13 in A. ¶ 22).

371. In spite of the simplicity of such demonstrations and their undeniable plausibility, we wish to prefer *thermodynamic* parameters to illustrate the energy relations in the formation of compound-like groupings in the constitution of glass such as were investigated by L. Shartsis and W. Capps[627] for the enthalpies of dissolution of alkali borate glasses in a 2-*N* nitric acid solution in a vacuum-bottle calorimeter. Combined with previously measured calorimetric data, the enthalpies of formation for crystalline and glassy forms of alkali borates could be calculated with an accuracy of some per cent. The curves of these parameters as a function of composition show uniformly (Fig. A. 218) minima at 20 molecular per cent R_2O in the glasses but for potassium and sodium borates only the same minima for the crystalline phases, whereas the lithium borate glasses have the minima at 25 per cent Li_2O. It is, therefore, assured that a compound $Li_2O.4B_2O_3$ does not exist whereas $Li_2O.3B_2O_3$ is stable; on the other hand, $K_2O.4B_2O_3$ and $Na_2O.4B_2O_3$ do exist.

372. For borosilicate melts, W. J. Knapp and H. Flood[628] made a most remarkable

[623] "Chemistry of Silicon and Physical Chemistry of the Silicates," Promistroĭ Edit. 1950; "Einführung in die Silikatchemie," German Edition by W. Hinz, VEB Verlag Technik, Berlin 1958, pp. 42 f.

[624] Cf. *Steklo i Keram.*, **11**, 1954 (2) 10-15; *Silikat Tech.*, **6**, 1955, 51-54; *Stroenie Stekla, Inst. Khim. Silikatov Akad. Nauk S.S.S.R., Trudy Soveshchaniya, Leningrad*, 1953 (Pub. 1955), 107-119, discussion remarks of A. G. Bergman on pp. 311-313.

[625] *Izvest. Akad. Nauk Armyan. S.S.R., Fiz. Mat. Estestven. i. Tekh. Nauki*, **8**, 1955, 49-63.

[626] *Radex Rundschau*, 1960, 230-238.

[627] *J. Am. Ceram. Soc.*, **37**, 1954, 27-32.

[628] *J. Am. Ceram. Soc.*, **40**, 1957, 246-249; 262-268; cf. as a theoretical basis M. Temkin, *Acta Physicochim. U.R.S.S.*, **20**, 1945, 411-420. Cf. H. Flood, T. Förland, and K. Grjotheim, *Z. anorg. u. allgem. Chem.*, **276**, 1954, 289-315; H. Flood and S. Urnes, *Z. Elektrochem.*, **59**, 1955, 364-370.

attempt to develop structural models of calcium borate melts by calculating *melt activities* and comparing these with the enthalpies of fusion. Calcium borates in this connection are most instructive because it is known from the crystal structure that they contain polymeric annular-shaped units $(BO_2)_3^{3-}$ whereas the pyroborate anion

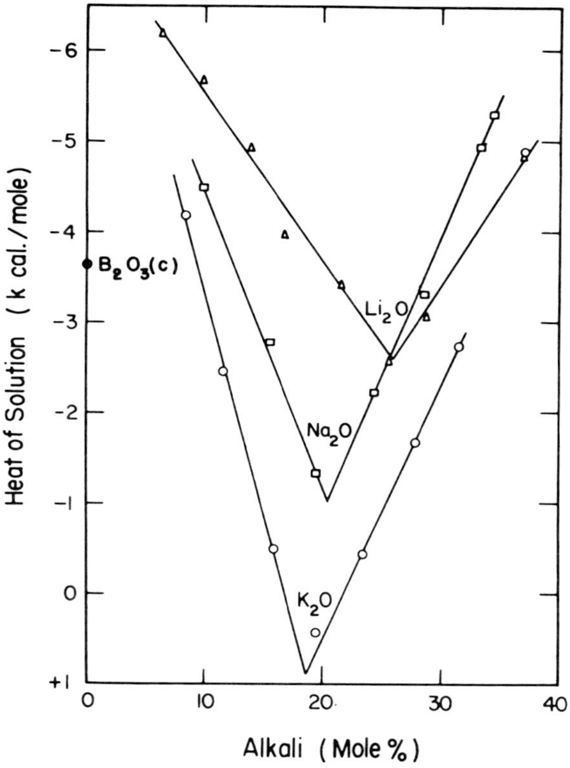

Fig. A. 218. Heats of dissolution of devitrified alkali borate glasses in 2 *N* nitric acid solution. (Shartsis and Capps).

$(B_2O_5)^{4-}$ is not yet certain, and also $(B_4O_7)^{2-}$ ions may exist in melts of $CaO.2B_2O_3$. The fundamental equation for calculation of the activities is $-RT \cdot \ln a_A = \Delta H_f(1 - T/T_0)$ for constituent A in the system A — B. A tabulation of the group types possible in borate and silicate melts lists the following:

Group Types for Borates and Silicates (Knapp and Flood):

General Group Family	Group Types
$BO_m^{(3-2m)}$	$(BO_3)^{3-}$ $(BO_{2.5})^{2-}$ $(BO_2)^{-1}$ $(BO_{1.5})^0$
$SiO_n^{(4-2n)}$	$(SiO_4)^{4-}$ $(SiO_{3.5})^{3-}$ $(SiO_3)^{2-}$ $(SiO_{2.5})^{1-}$ $(SiO_2)^0$

373. In connection with silicates in which several polymeric groups are sufficiently known, Knapp and Flood built up structural models of borosilicates, based on the calculation of activities as functions of composition and on measured enthalpies of fusion, and then for the systems $3CaO.B_2O_3$—$2CaO.SiO_2$, $2CaO.B_2O_3$—$2CaO.SiO_2$, $3CaO.B_2O_3$—$CaO.SiO_2$, $CaO.B_2O_3$—$CaO.SiO_2$, and $2CaO.B_2O_3$—$CaO.SiO_2$. Simple ionic models can be created in most cases but there are doubtless also highly polymerized mixed ions of the type metaborate-metasilicate for which group models are applicable, of the type

```
                            O
                            |
           O                B       O       O
           |               / \      |       |
    ···O—Si—O       O        O—Si—O—Si—O···
           |      \ /               |       |
           O       B                O       O
                   |
                   O
```

```
      O      O              O     O             O      O      O
       \     |              |    /               \     |     /
or      B—O—Si—O—···—O—Si—O—B          or        B—O—Si—O—B
       /     |              |    \               /     |     \
      O      O              O     O             O      O      O
```

and so on, to fit the activities determined from calorimetric measurements of ΔH_f. Most of the melts contain mixed pyro-borate-silicate ions $(BSiO_6)^{5-}$, with the exception of melts of the systems $CaO.B_2O_3$—$CaO.SiO_2$ and $2CaO.B_2O_3$—$CaO.SiO_2$ for which the assumption of polymeric mixed ions is indispensable. We discussed in A. ¶ 200 the structural group models used by W. J. Knapp and W. D. Van Vorst[629] as melts between $Na_2O.SiO_2$ and $Na_2O.2SiO_2$, on the basis of a comparison of the activities calculated from ΔH_f and different model concepts, using the ring or chain structures $(SiO_3)^{2x-}$ with a random distribution of $(SiO_{2.5})_2$ pairs bridging between rings and chains (cf. Figs. A. 144, 145) in good agreement with the calculations and calorimetric data.

374. How different the observations of authors on the appearance of break-points on curves, showing a given physical parameter as a function of composition, may be interpreted, is quite interesting to note even in the recent literature. J. O'M. Bockris and D. C. Lowe[630] gave in their studies on viscosity of fused glasses (cf. A. ¶ 22) no indication of any inflection points but found only smooth curves for the system CaO—SiO_2 although there are distinct changes of structure in the glass below, and above, 12 molecular per cent of the metal oxide. Below this concentration SiO_2 is the flowing unit, above it is a definite silicate anion the size of which increases

[629] *J. A. Ceram. Soc.*, **42**, 1959, 559-562.

[630] *Proc. Roy. Soc.* (*London*), A, **226**, 1954, 423-435.

with increasing concentration in silica. Also A. Dietzel and L. Merker[631] failed to establish the characteristic break-points, which were so impressively demonstrated before by W. E. S. Turner and E. Preston (1932/33) from alkali silicate glasses by different rates of *alkali vaporization* from the surface of the melts, combined with strong migration of Na^+ ions in the melt to the surface, and a corresponding ion exchange of Na^+ replacing Ca^{2+} and H^+ ions in the glass constitution. CaO is enriched in the alkali-impoverished surface layer (cf. B. ¶ 92, ¶ 168, on the role of water in such layers). Nothing from the volatilization curves speaks in favor of the existence of $Na_2O.2SiO_2$ or other compounds in the glass. On the other hand, we mentioned in A. ¶ 139 that J. M. Stevels[632] interpreted the break-points in curves representing the logarithm of viscosity (cf. A. ¶ 74,), and the activation energy of electric conductance versus composition for alkali and lead silicate glasses (expressed by the parameter Y equal to the number of O^{2-} anions per polyhedron which participate in the framework). In the system $Na_2O—SiO_2$ he found a break-point for $Na_2O.2SiO_2$ ($Y = 3.0$), for the system $K_2O—SiO_2$ one at $K_2O.4SiO_2$ ($Y = 4.0$) and in the system $PbO—SiO_2$ one at $PbO.SiO_2$ ($Y = 2.0$).

375. Very sharp break-points are characteristic of the "anomalous" phosphate glasses which E. Kordes, W. Vogel, and R. Feterowsky[633] studied in the systems $BeO—P_2O_5$, $MgO—P_2O_5$, and $ZnO—P_2O_5$. The discontinuities at 50 molecular per cent P_2O_5 could be explained by a sudden change of the atomic-structural arrangements as a function of the glass composition. It is particularly interesting that the discontinuity does not appear (cf. Fig. A. 219) in the curves for molecular refraction, which is a measure for the volume proper of the molecules, not for their state of *packing* in the glass structure. Sharp breaks on physical parameter versus composition curves play a particularly important role in the development of M. L. Huggins's "Structon" theory[634] discussed in A. ¶ 242 f. These Structons undergo changes in specific arrangements of atoms surrounded by close neighbors, with sharp boundaries of composition, and straight-line segments of those curves between different break-points, especially for volume versus composition. Huggins and T. Abe[635] applied the Structon theory to borosilicate glasses with sharp break-points in diagrams of the expansion coefficient ($\delta V/\delta T$) as a function of the Structon numbers for the cations, as in alkali and alkaline earth borate glasses.

[631] *Trav. Congr. Intern. du Verre*, IV, Paris 1956, II. 3; *Veröffentl. Max-Planck Inst. Silikatforsch.*, **16**, 1956, 105-110; more explicitly in *Glastech. Ber.*, **30**, 1957, 134-138.

[632] *Silicates inds.*, **22**, 1957, 325-335.

[633] *Z. Elektrochem.*, **57**, 1953, 282-288. On the system $ZnO-P_2O_5$ with interesting fluorescence and phosphorescence properties (Mn as activator) under cathode ray excitation, cf. F. A. Hummel and F. L. Katnack, *J. Electrochem. Soc.*, **105**, 1958, 125-133; 528-533.

[634] *J. Am. Ceram. Soc.*, **38**, 1955, 172-175; *J. Phys. Chem.*, **58**, 1954, 1141-1146.

[635] *J. Am. Ceram. Soc.*, **40**, 1957, 287-292.

376. Going back to the classical work of W. E. S. Turner and E. Preston on the break-points of the alkali and lead oxide volatilization of glasses, N. V. Solomin[636] reexamined these conditions in alkali borate melts and gave a valuable tabulation of quantitative data of evaporation, and the alkali contents in vapors of the compounds $2R_2O.B_2O_3$, $R_2O.B_2O_3$, and $R_2O.2B_2O_3$, over the range from 1100° to 1400°C.

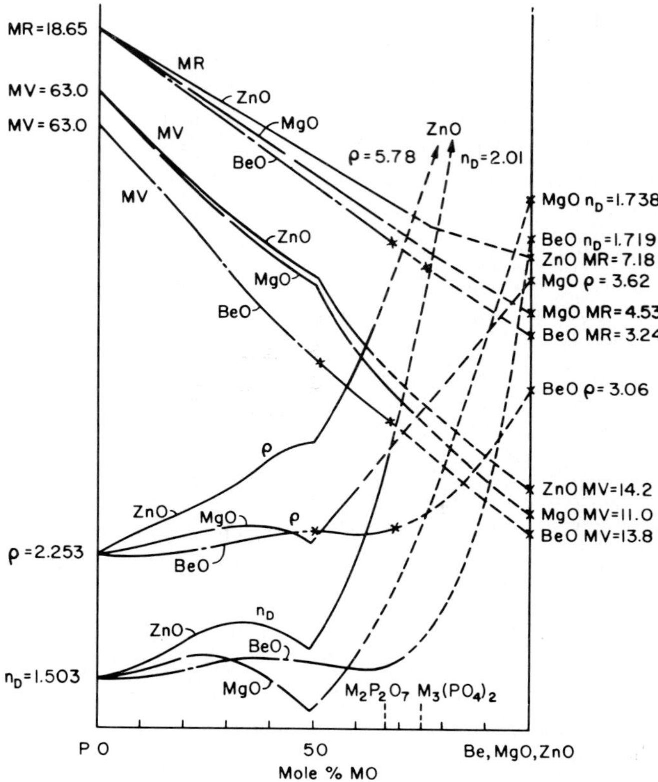

FIG. A. 219. Refractive index, density, molecular volume, and molecular refraction of anomalous phosphate glasses. (Kordes, Vogel, and Feterovsky).

N. V. Solomin and L. V. Potemkina[637] concluded that $Na_2O.B_2O_3$ (at 1400°C.), $K_2O.B_2O_3$ and $2K_2O.B_2O_3$ (at 1300°C.), $Li_2O.B_2O_3$ and $Li_2O.2B_2O_3$ (at 1400°C.) vaporize completely without perceptible changes in composition, but $2Li_2O.B_2O_3$ is decomposed in vapor to $Li_2O.B_2O_3$ and free alkali. $Na_2O.2B_2O_3$ is partially decom-

[636] *Stroenie Stekla, Inst. Khim. Silikatov Akad. Nauk S.S.S.R., Trudy Soveshchaniya, Leningrad*, 1953 (Pub. 1955), 230-233.

[637] *Doklady Akad. Nauk S.S.S.R.*, **96**, 1954, 91-93.

posed above 1200°C., while at 1400°C. the vapor phase contains 30 per cent $Na_2O.B_2O_3$. The maximum in the fusion equilibrium diagram is rather sharp for $Na_2O.B_2O_3$, but very flat for $Na_2O.2B_2O_3$. The disodium borate $2Na_2O.B_2O_3$ is not decomposed in its vapor at 1300°C., but at 1400°C. $Na_2O.B_2O_3$ is formed in the condensate. $K_2O.B_2O_3$ and $2K_2O.B_2O_3$ are stable below 1300°C. but $K_2O.2B_2O_3$ is changed to $K_2O.B_2O_3$ in the condensate. Metaborates are the most stable borates. The vapor pressure of B_2O_3 is much lower than that of free alkali and of the metaborates. For the manufacturing of borosilicate glasses it is important that $K_2O.B_2O_3$ is the most volatile constitutent. With increasing atomic weight and cationic radius of the alkali metals an approximately linear function is plotted for volatilization of the metaborates, determined in gm./cm.^{2} sec. 10^6 at constant temperature.

377. For volatilization from melts of the system $Na_2O—B_2O_3$, G. A. Kolykov[638]

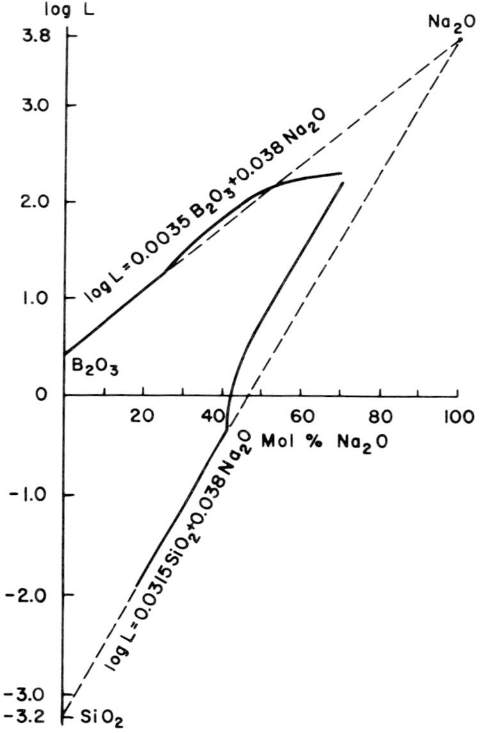

FIG. A. 220. Volatility, L, in logarithmic scale, in system $SiO_2—B_2O_3$, and $Na_2O—SiO_2$, as a function of concentration in Na_2O, at 1200°C. (Kolykov).

[638] *Stroenie Stekla, Inst. Khim. Silikatov Akad. Nauk S.S.S.R., Trudy Soveshchaniya, Leningrad*, 1953 (Pub. 1955), 234-244. On volatilization from borosilicate glasses, see recently L. F. Oldfield and R. D. Wright, *Glass Technol.*, **3**, 1962, 59-68.

determined systematically the rates of volatilization at different temperatures and as a function of time, expressed in diagrams of log L (L the volatility) versus the concentration in Na_2O, as seen in Figs. A. 220 and 221. The curves for L in the ternary glasses show a systematic shifting of the maxima from 80 per cent B_2O_3 for

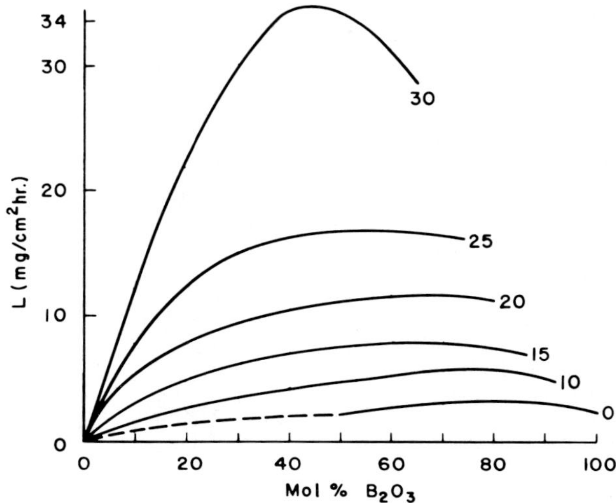

FIG. A. 221. Volatility, L, in system $Na_2O—B_2O_3—SiO_2$, as a function of the content in B_2O_3, at 1200°C., and for different Na_2O concentrations. (Kolykov).

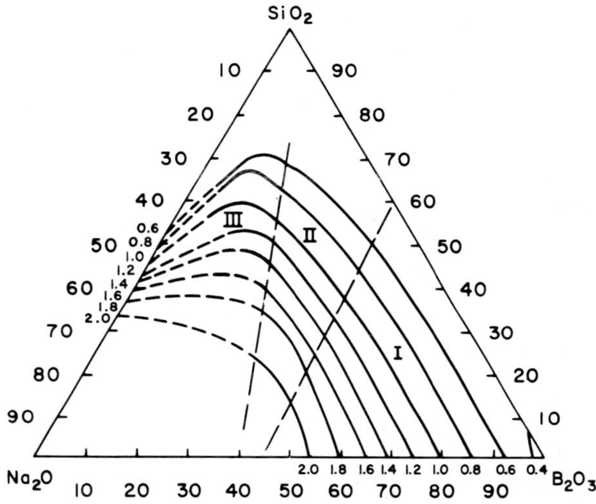

FIG. A. 222. Curves of constant volatility, L, at 1200°C., for glasses of system $Na_2O—B_2O_3—SiO_2$. (Kolykov).

pure $Na_2O—B_2O_3$ glasses, to 40 per cent B_2O_3 for glasses with constant 30 per cent Na_2O. In triangular projection, the curves of constant L are seen from Fig. A. 222, indicating the area *I* in which L is only determined by Na_2O, whereas in the fields *II* and *III* all three components influence it in different degrees. Comparing these curves with those known from the general discussion of D. P. Konovalov (1884) on the vapor pressure curves in binary and ternary systems, one must conclude that the sodium borate and silicate melts correspond to imperfect liquids with a negative deviation from Raoult's law. They are solutions of different chemical compounds and are disposed to form chemical compounds. The melts of the system $B_2O_3—SiO_2$ and the ternary melts, however, show positive deviations from the Raoult law, namely they are liquids disposed to unmix and are too similar in their chemical nature to form compounds. On the unmixing behavior and characteristics of porous glasses prepared from sodium borosilicates see Vol. III, B. ¶ 425, and Vol. V, Section B. (Vycor type glasses).

378. The volatilization of *lead oxide* from glasses and ceramic glazes is of high technological importance and, therefore, was systematically studied by W. Kerstan[639] with gravimetric determinations of the evaporation losses. The influence of B_2O_3 and Al_2O_3 on this volatility is relatively unimportant whereas CaO expels the lead oxide in a drastic way. Ch. Hirayama[640] quite recently measured the volatility from binary melts of $PbO—SiO_2$ and $PbO—B_2O_3$ (with more than 71 per cent PbO) at reduced pressures. At comparable temperatures, the volatilization from binary lead silicate melts is greater than from melts with a corresponding ratio PbO : B_2O_3. The diffusion of PbO is the driving force in the volatilization process, the pure vapor pressure is even at 900°C. relatively low (0.2 mm. Hg). A true "index of volatility" of PbO can be derived from relations which connect the rate, v_i, of loss after heating time, t, with changes in time corresponding to the equation $v_i = k \cdot e^{-at}$, if extrapolation is made to $t = 0$.

NONSILICATE GLASSES [641]

379. A comparison of the physicochemical characteristics of silicate glasses with those which do, or even do not, contain other related glass-forming anions is most

[639] *Ber. deut. keram. Ges.*, **30**, 1953, 254-256.

[640] *J. Am. Ceram. Soc.*, **43**, 1960, 505-509. We refer to the correlation of PbO volatilization with surface tension of lead oxide melts with B_2O_3 and SiO_2 as measured by L. Shartsis, S. Spinner, and A. W. Smock (1948). On break-points in the conductivity curves as a function of temperature in melts of the system $PbO—B_2O_3—SiO_2$ connected with unmixing phenomena cf. C. Kröger and Kl. Lieck, *Z. anorg. u. allgem. Chem.*, **279**, 1955, 300-312.

[641] A very good introduction into this field see W. A. Weyl, *J. Chem. Educ.*, **27**, 1950, 520-524, and generals also in *Central Glass & Ceram. Research Inst. Bull.* (India), **4**, 1957, 121-139; *Glastech.*

instructive from the theoretical viewpoint concerning their similar and more or less specific constitution. From the practical-technological side they are also informative since it was possible to develop beyond the general principles valid for silicate glasses, new types of other glasses which opened quite new perspectives in application, such as in the optical industry. The classical methods starting from V. M. Goldschmidt's ideas on the application of crystallochemical laws on glasses in this new field of nonsilicate glasses may be emphasized again.

380. We do not wish to repeat too much of what was discussed in previous sections, especially not concerning the *borate* glasses, which have been so often used as "model" systems for glass-theoretical studies. To a certain degree a similar parallelism is valid for *aluminate* and especially for *phosphate* glasses, although it is advisable to consider the particular physicochemical nature of the latter, as in the chain arrangements of the metaphosphate glasses. In this respect a comparison must be made not only with the silicates but also with high-polymers in general. The application of modern chromatographic methods making possible the isolation of different types of soluble polymeric phosphate anions became of high importance for knowledge of the contitution of such inorganic glasses, as a basis of comparison with the rules of polymerization in iso- and heteropolyacids or even in organic "model" systems. We mention here only the excellent work done by E. Thilo[642] and his school, and that of A.E.R. Westman[643] (cf. A. ¶ 311). We have often mentioned before the studies of E. Kordes[644] in connection with theoretical problems of phosphate glass constitution. As of great practical significance we also mention the devitrification studies of W. Vogel[645] in optical phosphate glasses, with the crystallization of $Al(PO_3)_3$ as a most characteristic product.

381. From the complete crystallochemical isotypism analogy existing between silica and $AlPO_4$ (cf. Vol. I, A. ¶ 20) it was expected that silica glass would have a

Ber., **30**, 1957, 269-282. We owe to Z. M. Syritskaya, *Structure of Glass. Proc. All-Union Conf. Glassy State, 3rd, Leningrad* (English Translation), 1959/60, 295-298, a systematic study of aluminophosphate glasses which have some importance as lubricants for metal extrusion processing (cf. A. ¶ 30). On the glass formation range of borate systems see the extensive discussions by M. Imaoka, *Yogyo Kyokai Shi*, **69**, 1961, 282-306.

[642] *Silikat Tech.*, **6**, 1955, 278-280; *Naturwissenschaften*, **46**, 1959, 367-373, with extensive literature references, further I. Schulze and W. Hinz, *Silikat Tech.*, **6**, 1955, 235-241, who emphasized the usefulness of G. Hägg's ideas on glass constitution; *Glastech. Ber.*, **29**, 1956, 319-323; H. Grunze, *Silikat Tech.*, **7**, 1956, 134-138. We finally mention W. Dewald, and H. Schmidt, *Z. anorg. u. allgem. Chem.*, **272**, 1953, 253-264; *Angew. Chem.*, **65**, 1953, 78-81, also on technical use of phosphates as detergents, etc. On sodium metaphosphate glass see X-ray investigations of G. W. Brady, *J. Chem. Phys.*, **28**, 1958, 48-50.

[643] *J. Am. Ceram. Soc.*, **37**, 1954, 420-427 (with J. Crowther), and *Ibid.*, **40**, 1957, 293-299 (with P. A. Gartaganis).

[644] Chiefly *Z. Elektrochem.*, **57**, 1953, 282-288 (with W. Vogel and R. Feterowsky).

[645] *Silikat Tech.*, **6**, 1955, 510-517.

model analogue in $AlPO_4$ glass with most desirable technological properties. These expectations, however, soon fade, since $AlPO_4$ is an excellently crystallizing compound, and in this point much different from SiO_2 as A. Dietzel and H. J. Poegel[646] observed. The fusion point of $AlPO_4$ is about 2000°C. and its melt is strikingly fluid in contrast with the high viscosity of silica melts. It is, therefore, concluded that the analogy between silica and $AlPO_4$ does not extend to the constitution of their melts probably because the fused $AlPO_4$ is a true salt with $[AlO_6]$ groups, with widened distances Al—O because of the strong field action of the P^{5+} cations, making the character of $AlPO_4$ highly ionic and unsuitable to form a glass. For these reasons there is also no analogy between the systems Na_2O—Al_2O_3—SiO_2 and Na_2O—Al_2O_3—P_2O_5, or in the partial system Na_2O—$AlPO_4$, in relation to Na_2O—SiO_2. The system Na_2O—$AlPO_4$ not only shows highly fluid melts and high Liquidus temperatures but is pseudobinary by the primary crystallization of corundum and by unmixing and opalescence phenomena. Glasses appear only in a narrow range between 17 and 20 per cent Na_2O (cf. Vol. III, B. ¶ 440).

382. The requirements of model glasses are much better fulfilled in the system *GeO_2* and its derivatives. We repeatedly mentioned the optical properties of pure germanium dioxide glass, such as in its discoloring after irradiation with X-rays (cf. A. ¶ 357*). The density and expansivity of this glass was recently measured by J. D. Mackenzie.[647] In strain-free glass the density was obtained to be 3.667 ± 0.002, or considerably higher than the generally accepted value of 3.638 ± 0.001. In air-quenched samples (from 600°C.) it is 3.653 ± 0.002, no devitrification being observed at 600°C. over 80 hours. The linear expansion coefficient between room temperature and 400°C. is 7.5×10^{-6}, or in other words, higher than for silica glass by more than one order of magnitude.

383. Concerning the analogy between silicates and *germanates*, W. Schütz (1933) described a Pb_2GeO_4 glass (molten in an alumina crucible); L. Merker and H. Wondratschek[648] demonstrated that *pure* Pb_2GeO_4 crystallizes very promptly (if molten in a platinum crucible). The data of Schütz are, therefore, only characteristic of a glass slightly contaminated by dissolution of the refractory wall material in which it was prepared and stabilized by the formation of $[AlO_4]$ groups. In no case are Pb_2GeO_4 and Pb_2SiO_4 isostructural. They are entirely different in their X-ray diffraction diagrams. Also a small addition of BeO or TiO_2 (2 per cent in maximum) act in Pb_2GeO_4 melts stabilizing the glass formation. No glasses were observed by

[646] *Naturwissenschaften*, **40**, 1953, 604; *Veröffentl. Max-Planck-Inst. Silikatforsch.*, **13**, 1953, 131; H. J. Poegel, Inaug.-Dissert. Techn. Univ. Berlin, 1959 (manuscript), abstract in *Glastech. Ber.*, **29**, 1956, 266. See also recent experiments on fusion of $AlPO_4$, by a A. Dietzel and I. Hinz, *Ber. deut. keram. Ges.* **39**, 1962, 569-572.

[647] *J. Am. Ceram. Soc.*, **42**, 1959, 310.

[648] *Glastech. Ber.*, **30**, 1957, 471-473.

P. P. Budnikov and S. G. Tresvyatskiĭ[649] in the system GeO_2—Li_2O, with eutectic melts between $Li_2O.GeO_2$ and GeO_2 at 920° to 930°C., and $Li_2O.GeO_2$ and $2Li_2O.GeO_2$ at 1113°C. Industrial uses of *germanate* glasses are repeatedly claimed. We note as a stable germanate glass of a highly complex type that described by H. C. Dudley[650] in the system GeO_2—$Ti_3(PO_4)_4$, of highly interesting transmittance properties in the near infrared spectrum (see the curve in Fig. A. 223), for optical purposes.

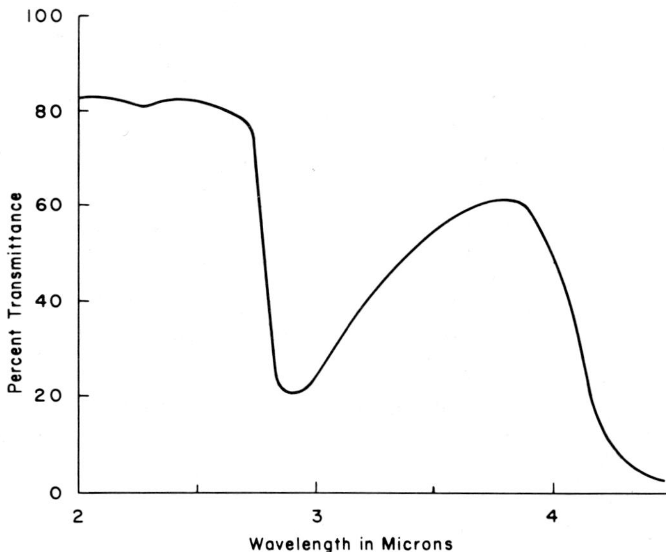

FIG. A. 223. Transmittance of glass 50 GeO_2—50 $Ti_3(PO_4)_4$, in infrared. (Dudley).

384. *Beryllium fluoride* as the best model glass in the definition of V. M. Goldschmidt has in spite of its highly interesting physical properties many difficulties in being directly applied, such as for optical purposes (see A. ¶ 387).[651] Many attempts have been made for the study of complex beryllofluorate glasses, namely by W. Vogel and Kl. Gerth[652] in the systems of BeF_2 with alkali-fluorides, molten in an atmosphere of anhydrous HF and followed by a rapid quenching of the melts between two silica glass plates, and annealed at nearly 110°C. There is a surprising analogy in the density versus composition curves with those of alkali-silicate glasses (cf. A. Dietzel and H. Sheybany, 1948/49). The refractive indices increase in the anomalous series $Na^+ < Li^+ < K^+ < Rb^+$ but in agreement with Dietzel's theory

[649] *Doklady Akad. Nauk S.S.S.R.*, **99**, 1954, 761-763.

[650] *Science*, **119**, 1954, 478-479.

[651] We note as a method for safe production of pure BeF_2 the U.S. Patent No. 2,653,856 (Sept. 1953) of G. Barr.

[652] *Glastech. Ber.*, **31**, 1958, 15-28.

of framework shrinkage in lithium containing glasses and increased electron density as seen in the rapidly increasing branch of the curve for the optical "space filling degree," $(n^2 - 1)/(n^2 + 2)$, as a function of the composition (Fig. A. 224). Polarization effects are indicated by deviations from a linear additivity of molecular volume and molecular refractivity in the glasses of potassium and rubidium beryllofluorates, whereas in the corresponding linear relations for sodium and lithium beryllofluorate glasses the non-deformable, "rigid sphere" behavior of the Na^+ and Li^+ ions is

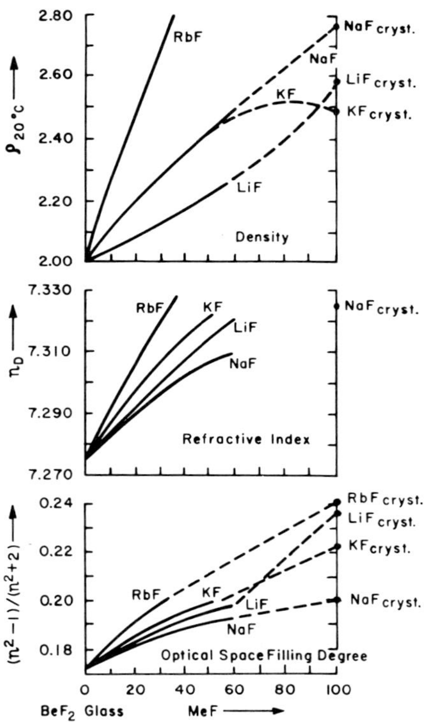

Fig. A. 224. Density, refractive index, n_D, and optical space filling degree, $(n^2 - 1)/(n^2 + 2)$, of alkali beryllofluorate glasses, as a function of composition. (Vogel and Gerth).

evident. The optically clear and Tyndall effect-free glasses show in the electron microscope the typical unmixing phenomenon described in Vol. I, C. ¶ 126, analogous to those in sodium borosilicate glasses. Turbid-opalescent Li-beryllofluorate glasses show particularly impressive electron micrographs, the other glasses the typical "droplets." The appearance of such droplets is very sensitive, contaminations of BeF_2 by only 1 per cent RbF are immediately detected in the electron microscope, and only purest BeF_2 glass is electron-optically "empty." The phase separation mechanism is the same as outlined in Dietzel's theories based on electrostatic field-strengths z/a^2, of the cations in the glass structure. "Swarm" development of molecular aggregates is the first step to phase separation, followed by formation of droplets under the action of surface tension. Corresponding to weaker bonds in beryllofluorate glasses the phase separation areas are larger in these (300 to 600 Å. in diameter) than in silicate glasses (100 to 300 Å.).

385. *Ternary* alkali beryllofluorate glasses were studied in their optical qualities by M. Imaoka and Sh. Mizusawa.[653] W. Vogel and Kl. Gerth,[654] on the other hand, extended their investigations to beryllofluorate glasses of the systems MgF_2, CaF_2, and SrF_2—BeF_2 and also to ternary compositions of these with alkali fluorides in addition. The glass formation is restricted to specific concentration fields. The physical properties of the glasses resulting show monotonous functions of composition. Beryllofluorate glasses containing MgF_2, however, indicate the specific role of the $[MgF_4]$ coordination groups which enter the framework skeleton of the structure, whereas the higher coordination of the cations Ca^{2+} or Sr^{2+} with F^- requires a distribution of these cations over the structural cavities in the same manner as is the case with alkali cations in beryllofluorate glasses. The electron-microscopic examination showed again the two-phase unmixing droplets which cannot be detected by optical Tyndall effects. The droplets are always enriched in BeF_2 and have a relatively higher surface tension. The swarm formation of molecular aggregates as microheterogeneities is not compatible with the Zachariasen framework theory. Vogel and Gerth discuss the idea that the unmixing reactions tend to the composition of distinct compounds known from the phase equilibrium diagrams of the systems concerned, with transitions from the droplets to crystallization nuclei with dendritic forms. The swarms may represent a precrystalline state in glass of a very low degree of crystallinity. The unmixing fields of the ternary beryllofluorate glasses are always in the neighborhood of the BeF_2 apex in the concentration diagrams. Pseudobinary sections show change in physical properties as a function of the addition of alkaline earth fluorides to constant binary RF—BeF_2 glass compositions.

386. In the same way as seen for silicate and borate glasses, the complex beryllofluorate glasses are intermediate between aspects of the Zachariasen framework and the crystallite theories. The formation of NaF.MgF_2 which is also observed in optical glass of the fluor opal type, is a typical compound also formed from beryllofluorate glasses, whereas in LiF containing magnesium fluoride glasses only MgF_2 crystallizes. Such glasses, therefore, correspond better to the postulates of the crystallite theory. A slight breakpoint is found in the curve for molecular volume and the molecular

[653] *Yogyo Kyokai Shi*, **61**, 1953, 13-14.

[654] *Silikat Tech.*, **9**, 1958, 353-358; 495-501. Melts of the system BaF_2–BeF_2 crystallize immediately on cooling.

refractivity of LiF—MgF_2—BeF_2 glasses at about 7 molecular per cent MgF_2 indicating a characteristic contrapolarizing effect of the Li^+ ion to suppress the entrance of [MgF_4] groups into the skeleton and to separate as [MgF_6] in the rutile type structure of MgF_2. The curve for molecular refractivity for KF—MgF_2—BeF_2 glasses is practically a horizontal line.

387. An elaborate investigation of the systematic development of new optical glasses from fluoride systems was made by W. Jahn,[655] especially for the evolution of particular properties in the refractive index, n , and the value

$$\nu = \frac{n_D - 1}{n_F - n_C}$$

over the experimental variation field of these parameters. For beryllofluorate glasses n_D is between 1.33 and 1.42; $\nu = 90$ to 105. A great hindrance in industrial production of such glasses is still their low degree of chemical durability, their hygroscopicity, and especially the toxicity of beryllium compounds in general. A more useful group of industrially considered glasses, therefore, are *phosphate* containing fluoride glasses for which a great many compositions are given and their properties and production described (with n between 1.44 to 1.54; $\nu = 72$ to 89). So-called fluorphosphate glasses are still higher in [PO_4] and lower in F^- than the glasses mentioned before ($n_D = 1.52$ to 1.62; $\nu = 62$ to 77). This group is described with some detail in its technological-optical characteristics. A graphic representation of the position of the different groups of fluoride glasses is given in Fig. A. 225, with n_D and ν as coordinates. It is highly instructive to compare the new glass types with those of previous optical glasses contained in the well-known Schott Catalogue. All compositions in which fluorides are compatible with oxides in the constitution of glasses, with an improved chemical durability and better glass-forming qualities, are designated *oxyfluoride* glasses. Particularly interesting among the phosphate-fluoride glasses are those in which no more BeF_2 is contained.

388. T. Izumitani and R. Terai[656] investigated the systematic development of new types of optical glasses on the basis either of *fluoroborates* in the systems BaF_2—B_2O_3 and PbF_2—B_2O_3, or of pure fluoride systems with complex *aluminofluorates*, as in the system BaF_2—PbF_2—AlF_3, analogous to the classical beryllofluorates, and finally of *rare element borates*, say of lanthanum, titanium, zirconium, thorium, tantalum, and wolfram, with metal oxides RO, or $R_2O_3 + B_2O_3$. The limits of glass formation areas for such compositions are very critical, especially in the pure fluoride

[655] *Glastech. Ber.*, 34, 1961, 107-120. On the general principles for development of new optical glasses, see W. Geffcken, ibidem 91-101, and M. Faulstich, *Ibid.*, 102-107, especially on La_2O_3 containing glasses, with rare elements in the basic system La_2O_3—ThO_2—B_2O_3—Ta_2O_5—Nb_2O_5.

[656] *Osaka Kogyo Gijutsu Shikenjo Kiho*, 3, 1952, 21-24, 25-28, 104-108; 4, 1953, 119-122, 123-127 (with H. Hamamura); see above work of Geffcken and Faulstich.

systems, although the crystallization tendency of PbF_2 can be kept low. The chemical durability can be increased by additions of TiO_2 or K_2ZrF_6, with very low softening points (330° to 340°C.). Among fluorophosphate glasses (see above) those are interesting which belong to the system $BaO—Al_2O_3—P_2O_5$, with additions of La_2O_3, and/or ZrO_2 to keep the crystallization tendency low. The systems of $AlPO_4$ with LiF, BaF_2, and PbF_2 also form stable glasses. Interesting *fluorosilicate* glasses were observed in the systems $Na_2O—PbF_2—SiO_2$, $NaF—SiO_2—TiO_2$ and furthermore a "Fluoro-Flint Glass" in the system $Na_2O—PbO—PbF_2—SiO_2—TiO_2$.

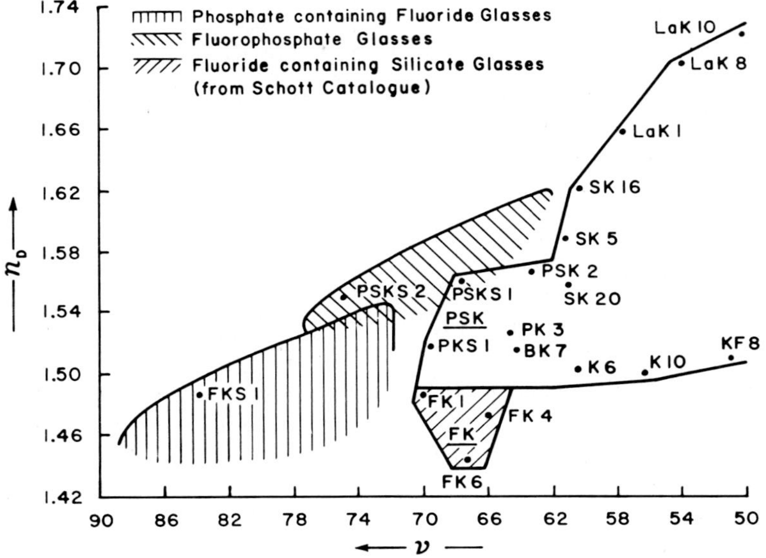

FIG. A. 225. Fields representing different fluoride containing glasses in optical diagram, with n_D and ν as coordinates. (Jahn).

389. As model glasses in a wider sphere of definition one may also accept the group of *sulfides*, *selenides* and *tellurides* which in the glassy state have attracted the attention of industrial research because of their infrared transmittance properties, known since 1870 by C. Schultz-Sellack as "diathermancy" such as for As_2S_3. We may emphasize that the presence of natural minerals of the As_2S_3 group in glass-form was also more recently described by G. C. Amstutz,[657] in very complex compositions e.g. from the Cerro de Pasco (Peru), with a remarkable isotope separation

[657] *Bol. soc. geol. Peru*, 1957, II, 25-53 (with P. Ramdohr and F. de las Casas). The name "revoredite" was proposed by these authors but it was refuted by Ch. Milton and Bl. Ingram, *Am. Mineralogist*, **44**, 1959, 1070-1076. See also G. C. Amstutz, *Bull. Geol. Soc. Am.*, **68**, 1957, 1696. Cf. C. Schultz-Sellack, *Ann. Physik*, **139**, 1870, 182.

(enrichment of S^{32} relative to S^{34}, from the normal ratio 22.23 or 22.31 to 22.96 : 1 in this special As_2S_3 mineral). A very accurate examination of physical and chemical properties of pure As_2S_3 glass was made by Fr. W. Glaze, D. H. Blackburn, J. S. Osmalov, D. Hubbard, and M. H. Black[658] chiefly because of its high transmittance in the infrared up to 13 μ. Its immediate use in optical instrumentation is therefore evident. W. A. Fraser and D. M. Fraser[659] studied the physical properties of synthetic As_2S_3 glass and G. Dewulf[660] described the manufacturing of As_2S_5 and As_2Se_5 glasses, with a high transmittance of infrared between 0.6 and 13 μ, and between 2.0 and 15 μ, respectively. A systematic determination of the position of *ternary sulfide glass* formation was given by S. S. Flaschen, A. D. Pearson, and W. R. Northover[661] in the systems As—Tl—S, As—Tl—Se, and As—S—Se, in the latter system with a wide field of glass-forming compositions extending to a small unmixing field near the S apex and down to 40 per cent Se around the As apex. The role of thallium in such glasses is quite specific in contrast with Sb, Bi, Cd, and Pb. All these glasses are low-softening (about 200°C.) but still highly viscous at 400°C. It is further interesting that GeS_2 also forms a regular glass, known since 1930.[662]

390. A recent extensive report on infrared transmitting glasses mostly on the basis of As_2S_3, As_2Se_3, and the like, was given by H. H. Blau,[663] in comparison with simple silicate, borate, aluminate, germanate, phosphate, and tellurite glasses, with their specific absorption curves over the range from 2 to 8 μ, for sulfide glasses up to 14 μ. We may also add gallate and thiogallate glasses discussed by P. L. Baynton, H. Rawson, and J. E. Stanworth,[664] and by W. A. Fraser and J. E. Brugger.[665] The *semiconductor* properties of glasses in the system As_2Se_3—As_2Te_3 were described by T. N. Vengel and B. T. Kolomiets.[666] The conduction mechanism is *electronic* in this case.

391. Coming back once again to the basic theories of V. M. Goldschmidt on model glasses, we must admit that his ideas on possibilities for a development of "strengthened" glasses of the group of nitrides, carbides, and silicides have not been fulfilled. At his time he could not be acquainted enough with the excellent crystallinity of

658 *J. Research Natl. Bur. Standards*, **59**, 1957, 82-92.

659 *J. Opt. Soc. Am.*, **49**, 1959, 497, TA 21. On the crystallization tendency and the electrical properties of such chalcogenide glasses see recently B. I. Kolomiets, T. N. Mamontova, and T. F. Nazarova, *Structure of Glass. Proc. All-Union Conf. Glassy State, 3rd, Leningrad* (English Translation), 1959/60, 418-422.

660 *Rev. opt.*, **33**, 1954, 513-518.

661 *J. Am. Ceram. Soc.*, **42**, 1959, 450; **43**, 1960, 274-278.

662 Cf. W. J. Pugh, *J. Chem. Soc.*, 1930, 2369-2373.

663 *Glastech. Ber.*, **32**, K, 1959, VII. 6 - VII. 10.

664 *Nature*, **179**, 1957, 434-435.

665 *J. Opt. Soc. Am.*, **49**, 1959, 497, TA 20.

666 *Zhur. Tekh. Fiz.*, **27**, 1957, 2484-2491.

those compounds which do not show any tendency to form glasses. Concerning their hardness, Goldschmidt's expectations were entirely confirmed and even excelled, some silicides also show valuable properties as refractory construction materials,[667] and others more. The question of why Goldschmidt's ideas on "strengthened" glasses were not successful was discussed by W. A. Weyl.[668] The fusion points of nitrides, carbides, and the like, are so high that in their melts the thermal energy is sufficient to overcome even the high activation energies of nucleation. On the other hand, in organic compounds fusion points and, therefore, the thermal energies are so low that even the low energy barrier of hydrogen bonding may be a hindrance to nucleation, and glass formation ensues. It is the relation between the energy barrier of a structural change and the value of $k \cdot T$ factors on the crystallization which determines the rate of reaction for nucleation and, therefore, glass formation. In these words, Weyl has outlined to the point the reason why Goldschmidt's purely crystallochemical treatment of the glass constitution could not fulfill his predictions except in the few real "model glasses" of which BeF_2 is the most impressive one.

SURFACE STRUCTURE OF GLASS AND ITS REACTIONS

392. The programmatic work done by W. A. Weyl since 1947 for an understanding of the chemical reactivity of glass surfaces from structural viewpoints conducted him to the evolution of the *screening theory* and successful application of the polarization-polarizability concepts for fundamental wetting properties of glass surfaces.[669] The screening of cations means the surrounding of positively charged atomic centers either by free electrons, or by polarizable anions which act on the coordination group thus formed like an electrostatic "Faraday cage." If the positive charge of the atom or its immediate surrounding is increased, a "tightening" effect is observed by an increased attraction between the central atom and the surrounding electrons. The electron cloud is "stiffened" and is much less influenced by exterior fields. Both processes, screening and tightening, are of particular importance for the interactions on the surface of a solid with surrounding solutions (or melts) from

[667] See e.g. R. Kieffer, F. Benesowsky and K. Konopicky, *Ber. deut. keram. Ges.*, **31**, 1954, 223-230.

[668] *Glastech. Ber.*, **30**, 1957, 269-282; *Central Glass & Ceram. Research Inst. Bull.* (*India*), **4**, 1957, 121-139.

[669] Cf. *J. Colloid Sci.*, **6**, 1951, 389-405, on the surface structure of water; Chapter IV in the book of R. Gomer and C. St. Smith, "Structure and Properties of Solid Surfaces," Univ. of Chicago Press 1954, 491 pp., especially pp. 147-184, on wetting as influenced by the polarizability of surface ions; *Glass Ind.*, **37**, 1956, 264-269, 286, 288; 325-331, 336, 344, 346, 350, on acid-base concepts; E. Plumat, *Silicates inds.*, **19**, 1954, 97-107; **21**, 1956, 259-264, gave a fascinating discussion on screening theory with applications. We will follow in many points this interpretation in our text.

which anions are attracted by a multiple charged cation like Si^{4+}, Al^{3+}, or B^{3+}. The screening is the consequence of the tendency of the cation to neutralize its exterior field in the smallest possible volume element. Screening directs simple surface reactions, like the aging of monomeric silica solutions, or acid-base reactions, but also oxidation and reduction as in the interaction of glass surfaces with gases like SO_2 and SO_3, the crystallization and epitaxis phenomena, and so on (cf. Vol. I, A. ¶ 248 ff.). Thus, a glass surface is in a pronounced state of non-equilibrium, and its disordered asymmetrical structure is the source for all solid state surface reactions like epitactic intergrowths and adsorption phenomena. The particular structure of water and interactions with HF, NH_3 solutions and others are ruled by cation screen-

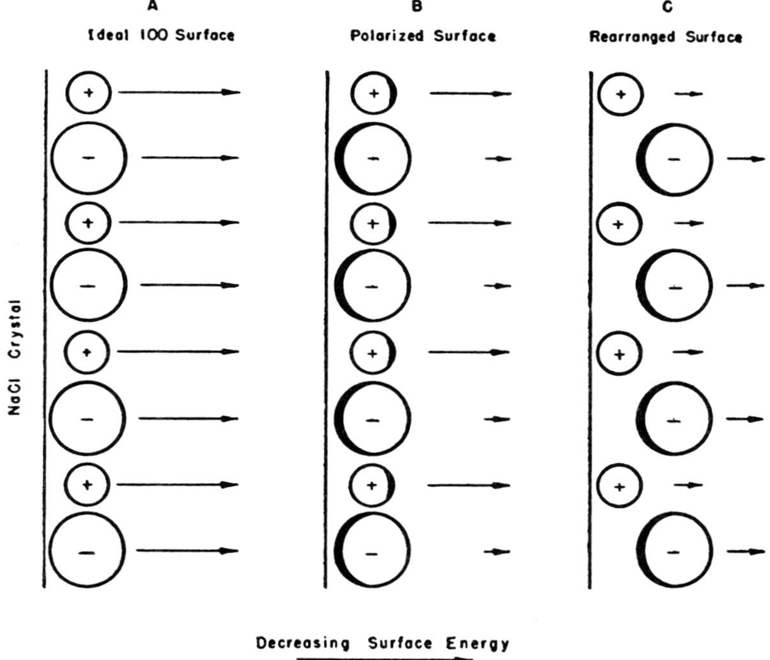

FIG. A. 226. Development of an equilibrium surface from nascent surface, *A*, through electronic polarization, *B*, to rearrangement of the surface ions, *C*. (Weyl).

ing tendencies in their interchanges with glass surfaces. The polarizability of anions in the surroundings follows the general laws previously formulated by K. Fajans and W. A. Weyl (since 1947). Also characteristic differences in the interactions of solutions containing cations of noble-gas configuration, or of nonnoble gas character, especially of cations with incomplete orbits, are discussed by Weyl from the viewpoints of their adsorption on glass surfaces.

393. Weyl[670] illustrated the effects of polarization on the surface of solids by a few impressive schemes of which Fig. A. 226 gives the change from the ideal (100) face of a NaCl crystal by the proceeding polarization of the oxygen anions to the asymmetric "rearranged" surface, and Fig. A. 227 the polarization of heavy metal

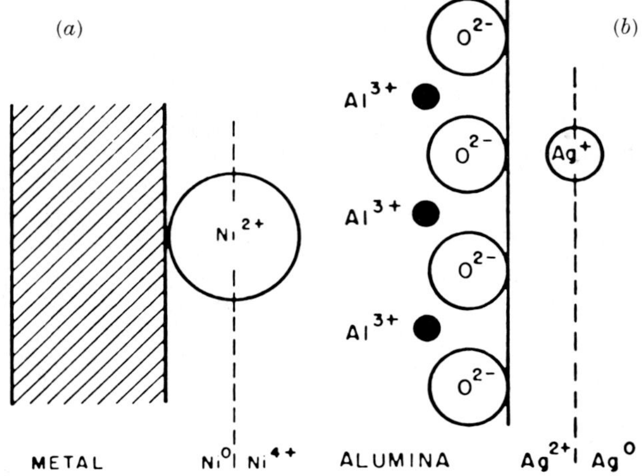

FIG. A. 227. Polarization of Ni^{2+} cations adsorbed on metal surface (*a*), and of Ag^+ ions on alumina surface. (Weyl).

cations (Ag^+, Ni^{2+} and Pb^{2+}) on silica glass, alumina, and platinum metal, respectively. The polarization effect of the noble metal on Ni^{2+} ions brings about a shifting of the electron cloud towards contact with the platinum making Ni^{2+} on the contact side conforming to a metallic bond, on the opposite side giving to the Ni^{2+} ion the quality of an ion of the type Ni^{4+}. Alumina reacts on the silver ion polarizing inversely, more explicitly forcing the electron cloud to be repulsed from the contact, and the exterior surface will therefore behave as if the Ag^+ ion would have been reduced to the metal atom Ag^0. Similarly silica glass acts on Pb^{2+} ions making the contact more intimate by states of Pb^{4+} ions, the other side more "metallophilic," or rather more metallike with Pb^0 atoms. These reactions and different behaviors in the exterior surface layers are of great importance to understand specific wetting and other surface reactions of solids with adsorbed cations. One may even imitate the surface of a real metal by adsorbed layers of silver ions on glass or on alumina, or of Pb^{2+} on common glass and silica glass, to a metallophilic behavior and bridging the gap of the ionic structure of the glass with the structure of the metal.

394. The significance of Weyl's development of the screening theory for glass surfaces is very far-reaching and has proved to be a principle of great multiplicity of

[670] *Penna. State Univ. Mineral Inds. Expt. Sta. Bull.*, 1951, No. 57, 118 pp.

application, as well in general silicate science as in glass technology. We may mention R. M. Gruver's ideas[671] to give an atomistic interpretation of *adhesion* phenomena on glass surfaces by using methods recommended by the screening theory to lower the surface free energy of glass and solids in general. In the field of glass/metal contacts this problem is of great practical importance in modern glass manufacturing. The solid (glass) surface acts as an acid if it contains poorly screened cations, and as a base if it contains better screened ones. On interfaces of acid/basic solids the protons move from the acid to the basic portion. The proton transfer takes place between the electron clouds of the "reactants." Electromotoric forces are the consequence of such moving electric charges into solids for better screening effects. The screening is improved by the electron shift and the sharing of anions, especially well in such amorphous solids as glass, that is, with a more efficient adjustment to the surfaces. Together with E. Ch. Marboe, W. A. Weyl[672] discussed extensively, on the enlarged basis of his theories the mechanical properties of glasses and their dependence on changes in chemical binding forces, to explain the action of glass surfaces on the mechanical strength. To this group of reactions also belongs hydrolysis under the action of strains or in the polishing process. Especially emphasized is not only the surface as the seat of asymmetric environments but resulting increased binding forces within the surface film, relative to the mechanical state of the bulk of the glass material. The chemical binding, influenced by these asymmetric environments is not restricted to a few atomic layers but has a certain action on depth, in a measure even beyond the sizes of the mosaic structure in crystals or of the blocks in common solids, namely extending over several 1000 Å. In undergoing such states of tension and strains in the surface, the glass explains the high rate of hydrolysis in exterior layers. That such hydrolysis actions under stress are of great importance for the interpretation of the practical polishing process is most obvious.

395. In an entirely different field of silicate investigation, namely in the study of silicate mineral associations in geological-hydrothermal units, Weyl's theory of surface reactions was successfully applied in a discussion of the important hydrothermal process of silicate polymerization, as in the feldspar group. G. W. De Vore[673] demonstrated how such a hydration leads to a decrease in the degree of polymerization on surface regions of such a silicate mineral with a framework structure. What he called the degree of "polarization interruption" rules the physicochemical reactions playing the basic role in the magmatic and postmagmatic mineral associations under nonequilibrium and under equilibrium conditions which are discussed from the thermodynamic viewpoint of effects of interface energies (cf. Vol. IV, Section A).

[671] *Glass Ind.*, **37**, 1956, 77-80, 94, 100-101.

[672] *Glastech. Ber.*, **32**, K, 1959, VI. 1 - VI. 18. On the mechanism of polishing combined with hydrolysis cf. especially pp. VI. 12 - VI. 14.

[673] *J. Geol.*, **64**, 1956, 31-55.

There is no reason why the same viewpoints of interruption in polymerization should not also be considered as a leading principle in the surface reactions of glasses as similar framework structures, their chemisorption, hydrolysis, and replacement behavior, in terms of ionic polarizabilities and polarizing powers.[674]

396. The strongly oxidizing properties of fresh-broken surfaces of quartz, compared by Weyl (1950) to that of ozone, was fully confirmed, and also for the surface of dehydrated silica gels, by K. G. Krasil'nikov, V. F. Kiselev, and E. A. Sysoev.[675] Organic dyestuffs which form colored products after oxidation stain silica gel more intensely the higher the calcination temperature. Silica gels with a specific surface of 400 and 695 m.^{2}/gm. (cf. Vol. I, C. ¶ 322) were used, and oxidizing properties could be quantitatively determined by iodometric titration with 0.01 *N* solution of sodium hyposulfite. The data show oxygen equivalents smaller by three orders of magnitude than for the hydroxyl groups adsorbed on the same gel surface. E. Richardson and C. D. Poucher[676] examined the time of declining oxidation power of fresh-broken quartz surfaces and found even after 25 weeks a relatively marked oxidizing activity by reaction with $KMnO_4$ solution. A. V. Bondarenko[677] on the other hand, used sensitive mass-spectrometric methods in the study of surface reactions on crystalline and amorphous modifications of silica (quartz powder of 35.8 and 8.0 m.^{2}/gm.; silica gels between 700 and 275 m.^{2}/gm. as specific surface, respectively) but did not find any charged ions (H^+; H_3O^+) in the dehydration products, thus speaking decidedly against Weyl's theory of surface dehydration, and no change of surface structure could be concluded. Nevertheless, some molecular oxygen was observed among the dehydration products which were evidently weakly bound to the surface, perhaps coupled to the adsorption of NO. Bondarenko is of the opinion that the oxygen is *not* derived from a destruction of the silica surface but by a decomposition of nitrogen oxides, and that "structural oxygen" is evolved above 900°C.

397. During intense grinding, quartz powder is able to react rather extensively with organic liquids. This fact is of great general interest because H. Deuel and R. Gentili[678] also found, e.g., traces of such reactions with butyl alcohol. More extensively, W. Hinz[679] applied sensitive radioactive tracer methods for observing *ion exchange* reactions on glass surfaces, say for Sr^{89} and Ca^{45} as indicators. These reactions with glass are not different in any way from the corresponding exchange

[674] In this respect see the thermodynamic calculations of H. Ramberg, *Am. Mineralogist*, **39**, 1954, 256-271, concerning polarization and contrapolarization of oxygen anions as an important factor in problems of the stability of silicate minerals (cf. Vol. I, A. ¶ 12).

[675] *Doklady Akad. Nauk S.S.S.R.*, **116**, 1957, 990-993.

[676] *Research (London)*, **11**, 1958, 247-248.

[677] *Doklady Akad. Nauk S.S.S.R.*, **125**, 1959, 573-576.

[678] *Helv. Chim. Acta*, **39**, 1956, 1586-1589.

[679] *Silikat Tech.*, **8**, 1957, 136-139. On the use of radioisotopes in glass investigations see the review by A. Turco and G. Sordello, *Vetro*, **1**, 1956 (1) 7-12, with valuable bibliography.

phenomena known from crystalline zeolites, glauconite, or on organic exchange resins. For Sr^{89} the adsorption isotherms show a saturation of exchange after about 19 hours; the strontium ions do not appreciably diffuse into the bulk of the glass and are promptly desorbed reversibly. Common glasses and silica glass behave in the same way. The density of the Sr^{89} population was about 13 μmoles/m.^{2}, equivalent to 26 μmoles/m.^{2} of hydroxyl ions, and somewhat more than a monomolecular layer of hydroxyl ions on the Si atom of a quartz-like surface structure. The existence of a swelling layer of variable thickness is most probable. A pretreatment of the glass surface strongly affects ionic adsorption. Very impressive effects are observed, for example, after treatment for 12 hours with a CrO_3–H_2SO_4 mix with $Ca(OH)_2$, i.e., "degreasing." The adsorption is increased in the sequence here given. The same is valid for the series of aqueous solutions of HCl, LiOH, NaCl, KNO_3, $MgCl_2$, and $CaCl_2$. Adsorption rapidly decreases when pH is below 5.1; it is = 0 for $pH = 1.8$. The intensity of the adhesion of cations on the glass surface is evidently different with the exchangeability for Sr^{89} in the sequence $H^+ < Li^+ < Na^+ < K^+$, and $Mg^{2+} < Ca^{2+} < Sr^{2+} < Al^{3+}$.

398. The exchange reactions with *para-hydrogen* and *deuterium* on borosilicate glass surfaces at 350°C. were studied by G. S. Annis, H. Clough, and D. D. Eley.[680] The activation energy of the deuterium exchange as a reaction of first order is 21,550 to 23,600 cal./mole, corresponding to the mechanism of a reaction $D_2 + 1(\ldots Si–OH) \rightarrow (\ldots Si–OD) + HD$. The conversion of parahydrogen at 350°C. which is also of the first order, has an enthalpy of reaction of 6,500 to 13,100 cal./mole for the process $p\text{-}H_2 + 2(\ldots Si–O–)^*$ (as active sites) $\rightarrow 2(\ldots Si –OH) \rightarrow 2(\ldots Si –O–)^* + o\text{-}H_2$, with a population of one active site per 10^7 sites in the surface structure.

399. The old problems of how to create an ideally "clean" glass surface was studied by L. Holland[681] by using ionic bombardment from a glow discharge in the range of 1000 to 5000 volts. The coefficients of (exterior) friction and of wetting characteristics are sensitive indicators of progressing cleaning. If an aluminum cathode is used the glass surface is coated by a hard, discolored layer, chiefly within the dark cathode space. Among different possibilities on the nature of this layer, the most probable one is the decomposition of hydrocarbons on the glass surface by the electrons evolved by the glow discharge. The bombardment with positive ions, on the other hand, is efficient enough to remove contaminations from the glass surface, although at high density of the current, components of the glass may be removed also at the same time.

400. Which powerful effects of "molecular attraction" are exerted by common glass surfaces was made evident by H. Elsner von Gronow's (1931) experiment of

[680] *Trans. Faraday Soc.*, **54**, 1958, 395-399.

[681] *Nature*, **181**, 1958, 1451-1452.

an oriented overgrowth of organic dyestuffs on one-directionally rubbed glass plates. These phenomena which are entirely analogous to the epitactic overgrowths of two different compounds by crystallographic orientation (cf. Vol. I, A. ¶ 248 ff.) were discussed by A. Neuhaus[682] not as a structure-specific influence of nucleation (seeding) of the supporting glass surface, but as a consequence of the tendency of crystallites of the dyestuff to develop a network of minimum free surface energy on contact with the glass. They are not azimuthal orientation effects but only textural orientation phenomena. There are also remarkable transitions between both forms observed, as on fibrous textures, or in O. Lehmann's classical "liquid crystals" (pseudoisotropic or paracrystalline phases), further with analogies to flotation and lubrication films.

401. To a certain degree we may include in this group of phenomena also the thin films namely of deposits of silica hydrogels on leached glass surfaces which can be made visible by TiO_2 vapor deposits and condensation of cadmium metal vapors by dark-field microscopy as recommended by H. König, H.-J. Löffler, and F. Lappe.[683] The adsorption layers show a characteristic pattern of network cracks (like the "craquelé" of ceramic glazes). On fire-polished glass surfaces the appearance of microcrack systems starting from distinct centers with concentric terrace-like structures is observed. Although highly purified glass surfaces do not show any condensates of this kind, seeds of condensation are immediately made visible after a steam treatment of the glass for some hours. The seeds are the alkalies; if they are entirely removed no cadmium condensates are seeded. Deposits of silicon monoxide on glass also strongly seed cadmium condensation. On aging the phenomenon is gradually lost, practically extinguished in the open air after one week. The seeding effects for cadmium vapor are not specific for glass surfaces. They are also observed on copper sheets with a temperature maximum at $+ 30°C$. Platinum cathode-vacuum deposits on glass show a nearly constant optical transmittance over the range of $\lambda = 200$ to 1,000 mμ, but in oxygen atmosphere the transmittance decreases with increasing wavelength by the formation of a low platinum oxide. The electrical conductivity of such platinum layers on silica glass is largest, and very low on common glass, probably because the silica gel film on the latter glass promotes oxidation of the metal.

402. Very important calculations of the *surface energies* of crystalline solid oxides by D. T. Livey and P. Murray[684] in relation to lattice energy, molecular volume, and enthalpy of formation, ΔH_r, are unfortunately not yet sufficient to give an accurate basis for a better quantitative treatment of surface reactions on glass. In spite of that, the range from 250 to 1420 erg./cm.^{2} is given for the oxide series from PbO to BeO. The stronger the polarization of the metal ion, as in Cd^{2+} or Pb^{2+} as

[682] *Fortschr. Mineral.*, **29/30**, 1952, 136-296, especially p. 140-141.
[683] *Glastech. Ber.*, **28**, 1955, 131-136.
[684] *J. Am. Ceram. Soc.*, **39**, 1956, 363-372.

drastic examples, the lower the surface energy of the oxides. Those containing the strongly polarizing cations Si^{4+} and B^{3+}, which are of interest for glass act in the same direction because of the strong deformability of the O^{2-} anion on the surface with an improved screening effect. Unscreened cations at the surface contribute to a higher free energy, and the surface cations are easily retracted, so that anions are present in the surface layer, as Weyl had postulated (cf. Fig. A. 228). The possibility of efficient screening is greater in oxides of the type RO_2 than in RO and R_2O_3, but for

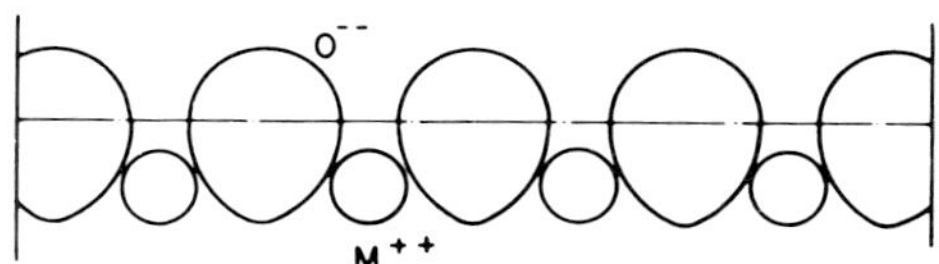

FIG. A. 228. Retraction of strongly polarizing cations M^{2+}, and strongly polarized oxygen anions, on the surface of an oxide RO_2, (Livey and Murray).

SiO_2 the data resulting from extrapolations made from measurements in the slag systems FeO—SiO_2 and PbO–SiO_2[685] are most unreliable and evidently much too low. At 1400°C. they are given as only 200 to 260 ergs/cm.², whereas SiO_2 and TiO_2 should have the highest surface energies in the series SiO_2, TiO_2 > ZrO_2(800) > UO_2(640) and > ThO_2(530) ± 20 per cent.

403. The *wetting of glass* by hot moisture, as by combustion gases in glass-melting furnaces creating the so-called firepolished surface is chiefly characterized by the introduction of hydroxyl ions into the structure. Hydrogen ions replace Na^+ by contact of the glass with liquid water; the resulting "water films" which were already known to E. Warburg (1886) and R. G. Shwerwood (1915, 1918), are highly important for the high-vacuum techniques due to their detrimental effects. The determination of the contact angle Θ on glass surfaces, as with small water droplets applied on it, is according to H. Wessel[686] (cf. A. ¶ 103 ff.) a very sensitive indicator for the presence of slight contaminations of the surface. If (1), (2), and (3) represent the gaseous, the liquid, and the solid (glass) phases, the surface tensions, σ, on the contacts of these phases have reached equilibrium when $\sigma_{13} = \sigma_{23} + \sigma_{12} \cdot \cos\Theta$. A hysteresis of the contact angle indicates the presence of highly polarizable ions, but particularly well-cleaned glass surfaces do not show this phenomenon. Such a cleaning can be done with trichloro-ethylene, CCl_2–CHCl, and a thermal after-treatment at 500°C. Wessel's experiments are in good agreement with Weyl's theoretical discussions. The best surface testing method is that of the "captured bubble" (cf. Fig. A. 229);

[685] Cf. P. Kozakevitch, 1949, and L. Shartsis, S. Spinner and A. W. Smock, 1948.

[686] *Silikat Tech.*, 4, 1953, 59-63.

if the angle Θ is larger than 0^0 gas bubbles adhere on the glass surface and are deformed by a slight shifting of the bubble holder. The effects of cations are made evident if the glass surface has been made water-repellent, say by a trace of laurylmethyl bromide.

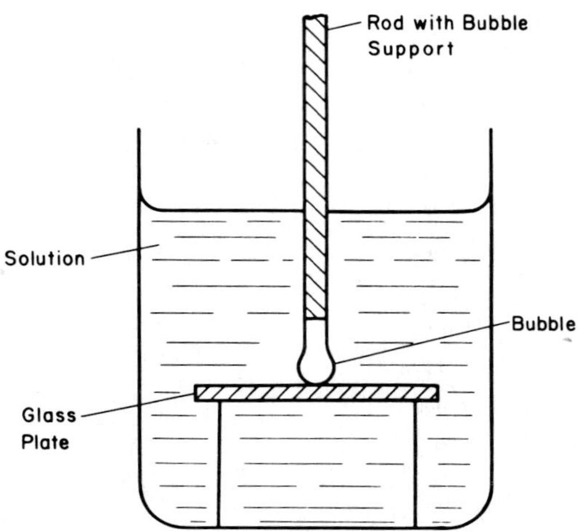

FIG. A. 229. Measurement of the contact angle, Θ, of an aqueous solution on a glass surface by the method of "captured bubbles." (Wessel).

404. S. P. Zhdanov[687] observed the very particular behavior of porous glasses (with 270 and 80 m.^{2}/gm. specific surface, and a diameter of the pores of 10 and 30 μ, respectively) in its wetting by water. The adsorption and desorption isotherms show the same hysteresis phenomena as do silica gels. Curves, plotted with the relative pressure p/p_s as abscissae and the water amount in millimoles/gm. substance as ordinates, show different branches for adsorption and desorption, and changes after a repeated hydration and dehydration cycling. The mechanism of dehydration is explained by the formation of unstable cyclic bonds of the type

$$\begin{array}{c} SiO_2\text{—}O\text{—}SiO_2 \\ \diagdown \quad \diagup \\ O \end{array}$$

which are usually forbidden in crystallochemistry and are changed in hydration to normal

$$\begin{array}{ccc} SiO_2 & \text{—}O\text{—} & SiO_2 \\ | & & | \\ OH & & OH \end{array}$$

[687] *Doklady Akad. Nauk S.S.S.R.*, **100**, 1955, 1115-1118; see also R. K. Iler, "The Colloid-Chemistry of Silica and Silicates," Cornell Univ. Press, 1955, 324 pp., especially pp. 239-240.

Weyl contested the possibility of cyclic bonds in general; Zhdanov admits that if they exist they must be highly unstable and deformed. The negative electrostatic charges postulated by Weyl for the dehydrated surface of silica gel have not been realized. On the contrary, the calcination of porous glass to 500° to 600°C. does *not* raise the adsorption capacity for water as Weyl's theory would have postulated.

405. A recent study of Sc. Anderson and D. D. Kimpton[688] investigated the conditions of non-wetted glass surfaces, among which the "cleanness" not only can mean the absence of foreign solids, semisolids, liquids, and gases but also may be caused by suitable adsorption layers of repellent anions. As such a layer, semi-chemisorbed oxygen which by "steric hindrance" inhibits the formation of the wetting causing hydrogen bonds, was studied by Anderson and Kimpton (cf. Weyl's theories with E. C. Marboe 1947, with L. R. Sonders and D. P. Enright 1950). The oxygen is held in deep potential wells associated with active centers in the glass surface. The contact angle with water then depends on the density of these active sites, and on how well the adsorbed oxygen covers the oxygen anions of the glass framework. A lead silicate glass with 90 per cent PbO supported a drop of water with a contact angle of about 90° when its surface was rubbed with pure cellulose and wiped dry. The temporary nonwetting is lost after a short heating to 275°C. The angle Θ was much lower when the ratio $PbO : SiO_2$ was below 70 : 30 (cation ratio $Pb^{2+} : Si^{4+} =$ 1 : 1), illustrating the high polarizability of the lead ions. A glass coated with titania has a contact angle with water of 90°, although the Ti^{4+} is not highly polarizable. It is probable, therefore, that non-wetting properties are related to the presence of octahedral coordinations in the glass itself whereas glasses with only tetrahedral bindings are characterized by angles Θ of 0° (e.g., on quartz, or orthoclase).

406. Anderson and Kimpton experimentally examined the correlation of octahedral coordinations with oxygen adsorption on glass surfaces making use of the sensitive phosphorescence behavior of certain organic dyestuffs (e.g., trypaflavine) when quenched by oxygen. After heating to 300°C. (see above) this adsorbed oxygen layer can be driven out and the oxygen amount determined by the phosphorescence test. Oxygen released by heating to only 175°C. is not related to the wetting of glass and the hydrophobicity of lead silicate glass was not destroyed. If, however, the heat exposure was above 270°C. further oxygen is given off and the surface loses its hydrophobic property, the phosphorescence test indicates an oxygen amount (measured by McLeod manometry, sensitive up to 10^{-6} mm. Hg) exactly corresponding to a monolayer of oxygen. Other experiments showed that in the wet-rubbing cycle oxygen adhering to the glass comes from the water and that the rubbing operation

[688] *J. Am. Ceram. Soc.*, **43**, 1960, 484-488. On the phosphorescence test for oxygen cf. J. Franck and P. Pringsheim, *J. Chem. Phys.*, **11**, 1954, 21-27; **12**, 1944, 295-299 (with M. Pollack and D. Terwoord).

is an essential part of the cycle. A zinc aluminophosphate glass and a titania-covered sodium-calcium silicate glass with 30 per cent TiO_2 also behaved very similar to the lead silicate glass. Heavy metals, therefore, are *not* required to develop water-repellent glass surfaces. Adsorption isotherms of water on powdered silica gel and on powder of a glass with 30 per cent TiO_2 show an isosteric heat of adsorption of 6000 to 7000 cal./mole for the gel, but of 19,000 cal./mole for the TiO_2 containing glass. A similar value is valid for TiO_2 powder and an X-ray protection glass shows a heat of adsorption of 5000 cal./mole. For the water-repellent TiO_2-silicate glass it is therefore much higher than that commonly observed for van der Waals and hydrogen bondings. It is of the rank of semi-chemical bonding strengths. The monolayer of oxygen is actually molecular O_2, comparable to the oxygen layers on silica, platinum black, or charcoal as examined by R. Juza and Fr. Grasenick[689] in magnetic susceptibility measurements at — 183°C. The Si^{4+} and B^{3+} cations near the surface of silicate and borosilicate glasses are completely pulled below it by the asymmetric fields, as Marboe and Weyl postulated, leaving the oxygen anions of $[SiO_4]$ and $[BO_3]$ configurations in the surface.

407. Very interesting studies were made by M. S. Aslanova and P. A. Rebinder[690] on the influence of atmospheric humidity on the creeping behavior of glass fibers and of their elastic fatigue. They also studied the fibers' behavior after immersion in different polar and nonpolar liquids (water, isoamyl alcohol, kerosene, and others), or in solutions of potassium oleate, sodium dioctyl sulfosuccinate, and "Velan" (a cationic-active compound in which the quaternary nitrogen atom of the pyridine group is bound with an amido and a complex ether group: $C_{17}H_{35}CO.NH.NH_3.\overset{+}{N}C_5H_5\overline{Cl}$). The tensile strength of aluminum borate and sodium-calcium silicate glasses is considerably reduced by water and active hydrophilic compounds as compared with the strength in dry air or in kerosene. This decrease reaching 40 per cent is due to the adsorption of polar groups like OH, NH_2, COOH, the CH_3 groups always pointing outward.

408. The *adhesion of organic polymers* on glass surfaces either highly cleaned, or modified by a treatment with alkyl chlorosilanes, was studied by N. A. Krotova, L. P. Morozeva, and B. V. Deryagin,[691] especially with organosilanes of the series $(CH_3)_{4-n}SiCl_n$, and further with layers of ethyl-, benzyl-, or nitro-cellulose, polyvinyl alcohol, perchlorovinyl, gutta percha, carboxylated rubber, and copolymers of methacrylic acid, styrene, and others. The adhesion work can be measured by tearing

[689] *Z. Elektrochem.*, **54**, 1950, 145-152.

[690] *Doklady Akad. Nauk S.S.S.R.*, **95**, 1954, 1215-1218; **96**, 1954, 299-302; a review cf. *Silikat Tech.*, **12**, 1961, 528-531. See also W. Haller and H. C. Duecker, *Nature*, **178**, 1956, 376-377, who used the contact angle method for a study of hydrophobic glass surfaces after treatment with *p*-nitrobenzyl bromide (cf. A. ¶ 324).

[691] *Doklady Akad. Nauk S.S.S.R.*, **129**, 1959, 149-152.

the films of the polymers from the glass surface with a standard speed of 1 cm./sec. Electrostatic charges developed in layers applied on non-modified glass. If modified by alkyl chlorosilanes, no charges were observed for films of nitrocellulose and perchlorovinyl; all the other polymers show a positive charge developed. A change from negative to positive charges is observed at 10 molecules CH_3SiCl_3, in a series with variations of the molecular ratio of chlorosilanes with increasing numbers of methyl groups applied on the modified glass surface, and a corresponding reduction of hydroxyl groups. In the double layer formed on the glass surface the role of an electron donor and acceptor in the modifying surface and in the polymer film material is evident. A reversal of these actions with the characteristic inversion of charges on the film is possible for a definite ratio of CH_3 : OH groups on the surface.

409. M. S. Aslanova[692] studied the influence of the chemical composition of the glass on the adhesion of polymers by a very sensitive "crossed-fiber contact" method illustrated by Fig. A. 230. Two crossed filaments have contact with each other in

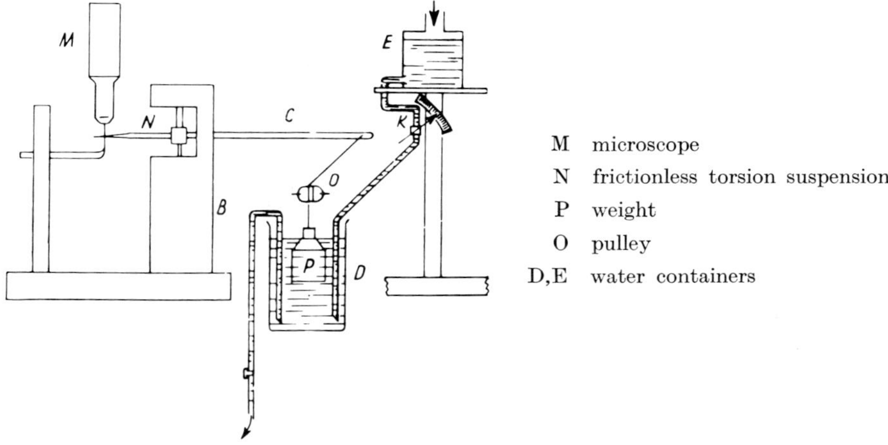

Fig. A. 230. Experimental device for measurement of adhesion forces of polymer films on glass fibers ("crossed filament method"). (Aslanova).

one point, and this contact is loosened mechanically, the adherence energy being by a dynamometric effect observed under the microscope. On one of the fibers a polymer is applied from a solution, in a thickness of 0.2 to 0.3 μ, the other filament remains clean glass. Glyptal (glycerine phtalate alkyd resin), phenol-formaldehyde and polyester resins, and silicone varnishes were used as adhesives on filaments of different composition, namely silica glass, alkali-free borosilicate, and heavy metal borate glasses. The lowest adhesion is shown by a silicone varnish containing no

[692] *Glastech. Ber.*, 32, 1959, 459-463.

polar groups. The adhesion is increased by 40 to 50 per cent for silicones if polar groups are introduced, and doubled for Glyptal. The adhesion energy for the non-polar poly-isobutylene and for paraffin is only 30 to 40 ergs/cm.². For Glyptal there is a strong dependence on the glass composition: on silica glass the adhesion energy is 30 ergs/cm.², for aluminum borosilicate glass 250 ergs/cm.², for sodium-calcium silicate glass 150 ergs/cm.², and on cadmium borate glass only 50 to 60 ergs/cm.². For alkali-free glass there is no direct relation to the cation polarizability which is known to be particularly high for Pb^{2+} and Cd^{2+}. If PbO is introduced into sodium-calcium silicate glass, the high polarizability of the Pb^{2+} ion raises the adhesion energy considerably, just as the treatment of the glass surface with lead acetate solution makes the glass hydrophobic. The organo-chlorosilanes and industrial "Volan" (see below) act in the same direction. Radioactive irradiation reduces the adhesion of polymers on glass by 60 to 70 per cent, ultraviolet irradiation acts (for Glyptal) by about 30 to 50 per cent and irreversibly, whereas silicone films remain indifferent to ultraviolet light (cf. Fig. A. 231).

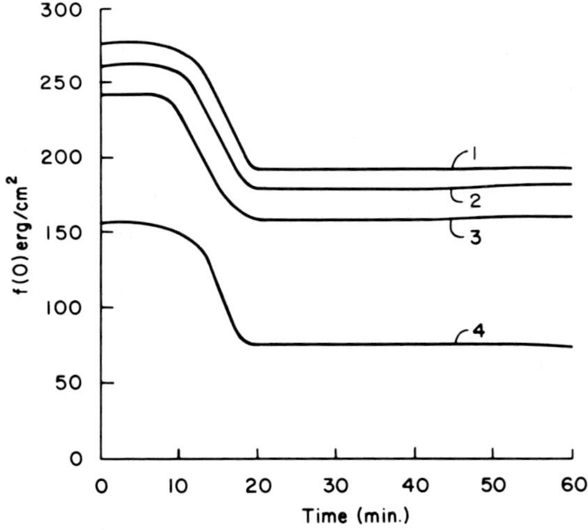

Fig. A. 231. Effect of ultraviolet irradiation on the specific adhesion energy of Glyptal resin film on glass fibers of different chemical composition. (Aslanova). (*1*) on silica glass; (*2*) on aluminum borosilicate glass; (*3*) on sodium-lead silicate glass; (*4*) on sodium-calcium silicate glass.

410. W. Hinz and G. Solow[693] compared for glass silk cloth the cohesion strength of inorganic binders (e.g., plaster of Paris) with the much higher adhesion of organic polyester resins (in the ratio of 2 to 45 kg,/cm.², respectively). These correlations which are of the highest importance for industrial manufacturing of reinforced plastics

[693] *Silikat Tech.*, **8**, 1957, 178-185.

were experimentally studied in tensile strength measurements of cylindrical specimens of glass fiber cloth embedded in hexane diol-maleic acid polycondensates dissolved in styrene and hardened with a peroxide catalyst. Overhardening makes the products brittle by tension strains. Specific effects are exerted by cation-active agents on the glass as a "finish" influencing the phase boundary conditions between the active groups in the finish (e.g., the NH group) and the polar CO groups of the polyester. A typical finish agent is "Volan," a copolymerisate on the basis of $CrCl_3$-methacrylate. As a function of the chemical composition of the glass and of its surface treatment, cation-active textile improving agents in combination with binders of the resin type are *not* to be recommended for a good cohesion of glass cloth with polyesters. Strikingly increased cohesion, however, is observed if unsaturated chlorosilanes are used, such as vinyl trichlorosilane, which bring about a valence-chemical binding between the glass textile and the plastic, even if the glass is high in alkalies, and in the presence of capillary moisture. The best cohesion on alkali-poor glass fibers is observed if cation-active agents are avoided. Moisture reduces the cohesion by the formation of swelling layers on the contacts between glass and plastics. The polarity of the glass surface can be reduced by a treatment with metal salt solutions, as shown by changes in the contact angles measured by Fr. Moser.[694] The cohesion is improved by an exchange of sodium with lead or zinc ions in alkali-poor glass, whereas in alkali-rich glass a rapid decrease in cohesion with the plastics occurs, especially if treated with bismuth salt solutions.

411. The hydrophobization of glass surfaces by organic agents has its parallel in the modern methods of preparing hydrophobic hydraulic binders, the theory of which was given by M. I. Khigerovich.[695] Surfactants (cf. Vol. V, Section C., with an air-entraining or plastifier effect) in concentrations of 0.10 to 0.25 per cent added to Portland cement clinker during grinding, cover the surface of the basic calcium silicates and aluminates (in addition the glass matrix) with a hydrophobic adsorption film. Especially suitable are oleinic acid, naphthene acids or higher fatty acids, their water-soluble salts, oxidized petroleum, and others. Naphthene acids, on the other hand, decrease the surface tension on the boundary air/water and bring about *air emulsion* in the mortar mix, but the calcium salts later formed do not have this specific property. The polar-active groups of the hydrophobic agents are strongly sorbed on the clinker grains, in contrast with the aliphatic water-repellent chains at the ends of their chain-shaped molecules, and make the systems thixotropic. In this respect we finally recall the experiments of H. Deuel *ct al.* (cf. A. ¶ 325 ff.) on the introduction of aliphatic and aromatic groups on the surface of clay minerals with hydroxyl groups containing silicates.[696]

694 *Glass Ind.*, **37**, 1956, 201-203, 228, 232.
695 *Silikat Tech.*, **8**, 1957, 534-536.
696 Cf. *Ber. deut. keram. Ges.*, **31**, 1954, 1-7.

412. Based on Weyl's previous experiments on the water-proofing of glass surfaces by an exchange of hydroxyl groups with fluorine anions, Sh. Nagai and M. Yamaguchi[697] improved the moisture stability of glass fibers by immersion in a solution of magnesium silicofluoride at 20° to 90°C. for 2 minutes up to 2 hours. The tensile strength, however, was reduced by about 10 to 20 per cent at the same time. The exchangeability of hydroxyl and fluorine anions as an analytical problem by F^- losses from solutions of fluorides in soft glass or borosilicate glass containers was studied by R. C. Specht.[698] Probably, the F^- ions are not strongly fixed on the glass structure since dilute acids containing Fe^{3+} ions are able partially to release them from the glass surface and a complete removal is possible by dilute solution of NaOH. The exchange of hydroxyl by fluorine anions on the surface of clay minerals has great theoretical significance as was especially emphasized in the investigations of Arm. Weiss, A. Mehler, G. Koch, and U. Hofmann,[699] and by L. A. Romo and R. Roy[700] (cf. Vol. IV, Section B, and Vol. I, C. ¶ 183).

413. We mentioned already in A. ¶ 394 how important W. A. Weyl estimates to be chemical-physical surface reactions of glass with aqueous suspensions and solutions in the higly complex process of *polishing*. From a practical viewpoint also A. Kaller[701] discussed the theoretical premises of polishing, with systematic experimental studies under well-defined conditions, chiefly on the preparation and grain size composition of the polishing "rouge" (iron oxide in the γ-modification, prepared from iron oxalate). The highly defective lattice of this agent is of great importance because of the higher binding energy of chemically sorbed and bound ions from the surrounding solutions. Kaller's theory is especially valid for Fe_2O_3, Cr_2O_3, Al_2O_3, CeO_2, and ThO_2 as polishing powders and entirely eliminates so-called thermoplastic deformation as an essential factor in the material removal from the glass surface. E. Brüche *et al.*[702] conclude from electron-microscopic examination of the behavior of scratches on glass during the polishing process, that the smooth walls of a scratch behave plastically, equally if produced in the open atmosphere, or under water, alcohol, benzene, trichloroethylene, or sodium-dried petroleum. There is no specific absorption of water or any other liquid in the glass surface needed for developing the type of smooth scratches, only a certain lubrication action of the liquid during

[697] *Yogyo Kyokai Shi*, **62**, 1954, 469-473.

[698] *Anal. Chem.*, **28**, 1956, 1015-1017.

[699] *Z. anorg. u. allgem. Chem.*, **284**, 1956, 247-271.

[700] *Am. Mineralogist*, **42**, 1957, 165-177.

[701] *Silikat Tech.*, **7**, 1956, 380-390, 399; *Brit. Chem. Eng.*, 1957/58, 145-167. Observations of silica crystallization on polished glass surfaces were described by P. de la Cierva and Fr. de Andrés, *J. Am. Ceram. Soc.*, **43**, 1960, 306-313.

[702] Cf. Vol. I, C. ¶ 117 ff., and extensive studies of E. Brüche, Kl. Peter, and H. Poppa, *Glastech. Ber.*, **30**, 1957, 387-393; **31**, 1958, 341-348. See also previously A. Smekal, *Ibid.*, **28**, 1955, 81 (abstract from Fachausschuss-Sitzung).

the scratching mechanism, as for the diamond pointer producing the scratch. Microplasticity, in the understanding of A. Smekal, is confirmed.

414. Most fascinating ideas on mechanical-chemical factors acting in the polishing process were recently discussed by W. A. Weyl and E. Ch. Marboe[703] on glass hydrolysis and the *establishment of bonds* between the glass surface and the polishing agent ("rouge") recalling the metallophilic properties of heavy metal cations in their affinity to the steel tool producing scratches, and a fixation of rouge particles on a clean glass surface by the high pressure exerted under a steel ball. Weyl and Marboe assume that during polishing the rouge combines temporarily with the glass surface thus increasing the exterior friction and establishing shear stresses, with water playing an important role in this bonding process. The process has some similarity with the formation of "iron (oxide) stains" on glass in contact with metallic iron described by Weyl (1947), as a consequence of hydrogen-bonding and screening. Even the Al^{3+} ion can enter the glass surface during polishing if an aqueous suspension (slurry) of meta-kaolin is applied as a (temporarily) effective polishing agent on a fresh glass surface. The kaolin-polished surface is hardened and even CeO_2, as well as rouge, applied after this polish, are less effective in turn. It is assumed that the application of shear stress during the polishing process promotes the formation of an aluminosilicate in the glass surface. This also explains why in polishing with rouge slurries the application of a 1 per cent solution of aluminum sulfate reduces glass removal by nearly 50 per cent. The breaking of major bonds of Si—O—Si in the glass surface and its re-establishment in corresponding bonds like Al^{3+}—O—Si^{4+} is a positive possibility in polishing.

[703] *Glastech. Ber.*, 32, K, 1959, especially pp. VI. 12 ff.

Section B

Industrial Glass and Enamels

REACTIONS IN THE GLASS BATCH

1. In a most elaborate series of investigations, C. Kröger and his coworkers have given a complete report on batch reactions for glass manufacturing, based on the simple systems $Na_2O—SiO_2—CO_2$ and $Na_2O—CaO—SiO_2—CO_2$, further those with SO_3 in the place of CO_2 (batches with soda or sodium sulfate mixtures). Kröger summarized his experience in a comprehensive book[1] from numerous single periodical papers. First, we wish to discuss those batch reactions in which only soda and limestone are the reaction raw materials of glass making, specifically the *carbonate* batch reactions. Concerning the basic heterogeneous gas reactions occurring in those systems we refer to Vol. IV, Section A. wherein we will treat these from the view-point of general equilibria in characteristic pressure-temperature diagrams developed by Kröger, which will then in Vol. V, Section A. be also discussed in connection with the equation of W. Jander for the rates of reactions in the "solid state" (in the broadest understanding of this word). We have to consider that in glass melting ideal equilibrium conditions, say of the system $CaSiO_3—Na_2SiO_3—SiO_2$, as given by N. L. Bowen and G. W. Morey (1925), are only of restricted helpfulness for an understanding of what happens during the melting-down of the batch. They are better applicable to describe the mechanisms of devitrification (B. ¶ 203 ff.).

[1] In his book "Glasgemengereaktionen und Glasschmelze," Westdeutscher Verlag, Köln, 1957, 110 pp. The most important single publications are: *Glastech. Ber.*, **25**, 1952, 307-324; **26**, 1953, 346-353 (with G. Ziegler); *Ibid.*, **27**, 1954, 199-212 (with G. Ziegler); *Ibid.*, **28**, 1955, 51-57, 89-98 (with Fr. Marwan); *Ibid.*, **29**, 1956, 275-289 (with Fr. Marwan); *Ibid.*, **30**, 1957, 42-52. Most recently, R. V. Harrington, J. R. Hutchins, and J. D. Sherman applied the theory of reactions of spherical particles, as developed by A. M. Ginstling and B. I. Brounshteĭn, *J. Appl. Chem.* (*U.S.S.R.*), English translation, **23**, 1950, 1327-1338, and T. P. Fradkina, *ibid.*, **25**, 1952, 1199-1205 (cf. Vol. V. ¶ A. 79, Footnote 159) to the kinetics of Subliquidus alkali carbonate-silica reactions. Cf. "Advances in Glass Technology", Internation. Congress on Glass, Washington D. C. 1962, Plenum Press, New York, 1962, pp. 75-92.

2. For mixes of the simplest model systems soda-quartz and soda-limestone-quartz (sand) the ideal equilibrium conditions of reaction with the participation of CO_2 as a volatile reagent are also only of theoretical interest. Kröger recognized that *kinetic* studies in such systems, however, with solid state reactions in the first stages, followed by a "polyeutectic" melt phase formation, even in subordinate amounts, are essential for the technological process involved. In experimental studies of batch reactions it is advisable to treat both cases not as separate phenomena but as intimately connected to each other in their evolution to glass formation in the industrial furnace.

3. The experimental methods for the study of the kinetic curves of CO_2 evolution as a function of time and temperature are very elementary as described in Vol. V, Section A. following the kinetics not only in loose powder batches but also in pressed pellets ("briquettes") of the batch mixtures. In initial stages the reaction rate is the same for both kinds of batch samples, but it becomes evident later that highly pressed pellets show a relatively reduced reaction rate. More residual carbonate remains in the final products and the output in CO_2 is reduced. Obviously, the diffusion of the gas from the inner portions of the pellets is impeded by the compacting process as the batches are "cemented" to a less permeable body. The volatility of the evolved CO_2 is a function of the logarithm of the external pressure, with the

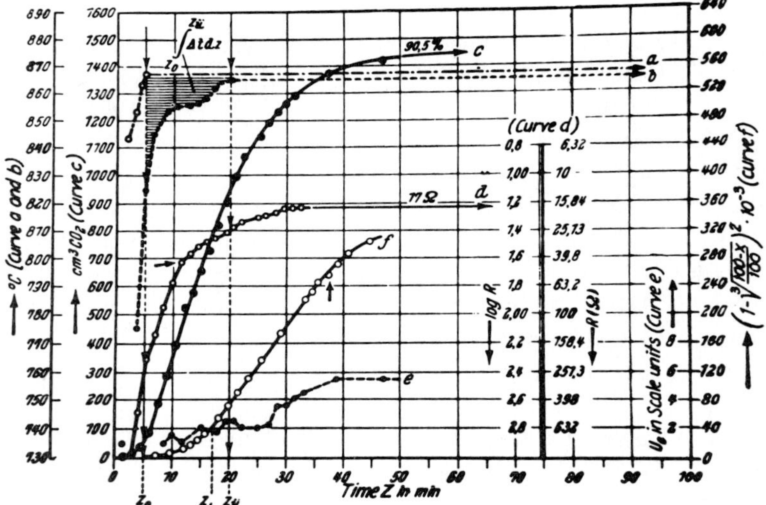

FIG. B. 1. Curves for the reaction of soda with quartz (molecular ratio 1 : 2), at a constant surface temperature of 865°C. (Kröger). Curve (*a*) shows the surface temperature of the mixture, (*b*) the central temperature, (*c*) the carbon dioxide evolved, (*d*) the resistance (log R), (*e*) data for the zero point displacement, U_0, as a function of melt formation, (*f*) the "ideal" curve for the Jander equation.

reaction constant $k = B - a \cdot \log(p)$. The important effects of the heating period and an example of the quantitative evaluation of the CO_2 evolution curves as a function of time is given in Fig. B. 1, for the constant temperature of 865°C. on the surface of the batch, and a composition 1 Na_2CO_3 : 2 SiO_2 (quartz). For the control of polyeutectic melts formed in the batch after the first solid state reactions, the curve (*d*) for log *R* (the electrical resistance) is very characteristic in this Figure. With increasing conversion of the batch, the initial liquid phase is resorbed again, but increasing amounts of the fused reaction products are formed which either compensate for the conductance effects of the liquid-consuming reaction or will continuously increase to a total melting. The first case is shown in Fig. B. 2, for a reaction mix of

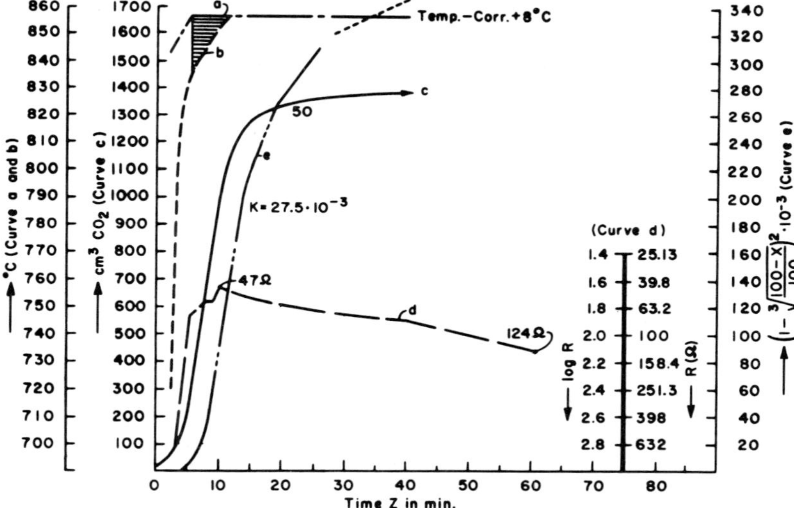

FIG. B. 2. Curves for the conversion of soda with sodium disilicate in the molecular ratio 1 : 1, at a constant temperature of the surface of the mixture of 864°C. (Kröger and Ziegler).

soda + quartz in the molecular ratio 1 : 1, or a mix of soda with $Na_2Si_2O_5$, also at 864°C. The initial polyeutectic melt is soon saturated with reaction products and these crystallize. In this case the conductance curve shows a maximum (for the resistance a minimum), and decreases with increasing time. The second case is that shown in Fig. B. 1, in which the curve (*d*) reaches a constant end value.

4. The simple Jander equations are not sufficient to describe the reactions in the quaternary system Na_2O—CaO—SiO_2—CO_2 when the dissociation of $CaCO_3$ (in limestone) is coupled with reactions of the type $Na_2CO_3 + Na_2Si_2O_5 \rightarrow 2\, Na_2SiO_3 + CO_2$. The reason for the anomalies here observed are to be found in an oversimplification in G. Tammann's and W. Jander's theory, wihch only considered a non-variable diffusion constant of the reacting ingredients in the reaction layers. Kröger previously had often observed structural changes in these layers during the process of molecular rearrangements by which "anomalous solid solutions" would be formed,[2] or similar "hermaphroditic" intermediate stages of crystalline products. By an extensively increased thermal treatment these defective crystal phases are changed into homogeneous nuclei of the final crystallizations. The modified Jander equation of the type $[1 - \sqrt[3]{(100 - x)/100}]^2 = k' \cdot \ln t$ (t = time) is then able to describe correctly also the kinetics of batch reactions with the particularly important composition $Na_2O,CaO,6SiO_2$ or of other mixtures derived from it.

5. On this basis, Kröger came to the definite conclusion that ternary sodium-calcium silicates are formed in very early stages of batch conversions, in combination with the electric conductance method for the detection of polyeutectic liquid phases above 760°C. The glass batch mixes $Na_2O,CaO,6SiO_2$, $Na_2O,2CaO,6SiO_2$, $Na_2O,2CaO,8SiO_2$, $Na_2O,2.7CaO,9.3SiO_2$, show continuous increase in conductance, and therefore also in melt phase as discussed above. $Na_2O,CaO,6SiO_2$ is characterized as presenting the most favorable conditions for an accelerated formation of a high share in fused products. This fact is remarkable in glass technology because A. E. Badger and L. G. Farber[3] observed practically the optimum of fusion speed conditions (and of fining, too) in pressed pellet samples. Kröger only confirmed this observation, say for $Na_2O,2.7CaO,9.3SiO_2$ batches, although the kinetic curve of this composition is nearly coincident with that of $Na_2O,CaO,6SiO_2$ in the early stages, and it is by far inferior to the latter batch composition in output of CO_2.

6. The reaction constants k' of the conversion curves are derived from modified Jander lines which correspond to the improved logarithmic time relation. These linear relations are valid over a certain field of variations of time, temperature and composition. For the mix $Na_2O,CaO,6SiO$, however, a distinct break point is observed after 50 minutes of heating and other slight anomalies occur in lime-rich mixes. From the inclination of the lines, the reaction constants were derived which show the very characteristic logarithmic time functions in Figs. B. 3 and 4, obeying the Arrhenius equation $k' = C \cdot e^{-E/RT}$. The activation energies of the reactions in the range of true solid state conversions below 780°C., are high (about 144,500 cal./mole), and 780°C. is the fusion point of the eutectic of the double-carbonate $Na_2Ca(CO_3)_2$ with crystalline solutions enriched in Na_2CO_3 ("double-carbonate eutectic"). At this temperature the reaction constants show a sudden change and activation energies of the next-following reactions are considerably lowered (about 78,500 cal./mole,

[2] On the disturbed structure of primary crystalline phases participating in the reactions cf. C. Kröger and J. Blömer, *Z. anorg. u. allgem. Chem.*, **280**, 1955, 51-64; $Na_2O.2CaO.3SiO_2$ crystallizes less well than $2Na_2O.CaO.3SiO_2$.

[3] *Glass Ind.*, **24**, 1943, 63-64.

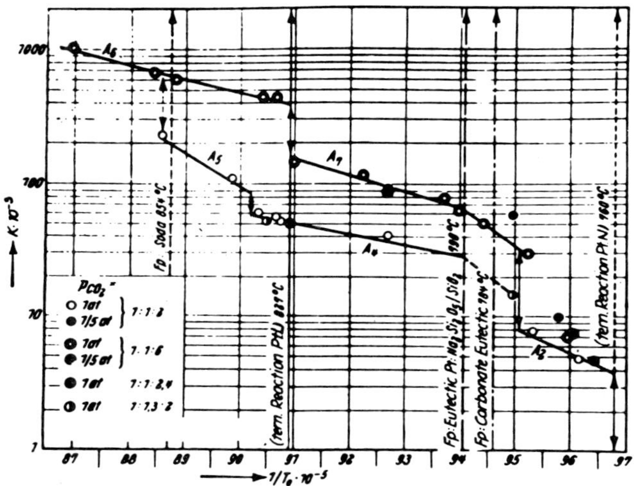

FIG. B. 3. Rate of reaction constants of the initial reactions, as a function of temperature. (Kröger and Marwan).

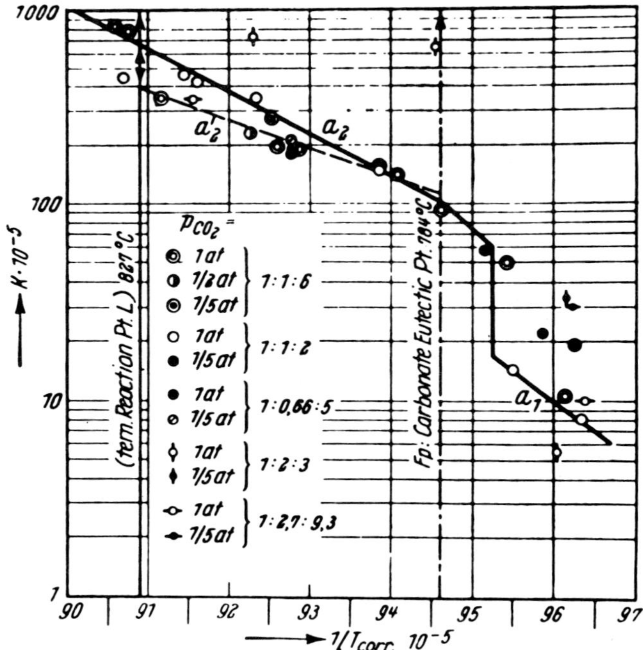

FIG. B. 4. Rate constants derived from the revised logarithmic equation, as a function of temperature. (Kröger and Marwan).

cf. curve A_2 in Fig. B. 3, between 760° and 780°C.; 33,600 cal./mole for 790° to 830°C. curve A_4). Then again, activation energies are raised to 109,000 cal./mole between 830° and 850°C. (curve A_5), and lowered to 33,600 cal./mole above 850°C. (curve A_6). For $Na_2O,CaO,6SiO_2$ mixes between 827° and 850°C. the apparent activation energy is 55,700 cal./mole (curve A_7). The accelerating effect of excess silica in this composition is most evident by the increased k' data. The sharp break point at 760°C. corresponds to the reaction equilibrium

$$Na_2O.2CaO.3SiO_2 + \text{liquid} \rightleftharpoons Na_2O.2SiO_2 + Na_2O.3CaO.6SiO_2,$$

but the ternary eutectic for

$$Na_2O.2SiO_2 + Na_2O.3CaO.6SiO_2 + \text{silica (quartz)} \rightleftharpoons \text{liquid}$$

is apparently not decisive for the formation of the first melt phase.

7. Instructive are further projections of the reaction constants k' derived from the modified Jander lines as functions of the ratios $Na_2CO_3 : CaCO_3$ and of the silica contents of the batches (Fig. B. 5). Up to 850°C. these reaction rates have a maximum

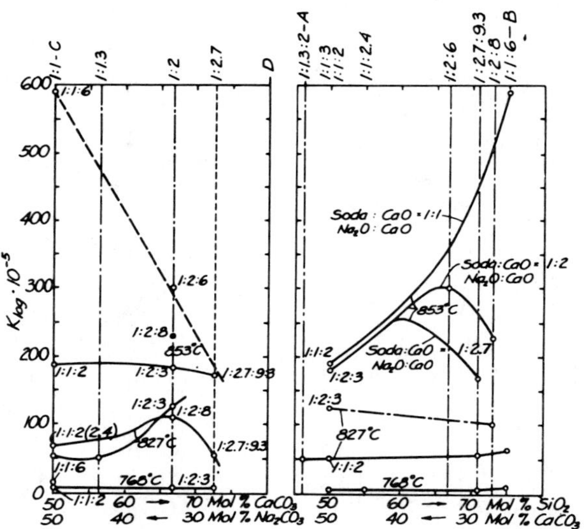

FIG. B. 5. Reaction rate constants as functions of the ratios $Na_2CO_3 : CaCO_3$ (left), and of the silica contents in the batch mixes (right). (Kröger and Marwan).

for the $Na_2CO_3/CaCO_3$ ratio of 1 : 2. The effects of changes in silica, however, are relatively low below 850°C. As soon as the fusion temperature of soda is reached the reaction rates in the batches with the $Na_2CO_3/CaCO_3$ ratio 1 : 1 are considerably increased with increasing silica contents, and the highest rate is that for Na_2O, $CaO,6SiO_2$ mixes. The higher the $CaCO_3$ content the more the maximum on the curves is lowered and shifted towards lower silica contents. The mixes $Na_2O,CaO,2SiO_2$, $Na_2O,2CaO,3SiO_2$, and $Na_2O,2.7CaO,9.3SiO_2$ have about the same reaction rates at 883°C. If primary $Na_2O.2SiO_2$ (as crystalline phase or as a glass) reacts with limestone, or $Na_2O.2SiO_2$ with quartz and $CaCO_3$ and presintered products of the compositions $Na_2O.2SiO_2 + SiO_2$, or $Na_2O.2SiO_2 + 4SiO_2$ with $CaCO_3$, it is somewhat unexpected to observe that $Na_2O.2SiO_2$ glass is, in final effect, somewhat *less* active than the crystalline form. The X-ray examination of products with calcium carbonate showed in those of $Na_2O.2SiO_2$ glass only the ternary silicate $2Na_2O.CaO.3SiO_2$, whereas in those with crystalline $Na_2O.2SiO_2$ this silicate is associated with $Na_2O.2CaO.3SiO_2$, Evidently, the "anomalous solid solutions" xNa_2O, yCaO, $zSiO_2$ are highly disordered intermediate products of different reactivity in both cases. The most rapid conversion was observed when the sintered product of the composition $Na_2O.2SiO_2$ + SiO_2 was used.

8. Concerning the amounts of liquid phase formed in the sodium silicates and the more siliceous products reacting with calcium carbonate, Kröger came to the important conclusion that in these more liquid is active than in soda-limestone-quartz mixes, say of the $Na_2O,CaO,2SiO_2$ type. The reaction characteristics of both groups of experiments are evidently different. The framework structure of quartz is changed into $[Si_3O_9]$ groups in the ternary sodium-calcium silicates in mixes of the soda-$CaCO_3$-quartz type by diffusion of sodium and calcium ions, whereas in the disilicate mixes the structure is already reduced to a simpler type and only the calcium ions must enter by diffusion. This fact is illustrated by the particularly low activation energy for reactions of $Na_2O,3SiO_2$ glass + $2CaCO_3$ (29,000 cal./mole), and Na_2O, $6SiO_2$ sinter + $CaCO_3$ (35,300 cal./mole).

9. The practical importance of Kröger's investigations is particularly evident in a discussion of one or the other industrial attempts for "accelerated" glass batch melting. The first fact is the rather unexpected impeding effect of the high pressure process of briquetting (pelleting) for the batch mixes.[4] The explanation is convincingly given by reduction in diffusion speed of the evolved CO_2 in the compacts during sintering. There are many other proposals made for acceleration of glass formation from batch mixes by special methods which deserve a detailed examination. We have spoken in B. ¶ 5 of the experiments of Badger and Farber concerning pellets of mixes with compositions contrasting to that of the final glass composition. Also P. Baldermann[5] recommended the use of premelted sodium waterglass to finish the glass in a second step in furnace processing. Very attractive are proposals

[4] We must not overlook that N. S. Kazanskiĭ and S. G. Klochkov, *Sprechsaal*, **74**, 1941, 340-342, observed a considerable reduction in time required for the refining process when the batch was briquetted.

[5] *Glass Ind.*, **29**, 1948, 447. See also I. I. Kitaĭgorodskiĭ and N. Solomin, *Glastech. Ber.*, **9**, 1931, 349-354.

of using the effects of reduced CO_2 partial pressures in the gases surrounding the batch. Kröger's experiments show convincingly how efficient must be such a reduction in pressure for the output in silicate formation from the mix according to the equation $k = B - a \cdot \log (p_{CO_2})$.

10. The granulometric composition of the batch ingredients is another factor of great significance. Kröger worked under standard conditions of grain size for the quartz sand (0.15 to 0.25 mm. fraction) as commonly used in glass manufacturing pratice and also soda and limestone in equal grain size classes. Industrially it is further important to consider the introduction of the batch into the melting section of the furnace either as a heap or deliberately spread out to a thin layer with a relatively large surface open to heat radiation from the burners and the roof. One may also calcinate the batch in a separate preheating chamber, or on a calcination belt as is done in the metallurgical practice of the Dwight-Lloyd process, or in the Lepol kiln of Portland cement manufacturing,[6] to abbreviate the time of the fusion process. Even the possibility of using a rotary kiln for calcination and fusion has been repeatedly proposed,[7] and special accelerating agents (ammonium sulfate, titanosilicates, and the like) were recommended.[8] Other ideas concern a separation of the primary formation of the glass-forming silicates and the final dissolution of the quartz sand in the excess, as in the erosion process of G. E. Howard,[9] and E. P. Danil'chenko and A. G. Repa.[10] All these many proposals, of which we give here only a selection, should be carefully reexamined under the aspects of reaction rates given in the rich data of Kröger's investigations.

11. With the same experimental methods, C. Kröger and Fr. Marwan[11] studied the effects of different *additions* to the ternary sodium-calcium silicate batches, such as the $Na_2O,CaO,6SiO_2$ type. Glass cullet cannot essentially affect the reaction course and the kinetics. If, however, limestone as a raw material is partly or totally replaced by dolomite, soda by potash or Li_2CO_3, considerable changes in the reac-

[6] Cf. the report in H. Kühl's book "Zementchemie," VEB Verlag Technik, Berlin 1951, Vol. 2, 439-442, and 465-473.

[7] E. G. in the Zotos kiln, cf. G. A. Zotos, *Glastech. Ber.*, **12**, 1934, 423-426; *Glass Ind.*, **22**, 1942, 393-396; W. Giegerich, *Glastech. Ber.*, **24**, 1954, 297 ff.; W. Küster, *Silikat Tech.*, **3**, 1952, 438-441.

[8] Cf. J. Enss, M. Thomas, and H. Kalsing, *Glastech. Ber.*, **18**, 1940, 213; A. Dietzel, *Ibid.*, **21**, 1943, 198-204.

[9] *Glass Ind.*, **29**, 1948, 615-620; 656-658; 694-697; 722; *Ibid.*, **30**, 1949, 492-496.

[10] *Steklo i Keram.*, **7**, 1950 (8) 10-12; *Doklady Akad. Nauk S.S.S.R.*, **86**, 1952, 1175-1178; *Zhur. Priklad. Khim.*, **25**, 1952, 740-744. See more recently, M. Boffé and G. Letocart, *Glass Technol.*, **3**, 1962, 117-123.

[11] *Glastech. Ber.*, **29**, 1956, 275-289; **30**, 1957, 42-52; a summary of experience on reaction rates, phase formations and solid state, with subsequent liquefaction reactions. In the systems $Na_2O–B_2O_3$ and $Na_2O–B_2O_3–SiO_2$, V. Gottardi and B. Locardi, *ibid.*, *Sonderband* **32**, K., 1959, V/8–V/15, studied the rates of the batch reactions.

tion rates are observed. Li_2CO_3 accelerates the reactions particularly intensively, K_2CO_3 somewhat less. Evidently, the conditions governing the formation of a primary melt phase are influenced by these agents (fluxing effects). Also B_2O_3, boric acid, and borax accelerate reaction rates of batch mixes but not so intensely as K_2CO_3 and also H_2O. Experiments with *moist* batches showed a slight increase in reaction rates (by 5 to 6 per cent) which is explained by a structural effect on the batch ("impregnation effect"). By a systematic application in glass manufacturing practice the output of the reaction of the basic batch might be increased by not less than 20 per cent.

12. It would be beyond the limits of our report if we would discuss in detail the wealth of valuable thermochemical data collected by Kröger and his coworkers in their investigations. For the reader it is particularly recommended that he become familiar with the excellent tabulations of the fundamental heats of formation of the compounds concerned in this work, further the heats of reactions, activation enthalpies of the different reaction steps, especially presented by Kröger in separate publications.[12] We may only mention the endothermic character of the formation of $Na_2O.3CaO.6SiO_2$ and $Na_2O.2SiO_2$ from the ternary metasilicates $Na_2O.2CaO.3SiO_2$ and $2Na_2O.CaO.3SiO_2$ + silica, and likewise the formation of $Na_2O.3CaO.6SiO_2$ from $CaSiO_3$, $Na_2O.2SiO_2$ and silica (quartz) (with 22,770, and 47,270) cal./mole, respectively). Exothermic are of course the formations of the ternary silicates from the oxides and the formation of $2Na_2O.CaO.3SiO_2$ from 2 mol. Na_2SiO_3 + $CaSiO_3$ (— 4,000 cal./mole), but is again endothermic that of $Na_2O.2CaO.3SiO_2$ from Na_2SiO_3 and 2 mols. $CaSiO_3$ (+ 21,430 cal./mole). As a summary of the practical deductions of the heat input in glass manufacturing, Kröger gives examples for different industrial glass types (calculated for 100 kg. glass):

Feeder container glass	62,400 kcal.	(relative to the final temperature of 1500°C.)
Fourcault sheet glass	70,050 do.	
Apparatus glass	53,650 do.	
Lead crystal glass	53,800 do.	

13. For examination of the complex thermochemical calculations C. Kröger and W. Janetzko[13] developed a special *direct-calorimetric* method for study of the enthalpies of sluggish reactions at high temperatures, based on the observation of the reaction in a calorimeter bomb containing a small electric resistance heater, as shown in Fig. B. 6. The energy imput is measured by a Watt-second meter. The accuracy of such a method is at least equivalent to that of the indirect, or heat-of-solution methods commonly used, and is particularly adequate to the measurement of the enthalpies of the carbonate and hydroxide melt syntheses of Na_2SiO_3,$Na_2O.2SiO_2$ or for devitrification and inversion effects. For different industrial glass batches C.

[12] *Glastech. Ber.*, **26**, 1953, 171-174; 202-214.

[13] *Z. anorg. u. allgem. Chem.*, **284**, 1956, 83-96.

Kröger, W. Janetzko, and G. Kreitlow[14] rediscussed the theoretical heat imputs based on direct-calorimetry and in comparison with the previously known input data. The results show the heat input for the *fusion* alone in the order of magnitude from 9,000 (for lead crystal glass) to 21,000 kcal./100 kg. glass (for sulfate glass).

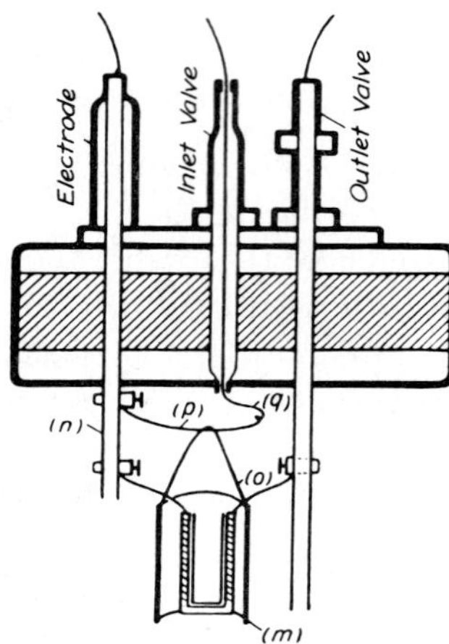

FIG. B. 6. Suspension of the heating furnace below the cover of a calorimeter bomb, for the direct-calorimetric method. (Kröger and Janetzko).

Carbonate glasses show input data from 12 to 13,000 kcal. It must be concluded that preference should be given in such calculations to the results from direct-calorimetry.[15]

[14] *Glastech. Ber.*, **31**, 1958, 221-228.

[15] It is not our intention to criticize the direct-calorimetry in comparison with the heat-of-solution method in general. We may only observe from a recent investigation of Chr. Hummel and H. E. Schwiete, *Glastech. Ber.*, **32**, 1959, 327-335, for the fundamental system Na_2O—SiO_2 that these authors established very accurate data for the sodium silicates and did not find a too good agreement with the direct-calorimetric data, as for the heats of fusion of the disilicate modifications α and β, namely:

$\alpha - Na_2O.2SiO_2$	Enthalpy of Fusion =	$6{,}700 \pm 400$	cal./mole	(Kröger and Janetzko: $8{,}810 \pm 120$ cal./mole)
$\beta - Na_2O.2SiO_2$	do.	$8{,}080 \pm 230$	do.	(Kröger and Janetzko: $7{,}220 \pm 140$ cal./mole)

Concerning enthalpy of inversion Kröger and Janetzko came to the improbable result that reaction α–$Na_2O.2SiO_2 \rightarrow \beta$–$Na_2O.2SiO_2$ would be endothermic (with + 1,590 cal./mole), whereas Hummel and Schwiete found the plausible value — $1{,}440 \pm 650$ cal./mole. Important data on the dissociation pressures of alkali carbonates participating in glass batch reactions were recently given by C. Kröger and J. Stratmann, *Ibid.*, **34**, 1961, 311-320, furthermore measurements

14. In a certain parallel to Kröger's investigations, A. G. Repa[16] and his coworkers made extensive studies on the glass formation process. With E. P. Danil'chenko he first examined the model system Na_2CO_3—SiO_2, with quenching experiments and determinations of the dissolution process of quartz sand particles in batches heated to different temperatures for variable exposure periods. The first reaction product observed is Na_2SiO_3 which reacts with excess soda to form the orthosilicate, with excess silica to form $Na_2O.2SiO_2$. The glass formation process is connected with dissolution of excess silica in the disilicate, in good agreement with Kröger's conclusions. Curves showing increasing saturation of the melts in silica are given for 1050°, 1250°, 1350°, and 1450°C. which are the basis for the kinetic calculations. They all show the type of a steep rise in the initial period of batch reaction but soon flatten out. The increasing dissolution of silica in the alkali silicates causes a continuous increase in viscosity and, therefore, a corresponding decrease in diffusion coefficients. Repa develops an equation $d[SiO_2]$ sol.$/dt = K_s[SiO_2/Na_2O]/\eta \cdot t = k[Na_2O/SiO_2]^2$, K_s being a coefficient determined by the surface area of the reacting quartz grains. The final dissolution time for the quartz grains is given by $t_f = k_0[SiO_2]^3/[Na_2O]^2$. Repa extended these studies to the reaction of quartz with lithium, sodium, potassium, calcium, magnesium, strontium, and barium carbonates, of which only $MgCO_3$ and $CaCO_3$ are easily dissociated at operating temperatures, and the remaining oxide reacts directly as such with the silica. In glass formation silica is dissolved in the metasilicate melt solutions whereas the reaction in the solid state is very sluggish. Diffusion of the alkali and alkaline earth ions to the surface of the sand grains accelerates the process with an intermediate formation of crusts of basic silicates on these. In agreement with Kröger's results great emphasis is also given by Repa to the secondary reactions, such as that of $Na_2O.2SiO_2$ with soda on one hand, and with silica on the other hand, to advance the interchange between the extremes of the alkaline melt and the slowly reacting quartz grains.

15. Repa[17] combined glass formation mechanisms with what he called the "oxygen potential," namely the molecular ratio of the total oxygen anions to the total number of cations $\Sigma O^{2-}/\Sigma$ (cations $+ Si^{4+}$). This means that in glasses of the common type Na_2O—RO—SiO_2 (R = Ca,Mg,Ba) the fusion times as the most striking technological property of the melting batch to produce glass, are variable if an exchange of R_2O by RO takes place. Glasses with the same oxygen potential (say 1.41 or 1.51) have the same fusion time at 1350°C. (a grain size of the quartz particles

of vapor pressures of such salts, using a modified Knudsen "vapor jet" method, modified by A. Herlet and G. Reich, *Z. angew. Phys.*, **9**, 1957, 14-23. On the thermodynamic constants of the reaction $Na_2SO_4 \rightleftharpoons Na_2O + SO_3$ see R. Brückner, *Glast. Ber.*, **35**, 1962, especially pp. 96-100.

[16] *Zhur. Priklad. Khim.*, **25**, 1952, 740-744; **26**, 1953, 475-481; **27**, 1954. 1184-1193.

[17] *Zhur. Priklad. Khim.*, **28**, 1955, 694-699. Repa called the oxygen potential coefficient "KP" or "η". We want to avoid this symbol in order to avoid confusion with viscosity which is commonly symbolized by η.

of 0.3 mm. in maximum) when 8.1 and 49.6 molecular per cent of RO replace 3.3 Na_2O + 4.8 SiO_2 and 20.3 Na_2O, + 29.3 per cent SiO_2, respectively. Also the viscosity at 1350° and 1450°C. of soda-lime silicate glasses with the same oxygen potential are the same, and one is able to compare different glass compositions over a wide range by plotting the properties as a function of the oxygen potential (cf. B. ¶ 119). In connection with the parallelism of oxygen potential and viscosity of glasses, Repa *et al.*[18] also give an instructive analogy of this property with the mineralizing agents which act as accelerators of glass melting, especially fluorides, chlorides, borates, phosphates, titanosilicates, but also oxides of barium and manganese. The tests of the rate of dissolution of the batch ingredients in the course of the vitrification process are based on the appearance of ultraviolet-luminescence if traces of uranyl compounds are added to the batch (cf. the method recommended by W. Weyl in 1937).

16. O. K. Botvinkin and E. M. Shpil'kov[19] determined the dissolution process of quartz grains in a series of ternary sodium-calcium silicate glass batches by the thin section method. The inversion of quartz to cristobalite is easily observed by reaction rim structures in the glass melt surrounding the quartz grains. This intermediate formation of cristobalite is essential for the dissolution mechanism, together with the mosaic structure of the defective quartz crystals controlling the rate of inversion. By measuring the refractive index of the glass in different distances from the surface of the dissolving quartz-cristobalite grains a diagram was developed for concentration in silica as a function of the distances which shows a smooth curve of decreasing silica contents from the saturation concentration (81 per cent) to the final silica concentration of the experimental glass (74 per cent) in a distance of 60 μ. The saturation degree, ΔC, is proportional to the logarithm of the viscosity of the reaction melts in all the intermediate transition stages of the quartz dissolution. As a function of time the percentage of dissolved silica is in a simple relation from the initial rapid process of assimilation to the asymptotically reached final state of the homogenized glass. For soda glasses this function is particularly steep in the first branch; no break point indicates the formation either of $Na_2O.2SiO_2$ or of $CaSiO_3$. Three-dimensional projections are given by Botvinkin and Shpil'kov for functional relations of the percentage of dissolved quartz to time and the amounts of sodium-calcium silicates in the glass for the constant experimental temperature of 1320°C. The rates of reaction are considerably reduced if, at constant lime content, soda is replaced by silica, or at constant silica (74 per cent), soda is replaced by lime, especially if the soda content

[18] With I. A. Tykachinskiĭ, L. I. Buneeva, V. V. Pollyak, E. P. Danil'chenko, and A. N. Afanas'ev, *Trudy Vsesoyuz. Nauch. Issledovatel. Inst. Stekla*, 1953, (32) 3-16.

[19] *Steklo i Keram.*, **13**, 1956 (10) 1-5; *Izvest. Akad. Nauk Kazakh. S.S.R. Ser. Gornogo Dela, Met. i Stroimaterial.*, **3**, 1957 (14) 86-102. With advanced microscopic-interferometric methods, and by radioautography, L. Leger, J. Bray, and E. Plumat studied the dissolution process of quartz grains and the diffusion on their peripheries; cf. "Advances in Glass Technology." Plenum Press, New York, 1962, 175-189.

is below 16 per cent. The relation of these facts with the viscosity and the diffusion coefficients makes possible a formulation of the reaction rate of quartz dissolution nemely $\omega = k \cdot C/\eta$, in excellent agreement with the experimental data.

17. Analogous to Kröger's investigations, Abd-El-Moneim Abou-El-Azm and H. Moore[20] extensively studied the rates of reaction between silica and other oxides at variable temperatures in binary and ternary mixtures, the influence of the form in which the nonsilica constituents are introduced upon their rates of reaction with silica, and a general theory of the reaction rates. The methods applied in the fundamental experiments in the temperature range from 700° to 1400°C. include especially determination of the amounts of crystalline silica left after the reaction in the mixes, by estimating the intensities of the X-ray diffraction interferences of quartz, cristobalite, and tridymite. The reaction rates in the mixes were arranged in the series with decreasing rates: $K_2O > Na_2O > B_2O_3 > CaO > MgO$ and Al_2O_3, as well for binary as for ternary mixes. Fe_2O_3 and TiO_2 are slightly accelerating agents in small amounts replacing CaO. For complex glass compositions the rates of reaction decrease in the order $K_2O > Na_2O > Li_2O > PbO > B_2O_3 > BaO > CaO > ZnO > MgO > TiO_2 > Al_2O_3 > ZrO_2$. Among fluxing additions, the fluorides are the most efficient accelerators and also the mineralizing influence of water is emphasized, if it is present in statu nascendi. If the glass-forming reaction between quartz grains and other oxides produces a highly viscous glass coating around the grains, progress of the reaction is formulated by the equation $R[1 - (m/M)^{1/3}] \cdot \log a = \log t +$ const. (R being the initial radius of the quartz grains, m the weight of the unreacted silica after t seconds, M the weight of silica originally introduced, a a factor depending on temperature and composition of the batch). If, on the other hand, a very fluid glass melt is formed which easily flows away from the quartz grains the formula $[1 - (m/M)^{1/3}] = B \cdot t$ (B a constant) is valid, in agreement with experimental results. For intermediate cases with a very viscous glass first formed, which becomes more fluid as the reaction proceeds, both equations must be applied one after the other. The time required for the change in reaction rates depends on the composition of the mixture and on the temperature. The quotient $1/\log a$ is a measure of the rate of reaction used to place the oxides in the series of those rates, as given above. Down to 600°C., $1/\log a$ increases approximately linearly with the rise in temperature over 600° to 850°C. (cf. Fig. B. 7). A sharp increase of $1/\log a$ is found at 850° to 900°C. (cf. Fig. B. 8, and Kröger's results), interpreted, however, by Abd-El-Moneim Abou-El-Azm and Moore as an inversion effect of quartz to tridymite at 870°C. The lower limit of reactivity between soda and quartz could be given by extrapolation of the curve for $1/\log a$ versus temperature at about 400°C.

[20] *J. Soc. Glass Technol.*, **37**, 1953, 129-154; 155-167; 168-181; 182-189; 190-210. On reactions in batches with red lead, potash, and sand see recent studies by B. Rosenkrands and B. Simmingsköld, *Glass Technol.* **3**, 1962, 46-51.

18. Thermogravimetric and differential-thermal analysis were applied to glass batch studies by F. W. Wilburn and C. V. Thomasson[21] in the model system Na_2CO_3—SiO_2, especially to study the influence of grain size variations in the reactants. In more finely ground batches the temperatures of the starting and of the completed

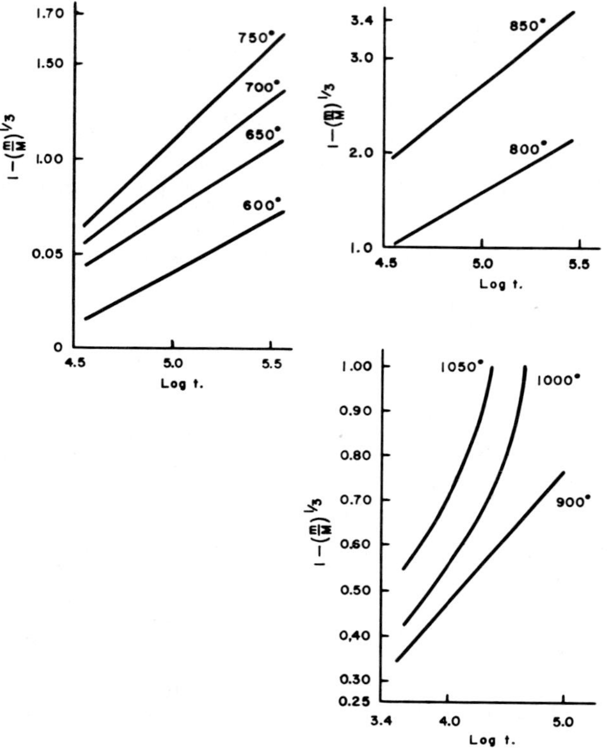

FIG. B. 7. Reaction rates of a mixture of 75 per cent silica (quartz sand) and 25 per cent Na_2O (introduced as soda), exposed to temperatures between 600° and 1050°C., as a function of time (in seconds). (Abd-El-Moneim Abou-El-Azm).

reactions, and those of the first perceptible glass formation are lower than with coarser mixes as seen from the instructive curves in Figs. B. 9 and 10. The effects observed at 700°C. (exothermic, by the formation of Na_2SiO_3) and at 830°C. (en-

[21] *J. Soc. Glass Technol.*, **42**, 1958, 158-175 T. We mention, in addition, the use of differential-thermal analysis for determination of residual quartz in silicate batches after a solid state reaction, as demonstrated by C. Kröger and J. Stratmann, *Glastech. Ber.*, **33**, 1960, 250-252, using the heat effect of the $\alpha \rightleftharpoons \beta$ inversion. For recent studies by C. V. Thomasson and F. W. Wilburn on the use of differential-thermal and thermogravimetric analytical methods, see *Phys. & Chem. of Glasses*, **1**, 1960, 52-69; **2**, 1961, 126-131.

dothermic) are particularly impressive and are to be compared with data given by Kröger in his investigations.

19. Starting from *structural* aspects of glass formation, C. L. McKinnis and J. W. Sutton[22] recently developed a series of ideas on the effects of "degradation" and "reorganization," as in the change of the ionic arrangements of Na_2SiO_3 as the simplest

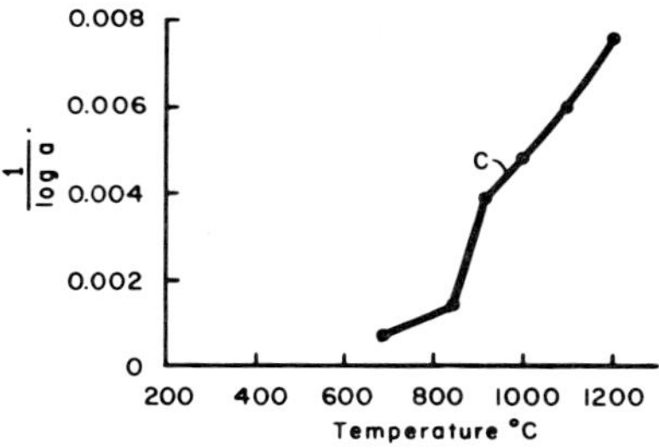

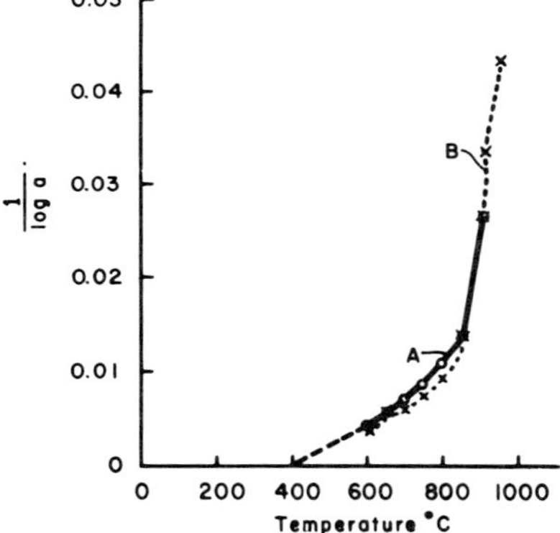

FIG. B. 8. Variation of the factor 1/log a, with temperature, for a mixture of (A) 75 per cent quartz sand, and 25 per cent Na_2O (introduced as soda); (B) 66.4 per cent quartz sand, and 33.6 per cent K_2O (introduced as potash); (C) 75 per cent quartz sand, and 25 per cent anhydrous B_2O_3. (Abd-El-Moneim Abou-El-Azm).

model in two limiting states (Figs. B. 11 and 12). A silicate melt at practical glass melting temperatures contains a multiplicity of anionic units in which there is a distribution of bridging oxygen function. In practical glass manufacturing the distributed silicate structure established at the melting temperature is "carried over" into

[22] *J. Am. Ceram. Soc.*, **42**, 1959, 194-199; 250-253. We, however, emphasize the critical remarks of R. Brückner, *Glastech. Ber.*, **35**, 1962, especially pp. 50-52, on the importance of melting history effects, and his careful viscosity and homogenization experiments in glass melts.

the glassy state for the most part. The significant steps in the dissolution process of silicate in a silicate melt are: the diffusion of a highly mobile cation, say of Na^+, into an interstitial position within the silica framework, the rupture of Si—O bonds in the framework near the "entrapped" ion, the coupling of the resulting unsaturated

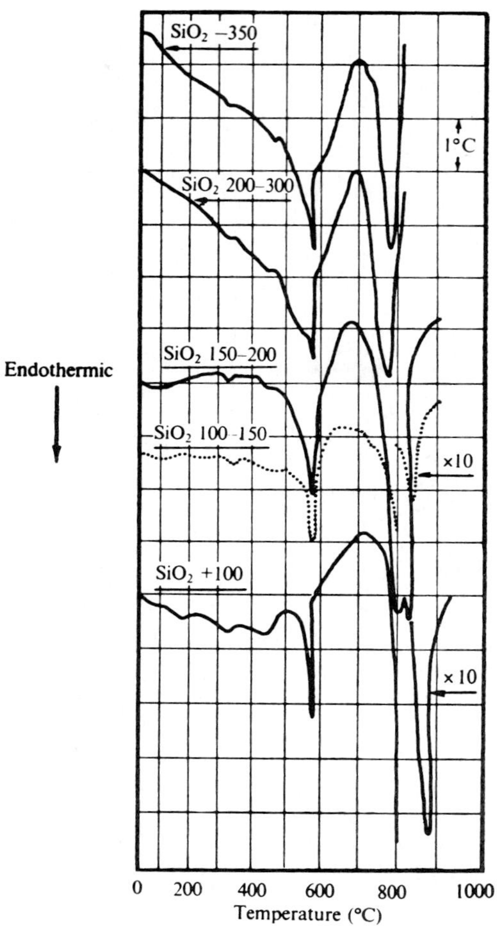

FIG. B. 9. Differential-thermal analysis curves of a mixture of 85 per cent silica (quartz sand, variable in grain size from 100 to 350 BSS), and 15 per cent soda (grain size constant below 350 BSS). (Wilburn and Thomasson).

silicon ions in the framework with a previously nonbridging oxygen anion of a unit in the melt, and the multiple repetition of this sequence of neighboring sites, yielding a silicate unit from the structure assimilated in the melt.

20. To the observations of Abd-El-Moneim Abou-El-Azm, McKinnis and Sutton supplement their own results that the temperature range of the transition from the equation $R[1-(m/M)^{1/3}] \log a = \log t + \text{const.}$ (for lower temperatures), to the high-temperature form $[1-(m/M)^{1/3}] = B \cdot t$ is by 120° to 220°C. higher than the Liquidus temperature (780°) of the system under investigation. Further, there is no discontinuity in applicability of the low-temperature equation in passing from the

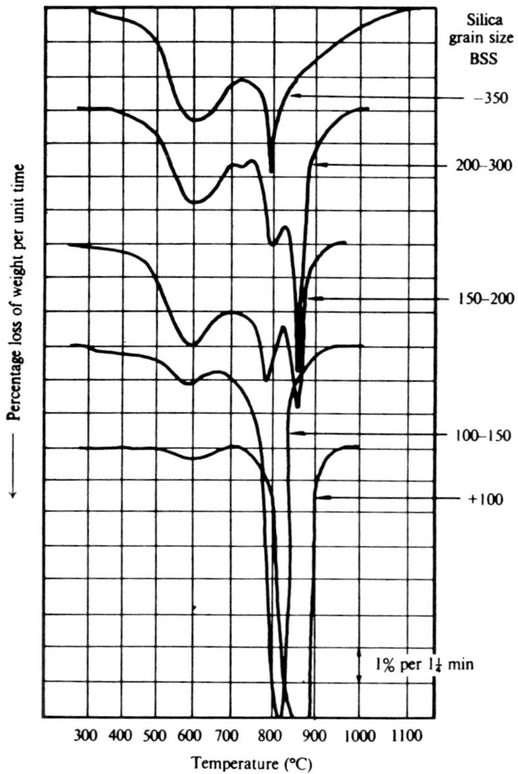

FIG. B. 10. Rate of loss in weight as a function of temperature, for mixtures of 85 per cent silica (grain size variable between 100 and 350 BSS), and 15 per cent soda (grain size constant below 350 BSS). (Wilburn and Thomasson).

solid state reaction at 600° to 750°C. to those involving melt-silica interactions from 800° to 900°C. The data of Abd-Al-Moneim Abou-El-Azm are in good agreement with the postulates of McKinnis and Sutton, especially for B_2O_3 containing batches, as striking evidence in support of their theoretical aspects, including viscosity effects and the possibility of a radical structural change at a ratio of Si/O of 0.444 (observed by H. H. Blau, 1951, in his studies on thermal expansion). It is important to compare also the *practical* experience that the temperature of economical glass melting is usually by more than 300°C. above the Liquidus temperature, i.e., in the range of the high-temperature mechanisms discussed above. For borosilicate glasses an example is discussed from Abd-El-Moneim Abou-El-Azm's results with a Liquidus

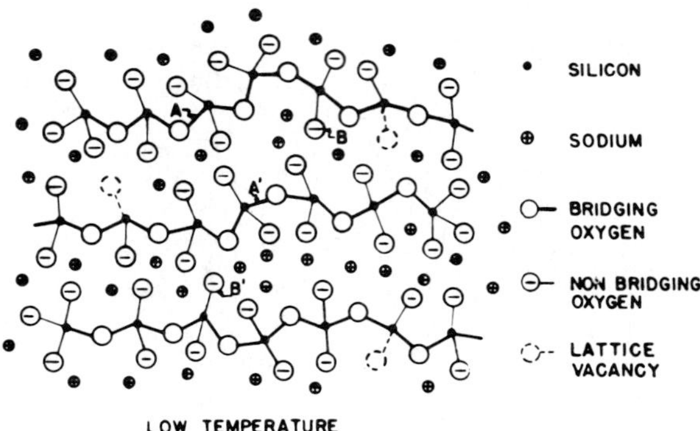

Fig. B. 11. Hypothetical structure of sodium metasilicate just above the Liquidus temperature. (McKinnis and Sutton).

temperature of 1080°C. (an alkali-free glass; the commercial melting temperature is 1580°C.), but if only 3 per cent Na_2O substitute for B_2O_3 the Liquidus temperature is raised to 1180°C., and the industrial melting temperature is only 1480°C. The difference between the Liquidus and practical glass fusion temperatures is thus decreased by 200°C. Such effects of a "melting history" illustrate how just in the

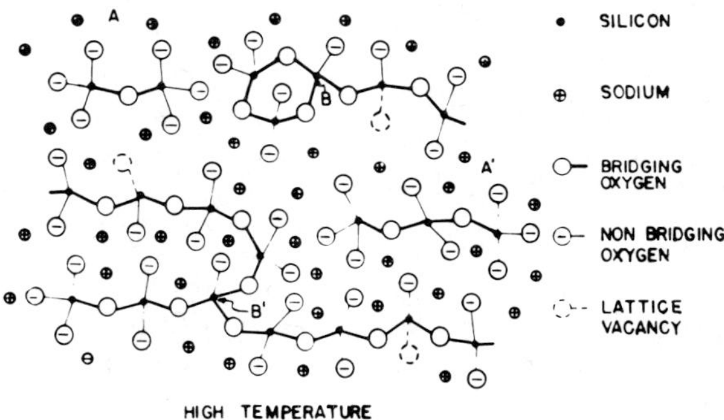

Fig. B. 12. Hypothetical structure of sodium metasilicate at higher temperatures, showing the effects of "degradation" and "reorganization." (McKinnis and Sutton).

neighborhood of the Liquidus temperatures a structural unit characteristic of the primary crystallization phase also persists in the glass which, however, is "degraded' and then "reorganized" at higher temperatures, with all the important consequences for the constitution of the glass and its sensitive physical properties. Reversely, the degraded structure can be "frozen-in" by quenching as is generally observed for glasses. Viscosity is one of the physical properties of the glass which excellently reflect structural changes in the melts. Tensile strengths of glass fibers proved to corroborate these conclusions in precision measurements (cf. A. ¶ 301 f.).

21. Speaking of *sulfate* glass batches, we have to discuss a series of investigations by C. Kröger and E. Vogel[23] who applied the manometric method, combined with detailed analyses of the gases evolved from *carbon* containing sulfate-silica mixes. The lowest reaction temperature observed was 580°C. the reaction rate being a function of temperature and of the ratios in the amounts of the initial raw materials [sodium sulfate and quartz (sand)], proportional to time in the degree of the chemical conversions. Above 50 per cent conversion, however, this proportionality is no longer valid and true equilibrium pressures are measured. Na_2S and COS are intermediate products of heterogeneous gas reactions of the summary type $7Na_2SO_4 + 13C \rightarrow 4Na_2CO_3 + 3Na_2S_2 + 7CO_2 + CO + COS$. No sodium sulfite was observed because this compound is disintegrated at about 530°C., or even at lower temperatures in the presence of carbon and quartz powder. The monovariant equilibria for the reaction $2Na_2CO_3 + 2Na_2S_2 + C \rightarrow 4Na_2S + 3CO_2$ is accompanied by the other reaction $6Na_2CO_3 + 7Na_2S_2 + 4C \rightarrow 13Na_2S + 8CO_2 + CO + COS$, which has an equilibrium pressure of 1 atm. at 770°C. In the temperature range from 530° to 640°C. anomalous higher pressures were observed which are non-equilibria. Quartz reduces the rate of the sulfate reduction, but the lowest temperature observed is the same for mixes with, and without, quartz. The rate of reaction is increased above 655°C., as a true sulfosilicate, $Na_2S.2SiO_2$, is formed which is also observed if Na_2SO_4 is reduced by 2 Si.

22. There is a sextuple point in the system $Na_2O—SiO_2—C—S$ (Fig. B. 13), with the phases Na_2SO_4, C, Na_2CO_3, Na_2S_2, Na_2S, and gas, corresponding to a nonvariant equilibrium with the coordinates 630°C. and 110 mm. Hg pressure, as the upper limit of the coexistence of Na_2SO_4 and C. Important is the strongly endothermic character of the reaction $Na_2S + 3Na_2SO_4 \rightarrow 4Na_2O + 4SO_2 + 399{,}870$ cal./mole. It starts in batches with SiO_2 at about 500°C. but the equilibrium is monovariant only up to 780°C. with an equilibrium pressure of 110 mm. Hg. Above 780°C. it

[23] *Glastech. Ber.*, **28**, 1955, 426-437; 468-474. From the practical viewpoint, studies on heat conductance of melting glass batches made by C. Kröger and H. Eligehausen, *Ibid.*, **32**, 1959, 362-373, supplement these chemical investigations. They show the influence of compaction of loose batch materials, typical increase in heat conductance at the start of solid state reactions (above 400°C.) and then in beginning melt reactions.

becomes divariant, and pressures are strongly decreased with increased reaction output, forming unsaturated melts, and *unmixing* phenomena occur. If lime is added the monovariant character of the reaction is preserved over a wider range, as the equilibrium temperatures are raised by about 100°C. If quartz is added the disproportionation of Na_2SO_4 is catalyzed and $Na_2O.2SiO_2$ is formed. The reaction $Na_2SO_4 + Na_2O.2SiO_2 + SiO_2 \rightarrow 4Na_2SiO_3 + 4SO_2 + 182{,}000$ cal./mole is much less endothermic than that in the absence of SiO_2 which, therefore, can not be easily verified. Simultaneous additions of lime and quartz impede the sulfate reduction by carbon and raise the equilibrium temperature for the reduction (mixing ratios $Na_2SO_4 : C$ tested were 1 : 2, and 2 : 1). Both reactions occur side by side on the temperature range from 700° to 900°C. The equilibrium pressures of some experimental batches as functions of temperature are given in Fig. B. 14.

23. The *industrial* sulfate glass process is sufficiently described by both basic reactions in the system Na—Si—O—C—S. The furnace atmosphere interacts in this case and open system conditions must also be considered, especially the withdrawal of gaseous reaction products by the combustion gases in which the partial pressures of CO_2 and water are particularly important because of reactions with Na_2S in the melt, much less the actions of CO and SO_2. Unmixing occurs in the temperature range between 770° and 1100°C. forming sulfate-enriched "gall" and silicate-enriched separate liquid phases. The amount of coke to be applied to industrial batches is nearly twice that for C required in formation of the sulfide in the ratio $3Na_2SO_4 : Na_2S$.

24. By the direct-calorimetric method described above, C. Kröger, W. Janetzko and G. Kreitlow[24] made possible a complete thermochemical calculation of the reactions in the system Na—Si—O—S—C, especially concerning the heats of fusion and the heats of formation of the interesting sulfosilicates like $Na_2S.SiO_2$, and $Na_2S.2SiO_2$, and further the supplementing heat of solution of the sulfide containing glasses produced by such reactions. For stability conditions of the sulfosilicates it is characteristic that the heats of the addition of silica to the Na_2S molecule in different stoichiometric ratios is much different from that of the binding of silica in the system $Na_2O—SiO_2$. We give from Kröger's and his coworkers' data the following comparison:

Reaction	ΔH in calories/mole
$Na_2S + SiO_2 \rightarrow Na_2SiO_2S$	— 9,000
$Na_2SiO_2S + SiO_2 \rightarrow Na_2Si_2O_4S$	— 31,400
$Na_2O + SiO_2 \rightarrow Na_2SiO_3$	— 54,680
$Na_2SiO_3 + SiO_2 \rightarrow Na_2Si_2O_5$	— 5,150

It would be particularly interesting to examine the possibilities of the existence of calcium containing sulfosilicates in glass batches with sulfates, a question which is briefly discussed by Kröger and Vogel, but not yet experimentally established.

[24] *Z. anorg. u. allgem. Chem.*, **287**, 1956, 33-44.

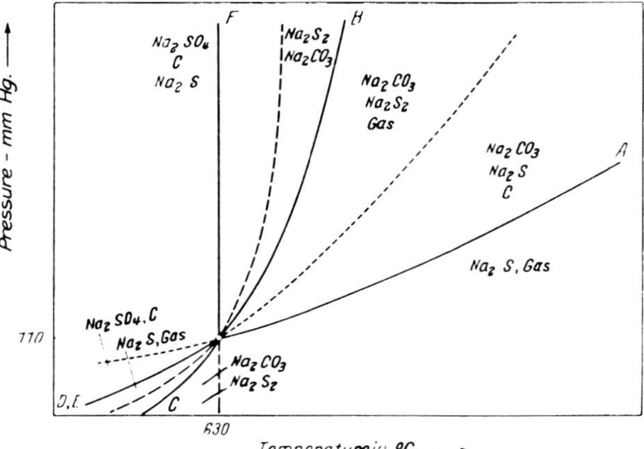

FIG. B. 13. Monovariant equilibrium curves in the system Na—O—S—C, in the neighborhood of the invariant sextuple point. (Kröger and Vogel).

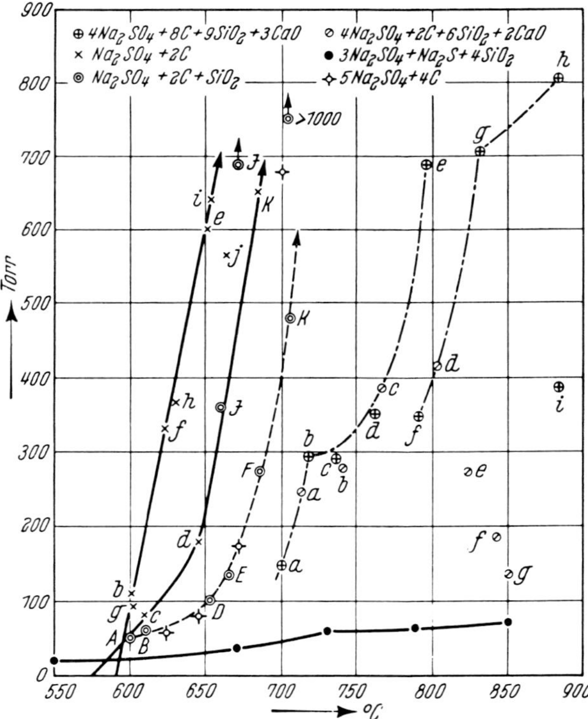

FIG. B. 14. Reaction equilibrium pressures for different sulfate glass batch compositions, as a function of temperature. (Kröger and Vogel).

25. Sulfo- or better said, *thiosilicate* glasses with bindings Si—S in the place of the common Si—O of silicate glasses were investigated by C. Kröger and J. J. Blömer[25] in the systems $Na_2S—SiO_2$, $Na_2S—CaO—SiO_2$, $Na_2S—B_2O_3$, $Na_2S—CaO—B_2O_3$, $Na_2S—As_2O_3$, and $Na_2S—CaO—As_2O_3$. The characteristic diagrams Figs. B. 15-17.

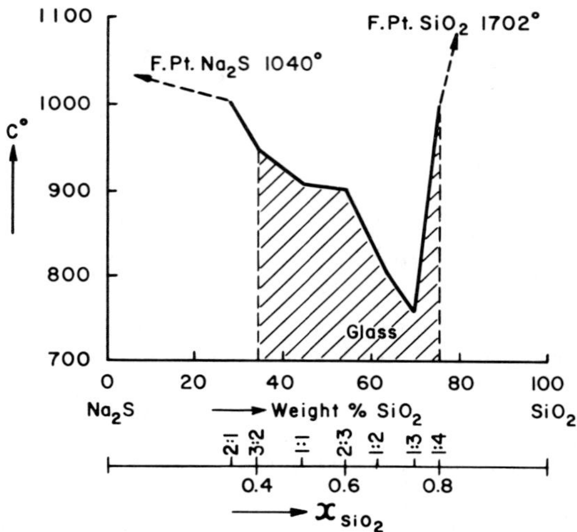

FIG. B. 15. Fusion characteristics and glass formation in the system $Na_2S—SiO_2$. (Kröger and Blömer).

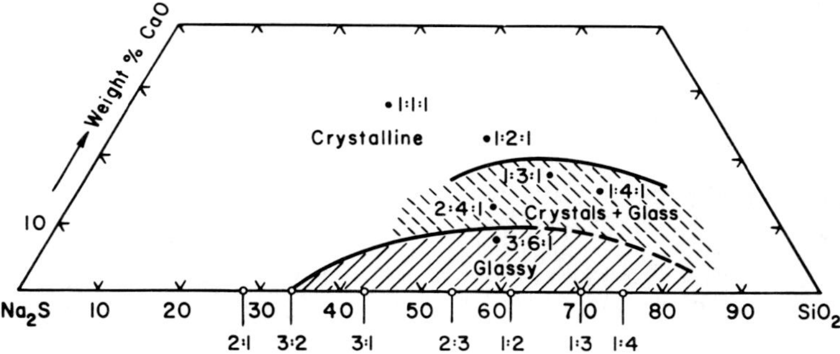

FIG. B. 16. Partial diagram of the system $Na_2S—CaO—SiO_2$, indicating the limits of homogeneous glass formation and characteristic molecular ratios. (Kröger and Blömer).

[25] *Glastech. Ber.*, **31**, 1958, 302-311. We may emphasize that E. Kohlmeyer and G. Lohrke, *Z. anorg. u. allgem. Chem.*, **281**, 1955, 54-63, reexamined previous data of G. Tammann and W. Oelsen (1930) on the system $Na_2O—Na_2S—Na_2SO_4$, and in general confirmed these. The eutectic in the ternary system was found at 620°C. (with 18 per cent Na_2S, 53 per cent Na_2O, and 29 per cent Na_2SO_4). In spite of their low fusion temperatures the melts have high boiling points and dissociation temperatures; no evolution of SO_2 was observed.

illustrate the restricted ranges of glass formation, more or less combined with crystallization of Na_2SiO_3 and $Na_2Si_2O_5$. The thiosilicate glasses are intensely colored, orange, red, or even deep-black. Above 700°C. a strong volatilization of SiS_2 is observed. The constitution of thiosilicate glasses is quite analogous to that of silicate glasses, but with widened near-order distances, as observed in increasing X-ray diffraction interferences with $d = 7.1$ Å. raised to 7.5 Å. None of the thiosilicates could be observed in the crystalline state. Thioborate glasses (with the compound

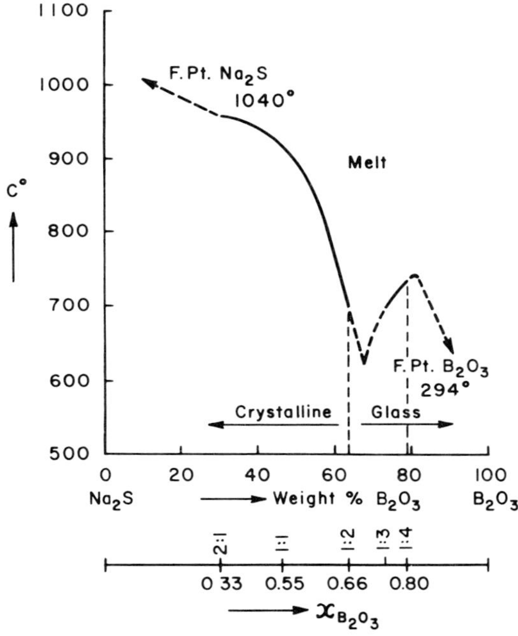

FIG. B. 17. Fusion characteristics and glass formation in the system Na_2S—B_2O_3. (Kröger and Blömer).

$Na_2S.B_2O_3$) are much less intensely colored (brownish-yellow, or bluish). No crystalline thioborates were observed, also no thioarsenites. If the systems contain CaO, CaS is always precipitated as a side-reaction product. The SiS_2 vapor pressure in the system Na_2S—SiO_2 shows a strictly linear function of $1/T$ for the logarithm of pressure. Thiosilicates, and the like, must be strongly dissociated in their melts although the compounds $Na_2S.2SiO_2$, $Na_2S.2B_2O_3$, $Na_2S.3B_2O_3$, $Na_2S.4B_2O_3$, and $3Na_2S.As_2O_3$ are rather probable to exist.

26. Previous to Kröger's investigations, E. Kordes[26] and his coworkers studied

[26] *Z. anorg. u. allgem. Chem.*, **264**, 1951, 255-271 (with B. Zöfelt and H. J. Pröger); W. Sack,

the large immiscibility gap between Na_2SO_4 (in the gall melt) and the liquid sodium-calcium silicates, with a mutual solubility for the contrasting liquid phase. The Na_2SO_4 contents in the resulting solidified glass are the greater the more basic it is in its composition (cf. Fig. B. 18). Industrial glasses of the so-called "tri-silicate"

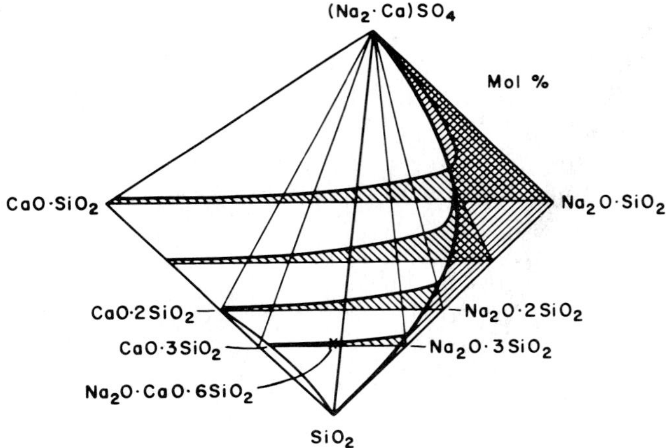

FIG. B. 18. Perspectivic sketch to show the solubility of liquid sodium sulfate in fused sodium-calcium silicates. (Kordes, Zöfelt, and Pröger).

type $1(Na_2O,CaO),3SiO_2$, in comparison with metasilicate glasses with $1(Na_2O,CaO)$, $1SiO_2$, show at 1200°C. a relatively low melt solubility for Na_2SO_4. As long as a gall is present, a certain amount of Na_2SO_4 always remains in the liquid glass and acts unfavorably on the properties of the finished glass. The molecular exchange of the two liquid phases in the immiscibility range is characterized by the transition of SO_3 into the silicate melt and the transition of soda, lime, and silica into the gall as shown by Fig. B. 19 particularly well for the metasilicate glass in the form of a reciprocal diagram, but is equally observed in the disilicate and trisilicate glasses (Figs. B. 20 and 21).

27. W. Sack studied especially the behavior of sodium *sulfite* in air and CO or CO_2 atmospheres, with its well-known disproportionation at 675°C. to form sulfate and sulfide together. With increasing temperature Na_2S can be oxidized to sulfate. CO plays a peculiar role in these reactions if catalyzed, showing its decomposition to CO_2 and a carbon deposit at 400°C. as minimum temperature. The sulfite reacts with carbon to form soda and sulfur (volatilized), or to form Na_2S and CO_2 with CO,

Ibid., **264**, 1951, 272-284; **266**, 1951, 105-117. See also older data given by F. Foerster and K. Kubel, *Ibid.*, **139**, 1925, 261-292, on sulfite reactions, and E. P. Danil'chenko, *Doklady Akad. Nauk S.S.S.R.*, **86**, 1954, 1175-1178.

finally to form sulfate, Na_2S_2, CO, and COS (cf. Kröger and Vogel). Experiments in a current of CO_2 demonstrate that the reaction of CO_2 with sulfite is most important for the change to sulfate. Kordes reduced the many possible reactions in sulfate batch glasses to a few principal types which are adequate for calculations with ther-

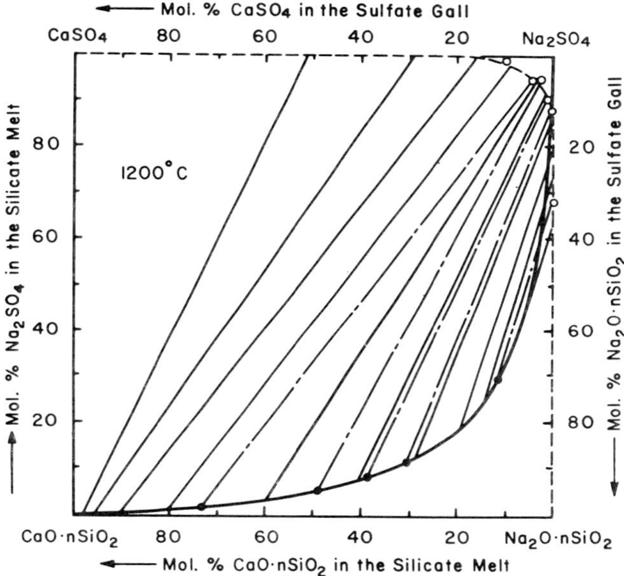

FIG. B. 19. Reciprocal transition equilibria between sodium-calcium metasilicate melts, and sulfate gall. (Kordes, Zöfelt, and Pröger).

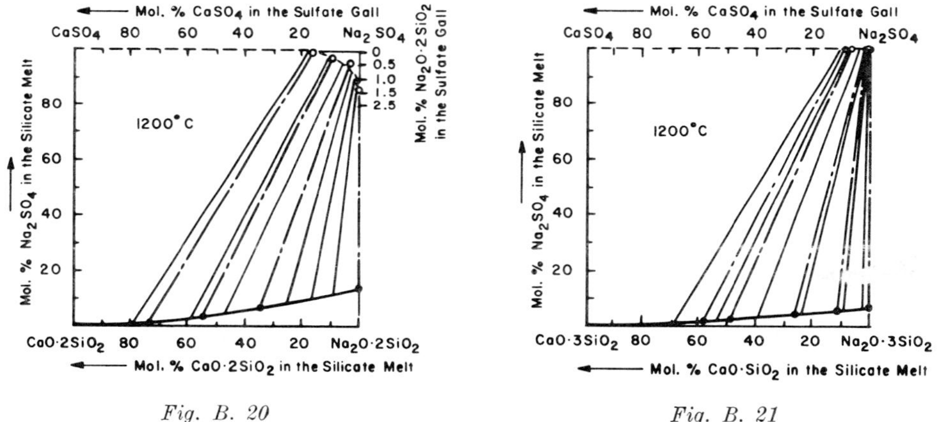

Fig. B. 20　　*Fig. B. 21*

FIG. B. 20 and 21. Reciprocal transition equilibria between disilicate, and trisilicate melts of sodium and calcium, and sulfate gall. (Kordes).

mochemical data and the ideal mass action law, including the distribution of the constituents of such glasses in the different phases:

For the Silicate Melt: $$(Na_2SO_4)_{Sil.} = c \cdot \left[\frac{(Na_2O,\, n\, SiO_2)}{(CaO,\, n\, SiO_2)}\right]_{Sil.}$$

For the Sulfate Gall: $$(Na_2O \text{ and } SiO_2)_{Gall} = 1/c' \cdot \left[\frac{(Na_2SO_4)}{(CaSO_4)}\right]_{Gall} = k \cdot \left[\frac{(Na_2SO_4)}{(CaSO_4)}\right]_{Gall}$$

with the mole fractions of the compounds indicated by rounded parentheses. The interchanges between both melts are particularly well illustrated by considering the distribution of CaO in the liquids:

For the Silicate Melt: $$\left[\frac{(CaO,\, n\, SiO_2)}{(Na_2O,\, n\, SiO_2) + (Na_2SO_4)}\right]_{Sil.} = \left[\frac{CaO}{\sum Na_2O}\right]_{Sil.}$$

For the Gall Melt: $$\left[\frac{(CaSO_4)}{(Na_2SO_4) + (Na_2O,\, n\, SiO_2)}\right]_{Gall} = \left[\frac{CaO}{\sum Na_2O}\right]_{Gall}$$

The ratio of the summarized expressions is

$$\left[\frac{CaO}{\sum Na_2O}\right]_{Gall} : \left[\frac{CaO}{\sum Na_2O}\right]_{Sil.} = K = k\left[\frac{1+c}{c}\right]$$

with $K = 0.0226$ to 0.0155 for the metasilicates, and then for the unmixing liquids in the disilicates with $K = 0.060$ and 0.00126; in the trisilicates with $K = 0.040$ and 0.060 for the Na_2SO_4 contents in the silicate, and in the gall melts, respectively, as seen from Figs. B. 19, 20, 21. The reactions between Na_2S and SiO_2, or of Na_2SO_4 and SiO_2 were discussed further by I. S. Lileev and T. I. Avdeeva,[27] using as silica raw material the fine-granular quartz powder occurring as the Russian mineral marshallite. Also these authors observed formation of thiosilicates of the type $Na_2S.nSiO_2$ as Kröger and Vogel described them. In the reactions of marshallite and Na_2SO_3 the formation of a homogeneous liquid glass at 900°C. after a rich evolution of SO_2 was observed, the gas evolution being extended over a rather long time period. The formation of sulfur is increased with the ratio Na_2SO_3/SiO_2, the reaction starting at about 500°C.

28. The problems of glass galls (sulfate "salt water") seen from the practical viewpoints were also discussed by P. Beyersdorfer[28] in connection with the evolution of SO_2 and sulfur vapors from sulfate-silica mixes, in oxidizing and reducing furnace atmospheres, either at 1400°C. or at 900°C. With lead crystal glass batches the pres-

[27] *Trudy Khim. Met. Inst. Akad. Nauk S.S.S.R., Zapadno-Sibir. Filial,* **3**, 1949, 35-43; 45-54; *Ceram. Abstr.*, 1954, 135, a, c.

[28] *Silikat Tech.*, **3**, 1952, 11-12.

ence of lead sulfate formed by a reaction of Na_2SO_4 is particularly troublesome,[29] and it is sufficient that SO_2 is taken up from the furnace atmosphere, namely the combustion products of sulfur containing producer gas, into the lead silicate melt to develop this harmful type of gall. To avoid this undesirable reaction it is advisable to keep the furnace atmosphere as neutral as possible so that SO_3 is not formed and absorbed in the melt. Such practical furnace control measures were further discussed by E. O. Schulz[30] citing the optimum temperature of introducing sulfate batches at 1430° to 1480°C. in maximum. Concerning the decomposition of calcium sulfate (introduced as gypsum rock) by reaction with carbon, R. R. West and W. J. Sutton[31] used the differential-thermal analysis technique in different gas atmospheres for experimental batches during heating. Analyses of the gases evolved by these reactions and X-ray diffraction identification of the reaction products confirm the observations of Kröger and Vogel in essential points and illustrate the sluggishness and incompleteness of the conversions even after a fusion started in the batch. The lowest temperature for the perceptible evolution of CO from a 20 per cent carbon and gypsum mix was 615°C. Characteristic of the inversion $\alpha \rightleftharpoons \beta$ $CaSO_4$ is the small endothermic peak at 1225°C., and a simultaneous slight dissociation of $CaSO_4$ to CaO,SO_2 and oxygen occurs. The eutectic of CaO and $CaSO_4$ melts at 1385°C. causing another endothermic peak. The formation of calcium sulfide is typical of gypsum heated in a CO atmosphere above 910°C., by the reaction $2CaSO_4 + 4CO \rightarrow CaO + CaS + SO_3 + 4CO_2$, and with more than 6 per cent carbon the reaction $CaSO_4 + 4C \rightarrow CaS + 4CO$ occurs, in addition to the reaction $CaS + 3CaSO_4 \rightarrow 4CaO + 4SO_2$.

29. By the presence of K_2SO_4 in glass galls a crystallization of *glaserite*, $(K,Na)_2SO_4$ (usually with the ratio 2 to 3 K : Na) is possible from a striaceous lead crystal glass as was described by N. Köppen and K. Wickert.[32] Such formations should not be misinterpreted as "stone" inclusions (cf. B. ¶ 212 ff.), they are true gall formations. Interesting also is the appearance of metastable (undercooled) high-temperature modifications (hexagonal, orthorhombic, or monoclinic) of sodium sulfate in place of thenardite which is stable at room temperature.

HOMOGENIZATION OF GLASS MELTS FROM BATCHES; FINING PROCESS

30. After the melting period of batch to crude glass, the processes of *homogenization and "fining"* are the most important stages in glass manufacturing, followed by a

[29] *Silikat Tech.*, **2**, 1951, 74.
[30] *Sprechsaal*, **83**, 1950, 213-216; 233-234; 258-259.
[31] *J. Am. Ceram. Soc.*, **37**, 1954, 221-224.
[32] *Glastech. Ber.*, **27**, 1954, 17-18.

period which is usually called "heat conditioning" for the last stage of making the glass the high-grade homogeneous material to be shaped. We wish to discuss these important steps together from the same viewpoint of physical-chemical processes, and refer to the report given by the German Society of Glass Technology, prepared by G. E. Rindone[33] in English translation, as one of the most essential recent investigations in this field. The report discusses in detail experience with gas content of the glass during the fusion, the development of bubbles and their ascension, growth, and bursting on the surface of the glass bath. In these prevailingly physical aspects of the fining process, the importance of the application of Stokes' law for the ascending gas bubbles is most obvious. It determines the speed of degassing of the melt as the phenomenon of primordial practical interest in our discussion. It is then more of secondary importance to consider how the ascension of the bubbles may be accelerated, e.g., by introduction of a current of an inert gas into the fining glass melt to accelerate the growth of the bubbles spontaneously developed, or by other mechanical or thermal means discussed in the report mentioned above.

31. J. Widtmann[34] used the "washing effect" of inert gases introduced through water-cooled nozzles of non-scaling steel, with a theoretical discussion of the gas pressure required for this purpose in equilibrium with the viscosity and surface tension of the glass melt. An interesting combination of the purely mechanical "washing" effect with a chemical fining (cf. below) is then possible, if the gas is not completely inert, but is absorbed by "acceptor agents," oxygen say by As_2O_3 and/or BaO present in the glass composition. These acceptors then form higher oxides and, by a "reboiling" effect at the higher temperatures of the fining chamber, help the homogenization by degassing. In place of a constant gas flow it was found profitable to apply the gas under repeated increased pressure impulses (50 to 100 pulsations per second).

32. G. Schilling and H. H. Franck[35] recommended a simple method of determining the degree of fining, to replace the microscopic enumeration method previously used by G. Gehlhoff, H. Kalsing, and M. Thomas (1931), which is best suitable for optical glasses containing only a low number of minute "seeds." Schilling and Franck base their determinations on routine (float) measurements of average density of the glass by suspension in liquids of accurately known density as recommended by M. A. Knight (1945), with an absolute accuracy of $\pm$ 0.001 for thoroughly annealed

[33] The original report prepared for a "Symposium on Fining," held in Paris in June 1955, was published in *Verres et réfractaires*, **9**, 1955, 82-97, and translated by G. E. Rindone in *Glass Ind.*, **38**, 1957, 489-493, 516, 526, 528, 561-565, 576-577, with a valuable bibliography of 66 references. On the formation and growth of bubbles in the glass fusion process, see also T. Izumitani and R. Terai, "Advances in Glass Technology," Plenum Press, New York, 1962, 205-216.

[34] *Glastech. Ber.*, **29**, 1956, 37-42.

[35] *Silikat Tech.*, **6**, 1955, 505-509.

glass samples, as in mixtures of toluene and bromoform. With increasing temperature of fusion and heat-conditioning, the densities increase, the rate of this increase being a function of temperature. An increase by 50°C. brings about the same seed volume in half the time needed for the initial temperature.

33. The problem of whether additions of *cullet* act as an improvement or as a hindrance to good homogeneity of the glass melt was extensively studied by W. Dietrichs, J. Löffler, H. Jebsen-Marwedel, and F. Wefer.[36] Cullet can influence the fining, especially by reactions of its contents in absorbed moisture and by the presence of sulfate (see later on sulfates as fining agents), and in a secondary measure by contaminations taken up from refractory material (indicated by raised alumina and Fe_2O_3 contents). A complete heat balance computation for cullet glass melts shows a distinct decrease in the specific input of thermal energy and of the total input in the charge. It requires, however, much technological experience to control the optimum, and permissible amounts of cullet additions to batches, the size of the cullet fragments, and "working" properties of the cullet glass. It must not be overlooked that the refractory walls of a glass tank furnace are much more seriously corroded if a pure batch is fused to prepare a glass of excellent quality, whereas in the fusion process with cullet glass this corrosion effects is usually rather low.

34. We have already learned in the discussion of the results of E. Kordes, B. Zöfelt, and H. J. Pröger[37] on sulfate batch reactions that, in the use of sodium sulfate as a raw material, the resulting glass of necessity must contain residual sulfate which therefore, fundamentally influences the fining mechanisms. A. Dietzel and O. W. Flörke[38] ascribed to the residual sulfate content of the melts a very favorable effect in the final dissolution of quartz grains suspended in the "pelt" or "skin" formed on the surface of the melt, because the surface tension is considerably reduced in the presence of sulfate anions, and in this way helps the accelerated finishing. The conditions of pelt formation may be compared with the flotation effects of a foam carrying solid particles adhering to gas bubbles. The sulfate containing melt, on the other hand, acts like a liquid with the addition of a depressor agent which impedes the adhesion of solid grains on the foam bubbles.

[36] *Glastech. Ber.*, **27**, 1954, 221-239; *Fachausschussber. deut. glastech. Ges.*, Fachausschuss III, No. 52, 1954. See also S. Kumar, *Central Glass & Ceram. Research Inst. Bull. (India)*, **8**, 1961, 23-29. The suspension method was considerably improved for glass routine density examinations by E. Plumat, *Verres et réfractaires*, **6**, 1952, 356-361, by centrifugation in heavy liquids. A very ingeneous method for measuring densities of glass samples is that of E. Plumat, *Silicates inds.*, **17**, 1952, 371-373, based on the different compressibilities of the glass and of the suspension liquids (mixtures of α-monobromonaphtalene and acetylene tetrabromide).

[37] *Z. anorg. u. allgem. Chem.*, **264**, 1951, 255-271, especially page 271.

[38] *Glastech. Ber.*, **32**, 1959, 181-185. See also "Comptes rendus du Symposium sur la fusion du verre," Bruxelles 1958, 465-474; *Veröffentl. Max Planck- Inst. Silikatforsch. Würzburg*, **19**, 1959, 107-114.

35. The German report on fining, mentioned above, enumerates in the section on *chemical* procedures the typical fining agents as previously classified in their chemical composition by R. Schmidt (1935),[39] with an excellent tabulation of the typical reactions of such agents in the degassing process. For the action of *sulfate* added only *for fining* (with the necessarily coupled addition of carbon) many of the reactions mentioned before are also useful here, especially in view of the possibilities of absorption of SO_2 from the furnace atmosphere. Ts. Okamura[40] gave a remarkable discussion of this phenomenon in experimental melts in SO_2 atmosphere at 1100° to 1400°C. to demonstrate the direct formation of a sulfate scum on the glass surface. Gas analyses over the whole thermal exposure of the melts showed a very complex interrelation of the amounts in SO_2 resorbed with those of the combustion gases (CO_2,H_2O). Glass which has absorbed SO_2 at 1100°C. presented a remarkable blue color (comparable with that of lapis lazuli or turquoise) and became paler in color with rising temperature. In every case, this typical bluish color is more intense than that observed in the interchange with oxygen and combustion gases. Also for the density of the SO_2 resorbing glass in the range from 1100° to 1200°C. complex relations were observed, as well as for the Liquidus temperature. For the particular role of SO_2 in comparison with CO_2 and oxygen, Okamura makes responsible the high dipole moment of SO_2 (1.69) applying the ideas of W. A. Weyl (1951) on the physical-chemical behavior of a glass surface.

36. The use of ammonium sulfate (about 3 weight per cent to be added) in place of common sodium sulfate as a fining agent was repeatedly proposed, by I. D. Tykachinskiĭ and G. V. Kataeva, later[41] by E. A. Engver, G. V. Kataeva, and M. F. Orlova, for a normal window glass melt. The most efficient fining agent was observed, when only a little Na_2SiF_6(0.18 F^-) were added to the ammonium sulfate containing mix, with a very remarkable increase in output and quality of the glass. The specificity of ammonium sulfate is explained by its melting effect (fusion point 350°C., decomposition to ammonia, water and SO_3 at 530°C.), in combination with "washing" by the evolved gases. With CaO in the batch, it reacts to form $CaSO_4$. Then in combination with the fluoride, the low-melting eutectic CaF_2—$CaSO_4$ (960°C.) of high fluidity thus intensifies glass formation and homogenization at the same time. G.

[39] "Die Rohstoffe zur Glas-Erzeugung," Akademische Verlags-Gesellschaft Becker & Erler, Leipzig 1943, 472 pp., especially pp. 427-466, with a very complete bibliography up to 1940. See also F. V. Tooley, "Handbook of Glass Manufacture," Ogden Publishing Company, New York, 1953, 506 pp., especially pp. 252-254.

[40] *Asahi Garasu Kenkyu Hokoku*, **5**, 1955, 113-125. On the effect of small additions of Na_2SO_4 on the fining process, see also C. Thorpe and F. Shaw, *Glass Technol.*, **3**, 1962, 135-139, further detailed studies of refining by M. Cable, *ibid.*, **1**, 1960, 144-154; **2**, 1961, 60-70, 151-158, and on the effects of the sand grain size, *ibid.*, **1**, 1960, 139-143.

[41] *Steklo i Keram.*, **15**, 1958 (2) 1-4, 4-5, 6-7. See also B. Simmingsköld and B. R. Jönsson, *Glastek. Tidskr.*, **10**, 1955, 151-159, 162-168.

Schilling and H. H. Franck[42] observed in the use of ammonium salts in general a particular red-brown coloring of the glass, especially for ammonium sulfate, chloride, and carbonate, but not with NH_4NO_3. This color effect is strikingly analogous to the well-known "coal-brown" (in reality FeS) discoloration in reducing atmosphere. In distinction from the coal-brown glass, however, the ammonium salts do not decolorize if the sodium-calcium glasses contain more than ten per cent CaO, nor with common Fourcault glass. Evidently, the ammonia in the decomposition products of ammonium salts dissociates partially, and then acts as a reducing gas evolving traces of intensely coloring sulfides. The range of chemical composition of glasses sensitive to this brown coloring is much smaller than it is for the typical coal-brown phenomenon (cf. B. ¶ 108).

37. E. Eipeltauer[43] studied the use of simple rock salt as a constituent of glass batches, not only for production of waterglass as batch material, or reacting with steam at high temperatures (as in the J. Kersten process, 1935), but also for fining, as V. P. Smirnov[44] demonstrated, especially for sulfide containing glasses, in combination with sulfate and coke powder.

38. Incomparably more important is the role of *oxygen acceptors* like As_2O_3, Sb_2O_3, and BaO for fining agents, as we observed above in the attempts of J. Widtmann on gas "washing effects." For the use of arsenic, Tr. Baak[45] gave recent data (including free energy computations) of the vapor pressure of As_2O_3 over glass melts and by combining them with analytical determinations of the $As_2O_3 \rightleftharpoons As_2O_5$ equilibrium as a function of the oxygen partial pressure. The free energy of the solution of both oxides could be calculated for the equation $\frac{1}{2}\,As_4O_6$ (gas) $+\ O_2$ (gas) $\rightleftharpoons$ As_2O_5 (gas), and ΔF as a function of temperature ($\Delta F = 0$ near 1000°K.). The much higher stability of As_2O_5 than that of As_2O_3 at high temperatures is evident. Losses of arsenic from the glass can be considerably reduced by oxidizing agents, say nitrates, or by introducing As in the pentavalent state to the batch. For determination of the vapor pressure of As_2O_3 over the molten glass, Baak used the dew-point method previously developed by R. Hargreaves, with an accurately known temperature difference between a high-temperature, and a low-temperature end of an evacuated tube, with the sample on the hot side (Fig. B. 22). The deposition of solid As_2O_3 (in the cubic form) on the tip of the thermocouple indicates exactly the equilibrium phenomenon at a measured temperature.

[42] *Silikat Tech.*, 7, 1956, 46-48. See also M. A. Bezborodov, *Central Glass & Ceram. Research Inst. Bull.* (*India*), 7, 1960, 54-57.

[43] *Silikat Tech.*, 9, 1958, 99-105; cf. A. Dietzel, *Glastech. Ber.*, 22, 1949, 246-248.

[44] *Steklo i Keram.*, 15, 1958 (7) 40-41.

[45] *J. Am. Ceram. Soc.*, 42, 1959, 27-29; on the Hargreaves method of vapor pressure measurements cf. *J. Inst. Metals*, 64, 1959, 115-125, for zinc vapors in brass alloys with 11 per cent zinc.

39. The important observation that gas bubbles in glass can be contracted, even completely disappear, by a reversal of heterogeneous gas reactions in the firing range, if on cooling a *resorption* of their gas contents takes place, was carefully studied by Ch. H. Greene and R. F. Gaffney,[46] further by Ch. H. Greene and I. Kitano,[47] for a

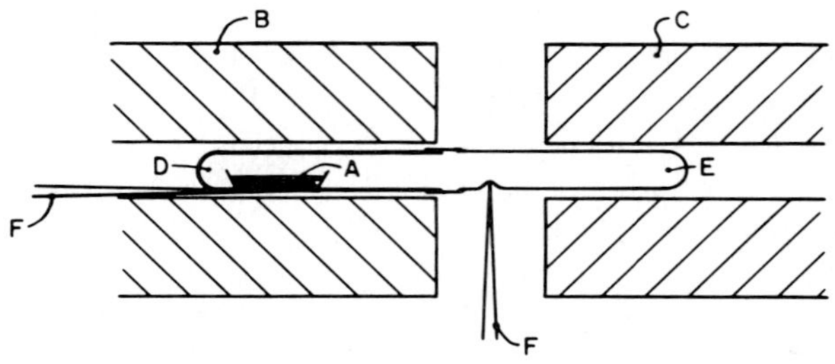

FIG. B. 22. Apparatus for the determination of the vapor pressure of arsenic over molten glass. (Baak). (*A*) Sample in porcelain boat; (*B*) high-temperature furnace; (*C*) low-temperature furnace; (*D*) and (*E*) evacuated tube; (*F*) thermocouple.

glass with a substantial amount of BaO (No. 9010, Corning Glass Works), with an artificial bubble drilled as a small depression in two nearly equal glass pieces and, therefore, nearly completely fitting together, then making a thermal exposure of it in a closed container as seen in Fig. B. 23, which was filled with oxygen and sealed. The resorption of the gas in the "bubble" walls is observed microscopically through a sapphire window in the furnace, as shown in Fig. B. 24, for temperature ranges

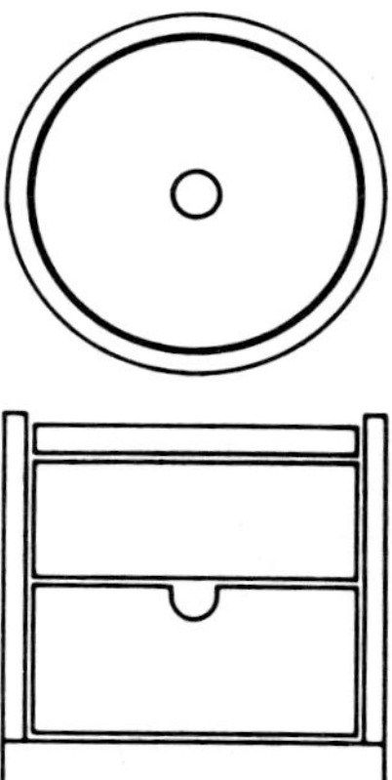

FIG. B. 23. Glass sample with artificial bubble in container. (Greene and Gaffney).

[46] *J. Am. Ceram. Soc.*, **42**, 1959, 271-275.
[47] *Glastech. Ber.*, **32**, K, 1959, V. 44-48.

from 1000° to 1300°C. The differential equation for the diffusion of oxygen away from the bubble (of the initial radius r_0) is given by Fick's law

$$\delta c/\delta t = a \nabla^2 \cdot c$$

or for spherical coordinates with the origin in the center of the bubble:

$$\frac{\delta c}{\delta t} = a \cdot \frac{\delta^2(cr)}{\delta r^2}$$

with the boundary condition $c = c_0$ on the surface. The calculation is made (a is the diffusion coefficient) more difficult by the so-called Stefan problem that the radius

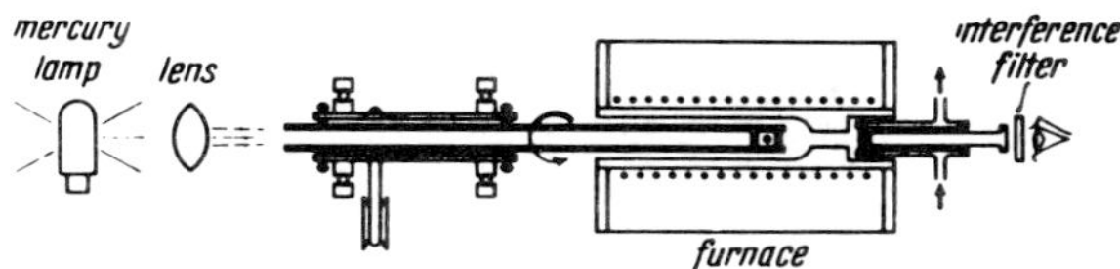

FIG. B. 24. Apparatus for observing bubbles in glass at high temperatures. (Greene and Kitano).

of the bubbles, r_0, on which the boundary condition $c = c_0$ holds, decreases with time in a way depending on the solution of the diffusion problem itself. An approximation is possible for solution of the Stefan problem by the equation

$$\frac{dr_0}{dt} = a \cdot c_0 \left(\frac{1}{r_0} + \frac{1}{\sqrt{\pi a t}} \right)$$

and finally for the increasing diameter of the bubble:

$$\frac{1}{r_0} - \frac{1}{r_1} = \frac{c_0}{r_1^2} \left(\frac{at}{r_1} + 2\sqrt{\frac{at}{\pi}} \right)$$

taking into account the decrease in concentration gradient at the surface of the bubble which also results from its contraction but which causes the shrinking to become slower. In reality, the change of r_0 as a function of $\sqrt{at}$ is less curved than shown by this equation alone. Approximation is then better (for t small so that $r \gg \sqrt{at}$) by

$$r_0 = r_1 - \frac{2\, i_0}{\sqrt{\pi}} \sqrt{at}$$

for a large portion of the bubble. In the initial stages of shrinking, the experimental results plotted as the bubble diameters in function of the square root of time show really such linear branches in the curves (Fig. B. 25) but are complicated by an abrupt change in direction after say 3 hours, to a nearly constant diameter. A very slow further shrinking occurred when its diameter was below 0.1 mm. perhaps due to an increasing internal pressure under the action of surface tension of the glass (pressure about 0.12 atm. to 1 atm.) with increasing temperature. The rates of decrease in bubble diameter are sligthly increased (Fig. B. 26).

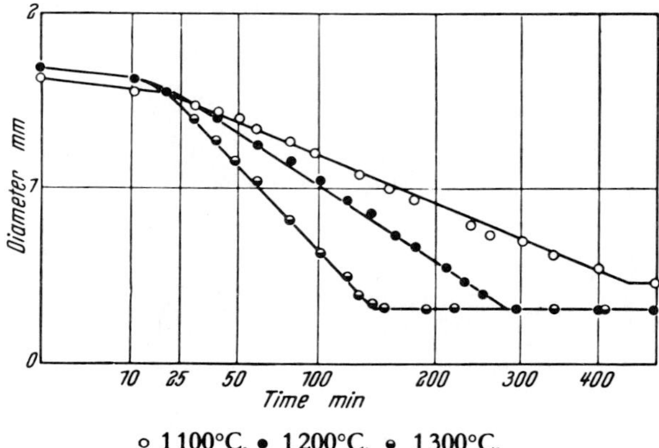

FIG. B. 25. Solution of oxygen in barium-aluminum-alkali silicate glass. (Greene and Kitano).

40. The diffusion of oxygen from contracting bubbles in molten glass was also investigated by R. H. Doremus,[48] and an equation was developed which is valid until about 80 per cent of the oxygen content of the bubble have diffused outside. The expansion or contraction of a bubble is given by the relation of solid flow, J, to initial

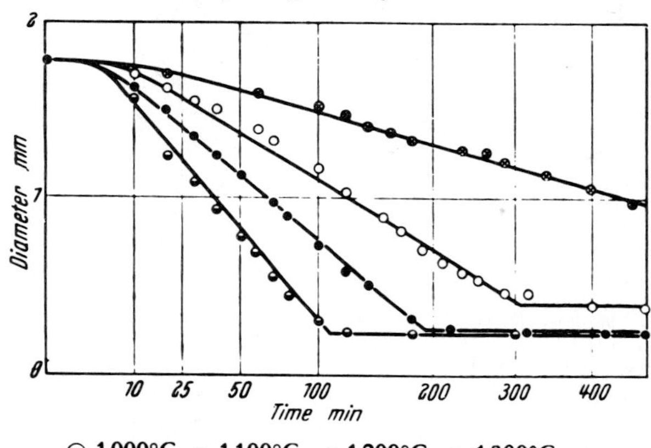

FIG. B. 26. Solution of oxygen in borosilicate glass, at 1000° to 1300°C. (Greene and Kitano).

[48] *J. Am. Ceram. Soc.*, **43**, 1960, 655-661; more recently, *Phys. & Chem. of Glass*, **3**, 1962, 127-128.

radius, R_0, and instantaneous radius, R, by

$$J = \frac{(c_\infty - c_i) \cdot D}{R} \left(1 + \frac{R_0}{\sqrt{\pi Dt}}\right)$$

as a satisfactory approximation of the not exactly soluble mathematical differential equation for shrinking bubbles. The equations are referred to the integrated form

$$R_0^2 - R^2 = 2D\beta t \left(1 + \frac{2R_0}{\sqrt{\pi Dt}}\right)$$

in which $\beta = (c_i - c)/(c_s - c_i)$, and in the case of the reaction on the surface of the bubble:

$$J = G(c_e - c_m)^n = (c_e - c_m) \frac{dR}{dt}$$

with the growth constant, G, c_m the concentration of the solute in the solution, c_e the solute concentration in the glass in equilibrium with oxygen at the pressure in the bubble. From this latter equation a linear relation between R and time is found, although only in the ideal case of the validity after an initial incubation time in which contraction is not ruled by an interfacial reaction.

41. The surface tension exerts an influence on the pressure in the bubble only for bubbles smaller than 0.5 mm. in diameter when $c_i = c_e$, being directly propor-

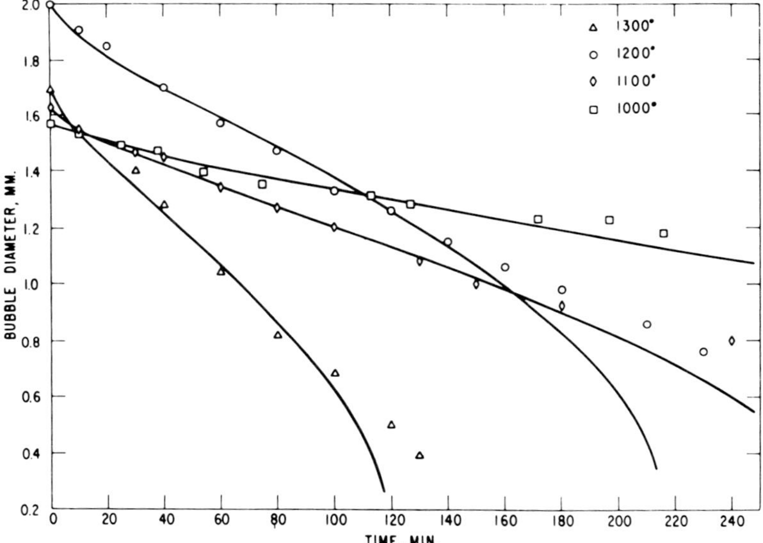

FIG. B. 27. Contraction of bubbles containing oxygen, in molten barium-aluminum-alkali silicate glass (Doremus).

tional to the oxygen pressure in the bubble as observed in special experiments of Greene and Kitano from which the Figs. B. 27 and 28 are taken. The rates of contraction of the bubbles observed slowed down sharply at a diameter between 0.3 and 0.5 mm. The assumption that the presence of foreign gases in the bubble (less than 1 per cent of the initial oxygen volume) influences the contraction effect is

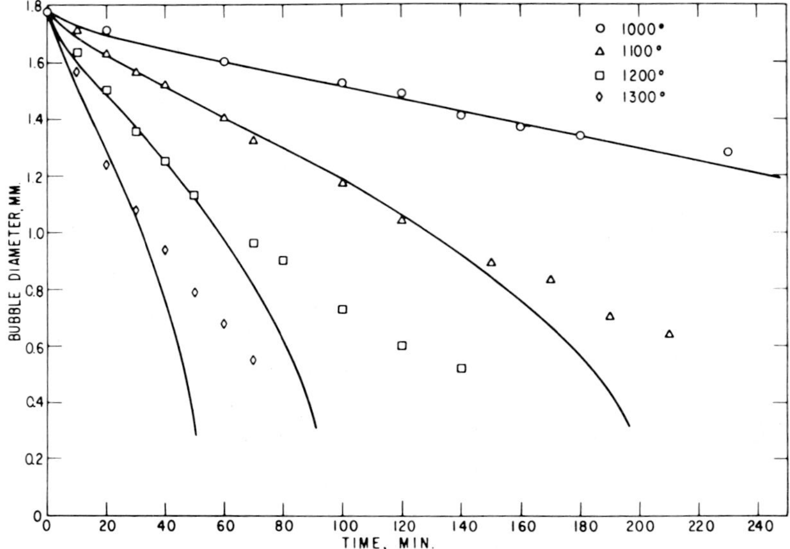

FIG. B. 28. Contraction of bubbles containing oxygen, in molten borosilicate glass. (Doremus).

only slightly plausible. If $\sqrt{\pi Dt}$ is much larger than R_0, the relation $R_0^2 - r^2 = D\beta t$ is useful for a certain temperature range and straight lines are found for the diffusion coefficients obeying the exponential law $D = D_0 \cdot e^{-Q/RT}$ (Fig. B. 29). It is difficult to see any direct relation of viscosity of the glass to the diffusion coefficient of oxygen in it, as the diffusion coefficient should be determined by the Stokes-Einstein equation $D = kT/6\,\pi\eta r$ (r being the radius of the diffusing molecules). If $\eta = 1000$ poises at 1200°C., r would be 2×10^{-11} or 10^{-12} cm., an impossible result which indicates the possibility of a formation of higher-associated complexes than the single molecules assumed. The movement of the oxygen molecules would be better explained by a kind of movement through "open channels" than by that of spherical particles through the viscous liquid, in analogy with the diffusion of helium through glass (cf. F. J. Norton[49]); (Vol. IV, Section A).

[49] *J. Am. Ceram. Soc.*, **36**, 1953, 90-96; see also T. B. King and P. J. Koros, Chapter 12, pp. 80-85 in "Diffusion in Liquid Silicates," Editor W. D. Kingery, Technology Press, Massachusetts Institute of Technology, J. Wiley & Sons, Inc., New York, 1959, 326 pp. King and Koros work-

42. It is probable that in experiments of Greene *et al.* the temperatures were so high that oxygen was already dissociated to the atoms which then easily migrate through open spaces in the glass. The "foreign gases" had much larger molecules than oxygen. They may have been CO_2 and/or H_2O, with diffusion constants of

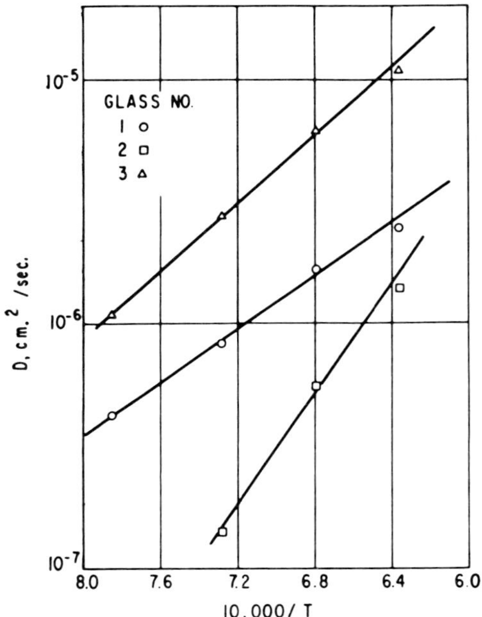

FIG. B. 29. Diffusion coefficients for oxygen in different molten glasses, at different temperatures. (Doremus). Glass No. 1 is a barium-aluminum-alkali silicate glass; No. 2 a sodium-calcium silicate glass; No. 3 a borosilicate glass.

2×10^{-9} cm.²/sec. at 1200°C. (cf. A. J. Moulson and J. P. Roberts; Vol. V, Section B).[50] As Greene, Kitano, and Doremus emphasize that in fining process not only a removal of bubbles takes place by rising to the surface, but also by the "resorption" of seeds due to diffusion of their gas contents into the glass, namely if there is an acceptor mechanism by As_2O_3 or Sb_2O_3, the latter (see above) increasing the oxygen pressure diffusion into the glass, thus accelerating the rate of fining. Doremus is of the opinion that the contents in As_2O_5 of the glass are fixed early in the fusion process and remain unchanged during the fining period. The ability of As_2O_3 to

ed with an isotope tracer diffusing through a melt of calcium aluminosilicate. W. D. Kingery and J. A. Lecron, *Phys. & Chem. of Glasses*, **1**, 1960, 87-89, determined the diffusion of O^{18} in glass spheres to study the oxygen mobility in silicate glasses.

[50] *Nature*, **182**, 1958, 200-201.

reduce the number of bubbles in the glass cannot be ascribed to its decomposition during melting. The alkali nitrates if present are aids in removing the bubbles (see above) but even this assumption is rather doubtful in view of the rapid decomposition of the nitrates above 500°C., to form with arsenic alkali arsenates, such as $K_2As_2O_7$. The real accelerating action of fining exerted by nitrates, however, will most probably consist in the increase in oxygen amount in the glass and a stimulation of the acceptor effects of As_2O_3. It is, therefore, this removal of oxygen dissolved in the glass more than the increase in gas volume evolved which characterizes the "fining aid" action of As_2O_3 and Sb_2O_3. If NaCl is used as a fining agent in the place of As_2O_3, the resorption mechanism from the bubbles is much different, as seen in the curves of Fig. B. 30.

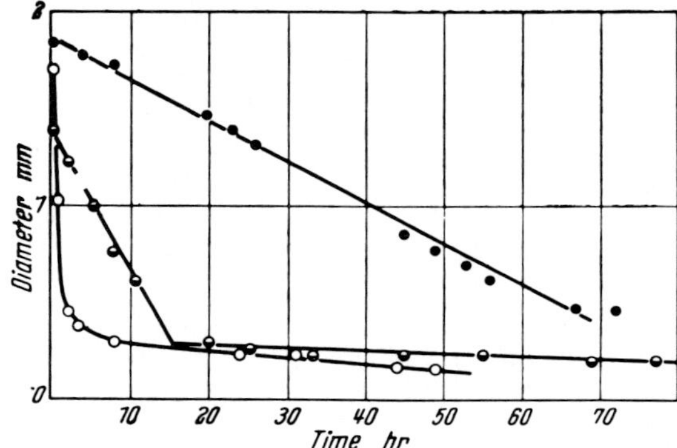

Glass 3: ○ 1200°C, glass 4 ● 1200°C, ◐ 1300°C.

FIG. B. 30. Solution of oxygen containing bubbles, in molten borosilicate glass. The time scale is linear. (Greene and Kitano). The glass was fined with sodium chloride.

43. From the merely practical viewpoint, studies on the *flow* of the glass in tanks through the fusion, fining, and conditioning sections are of high importance for localization of the fining reactions determining the presence or absence of the seeds suspended. In this field, model esperiments have often been applied. We mention, in this connection, the studies of H. Shimada,[51] with complete analyses of the local distribution of striae heterogeneities in the single sections, and a calculation of the apparent diffusion coefficient of CaO and ZnO, to illustrate the real diffusion phenomena and mixing effects (homogenization) by convection currents. This was done

[51] *Glastech. Ber.*, **27**, 1954, 151-159. Shimada's calculations concern electric-heated tanks, with specifically varied potential fields. For the theory of model experiments for tank flow conditions see the previous calculations and the excellent experiments of I. Sawai, *Bull. Inst. Chem. Research, Kyoto Univ.*, **30**, 1952, Sept., 1-14. See also M. Kunigi, K. Takahashi, and I. Sawai, "Advances in Glass Technology," Plenum Press, New York, 1962, 165-174.

under the assumption of quite definite boundary conditions. Also S. Kruszewskiĭ[52] developed such convection patterns in glass tank models in the scale of 1 : 20, determining in a "fining factor" a combination of viscosity and time effects to characterize the design and operation of the sidewall, bottom, and surface currents and their speeds. He also described the thermal behavior and optimum conditions for applying the "throat." The use of Sr^{89} as radioactive tracer isotope element in a large tank furnace was performed in convection current studies by V. A. Dubrovskiĭ,[53] the tracer being added in quantities of 50 mc. activity in $SrCO_3$. Samples from the glass bath were systematically drawn. This is doubtless the most elegant and conclusive method of marking atoms without any intermediate agents for a study of behavior in any part of the glass bath, from the feeding of the batch to the machining portion. The hydrodynamic conditions of glass flow in the bath were recently investigated by A. R. Cooper,[54] especially in the interactions of laminar flow with secondary flow. If the flow rates are high, laminar flow transfer provides the dominant influence. With decreasing flow rates, the velocity gradients of laminar flow are diminished and modified until the parallel components of secondary flow prevail, which then increase the velocity gradients and by this effect accelerate the mixing and homogenization.

COLOR CORRECTION AND DECOLORIZING OF GLASSES

44. In his masterly monograph on colored glasses (1943) and his more recent discussions on this matter, W. A. Weyl[55] emphasized how much the colors in silicate *minerals* are analogous to those observed in artificial glasses. As a matter of fact, in their absorption spectra both groups of silicates show most impressive parallelisms. From the side of mineralogists, we note especially from the studies of S. V. Grum-Grzhimaĭlo[56] on rose-colored and yellow topaz from the Southern Ural, or of multicolored beryls, tourmalines, and the like, how much these have in common with some industrial and laboratory-made glass melts, notably the role of manganese in the phosphates of the apatite group, as Z. V. Vasil'eva[57] demonstrated. The detailed analysis of spectrophotometric absorption curves in the range of wavelengths from

[52] *J. Soc. Glass Technol.*, **41**, 1957, 259-275.

[53] *Steklo i Keram.*, **14**, 1957 (6) 1-6.

[54] *J. Am. Ceram. Soc.*, **42**, 1959, 93-101. See also " Advances in Glass Technology," Plenum Press, New York, 1962, 217-229.

[55] *Penna. State Univ. Mineral Inds. Expt. Sta. Tech. Paper*, Final Report on "Color of Minerals," January 1954, 31 pp. Cf. Vol. II, Par. A. 347.

[56] *Zapiski Vsesoyuz. Mineral. Obshchestva*, **82**, 1953, 142-146; *Trudy Inst. Krist. Akad. Nauk S.S.S.R.*, **12**, 1956, 85-92 (with L. A. Pevneva); *Ibid.*, **12**, 1956, 79-84.

[57] *Zapiski Vsesoyuz. Mineral. Obshchestva*, **87**, 1958, 454-468.

200 to 1200 mμ is here particularly instructive and gives clear indications for the chromophoric groups [MnO_6], [CrO_6] or [FeO_6], and their influence on the position of the sharp near-ultraviolet absorption edge. Grum-Grzhimaĭlo emphasizes for such minerals application of polarized ultraviolet light to detect zonal structures and absorption anisotropies in crystalline phases like tourmaline, which in some cases are "colorless" but have a strong ultraviolet absorption $\omega > \varepsilon$. It is most characteristic that in some rose-colored and colorless zones in tourmalines, intensity ratios in the visible range and those in the ultraviolet are just reversed. Such phenomena are also particularly important for the study of changes in valencies which are so characteristic of iron, manganese, vanadium, cerium, and the like, either independently, or coupled with other "sensitizing" elements (cf. B. ¶ 59 f.). The absorption bands for Mn^{2+} (630 mμ) and Mn^{7+} (540 mμ) in the apatites studied by Vasil'eva are a particularly characteristic example for the first group of phenomena here to be mentioned.

45. Another highly variable group of mineral colors is described by N. M. Melankholin[58] in the emphiboles, with superposition spectra for the cations Fe^{2+} and Fe^{3+}, similar to common iron-colored glasses (see below), or the polarization effects exerted by alkalies (specifically sodium) on the Fe^{2+} centers in glaucophane, with a deep minimum of absorption at 500 mμ and a shifting of the maximum in the infrared and the red towards the violet end of the visible spectrum. Evidently, the position of the Fe^{2+} ions in [FeO_6] is different from that of Fe^{3+} in the same amphiboles.

46. For a correct characterization of mineral and glass colors the tricolorimetric method of spectrophotometric measurements and the plotting in corresponding diagrams (CIE type, based on Helmholtz' tricoloric theory) is of fundamental value, as it was used by B. Simmingsköld and B. R. Jönsson,[59] and more recently by M. Ihara and T. Yamamoto[60] for iron-manganese colored glasses. The diagrams represent very clearly the color change from blue to green with increasing total iron contents. The results illustrate also very well the influence of the partial pressure of oxygen with the constant $K[O]^{1/2} = [FeO]^2/[Fe_2O_3]$. With increasing total iron the ratio $[FeO]/[Fe_2O_3]$ decreases, and the yellowish tint of Fe_2O_3 becomes stronger than the blue tint of FeO. Also the black-white contents of the color are influenced by these changes: FeO darkens the glass relatively to Fe_2O_3 by a 9-times stronger effect.

[58] *Zapiski Vsesoyuz. Mineral. Obshchestva*, **85**, 1956, 218-223.

[59] *Glastek. Tidskr.*, **10**, 1955, 151-159, 162-168.

[60] *Yogyo Kyokai Shi*, **66**, 1958, 144-152. See also H. E. von Steinwehr, *Fortschr. Mineral.*, **33**, 1955, 150-151. On the 454 mμ absorption band see E. Kolbe, *Neues Jahrb. Mineral. Geol. Beilage-Bd.*, **69**, A, 1935, 183-254, especially pp. 232 ff. The oxygen potentials in colored glasses are extensivelly discussed by H. J. Tress, *Phys. & Chem. of Glasses*, **3**, 1962, 28-36; *Glass Technol.*, **3**, 1962, 95-106, 176-181, also for gold and copper glasses, or for cadmium sulfide-selenide glasses. On Redox equilibria in glasses see H. J. Tress, *Phys. & Chem. of Glasses*, **1**, 1960, 196-197.

The light transmittance is a linear function of the FeO content. This makes a relatively accurate determination of Fe_2O_3 in the range from 0.5 to 2.5 per cent possible, using linear changes in light transmittance. If iron *and* manganese are present at the same time, the ratio Mn/Fe rules the color tint. One may express the interactions of their oxides by the reactions

$$3\,MnO_2 + Fe_2O_3 \rightarrow Fe_2O_3 + Mn_2O_3 + MnO + O_2 \quad \text{for low ratios,}$$
$$2\,MnO_2 + 2\,Fe_2O_3 \rightarrow 2\,Fe_2O_3 + 2\,MnO + O_2 \quad \text{for medium ratios,}$$
$$MnO_2 + 3\,Fe_2O_3 \rightarrow 2\,Fe_2O_3 + 2\,FeO + MnO + O_2 \quad \text{for high Fe-ratios.}$$

47. The simultaneous presence of Fe^{2+} and Fe^{3+} ions which is so characteristic of many colored silicate minerals and industrial glasses (cf. Chr. Andresen-Kraft, 1931) was studied by E. Baier, W. Schefer, and H. E. von Steinwehr[61] in iron-colored

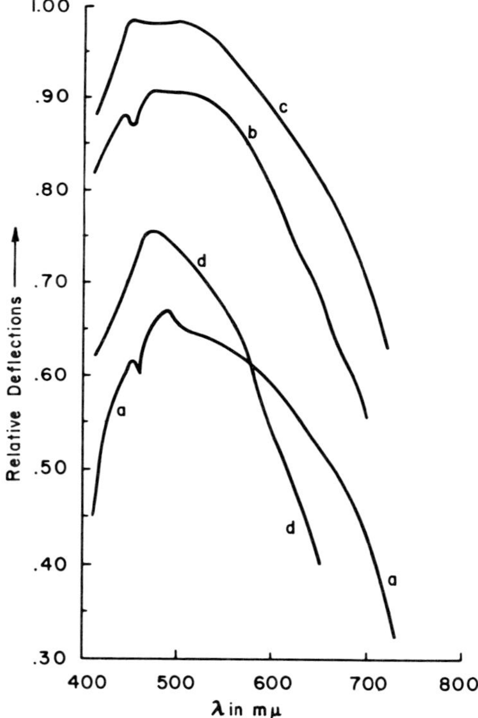

FIG. B. 31. Transmittance curve of iron-colored borax glass. (Baier, Schefer, and von Steinwehr). Glass (*a*) and (*b*) contain 20.6, and 17.6 per cent of the total iron present as Fe^{3+} cations. Glass (*c*) and (*d*) contain 1.8, and 3.0 per cent of these, respectively.

[61] *Glastech. Ber.*, **29**, 1956, 247-251. It is interesting to mention here that the iron colors of *antimony* glasses, this element containing as a framework forming constituent, are much different

borax glasses (molten in pure Argon, and by dissolving purest iron wire in the borax) to explain the deep-blue colors of minerals like vivianite, $Fe_3(PO_4)_2.8H_2O$. The spectrophotometric examination was made over the range from 400 to 750 mμ in which the sharp absorption band at 454 mμ is most characteristic (cf. Fig. B. 31). This line is always present in borax glasses with slight contents in trivalent iron, but absent in the purest FeO-borate glasses of deep-blue color. The more viscous the melts the more stubborn are non-equilibria of iron metal with divalent and trivalent iron cations. Only in highly fluid melts are they eliminated and typical blue glasses appear which do not show the 454 mμ band with considerably more than 95 per cent Fe^{2+}. It is therefore definitely established that the well-known greenish iron colors are mixed colors. These conclusions are not only in agreement with older data of Andresen-Kraft but also with those of H. Moore and S. N. Prasad (1950) as A. G. F. Dingwall and H. Moore[62] showed in their extensive measurements of viscosity, with important conclusions on the coordination of iron in glasses of different colors (blue with Fe^{2+}, amber to brown with Fe^{3+}, green as mixed colors of glasses with both cations, and gray with magnetite, colorless with $[FeO_4]$ coordinations.

48. Andresen-Kraft (1931) emphasized the correlation of absorption characteristics in the visible range of iron-colored silicate glasses with their *magnetic susceptibility*. These relations were extensively studied by J. de Jong[63] and by Abd-El-Moneim-Abou-El-Azm.[64] We can here present only a very brief recapitulation of their many results in ultraviolet, visible, and infrared transmission curves, from which de Jong concluded that trivalent iron is a typical framework former in the coordination

from those in silicates, phosphates, etc., but are deep-red, with a fairly high transmittance in the infrared, but a nearly complete absorption below 650 mμ; cf. W. A. Hedden and B. W. King, *J. Am. Ceram. Soc.*, **39**, 1956, 218-222; cf. A. ¶ 222.

[62] *J. Soc. Glass Technol.*, **37**, 1953, 316-372 T, especially pp. 364-372; see also H. E. Simpson, *Glass Ind.*, **37**, 1956, 257-261, 280, on experimental window and cullet glasses with Fe colors.

[63] *J. Soc. Glass Technol.*, **38**, 1954, 57-83 T. On titanium and chromium colors see *Ibid.*, **38**, 1954, 84-88 T. The iron-oxygen equilibria in glass, including the experimentally important effects of platinum on the Fe^{2+}/Fe^{3+} equilibrium were recently discussed on a thermodynamic basis by T. Bååk and E. J. Hornyak, *J. Am. Ceram. Soc.*, **44**, 1961, 541-544 (cf. Vol. III, B. ¶ 315, footnote 492). On magnetic resonance spectroscopy for the study of electronic and structural properties of glasses see R. F. Tucker, "Advances in Glass Technology," Plenum Press, New York, 1962, 103-114. Magnetic susceptibility measurements up to annealing temperatures were made by C. R. Bamford, *Phys. & Chem. of Glasses*, **1**, 1960, 143-147 (with H. Charnock), 159-164, 165-169; **2**, 1961, 163-168; **3**, 1962, 54-57, 189-202.

[64] *J. Soc. Glass Technol.*, **38**, 1954, 101-145, 146-196, 197-243, 244-270, 271-276. The influence of valence changes in iron glasses has its parallels in *copper*-colored glasses, investigated by A. Ram, S. N. Prasad, and V. K. Vaish, *Central Glass & Ceram. Research Inst. Bull.* (*India*), **6**, 1959, 86-88. Glasses with Cu^{2+} ions are deep-blue, those with Cu^{+} nearly colorless. It is erroneous to assume that the entire amount of CuO added to a glass batch is present (under normal melting conditions) as Cu^{2+} ions. See also A. Ram and S. N. Prasad, "Advances in Glass Technology," Plenum Press, New York, 1962, 257-269.

$[FeO_4]$ but that even divalent iron (besides the normal $[FeO_6]$) may also occur in tetrahedral coordination with oxygen. The measurements of magnetic susceptibility make most probable from the analogy with cobalt-containing glasses that the bonding mechanisms both for divalent and trivalent iron ions are mainly ionic. Most interesting is the distinct *solarization* effect observed by de Jong for a glass with only 0.1 per cent Fe_2O_3 as seen from Fig. B. 32. The even more elaborate studies of Abd-

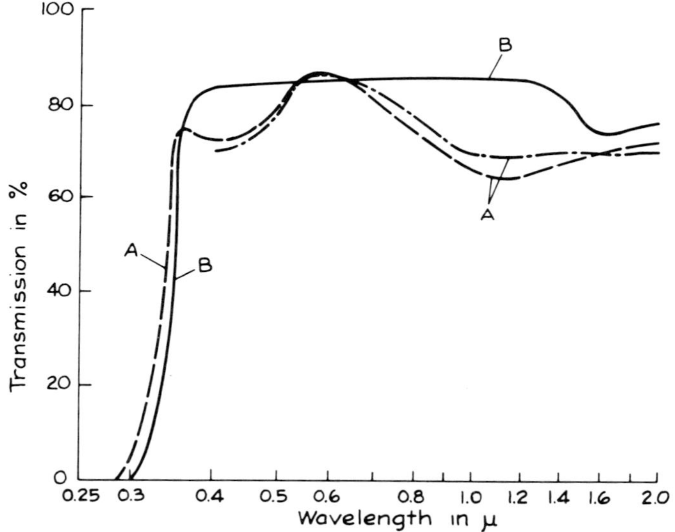

FIG. B. 32. Transmittance curve of a sodium-barium borosilicate glass, with the addition of 0.1 per cent Fe_2O_3. (*A*) is 2.4 mm. thick, and was melted under normal furnace conditions, (*B*) is 2.8 mm. thick, and was melted in oxidizing atmosphere, for comparison (dash-dot curve), with the same glass after solarization. (de Jong).

El-Moneim Abou-El-Azm concern the different forms in which iron exists in glass, based on the colorimetric analysis as Moore and Prasad had done before, and a redefinition of the specific optical density for pure Fe^{2+} or Fe^{3+} containing glasses.

49. To explain the colors of mixed glasses, the ferric oxide giving the glass a brown color, as well as magnetite which colors it gray, are colloid-dispersed suspensions, with a paramagnetic mass susceptibility of 425×10^{-6} at 20°C. for low concentrations, but a ferromagnetic susceptibility in higher amounts of iron after heating to 550°C. for 200 hours. The colorless glass with $[FeO_4]$ and the possibility of having ferrate ions $[FeO_4]^{2-}$ in highly alkaline fox-red colored glasses are also of basic importance. Many details were given to demonstrate the very characteristic influence of alkalies and alkaline earth ions present in the glass, on the variability in the contents of different forms of iron ions, especially the great influence of the introduction of soda, as carbonate, sulfate, oxalate, fluoride, or chloride, the effects of raised temperatures, the rates of heating, and of thermal exposure over longer periods, and so on. Concerning ferromagnetism of glasses containing an iron equivalent of more than 4 per cent Fe_2O_3, it is remarkable that no discrete suspended particles of magnetite could be detected. Further studies extended to the influence of trivalent and tetravalent metal oxides present at the same time on the different forms in which iron occurs in glass. In all these diversified compositions a "critical" one was found to exist, in which the amounts of Fe^{2+} cations are at a maximum, that of "colorless" iron a minimum. In this critical composition the ratio between the number of framework-forming and framework-modifying cations is somewhat above 1.5. In other words this corresponds to an arrangement in which the interstitial positions between six or more structural units are occupied by framework-modifying cations (of calcium, strontium, barium, cadmium), although some of the interstices may be too small to permit entrance of such cations. On the other hand, magnesium, zinc, and lead may participate in framework forming positions.

50. We cannot list here more details on iron-colored borate, boro-aluminate, and phosphate glasses, except to mention that, in principle, analogous regularities exist as observed in silicate glasses. Some phosphate glasses, however, behave somewhat particular in showing considerably different absorption curves which cannot be evaluated on the basis of those optical densities which were valid for the presence of iron in silicate or borate glasses. Iron can exist in phosphate glasses, as in a form of trivalent iron, covalently bonded with three oxygen atoms (phosphoferrate type, given in its spectral reflection curve). In Fe^{3+} phosphate glasses magnetic susceptibility is $= 180 \times 10^{-6}$, for the colorless ion it is 182.5×10^{-6}. This close analogy speaks for the presence of $[FeO_4]$ groups in such glasses of the ferric phosphate type.

51. The effects of increased alkali contents in silicate glasses, and of the replacement of sodium by potassium and lithium ions, were investigated in Fe^{2+} containing glasses, prepared under strongly reducing conditions, by A. Bishay.[65] In general, absorption over the whole range of the visible spectrum is increased with increasing alkali contents. This distinct effect is explained from results of magnetic susceptibility measurements in combination with the optical data by formation of groups of $[Fe^{2+}O_4]Na_2$, the alkali ions of which are held in close proximity to the central group. As such, they are framework participants and analogous to the $[Fe^{3+}O_4]$ groups or to $[Co^{2+}O_4]$ in deep-blue alkali silicate glasses. The magnetic susceptibility of such glasses is relatively very low and may be compared to that of hercynite-type spinels ($FeO.Al_2O_3$). The grayish tint of alkali-rich glasses with $[Fe^{2+}O_4]$ groups postulates

[65] *J. Am. Ceram. Soc.*, **42**, 1959, 403-407. A very recent investigation on iron containing Cabal (Calcium-boroaluminate) glasses in their transmission and magnetic susceptibility behavior see by A. Bishay, *Ibid.*, **44**, 1961, 16-21. Also here the conclusion is important that ferrous iron favors network-forming positions.

specific magnetic susceptibility properties, either in covalent, or in ionic-coordinate form:

	3d	4s	4p	Spin moment	Type of bonding
(*a*)	•• | • | • | • | •	××	×× | ×× | ××	4.9	coordinate
(*b*)	•• | •× | •× | •× | •×		| |	0.0	covalent

with basically different spin moments and responses. The extensive discussion of these conditions makes evident that as the alkali contents are increased (say from 0.258 to 0.516 moles R_2O) in fully reduced silicate glasses, about nine tenths of the iron remains in the coordination $[FeO_6]$ whereas one-tenth of the Fe^{2+} cations enter the framework as $[Fe^{2+}O_4]$ groups, with *covalent* bonding of O to Fe, with a zero moment. The absorption coefficients of blue FeO glasses with high alkali contents do not obey the Lambert-Beer law, as little as it does hold for the Fe^{3+} ions in glasses or in aqueous solutions of ferric salts. This is also the reason why Moore's and Prasad's calculations of optical density cannot be applied to iron in glasses of highly alkaline composition.

52. As a supplement we may mention the interesting observation of H. Hartmann and H. Brand[66] that for deep-brown colored iron-manganese glasses which are paramagnetic a particular change of the susceptibility with temperature is found; at 650°C. the susceptibility converges to zero and this effect is reversible. This phenomenon is discussed as indicating a change of the Fe^{3+} ions to complex $[FeO_4]^{2-}$ anions which are diamagnetic. At the same time there is a distinct increase in light transmittance for the brown iron glass.

53. A careful study on the color of *nickel*-containing glasses was made by H. Moore and H. Winkelmann.[67] The great contrasts of glass colors with those of aqueous solutions of nickel salts are striking (cf. A. ¶ 348), as observed in different binary alkali silicate, ternary alkali-lime silicate, borate, and borosilicate glass, and further in some non-silica, non alkali glasses with CaO, B_2O_3, and Al_2O_3 (called "Cabal" glasses). The transmission curves over the range from 300 to 2300 mμ show three very characteristic types, the simplest (Fig. B. 33) of which is that of the Cabal compositions, generally called the "brown" nickel type. Borates and borosilicate glasses belong in their transmission curves to the "green" type (Fig. B. 34), and alkali and alkali-lime silicate glasses with much variant color tints (brown to blue) to the "undulatory" type (Fig. B. 35). There are transitions from one type to the other,

[66] *Glastech. Ber.*, **27**, 1954, 12-15. See also P. W. McMillan, "Advances in Glass Technology," Plenum Press, New York, 1962, 333-347.

[67] *J. Soc. Glass Technol.*, **39**, 1955, 215-249, 250-286, 287-313 T.

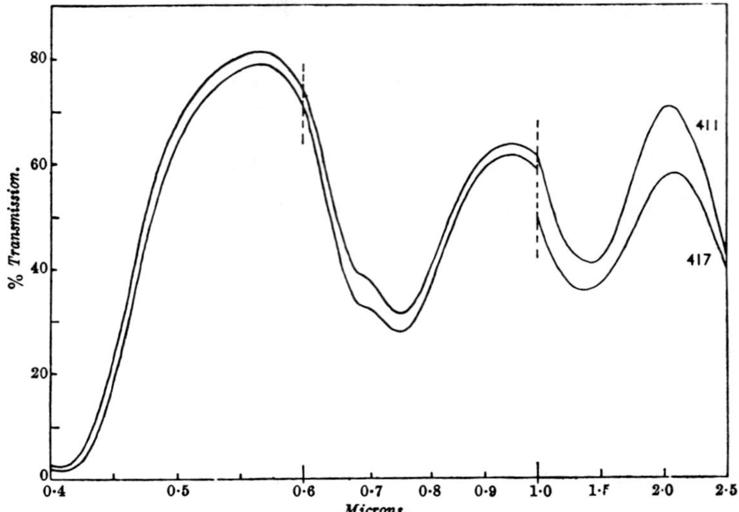

Fig. B. 33. Transmittance curve of a typical alkali borate glass ("green" type), with characteristic absorption bands at 450, 690/770, and 1200 mμ. (Moore and Winkelmann). No. 411 with 5 per cent Li_2O, No. 417 with 15 per cent K_2O, both glasses with 1 per cent NiO.

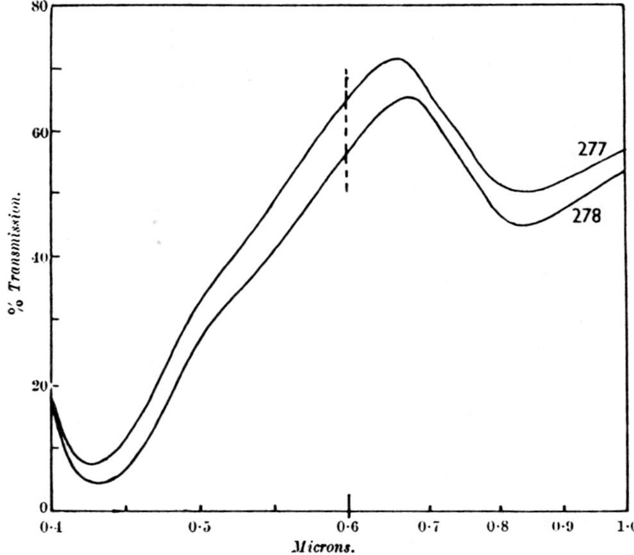

Fig. B. 34. Transmittance curve of two "Cabal" glasses, No. 277 with 54.8, No. 278 with 34.6 per cent B_2O_3, both glasses colored with 0.2 per cent NiO, constant ratio CaO/Al_2O_3 ("brown" type). (Moore and Winkelmann).

dependent chiefly on the alkali contents, but with highly complex relations between absorption characteristics and chemical composition. It is concluded that the green type of transmission curves corresponds to the simple $[NiO_6]$ coordinations, within the interstitial positions of the glass framework structure, whereas in the brown type the nickel ions are amidst two single-bonded oxygen anions between the structural framework units. For alkali silicate glasses with their undulatory color tints, Ni^{2+}

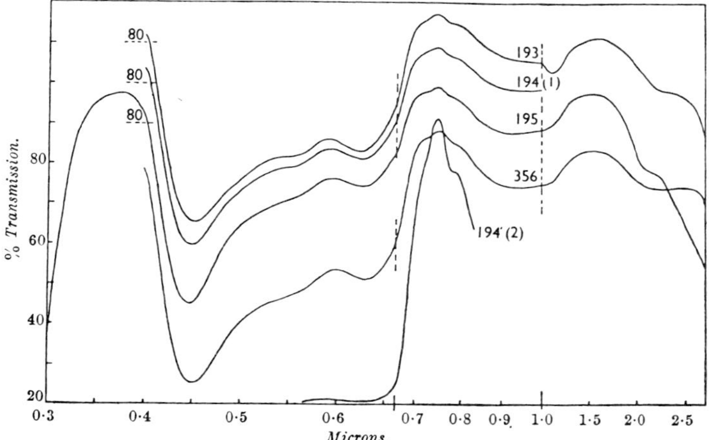

FIG. B. 35. Transmittance curve of five sodium silicate glasses, colored with NiO (0.05 per cent, but in No. 194, (2) 0.56 per cent). "Undulatory" type. (Moore and Winkelmann).

is associated with the alkali ions in complexes of the type $[NiO_4]^{2-}Na_4$ of tetrahedral configuration, with the alkali ions outside the faces of the tetrahedra. This complex may be built between four $[SiO_4]$ groups in its neighborhood in alternative configurations which may form a resonance system, the resonance point depending on the relative polarizing effects of the alkali and nickel ions on the electron systems of the four oxygen anions associated with these.

54. Interesting alkali-free, brown-colored *manganese* glasses were prepared by I. I. Kitaĭgorodskiĭ and V. A. Rishina[68] in compositions of the system anorthite-diopside-rhodonite (cf. Vol. III, B. ¶ 313, Footnote 488). The particular role which *titanium* plays in the coloring of glass was studied by V. V. Vargin,[69] especially in combination with iron colors (for 0.2 per cent of Fe_2O_3) in yellow-orange and bluish-green tints, if molten in oxidizing and reducing atmospheres, respectively. In a 20 per cent TiO_2

[68] *Trudy Moskov. Khim. Tekhnol. Inst. im. D. I. Mendeleeva*, 1956, No. 21, 61-66. On the influence of thermal treatment of Fe–Mn colored container glasses see G. Paoletti, *Vetro e Silicati*, **5**, 1961, (30) 14-19.

[69] *Doklady Akad. Nauk S.S.S.R.*, **103**, 1955, 105-106.

containing sodium-zinc silicate glass, yellow-brown colors appear which are reduced in their intensity, however, by the use of KNO_3 as an oxidizing agent. The more Fe^{2+} ions are present the deeper are the colors. If 0.5 per cent carbon is added, the spectral-photometric effect is similar to that observed in Ti_2O_3-phosphate glasses with a blue-violet color. Ti^{4+} exerts a strongly polarizing effect on iron ions in the equilibrium $[FeO_6]$ (blue) $\rightleftharpoons$ $[Fe^{3+}O_4]$ (yellow). Also G. S. Bogdanova and E. M. Orlova[70] studied the production of yellow-orange colored glasses on the basis of titanium-*cerium* compositions. Calcium- and borosilicate glasses are not suitable for this purpose, whereas compositions with zinc, lead, and barium gave very satisfactory results in the production of pressed filter glasses, for instance with 4 per cent TiO_2 and 2 per cent CeO_2. Cerium glasses, and others colored with rare earths, especially with neodymium and the like have an increasing usefulness as decolorizers and are in many respects superior to the common ones, such as As_2O_3 and Sb_2O_3, as K. T. Bondarev and V. A. Dubrovskiĭ[71] emphasized, not to speak of the excellent suitability of CeO_2 as a polishing powder for optical glass (CeO_2 in mixture with rare earths is the commercial "Polyrite").

55. The phenomena of *solarization*, glass coloring by solar irradiation, is an elementary case of the more general problem of high-energy radiation effects on glass, and was recently extensively studied especially because of its high significance in nuclear energy uses, for dosimetry, and so on. We discussed in Vol. I, C. ¶ 89 ff. the elegant method of making glass photosensitive by sensitizers among which CeO_2 is the most commonly used material. S. D. Stookey gave an excellent report in this field[72] and also N. J. Kreidl[73] presented a corresponding report on high-energy radiation in its effects on glass, including principles of radiation dosimetry by using color changes in glasses as indicators for the action and intensity of such radiations. Cerium inhibits in a high degree these coloration effects so as to develop "nonbrowning" or

[70] *Steklo i Keram.*, **15**, 1958 (5) 21-25; **16**, 1959 (3) 13-16.

[71] *Steklo i Keram.*, **15**, 1958 (2) 21-24. New experiments using selenium as decolorizing agent in container glass are described by H. R. Persson, *Central Glass & Ceram. Research. Inst. Bull. (India)*, **6**, 1959, 127-135, with the remarkable result that adequate annealing may be very efficient in color corrections.

[72] *Ind. Eng. Chem.*, **46**, 1954, 174-176. On glass scintillators for the detection of neutrons see A. Bishay, *J. Amer. Ceram. Soc.*, **44**, 1961, 231-233; D. G. Anderson, J. Dracass, T. P. Flanagan, and E. H. Noé, "Advances in Glass Technology," Plenum Press, New York, 1962, 429-441. On the ratio Ce^{3+}/Ce^{4+} in glasses as influenced by fusion conditions see V. Gottardi, G. Paoletti, and M. Tornati, *ibid.*, 412-423; *Vetro e Silicati*, **6**, 1962 (35) 12-16. On the role of Ce in suppressing γ-ray-induced coloration of borate glasses see A. Bishay, *J. Am. Ceram. Soc.*, **45**, 1962, 389-393.

[73] *Ind. Eng. Chem.*, **46**, 1954, 170-171. See also J. Paymal, *Verres et réfractaires*, **15**, 1961, 259-269, an excellent review on the n,α radiation reactions in glasses. Temperature effects during ultraviolet exposure of photosensitive glasses were recently studied by R. Yokota, "Advances in Glass Technology," Plenum Press, New York, 1962, 424-428.

"protected" glass especially shielding from γ-radiation. G. E. Rindone[74] comes back to the development of color centers by solar radiation in its specific difference from ultraviolet irradiation effects. In sunlight short-waved radiation causes the initial electron transfer, whereas the longer-waved rays promote the return of electrons to their original positions. These transfer phenomena depend very characteristically on the chemical composition of the glass, especially if photosensitive combinations are present as Ce—As in silicates.

56. The speed of the self-extinction, or the disappearance of the color centers if preserved in the dark, depends on the rigidity ("tightening") of the oxygen packing in the glass structure. Rindone presents a rich exhibit of curves to illustrate these general outlines of the solarization phenomenon in which the "oscillation of valencies," as between Fe^{2+} and Fe^{3+}, plays a particularly important role. Solarization cannot be explained alone on the basis of electron transfer; it is the firmness with which the electron is held in "trapped" positions which ultimately determines the changes observed in optical transmission curves and the activation energy required for restoring the electrons. If this activation energy is very low, little or even no change at all can occur. Schematically, Rindone explains the solarization in Ce^{3+} and As^{5+} containing glasses by addition of energy quanta, $h\nu_1$. The discoloring effect in the $Ce^{4+} + As^{3+}$ state of the color centers is most striking in the loss of blue fluorescence in long-waved ultraviolet, if by the energy addition $h\nu_2$ the return of the electrons to their original positions is indicated. The degree to which this reversal occurs in glasses of the type $Na_2O,RO,4SiO_2$ is different for variable R^{2+} cations. Little or no tendency for the electrons to return through the glass structure and a high activation energy is indicated in the following scheme by a short arrow, whereas a low activation energy of this restoring reaction corresponds to a long arrow:

Possible Solarization Process in $Na_2O,RO,4SiO_2$ (Rindone)

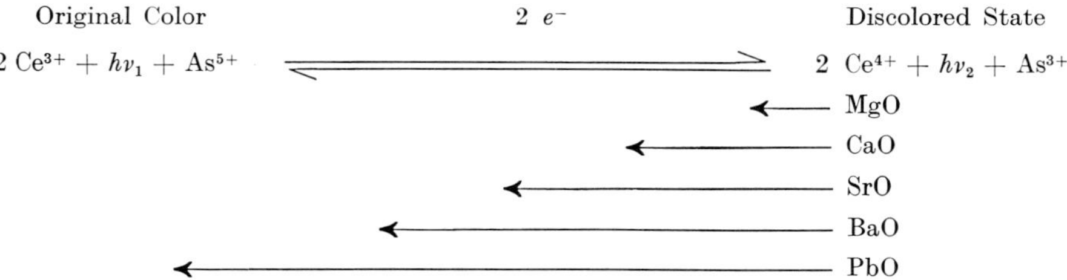

Iron and titanium, which damp the solarization effects of cerium + arsenic, are themselves able to trap electrons and thus interfere in their transfer between both photosensitive ions, decreasing the number of color centers in this way.

[74] *Trav. Congr. Intern. du Verre,* 4, Paris 1956, VII. 4, 373-389.

57. From practical viewpoints, B. Simmingsköld and B. R. Jönsson[75] made spectrophotometric measurements of solarized industrial glasses (after exposure in full sunlight for 60 days) and by plotting the results in tricoloric CIE diagrams. The most important changes were observed in glass containing arsenic or antimony, with the tendency to develop an equilibrium tint characterized by the optical "center of gravity" at $\lambda = 545$ mμ. CeO_2 caused much weaker color changes and the final equilibrium wavelength was shifted towards the ultraviolet. Additions of PbO to common soda-lime silicate glass compositions reduced the intensity of solarization, and above 3 per cent PbO no more trace of it was observed even when arsenic was added as a stabilizing agent. The presence of pentavalent As and Sb, even in very low amounts in weakly iron-colored glasses make it practically impossible to produce a pure soda-lime silicate glass which would not show solarization effects.

58. Before we enter the wide field of glass coloration by *high-energy radiation* we wish to emphasize, as an eminent introduction, the reports of Kuan-Han Sun and N. J. Kreidl[76] which extend over all types of radiation, namely atomic radiation with either neutral or electrostatic-charged particles, with a mass from zero (in the photon) to that of large fission fragments from uranium, and with energies of practically nil (in slow neutrons at low temperatures) to 10^{16} mev. (in cosmic ray particles). The basic physical-chemical effects of such radiation may become evident either in the creation of free electrons in the glass with their multiple reactions with the structural details in it, the excitation of luminescence, or the displacement of atomic nuclei and formation of vacancies. Since glass usually becomes dark-colored and opaque, then eventually ruptures under the action of irradiation intensities above 10^{10} Roentgen units, most investigations on glass color changes are made in the range below this intensity. Most conclusive have been the many parallels existing between the coloration of glasses and that of crystals such as of alkali halides, especially in development of color saturation by long irradiation, the disappearance of color by illumination with light, the self-extinction (fading) effects if preserved in the dark, or due to heating, and further in the phenomena of thermal luminescence and photoconductivity. Glasses containing cerium, thorium, silver, cobalt, manganese, or chromium are especially capable of thermoluminescence which is much more sensitive than the coloration process.

59. The specific action of *gamma-radiation* on sodium aluminosilicate glasses was described by M. Levy[77] in the superposition of several absorption bands, each one of

[75] *Glastek. Tidskr.*, **10**, 1955, 162-168. On the absorption spectra of solarized Mn^{3+}–V^{2+} glasses, see S. Kumar and P. Sen, *Phys. & Chem. of Glasses*, **1**, 1960, 175-180.

[76] *Glass Ind.*, **33**, 1952, (10) 511-514, 546; (11) 589-594, 614; (12) 651-653, 674, with extensive bibliography (170 references). See also G. E. Blair, *J. Am. Ceram. Soc.*, **43**, 1960, 426-429, on glasses for low- and high-level dosimetry.

[77] *J. Soc. Glass Technol.*, **40**, 1956, 18-24; 462-469 T. See also R. Yokota, *J. Phys. Soc. Japan*, **7**, 1952, 222-223, 316-321; *Phys. Rev.*, **53**, 896-897; **59**, 1954, 1145-1148.

these being analogous to those observed in the color centers corresponding to the F bands of alkali halides. The analogous centers, G, in glass show the same electron trapping mechanisms, namely in oxygen anion vacancies. Analogous inhibitions and sensitization effects are also observed. For the absorption bands Levy formulates the approximation function

$$\frac{e^{\mu d} - 1}{e^{\mu_0 d} - 1} = e^{-a(h\nu - h\nu_0)}$$

(with μ_0 = the absorption coefficient at the maximum of the band, $h\nu_0$ the photon energy at this maximum, d the thickness of the glass sample). This formulation was derived from the observation that absorption spectra recorded in terms of transmission, as a function of the photon energy, consist of bands each approximating closely to a Gauss distribution. This form of the equation is preferred since the method of plotting absorption spectra in terms of μ as a function of $h\nu$ gives a more direct measurement of the number of color states involved. For glasses on the basis of the composition $Na_2O,4SiO_2$, with variable contents in alumina, the absorption spectra are of the type shown in Fig. B. 36, with the bands A, B, C to the referred in A to G centers associated with Na_2O, in B to those associated with alumina, and in C to those associated with SiO_2.

60. We also mention the detailed studies of P. L. Baynton and H. Moore[78] in this connection. They referred to ultraviolet and X-ray radiation distinct effects

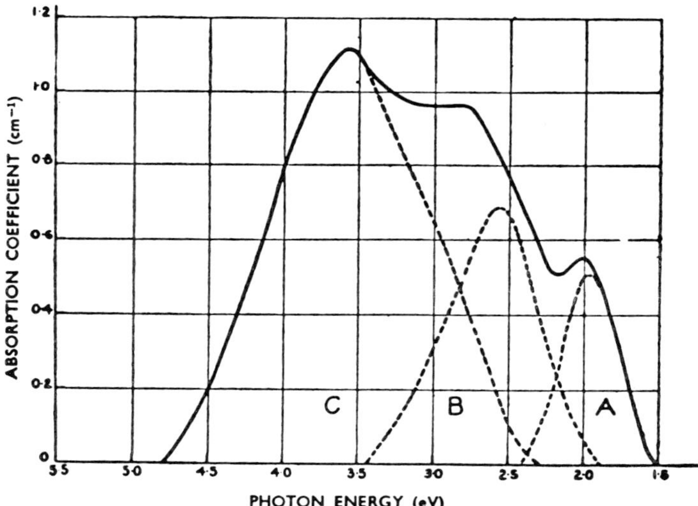

FIG. B. 36. Absorption spectrum of a sodium aluminosilicate glass, with three characteristic bands, A, B, C. (Levy).

[78] *J. Soc. Glass Technol.*, 40, 1956, 187-251.

in iron and manganese containing glasses, with the sensitizing agents Ti, Ce, and As, alone, or together, including also the characteristic fading of color in the dark (self-extinction, see above). The durability of industrial glasses under the action of higher energy gamma-radiation from Co^{60} was further discussed by S. M. Brekhovskikh,[79] by simple measurements of transmittance before and after the irradiation, with corresponding spectrophotometric evaluation of absorption curves. Colored glasses stained with selenium-cadmium, then with Co, Cu, or chromium did not change their absorption characteristics, and the transparency was not strongly affected. The energy of the free or excited electrons arising during the γ-action on the colored glass is expended in the Se-Cd glass, not in the reduction of Si^{4+} cations to Si atoms. The ions Co^{2+} and Cu^{2+} or Cr^{3+} change their valencies under the action, but it is restricted to a relatively small number of ions which cannot much change the color and absorption characteristics. Glasses with only 0.5 to 2 per cent CeO_2 (as they occur in the Corning glasses No. 8362, 8363, and 8365) are only slightly discolored by gamma irradiation, and may, therefore, be used in optical instruments for nuclear applications. Mass absorption of γ-rays below 1 mev. according to E. Umblia[80] are highest for PbO,Bi_2O_3, Ta_2O_5,WoO_3; high for BaO, and CdO. An efficient γ-protection glass is, for example, one with 60 per cent Bi_2O_3, 30 per cent PbO, and 10 per cent $B_2O_3 + SiO_2$. Thermal neutrons are most efficiently absorbed by glasses containing cadmium, boron, and lithium, with particularly large cross section values (see below), and also by the rare earth elements Sm, Gd, and Eu. Cadmium, on the other hand, shows a strong induced radio-activity. Glasses of the system PbO—CdO—B_2O_3 efficiently protect against mixed neutron radiation.

61. A complete discussion of the theory and specific action of α, β, and γ radiation from radioactive material, and that of X-rays and neutron radiation in interaction, as with lead glass, was given by W. Jahn[81] in a complex of ionization and induction effects, total absorption, Compton diffraction, photo-effects, and so on, especially for corpuscular electron radiation on phosphate glasses, and of γ-radiation from Co^{60} on optical (Baryte Crown), phosphate, and lead silicate glass. Particularly characteristic is the spontaneous bleaching of radiation-discolored glass in the course

[79] *Steklo i Keram.*, **15**, 1958 (1) 10-14; *Structure of Glass. Proc. All-Union Conf. Glassy State, 3rd, Leningrad* (English Translation), 1959/60, 314-318.

[80] *Glastek. Tidskr.*, **13**, 1958 (3) 67-70. See also studies of A. Bishay on γ-ray-induced absorption bands in lead borate glasses, *J. Am. Ceram. Soc.*, **43**, 1960, 417-421, and on anomalous γ-ray coloring effects in Ce-containing glasses, *Phys. & Chem. of Glasses*, **2**, 1961, 169-175.

[81] *Glastech. Ber.*, **31**, 1958, 41-54. See on the effects of impurities in silica glass determining its sensitivity to γ-radiation the recent studies of N. F. Orlov, *Structure of Glass. Proc. All-Union Conf. Glassy State, 3rd, Leningrad* (English Translation), 1959/60, 306-309. Silica glass prepared from synthetic quartz has a higher resistance to γ-irradiation than glass from melts of natural quartz. A certain residual "crystallinity" of the silica glass as dependent on the raw material processing and contaminations is made probable by these effects.

of time. The multivalent cations of lead and cerium (in amounts of 0.8 to 2 per cent) are important because of their electron-catching properties, and cerium as a sensitizer in photosensitive silver glasses (cf. Vol. I, C. ¶ 90). Both for development of radiation-protecting glass shields, and for *dosimetry* of high-energy radiation these facts are of the highest significance. They apply also to the fundamentally different mechanisms of neutron dissipation by B^{10} and Cd^{113}, which have the highest known cross section values, as applied to development of neutron-absorbing glasses, on the basis of beryllium-lithium borate compositions. Cerium may be added to eliminate the so-called "edge effect" caused by α-particles emitted during the capture of neutrons by boron, a process bringing about intense coloring of the edges and surfaces.

62. J. Paymal, M. Bonnaud, and P. Leclerc[82] especially described the application of element combinations like manganese-iron, manganese-vanadium and manganese-vanadium-iron for the systematic development of radiation *dosimeter glasses* with a sufficient color stability over the range from 10^5 to 2×10^7 Roentgen units. As an

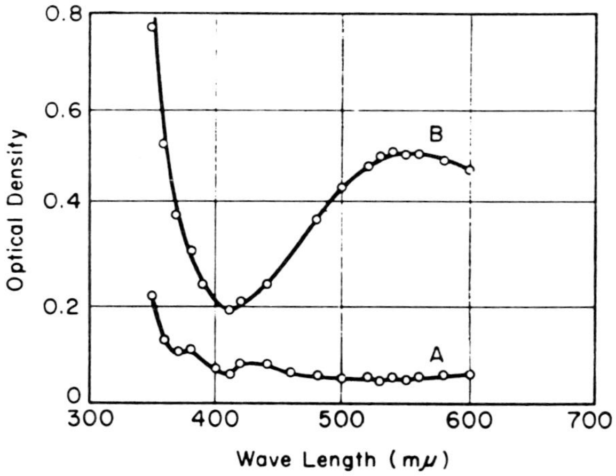

A unradiated B irradiated

FIG. B. 37. Absorption curves of a dosimeter glass with the sensitive combination Fe + Mn. (Paymal, Bonnaud, and Le Clerc). Irradiation intensity 0.9 megaroentgen units.

example, we note here in Fig. B. 37 the absorption curves of a dosimeter glass of the Mn-Fe silicate type, and in Fig. B. 38 its recovery at 20°C. after a strong irradiation, in Fig. B. 39 the absorption curve of a Mn—V glass with a characteristic

[82] *J. Am. Ceram. Soc.*, **43**, 1960, 430-436, and W. A. Hedden, J. F. Kircher, and B. W. King, *ibid.*, pp. 413-415. A. Bishay, *Phys. & Chem. of Glasses*, **2**, 1961, 33-38, developed a Bi–Pb borate glass for high-level γ-dosimetry. See also J. Paymal, *Verres et réfractaires*, **15**, 1961, 259-269, 341-351, and *Compt. rend.*, **252**, 1961, 1939-1941 (with B. Savouret and Kl. Leibrecht).

increase of the optical density at $\lambda = 540$ mμ as a function of the doses applied in Roentgen units.

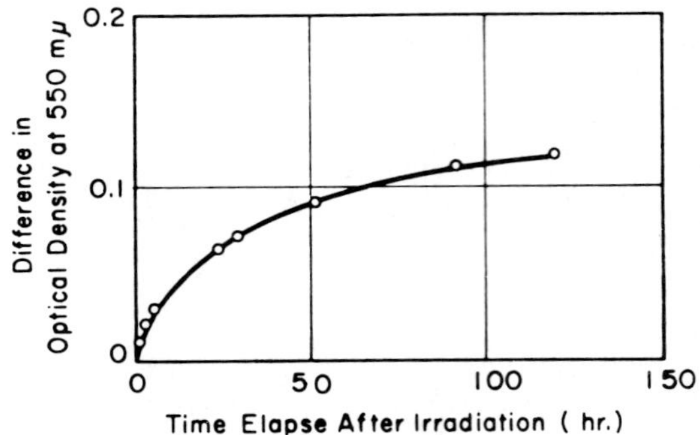

FIG. B. 38. Recovery at 20°C. of the same dosimeter glass, after irradiation by 3 megaroentgen units. (Paymal, Bonnaud, and Le Clerc).

63. For the fundamental problems of the *basicity-acidity* behavior of glass melts the use of coloring cations as *indicators* was previously found by W. Weyl and E. Thümen (1933), A. Dietzel and W. Stegmaier (1940), and especially by H. Lux and his coworkers (1942) to be an excellent experimental method. More recently, Fr.

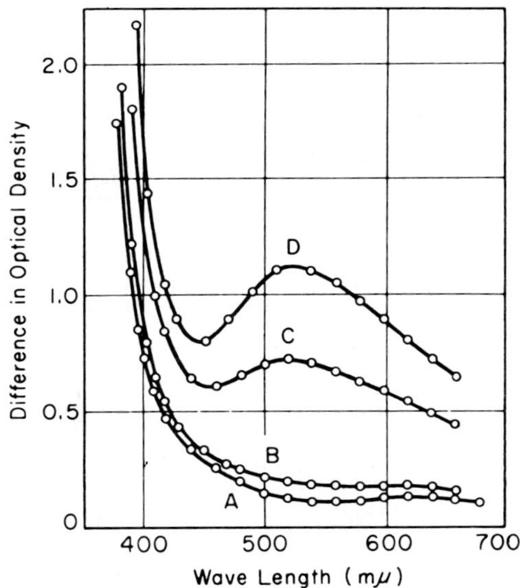

FIG. B. 39. Absorption curves for a dosimeter glass with the sensitive combination Mn + V. (Paymal, Bonnaud, and Le Clerc). Curve *A* for nonradiated glass, *B* for an irradiation by 1×10^5 Roentgen units, *C* for 1.48×10^6, *D* for 5.6×10^6 Roentgen units.

Irmann[83] investigated molten alkali borates of the systems of B_2O_3 with Li_2O, BaO, ZnO, and the Cr(VI)—Cr(III) indicator at 1200°C. The two characteristic cations Cr^{6+} and Cr^{3+} were finally determined by wet-analytical methods and their equilibrium concentrations as a function of the oxygen pressure calculated by the equation

$$K' = \frac{(c_6)}{(c_3)} \cdot p_{\mathrm{O}}^{2-3/4}$$

(from the reaction equilibrium of Cr^{6+} (melt) $\rightleftharpoons$ Cr^{3+} (melt + 3/4 O_2). The constant K' is an analogue to the H-function of protonic solids, as in cracking catalysts of the alumina-silica gel group.[84] If the equivalent function of the metal oxide RO in

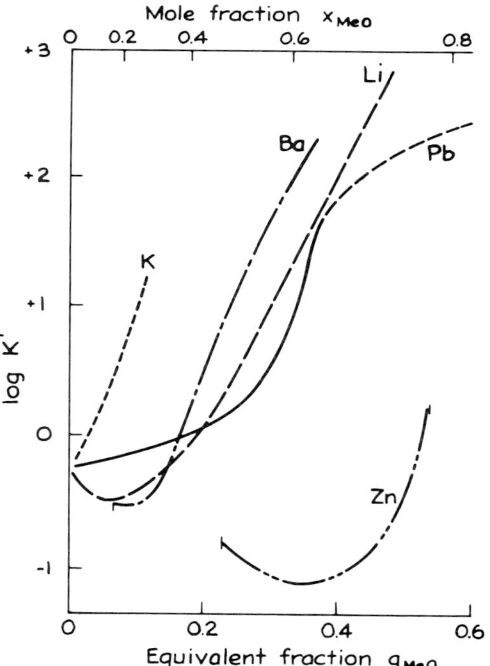

FIG. B. 40. Cr(VI) - Cr(III) indicator equilibrium constant, K', as a function of the equivalent fraction of metal oxides in binary borate melts at 1200°C. (Irmann).

[83] *J. Am. Chem. Soc.*, **74**, 1952, 4767-4770. N. I. Vlasova, E. I. Galant, and A. A. Kefeli, *Structure of Glass. Proc. All-Union Conf. Glassy State, 3rd, Leningrad* (English Translation), 1959/60, 327-330, used Co^{2+} as a color indicator for the determination of the coordination of B^{3+} and Al^{4+} in borosilicate and alumina-borosilicate glasses. The absorption spectrometric reactions are made in presence of Cl in the glasses (halide effect).

[84] On the so-called Hammett indicators and their general principles cf. L. P. Hammett and A. J. Deyrup, *J. Am. Chem. Soc.*, **54**, 1932, 2721-2739.

the borate melts is expressed by $q_{RO} = 2n_{MeO}/(6n_{B_2O_3} + 2n_{MeO})$, n denoting the number of oxide moles, for the five borate systems studied, the functional relation of log K' versus q_{MeO} is given in Fig. B. 40, with characteristic minima in the borate system with the highly polarizable Zn^{2+} cation at q_{ZnO} of about 0.35, for Ba^{2+} at about 0.15. PbO behaves quite particularly; up to q_{PbO} of about 0.2 there is only a very low change in log K'. A similarity is indicated for lead, boron, and silicon in their oxygen availability for bond formation, observed also in their magnetooptical behavior, from measurements of H. Cole[85] who concludes the presence of tetra-covalent lead in lead silicate glasses. The striking steep increase in the curve at q_{PbO} = 0.3 corresponds to the anomaly of surface tension in melts of lead borates as observed by L. Shartsis, S. Spinner, and A. W. Smock.[86]

64. Also the determination of the electromotive force (EMF) of galvanic concentration cells of the type Pt in O_2/(silicate solvent)/(silicate solvent + different oxides)/Pt in O_2 (cf. A. ¶ 171) was applied by R. Didtschenko and E. G. Rochow,[87] for the determination of the oxygen anion activity, as in $PbSiO_3$ melts as solvents, to characterize the acidity and basicity of the different oxides. The negative potentials of the alkali oxides, MgO, CaO, SrO, BaO (in increasing order) are contrasted with the positive potentials of the acid oxides $BeO < Al_2O_3 < B_2O_3 < SiO_2$, both groups on one smooth curve (Fig. B. 41). The alkali oxides R_2O lie on another curve shifted more to the negative potentials and show a distinct minimum for K_2O, whereas cations with 18-electron outer shells lie on a third hyperbolic curve, displaced to positive potentials, presumably because of greater polarizing actions.

65. In a fundamental investigation, W. A. Weyl[88] together with T. Förland and M. Tashiro, and later with G. S. Smith and G. E. Rindone (see below), discussed the acid-base relationships in glass systems to find a concept equally applicable to aqueous systems, fused salts, glasses, and solids, further to develop and to use oxygen electrodes for the measurement of oxygen activities in these systems for an exact basis of such basicity-acidity phenomena. Weyl discusses chiefly previous results of H. Lux and his coworkers on the color indicator method with Cr(VI)—Cr(III), or Mn(III)—Mn(VI), or V(V)—V(III) color change reactions, but also the important question of the stability of isopolyacids in fused salts and the consequences of the possibility of an atomistic interpretation of the acidity in glass from the viewpoint of the *screening theory* (cf. A. ¶ 392 ff.). This defines acids as molecules containing incompletely screened cations, the acidity being inversely proportional to the degree of screening. The arbitrary selection of a definite composition (in aqueous solutions

[85] *J. Soc. Glass Technol.*, **34**, 1950, 220-237.

[86] *J. Research Natl. Bur. Standards*, **40**, 1948, 61-66.

[87] *J. Am. Chem. Soc.*, **76**, 1954, 3291-3294.

[88] *Glass Ind.*, **37**, 1956 (5) 264-269, 286, 288; (6) 325-331, 336, 344, 346, 350.

it is H_2O) for the "neutral point" to distinguish acids and bases is in every case no more suitable for glass systems in which the mobility of the protons equalizes the degree of screening within the whole system and there is no parameter equivalent to *pH*. In this respect the glasses may show some parallels with alumina-silica gel catalysts for cracking hydrocarbons (cf. Vol. I, C. ¶ 211 ff.) which behave as highly

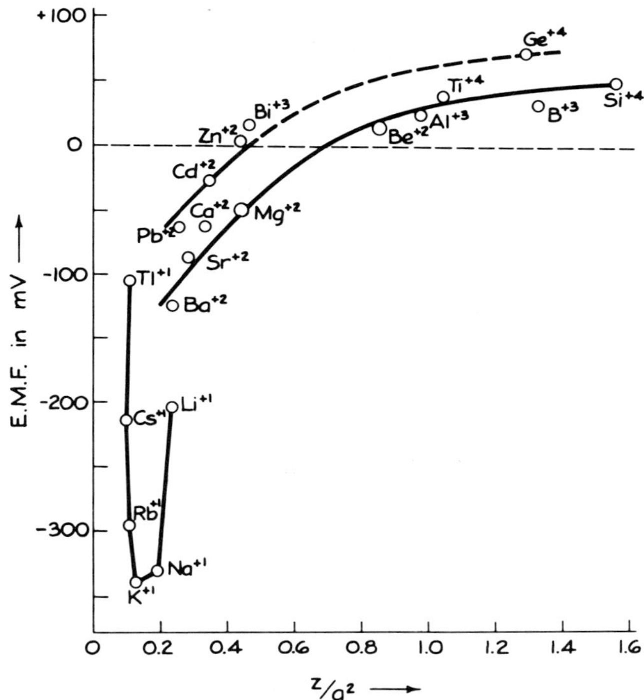

FIG. B. 41. Potentials as a function of the z/a^2 values for different cations, at constant concentration of 0.2 mol. and temperature of 900°C. (Didtschenko and Rochow).

acidic systems due to their incompletely screened cations. In the same way, also "intermediate states" in incomplete solid state reactions of silica with oxides, and "acids," although of indefinite acidity because of the variability of local positions concerned, are highly unscreened.

66. T. Förland and M. Tashiro[89] constructed a galvanic concentration cell with two oxygen electrodes, and a combination of two glasses as electrolytes, of the type

Ag (in air) / Glass I / Glass II / Ag in air

[89] *Glass Ind.*, 37, 1956 (7) 381-385, 399, 402.

as schematically shown in Fig. B. 42. The experimental glasses contained only one kind of alkali ions but in different concentrations. If an alkali earth cation is added the basicity of the glass is strongly reduced. Taking the measured EMF as a measure of basicity (if the potentials are negative), and of acidity (if they are positive)

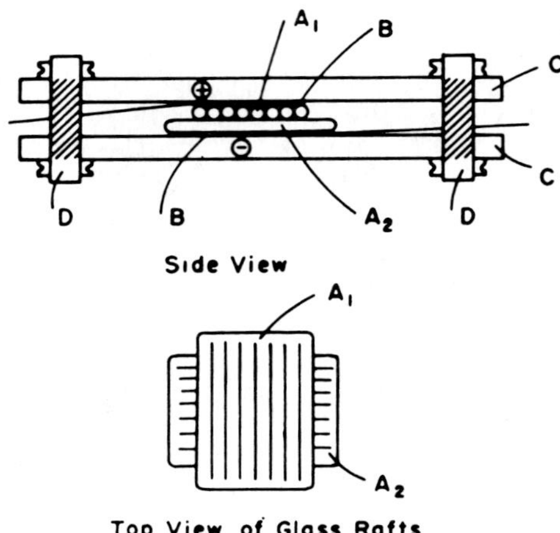

Fig. B. 42. Galvanic concentration cell with oxygen electrodes, for measurements of the acid/base properties of glasses. (Förland and Tashiro).

(see above), the characteristic diagrams of basicity as a function of the moles silica in $Na_2O,xSiO_2$, as shown in Figs. B. 43 to 45, result. In Fig. 45 particularly characteristic is the maximum with a relative concentration of $x = 0.5$ for K_2O in mixed sodium-potassium silicate glasses of the type $(1 - x)Na_2O,xK_2O,3SiO_2$. The interpretation of the basicity concept is given by the partial free energy, $\bar{F}$, of Na_2O in the glass, implying also the entropy of mixing. Arbitrarily, the composition Na_2O, $3SiO_2$ may be used as the "standard state." The different EMFs measured in concentration cells with increasing silica contents the difference $\Delta\bar{F}_{Na_2O} = \bar{F}_{Na_2O(II)} - \bar{F}_{Na_2O(I)} = -E.2F$ (E the EMF of the cell, defined by the activities of Na_2O in the electrolytes

$$a_{Na_2O} = -\frac{RT}{F} \cdot \ln\left(\frac{a_{Na_2O(II)}}{a_{Na_2O(I)}}\right)$$

F = Faraday's number), ΔF being a function of temperature make evident that the liquid junction potentials are insignificant (less than 5 mv.).

67. If the glass is considered as a mixture of bridging oxygen atoms in $Si^{4+} - O^{2-} - Si^{4+}$, and nonbridging groups $Si^{4+}—O^{2-}—Na^{+}$, the change of relative basicity as a function of temperature corresponds to an entropy of mixing of these two

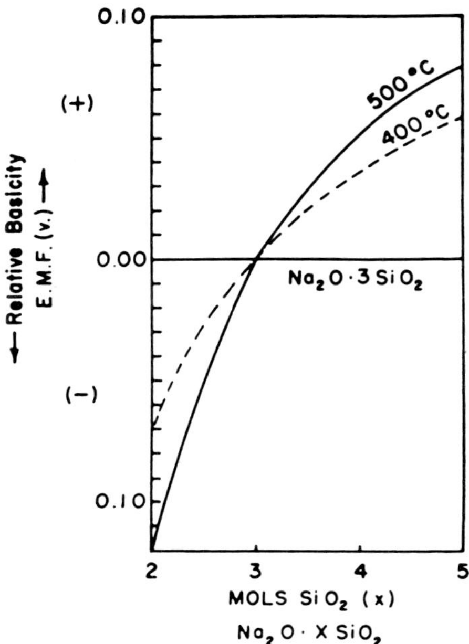

FIG. B. 43. Relative basicity, determined in EMF, of glasses $Na_2O,xSiO_2$, with the trisilicate glass taken as a standard. (Förland and Tashiro).

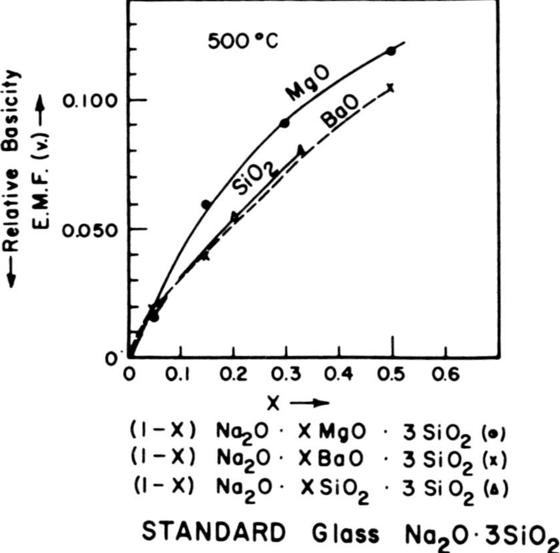

FIG. B. 44. Relative basicity, determined in EMF, of glasses $(1 - x)Na_2O,x(MgO,BaO,SiO_2)$ $3SiO_2$. (Förland and Tashiro).

kinds of oxygen anions, strongly deviating from an ideal mixture. If magnesia is introduced into both glasses of the galvanic cell, the basicity is practically independent of the temperature, the entropy of mixing for the bridging and nonbridging oxygen shares, therefore, close to zero. Clusters are formed of non-bridging O^{2-} anions with Mg^{2+} cations, as they are also known from the immiscibility region in the system SiO_2—MgO. The Na^+ ions, however, impede a coalescence of the clusters to a separate phase. The basicity of the sodium silicate glasses is more strongly decreased by MgO than by BaO, or by SiO_2 in analogy to previous color indicator observations of Lux and Rogler (1942) who examined additions of MgO and B_2O_3. It is evident, as Weyl, Flood and Förland (1947) concluded, that the Mg^{2+} and Ba^{2+} ions surround themselves with a number of nonbridging oxygen anions which are strongly *tightened* by the alkaline earth ions and thus decrease the basicity. In the glasses $Na_2O,RO,5SiO_2$ the decrease in basicity is in the order BaO $>$ CaO $>$ MgO. The opposite was found by Lux and Rogler for corresponding borate glasses. The maximum in EMF in Fig. B. 45 may be caused by different transfer numbers of the Na^+ and K^+ ions in the mixed glasses (on these numbers cf. B. Lengyel, 1940, for a disilicate glass).

68. In another experimental series, G. S. Smith and G. E. Rindone[90] used corresponding galvanic cells with two alkali silicate glasses of different concentrations,

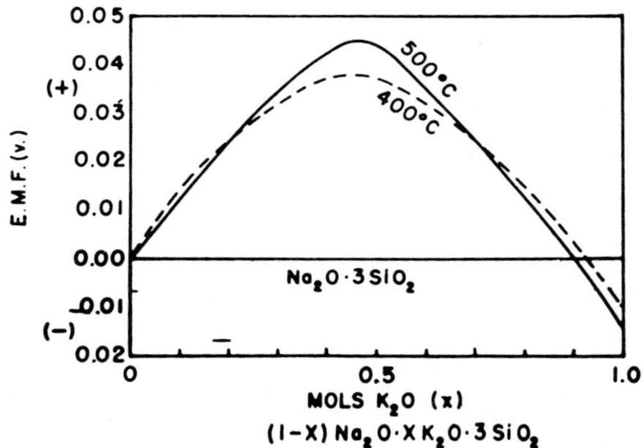

FIG. B. 45. EMF as a function of the molar concentration of K_2O in glasses $(1-x)Na_2O$, $xK_2O,3SiO_2$, at 500°C. (Förland and Tashiro).

[90] *Glass Ind.*, 37, 1956 (8) 437-443, 454, 458. Most recently G. E. Rindone, D. E. Day and R. Caporali, *J. Am. Ceram. Soc.*, 43, 1960, 571-577, used oxygen electrode methods for an investigation of sodium and potassium silicate glasses containing Al_2O_3 and TiO_2 (cf. A. ¶ 177). Alumina increases the acidity of the potassium glasses, whereas in sodium silicate glasses the first additions of alumina are more efficient in increasing the acidity than are larger additions. The phenomena indicate the formation of $[AlO_4]$ groups. Ti^{4+} is present in both tetrahedral and octahedral coordinations in the potassium and sodium silicate glasses.

and silver electrodes, in air. Simple alkali trisilicate glasses are used as reference glasses in contact with similar glasses into which MgO, BaO, and SiO_2 were substituting for an equivalent amount of Na_2O. The acidity relationships in the alkali silicate systems are determined for the Li, — Na, — K series, with varying amounts of MgO, BaO, and SiO_2. The alkali earth oxides cause a larger decrease in basicity (increase in acidity) than an equivalent amount in silica (cf. Fig. B. 46 for lithium silicate

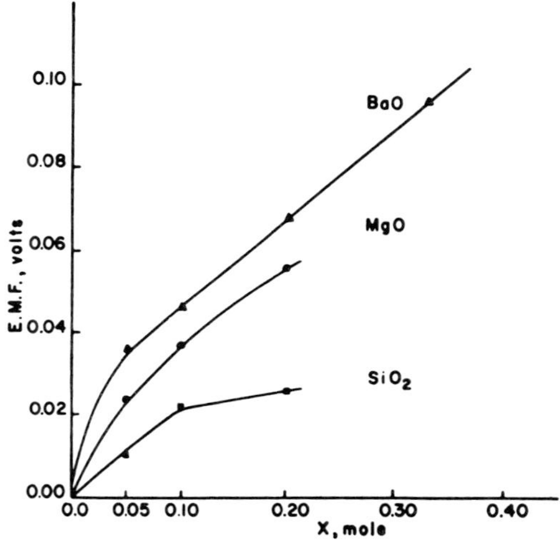

FIG. B. 46. EMF as a function of the molar substitution of Li_2O in glasses $Li_2O,3SiO_2$, by MgO,BaO, and SiO_2, at 500°C. (Smith and Rindone).

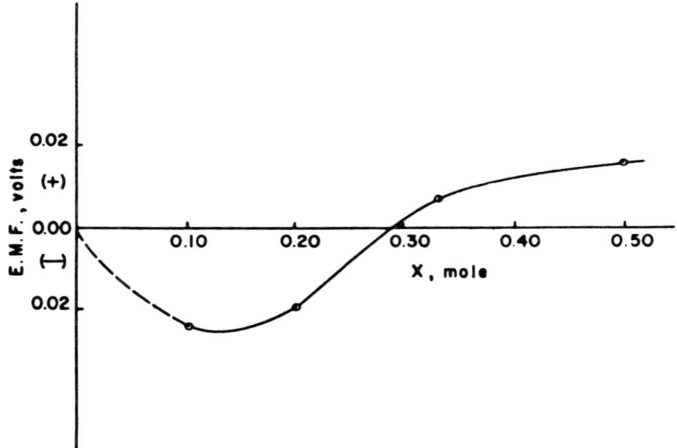

FIG. B. 47. EMF developed at 500°C. by the galvanic cell $Ag/(1 - x)K_2O,xMgO,3SiO_2/(1 - x)K_2O,xBaO,3SiO_2/Ag$. The potential of the electrode in contact with the BaO-containing glass is taken as zero. (Smith and Rindone).

glasses). This effect is a consequence of the field strength and coordination number of the ions introduced, as particularly evident in Fig. B. 46, with Ba > Mg > Si. Very interesting also is the reversal of basicity-acidity relations in magnesia and BaO containing glasses $K_2O,3SiO_2$ at 500°C., as observed from Fig. B. 47, as a function of temperature.

PROBLEMS OF REACTION RESIDUES IN GLASS MELTS

69. The process of fining is evidence that, with apparent homogenization of the glass from the molten batch, inner equilibria are by no means finally established. We emphasized repeatedly that there are nonequilibria especially with the residual carbonates, sulfates, and other salts, in the glass as it leaves the fining zone and remains stagnant in the thermal conditioning chamber of a tank furnace. In other words, the residual contents of the glass in CO_2,SO_3, and especially in H_2O, show that there are those nonequilibria which influence the physical-chemical properties of the glass and bring about reboiling and secondary degassing phenomena of a highly undesirable character. We must understand that the "residual gases," including water, are not only physically absorbed but real chemical reaction residues, or chemically bound in the form of the anions CO_3^{2-}, SO_4^{2-}, OH^-, and so on. H. Scholze and A. Dietzel[91] used the infrared absorption technique to investigate the presence of "water" in glass by a study of the absorption in the range from 1 to 5 μ wavelength, to make clear the distinction between the presence of free H_2O, CO_2, and other compounds, and of bound anions OH^-, CO_3^{2-}, and the like. For residual CO_2 the band at 4.25 μ would be conclusive, for CO_3^{2-} that at 3.5 to 3.6 μ. As a matter of fact, only the CO_2 band at 4.25 μ is indicated in most cases, but the intensity of the 3.5/3.6 μ band systematically increases with an increasing alkalinity of the glass, in agreement with the previous experimental observations of W. Weyl (1931/32) for the presence of carbonates in equilibrium with alkali silicate melts. Only when such melts are heated above 1000°C. the CO_3^{2-} band slowly disappears by progressive dissociation of the carbonate and its reaction with the siliceous medium.

70. For discrimination of the OH^- and the H_2O bands it is important to know the influence of the dielectric constant of the glass on the position and the frequency of the absorption band, ν. The relative shift of frequency as a function of the dielectric

[91] *Glastech. Ber.*, **28**, 1955, 375-380. On the different infrared bands important for glass in this connection see J. M. Florence, F. W. Glaze, E. H. Hahner, and R. Stair, *J. Am. Ceram. Soc.*, **31**, 1948, 328-331. Concerning the possibility of carbonate residues in glass see also, in continuation of the previous data given by T. J. Howarth and W. E. S. Turner, (1930/31), the discussions mentioned more in detail in Vol. V, Section A, by G. Batta, P. Gilard and L. Scheepers, *Proc. Intern. Symposium on Reactivity of Solids. Gothenburg*, 1952, Vol. II, 1954, 801-821.

constant of the medium, ε, is given by the relation

$$\Delta\nu/\nu = C \cdot \left(\frac{\varepsilon - 1}{2\varepsilon + 1}\right)$$

(C a constant). With ε for silica glass $= 3.75$, and another glass with $\varepsilon = 8$, the shift is calculated from 2.74 to 2.76 μ. The observations showed bands at 2.75 μ and 2.95 μ for both glasses, respectively, definitely indicating an association of OH^- with hydrogen bondings in the latter result. The presence of water in glass depends very much on the partial pressure of water vapor in the furnace atmosphere, and also on the form in which the batch contained it, say whether $Na_2CO_3.10H_2O$ or anhydrous soda was used in the batch composition. The total transmittance in the infrared range up to 4 μ is particularly high, or in other words, the water content low, in glasses molten from anhydrous batches and *in vacuo*. In every case, the majority of the water in glass is firmly bound and cannot easily be driven out quantitatively by vacuum fusion (cf. B. ¶ 89-97).

71. The sluggishness of the disappearance of quartz sand particles in the glass melt is one of the important factors favoring the development of reaction residues (relics). It was, therefore, extensively studied by many technologists with many different methods. Simple microscopic observations of these processes were chiefly made by H. Jebsen-Marwedel,[92] and by H. Reumuth and W. Buss[93] which well illustrate the last stages of the reactions, according to results of C. Kröger *et al.* (cf. B. ¶ 2 ff.). Also the formation of well-known dissolution striae in the structure of the glass, is best made visible by the Toepler schlieren method. By interferometric-microscopic observation, J. Löffler[94] contributed many details on the formation of the striae around the sand grains, with their characteristic zoning. There is one zone in which Na_2O is enriched by diffusion followed by another zone in which the soda is much impoverished by the migration of the Na^+ ions towards the sand grain. Measurements of the refractive index and specific HF—HBF_4 etching methods developed in the structure of such striae the finest details which are useful for a quantitative determination of their composition from spot to spot. Thus it could be made evident that the quartz grains are suspended in a peripheral "swimming" layer

[92] *Glastech. Ber.*, **30**, 1957, 122-129. Excellent micrographs are also given by Jebsen-Marwedel in the particularly recommended book "Microscopic Examination of Glass Melting," Verlag der Deutschen Glastechnischen Gesellschaft, Frankfurt am Main 1951, 151 pp., and in the book "Glastechnische Fabrikationsfehler," Springer-Verlag, Berlin, Göttingen, and Heidelberg, 2nd edition 1959, 529 pp., especially pp. 83-181, on schlieren (striae), bubbles, etc.

[93] *Glastech. Ber.*, **30**, 1957, 145-149; **32**, 1959, 89-95.

[94] *Glastech. Ber.*, **30**, 1957, 129-133; **31**, 1958, 268-269. In *Ibid.*, **30**, 1957, 117-121, Löffler demonstrated that no gravitative differentiation of polyeutectic melts can normally occur separating these from the sand grains.

relatively enriched in Na_2O. The ratios Na_2O/CaO and Na_2O/MgO are markedly higher in this layer than in the bulk of the glass.

72. Measures to accelerate the last reactions before the final homogenization of the glass melt often have been proposed. We here mention only the elementary recommendation of B_2O_3, borax, and especially of H_3BO_3, which was made by R. S. Allison and W. E. S. Turner[95] for the practitioner, or of BaO and fluorides for the same purpose, and the great importance of correct *moisture* content of the batches to control a uniform melting and homogenization, as studied by A. Dietzel and H. Wickert[96] to accelerate assimilation of quartz when applied in the form of a very fine powder. In this latter case, high additions of water must be made (20 per cent and more) to avoid an insufficient soaking of the batch. The mixture becomes, in this latter case, dry-friable and crumbling and "stone" formation is unavoidable. In this direction are also recalled the studies of C. Kröger and Fr. Marwan[97] concerning rates of the total melting process for batches which were distinctly accelerated by "catalyst" additions of B_2O_3, Na_2SO_4, H_3BO_3, $Na_2B_4O_7.10H_2O$, and especially $K_2CO_3.H_2O$. Such additions combined with moisture develop what those authors called "impregnation" effects by the moisture-treated batch which may bring about a considerable increase in the industrial glass output (Fig. B. 48).

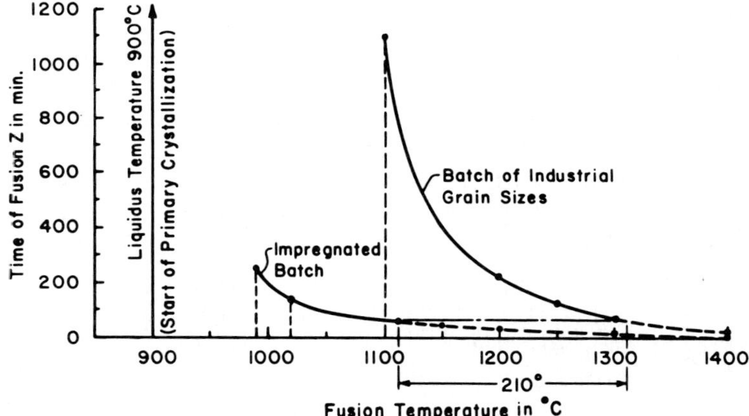

FIG. B. 48. Total time of glass melting as a function of temperature, (*a*) starting from an industrial grain size classification of the batch, (*b*) starting from the same batch, but impregnated by a moisture treatment. (Kröger and Marwan).

95 *J. Soc. Glass Technol.*, **38**, 1954, 297-364 T.

96 *Glastech. Ber.*, **27**, 1954, 170-172. See also theoretical discussions of I. D. Tykachinskiĭ, *Khim. Nauka i Prom.*, **3**, 1958 (1) 65-71; *Ceram. Abstr.*, 1958, 308g, who urgently recommends the use of accelerating agents and fine dispersions of reagents for glass fusion.

97 *Glastech. Ber.*, **29**, 1956, 275-289; **30**, 1957, 222-229; cf. older data given by W. E. S. Turner and E. Preston, *J. Soc. Glass Technol.*, **24**, 1940, 124-138. See further specifications for accelerat-

73. Whereas the common fining reactions are so conducted that the resulting glass melt becomes "fined," or free from bubbles and seeds, on the contrary, one found ways to make the glass *foamy*, in other words, interspersed with more or less fine gas bubbles and closed pores. The production of foamy glass which rapidly becomes of great importance for heat and sound insulating purposes in building industry, was described in certain special reports. We mention here only those of M. Hübscher,[98] H. J. Karmaus,[99] and E. O. Schulz.[100] Hübscher defines the physical structure behavior of foamy glass on the fundamental classification principles given by E. Manegold,[101] most of the glasses being of the polyhedral type, with closed cells. It is undesirable to have spongelike structures with disrupted cell walls. The expansion of the glasses to foams, under the action of a bloating agent (of multiply varied chemical character), is in principle nothing more than rapid development of gas bubbles of optimum sizes, practically controlled, in a functional relation to temperature and viscosity of the glass melt (cf. B. ¶ 55 ff.). Schulz emphasizes the variable anomalies in the isotropy of foam glass structures which occur as a consequence of the low heat conductance of the material between sintering and bloating, and the necessity of a careful thermal treatment (annealing) especially for materials having a volume-weight near 0.13 gm./cm.^{3}.

74. Interesting is the information given by R. L. Shuster and L. K. Kovalov[102] that foam glass building blocks have a high importance in construction of earthquake-proof housing in Kazakhstan. Many details on the optimum granulometric composition of the glass and the gas-evolving addition (limestone or dolomite preferably, added in the order of magnitude of 3 per cent) are given by the same authors to develop closest packing of gas bubbles in the foam structure. More recently, I. I. Kitaĭgorodskiĭ and N. V. Artamonova[103] described interesting highly siliceous glasses, with 90 to 94 per cent silica, and 6 to 10 per cent $B_2O_3 + Sb_2O_3$, the latter oxide forming the bloating-activating agent. The finely powdered glass is sintered in pressed pellets or bricks, heated up to 1400°C., soaked for about half an hour, and slowly cooled. Such a highly porous glass (called "Penosil") is mechanically stable up to 1300°C. in its foam structure (common glasses collapse at only 500°C., alkali-free foam glass perhaps at 1050° or even 1100°C.), and of course, acid-proof.

ing agents like H_3BO_3, and CaF_2, etc. by L. G. Mel'nichenko and F. Ya. Kharitonov, *Steklo i Keram.*, **15**, 1958 (5) 18-21, especially for briquetted batches.

[98] *Silikat Tech.*, **5**, 1954, 243-247, chiefly based on Russian literature in this field.

[99] *Glas-Email-Keramo-Tech.*, **8**, 1957, 172-176; 224-229, an excellent Patent report, including chiefly European Patent literature.

[100] *Silikat Tech.*, **5**, 1954, 343-346, on manufacturing methods in details.

[101] See the book "Schaum," Verlag Chemie und Technik, Heidelberg ,1953, 520 pp.

[102] *Izvest. Akad. Nauk Kazakh. S.S.R., Ser. Gornogo Dela, Met. i Stroimaterial.*, **3**, 1957, 54-69.

[103] *Doklady Akad. Nauk S.S.S.R.*, **130**, 1960, 377-378.

75. For the formation "*cords*" or "*schlieren*" (striae) the dissolution of refractory material from the walls of the glass tank is of particular importance. This process emits continuously from the walls into the glass a current of alumina-enriched heterogeneity "threads," in a process which is similar to the spinning of glass fibers from the melt into the open air. M. Jaupain[104] used a simple "staining" method for identification of such threads by adding some CuO to the refractory material from which the glass striae transfer the Cu^{2+} tracer into the glass. They are then easily recognized by the ultraviolet fluorescence test of samples taken from the glass. In principle the same effect is produced with radioactive tracers as shown by W. Jahn,[105] for instance with β-radiating elements of V^{57}, Cu^{63}, Mn^{55} and especially rare earth elements activated by irradiation in a nuclear pile. Some of these elements with a high activation cross section are very sensitive as tracers for the purpose here mentioned, whereas unfortunately Si^{31}, Al^{28}, Na^{24}, K^{42}, Mg^{27}, Ca^{45}, Ca^{49}, Zn^{69}, Ba^{139}, As^{75} have far too short half-life times. Jahn is of the opinion that tracing of the tank blocks with Li, Be, Sc, or Ti, V, Mn, Sr, Y, Zr, Mo, Ba, La, Re would be sufficient in combination with flame-spectrometric determinations, with sensitivities of 0.002 γ/cm.^{3} in arc emission spectrometry. Be is easily enriched from the cords and this could even increase the sensitivity of their identification. For practical purposes, J. Löffler's interferometric-microscopic method, combined with HF—HBF_4 etching (cf. B. ¶ 71), is doubtless the best quantitative method, also for determination of differences between the chemical composition of the striae relative to that of the basic glass, as L. Riedel[106] outlined.

76. The changes of the shapes of striae (cords) as a function of time under the action of the flow gradient ("pull") in the glass and the diffusion problems involved are mathematically treated by W. Geffcken,[107] for low Reynolds numbers as a creeping motion phenomenon, or in a stagnant glass mass with tremendously long periods of self-homogenization. Nothing can better illustrate the significance of glass flow in the homogenization process than this result of the mathematical deductions. Also the conditions of circular stirring with a spiral motion of the glass are discussed by Geffcken. On the other hand, changes in shape of cords as heterogeneities in the glass mass under the influence of *surface tension* phenomena (cf. A. 93 to ¶ 117) were extensively described in the rich practical experience systematically collected by H. Jebsen-Marwedel since 1936. We wish to emphasize here chiefly observations on cell-shaped network structures in heterogenous liquids, comparable to industrial

[104] *Verres et réfractaires*, **6**, 1952, 356-361. On the effects of material heterogeneities by striae in glass and the problem of its homogenization by laminar convection currents in the melt, experimentally verified by Couette flow phenomena in a rotation viscosimeter, see R. Brückner, *Glastech. Ber.*, **35**, 1962, 43-52.

[105] *Glastech. Ber.*, **32**, 1959, 103-106.

[106] *Fachausschussber. deut. glastech. Ges.*, No. **56**, 1958, 526-531.

[107] *Glastech. Ber.*, **30**, 1957, 143-145.

glass melts in their state after fining[108] on whirl (vortex) striae,[109] particularly on what he called "dynactive pairs" of two liquid phases. There are typical vortex motion effects if one liquid of a higher, or lower, surface tension falls onto and through the surface of a second liquid. This effect known in the hydrodynamic literature[110] as "spreading" is as a matter of fact most striking, as in the model system of water on glacial acetic acid, when a water drop is rapidly propelled toward the periphery displacing the acetic acid which in turn approaches the center. In this way many complex structures were fixed in masterly reproduced instantaneous photographs by Jebsen-Marwedel.[111]

77. Technologically most characteristic property of dynactive phenomena is the possibility that even a liquid portion of higher density is driven to the surface of a lighter melt, if the first glass has a particularly low surface tension. Glasses of low surface tension tend to subdivide the glass with the higher one to the cellular-

[108] E.g. in *Naturwissenschaften*, **84**, 1954, 1-5, 22-27.

[109] *Sprechsaal*, **84**, 1954, 1-5, 22-27.

[110] Cf. K. L. Wolf: in "Physik und Chemie der Grenzflächen," Springer-Verlag, Berlin, Göttingen, Heidelberg, Vol. I, 1957, 262 pp., especially p. 184. Cf. H. Kroepelin and E. Prött, *Naturwissenschaften*, **45**, 1958, 333-334, for absorption of acetic acid on a water surface, with turbulent schlieren phenomena, further R. Brückner, *Ibid.*, **47**, 1960, 371-372, and very extensively in *Glastech. Ber.*, **34**, 1961, 439-456, 515-528, on non-stationary surface processes between miscible liquids, who, however, does *not* confirm Jebsen-Marwedel's conclusions that the condition $\Delta\varrho \cdot \Delta\sigma < 0$ must be fulfilled (see below). Intense vortex formation is observed as well for liquid combinations with a positive, as with a negative value of that product, provided the differences in concentration and surface tension before the mixing are large enough. For water-acetic acid, and aniline-CCl_4 is the product below 0, but for water-acetone, aniline-benzene, and nitrobenzene-dioxan it is positive. The mathematical discussion makes possible a more generally applicable theory of "dynactivism" than previously given by Jebsen-Marwedel. The high importance of his observations and conclusions for glass is not affected by the critical remarks of Brückner.

[111] Cf. *Kolloid-Z.*, **137**, 1954, 118-120; *Glastech. Ber.*, **27**, 1954, 357-374; *Ind. chim. belge*, **20**, Special Number, 1955, 98-104; *Compt. rend. congr. intern. chim. ind.*, **27**, Bruxelles 1954, No. 3; *Glastech. Ber.*, **29**, 1956, 233-238, the whirl phenomenon described as of a high technological significance for development of characteristic spherical *holes* of erosion in the refractory wall of glass tanks, often combined with bubbles in the whirling liquid. In *Ibid.*, **29**, 1956, 269-275, the dynactive splitting away of silica- or alumina-enriched threads from the melt of lower density, but with a higher surface tension, is described. They especially develop on the "metal line" as a consequence of dynactive rotations in corrosion grooves of the refractory wall. Such threads are particularly well identified by Löffler's HF—HFB_4 etching method and interferometric-microscopic examination. Very interesting microscopic studies by W. Buss, *Ibid.*, **32**, 1959, 89-95, especially pp. 94 f. demonstrate dynactive spreading phenomena in experiments with glass spheres stained by CdS-CdSe, of variable chemical composition, with the characteristic spreading networks, funnel-shaped whirling, etc. Concerning the influence of CaO, MgO, and especially of SO_3 on the surface tension of highly viscous glasses, we recall A. ¶ 113 f., and studies of M. V. Okhotin and I. G. Bazhbeuk-Melikova, *Steklo i Keram.*, **9**, 1952 (4) 12-14; (6) 3-4; *Zhur. Priklad. Khim.*, **26**, 1953, 1320-1322, and of A. Dietzel and E. Wegner, *Veröffentl. Max-Planck-Inst. Silikatforsch.*, **13**, 1953, 120-130; *Atti congr. intern. vetro, 3rd Congr. Venice* 1953, 354-363.

network structure mentioned above. A difference of only one per cent (equal to about 3 dynes/cm.) in surface tension is sufficient to make this effect evident, followed by a "whirling" because of density disequilibrium attained by the cellular arrangements. The mixing by flow of glasses of different density thus efficiently enhances the diffusion process. This is the principal importance of dynactivity in glass homogenization. Jebsen-Marwedel applied to the explanation of these phenomena the previous theory of H. Jettmar and F. C. Roesler[112] that the condition $(\varrho_1 - \varrho_2) \times (\sigma_1 - \sigma_2) < 0$ must be fulfilled (ϱ the density, σ the surface tension) for its action. The alumina-enriched reaction glass derived from dissolution of tank block material is usually lower in density, but higher in viscosity and surface tension, relative to the bulk of the glass melt. Therefore, it brings about intense enveloping and dynactive rotation currents in the liquid which are so well described by Jebsen-Marwedel. He gave further a tabulation to demonstrate the effects of different oxides on density and surface tension.[113] SO_3, V_2O_5, and CrO_3 have particularly strong negative decrements in this table. Intense dislocations of layers in the melts and spreading phenomena are the immediate consequence made visible by the striae-microscopic examination. Oxidation and reduction reactions further play an important role not only in glass melts but also in ceramic glazes and enamels which often show a well developed network-cellular structure or crazing, curling, craquelé patterns, or a "breaking-through" of the ground enamel through the cover enamel by dynactivity.

78. We mentioned in A. ¶ 100 that the particularly strong effect of V_2O_5 on lowering the surface tension of silicate melts was made useful in A. Staerker's "Vanal" method for protecting refractory wall surfaces from corrosion by slags. The repellent effect shown on such surfaces for the V_2O_5 containing reaction melts was very impressively demonstrated by H. Lehmann's and U. S. Singh's photographs of Vanal-treated refractories and their contact angles, with an extensive discussion of the theoretical basis of that method,[114] applying the decrements (oxide factors) for the calculation of surface tensions based on A. Dietzel's empirical coefficients (1942). Vanal increases the contact angles up to 1000°C., but not above this temperature.

[112] *Kolloid-Z.*, **123**, 1957, 2-11. The conditions as formulated by Jettmar and Roesler are too narrow; cf. R. Brückner, Footnote 110. This author especially studied the behavior of Na_2SO_4 and of its decomposition products, and of the exchange of sulfate anions between the two liquid phases, gall/glass melt.

[113] *Glastech. Ber.*, **31**, 1958, 431-438. On the particular sensitivity of surface tension conditions in SO_3 containing glasses, and consequences in leachability behavior of SO_3 from glass see I. Hilgenfeldt and H. Jebsen-Marwedel, *Ibid.*, **31**, 1958, 161-170 (cf. B. ¶ 194).

[114] *Tonind.-Ztg. u. Keram. Rundschau*, **79**, 1955 (1/2) 2-3; *Ber. deut. keram. Ges.*, **34**, 1957, 353-362. For a particularly impressive photograph of the Vanal effect see H. Lehmann and Kl. H. Müller, *Tonind.-Ztg. u. Keram. Rundschau*, **79**, 1955, p. 311, Figures 3 and 4. See also U. S. Singh, *Glas-Email-Keramo-Tech.*, **10**, 1959, 165-169. See further A. Staerker, *Silicates inds.*, **21**, 1956, 509-512; *Ber. deut. keram. Ges.*, **34**, 1957, 329-334; *Silicates inds.*, **23**, 1958, 650-661. with practical experience reports and impressive photographs.

An explanation of the Vanal effect is possible if one considers the specific polarization effects of the ions on the surface associated with V^{5+} ions. Other methods for calculation of surface tension, but in principle also based on oxide decrements, were shown by A. A. Appen, K. A. Shishov, and S. S. Kayalova[115] (cf. A. ¶ 95).

79. An interesting correlation between surface tension effects and the stability of seeds (gas bubbles of small dimensions), with their interior gas pressure was used by G. Koranyi[116] for approximate determination of surface tension in glass melts. The pressure in the seeds rising in the viscous liquid does not differ much from the sum of the atmospheric and the hydrostatic pressures. In the correct order of magnitude, the surface tension, therefore, can be calculated from the minimum diameter of the seeds in the glass, frozen during the fining and conditioning periods. In a simple window glass of the soda-lime-magnesia silicate type surface tension was calculated to be 382 dynes/cm. at 1375°C.

EVOLUTION OF ABSORBED GASES FROM GLASS BY FUSION

80. In close relation to problems of residual carbonates, sulfates, and other compounds in glass melts is the phenomenon of the so-called release of "absorbed" gases, commonly assumed to be physically or chemically "dissolved" in the glass as the liquid and undercooled solvent. We owe to Fr. Mahieux[117] a comprehensive investigation on this important problem based on H. Jebsen-Marwedel's "fining diagrams" (1949). Beyond that, special radioactive tracer methods were developed by Mahieux using C^{14} in the carbonate and CO_2 evolved, and S^{35} in the sulfates and the gaseous SO_2 or SO_3, released from the glass during heating in a microfurnace (Vol. V, Section A) (Fig. B. 49). Inversely Mahieux studied absorption of the labeled gases into a synthetic soda-lime silicate melt, or introduced labeled carbonates and sulfates applied to the batches, in exchange with the furnace gas atmosphere. From tracer activity it was concluded that a glass with 75 per cent silica fixes at 1315°C about 0.5 per cent of its weight in SO_3. Another glass with 56 to 57 per cent silica, however, fixes 0.8 per cent SO_3, in good agreement with the previous data given by Jebsen-Marwedel (1940), J. E. Stansworth and W. E. S. Turner (1937), and the older vacuum expulsion data of H. Salmang and A. Becker (1930), or gas microanalyses from the bubble contents, by H. A. Shadduck and A. Van Zee (1942). The "solu-

[115] *Zhur. Fiz. Khim.*, **26**, 1952, 1131-1138, 1399-1404; *Silikat Tech.*, **4**, 1953, 104-105; see also A. Petzold, *Ibid.*, **5**, 1954, 11-12.

[116] *Acta Tech. Acad. Sci. Hung.*, **11**, 1955, 155-159; *Ceram. Abstr.*, 1955, 176 h.

[117] *Verres et réfractaires*, **10**, 1956, 277-298, 342-356; **11**, 1957, 9-17. See also the vary recent results of R. Brückner, *Glast. Ber.*, **35**, 1962, 93-105, especially on the exchange of glass melts with a SO_3 (and water) containing furnace atmosphere. A review of 40 publications on gas solubility in glass was presented by T. J. Harper, *Glass Technol.*, **3**, 1962, 171-175.

bility " of CO_2 (as CO_3^{2-}) in glass melts is absolutely much lower than that of SO_3 (as the SO_4^{2-} anion), but follows the same functional relations to acidity of the glass composition. The CO_2 "retaining" effects of bases are illustrated by the series of homologous dissociation temperatures of the carbonates, namely by $T^0(BaCO_3) > T^0(Na_2CO_3) > T^0(CaCO_3)$.

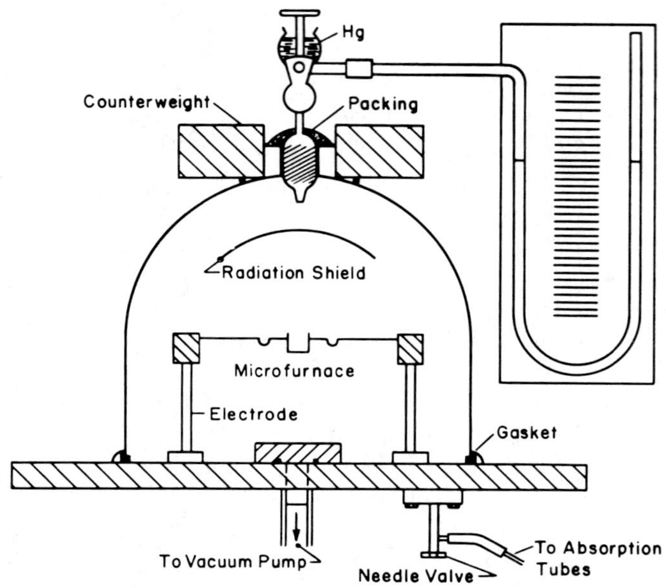

FIG. B. 49. Apparatus for fusion of experimental glasses, labeled with radioactive tracers, *in vacuo*, and for the determination of absorbed gases released from the glass. (Mahieux).

81. Desorption and spontaneous release of CO_2 and SO_3 under conditions of supersaturation from the glass surface indicates the tendency to an establishment of equilibria of the melts with the surrounding atmosphere (cf. Fig. B. 50). The desorption equilibria are especially evident in a series with the introduction of $C^{14}O_2$ and $S^{35}O_3$ from the furnace atmosphere into the glass melt. The ideal equilibrium concentration of SO_3 at 1315°C. is $95 \pm 15 \times 10^{-3}$ mg./gm. glass (with 75 per cent SiO_2). The accuracy of the tracer method in this case is only $\pm$ 10 to 15 per cent, similar to that of the measurements of the solubility of water in silica. In every case, the absorbed gases are preferably enriched in the exterior layers of the glass melts and distribution of the marked atoms in the glass is not homogeneous. The penetration of the "absorbed" CO_2 and SO_3 into the bulk of the glass is not only a diffusion problem but also is influenced by "dynactivity" (see B. ¶ 76 f.), that is surface tension effects and thermal convections made evident by contact radiographs, with typical "whirls" near the borders of the crucible melts.

For pure B_2O_3 glass the absorption of labeled SO_3 is restricted to a slight surface fixation. In GeO_2 glass, however, SO_3 is much better fixed if traces of water are present, whereas entirely anhydrous GeO_2 fused in a totally dry furnace atmosphere causes only an extremely low absorption effect. A given glass is much more able to evolve gases when it contains more of the anions CO_3^{2-}, SO_4^{2-}, or O_2^{2-} (peroxides),

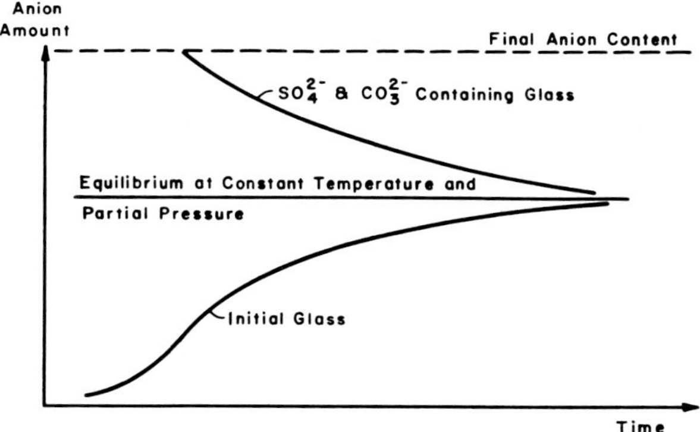

FIG. B. 50. Approach to equilibrium concentration in CO_2 and SO_3 in an experimental glass, from both sides, by desorption and absorption (at 1315°C.). (Mahieux).

at a constant temperature (1315°C. in Mahieux's experiments) and variable pressures (down to 10^{-3} mm. Hg). Sulfate glass bubbles show a greater tendency to evolve gas than carbonate or peroxide batches. Ba^{2+} in glass melts retains more $CO_3{}^{2-}$ than Na^+ and Ca^{2+}. A typical secondary seeding (reboiling, cf. B. ¶ 69) is often observed, even during decreasing temperatures when a state of supersaturation has developed.

82. Concerning the so-called "fining" reactions as heterogeneous gas equilibria we have only to recall of our discussions in B. ¶ 30 ff. especially the report given by the German Society of Glass Technology and the translation by G. E. Rindone.[118] The same author[119] explained reboiling or "secondary seeding" of SO_3-containing optical glass of the high-in-barium type by experiments demonstrating that the introduction of sulfate into batches of sodium-barium silicates causes a pronounced change in the structure of the glass melt developing a high interfacial tension between it and a similar glass which is sulfate-free. There is also a distinct tendency to unmix a separate liquid phase of alkali-barium sulfate composition. In industrial barium

[118] *Glass Ind.*, **38**, 1956, 489-493, 516, 526, 528; 561-565, 576-577.

[119] "Symposium sur l'affinage du verre," Paris 21-23 Juin 1955 (Union scientifique continentale du verre, Paris) pp. 399-421. On the formation of gas bubbles in glass melts see recently S. M. Budd, V. H. Exelby, and J. J. Kirwan, *Glass Technol.*, **3**, 1962, 124-129.

glasses this tendency is generally more latent but gas evolution takes place with "reboiling" when nuclei are present around which the liquid structure is "broken." This same effect was previously emphasized in its technological significance by W. A. Weyl and E. C. Marboe (1948). In every case, "delayed seeds" formation can be traced backward to reactions between sulfate containing portions of the melt with other parts of the heterogeneous glass. The great difficulties in a satisfactory fining of sulfate glass batches are also projected into gas content problems as shown by V. V. Pollyak,[120] and their optimum temperature cycling to avoid troublesome opaque inclusions of alkali sulfates together with the SO_3 "solubility" in the range from 1000° to 1400° as a function of the silica contents.

83. *Microanalytical* methods for determination of the composition of gases evolved from glass melts *in vacuo* were developed by P. K. Chu, J. H. Keeler, and H. M. Davis[121] and applied to enamels (cf. B. ¶ 249), combined with methods for analyzing tiny gas bubbles from reboiling blisters, consisting of a pricking device for releasing the confined gas, absorption-observation cells for use on the microscope stage, as it was first used by J. Enss (1933). V. T. Slavyanskiĭ and E. N. Krestnikova[122] for the same purpose crushed the glass sample between a convex and a concave lens, and absorbed the gas in glycerine from which it is then released *in vacuo* and gathered in a microburette for standard gas analysis. A more elaborate vacuum apparatus was used by Slavyanskiĭ[123] for extracting the gas and measuring its volume by an absolute manometer, completely separating the reaction chamber from any rubber or lubricant contacts securing the vacuum pump with a mercury seal. The heating was made stepwise to 400° and 900°C. when steady degassing began. Water vapor and CO_2 are immediately condensed in the liquid air traps, then the flask is filled with mercury and the gas released pumped over to the analysis bulbs.

84. A most powerful modern tool in such microanalytical investigations of the contents of gas bubbles in glass (or related materials) is *mass spectrometry*. B. J. Todd[124] applied this instrumentation successfully to analysis of the contents of single gas bubbles and condensation products of the gases in the sampling system. The bubbles were broken up in vacuo before introduction into the mass spectrometer (Fig. B. 51). The pressure in the blisters before they were opened is calculated

[120] *Steklo i Keram.*, **15**, 1958 (6) 1-4. See also more recently H. Krause, W. Vogel, and H. Wessel, *Silikat Tech.*, **10**, 1959, 401-404, with interesting observations of condensation of the metastable monoclinic modification of α-As_2O_3 in sulfate bubbles. By scratching, the monoclinic needles are spontaneously inverted to stable cubic As_2O_3.

[121] *J. Am. Ceram. Soc.*, **36**, 1953, 48-59.

[122] *Steklo i Keram.*, **10**, 1953 (11) 11-15.

[123] *Steklo i Keram.*, **14**, 1957 (2) 11-17. The apparatus and operation is similar to that previously used by E. W. Washburn, F. F. Footitt, and E. N. Bunting, (1920).

[124] *J. Soc. Glass Technol.*, **40**, 1956, 32-38 T.

from the volume of evolved gas and the volume of the blisters themselves weighed after being filled with mercury. The pressure was determined to have been between 0.26 and 0.29 atm. (for a Corning glass No. 9010). In this case, oxygen in some seeds totalled 98 per cent, in some of the seeds only 1 to 8 per cent, in others 92 to 98 per cent. How far accuracy of mass-spectrometric determinations of gases

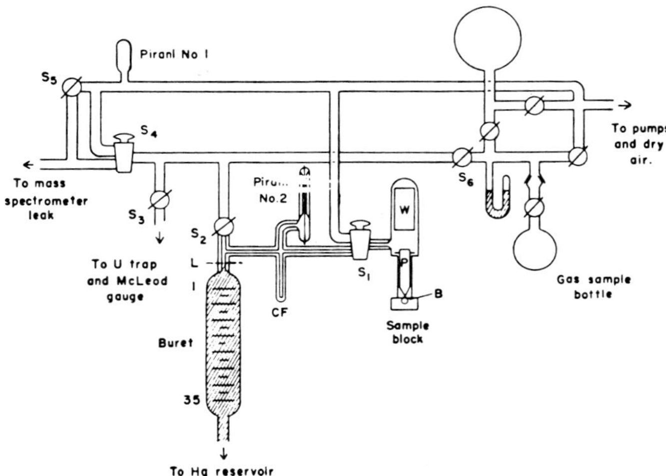

FIG. B. 51. Sampling apparatus for mass-spectrometric analysis of gases in glass blisters. (Todd).

can be driven was demonstrated by J. F. Martin, J. E. Friedline, L. M. Nelnick, and G. E. Pellissier[125] in steel inclusions, by a vacuum fusion technique: for oxygen and nitrogen to 0.0001 per cent, with an average deviation of 0.0001 per cent for oxygen, and 0.00008 per cent for nitrogen.

85. Concerning the early ideas of E. Zschimmer (1924) to remove seeds by the application of high external pressures in order to collapse the tiny bubbles, F. A. Kurlyankin[126] gave interesting details of calculated pressure effects for silica glass by charging with nitrogen or carbon dioxide: about 100 atm. are required for making the melts bubble-free. Similar pressures should probably be applied to common glass melts. In the inverse case of reducing the external pressure large bubbles are evolved which bloat the glass up to produce a typical foamy glass as discussed in B. ¶ 73 f., and as it was impressively described by E. W. Washburn, F. F. Footitt, and E. N. Bunting in their classical investigation on gases dissolved in glass (1920).[127]

[125] *Trans. Met. Soc. AIME,* 1958 (August) 514-519.

[126] *Steklo i Keram.*, **8**, 1951 (12) 13-17.

[127] In the German DR Patent No. 492,004 (January 30, 1930) the vacuum-bloating process is used to counteract collapsing of foamed glass bubbles on cooling by a vacuum after-treatment.

During the well-known "out-gassing" of glass in a high vacuo at temperatures not far from the softening point water is found to be the most important constituent of the released gases. B. J. Todd[128] observed that there is a linear relation between the square root of the "bake-out" time and the gas volume evolved from a unit surface above 300°C. For common soda-lime silicate glasses one may calculate the diffusion constant of water and its concentration gradient after "bake-out" which is an exponential function of the reciprocal value of the absolute temperature of bake-out. The activation energy of the water diffusion process for different glasses varies between 36,000 cal./mole for lead-potassium silicate and borosilicate glasses up to 75,000 cal./mole for Vycor glass with 96 per cent silica. The diffusion constant for water between 430° and 530°C. in the glass increases from 6.1×10^{-13} to 6.1×10^{-11} cm.^{2}/sec.

86. The so-called "solubility" of elemental gases in Pyrex glass at 1170°C. was determined by J. J. Naughton[129] for helium, argon, hydrogen and oxygen. Only that of hydrogen is measurable in the amount of 0.060 ± 0.004 cm.^{3}/gm. glass (under standard conditions) measured under a pressure of 10 mm. Hg. The dissolution process is reversible and does increase slightly with increasing pressures. More from the glass-technological viewpoints, Ts. Okamura[130] studied the absorption of CO_2, SO_3, and H_2O in industrial soda-lime-magnesia silicate glasses, starting from analyses of the gas contents of bubbles in the glass, especially in that made from sulfate batches. In agreement with the results of J. B. Todd (see above) it is rather surprising that all of the bubbles examined were found to contain one single gas prevailing, but no mixtures of about equal parts of different gases. Air bubbles were observed more in the initial stages of fusion and SO_3 bubbles subsequently increase with gradually increasing temperatures and fining. The gas content of the molten glass also rapidly decreased from the initial stage of refining to the finished homogeneous glass, and no remarkable change was observed after this end state. Decreases chiefly concern CO_2 and SO_3, whereas water revealed little change from the fining to the end state. The moisture content of the batch (cf. B. ¶ 72) did not much affect the water content of the molten glass, but a replacement of carbonates in the batch by the hydroxides immediately raised the observed water contents and reduced the CO_2 present in the melt. It made, on the other hand, a considerable difference

[128] *J. Appl. Phys.*, **26**, 1955, 1238-1243.

[129] *J. Appl. Phys.*, **24**, 1953, 499-500. New data on the solubility of helium in Li-silicate melts are presented by H. Scholze and H.-O. Mulfinger, *Angew. Chem.*, **74**, 1962, 75-76; *Veröffentl. Max Planck-Inst. Silikatforschg.*, **22**, 1962, 243. More detailed discussions of methods and results of gas solubility in glass, and the problems of "water" in glass see by the same authors (and H. Franz), *ibid.*, pp. 251-269, and in "Advances in Glass Technology," Plenum Press, New York, 1962, 230-248.

[130] *Asahi Garasu Kenkyu Hokoku*, **2**, 1952, 1-10; **4**, 1954, 1-7, 8-18 (with T. Uno); *Ibid.*, **6**, 1956, 75-88, 89-97.

whether the melts were prepared in an open platinum crucible, in one of large, or another of small size, or in an open pot, or in the tank furnace.

87. Okamura also measured the absorption of CO in a soda-lime-magnesia silicate glass under 1 atm. pressure, at 1100° to 1400°C. None of the glasses really took up CO as such in perceptible amounts, but H_2O was abundantly evolved. CO acts only indirectly by a reaction with the iron oxide traces in the glass reducing it in the normal way to greenish-blue or yellowish-brown colored solutions by the formation of FeO and FeS (coal-brown) (cf. B. ¶ 36). CO also reduced the sulfate with the evolution of secondary SO_2 as the most characteristic gas constituent in the bubbles. A characteristic relation of the gas evolved *in vacuo* as a function of the temperature at which CO had reacted is given in Fig. B. 52; the corresponding Liquidus temperatures are shown in Fig. B. 53. If the melts were made in pure argon at 1 atm. pressure only water, but no argon, was subsequently evolved in a high *vacuo* regardless of temperatures at which the glass had been molten. Also the Liquidus temperatures were practically not changed at all. The method of microanalysis of gases included in bubbles in comparison with the gases evolved from the vacuum fusion, was applied by Ts. Okamura, T. Sasaki, and Sh. Uemura[131] for samples systematically taken from different portions of the glass bath in a tank furnace. The majority of these bubbles contained SO_2 as the main constituent, very few of them CO_2 in the place of SO_2, or air when there were only few seeds. The contact of the

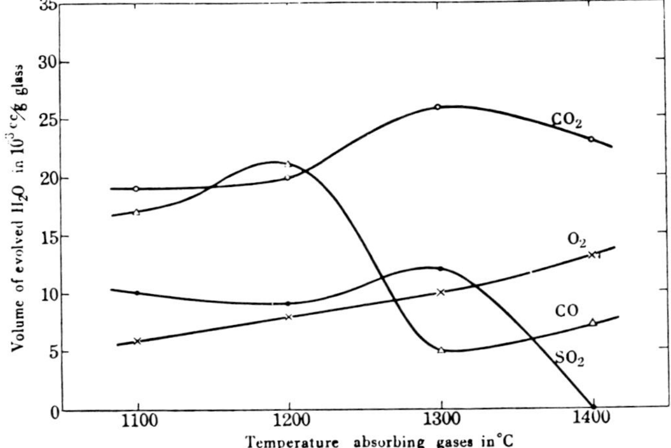

FIG. B. 52. Volume of water evolved from a soda-lime-magnesia silicate glass, as a function of temperature, and of gases which had reacted ("absorbed") by the glass. (Okamura).

[131] *Asahi Garasu Kenkyu Hokoku*, **9**, 1959, 1-12, 13-29.

glass bath with the refractory material of the walls develops air-filled bubbles in the first period of the service of the brick, but after longer periods the bubbles from the surface of the refractories were enriched prevailingly in SO_2, thus indicating considerable influence of the brick material on the gas contents of the glass in toto, a factor of great practical importance.

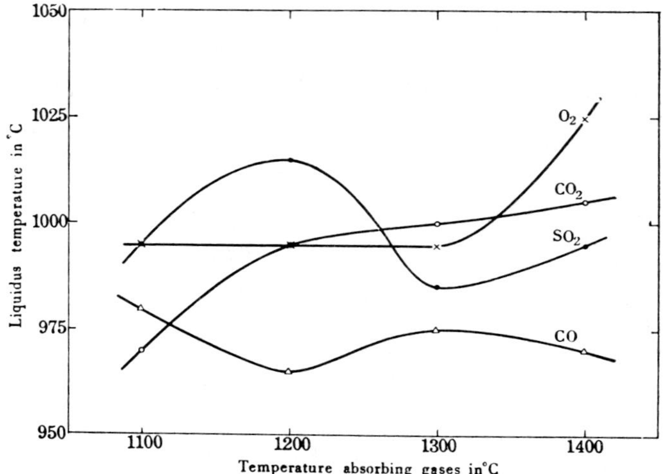

FIG. B. 53. Liquidus temperature of the same glass, after "absorption" of different gases. (Okamura).

88. The kinetics of reactions of gaseous HCl (or HBr) with the surface of Pyrex glass (No. 7220) in the form of wool (of about 10^4 cm.^{2} total surface in the experiments) and over temperature range between 295° and 385°C. was studied by J. E. Boggs and H. Mosher.[132] Changes in the HCl pressure were used as a measure for determination of rates of the reaction. Water formed was frozen out in a trap cooled down to — 78°C. The rate-determining reaction step is a diffusion process through the glass. Since it is rather improbable at the relatively low temperature range of the observed reaction that a diffusion of HCl through the glass bulk takes place, a dissolution of this gas in a layer of adsorbed water on the glass surface must be assumed, with a subsequent diffusion of protons into the glass, and of sodium ions to the surface. J. E. Boggs, L. L. Ryan, and L. L. Peck[133] studied the analogous reaction of HBr with Pyrex glass at 300° to 450°C. and gave an explanation of the observed reaction kinetics. They also considered the antagonistic diffusion of protons and alkali ions through the glass. This ionic counter-diffusion process is, in every

[132] *J. Am. Chem. Soc.*, **78**, 1956, 3901-3903.

[133] *J. Phys. Chem.*, **61**, 1957, 825-827.

case, the step controlling the reaction rate observed. The strictly linear function of the reaction constant, ln K, versus $1/T$ has for HCl and HBr practically the same slope, but the line for HCl is slightly below that for HBr.

89. Because of the high theoretical and practical importance of the problem involved with the *water* content of glass melts, J. W. Tomlinson[134] studied this phenomenon with alkali and alkaline earth silicate melts. He saturated these melts with water vapor under partial pressures of 0.064, 0.117, 0.238, and 1.0 atm. at 900° 1000°, 1100°C., and determined the "absorbed" water by evacuation at 1200°C. for a period of 3 to 10 hours. The amount of water evolved was measured by manometric methods. At 1100°C. and 1 atm. water vapor pressure about 3 cm.^{3}/gm. glass (standard conditions) were taken up. Over the range from 900° to 1100°C. the solubility decreases only a little (about 10 per cent). In every case, the solubility data are not proportional to $(p_{H_2O})^{1/2}$ as they should be if a reaction with the silicate of the type

$$H_2O + \equiv Si{-}O{-}Si \equiv \rightarrow \equiv Si{-}O{\cdot}H + H{\cdot}O{-}Si \equiv$$

would occur. It is below the amounts to be expected if the partial pressures are low (Fig. B. 54). Tomlinson assumes a partial dissociation of the alkali silicate melt in vacuo, a release of atomic sodium which is condensed on cooler parts of the reaction chamber, and thus withdraws water from being condensed in the cold trap.

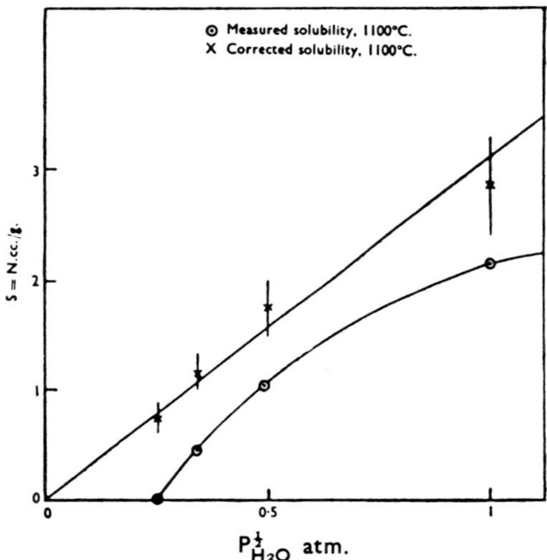

Fig. B. 54. Solubility of water vapor in sodium disilicate melt at 1100°C., as a function of the square root of vapor pressure, observed, and corrected. (Tomlinson).

[134] *J. Soc. Glass Technol.*, **40**, 1956, 25-31 T. On the removal of water from glasses see R. V. Adams, *Phys. & Chem. of Glasses*, **2**, 1961, 50-54.

90. L. E. Russell[135] applied the principle of expelling the water from alkali silicate melts by a current of a highly dried inert gas (Fig. B. 55 shows the apparatus for this operation). This water extraction from an initially water-saturated glass melt at a partial pressure of 1 atm. is of the order of 1 cm.^{3}/1 gm. glass at 1400°C. (meas-

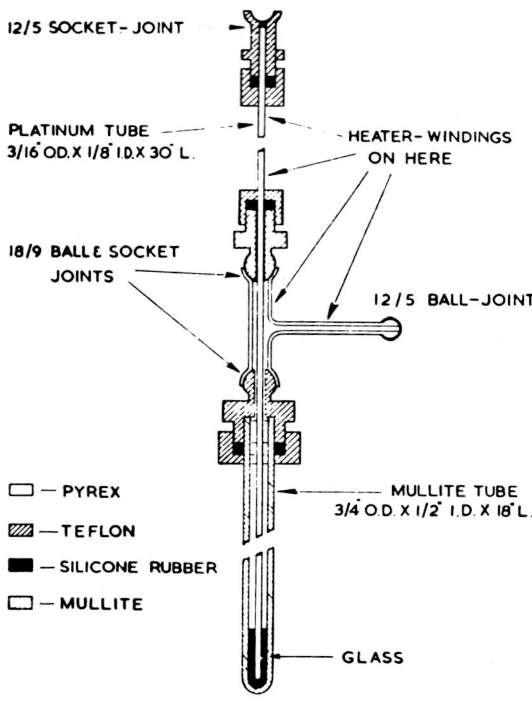

Fig. B. 55. Details of the apparatus for driving out water vapor absorbed in alkali silicate melts, by a dry current of inert gas. (Russell).

ured at standard conditions). The temperature coefficient of the water solubility in disilicate glasses (of alkalies, and alkaline earths) is distinctly a function of the field strengths, z/a^2, of the cations. If the field strength is high (as for Li^+, Sr^{2+}, Ba^{2+}) the solubility increases with rising temperature, but decreases if the field strengths are low (as for Na^+, K^+, Cs^+) (Fig. B. 56). The solubility of water as a function of the various partial pressures strictly follows in this case the square root law, namely in proportionality with $\sqrt{p_{H_2O}}$. The heat of gas evolution,

$$\Delta H_g = R \frac{T_1 T_2}{(T_2 - T_1)} \cdot \ln \left(\frac{S_2}{S_1} \right)^2$$

[135] *J. Soc. Glass Technol.*, **41**, 1957, 304-317 T.

(S the solubility $= k \cdot \sqrt{p}$), is also a linear function of the field strength, z/a^2. A negative value of ΔH_g corresponds to a positive temperature coefficient of the water solubility.

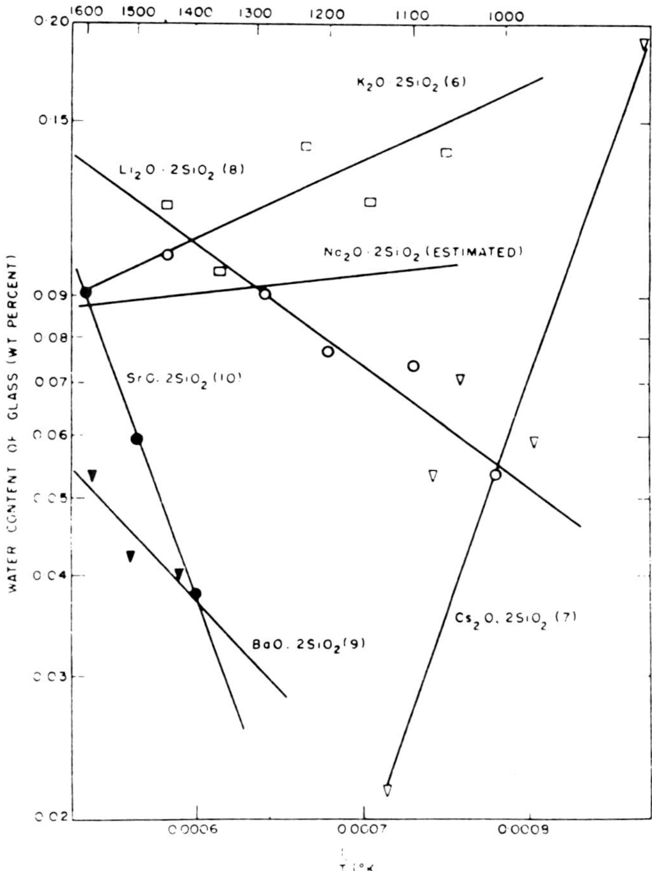

Fig. B. 56. Variation of the water vapor solubility in different disilicate glasses, as a function of temperature. (Russell).

91. With C. R. Kurkjian, Russell[136] also investigated the water vapor solubility in fused alkali silicates at equilibrium pressures, for variable alkali concentrations, and as a function of temperature (Fig. B. 57). The solubility–composition curves show a pronounced minimum (Fig. B. 58) for about 25 molecular per cent R_2O which is explained by two antagonistic mechanisms, side-by-side, of the water so-

[136] *J. Soc. Glass Technol.*, 42, 1958, 130-144 T.

lution reaction. For the first mechanism, solubility is increased with the R_2O contents, and the reaction is for high activities a_{Na_2O}:

$$\begin{array}{ccc} & \text{Na} & \\ \equiv \text{Si—O} & & \text{O—Si} \equiv + \text{H}_2\text{O(gas)} \rightarrow \equiv \text{Si—O—Si} \equiv + 2\,\text{Na}^+ + 2\,\text{OH}^- \\ & \text{Na} & \end{array}$$

For the other mechanism, with a solubility decreasing with increasing alkali contents and with high activities a_{SiO_2}, the reaction is:

$$\equiv \text{Si—O—Si} \equiv + \text{H}_2\text{O(gas)} \rightarrow \equiv \text{Si—O-H/H-O—Si} \equiv$$

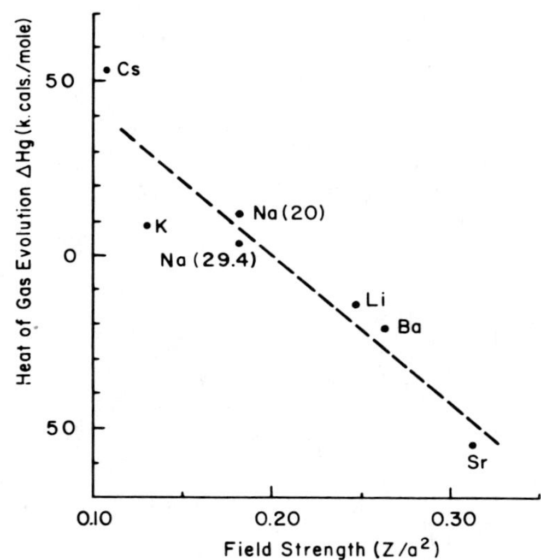

Fig. B. 57. Heat of water vapor evolution from disilicate glasses, as a function of field strengths, z/a^2, of the cations. (Russell).

The first (high-alkali) type of reaction is connected with *hydrogen bonding mechanisms* as they are confirmed by the infrared absorption data given by H. Scholze (see below). From this viewpoint the high-alkali side of the minimum in Fig. B. 58 follows the reaction with hydrogen bonding as written in the form

$$2\left\{\begin{array}{ccc} & \text{Na} & \\ \equiv \text{Si—O} & & \text{O—Si} \equiv \\ & \text{Na} & \end{array}\right\} + \text{H}_2\text{O(gas)} \rightarrow 2\left\{\begin{array}{c} \text{Na} \\ \equiv \text{Si—O-H-O—Si} \equiv \\ \text{Na} \end{array}\right\} + \text{O}^{2-}$$

92. The fundamental importance of *infrared transmission* measurements in the short-wave length region between 2.5 and 5.0 μ for a solution of the problem in which form "water" is absorbed in glass melts, will be discussed in detail in Vol. IV, Section B, in connection with the investigations of J. M. Florence, Fr. W. Glaze, and M. H. Black.[137] H. Scholze and A. Dietzel[138] emphasized the possibilities of distinguishing the types of binding mechanism of free OH^- groups by the absorption band at 2.75 μ, and of the hydroxyl group in the associated state at about 2.95 μ, in connection with the "moisture" content of gases in the furnace atmosphere of

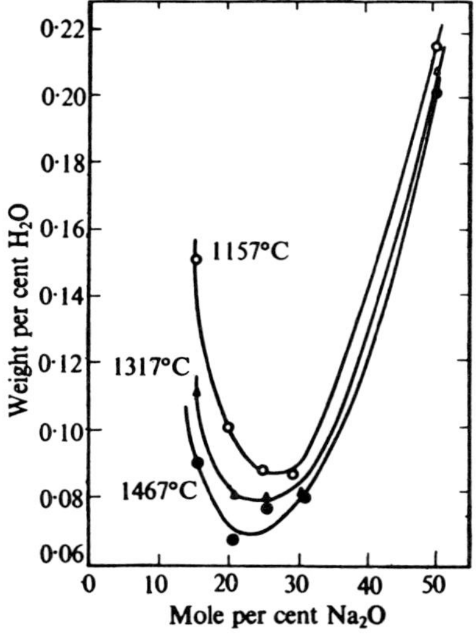

Fig. B. 58. Variation of water solubility as a function of the Na_2O contents in silicate glasses. (Russell and Kurkjian).

experimental glass melts (cf. A. ¶ 256 f.). The more elaborate measurements performed by both authors[139] (see also B. ¶ 69 on water as a "reaction residue" in glass) showed that an important influence on the position of characteristic infrared bands is exerted by the alkali ions of strong electrostatic field potentials (e.g., Li^+), or moderate effects (Na^+, K^+), by their polarizing action. The type of shiftings thus observed in alkali silicate glasses is shown in Fig. B. 59. Scholze comes to the important conclusion that the variability of the band from 2.75 to 2.95 μ makes

[137] *J. Research. Natl. Bur. Standards*, **50**, 1953, 187-196. See also the interesting infrared absorption method used by D. K. Priest and A. S. Levy *J. Am. Ceram. Soc.*, **43**, 1960, 356-358, to study the correlation of water content of a borosilicate glass with its chemical corrosion resistance.

[138] *Naturwissenschaften*, **42**, 1955, 342-343. See also R. S. McDonald, *J. Phys. Chem.* **62**, 1958, 1168-1178, on infrared spectroscopy applied to a study of amorphous silica.

[139] *Glastech. Ber.*, **28**, 1955, 375-380; *Veröffentl. Max-Planck-Inst. Silikatforsch.*, **16**, 1956, 135-140; *Atti congr. intern. vetro, Venice*, **4**, 1956, VIII. 3, pp. 1-6.

a clear decision impossible for distinguishing molecular water, or associated hydroxyl groups, in the cavities of the glass framework structure. In every case, it is also erroneous to assume that the band at 3.6 to 3.7 μ should belong to the CO_3^{2-} groups of carbonate residues, as was previously done by Florence *et al.* This latter band is certainly caused by hydroxyl groups the H^+ ion of which is in a strong bridge-bond-

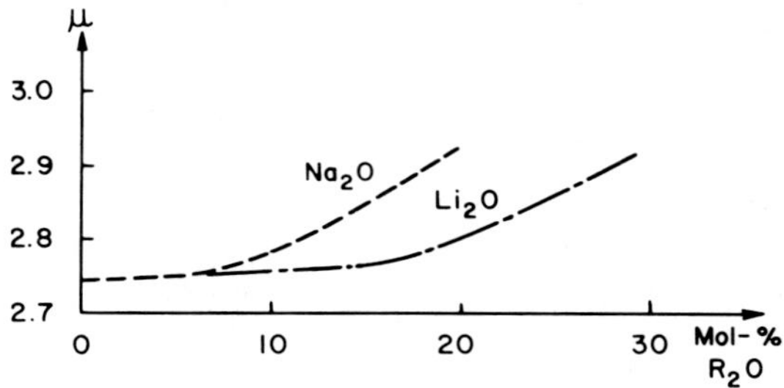

FIG. B. 59. Variation of the characteristic hydroxyl absorption band in the short-wave infrared range for binary silicate glasses with Li_2O and Na_2O. (Scholze).

ing with a simply-bound oxygen anion and a neighboring oxygen ion in a distance of 2.66 Å. In principle, the decision on the presence or absence of free water molecules could only be made in the positive if the deformation oscillation at 6.25 μ could be identified. Unfortunately, it lies in a wavelength region in which the strong silica absorption band entirely overlaps the H_2O effects.[140]

93. R. Brückner and H. Scholze[141] in a more detailed series of investigations first studied the behavior of infrared "water" bands in freshly prepared and in "weathered" B_2O_3 glass. The conditions for preparing anhydrous B_2O_3 melts were varied extensively, the most efficient method being the fusion in an atmosphere of BCl_3 at 1350°C. The "weathering" of rehydration was made as well by H_2O as by D_2O vapors for varying exposure times. Completely anhydrous B_2O_3 does not show any 2.8 μ bands; the weathered glass shows the very strong absorption of H_3BO_3 (sassoline) with hydroxyl valence oscillations at 3.1 μ, and 8.4 μ (OH deformation oscillation, parallel to the structure layers of H_3BO_3). The OH groups in the glass structure are made evident by the 2.8 μ band, the intensity of which is a function of the water content. But with increasing degrees of weathering this band is gradually overlapped by the 3.1 μ band of H_3BO_3. If OD is present in the place of OH, the characteristic band is displaced to 3.8 μ (cf. Fig. B. 60), those of H_3BO_3 to 4.15 and 11 μ, for D_3BO_3.

[140] On the Si—O oscillation band at 6.2 μ cf. D. G. Drummond, *Proc. Roy. Soc.* (*London*), A, **153**, 1936, 328-339.

[141] *Glastech. Ber.*, **31**, 1958, 417-422.

94. Corresponding studies of Scholze with sodium and sodium-calcium silicate glasses[142] concern the possibility of a quantitative determination of the "water" dissolved in glass from the extinction coefficient $\varepsilon = 1/cd \cdot \log D$ (c = the water concentration, d the thickness of the sample, D the transmissivity), and a determination of the integral absorption, $\int \varepsilon d\nu^*$ for the bands. The results of such deter-

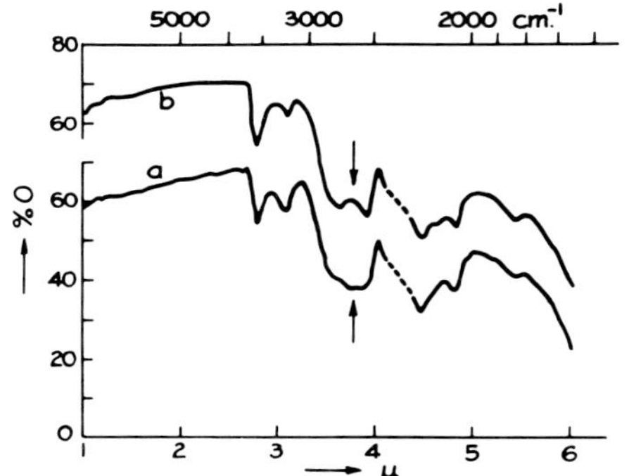

Fig. B. 60. Infrared absorption curve of B_2O_3 glass, (a) containing in the same time OH and OD absorption bands, in comparison with a similar B_2O_3 glass (curve b) which only shows the OH band. (Brückner and Scholze). Thickness of sample 50 to 60 μ.

minations vary between 0.02 and 0.06 per cent for silicate glasses. As a general rule, it is observed that the greater the wavelength of the absorption bands, the lower the wave number ν^*, the stronger the H bridge bonding to neighboring oxygen anions, the lower the distance O—O. The coefficient ε as a function of ν^*, and $\int \varepsilon d\nu^*$ as a function of ν^* are for the characteristic absorption bands about on one line with those determined for liquid water (dissolved in CCl_4) and in ice. The strong effect of washing-out of water from the glass by a current of a dry inert gas is evident from Fig. B. 61.

95. Systematic variations of the chemical composition of binary, ternary, and quaternary glasses in their effects on infrared absorption characteristics were made by H. Scholze.[143] Without going into a detailed discussion of the results here, it may be emphasized again that the interpretation of the OH bands cannot decide the problem of whether the 2.75 μ band corresponds to molecular water or to hydroxyl groups in —Si—OH bonding similar to that in organic silanols. The systematic shifting of the short-wave band with increasing alkali contents (see above) makes

[142] *Glastech. Ber.*, **32**, 1959, 81-88. On the infrared absorption of water in CCl_4, and of ice cf. J. J. Fox and A. E. Martin, *Proc. Roy. Soc.* (*London*), *A*, **174**, 1940, 234-262.

[143] *Glastech. Ber.*, **32**, 1959, 142-152, 278-281. On the hydroxyl groups in organosilanols cf. K. Damm and W. Noll, *Kolloid-Z.*, **158**, 1958, 97-108.

it *indirectly* evident that water is introduced into the glass structure really as hydroxyl groups. For the 2.85 μ band which is independent of temperature, the same binding mechanism is valid for all of the glasses investigated, and the 3.6 μ band is caused by hydroxyl groups in strong hydrogen bonding. Tha variability between 3.35 and 3.85 μ indicates the variability of the O—O distances in the OH—O bridges, between 2.6 and 2.7 Å. The groups O—Na favor the formation of such bridging bonds. Therefore, the higher the alkali contents the higher the intensity of the corresponding band near 3.6 μ, but the wave length is increased with decreasing field strength, z/a^2, of the alkali ions. On the contrary, alumina in the glass counteracts

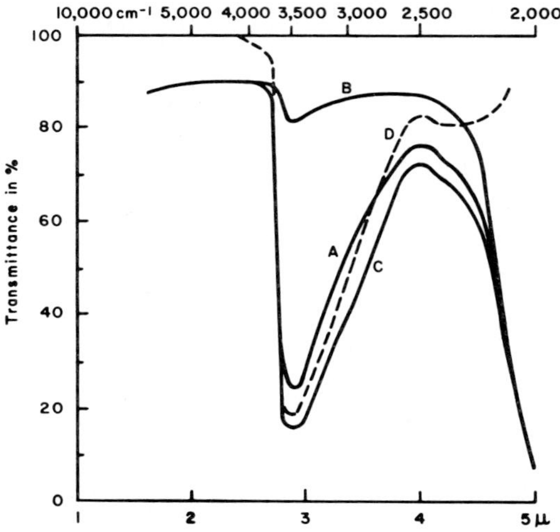

FIG. B. 61. Infrared absorption curves of a water containing sodium-calcium aluminum silicate glass, with 20 mol. per cent Na_2O, CaO, Al_2O_3, and 40 per cent SiO_2. (Scholze). (*A*) is the curve for the initial glass, (*B*) of the same glass, washed-out by a dry current of nitrogen, (*C*) washed-out with a mixture of 50 : 50 nitrogen and water vapor, (*D*) difference curve of (*B*) and (*C*).

those alkali effects, and a glass with 20 molecular per cent soda, 20 per cent alumina does not show any more hydroxyl bonding with hydrogen bridging. The band at 4.25 μ is rather invariable in its position, indicating an extremely strong hydrogen bridging bond with the O—O distance of 2.55 Å. It chiefly occurs in alkali-rich glasses. The introduction of CaO, MgO, and the network-forming cations Zn^{2+} and Be^{2+} increases the number of free hydroxyl groups (cf. Fig. B. 62). If titanium is introduced into soda and baryte silicate glasses, the band at 3.6 μ is reduced in favor of that at 2.9 μ. $[TiO_6]$ groups withdraw oxygen anions from the O—R bonds.

96. *Fluorine* containing silicate glasses, or free aluminate, and borate glasses also show an increase in free hydroxyl groups (corresponding to the 2.85 μ peak), relative to the firmly bound hydroxyl groups (bands at 3.6 μ) (cf. Fig. B. 63). In alkali borate glasses the free hydroxyl groups are even present alone (shifted in wavelength from 2.81 to 2.90 μ). The ions Li^+, Na^+, K^+ have specific effects on the final position of the bands (2.88 μ for Li and Na, 2.90 μ for K) (cf. Fig. B. 64). Concerning the influence of temperature changes, Scholze[144] finds a general trend to change the firmly

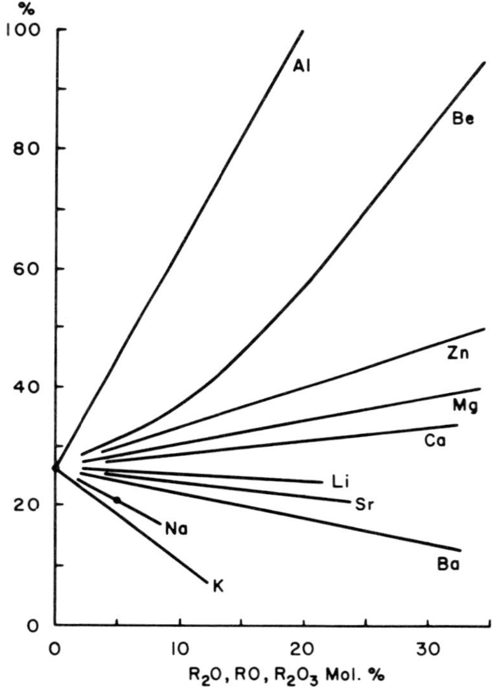

FIG. B. 62. Share in free hydroxyl groups in total water content in glasses with 20 mol. per cent Na_2O, if silica is systematically replaced by R_2O,RO, or R_2O_3 oxides. (Scholze).

bound hydroxyl groups into free OH with rising temperature. The intensity of the band at 3.6 μ decreases, its position is shifted to shorter wave lengths, indicating a loosening in the glass framework. The concentration of free hydroxyl groups is highest in lithium aluminosilicate glasses, somewhat less in sodium-calcium silicate glasses, and in potassium silicate glasses they are still relatively low even at 1300°C. (cf. Fig. B. 65).

97. The *diffusion coefficient of "water" in glass* was determined by H. Scholze

[144] *Glastech. Ber.*, **32**, 1959, 314-320. On transmission in the near infrared at temperatures up to the freezing point range (cf. A. ¶ 259), cf. *Ibid.*, **32**, K, 1959, VII. 5, 1-5.

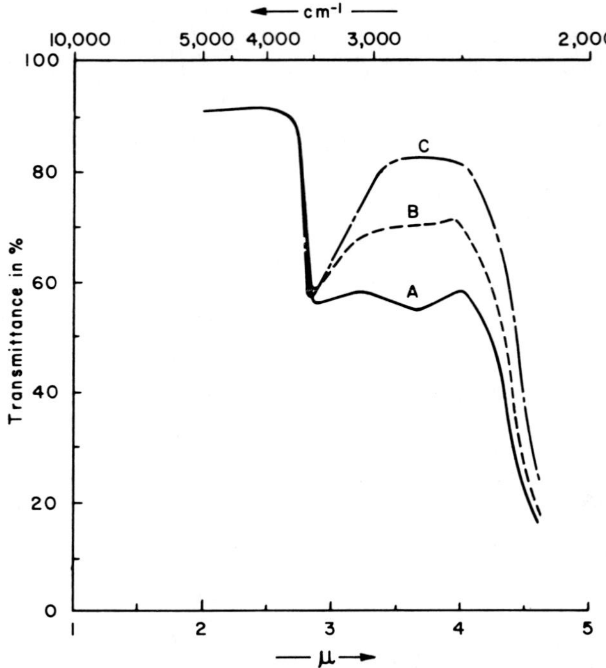

FIG. B. 63. Infrared spectrum of fluorine containing sodium-aluminosilicate glasses (20 mol. per cent Na_2O, 10 Al_2O_3, 70 SiO_2), (*A*) without fluorine, (*B*) with 3 per cent fluorine, (*C*) with 6.4 per cent fluorine. (Scholze).

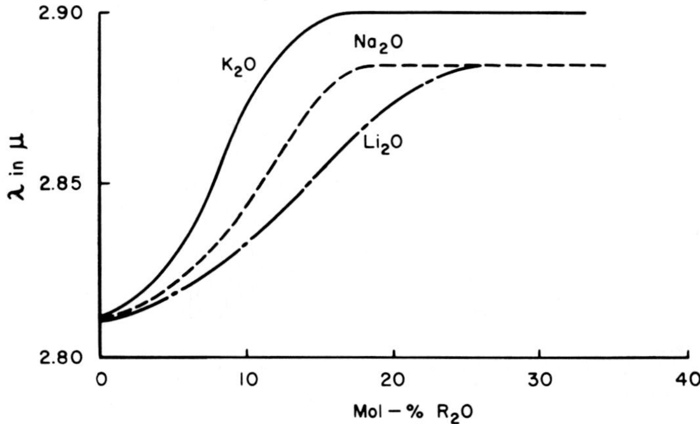

FIG. B. 64. Shifting of the hydroxyl bands in binary alkali borate glasses, as a function of the alkali concentrations. (Scholze).

and H.-O. Mulfinger[145] by the determination of changes in the transmissivities in the range from 2.8 to 3.8 μ by a unilateral exposure of the glass surface to a current of dry nitrogen and the ensuing transfer of water from the surface. The order of

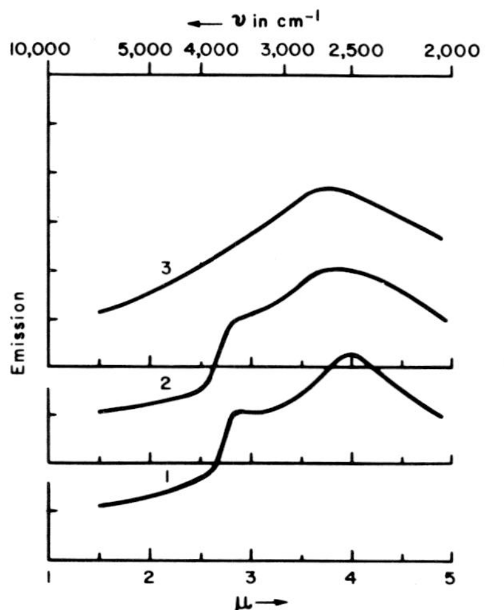

FIG. B. 65. Infrared emission of glass at 1300°C. (Scholze). Curve (1) for glass with 10 mol. per cent Li_2O, 20 Al_2O_3, 70 SiO_2, (2) for glass with 16 mol. per cent Na_2O, 10 CaO, 74 SiO_2; (3) for glass with 30 mol. per cent K_2O, 70 SiO_2.

magnitude for the diffusion coefficient, D, is 10^{-6} cm.^{2}/sec., increasing with temperature and the alkali content of the glass, and also with increasing field strength, z/a^2, of the alkali ion (cf. Fig. B. 66). The activation energy for the diffusion, AE_D is related to that for the viscous flow AE_η, by the equation

$$\log D = -\log \eta - \left[\frac{(AE)_D - (AE)_\eta}{4.57\ T}\right] + 5.82$$

The jumping of the protons from hydroxyl groups to neighboring single-bonded oxygen in the glass constitution determines the mechanism of the diffusion. The activation energy AE_η is of the order of magnitude between 35,000 and 49,000 cal., that for the diffusion AE_D for a counter-diffusion of alkali ions 19,000 to 31,000 cal.,

[145] *Glastech. Ber.*, **32**, 1959, 381-386. Influence of water contents on density and refractive indices, including molecular refraction, of the glasses see investigations of H. Scholze, H. Franz, and L. Merker, *Ibid.*, **32**, 1959, 421-426. See also recently S. Garbe, *Ibid.*, **34**, 1961, 413-417, on high-vacuum release of water from sodium silicate glasses, using for the measurement of partial pressures of gases evolved the Omegatron of A. Klopper and W. Schmidt, *Philips' tech. Rundschau*, **22**, 1960, 221-229.

to be compared with that of electrical conductance as measured by J. O'M. Bockris, J. A. Kitchener, S. Ignatowicz, and J. W. Tomlinson[146] which is only about 9,000 to 12,000 cal. (cf. A. ¶ 126).

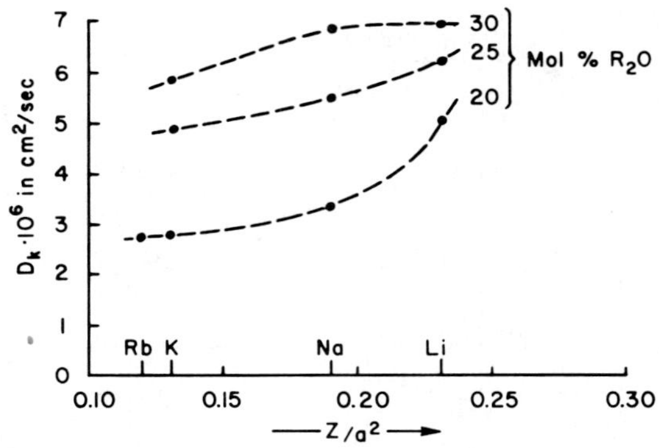

FIG. B. 66. Diffusion coefficient of water in binary alkali silicate glasses, as a function of cation field strength, at 1400°C. (Scholze and Mulfinger).

98. The water content of glasses, and especially its action on increased *volatility of alkalies* from glass melts was discussed by A. Dietzel and L. Merker[147] with relation to the chemical and ensuing optical heterogeneities developed in the final glass. By precision measurements of the refractive indices under varying furnace atmosphere conditions, as in pure nitrogen, or in a mixture of 75 per cent nitrogen and 25 per cent water vapor, or in pure water steam, the specific volatilization of alkalies was determined. It occurs in pure nitrogen between 0.02 and 0.04 mg./cm.²/hour, in the nitrogen-water vapor mixture between 0.08 and 0.15, in steam between 0.13 and 0.23 mg./cm.²/hour. Evaporation losses in alkali are compensated by a migration of the sodium ions from the bulk of the glass to the surface where an exchange of Na^+ and H^+ ions with Ca^{2+} ions takes place. This exchange is also analytically confirmed by an enrichment of Ca^{2+} in the alkali-impoverished surface. Small amounts of water dissolved (order of magnitude 0.1 per cent, determined by the infrared absorption band at 2.95 μ) increase the refractive index, a rather unexpected effect which is explained by the high electrostatic field strength of the protons.

[146] *Trans. Faraday Soc.*, **48**, 1952, 75-91.

[147] *Glastech. Ber.*, **30**, 1957, 134-138. See also *Trav. Congr. Intern. du Verre*, **4**, Paris 1956, II. 3, pp. 1-6, and *Veröffentl. Max-Planck-Inst. Silikatforsch.*, **16**, 1956, 105-110, especially in connection with technological problems of alkali volatilization from glass surfaces, cf. A. ¶ 374, 376. The considerable volatilization of sodium sulfate by interaction of the glass surface with combustion gases of the tank furnace atmosphere is described by J. E. Peña de Castro, *Verres et réfractaires*, **11**, 1957, 89-93.

VISCOSITY OF INDUSTRIAL GLASS MELTS

99. In view of the fundamental importance of viscosity characteristics of industrial glass melts for the technological manipulation of their shaping process, the correlation of this property with temperature and also with the chemical composition of the glass was extensively investigated. Of particular significance in this connection are the investigations of J. Boow and W. E. S. Turner,[148] extending over the temperature range between 500° and 1400°C., chiefly based on measurements with the Margules rotation viscosimeter or by the fiber elongation method. The results were used to develop the "oxide factors" (increments) of constituents of the complex composition of commercial glass types (cf. B. ¶ 114 ff.). For colored glasses their specific heat (radiation) conductance characteristics on cooling will be discussed in B. ¶ 107 ff. which are important for optimum machining rates or for manipulation by hand. Especially for soda-lime silicate glass, A. G. F. Dingwall and H. M. Moore[149] studied effects of the replacement of silica by different oxides on the basis of cation-for-cation exchanges, in which field strength parameters and sizes of the different cations are determining factors of the viscosity. If the temperature for $\eta = 10^{12}$ poises is plotted as a function of the radii of substituting monovalent and divalent cations, the resulting curve (Fig. B. 67) is subdivided in branches with break points

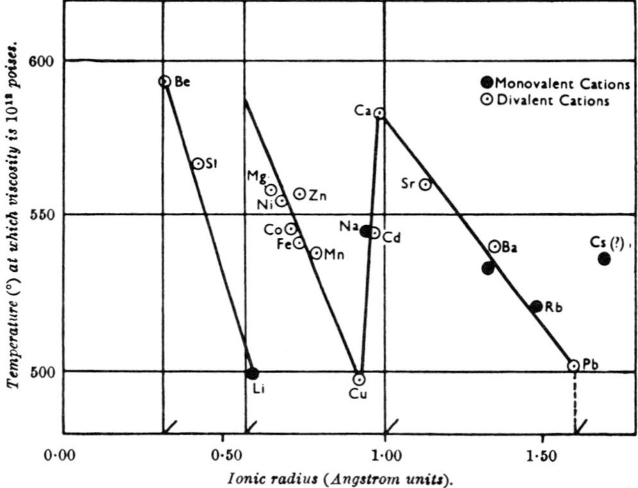

FIG. B. 67. Temperatures for $\eta = 10^{12}$ poises, for glasses containing oxides with monovalent and divalent cations, replacing cation-for-cation up to 8 per cent SiO_2, as a function of the radii of the cations participating. (Dingwall and Moore).

[148] *J. Soc. Glass Technol.*, **26**, 1942, 215-240; **27**, 1943, 94-112, 207-237; **29**, 1945, 199-232, 233-249.

[149] *J. Soc. Glass Technol.*, **37**, 1953, 316-372 T.

at radii of 0.31, 0.60, 1.00, and 1.60 Å., corresponding to ions with a tetrahedral, octahedral, cube-shaped, and higher ($n = 12$) coordination configuration. Viscosity is highest for the smallest cation on each branch of the curve, and systematically diminishes with increasing cation radius, but in the break points a sudden change takes place from one to the next coordination type with an abrupt increase in viscosity. Most interesting also are tables of *activation energies* of viscous flow at 1300°C., and for the temperature $t(10^{12})$, corresponding to $\eta = 10^{12}$ poises, for different glass-technologically important cations (see the accompanying tabulation).

Substituting cation	Log viscosity at 1300°C.	Temperature at which the viscosity is $= 10^{12}$ poises $[t(10^{12})]$	Activation energy of viscous flow in kcal./mole	
			at 1300°C.	at $t(10^{12})$
Base glass	2.34	566°C.	42.5	145
Monovalent cations				
Li^+	1.94	500	37.8	129
Na^+	1.99	544	39.4	155
K^+	2.04	533	39.7	139
Rb^+		522		128
Cs^+		536 (?)		142 (?)
Divalent cations				
Be^{2+}		593		162
Mg^{2+}	2.11_5	558	41.7	141
Ca^{2+}	1.92	583	41.4	157
Sr^{2+}	1.87	560	40.0	147
Ba^{2+}	1.84_5	540	38.0	142
Ni^{2+}	1.95	555	38.8	148
Co^{2+}	1.95	545	38.3	145
Mn^{2+}	1.91	538	39.4	138
Cu^{2+}	1.88	498	36.0	127
Zn^{2+}	2.00	557	41.4	147
Cd^{2+}	1.88	544	38.6	143
Pb^{2+}	1.73	502	37.2	126
Fe^{2+}		541		137

100. For glasses of the system $Na_2O—PbO—SiO_2$, M. N. Skornyakov,[150] and B. A. Pospelov and K. S. Evstrop'ev[151] presented detailed data at high temperatures and near the softening range, with nomograms (cf. A. ¶ 31). Direct relations between the chemical composition, viscosity, and the machining properties of industrial glasses were also tabulated with instructive graphs by A. K. Lyle,[152] including equations and graphs for the softening temperatures as a function of the annealing point and the relative machine speeds for commercial glasses (cf. A. ¶ 69). Such nomograms are of a high value practically for a rapid orientation on the properties to be expected if glass compositions must be changed. They offer the possibility of making sufficiently accurate interpolations in place of time-consuming measurements. Also M. V. Okhotin[153] developed similar binominal equations and nomograms for soda-lime-magnesia (3 per cent constant)-alumina silicate glasses, with correction tables for magnesia contents up to 5 per cent, for temperatures corresponding to viscosities between 10^3 and 10^{13} poises. Another investigation concerning the temperature range between the annealing and softening points (determined by the fiber elongation method) of different commercial glasses was made by M. V. Okhotin and R. I. Tsoĭ.[154] V. T. Slavyanskiĭ, M. P. Novikova, L. V. Isaeva, and E. N. Krestnikova[155] carefully determined the temperature for $\eta = 10^2$ and 10^4 poises, additionally the "length," $l = t(100) - t(10{,}000$ poises) of different glasses, and for the standard $Na_2O.2SiO_2$ glass up to 1380°C., to develop a functional scale for viscosity as a function of temperature. For $Na_2O.2SiO_2$ the data of viscosity fit well on lines for $t(100)$ and the length; on the average, linear functions also fit for optical glasses like several Flints, Baryte Flints, and heavy Crown glasses. A nomogram is given for glasses of the systems $Na_2O—PbO—SiO_2$, $K_2O—PbO—SiO_2$, and $Na_2O—K_2O—PbO—SiO_2$, in viscosity vs. $t(100$ poises) curves. Factors for potassium-lead silicate glasses and for quaternary lead glasses are also linear in the limits of $Na_2O + K_2O = 3$ to 10 mole. per cent, and PbO from 10 to 35 per cent. The accuracy of calculated viscosity data is 12° for $t(100)$ and 15° for $t(10{,}000)$.

101. J. Herbert,[156] from the practical viewpoint, developed a method for defining the working range in the viscosimetric-plastic behavior of industrial glasses by plotting $\log(10^{13}/\eta)$ as a function of temperature. This function shows two linear

[150] See in I. V. Grebenshchikov, "Physical-Chemical Properties of Glasses of the ternary System $Na_2O—PbO—SiO_2$," Symposium (Sbornik Stateĭ), under auspices of the Academy of Sciences, Moscow 1949, 219 pp., especially pp. 39-69.

[151] *Ibid.*, pp. 70-82.

[152] *Ind. Eng. Chem.*, **46**, 1954, 166-170.

[153] *Steklo i Keram.*, **11**, 1954 (1) 7-11.

[154] *Steklo i Keram.*, **10**, 1953 (6) 11-13. See also Anonymus, *Silikat Tech.*, **5**, 1954, 503-504, for $\eta = 10^3$ to 10^5 poises, with excellent nomograms and valuable literature references.

[155] *Optiko-Mekh. Prom.*, 1958 (1) 53-58.

[156] *Verres et réfractaires*, **8**, 1954, 74-88.

branches with an intersection point for $\log(10^{13}/\eta)$ = constant = 7.5 (for $\eta = 5.5$) as the transition phenomenon from plastic to fluid states, and vice-versa, for most of the commercial glasses. Only those with a composition high in soda and boric acid behave anomalously (cf. A. ¶ 74). The corresponding temperature called by Herbert "T_2" is near Lillie's "flow point," and practically coincides with the temperature of the maximum rate of devitrification. If T_1 is the temperature corresponding to $\eta = 10^{13}$ poises, then the range of plasticity is outlined by the empirical formula $\log(10^{13}/\eta) = A \cdot \log(T/T_1)$, the fluidity range by $\log(10^{13}/\eta) = B \cdot \log(T/T_2) + C$. If the Littleton point (cf. A. ¶ 30, 39) is known by the condition $\log \eta = 7.5$, the constant A is given by $A = (\log T_2 - \log T_1)/5.35$, and as an approximation can be used $\log T_2 = (A \cdot \log T_1 + 7.5)/A$. The viscosity corresponding to the temperature T_2 is on an average = 8.2, or higher than for the Littleton point.

102. Particular requirements in viscosity must be fulfilled by glass melts which are used as a *lubricant* in the high-pressure metal extrusion process in the patented shaping method after L. Bataille and J. Séjournet.[157] I. Peychès and J. Séjournet[158] described the process applied especially for copper, brass, aluminum, but also for molybdenum and related metals, in extruding through a nozzle. Even steel can be extruded in this way under a pressure of 10,000 kg./cm.^{2} at 1200°C. to a rod, using a prop of compacted glass fibers, or fritted glass powder, which must have at operation temperature (determined by the plasticity of the metal to be extruded) a viscosity of the order of magnitude of 100 poises, sufficient to glide over the surface of the nozzle and along the walls of the extrusion chamber. The heat transfer from the pressing cylinder to the glass is a kind of diffusion problem, in the case of steel bar extrusion extending over a temperature range from 100° to 1150°C. The glass film after operation shows a characteristic zonal structure. The temperature near the nozzle was not above 450°C. followed by a fritted layer which underwent a thermal exposure to about 650°C. and another zone with gas bubbles evolved after release from the very high pressure exerted on the glass. The lubricant film of about 20 μ thickness was the only real liquid. The viscosity of the glass must fit the plasticity behavior of the metal extruded, the fusion point must, therefore, be low for copper-aluminum alloys, high for molybdenum. Quite particular is the *instantaneous* condition of the viscosity changes following "temperature jumps" which are not comparable with the ordinary conditions of viscosimetry.[159] The suitability of a glass

[157] U. S. Patent of L. Bataille and J. Séjournet, No. 2,538,917, (1951). Ordinary window glass is recommended, but also sodium borate or metaphosphate glasses may be used. See also J. Buffet and A. Collinet, *Glass Technol.*, **2**, 1961, 199-200.

[158] *Verres et réfractaires*, **8**, 1954, 131-135. Similar processes for shaping high-strength titanium alloys (with aluminum and vanadium), as for forming sheet metals with enamel type glasses as lubricants, were described by W. M. Sterry and Fr. H. Simpson, *Am. Ceram. Soc. Bull.*, **35**, 1956, April, Program p. 25; *Ceram. Ind.*, **68**, 1957 (5) 92-94, 113.

[159] On the physical meaning of instantaneous viscosity in chilled glasses cf. R. Guy, *Silicates*

as a lubricant is influenced by the molecular cavities in its structure; annealed glass fibers behave, therefore, better than quenched fibers of the same composition.

103. Also L. K. Kovalov and V. A. Ryabov[160] discussed the technological conditions of glass lubrication and the external friction of glass to metal in hot-pressing, which in the coefficient 0.027 to 0.033 is comparable to that of graphite mixed with petrolatum on metal (0.026 to 0.030). Important also are the wetting conditions of both phases which are favorable for nickel and chromium alloys but very difficult for molybdenum and wolfram alloys. Titanium alloys become wettable with glass at 1050° to 1100°C. if the glass contains Li_2O, BeO, and/or B_2O_3. The glass must keep its working viscosity above a critical time for giving optimum lubricant efficiency, when the metal is totally immersed in the glass from every side.

104. The practitioner often comes into contact with the problem of how viscosity of a given glass in a tank bath is affected by the more or less controlled dissolution of refractory material. O. Gott[161] investigated for a Fourcault crown glass the functional relation of the viscosity (and also of the thermal expansion coefficients) to the assimilation degree of silica brick, of fireclay material (with 25 per cent Al_2O_3 content), or even pure alumina, in pilot plant melts. For such a comparison of absolute viscosities the difference of temperatures for $\eta = 10^{4.5}$ and $10^{7.5}$ poises is important. The higher this temperature difference, the lower is the tendency to devitrification. For glass machining, the temperature for $\eta = 134$ poises, and a difference of this temperature determined for the base glass, and that for the glass containing dissolved material, is essential. Alumina added to the glass changes its viscosity (and thermal expansion coefficient) in a much higher degree than silica. The expansion coefficients of the derived glasses are lowered relative to that of the initial glass.

105. The fundamental conditions of the corrosion of refractories as of tank blocks in contact with glass melts concerning their constitution are discussed in Vol. V, Section B. No less important are in this highly complex phenomenon the viscosity and wetting characteristics of the glass melt as such on the solid surface. R. Rasch[162] developed a series of theoretical factors essential for loosening ("activation") of the crystal structure of refractories, with their more or less covalent bonding types in the crystalline phases and surface activities as a function of the composition, but on the other hand, due to the particularly high corrosion exerted by sulfate

inds., **31**, 1956, 69-73 (cf. A. ¶ 30).

[160] *Steklo i Keram.*, **15**, 1958 (7) 8-12. Russian lubrication glasses are relatively high in alkalies; they contain some BaO, and rather much CaF_2 and Fe_2O_3.

[161] *Glastech. Ber.*, **30**, 1957, 139-142.

[162] *Glas-Email-Keramo-Tech.*, **5**, 1954, 393-399. On the kinetic contact conditions between Na–Ca silicate melts and borides, nitrides, and other refractory alloys, see M. Boffé, M. Vantournhoudt, and E. Plumat, "Advances in Glass Technology," Plenum Press, New York, 1962, 580-602.

containing batches and partial glass melts. P. G. Herold and V. P. Maheshwary[163] directly measured changes of the viscosity of corrosive melts in correlation to the refractory material, following the progressive dissolution of the body. A most valuable collection of experience data from tank blocks in service was finally brought together by K. Konopicky.[164]

106. The *wetting* of refractories by glass melts was studied by K. P. Azarov[165] in model experiments for cylindrical shapes of pressed glass powder, high in CaF_2 (12.6 per cent) and alkalies (23.8 per cent Na_2O), at 950°C. on the solid support of CaO, Al_2O_3, quartz, and TiO_2. The wetting was increased when the A. F. Kapustinskiĭ electronegativity[166] increased. The best wetting was observed for glasses containing alkalies and alkaline earths, on SiO_2, TiO_2, Fe_2O_3, Cr_2O_3, and less complete with the basic oxides CaO, Al_2O_3. With increasing ionic radii and decreasing electronegativities, the wetting of acidic solids decreased, that of basic oxides was improved. Inverse wetting phenomena were observed for cations of the alkaline earth group. If the oxide, sulfide, or selenide of cadmium in amounts of only 0.009 gm.atoms of the anion per 100 gm. of the glass were added, acidic solids were better wetted, and wettability of basic oxides decreased as the radius of the anion increased. The inverse phenomena occurred when the electronegativity of the anion increased. For pressed cylindrical shapes of iron and nickel oxides the degree of wetting by glass is reduced with increasing oxidation of the iron, but for nickel the oxidation was related to better wetting. Such observations are particularly important for the explanation of the wetting phenomena in sheet metal enameling (cf. B. ¶ 255 ff.), especially if boron-free enamels are concerned.[167]

107. Besides the fundamental importance of viscosity in glass melting and shaping, a similar correlation with the *heat transfer* conditions in the whole process of the glass tank operation is no less essential for the technology of glass manufacturing*. Heat conductance in the melting glass batches was especially studied by C. Kröger and H. Eligehausen[168] making direct measurements with the classical "hollow-cylinder" method, with compacts of quartz sand, limestone, and soda, in technical standard grain sizes. The conductance, λ, up to the beginning chemical reactions

* On the practical importance of the heat transfer phenomena for glass furnace control see the recent symposium report in *J. Am. Ceram. Soc.*, **44**, 1961, 301-373, with contributions by C. L. Babcock, W. D. Kingery, R. Gardon, H. Charnock, F. J. Grove, van Laethem, L. Leger, M. Boffé, and E. Plumat, S. Kruszewski, W. Trier, W. Giegerich, D. A. McGraw, and P. Acloque.

[163] *Missouri Univ. School Mines and Met. Bull. Tech. Ser.*, 1952, No. 82, 28 pp.

[164] *Glastech. Ber.*, **27**, 1954, 65-70; *Fachausschussber. deut. glastech. Ges.*, No. 51, 1954.

[165] *Doklady Akad. Nauk S.S.S.R.*, **82**, 1952, 79-82.

[166] *Doklady Akad. Nauk S.S.S.R.*, **67**, 1949, 467-470, 663-666.

[167] Cf. K. P. Azarov, *Zhur. Priklad. Khim.*, **27**, 1954, 33-42; *Silikat Tech.*, **5**, 1954, 347-348.

[168] *Glastech. Ber.*, **32**, 1959, 362-373.

can be expressed by an elementary binominal equation as a function of temperature. Such relations are also valid for industrial mixes in manufacturing window glass, tube glass, or white container glass. As soon as *colored* glasses (as with Fe-Mn green or brown) are concerned, specific factors of the emission properties of coloring oxides become evident, and λ appears increased because of the superposition of *radiation conductance* effects. This latter factor in its connection with the working properties of the finished glass was the subject of investigations of J. Boow and W. E. S. Turner[169] on the cooling and setting characteristics of colorless and colored glasses during their shaping process. It became evident in these studies, for example in blue cobalt or green and brown iron manganese-colored glass, that the temperature (and with it also the viscosity) changes are influenced by infrared emission properties of the glass in the cooling range. Measurements of the temperature differences on

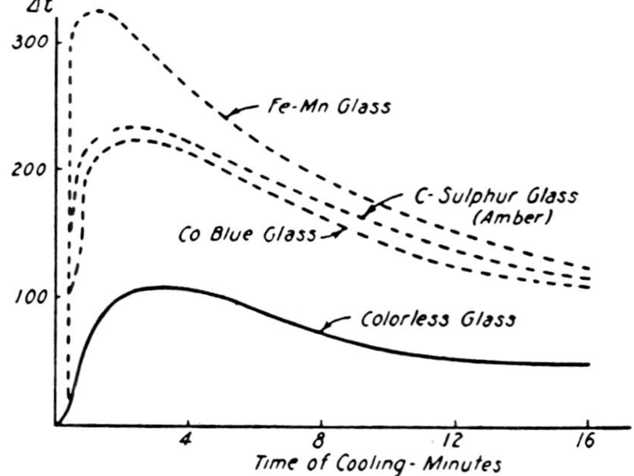

FIG. B. 68. Effects of infrared absorption in glass melts. (Boow and Turner). Temperature difference between the surface of the glass melt, and 2 cm. deeper.

the surface of the cooling glass bodies and in their interior (Fig. B. 68) made clearly evident the strong effects of the specific heat radiation qualities of such colored glasses. On the other hand, the heat radiation transmittance (diathermancy) of the colorless glasses creates in the cooling mass opposite temperature distribution conditions. The particularly marked differences of temperature within the first 2 to 3 mm. depth of the "surface layer" in the case of colored glass, gives rise to correspondingly high differences in viscosity and causes the "skin" effect which is very pronounced for all subsequent manipulative operation in glass shaping, whether for machine-operated glass or in manipulating it by hand.

[169] *J. Soc. Glass Technol.*, especially **27**, 1943, 94-112 T, 207-237 T; *Ibid.*, **29**, 1945, 199-232 T, 233-249 T.

108. The transfer of temperature distribution and heat radiation (diathermancy) effects to the energy balance calculation of a complete glass tank was made by R. Halle, R. S. Allison, and W. E. S. Turner[170] in a series of investigations of the highest value for glass technology operation. Although we cannot discuss here details of this eminently comprehensive work, we find in it the first successful attempt to derive from systematic measurements of temperature distribution and radiation energy input and output in the tank furnace (flame-filled heating chamber and glass bath), an accurate heat balance for the whole unit, with a calculation of over-all heat transmission through the glass bath to the tank bottom, as certain for a colorless glass as for a "coal-amber," Fe-Mn brown and deep-green glass. The heat transmissions were found characteristically different, namely about 60 per cent for colorless glass, 49 per cent for amber, and only 37 per cent for dark green glass. The depth of heat penetration for these glasses, at the same viscosity, was found characteristically different in the model tank used, as shown in Fig. B. 69. Further, the specific action

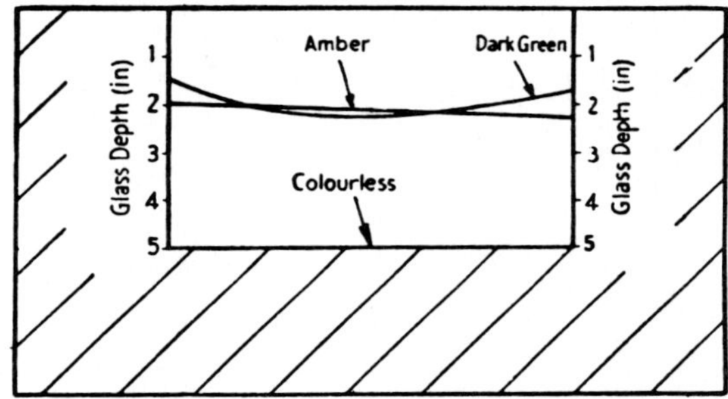

Fig. B. 69. Depth of heat penetration into glasses of different color type, but with the same viscosity, in a cross section through a model tank. (Halle, Allison, and Turner).

of the iron oxides on heat transmission was investigated in such model experiments, and clear evidence proved the tremendous influence of the infrared absorption and emission properties of such glasses in the heat distribution equilibrium of the glass bath, as a function of the ratio $FeO : Fe_2O_3$ in the glass composition. Very instructive are the comparative results of cobalt-blue glass with amber, or green-colored glass, as mentioned before. They show convincingly that color and transmission of cobalt-blue glass at room temperature *cannot* be taken as an indication of their heat-transmitting properties in the molten state. The temperature gradients measured for a commercial cobalt-blue glass and those of amber and green glass were 36° (per inch),

[170] *J. Soc. Glass Technol.*, **29**, 1945, 5-34, 170-191; **30**, 1946, 343-355, 356-363; **31**, 1947, 122-133 T. We add here immediately the discussions of F. W. Preston, *Ibid.*, **31**, 1947, 134-140, on the exact physical meaning of the term "diathermancy" as the transparency phenomenon for heat radiation.

and 45° and 85°, respectively, indicating how much more intense the latter two glasses absorb the heat radiation than the intensely colored cobalt glass as it appears at room temperature. Their light transmittances in the visible spectrum are (at 20°C.) 83.9 per cent for the colorless glass, but only 23.4 per cent for the cobalt glass, 42.4 per cent for the amber, and 56.8 for the green glass, whereas the heat radiation absorption of green glass is highest, followed by amber glass, cobalt glass, and least for colorless glass.

109. Direct measurements of heat radiation *emission* by infrared spectrometry were made more recently by M. Czerny and his coworkers,[171] based on a method developed by M. Czerny, W. Kofink, and W. Lippert,[172] as sketched in Fig. B. 70. The radiation proper of the glass melt as the absorbing medium is separated from the energy of a light source by modulating the latter periodically before it enters the

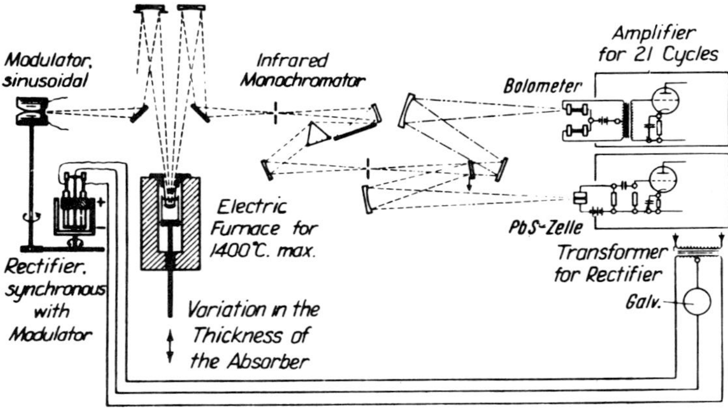

Fig. B. 70. Schematic drawing of infrared absorption apparatus, with sinusoidal modulator. (Genzel).

glass. This is done by a rotating slit device; the receiver of the radiation then records a mixture of a nonmodulated radiation which is transformed to electric energy as a direct current and a superimposed alternating current corresponding to the modulated energy. The separation of these two components is accomplished simply by a transformer and the measurement of the alternating component will almost completely

[171] Cf. N. Neuroth, *Glastech. Ber.*, **25**, 1952, 242-249; **26**, 1953, 66-69; M. Czerny and L. Genzel, *Ibid.*, **25**, 1952, 134-139, 387-392; *Ceram. Abstr.*, 1953, 117-118, 151-152; L. Genzel, *Glastech. Ber.*, **24**, 1951, 53-63; **26**, 1953, 69-71. On the total heat transmission coefficients of amber and green glasses in the molten state, see also S. Kruszewski, *Vetro e Silicati*, **4**, 1960 (24) 11-20; *J. Am. Ceram. Soc.*, **44**, 1961, 333-339.

[172] *Ann. Physik*, **8**, 1950, 65-86; see on older measurements for temperatures up to 500°C. W. Meyer, *Glastech. Ber.*, **14**, 1936, 305-321.

eliminate the effects of the radiation proper of the glass melt in the receiver. For infrared, a sensitive receiver is a bismuth bolometer of low inertia and a lead sulfide photoelectric cell whose response, by electronic amplifying, is measured with a considerable accuracy as well at high as low temperatures up to 3 μ wavelength. As a light source which must be rich in infrared radiation, a Nernst glower was used, or an electrically heated infrared radiator of silica glass. The photoelectric cell was sensitive enough to indicate radiation energy of 5×10^{-9} watts/cm.^{2} for $\lambda = 1.7\ \mu$. The infrared radiation of the glass melt is reflected from a platinum/30 per cent rhodium-platinum alloy mirror into a quartz spectrometer, here used as a monochromator. The mirror was submerged into the glass melt to a depth which could be accurately adjusted and varied in the electric furnace. The Lambert-Beer law is strictly fulfilled by the monochromatic radiation from room temperature up to 1300°C. for different optical glasses. L. Genzel's investigations were extended by N. Neuroth (cf. A. ¶ 246 f.) in measurements of commercial glasses. The great difference in the absorption coefficients below and above the transformation range are highly significant for window glass and a lead glass. Increasing temperatures cause a flattening of the absorption bands, as shown for a lead glass in Fig. B. 71 and corresponding data were given for different types of colored glasses.

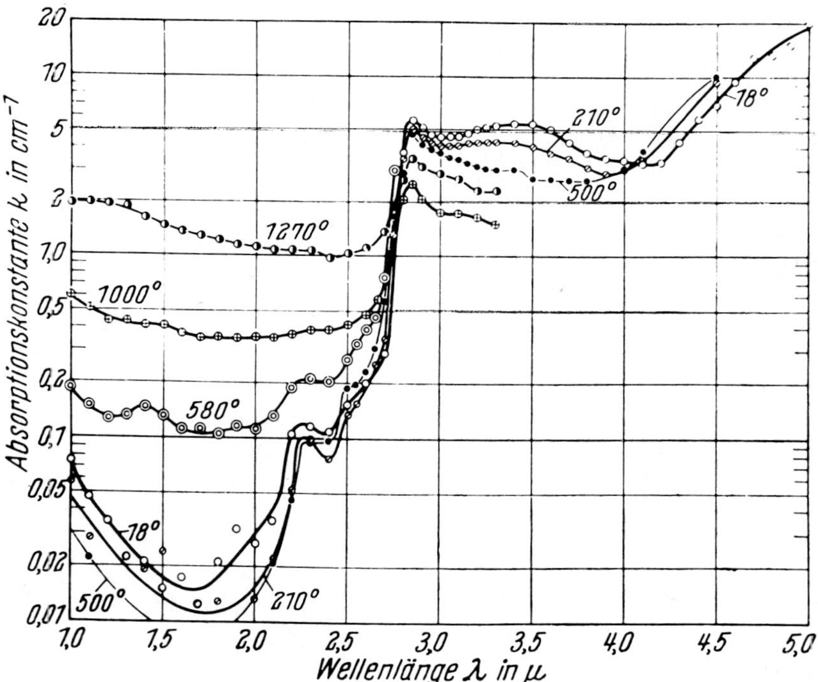

Fig. B. 71. Infrared absorption of a lead glass (for X-ray shielding), at variable temperatures, as a function of wavelengths. (Neuroth).

110. The particular importance of the investigations of M. Czerny and his school is seen in the ingenious combination of their results with the mathematical solution of the heat transfer problems in the glass tank, especially for the global transmittance, and its integration at high temperatures over the visible and the infrared

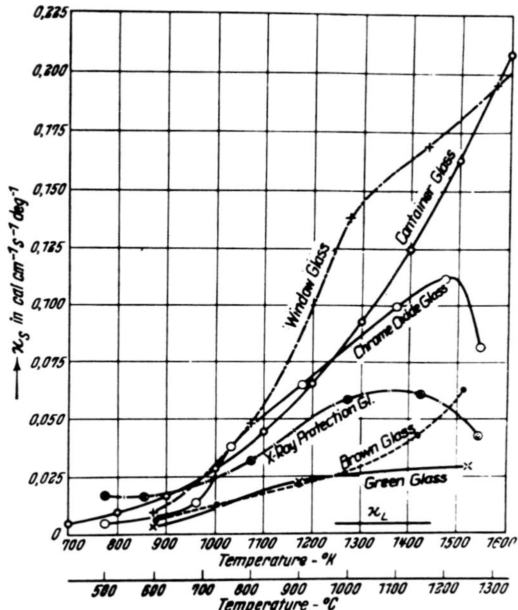

FIG. B. 72. Radiation heat conductance, $\varkappa_S$, of some typical glasses, as a function of temperature, in comparison with common heat conductance, $\varkappa_L$, as defined for low temperatures. (Genzel).

ranges. This is based on the radiation law of Stefan and Boltzmann[173] which also rules the problems of radiation transmittance and the depth to which irradiated heat energy permeates into the absorbing glass bath, as already discussed in the previous work of Halle, Allison, and Turner. We cannot here show the theoretical method for

[173] That part of the problems here involved which can only be treated very briefly, is contained in theoretical deductions made by W. Geffcken, *Glastech. Ber.*, **25**, 1952, 392-396; *Ceram. Abstr.*, 1954, 31-32; *Glastech. Ber.*, **29**, 1956, 42-49; for the mathematical performance of extensive calculations cf. A. Walther, J. Dörr, and E. Eller, *Ibid.*, **26**, 1953, 133-140, including the complications by the "wall and bottom" effects of the glass tank. See also W. Trier, "Advances in Glass Technology, Plenum Press, New York, 1962, 619-634, and J. Segall, *ibid.*, 635-639, on heat-technological problems of modern glass tank furnaces. M. Coenen, *ibid.*, 93-102; Veröffentl. Max Planck-Inst. Silikatforsch., **22**, 1962, 139-148, presented measurements and calculations on the time functions of temperature changes in boundary zones between glass melt and metal walls, and T. J. Naughton and D. A. McGraw, "Advances in Glass Technology," 1962, 603-618, analyzed the heat transfer in glass pressing processes.

solution of these calculation problems.[174] The calculation of the "radiation conductance" proper and the energy current is given in the form of an integral equation including not only the "gray" absorption but also the specific absorption characteristics of every given glass when its characteristic bands of absorption (or emission) in the infrared are known. A series of characteristic radiation conductance curves, $\varkappa_s$, are given in Fig. B. 72 as a function of temperature, in comparison with the common heat conductance as defined at low temperatures, $\varkappa_L$. The principal advantage of the manner in which Czerny and his group of investigators challenged the complex problems here discussed is that not only the radiation flow in one direction, namely the direction of the temperature gradient from the surface of the glass bath to the bottom of the tank, can be correctly calculated but the integral equations established also give the solution for radiation flow over the whole half-space.[175] The results confirm how important the iron content of commercial glasses is for heat flow characteristics; even a small amount of FeO may change the temperature distribution in the glass bath of a tank and even more in a pot-crucible.

111. Since the theory of the methods for a calculation of the radiation current in glass tanks is based on Planck's radiation law, M. Czerny, L. Genzel, and G. Heilmann[176] formulated integration rules of a particular mathematical convenience, with a graphic evaluation of infrared absorption curves, to determine the "effective mean free path" for the absorption-controlled polychromatic radiation. The importance of the heat conductance of the bottom lining of the glass tank is emphasized, as well as a current control of infrared absorption of the glass at service temperature, in connection with the examination of its FeO contents. We further mention the compilation of valuable heat conductance data for industrial glass melts as recently presented by N. Neuroth,[177] in comparison with earlier data of A. F. Van Zee and C. L. Babcock (1951) for the total heat transferred by both, common conduction, and radiation. Contradictory results at high temperatures, even for glasses of nearly identical composition, are explained by the fact that in a transparent glass the conductance cannot be an individual material constant of the heat transfer, but that

[174] A very brief introduction was made by W. Eitel, *Glass Ind.*, **36**, 1955, 575-581, 592-593.

[175] We mention here as an excellent solution of the one-directional flow problem the calculations of B. S. Kellett, *J. Opt. Soc. Am.*, **42**, 1952, 339-343; *J. Soc. Glass Technol.*, **36**, 1952, 115-123 T; *Glass Ind.*, **34**, 1953, 73-76; **35**, 1954, 131-134; see also *Ceram. Abstr.*, 1953, 117-118.

[176] *Z. Physik*, **139**, 1954, 302-308; *Glastech. Ber.*, **28**, 1955, 185-190; **30**, 1957, 1-7. See in addition the studies of Ilse Huhmann-Kotz, *Ibid.*, **32**, 1959, 47-52, 189-197. The use of the equations in the calculation of the radiation current can be much facilitated by Czerny's fractional-function tables and auxiliary functions tabulated by M. Czerny and A. Walther in "Tables of the Fractional Functions for the Planck Radiation Law," Springer-Verlag, Berlin, Göttingen, Heidelberg, 1961, 59 pp.

[177] *Glastech. Ber.*, **32**, 1959, 197-198; cf. A. F. Van Zee and C. L. Babcock, *J. Am. Ceram. Soc.*, **34**, 1951, 244-250; see also Gisela Eckhardt, *Glastech. Ber.*, **32**, 1959, 373-380.

the radiation conductance (with an effective mean free-path) is superimposed on common heat conductance, combined with wall effects, and other factors.

112. In conclusion of this most fascinating field of investigation we may emphasize the interesting measurements of infrared transmission of solid and molten glass in the range up to 1400°C. made by F. J. Grove and P. E. Jellyman,[178] who used a horizontal beam of radiation entering the glass melt through a window of synthetic sapphire in a refractory crucible cell. The measurements concerned different commercial glasses, some of them with a relatively high iron content (the bluish-green heat-absorbing "Calorex" glass containing 0.86 per cent Fe_2O_3), confirming the rule that radiation conductance is strongly influenced by the content in FeO. Further, a 2 per cent nickel glass was investigated up to 1400°C. and 5 μ wavelength. The emissivity phenomenon of transparent materials in general was rediscussed by R. Gardon[179] and illustrated by the hemispherical emissivity of window glass sheets at various temperatures. The radiation flux within a transparent radiator exceeds that emitted in air by a factor of approximately the square of the refractive index.

113. On the emissivity of refractories cf. Vol. V, Section B. Interesting discussions on the radiation heat transfer as an important geophysical factor governing thermal equilibria in the inner earth are made by F. W. Preston[180] and S. P. Clark.[181]

PHYSICAL PROPERTIES OF GLASS AS A FUNCTION OF CHEMICAL COMPOSITION

114. In continuation of the principle of additivity of physical properties as a function of oxide increments of glass composition which was classically formulated in the rule of A. Winkelmann and O. Schott (1901), many attempts have been made to improve this method for the calculation of material constants. H. H. Blau[182] started in his investigations on the additivity of the thermal expansion coefficients in soda-lime silicate glasses from a new type of equation, namely $\alpha = (1/3\ V_0)\ (\Delta b_{Si} + \Delta c_{Si} N_{Si} + \Delta c_M N_M)$ in which N_{Si} is the ratio of the number of the Si^{4+} ions to the number of O^{2-} anions, N_M the ratio of the number of all the framework-modifying (base) ions to the total number of O^{2-} anions, b and c constants, and V_0 the volume

[178] *J. Soc. Glass Technol.*, **39**, 1955, 3-15 T.

[179] *J. Am. Ceram. Soc.*, **39**, 1956, 278-287.

[180] *Am. J. Sci.*, **254**, 1956, 754-757.

[181] *Am. Mineralogist*, **42**, 1957, 732-742, especially pp. 737 f., and *Bull. Geol. Soc. Am.*, **67**, 1956, 1123-1124.

[182] *J. Soc. Glass Technol.*, **35**, 1951, 304-317. See also the fascinating article of S. Kumar, *Central Glass & Ceram. Research Inst. Bull. (India)*, **7**, 1960, 58-69, on a more theoretical background.

of the glass containing one gram-atom oxygen at temperature T, and the difference $\Delta c_{Si} N_{Si}$ an abbreviation for the difference product $(c'_{Si} - c_{Si})N_{Si}$ at temperatures T' and T, $(T' = T + 1°C.)$, respectively, etc. The calculation based on this type of equation shows distinct advantages over previous summation formulae based on "oxide increments," because it enables the operator to consider also structural implications, if breaks occur in the curves representing the property parameters as a function of composition. Non-bridging oxygen anions which do not share two neighboring Si^{4+} cations, tend to occupy positions in the framework as remote as possible from one another, or to repel one another at least within the scope of short-range distances. For soda-silica glasses the break points are observed for $N_{Si} = 0.333$, 0.400, and 0.500, and the following constants for the calculation are derived:

N_{Si}:	from 0.286 to 0.333	from 0.333 to 0.400	from 0.400 to 0.444	from 0.444 to 0.500
b_{Si}	5.69	4.21	2.66	...
c_{Si}	13.00	17.27	21.14	27.26
c_{Na}	difficult glass formation range	8.7	8.7	8.7
Δb_{Si}		− 0.001049	− 0.000720	...
Δc_{Si}		0.00240	0.00163	0.0000327
Δc_{Na}		0.00168	0.00168	0.00168

115. The structural meaning of these break points is:

at 0.286 for N_{Si} : per tetrahedron three nonbridging oxygen anions;
at 0.333 per tetrahedron two nonbridging oxygen anions;
at 0.400 per tetrahedron one nonbridging oxygen anion;
at 0.500 no more nonbridging oxygen anions.

In the range
from 0.286 to 0.333 are three, and two, non-bridging oxygen anion bonds side-by-side;
from 0.333 to 0.400 are two, and one, nonbridging oxygen anion bonds side-by-side;
from 0.400 to 0.444 are some nonbridging oxygen anions in adjacent positions;
from 0.444 to 0.500 are nonbridging oxygen anions in alternate, or more remote positions.

The agreements of the experimental with the theoretical values for the thermal expansion coefficients are very satisfactory, much better than with those calculated from the Winkelmann-Schott increments, and distinctly better than those of English and Turner, or with Gilard-Dubrul's increments.

116. A. A. Appen[183] developed fundamental ideas of combining such more practical viewpoints of a convenient calculation and prediction of property constants for glasses

[183] A valuable report on these investigations see the Symposium on "Structure of Glass," under auspices of the Academy of Sciences, Moscow-Leningrad, 1953/1955, pp. 96-106. See details, *Zhur. Priklad. Khim.*, **24**, 1951, 904-914, 1001-1009, 1122-1130; **27**, 1954, 121-126; *J. Appl. Chem. (U.S.S.R.)*, **25**, 1952, 1241-1248, 1273-1280, 1303-1309 (English Translation).

with our present ideas on the constitution of glass (cf. Par. A. 283). As the most characteristic parameters suitable for additive calculations Appen discussed density, d, molecular volume, V, index of refraction, n_D, average dispersion, $D = n_F - n_C$, and the coefficient of linear thermal expansion, $\alpha_{20°}^{400°}$, as a function of the mole fractions, γ_i, of the oxide constituents $(\Sigma \gamma_i = 1)$, and the intensive characteristic $\bar{g}_i = g + (1 - \gamma_i)dg/d\gamma_i$ approximated for practical calculations by $\bar{g}_i = g + (1 - \gamma_i) \times \Delta g/\Delta\gamma_i$ in which g and Δg are experimentally determined, γ_i in the original composition, $\Delta\gamma_i$ for the change in composition. In this way, the different properties with the "partial characteristics" $\bar{N}_{SiO_2}$, $\bar{V}_{SiO_2}$, $\bar{D}_{SiO_2}$, and $\bar{\alpha}_{SiO_2}$, etc., were calculated from experimental data of simple binary and ternary systems. $\bar{N}_{SiO_2}$ and $\bar{V}_{SiO_2}$ varied somewhat, decreasing with increasing contents in SiO_2, whereas $\bar{D}_{SiO_2}$ is practically constant. Average arithmetic values for $\bar{N}$, $\bar{V}$, $\bar{\alpha}$ are 1.475, 26.1, and 38×10^{-7}, respectively, and more explicitly $\bar{N}_{SiO_2} = 1.475 - 0.0005\ (a - 67)$; $\bar{V}_{SiO_2} = 26.1 + 0.035\ (a - 67)$; $\bar{\alpha}_{SiO_2} \cdot 10^7 = 38 - 1.0\ (a - 67)$.

117. The partial characteristics of the alkalies in glasses are relatively stable, but with a systematic substitution of Na_2O by K_2O the curves for the thermal expansion coefficient show maxima. For the alkaline earths the partial characteristics are practically constant over a rather wide range of compositions. Particularly the $\bar{N}_{Al_2O_3}$ and $\bar{V}_{Al_2O_3}$ data, and also those of B_2O_3 require a special consideration in alumino- and borosilicate glasses because of the change of coordination from $[RO_6]$ to $[RO_4]$, or of $[RO_3]$ to $[RO_4]$, respectively. $\bar{V}$, $\bar{D}$, and $\bar{\alpha}$ are rather constant for ZnO, vary somewhat for CdO, and considerably for PbO, chiefly depending on the alkali contents of the glasses. The limits of the applicability of binominal equations as those developed by Appen for the different properties are mostly above 15 per cent Na_2O and above 10 per cent K_2O. The lead glasses are subdivided into two classes, a "tight" and a "loose" structure type, the first including the nonalkaline glasses with constant $\bar{\alpha}$, the second class containing alkalies and having increased $\bar{\alpha}$ characteristics by a looser packing effect with increasing alkali contents.

118. Especially for calculation of thermal expansion coefficients, Appen[184] tabulated the $\bar{\alpha}$ parameters for SiO_2, PbO, TiO_2, and B_2O_3, in the latter case in the form $\bar{\alpha}_{B_2O_3} \times 10^7 = 12.5\ (4 - \psi) - 50$, to express the functional relation to the ratio, ψ, of the total number of the molecules of Li_2O, Na_2O, K_2O, CaO, BaO, and CdO to that of B_2O_3. A limit for the applicability of the equations is given by the condition $SiO_2 > 45$ to 50 molecular per cent in the glass composition, but they are also valid

[184] *Steklo i Keram.*, **10**, 1953, (1) 7-10; *Silikat Tech.*, **5**, 1954, 113-114. On borosilicate glasses see *Zhur. Priklad. Khim.*, **26**, 1953, 569-578, further the extensive contributions of Kl. Kühne, *Silikat Tech.*, **12**, 1961, 313-316, 368-376, 408-421, for a systematic development of adequate borosilicate glasses of definite physical properties, chiefly in the optical, and the electrotechnical industry, with many important data tables and diagrams.

for glazes and enamels in the same ranges of composition as for common glasses. Fluorine containing glasses, however, show anomalies with uncontrolled losses in fluorine. Particular care was given by Appen to the role of alumina and titania in silicate glasses,[185] in which even negative values for the partial expansion coefficient may occur indicating that alumina reduces the expansion coefficient more strongly than any other normal constituent, because of the change in coordination from $[AlO_6]$ to $[AlO_4]$. For titania similar anomalies are valid, caused by the transition of $[TiO_6]$ to $[TiO_4]$, as a function of the basicity of the glass, with considerable reductions of $\bar{n}_{TiO_2}$ and $\bar{D}_{TiO_2}$. Optimum properties of titania containing glasses cannot be exactly calculated because of this difficultly controlled variability of the "partial" properties. For borosilicate glasses an anomalous behavior is also characteristic. In the binary glasses SiO_2—B_2O_3 is $\bar{n}_{B_2O_3} = 1.461$; $\bar{V}_{B_2O_3} = 38.0$ cm.^{3}/mole. In ternary glasses with the ratio $\psi_B = Na_2O/B_2O_3$ below 1/3 they are the same, but for ψ_B above 1/3 the coordination $[BO_3]$ is changed to $[BO_4]$ and $\bar{n}_{B_2O_3}$ is increased with an end value of 1.710, $\bar{V}_{B_2O_3}$ decreased to 18.5 cm.^{3}/mole, as also observed in danburite. In quaternary glasses with alkaline earths, the cations of small radius retain their oxygen more firmly than cations of large radius, and the formation of $[BO_4]$ is suppressed. Alumina takes away all available oxygen to form the groups $[AlO_4]$ and $[BO_3]$ remains. The ratio $\psi_B = (a_{R_2O} + a_{RO} - a_{Al_2O_3})/a_{B_2O_3}$ determines the amount of coordinations changed from $[BO_3]$ into $[BO_4]$ configurations. Only by empirical formulae is it possible to calculate $\bar{V}_{2BO_3}$, $\bar{n}_{B_2O_3}$, $\bar{\alpha}_{B_2O_3}$, and $\bar{d}_{B_2O_3}$ in complex glasses as a function of ψ_B and the silica contents.

119. Considering the surface tension, σ, as a function of composition, we mentioned before the calculations of A. A. Appen, K. A. Shishov, and S. S. Kayalova[186] which were supplemented by Appen[187] with calculations of the "partial molar surface tension" $\sigma_1 = \sigma + (1 - \gamma_i)\,\Delta\sigma/\Delta\gamma_i$. On the basis of extensive measurement data given by Appen, A. Petzold[188] criticized the previous measurements and conclusions of K. C. Lyon (1944). Whereas one group of silicate glasses containing surface-inactive constituents make possible an approximative calculation of σ by the additivity rule, this is not the case for glasses with compositions of more complex functional relations to σ, or with glasses containing pronouncedly surface-active components (cf. A. ¶ 95). A. G. Repa[189] applied his theory of the "*oxygen potentials*" (KP) defined by

[185] *Zhur. Priklad. Khim.*, **26**, 1953, 9-17. On the considerable anomalies in fluorine containing glasses see also V. V. Vargin and N. I. Krasotkina, *Doklady Akad. Nauk S.S.S.R.*, **108**, 1956, 1133-1136, in which cristobalite and tridymite are assumed.

[186] *Silikat Tech.*, **4**, 1953, 104-105. On the properties of sodium aluminosilicate glasses see specifically D. E. Day and G. E. Rindone, *J. Am. Ceram. Soc.*, **45**, 1962, 489-496, 496-504.

[187] *Zhur. Fiz. Khim.*, **26**, 1952, 1399-1404.

[188] *Silikat Tech.*, **5**, 1954, 11-12; cf. K. C. Lyon, *J. Am. Ceram. Soc.*, **27**, 1944, 186-189.

[189] "Structure of Glass," Symposium Moscow-Leningrad, 1953/1955; pp. 276-279; *Zhur. Priklad. Khim.*, **28**, 1955, 694-699.

the ratio of the total sum of oxygen available, to the sum of the cations, including Si^{4+}, to problems of the additivity of physical properties of glass from the composition. He especially demonstrated for the thermal expansion coefficient (Fig. B. 73) how for different glasses a characteristic fundamental relation of this parameter to the KP value is quite regular, such as for soda and potash silicate glasses. This is

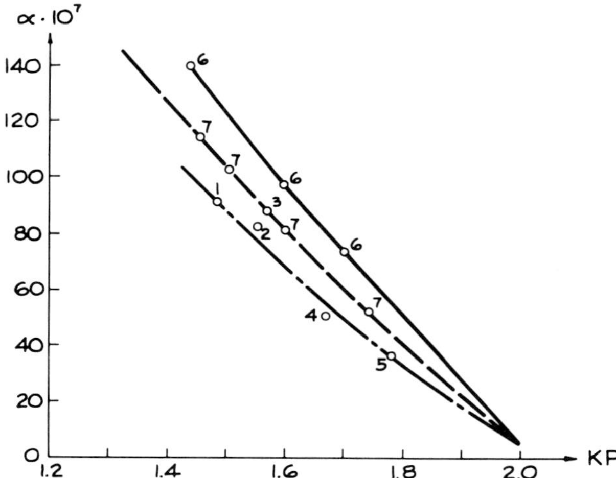

FIG. B. 73. Thermal expansion coefficient of different glasses, as a function of the "oxygen potential," *KP*, (Repa). (1) window glass, (2) plate glass, (3) lead glass for high vacuum-tubes, (4) Russian special glass *Zs* — 5, (5) Pyrex glass.

technologically remarkable, because among glasses with the same expansion coefficient those with the highest KP merit the preference in their weldability, workability in the flame, and other characteristic properties in shaping. Pyrex type glasses appear here as an optimum. For other properties, the projecton of their parameters as a function of KP is also very useful as a basis for comparison.

120. For the additivity calculation of the density of glasses, M. L. Huggins and J. M. Stevels[190] developed new equations either valid for glasses high in the framework-forming cations Si^{4+}, Al^{3+}, B^{3+}, and the like, or for glasses of more extensive ranges of composition with framework-modifying cations especially suitable for annealed glasses. The first equation (by Stevels) relates the specific volume $V = V_0/(1 - Rx')$ or $100/V = 100/V_0 \cdot (1 - Rx')$, if V_0 is the volume per gram atom of oxygen at the limit $R = 0$. The empirical constant V_0 is $= 8.224$ cm.^{3}, and $x' =$

[190] *J. Am. Ceram. Soc.*, **37**, 1954, 474-479. Concerning Huggins's "*Structon*" theory we refer to A. ¶ 242 f. and *Ibid.*, **38**, 1955, 172-175; *J. Phys. Chem.*, **58**, 1954, 1141-1146, especially on the structural meaning of the four ranges mentioned above.

0.19, R meaning the number of oxygen atoms per atom of the framework-forming cations, namely $R = 1/\Sigma (N_{\text{m}}) = 1/(N_{\text{Si}} + N_{\text{B}} + N_{\text{Al}})$. The applicability of this equation to glass compositions with higher contents in alkalies, alkaline earths, lead, and the like, is doubtful. The second equation (Huggins) is given by Huggins in the form $V = b_{\text{Si}} + \Sigma\, c_{\text{M}} N_{\text{M}}$, in which b and c_{M} are empirically determined constants for the component oxides, and the sum Σ is taken over all these. Four different ranges for applicability of this type of equation are given in the following:

Equations:

$$b_{\text{Si,C}} = (c_{\text{Si,D}} - c_{\text{Si,C}})\, N_{\text{Si,CD}}$$
$$b_{\text{Si,B}} - b_{\text{Si,C}} = (c_{\text{Si,C}} - c_{\text{Si,B}})\, N_{\text{Si,BC}}$$
$$b_{\text{Si,A}} - b_{\text{Si,B}} = (c_{\text{Si,B}} - c_{\text{Si,A}})\, N_{\text{Si,AB}}$$

Ranges:

	M	N_{M}	c_{M}	Range of N_{Si}	Range of R
A	Si	5.69	13.00	0.275 to 0.345	3.705 to 2.90
B	Si	4.21	17.17	0.345 to 0.400	2.90 to 2.50
C	Si	2.66	21.14	0.400 to 0.435	2.50 to 2.30
D	Si	0	27.26	0.435 to 0.500	2.30 to 2.00

121. For silica glass V is reduced to $V = c_{\text{Si, D}}/2 = 13.63$. The relation of both equations here discussed is shown graphically in Fig. B. 74 in which the first (Stevels)

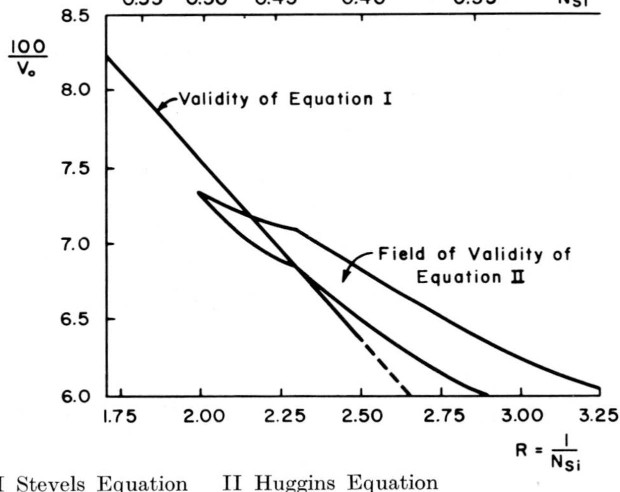

FIG. B. 74. Comparison of the validity of Stevels' equation for the calculation of glass density from composition (straight line), and of the field of validity of Huggins' equation, for silicate glasses. (Huggins and Stevels).

equation is the straight line extending from $100/V_0 = 6.00$ and $R = 2.70$ to $100/V_0 = 8.25$ and $R = 1.70$. The four ranges of the second (Huggins) equation are indicated by the smoothly curved branches of the other designed curves. The agreement of the calculations with the experimental data is excellent in the different ranges of composition. In an investigation of surface tension, density, viscosity, and electrical conductance of fused borates of calcium, barium, and strontium, by L. Shartsis and B. F. Shermer,[191] however, it was demonstrated that both equations given above are not able to represent the density properties of such glasses, evidently by large contraction effects caused by the addition of divalent cations, in the order of their ionic potentials Ca > Sr > Ba. How strong these deviations may be is seen in Fig. B. 75 in which the Huggins equation for alkali ions in borate glasses corresponds to dotted straight lines in comparison with the experimentally established curves for V_0 versus N_M (gram atom M/gm. atom O).

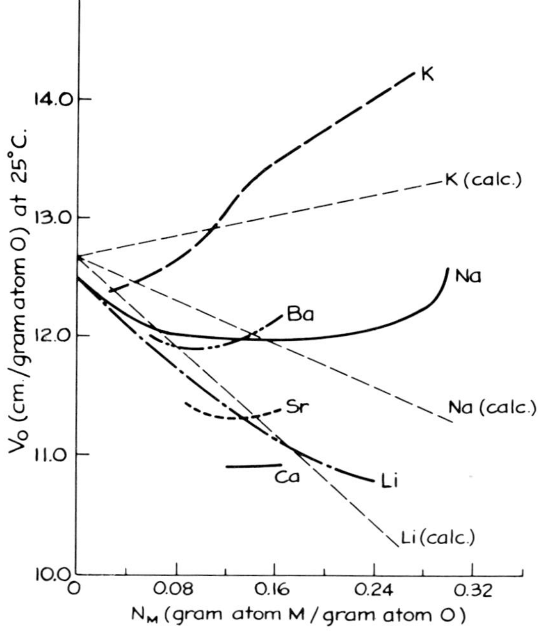

Fig. B. 75. Volume, V_0, of borate glasses at room temperature, containing 1 gm. atom of oxygen, versus N_M gm. atoms of cations per gm. atom oxygen. Curves for alkali, and alkaline earth borate glasses, straight line indicating the calculated data from Huggins' equation. (Shartsis and Shermer).

122. I. Náray-Szabó[192] discussed the volume of oxygen in the glass structure, v in Å^3., and its role in the thermal expansion process, as a function of the composition, in the linear relation $R = \mathrm{O}/(\mathrm{Si} + \mathrm{Al} + B..)$ or as a function of the concentration in framework-forming cations, with $v = aR + b$. In tabulated data for

[191] *J. Am. Ceram. Soc.*, **37**, 1954, 544-551. On density and thermal expansion of alkali borate glasses see the extensive and systematic studies of L. Shartsis, W. Capps, and S. Spinner, *Ibid.*, **36**, 1953, 35-43 (cf. A. ¶ 183).

[192] *Acta Phys. Acad. Sci. Hung.*, **8**, 1957, 37-64; **9**, 1958, 151-161, 403-421; *Ceram. Abstr.*, 1958, 145 b; *Glastech. Ber.*, **32**, 1959, 185-189.

binary glasses ($NaO—SiO_2$, $K_2O—SiO_2$, $SiO_2—B_2O_3$, and $Na_2O—P_2O_5$) an excellent agreement with experimental data was observed. Ternary and quaternary glass systems of additive arithmetic functions of composition were developed which bring about simple equations for the thermal expansion coefficient as a function of composition, with a clear grouping of framework-forming and framework-modifying cations, the forming cations exerting a negligible effect:

$$\alpha_{\text{linear}} \times 10^8 = 35(Na_2O) + 30(K_2O) - 8(MgO) + 10(CaO) + 10(BaO) + 6(PbO) + 10(Fe_2O_3) - 6(ZrO_2) + A,$$

the oxides (in parentheses) being introduced with their percentages, and with $A = 200$ for common glasses, $= 220$ for borosilicate glasses up to 30 B_2O_3, $= 250$ for zinc silicate glasses with less than 5 per cent ZnO, $= 220$ for barium glasses with more than 5 per cent BaO, and $= 280$ for lead glasses with more than 5 per cent PbO.

"Glasses proper" with R varying between 2.0 and 2.7 have gradual transitions to glasses with higher R values, but these latter glasses have low mechanical strength, devitrify easily, and are not water-stable. Amorphous boron, arsenic, and antimony trioxides are not "glasses proper"; they consist of discrete molecules. Borosilicate glasses may show R values down to 1.6.

123. In a method similar to what Appen had used, O. Knapp[193] developed additive rules for the special field of electronic vacuum tube glasses and so-called *solder glasses* used in modern radio techniques, especially for glass/metal or glass/ceramics sealing. Knapp started from the factor of viscosity for the transformation point at $\eta = 10^{13}$, of the softening point at $10^{7.6}$, flow point at 10^5 poises, and the electrical conductance defined by the temperature t_{100} for $100 \times 10^{-10}\ \Omega^{-1}$ cm.$^{-1}$. With these most important property data it is possible to classify bulb glasses, metal/glass sealing materials, and glass/ceramics sealings, besides some special (solder) glasses, as illustrated for typical vacuum tube glasses, many of which are Schott & Genossen products. Among the special glasses are those highly resistant to metal vapors (especially to Hg, Na, Cd, Mg, Cs vapors). In the glasses for sealing metal/glass are some special glasses for Fe—Ni alloys, for copper, Cr—Fe and Fe—Ni—Co alloys, for Mo, Wo, and the like also zinc-aluminum borophosphates, zinc borosilicates, alkali-free aluminum phosphates, and that glass which corresponds to the ternary eutectic of the system anorthite-clinoenstatite-SiO_2 (cf. Vol. III, B. ¶ 274). Other glasses must show adaptation to contrasting thermal expansion characteristics (a difference of more than 10 per cent), of so-called connecting glasses. Typical solder

[193] *Silikat Tech.*, **5**, 1954, 51-54; **6**, 1955, 99-104. On aluminoborophosphate glasses see recently S. Kumar, *Central Glass & Ceram. Research Inst. Bull.* (*India*), **7**, 1960, 117-122. Lead containing aluminoborophosphates are discussed by the same author in *Ibid.*, **6**, 1959, 13-22, for solder glasses. See also R. H. Dalton, *J. Am. Ceram. Soc.*, **39**, 1956, 109-112.

glasses must have[194] very low softening temperatures and a working viscosity of 10^4 to 10^6 poises. Glasses with 30 per cent PbO have these viscosities at 725° to 925°C., common soda-lime-magnesia silicate glasses at 800° to 1000°, Pyrex glass at 920° to 1200°C. They must be "short" in their temperature-viscosity characteristics and in the annealing range the thermal expansion coefficient of the glasses to be welded must be identical. Knapp demonstrates such glasses from the "plumbate" series of the system PbO—ZnO—SiO_2, with small additions of silica to reduce the expansion coefficient. An exceedingly steep viscosity versus temperature characteristic between the softening point and the lower annealing temperature is observed in glasses of the system PbO—B_2O_3—SiO_2. An addition of ZnO lowers the softening temperature but increases the devitrification tendency. The introduction of ZnF_2 in place of ZnO is sometimes profitable. Glasses of the system ZnO—B_2O_3—V_2O_5 were recommended similarly for molybdenum and Fenico (Fe-Ni-Co) metal soldering, as shown by E. P. Denton and H. Rawson.[195] The softening temperature is considerably lowered by introduction of 20 per cent V_2O_5 although the expansion coefficient is only little changed.

124. Systematic studies on changes in properties of sodium borosilicate glass (5 per cent Na_2O, 15 B_2O_3, and 80 SiO_2) by an introduction of CaO, MgO, BaO, ZnO, and PbO, were made by J. Yamamoto,[196] especially when silica is replaced by these oxides. The properties studied are thermal expansion, transformation range, softening temperature, viscosity, density, electrical conductance, and chemical durability, as the most important technological factors. As alumina is introduced transformation temperature, softening temperature, viscosity, and conductance are decreased, chemical durability and thermal expansion coefficient increased.

125. Of high practical importance are studies on *thermal shock-resistivity* of ceramics in general, and glasses in particular, as a function of chemical composition. Beyond the classical but empirical formulation of this resistivity as a function of thermal expansion, heat conductance, elasticity modulus, and others combined, by Winkelmann and Schott (1901), W. R. Buessem[197] made remarkable efforts to find

[194] *Silikat Tech.*, **9**, 1958, 153-155; see also German Patent 709,479 (DeVries). On aluminophosphate glasses see also Z. M. Syritskaya, *Structure of Glass. Proc. All-Union Conf. Glassy State, 3rd, Leningrad* (English Translation) 1959/60, 295-298. See recently J. Broukal. *Silikattechnik*, **13**, 1962, 428-433. On Na-vapor resistant glasses see C. A. Elyard and H. Rawson, "Advances in Glass Technology," Plenum Press, New York, 1962, 270-286, on B and F containing special flux compositions for glass/metal welding also B. Locardi, *Ibid.*, 566-579; *Vetro e Silicati*, **6**, 1962 (35), 5-11.

[195] *J. Soc. Glass Technol.*, **40**, 1956, 252-259; see also J. Gallup and A. G. T. Dingwall, *Am. Ceram. Soc. Bull.*, **36**, 1957, 47-51.

[196] *Yogyo Kyokai Shi*, **61**, 1953, 257-265, 491-494, 495-499; **62**, 1954, 125-128.

[197] *J. Am. Ceram. Soc.*, **38**, 1955, 15-17. In this connection the recent studies of E. I. Kozlovskaya, *Structure of Glass. Proc. All-Union Conf. Glassy State, 3rd, Leningrad* (English Translation),

a modern, mathematically better founded relation for this significant property, and to formulate an improved method for its testing. The commonly used shock methods are evidently too inconclusive. In their place, a determination of the five fundamental parameters given in this critical discussion is indispensable, to have a basis for thermal relation of optimum materials for a definite construction purpose, in the light of theoretical deductions developed by W. D. Kingery[198] on different factors affecting thermal stress resistance of ceramic materials in general. Analogous discussions of M. Mehmel[199] are based on the theories of Büssem, also with a serious warning against misinterpretation of thermal shock test results especially for multiphase ceramic bodies (cf. Vol. V, Section B). The method of preparing silica-enriched glass compositions as proposed by I. I. Kitaĭgorodskiĭ and V. A. Blinov[200] may be one promising way to develop glasses systematically with considerably improved thermal stabilities.

126. Amidst the rich literature on changes of physical properties by systematic changes in chemical composition of glasses we mention here further the studies of H. E. Schwiete and G. Ziegler[201] with binominal interpolation formulae for calculation of temperature function of enthalpy and derivation of partial increments for Na_2O, K_2O, CaO, MgO, and Al_2O_3, based on the assumption that the state of SiO_2 in silicate glasses is thermodynamically identical with that in pure silica glass. In general, Schwiete's and Ziegler's data agree very well with those previously calculated by D. E. Sharp and L. B. Ginther (1951), for the basic glasses K_2O, $2.5SiO_2$, $Na_2O,3.3SiO_2$, $Na_2O,CaO,6SiO_2$, but emphasis always is placed on the fact that even the best increment calculations can be valid only as more or less good approximations for the enthalpies. Particularly important is the observation that the calculations may be extended in their usefulness to strongly colored iron-manganese glasses with $[FeO_4]^{2-}$ anions which have a higher enthalpy contribution than Fe^{3+} cations. It is, therefore, in general not possible to treat Fe^{3+} in glasses like a substitution for Al^{3+} as did Sharp and Ginther. J. Moore and D. E. Sharp[202] corrected in their interpolation formulas

$$c_{\mathrm{m}} = \frac{at + c_0}{(0.00146\, t + 1)}$$

1959/60, 299-301, on the elastic properties of glasses as a function of chemical composition are particularly valuable.

[198] *J. Am. Ceram. Soc.*, **38**, 1955, 3-15; with en elaborate tabulation of symbols to be used for thermal stress analysis, on pp. 1 and 2.

[199] *Ber. deut. keram. Ges.*, **35**, 1958, 225-232.

[200] *Doklady Akad. Nauk S.S.S.R.*, **127**, 1959, 392-395.

[201] *Glastech. Ber.*, **28**, 1955, 137-146. Non-linear deviations in the enthalpy-temperature functions of simple soda-silicate glasses are discussed by H. E. Schwiete and Chr. Hummel, *Ibid.*, **32**, 1959, 413-420, and similar deviations for density (volume changes with composition) by W. Schwiecker, *Ibid.*, **30**, 1957, 379-386 (cf. A. ¶ 284), as near-order effects in the glass structure.

[202] *J. Am. Ceram. Soc.*, **41**, 1958, 461-463. Anomalies observed in those glasses which H.

for the average specific heat the previous increments for K_2O and B_2O_3, and detected a characteristic deviation of the enthalpies calculated for glasses with metals of an atomic weight above 50 (e.g., iron, zinc, lead) from the ideal course over a portion of the temperature ranges desired. Nevertheless, for practical application it is sufficient to introduce provisional increments for iron and manganese in the range from 0° to 600°C.

127. We further mention the investigations of L. W. Coughanour, L. Shartsis, and F. H. Shermer[203] on the systematic determination of viscosity, density, and electrical conductance of fused alkaline earth borates with a constant addition of 3 molec. per cent of K_2O, for the purpose of eliminating the liquid immiscibility gap which occurs in the system without this alkali addition. On a molecular per cent basis, densities decrease in the series of BaO > SrO > CaO for the borate glasses, whereas determined at room temperature and up to 1100°C. for viscosity the sequence is CaO > SrO > BaO. With increasing concentration in RO at 1000° to 1100°C., conductance decreases, whereas in the temperature range between 800° and 900°C. it increases. For lithium-aluminosilicate glasses, H. W. Rauch, C. H. Commons, and H. H. Blau[204] determined the thermal expansion coefficient, dielectric constant and power factor, optical constants, and chemical durability as a function of composition. In comparison with lithium aluminosilicate glasses, corresponding sodium aluminosilicates have by about 10 per cent higher thermal expansions and dielectric constants, and densities higher by about 2 per cent, but an alkali loss by leaching which is by 3 to 6 times larger, and a slightly lower refractive index.

128. For the thermal expansion of complex glasses, K. Takahashi[205] calculated a series of new increments of the cations participating in glass compositions, with particular emphasis given to structural effects of O^{2-} coordination in the network theory. The applicability of the new factors is claimed to be in a wider field of glass compositions and with a higher accuracy than for the previous factors given by Winkelmann and Schott (1901), English and Turner (1927), Gilard and Dubrul (1938), and Danzin (1949).

129. Very interesting studies on effects of the introduction of *rare earths* into different lead silicate and soda-lime silicate glasses, and further into phosphate glasses

Hartmann, H. Brand, and K.-H. Kiessling, *Glastech. Ber.*, **26**, 1953, 29-33; **27**, 1954, 12-15; **30**, 1957, 186-188, had found, and striking divergencies as well from the Sharp-Ginther equation as from the data of Schwiete and Ziegler (see above) are evidently caused by serious inaccuracies in the chemical analyses of glasses used by those authors (cf. A. ¶ 284).

[203] *J. Am. Ceram. Soc.*, **41**, 1958, 324-329.

[204] *J. Am. Ceram. Soc.*, **42**, 1959, 113-120.

[205] *Yogyo Kyokai Shi*, **63**, 1955, 143-147; tabulation of increments also in *Ceram. Abstr.*, 1955 200 d. On thermal expansion of borosilicate glasses see K. Takahashi, *Bull. Fac. Eng. Yokohama Nation. Univ.*, **3**, 1954, 201-213.

were performed by R. C. Vickery and R. Sedlacek.[206] From the quantitative data it is evident that lanthanum as a representative of the rare earth elements is exclusively a framework modifier, whereas scandium and yttrium behave as framework formers

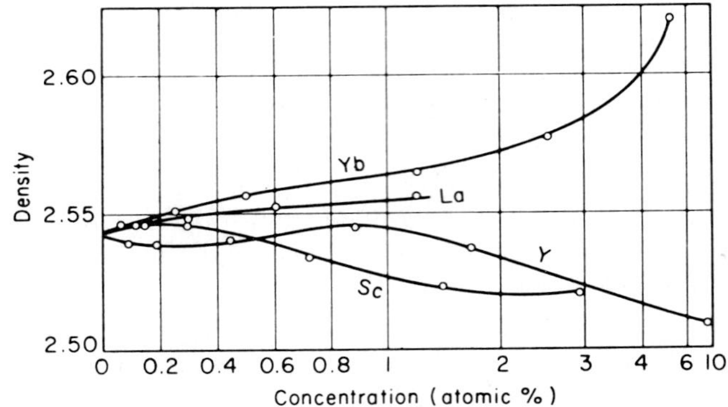

Fig. B. 76. Density of phosphate glasses containing different rare earth elements, as a function of concentration. (Vickery and Sedlacek).

in low concentrations (between 0.2 and 1 atomic per cent) (cf. Fig. B. 76). The most distinct effects are observed in phosphate glasses, the least characteristic ones in lead silicate glasses (cf. A. ¶ 352). The *microhardness* (determined by the Vickers diamond prism indenter test) of different glasses as a function of composition for silicate and borosilicate glasses was examined by L. Ainsworth,[207] taking the mechanical yield point of deformation in the glass structure as basis of the measurements with a plastic flow under a small load. The experiments extended over alkali silicate glasses, soda-lime silicate glasses, soda-lead silicates, and complex compositions. Fused silica glass has the highest yield point of all of these. An addition of alkali to silica glass causes an initial decrease in bond strength, followed by a slight increase until a critical concentration is reached beyond which another decrease occurs (Fig. B. 77). Mixed soda-potash silicate glasses are stronger than the two pure alkali glasses alone. Zinc, magnesium, and lead participate in the framework-forming mechanism to a considerable extent. This explains the apparently inconsistent be-

[206] *J. Am. Ceram. Soc.*, **41**, 1958, 422-426.

[207] *J. Soc. Glass Technol.*, **38**, 1954, 479-500, 501-535, 536-547. Very important connections between the microhardness and the atomistic constitution of glasses (as indicated e.g. in the molecular refraction) were quite recently demonstrated by A. Petzold, F. G. Wihsmann and H. v. Kamptz, *Glastech. Ber.*, **34**, 1961, 56-71, with an extensive bibliography. On technological grindability (abradibility) of glasses as a function of composition cf. J. Gypser, *Silikat Tech.*, **6**, 1955, 372-377, see also S. R. Scholes, *J. Am. Ceram. Soc.*, **28**, 1945, 133-136. Mechanical properties of ternary glasses were recently discussed by F. K. Aleĭnikov, R. B. Paulavichius, and V. A. Slizhis, *Lietuvos T. S. R. Mokslų Akademijos Darbai, Ser. B.*, **2** (29), 1962, 69-93.

havior of glasses containing these oxides, with exceptions to the general rule of sequence of decreasing effects, for $Al_2O_3 > CaO > SrO > BaO > ZnO > MgO > PbO > SiO_2 > Na_2O > K_2O$.

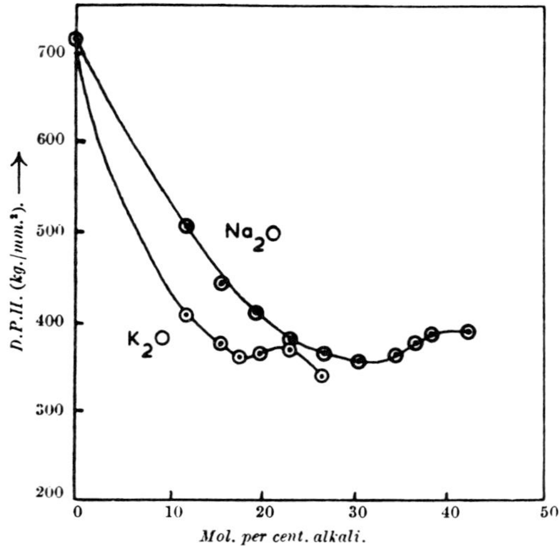

FIG. B. 77. Variation of the diamond pyramid hardness number, as a function of the alkali content of experimental glasses. (Ainsworth).

130. Densities and *optical* properties (refractive index, dispersion) of soda-lead silicate glasses were studied by K. A. Krakau.[208] There is a distinct deviation from the smooth curve of changes in optical constants from additivity on the join $Na_2O.2SiO_2$—$PbSiO_3$, near the composition of $Na_2O.2PbO.4SiO_2$, and on the join $Na_2O.2SiO_2$—$Na_2O.3PbO.7SiO_2$ near the latter composition. M. A. Bezborodov and A. I. Zelenskiĭ[209] demonstrated how zirconia introduced into a soda-lime silicate glass (replacing SiO_2, up to 3 per cent, and replacing CaO up to 10 per cent) affects the refractive index. The index is raised the more zirconia is introduced, but with dif-

[208] In the Symposium "Physical-Chemical Properties in the Ternary Soda-Lead Silicate System," edited by I. V. Grebenshchikov, Akad. Nauk S.S.S.R., Moscow-Leningrad 1949, pp. 123-131. See also N. M. Medvedev, *Structure of Glass. Proc. All-Union Conf. Glassy State, 3rd, Leningrad* (English Translation), 1959/60, 281-284.

[209] *Doklady Akad. Nauk S.S.S.R.*, **96**, 1954, 137-139, see also N. M. Verebeĭchik and V. I. Odelevskiĭ, *Structure of Glass. Proc. All-Union Conf. Glassy State, 3rd, Leningrad* (English Translation), 1959/60, 248-250. On *hafnium*-containing silicate glasses see recently S. M. Brekhovskikh and V. N. Sesorova, *Ibid.*, 1959/60, 396-398. M. A. Bezborodov and N. M. Bobkova, *Doklady Akad. Nauk S.S.S.R.*, **116**, 1957, 652-655, offer interesting data also for *cesium* silicate glasses in their optical properties, applying Appen's theories (see above). On zirconia and titania glasses, for a comparison see M. A. Bezborodov and I. S. Kachan, *Ibid.*, **115**, 1957, 1148-1151.

ferent slopes in the lines as a function of the replacements mentioned before. The refractive index properties are so favorable that it is possible to substitute zirconia glasses with very comparable properties for certain kinds of lead crystal glass.

131. *Lithium-beryllium borate* glasses, as "Lindemann Glasses," which have a great significance for X-ray diffraction practice as a container material of high *transmittance* were systematically studied by H. Menzel and S. Sliwinski,[210] followed by H. Menzel and J. Adam,[211] with an extensive examination of the X-ray transmittance. They also examined other complex borate systems. An improved glass of a similar composition type was recommended by A. Schleede and M. Wellmann.[212] Other investigations in this field were more recently made by L. Ya. Mazelev,[213] especially on the system $Li_2O—MgO—B_2O_3$. Density and refractive index of such glasses do not follow the additivity rule. On the other hand, S. M. Brekhovskich[214] developed special glasses of a very high X-ray and γ-ray *absorption* for irradiation protection purposes. In these ultraheavy glasses of the system $PbO—Bi_2O_3—SiO_2$ the lead and bismuth ions are framework formers. Also in their infrared transmittancy in the range from 1 to 4.5 μ these glasses show remarkable properties. Brekhovskich also studied the systems $PbO—Bi_2O_3—B_2O_3$, $PbO—Bi_2O_3—P_2O_5$, $PbO—CdO—B_2O_3$, $CdO—BaO—B_2O_3$, $PbO—CdO—SiO_2$, and the quaternary system $PbO—Al_2O_3—Bi_2O_3—SiO_2$, with nomograms for the curves of constant density and absorption for γ-radiation (of energy 1.326 mev.).

132. Lead silicate and phosphate glasses with the highly polarizable Pb^{2+} ion were the subject of theoretical studies of M. Tashiro[215] especially in connection with their *photoelastic birefringence* behavior under stress. This property is influenced quite specifically by the presence of alkali ions and by the substitution for O^{2-} anions by F^-, in the mutual polarization effects of anions and cations, as W. A. Weyl had postulated[216] (cf. A. ¶ 392 ff.). It was found that Pb^{2+}, Ba^{2+}, Sr^{2+}, and Tl^+ change the value of the birefringence of the glass of which they are a constituent, in the negative direction. The amount of PbO required to decrease the birefringence

[210] *Z. anorg. u. allgem. Chem.*, **249**, 1942, 357-385.

[211] *Glastech. Ber.*, **22**, 1949, 237-246; **25**, 1952, 143-152, 354-361.

[212] *Z. Krist.*, **83**, 1932, 148-149; the "Wellmann glass" with a low content in soda and potash, was e.g. used by I. N. Stranski, E. Kürbs, and K. Plieth, *Z. anorg. u. allgem. Chem.*, **258**, 1949, especially 240-241, in their studies on modifications of As_2O_3 (as claudetite and arsenolite).

[213] *Zhur. Priklad. Khim.*, **13**, 1940, 1288-1302; **26**, 1953, 286-297. See further the recent results of L. Ya. Mazelev, *Structure of Glass. Proc. All-Union Conf. Glassy State, 3rd, Leningrad* (English Translation), 1959/60, 390-392, on systematically varied borate glass compositions.

[214] *Steklo i Keram.*, **14**, 1957 (8) 1-4; *Glastech. Ber.*, **32**, 1959, 437-442.

[215] *J. Soc. Glass Technol.*, **40**, 1956, 353-362.

[216] *Trans. N.Y. Acad. Sci.*, [2] **12**, 1950, 245-257; *Glastech. Ber.*, **30**, 1957, 269-282, especially p. 276; **32**, K, 1959, especially p. VI. 4.

to zero depends on the base composition of the glass and on the ratio of Pb^{2+}/Si^{4+}, which is in every case higher for lead silicate glasses than for lead phosphates with the ratio Pb^{2+}/P^{5+}. The alkali ions tend to increase the birefringence effect in the series $Li^+ < Na^+ < K^+$. Mg^{2+} and Ca^{2+} act in the same direction. The negative effect of Ba^{2+} is stronger than that of Sr^{2+}. The sign and amount of photoelastic birefringence is the result of two antagonistic effects, namely that of a deformation of the polarizable anions (acting in the positive direction) and the counter-effect of deformation of the polarizable cations (acting negatively).

133. Concerning the effects of the composition on *molecular refraction*, R, we mention the extensive investigations of E. Kordes (1939, 1941), and a more recent study of his with H. Becker,[217] in the form of graphs showing the functional relations. In contrast with the glasses of the system $PbO—P_2O_5$, those of Na_2O, Li_2O, and CdO with P_2O_5 do *not* change their molecular volumes and molecular refractions as linear functions of composition, and apparent molecular refraction increments of the oxides as constituents of the glass are not constant, but approximately a linear function of the molecular oxide composition. The same is valid for the oxygen atoms, $O^{P,M}$, which are connected on one side with P^{5+}, on the other side with the metal atom M. The refraction of the $O^{P,Pb}$ is $R = 3.70$ cm.³, that of $O^{P,Cd}$ also $R = 3.70$, but for $O^{P,Na}$ it is $= 2.84$, and for $O^{P,Li} = 2.64$ cm.³. From the phosphate glasses and their molecular refraction the ion refractions are calculated; they are for $Pb^{2+} : R = 9.12$ cm.³, for $Cd^{2+} : R = 2.15$ cm.³, in good agreement with the experimental data. In glasses of the type $Na_2O—P_2O_5$, or $CdO—P_2O_5$, there are distinct negative packing effects. Negative ionic refractions for Na^+ and Li^+ which have no direct physical meaning, are only indications of mutual polarization effects and make a calculation of those parameters from phosphate glasses illusionary. On the calculation of ionic refractions for *crystalline* phosphates and other salts, cf. Vol. I, A. ¶ 224, and the recent investigations of G. B. Bokiĭ and E. S. Batsanov.[218]

134. Aniuta Winter-Klein[219] (cf. A. ¶ 2) in her fundamental investigations of the physical properties of glasses, as a consequence of their constitution, built up by simple additivity principles for the glass oxides, A_xB_y, and the atomic masses M_A and M_B formulated the refractive indices $n_{A_xB_y} = n_O + C \cdot y \cdot M_{A,B}$ with C a constant, and $n_O = 1.24$, the refractive index of glassy oxygen. The difference of the indices $n_{A_xB_y}$ and n_O divided by y is a linear function of the atomic weight of A, as seen in Fig. B. 78. In silicate glasses the optical dispersion $\nu = (n_D - 1)/$

[217] *Z. anorg. u. allgem. Chem.*, **260**, 1949, 185-207.

[218] *Kristallografiya*, **1**, 1956, 81-89.

[219] *Compt. rend.*, **242**, 1956, 2931-2933. A basic investigation of the optical properties of *borate glasses* was presented by M. Imaoka and T. Yamazaki, *J. Ceram. Assoc. Japan*, **70**, 1962, 89-100, and specifically for *lanthanum borate glasses*, their refractive indices and Abbe numbers, *ibid.*, pp. 115-123.

($n_F - {}_Cn$) is for cations of different groups of the Periodic System[220] a function of the molecular proportion of the oxide in the glass composition with one characteristic curve for each group, as shown for the oxides of groups I and II (alkalies and alkaline earths) in Fig. B. 79. The variation of ν in silicate glasses therefore

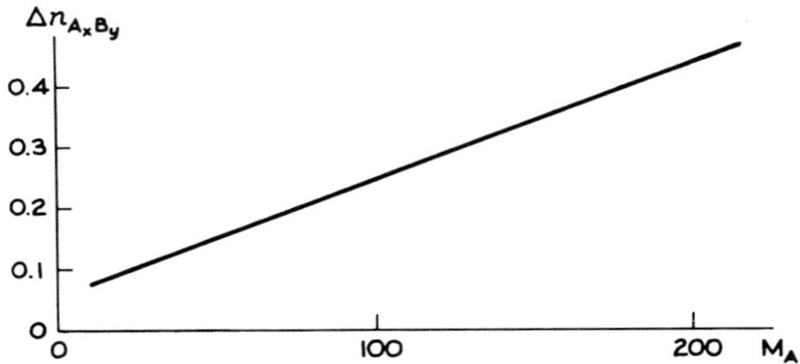

FIG. B. 78. Linear additivity of refractive indices of glass-forming oxides from the atomic mass, M_A, of the cation element, and the parameter $n_{A_xB_y}$. (Winter-Klein).

depends on the number of molecules modifying the structure ("molécules perturbatrices") and the group in the Periodic System, whereas the weight and sizes of the modifying elements are ineffective.

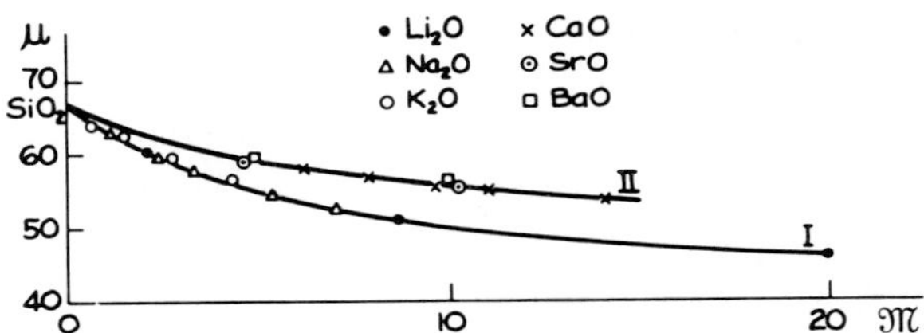

FIG. B. 79. Dispersion, ν, for silicate glasses with cations of groups I and II of the Periodic System, as a function of the fictive molecular mass, $\mathfrak{M}$, of the bases. (Winter-Klein).

135. Concerning a comparison of monoatomic, or diatomic glasses, as mixtures of the elements of the VIth group with elements of other groups, namely of oxides, sulfides, selenides, tellurides in glassy form, and their refractive indices we refer again to A. ¶ 2, according to A. Winter-Klein.[221] The validity of the Lorentz-Lorenz relation

$$\left(\frac{n^2 - 1}{n^2 + 2}\right) \cdot \frac{M}{\varrho} = \frac{4\pi}{3} N_0 \sum_i f_i \alpha_i$$

(with α the polarizability, ϱ the density, f_i the molar fraction of ions of type i, N_0

[220] *Compt. rend.*, **242**, 1956, 625-627.

Avogadro's number) for an accurate relation between density and refractive index, is for *glasses* of restricted reliability, as H. N. Ritland[222] made evident from a very large material of numerical data on optical glasses and other glasses with a widely varying thermal history (cf. A. ¶ 272 ff.). The increase in refractive index for a given change in density is always *smaller* than it should be as required by the Lorentz-Lorenz equation. The apparent contribution of induced polarization to the local electric field is very small as a consequence of the reduction in ionic polarizability when an increase in density occurs. The Newton-Drude relation $(n^2 - 1) \cdot M/\varrho = 4\pi N_0 \sum_i f_i \alpha_i$ fits somewhat better than the Lorentz-Lorenz relation although in the other direction of deviations.

136. An extensive investigation of the specific refraction of glasses,

$$R_s = \left(\frac{n^2 - 1}{n^2 + 2}\right) \cdot \frac{1}{\varrho}$$

and its temperature coefficient by L. Prod'homme[223] confirms the deviations from the Lorentz-Lorenz equation in so far as it would predict a negative temperature coefficient of n, whereas a rather regular *in*crease of n is observed up to the transformation range. The specific refractivity also varies with temperature. The Lorentz-Lorenz equation is well verified the less in glass the higher their contents in heavy atoms, which are sensitive to temperature changes by their electric polarization. The temperature coefficient dn/dT may be considered as being composed of two antagonistic factors of positive and negative effects on their relation to n. Thus it may happen, that silica glass, and a heavy Flint glass have the same temperature coefficient of n although in the silica glass the effect of a slight dilatation prevails, whereas in the Flint glass a large polarization of Pb^{2+} brings about the end effect. The temperature coefficient is negative only in glasses of large thermal expansion, as in phosphate glasses, or in other types with very light framework forming atoms like boron and even carbon. The coefficient dn/dT is a linear function of the "fictive" mass, $\mathfrak{M}$, of the glass in question, namely the mass of that volume of it which would correspond to a content of one gram-molecule oxygen.

137. The interesting effects of *titania* as an addition to silicate glasses were discussed by M. A. Bezborodov and I. I. Kisel'.[224] Titania glasses may substitute in many

[221] *Compt. rend.*, **244**, 1957, 750-752; *J. Am. Ceram. Soc.*, **41**, 1958, 464-466.

[222] *J. Am. Ceram. Soc.*, **38**, 1955, 86-88. The best agreement with the Lorentz-Lorenz relation is shown by alkali halide *crystals*, cf. J. R. Tessman, A. H. Kahn, and W. Shockley, *Phys. Rev.*, **92**, 1953, 890-895.

[223] *Compt. rend.*, **244**, 1957, 584-586; Thèse, Faculté des Sciences, Université de Paris, Série A, No. 3046, 1957, 74 pp.; *Verres et réfractaires*, **10**, 1956, 267-276; **11**, 1957, 351-360; **12**, 1958, 3-22.

[224] *Doklady Akad. Nauk S.S.S.R.*, **103**, 1955, 1073-1076. See also R. G. Tempel'man, *Legkaya Prom.*, **12**, 1952 (8) 34-36, especially on the fusion and the optical properties of TiO_2 containing

cases for lead crystal glasses, with good optical properties, and a satisfactory chemical durability. Barium-titanium silicate glasses were studied by G. W. Cleek and E. H. Hamilton,[225] in this connection, with excellent optical properties, infrared transmittance, and a high chemical resistivity to acidic and alkaline solutions. A broad minimum of leachability for increasing titania contents between 12 and 30 per cent was also observed by V. V. Vargin[226] in sodium titanosilicate glasses. He also noted nearly linear, smooth curves for refractive indices and densities with increasing TiO_2 contents. J. H. Strimple and E. A. Giess[227] studied glasses of the system $Na_2O—B_2O_3—SiO_2—TiO_2$, up to 45 molecular per cent TiO_2 and 30 per cent alkalies, with detailed measurements of the thermal expansion, density, refractive index, and molecular refraction as functions of the composition. The chemical durability is also in this case improved by addition of TiO_2, the thermal expansion characteristics comparable with those of *borosilicate* glasses.

138. This latter type of complex glasses is much elucidated by extensive investigations of T. Abe,[228] especially in connection with problems of the "boric acid anomaly," in other words the change from the coordination $[BO_3]$ to $[BO_4]$, as a function of the composition but also with rich numerical data of thermal expansion coefficients, densities, refractive indices, and molecular refractions in the system $Na_2O—B_2O_3—SiO_2$. E. H. Hamilton, G. W. Cleek, and O. H. Grauer[229] studied in approximately the same direction the properties of glasses of the system $BaO—B_2O_3—SiO_2$, with some typical anomalies for a continuous change in composition, characterized by maxima of the refractive index and density for compositions near distinct stoichiometric ratios corresponding to the compound $3BaO.3B_2O_3.2SiO_2$ (cf. Vol. III, B. ¶ 429). The most stable glasses of the series are located in the field of primary crystallization of this ternary compound which shows a very flat Liquidus surface indicating the high degree of thermal dissociation in the melt. There is an evident analogy with conditions in the system $BaO—TiO_2—SiO_2$, with the ternary compound $BaO.TiO_2.2SiO_2$ or with $ZnO.B_2O_3$ in the system $ZnO—B_2O_3—SiO_2$.

glasses. More recently A. Ram, S. V. Bhatye and K. D. Sharma, *Central Glass & Ceram. Research Inst. Bull. (India)*, **6**, 1959, 3-12, studied the influence of TiO_2 on viscosity and surface tension of soda-lime silicate glasses. On the dielectric properties of glasses in the system $Na_2O–SiO_2–TiO_2$, see recently Ch. Hirayama and D. Berg, *Phys. & Chem. of Glasses*, **2**, 1961, 145-151.

[225] *J. Research Natl. Bur. Standards*, **57**, 1956, 317-323.

[226] *Doklady Akad. Nauk S.S.S.R.*, **111**, 1956, 848-850.

[227] *J. Am. Ceram. Soc.*, **41**, 1958, 231-237.

[228] *J. Am. Ceram. Soc.*, **35**, 1952, 284-299. Concerning the boric acid anomaly cf. O. S. Molchanova and M. V. Serebryakova, *Trudy Gosudarst. Opt. Inst. Leningrad*, **23**, 1953, No. 141, 3-12, especially observed in changes of the refractive index and density in the annealing range (cf. A. ¶ 290). A. A. Appen and Kan' Fu-Si, *Structure of Glass. Proc. All-Union Conf. Glassy State, 3rd, Leningrad* (English Translation), 1959/60, 445-450, observed also aluminoborate anomalies in silicate glasses containing simultaneously Al_2O_3 and B_2O_3.

[229] *J. Am. Ceram. Soc.*, **41**, 1958, 209-215.

139. A very specific effect of *fluorides* in glasses containing *rare earths* was described by G. W. Cleek and Th. G. Scuderi[230] for the infrared transmittancy of lanthanum silicate glasses in the range from 2.75 to 4.25 μ. The absorption is considerably reduced by the presence of fluorine anions and many of the glasses do not show any turbidity or opalescence. The only serious inconvenience in manufacturing of such glasses is their tendency to form fluoride striae, or local heterogeneities if more than 4 molecular per cent of fluoride is introduced. The refractive indices are lowered by the fluorine content. At the same time the undesirable absorption band at 2.75 μ ("water band") is to a high degree eliminated. Particularly valuable are glasses of the kind here discussed on the basis of barium titanosilicates or of $BaO—La_2O_3—SiO_2$, with a substitution of BaO by BaF_2. On the other hand, AlF_3 could be introduced into quaternary glass compositions $BaO—La_2O_3—TiO_2—SiO_2$, although in this case the fluoride striations are very disturbing.

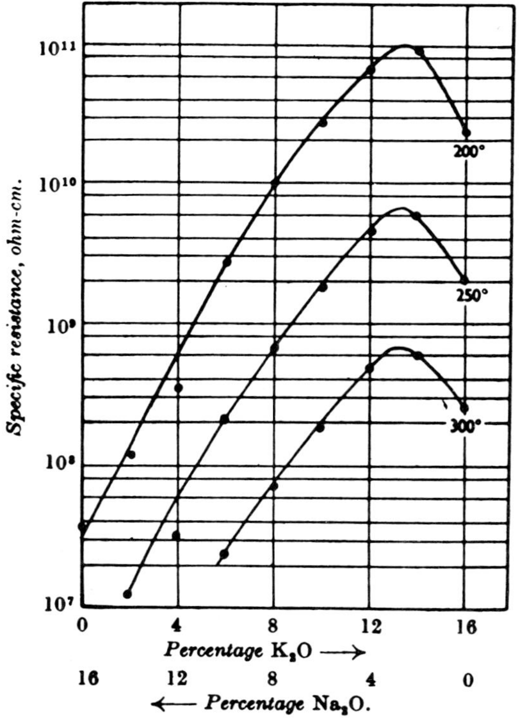

FIG. B. 80. Variation of specific resistance of mixed alkali-lime silicate glasses in the series 74 SiO_2, 10 CaO, 16 (Na_2O + K_2O), as a function of composition. (Moore and De Silva).

[230] *J. Am. Ceram. Soc.*, 42, 1959, 599-603.

140. Among investigations on the *dielectric* properties of systematically changed glasses, we mention those of A. A. Khar'kov[231] of dielectric losses in the system Na_2O—PbO—SiO_2 at high frequencies ($\nu = 1.715 \times 10^5$ to 1.3×10^7 cycles). Up to 300°C. the loss angle (tan δ) decreases with increasing ν. The binary section $Na_2O.2SiO_2$—$PbO.SiO_2$ shows a minimum for the loss angle at low temperatures, corresponding to the compound $Na_2O.2PbO.4SiO_2$. If PbO is kept constant (28.6 mol. per cent), and Na_2O substitutes SiO_2, tan δ suddenly increases as soon as the composition $Na_2O.2PbO.4SiO_2$ is passed. A most elaborate study of the electric properties of alkali-lime silicate glasses, some with B_2O_3 and Al_2O_3 contents, was performed by H. Moore and R. C. De Silva,[232] based on an extensive theoretical treatment of physical problems involved with conductance, dielectric constants and power factors, in relation to the glass structure and the alkali ion mobility (cf. A. ¶ 151). Divalent cations are in all cases held much more strongly than the alkali ions as a part of the glass structure. Alumina loosens the structure and facilitates the alkali ion migration, whereas B_2O_3 tightens the structure and impedes the alkali migration. In general, the potassium ion is kept firmer in the glass framework than sodium ions. Among the experimental results, given in Fig. B. 80, are typical variations of specific resistance as a function of composition in mixed glasses of the type 74 SiO_2, 10 CaO, 16 (Na_2O + K_2O). In Fig. B. 81 are noted the same results for the dielectric constant.

141. More recently, D. W. Rinehart[233] examined the dielectric losses for commercial plate glass as a function of temperature to establish irreversible increases in tan δ at quenching temperatures below the transformation range, or within it, evidently in consequence of changes in the distribution of Na^+ ions during cooling. The increases in tan δ are connected with changes in density indicating changes of configuration in the framework structure of the glass. Simple ternary systems of the type Na_2O, or K_2O—SiO_2 with MgO, CaO, SrO, BaO, CdO, ZnO, PbO, B_2O_3, Al_2O_3, SiO_2, TiO_2, ZrO_2, and SnO_2 (for constant $\nu = 1$ kc. at room temperature) show profound effects and changes of configuration by added oxides on the mobility of the Na^+ ions, decreasing tan δ, with amounts which bear some relation to the sizes of the introduced cations. Quite particular are the maxima and minima observed in the functional relation of tan δ to the contents of the glasses in MgO and ZnO (Fig. B. 82), and a characteristic minimum in the corresponding curves with additions of alumina, followed by a steep increase in tan δ with higher alumina

[231] "Physical-Chemical Properties of the Ternary System Soda-Lead Oxide-Silica," edited by I. V. Grebenshchikov, under auspices of the Akad. Nauk S.S.S.R., Moscow-Leningrad 1949, pp. 158-163. On lead borate and borosilicate glasses with dielectric constants up to about 80 see Fr. Bischoff, *Glastech. Ber.*, **28**, 1955, 98-100.

[232] *J. Soc. Glass Technol.*, **36**, 1952, 5-55 T.

[233] *J. Am. Ceram. Soc.*, **41**, 1958, 470-475; **42**, 1959, 107-112 (with J. J. Bonino).

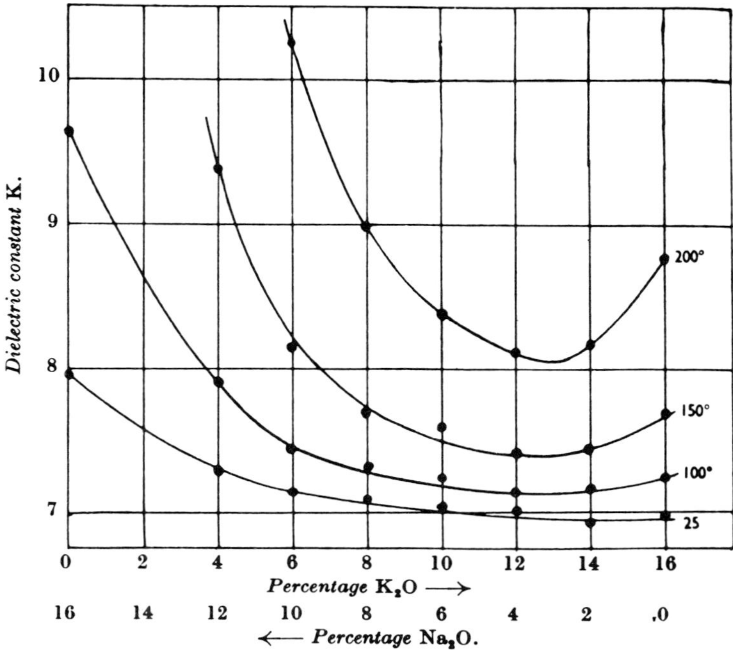

FIG. B. 81. Variation of dielectric constant of mixed alkali-lime silicate glasses of series 74 SiO_2, 10 CaO, 16 (Na_2O + K_2O), as a function of composition. (Moore and De Silva).

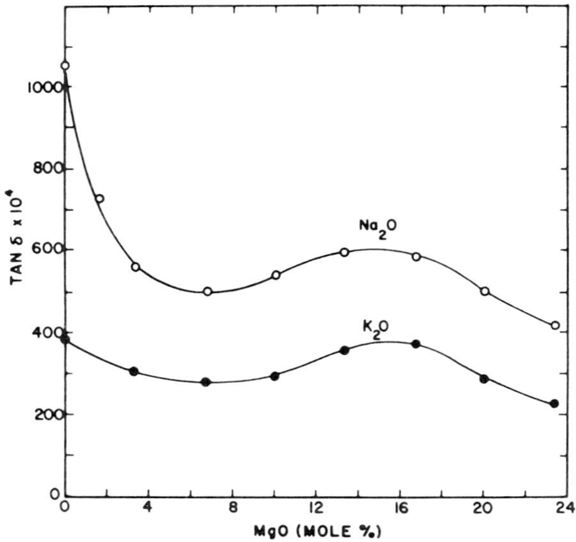

FIG. B. 82. Tan δ of glasses of the systems Na_2O—MgO—SiO_2, and K_2O—MgO—SiO_2, as a function of magnesia contents, for a frequency $\nu = 1$ kc., at room temperature. (Rinehart and Bonino).

contents (Fig. B. 83), and also the comparison with the behavior of additions of SiO_2, TiO_2, ZrO_2, and SnO_2 (Fig. B. 84). B_2O_3 decreases tan δ, sharply contrasting to the behavior of alumina (Fig. B. 85). The anomalous maxima and minima may be explained by the change of the cations added in their coordination properties and the changes from framework-modifying to framework-forming qualities.

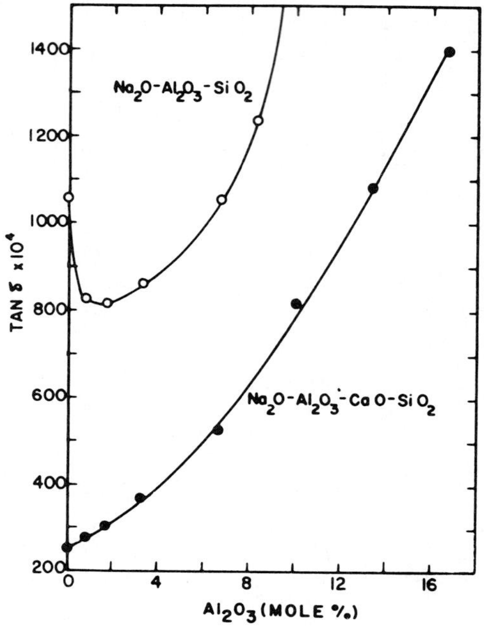

Fig. B. 83. Tan δ of glasses of the systems Na_2O—Al_2O_3—SiO_2, and K_2O—Al_2O_3—SiO_2, as a function of the contents in alumina, for a frequency of 1 kc., at room temperature. (Rinehart and Bonino).

142. The electrical properties of the system Li_2O—CaO—Al_2O_3—SiO_2, at different levels from 0 to 30 weight per cent CaO, were measured by H. W. Rauch, C. H. Commons, and H. H. Blau.[234] Some of the compositions in higher lime concentrations may find commercial application. The dielectric constants are moderately increased with Li_2O increasing up to 15 per cent, but then with an accelerated rate. Double break points appear on the curves of the ternary compositions, whereas those of the quaternary glasses exhibit a single change in slope. Also the power factor is increased with increasing Li_2O contents. The ternary glasses have higher power factors than quaternary types; break points are less distinct than they are for the dielectric constants. The surface resistances for the measured glasses are higher

[234] *J. Am. Ceram. Soc.*, 42, 1959, 113-120.

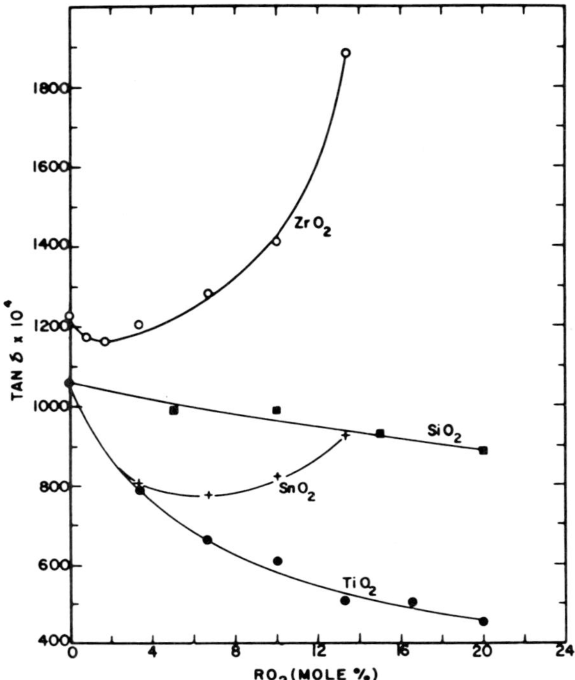

Fig. B. 84. Tan δ of glasses of the systems $Na_2O—RO_2—SiO_2$, with $RO_2 = SiO_2$, TiO_2, ZrO_2, and SnO_2, for a frequency of 1 kc., at room temperature. (Rinehart and Bonino).

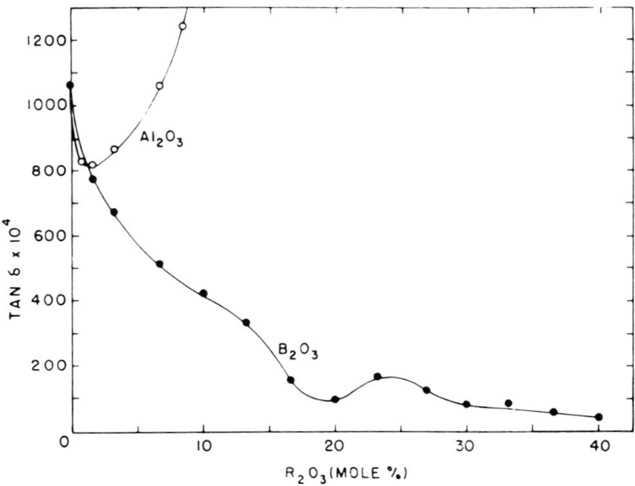

Fig. B. 85. Tan δ of glasses of the systems $Na_2O—R_2O_3—SiO_2$, with $R_2O_3 = B_2O_3$, and Al_2O_3 for a frequency of 1 kc., at room temperature. (Rinehart and Bonino).

than volume resistances, decreasing with increasing Li_2O contents, but in one of the series a maximum of volume resistance appears in glasses with 5 per cent CaO and 5 per cent alumina. In comparison with corresponding soda glasses, these have a thermal expansion coefficient and dielectric constant about ten per cent higher than the lithia glasses, and about 2 per cent higher density, but slightly lower refractive indices. The chemical durability of lithia-lime aluminosilicate glasses decreases with increasing lithia contents. For some lead borate and phosphate glass systems, with a high specific resistance and power factor, and with a yield point temperature (for a glass rod deforming under its own weight) of only 350° to 500°C. Ch. Hirayama and M. M. Rutter[235] studied the changes in their properties as a function of the systematic substitution of PbO by BaO, alumina, and silica. In general, the substitutions indicated improve the dielectric properties and raise the "sag" points. The dielectric constants (for $\nu = 60$ cycles, and 1 kc.) are linear functions of the densities at room temperature. If Pb^{2+} and Ba^{2+} ions are present simultaneously, there is a distinct minimum of the power factor. With particular respect to practical application, Hirayama and Rutter discuss the temperature effects on electric resistance and power factors of the different glass compositions.[236] For glasses of comparable softening temperatures the lead borate glasses are superior to the phosphate glasses.

ELECTRICAL PROPERTIES OF GLASSES AT LOW TEMPERATURES

143. The *theory of glass electrodes* was extensively discussed by N. A. Izmaĭlov and A. G. Vasil'ev.[237] M. Dole's hypotheses (1931) on anionic reactions of the electrode in strongly acidic solutions are evidently not sufficient to explain their particular behavior. Anionic volume and adsorption factors were previously considered by B. P. Nikol'skiĭ[238] under the assumption that anions as well as cations enter the sur-

[235] *J. Am. Ceram. Soc.*, **42**, 1959, 367-373. Interesting comparisons between aluminosilicate, and galliosilicate glasses were made by S. K. Dubrovo, *Structure of Glass. Proc. All-Union Conf. Glassy State, 3rd. Leningrad* (English Translation), 1959/60, 373-377, concerning their physical and chemical properties. See also E. I. Galant, *Doklady Akad. Nauk S.S.S.R.*, **141**, 1961, 417-419, on galliosilicate, and gallioborosilicate glasses.

[236] We may here expressively emphasize the important monographs on dielectric properties of glasses in Chapter III, by J. M. Stevels, in "Handbuch der Physik," edited by S. Flügge, Springer-Verlag, Berlin, Göttingen, Heidelberg, Vol. **20**, 1957, 1056 pp., and G. W. Morey, "Properties of Glass," American Chemical Society Monographs, No. 124, Reinhold Publishing Corp. New York, 2nd edition, 1954, 591 pp., especially Chapters 17 and 18. Ch. Hirayama and D. Berg, *Am. Ceram. Soc. Bull.*, **40**, 1961, 550-555, discussed the low-temperature resistance of commercial glasses.

[237] *Doklady Akad. Nauk S.S.S.R.*, **95**, 1954, 579-582.

[238] *Zhur. Fiz. Khim.*, **25**, 1951, 1335-1346 (with E. A. Materova); *Ibid.*, **27**, 1953, 724-743.

face layer of the glass electrode. Izmaĭlov and Vasil'ev come to the conclusion that this assumption is in contradiction with the fact that cations are *not* adsorbed at all in the highly acidic range and anions *not* adsorbed in the alkaline range of *pH*. A new theory is developed starting from the amphoteric character of silicon in the surface layers of glass electrodes. In the active state, Na^+ ions are exchanged for H^+, expressed by the reaction $\equiv$Si—O—Na $\rightleftarrows$ $\equiv$Si—O—H, or in other words, they function as a hydrogen electrode with base exchange capacity. In strongly alkaline solutions it is a cationic electrolyte. In strongly acidic solutions, as in dilute HCl, the anionic function is expressed by a group $\equiv$Si—Cl, formed in analogy to the formation of $\equiv$C—Cl from $\equiv$C—O—H in HCl on charcoal (cf. N. Shilov, H. Shatunovska and K. Chmutov[239]). When the electrochemical potentials $\bar{\mu}'$ and $\bar{\mu}$ for the glass and the solution are calculated, equations can be developed for the potential jump on the phase boundaries in relation to the constant of ionic exchange of anions in the acidic *pH* range in which the $\equiv$Si—OH groups start to show basic functions. The activity of H^+ and of the anions determines the potential jump. For lower acidities and neutral solutions, the calculation characterizes the glass electrode as a hydrogen electrode in its potential function. The calibration curve in the acidic range has a minimum for the function $E = f(pH)$.[240] The exchange constants of the anions in the acidic range, K_A, are calculated from experimental data. The minimum was found for $pH = -0.2$, the constant $K_A = 4.0 \times 10^{-15}$, whereas the Na^+ exchange gives the constant 6.3×10^{-13}. The calibration curve derived from Nikol'skiĭ's theory is in good agreement with previous data, showing in the alkaline range a characteristic maximum. A combination of Nikol'skiĭ's equations with those of Izmaĭlov and Vasil'ev provides a general solution of problems of the *pH* response of glass electrodes.

144. D. Hubbard and R. G. Goldman[241] studied the heterogeneous equilibria at the interface of the glass electrode with the solution in connection with its *pH* response (Fig. B. 86). Preferential leaching of alkali ions from the glass surface in acidic aqueous solutions and formation of a silica-enriched surface layer cause an uneven distribution of the migratory ions, corresponding to a mechanism in a Donnan membrane with "equation of products" equilibria. For Corning electrode glass No. 015 and a series of experimental sodium silicate and soda-lime silicate glasses the potentiometric method to determine voltage

$$E = \frac{RT}{nF} \cdot \log \left(\frac{M^{n+}_{\text{soln.}}}{M^{n+}_{\text{glass surf.}}} \right)$$

[239] *Z. physik. Chem. (Leipzig)*, **150**, A, 1930, 31-36.

[240] Cf. N. A. Izmaĭlov and A. M. Aleksandrova, *Sbornik Stateĭ Obshcheĭ Khim. Akad. Nauk S.S.S.R.*, **1**, 1953, 173-178.

[241] *J. Research Natl. Bur. Standards*, **48**, 1952, 428-437.

makes it possible to observe the uneven distribution, say of Br^- anions, and the complex ions $Ag(NH_3)^+$ (Fig. B. 87). The experimental data are compatible with requirements of the Donnan membrane theory, especially the swelling phenomenon of the glass surface, and repression of the swelling by high electrolyte concentra-

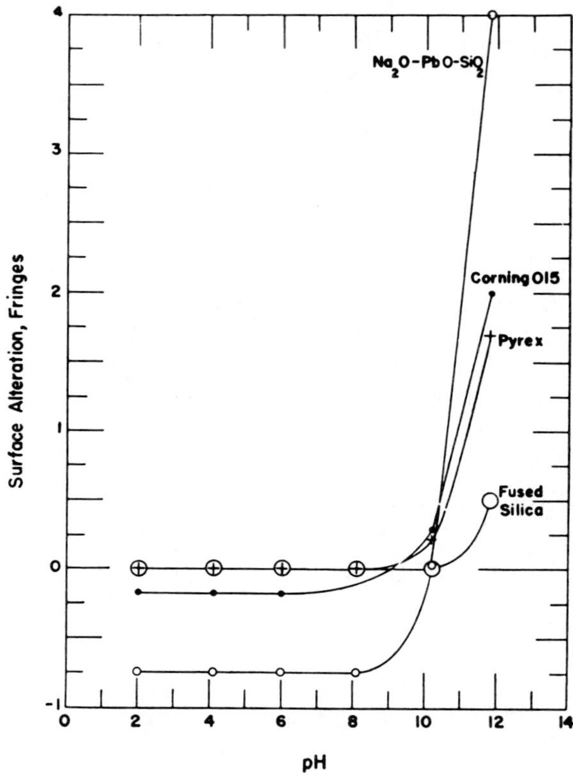

FIG. B. 86. Surface alteration — pH curves of some typical silicate glasses, exposed for 6 hours at 80°C. (Hubbard and Goldman).

tions. Further, the departures in voltage in alkaline and HF solutions (Fig. B. 88), in the "superacid" region, and asymmetric potentials are explained in the light of this theory. The Donnan membrane aspects, however, are not able to interpret the preferential voltage response of the glass electrode for H^+ ions, excluding all the other ions.

145. Lithium silicate glasses, in their suitability for hydrogen and deuterium electrode characteristics, were examined by D. Hubbard and G. W. Cleek.[242] These

[242] *J. Research Natl. Bur. Standards*, **49**, 1952, 267-272.

glasses show a distinct maximum of the *pH* and *pD* response curves at 82 per cent silica, with a potential of 0.59 mv. per *pH* unit, limited on the alkali-rich side of the compositions by the strong tendency to devitrify, and correspondingly by a very

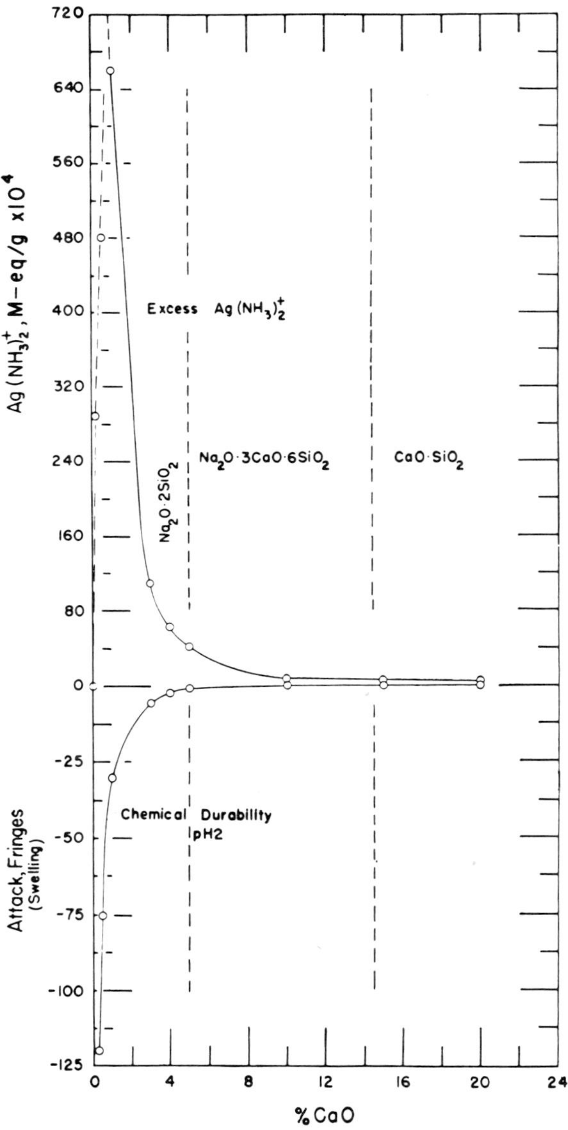

FIG. B. 87. Comparison of excess $Ag(NH_3)^+$ ions in glass surface, combined with chemical durability (swelling) of soda-lime silicate glasses, with constant silica contents of 70 per cent. (Hubbard and Goldman).

poor chemical durability, whereas on the siliceous side opalescence appears, but the hygroscopicity is decreased, although the response drops rapidly (cf. Fig. B. 89). The optimum composition for the usefulness of lithium silicate glasses for glass electrodes is in the field of primary crystallization of tridymite. Concerning the *swelling* behavior it is interesting that the glasses show a stronger swelling in acidic H_2O than in D_2O solutions. For D_2O, buffer solutions could be prepared with P_2O_5

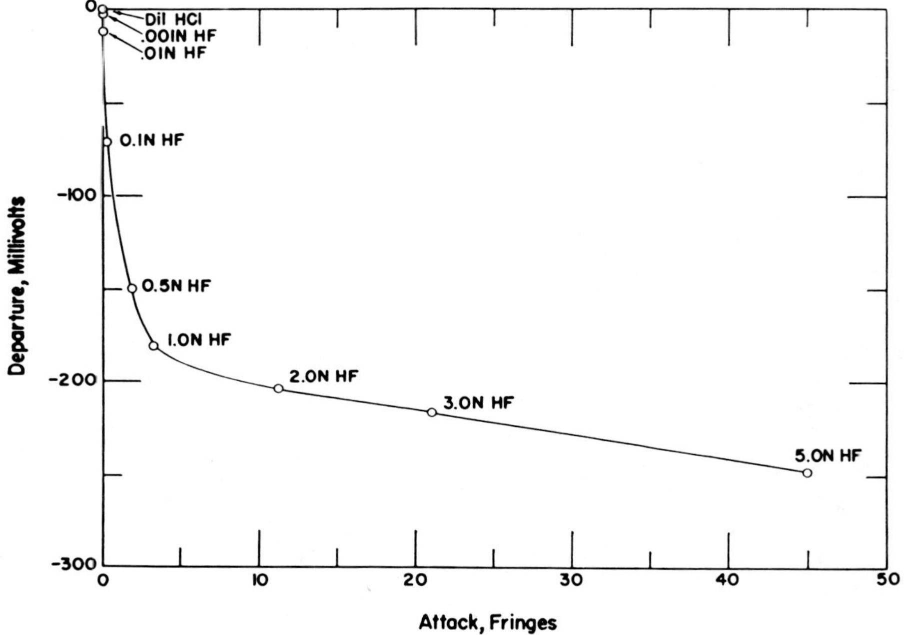

FIG. B. 88. Voltage departure of a glass electrode by partial removal of the silica-enriched surface layer. (Hubbard and Goldman).

(D_3PO_4, with $pD = 0.65$), and $Ca(OD)_2$ (with $pD = 12.4$). Hubbard[243] also examined Pyrex glass No. 7740 and Corning 0.15 concerning the temperature effect of their electric resistance and voltage deviations (errors). Pyrex does not give any significant pH response, even at resistances as low as 12 megohms (normally they are in the order of magnitude of 10^4 or 10^3 megohms). Glass electrodes with large errors in pH must fail if the hygroscopicity is very low, or the cell wall thickness[244] is too great, and the chemical durability of the glass is too poor, or if inhibiting films

[243] *J. Research Natl. Bur. Standards*, 50, 1953, 337-342.

[244] On a non-destructive (magnetic) method for determination of wall thicknesses, and especially of the optimum ("departure") thickness, cf. J. J. Diamond and D. Hubbard, *J. Research Natl. Bur. Standards*, 47, 1951, 443-448.

are deposited on the glass surface. The highly hygroscopic glass 015 (like common soft glasses) has a considerable increase in hygroscopicity between 20° and 100°C., whereas the decreasing resistance is of subordinate effect (cf. Fig. B. 90).

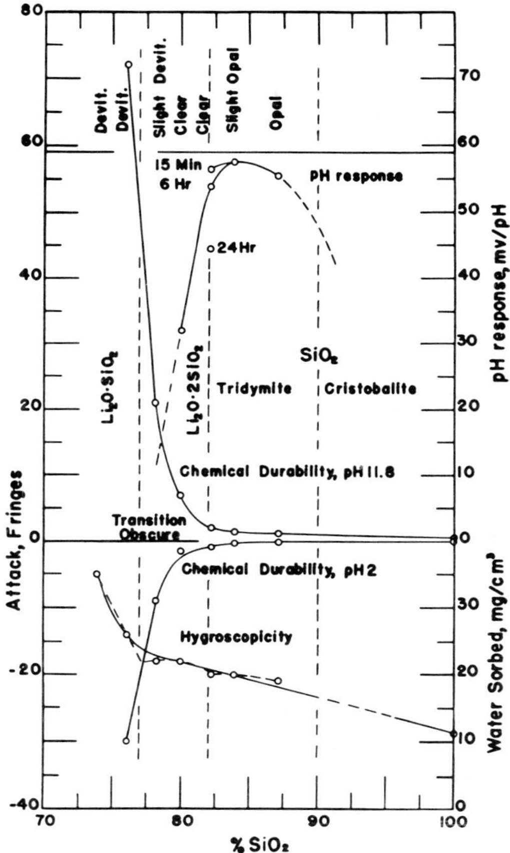

FIG. B. 89. Comparison of the pH response, the hygroscopicity, and chemical durability of lithium silicate glass, at pH = 11.8, as a function of the silica content. (Hubbard).

146. E. Budewski, D. Michailowa, and E. Pentschewa[245] investigated the glass electrode properties and resistance of glass containing 25 per cent Li_2O, and 8 per cent BaO, with improved devitrification stability, and with a resistance 6 to 8 times lower than that of the common "Perley" electrodes, combined with a high specific conduction similar to that of Corning 015 (with 3.3 to 4.8 $\times$ 10^{11} ohm^{-1} cm.^{2} at

[245] *Z. Elektrochem.*, 61, 1957, 158-162.

25°C.). V. V. Moĭseev, E. A. Materova, and A. A. Belyustin[246] made an interesting attempt to use rubidium and cesium silicate glasses as electrode material. Such compositions on the alkali-lime silicate basis are highly refractory (melting above 1600°C.), and even borosilicates with 11 per cent B_2O_3 melt above 1500°C. In the same way as proved for lithium and sodium silicate glasses, the hygroscopicity and chemical durability are factors determining usefulness as glass electrodes. If Na^+

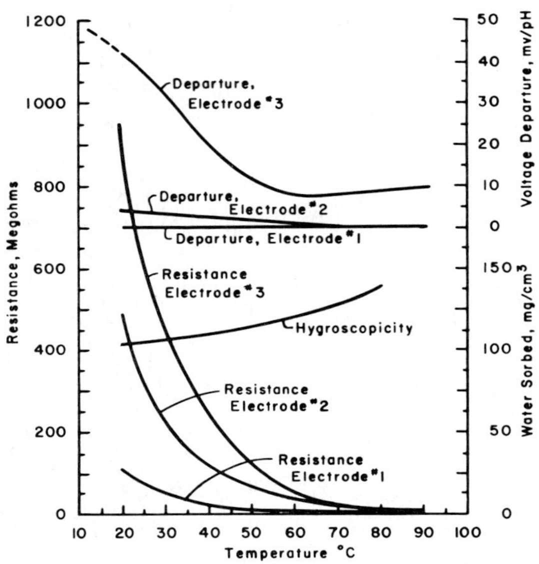

FIG. B. 90. Effect of temperature on electrical resistance, and hygroscopicity, and pH response of electrodes from glass Corning 015. (Hubbard).

is present the potential and resistance data clearly depend on the Na^+ contents, but the Na^+ function is only approximately fulfilled (80 to 90 per cent of the theory). Particularly important is the sorption behavior of sodium, potassium and rubidium cations as easily controlled by the base exchange of the glasses containing Rb^{86} as isotope indicator.

147. Modern glass electrodes for measuring sodium ions were discussed by G. Eisenman, D. O. Rudin, and J. K. Casby,[247] on the basis of sodium aluminosilicate

[246] *Doklady Akad. Nauk S.S.S.R.*, **113**, 1957, 824-827. On the possibilities of the introduction of rubidium in different glasses see H. E. Simpson, *Glass Ind.*, **42**, 1961, 222-223. See also B. P. Nikolskiĭ, M. M. Shul'ts, N. V. Peshekhonova, and A. A. Belyustin, *Doklady Akad. Nauk S.S.S.R.*, **140**, 1961, 641-643.

[247] *Science*, **126**, 1957, 831-834. Cf. previous data of B. von Lengyel and E. Blum (1931). New glass electrodes of high chemical resistivity and sensitive to divalent cations were recently studied by R. M. Garrels, M. Sato, M. E. Thomson, and A. H. Truesdell, *Ibid.*, **135**, 1962, 1045-1048.

compositions (with 11 per cent Na_2O, and 18 mol. per cent Al_2O_3) for the selective measurement of the Na^+ ion activity as a continuous function of time, as in extracellular biological fluids, with graphic contours of isosensitivity, interpolated from electric potentials in the siliceous portion of the system Na_2O—Al_2O_3—SiO_2. For the important $pH = 7.6$ a regular change in Na^+ to K^+ sensitivity is observed, and a remarkable reversal occurs for the contour line indicating the ratio 1 : 1 for the relative sensitivities for this ion pair (cf. Fig. B. 91). Small amounts of CaO and

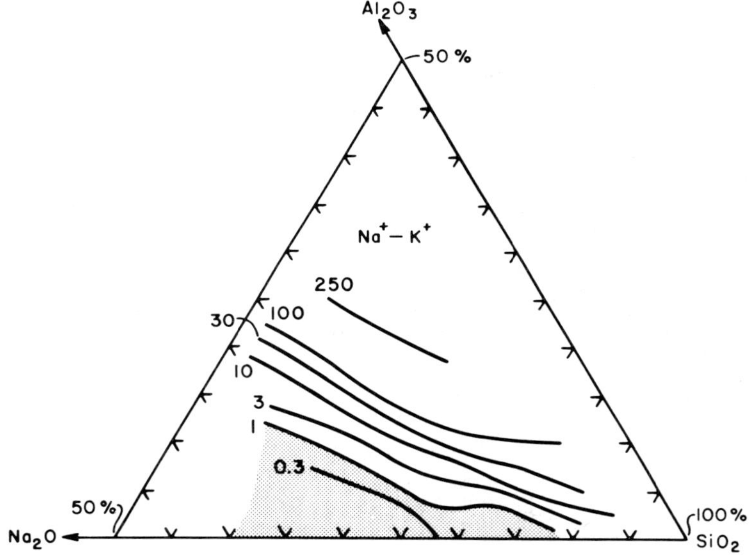

FIG. B. 91. Contour curves diagram for the Na^+ — K^+ ion pair in the sensitivity of a glass electrode with 11 Na_2O, 18 Al_2O_3, 71 SiO_2, for pH = 7.6. (Eisenman, Rudin, and Casby).

Fe_2O_3 do not influence significantly the electrode function. For similar biological uses in which Na^+- and K^+-sensitive glass electrodes are required for the control of circulating fluids (as in serum and blood) S. M. Friedman, J. D. Jamieson, M. Nakashima, and C. L. Friedman[248] used a sodium to silver glass electrode which is capable of discriminating $\Delta[Na^+]$ of less than 1 meq. per liter in 140 meq./liter. A corresponding potassium glass electrode discriminates less than 1 meq./liter, $\Delta[K^+]$ in the range of 1 to 10 meq./liter. The electrode may be used singly or pairwise in connection with a common calomel electrode for simultaneous monitoring $\Delta[Na^+]$ and $\Delta[K^+]$ in mixed solutions.

148. Glasses of the system Li_2O—La_2O_3—SiO_2 were studied by B. P. Nikol'skiĭ,

[248] *Science*, **130**, 1959, 1252-1254.

A. I. Parfenov, and M. M. Shul'ts[249] in their properties as glass electrodes, their conductance and chemical durability over pH = 0 to 14, under aspects similar to the investigations of commercial glasses by Hubbard *et al.* The introduction of lanthanum into lithium silicate glasses increases their electrical resistance and the activation energy of conduction. The $Li^+ \rightleftharpoons H^+$ exchange constant

$$K = \left(\frac{a_{H^+}^{soln.} \cdot N_{Li^+}^{gl.}}{a_{Li^+}^{soln.} \cdot N_{H^+}^{gl.}} \right)$$

varies from 0.12 to 4.1×10^{-11} for La_2O_3 concentrations of 0 to 9 molecular per cent. In a solution of lithium salts an increased content in La_2O_3 in the electrode glass impairs its *pH* function in an alkaline medium by lowering the upper limit of response. If lithium salt solutions with *pH* equal to, or above 12 are used, practically all of the H^+ ions are replaced by Li^+ on the glass surface. With Na^+ and K^+ in the solution, the influence of La_2O_3 in the electrode glass is just raised, that is the upper limit of the *pH* response is raised to the more alkaline side. Evidently the introduction of La_2O_3 tightens the structure and impedes the exchange of Li^+ by larger cations. The series Li > Na > K is valid for the *pH* function. The specific effects of La_2O_3 in solutions at 95°C. are less pronounced than they are at 25°C. Optimum responses are observed in glasses with 24 to 28 molecular per cent Li_2O, 4 to 7 per cent La_2O_3, and 65 to 70 per cent SiO_2.

149. The *alkali anomaly* in sodium silicate glass electrodes observed above *pH* = 10, which is characterized by systematic deviations of the potentials from theory, is caused by the exchange of alkali ions with hydrogen ions on the boundary surface of glass/solution. A. O. Long and J. A. Willard[250] used the radioactive tracer element Na^{24} as an indicator of the specific exchange reactions, and E. Trebge and R. Fischer[251] studied this problem with $Na^{24}Cl$ for labeled NaOH solutions in contact with a sodium silicate glass electrode. A glass of this kind with $SiO_2 : Na_2O = 2.2$, after exposure for up to 37 hours, had an activity about 10 times greater than that of a glass with the ratio 3.3. A saturation value was finally reached, increasing with increasing alkalinity of the solution above 10. Alkali anomalies in the potential indications are a function of the chemical durability of the glass. A HCl-glycol buffer

[249] *Doklady Akad. Nauk S.S.S.R.*, **127**, 1959, 599-601. See also *Structure of Glass. Proc. All-Union Conf. Glassy State, 3rd, Leningrad* (English Translation), 1959/60, 256-262.

[250] *Ind. Eng. Chem.*, **44**, 1952, 916-920. See also J. O. Isard, *Nature*, **184**, 1959, 1616-1618, and the extensive experiments of B. P. Nikol'skiĭ, E. A. Materova, and V. V. Moĭseev, *Structure of Glass. Proc. All-Union Conf. Glassy State, 3rd, Leningrad* (English Translation), 1959/60, 378-381. On sodium-barium silicate compositions for glass electrodes see G. T. Petrovskiĭ, ibid., pp. 263-265. See also J. A. Savage and J. O. Isard, *Phys. & Chem. of Glasses*, **3**, 1962, 147-156, on the electrode properties of cation-sensitive glasses.

[251] *Silikat Tech.*, **10**, 1959, 351, 385-389.

(with $pH = 2$), or an acetate buffer ($pH = 4.62$), phosphate buffer ($pH =$ of 5 to 8), or a NaOH-glycol buffer (pH between 8 and 13) were used. The millivolt per pH unit and the Na^+ activity versus pH curves (cf. Fig. B. 92) show the characteristic start of the anomaly at $pH = 10$. The anomaly, therefore, is caused by Na^+ ions entering the glass surface. It is expressed by an empirical formula for the measured potential $E' = 2.3\,(E'_a - E_c) \cdot \log\,(pH - pH_c) - E_c$, in which E_a is the potential for which the anomaly starts, E_c and pH_c the potential and pH of pure H_2O. Below $pH = 9.5$ to 10 this equation is not valid because of the exchange $Na^+_{glass} \rightleftharpoons H^+$ in an unbuffered solution or in water.

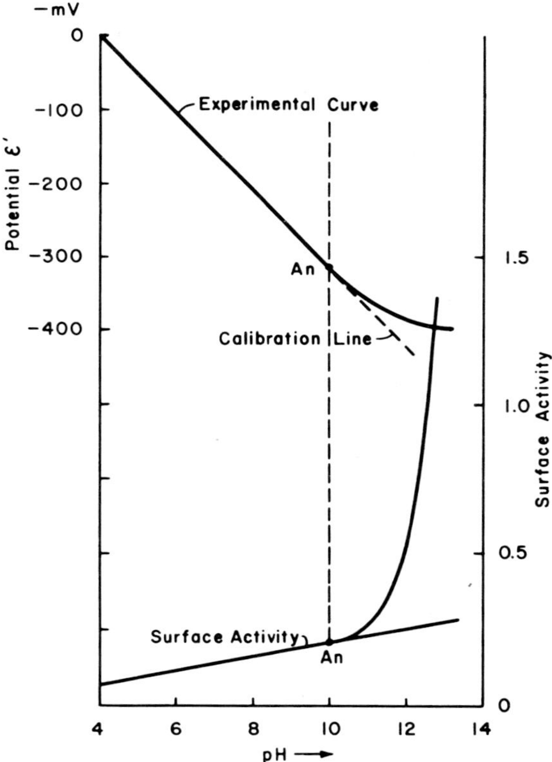

FIG. B. 92. Comparison of the alkali anomaly in pH response of a glass electrode, and of radioactivity of a Na^{24}- induced glass surface, as a function of pH. (Trebge and Fischer).

150. The method of *radioactive tracers* in labeled alkaline solutions in their interaction with the surface of a glass electrode was recently applied by K. Schwab and H. Dahms,[252] also with the activated glass phase, and by using tritium as the indi-

[252] *Naturwissenschaften*, 47, 1960, 351-352.

cator of hydrogen. The experiments with labeled solutions confirm the previous results, mentioned above, that in the range of *pH* between 1 and 8 only little Na^+ is taken into the glass surface. The ion exchange reaction at higher *pH* was confirmed also by the anion Br^- labeled with Br^{82}. For marking the glass itself the activation was made in a thermal neutron current of 8×10^{12} *n*/cm.²/sec. in the atomic reactor. The measurement of the release of Na^+ ions from the glass surface in the solutions of variable *pH* and alkali concentrations gave evidence that the rate of this release is influenced neither by *pH* nor by alkali concentration (*pH* = 13). As a matter of fact, H^+ ions in the glass are exchanged by Na^+ from the solution. The instantaneous observation of the alkali anomaly of the potential is a consequence of the very rapid establishment of equilibrium on the phase boundary, whereas the Na^+ equilibrium concentration in the whole swelling layer is a consequence of diffusion processes which require comparatively much more time. The determination of the H^+ content in the swollen layer by its marking with tritium makes evident that with 50 to 70 *n* gm. H/cm.^{2} enough hydrogen ions are present to make possible the exchange with the sodium ions.

151. Very important relations between *electric properties* of *glass/metal sealings* and their thermal expansion coefficient and adhesion characteristics must be observed for modern high-vacuum tube manufacturing. In the rich literature in this field we find an excellent report of H. Kalsing[253] on industrially developed special glasses of this kind, particularly for vacuum sealing and welding with wolfram, molybdenum, iron-nickel alloys, iron-nickel-chromium and iron-chromium, further wigh copper-, silver-, or gold-coated metal wires, as with a core of 43 per cent nickel, 57 per cent iron alloy, and pure iron. Highly variable conditions of thermal and electric properties must be mastered to avoid destructive formation of microcracks in the socket portion of vacuum tubes and to avoid electrolysis effects. W. Düsing[254] studied the electrolysis through the sealing glass by cinematography to determine the life-time of the sealings which is a very characteristic function of the temperature, say 7,000 hours at 200°C., and only 87.5 hours at 300°C. for a commercial glass type. Alkali metals and lead are the most destructive products of electrolysis, staining the glass intensively black, with serious damage done to the glass, and a loss of vacuum tightness. The reduced glass often shows lead metal in minute droplets showing a typical Tyndall effect.

152. A. Danzin[255] described the development of a lead glass with a low thermal expansion in metal seals, combined with "Kovar", and iron-nickel-cobalt alloys for vacuum tube manufacturing. This glass called "Neutrohm" at the same time,

[253] *Sprechsaal*, **86**, 1953, 212-216, 339-341, 363-365, 389-392, chiefly on composition of many of the industrial glasses used for sealing purposes.

[254] *Glastech. Ber.*, **26**, 1953, 232-238.

[255] *Silicates inds.*, **18**, 1953, 321-324; *Verres et réfractaires*, **7**, 1953, 371-378.

has excellent moisture and climatic durability. The dielectric properties (dielectric constant and tan δ) as a function of frequency follow the equation for the loss factor $\pi = \varepsilon \cdot \tan \delta/(1 + \tan^2 \delta)$. In relation to temperature, frequency, and resistance they are shown in a three-dimensional nomogram. The Gehlhoff-Thomas (1925) characteristic parameter T_{k100} gives sufficient information on conductance of the Neutrohm glass type and of another molybdenum sealing glass "Moly A 119" in the practically useful temperature range. T_{k100} is for Neutrohm = 340°C. However, this parameter alone does not provide distinct information on intensity of the current flowing through the polarized glass in a given time, its phase boundary with the metal, or its electrolysis.

153. Concerning the elementary phenomenon of electric conductivity as a function of composition, E. Plumat[256] recommended a method for control of the homogeneity of well-annealed glass samples based on previous measurements of E. Seddon, E. J. Tippett, and W. E. S. Turner (1932) for sodium silicate and sodium-calcium silicate glasses. A special furnace was constructed in which eight glass samples are exposed simultaneously at 200°C. and electrolysed by direct current (100 volts), with particular care given to control the polarization by reversal of the current. If the temperature is raised to 450°C. an alternating current of 1000 volts is preferred. For window glass characteristic diagrams were developed which show the electric resistance as a function of the contents in soda, lime and magnesia. Thus it is possible to detect under plant manufacturing conditions without any chemical-analytical methods every variation in the silica content of the glass or replacement of alkalies by lime and magnesia.

154. The systematic study of volume and surface electric conductivity of glasses in the system $Na_2O.2SiO_2$—$PbO.SiO_2$ at room temperature, 30° and 75°C. was made by L. Yu. Kurts.[257] For the measurement of surface conductance freshly fractured specimens were used and the temperature coefficient as well as the polarization voltage determined. The polarization curve as a function of composition indicates very distinctly the compound $Na_2O.2PbO.4SiO_2$. The surface conductance as a function of time also indicates this compound. P. Le Clerc[258] could observe surface conductance as a function of time, especially under moisture conditions (relative moisture 75

[256] *Silicates inds.*, **18**, 1953, 17-19. See recently O. V. Mazurin, *Structure of Glass. Proc. All-Union Conf. Glassy State, 3rd, Leningrad* (English Translation), 1959/60, 229-231. Especially on glasses of the system SiO_2—B_2O_3—Al_2O_3—BaO, see V. A. Khar'yuzov, O. V. Mazurin, and N. M. Zubkova, *ibid.*, 232-233. P. R. Segatto, *J. Am. Ceram. Soc.*, **45**, 1962, 102-104, developed a controlled-potential coulometry method of Pb, Cd, and Zn in glasses.

[257] "Physical-chemical Properties of the System Na_2O—PbO—SiO_2," edited by I. V. Grebenshchikov, under auspices of the Acad. Sci. U.S.S.R., Leningrad, 1949, pp. 110-112.

[258] *Silicates inds.*, **19**, 1954, 237-242. Practically important effects of leaching, before application of silicone films on glass, in its surface conductivity are discussed by J. Edge and L. F. Oldfield, *Glass Technol.*, **1**, 1960-69-79.

per cent constant, 25°C., automatically controlled), and measured the resistance by a quadrant electrometer method, polarization effects being eliminated by a frequent reversal of the poles. The measurements extended to Flint glass and a barium borosilicate composition. The order of magnitude of the surface resistance (10^8 to 10^9 ohms) for both glasses corresponds to the equivalent resistance of a NaOH solution with 0.3 equiv./liter in a layer of 1 cm.² area and a film of 10 Å. thickness. The observed decrease of surface conductance with time is evidently caused by a migration of the adsorbed water into the glass structure. This process is much slower than the ion migration of electrolytes from the interior of the glass to the surface in the first 2 to 3 hours. F. W. Dunmore[259] observed in experiments on "electric hygrometry" similar "aging" effects on the surface of Flint glass after exposure to hygroscopic salts.

155. The conductivity of *mixed* alkali glasses in general was discussed in a theoretical investigation by B. Lengyel and Z. Boksay[260] in continuation of previous studies of B. Lengyel (1940). The characteristic minimum of conductance (cf. G. Gehlhoff and M. Thomas, 1925) for mixed glass compositions is explained by aid of a model concept for the conductivity mechanism on the assumption that the glass would be a pure alkali ion conductor and that the ion migration would be caused by the transition from one structural hole to another in the neighborhood. The "geometric hindrance" can then be calculated for the transition of a larger ion into the structural hole emptied by a smaller ion. The conductance of mixed glass is then $\varkappa = A^r(n_1F_1\varkappa_1 + n_2F_2\varkappa_2)$ in which $\varkappa_1$ and $\varkappa_2$ are conductances of pure alkali glasses, n_1 and n_2 the molar fractions, A, F_1 and F_2 factors of the "hindrance" for the migrating ions, r a constant characteristic of the glass, indicating the number of structural holes before the current passed through the glass. An analogous equation is developed for the transfer numbers: $v_{1(2)} = n_{1(2)}F_{1(2)}/(n_1F_1 + n_2F_2)$, independent of temperature. There is a remarkably satisfactory agreement between calculated and observed data for lithium-sodium, sodium-potassium, and lithium-potassium silicate glasses.

156. A typical case of influencing what Lengyel called the hindrance of alkali migration in glasses is the effect of introduction of *fluorine* into the glass constitution, as O. V. Mazurin and E. V. Molchanova[261] observed for a series of alkali glass compositions in the range between 100° and 400°C. Potassium silicate glasses with about 2 per cent fluorine have a distinctly increased resistance, whereas in sodium silicate glasses this effect is much less pronounced. Mixed sodium-potassium silicate glasses

[259] *J. Research Natl. Bur. Standards,* **20,** 1938, 723-744, especially pp. 727-731.

[260] *Z. physik. Chem. (Leipzig),* **203,** 1954, 93-112; **204,** 1955, 157-164; *Silikat Tech.,* **7,** 1956, 391-394. On surface layers on electrode silicate glass see B. Lengyel and S. Dobos, *Proc. 6th Conf. Silicate Industry, Budapest, Oct.* 1961 (publ. 1963), 279-290.

[261] *Trudy Leningrad. Tekhnol. Inst. im. Lensoveta,* 1955, No. 34, pp. 48-52.

show intermediate effects, but a reduction of the alkali contents even reinforces them. This phenomenon of increased resistance for fluorine containing glasses is particularly important for the electrotechnical use of glasses as insulator materials.

157. Concerning the problems of the presence of *unipolar conduction* in glasses at low temperatures, Ya. I. Ol'shanskiĭ[262] conducted interesting experiments to examine conclusions of C. Tubandt (1920/21) in his classical studies on the conductivity of solids. At least in Cu_2S films creeping over a copper wire creating an "inner electrolysis" by the galvanic cell $Cu/Cu_2S/H_2S$ (atm.) the rate of material transfer M/t (time) in moles is proportional to the EMF, ΔF, and a *mixed* conductance. It is a maximum for equal shares of electronic and ionic migrations, as C. Wagner (1933) had demonstrated. Tubandt who had no knowledge of this mixed conductance phenomenon, therefore, came to the erroneous interpretation of a purely ionic conductivity of heavy metal sulfides at high temperatures and Ol'shanskiĭ could determine the share of ionic conductance in Cu_2S lower than 0.1 per cent of the total conductance, with an absence of electrolytic phenomena. Very pronounced mixed conductance, however, occurs in the system Fe—FeS—FeO—SiO_2, with a temperature coefficient and a coupling of these phenomena with electrochemical galvanic cell effects in copper ores, which may have a great importance in the formation of magmatic sulfide deposits under the action of "inner electrolysis." Direct current actions with electrolytic ion migration are known in the glass of porcelain up to 700°C., under low voltage potentials, and a freezing-in of the cations on cooling causing a polarization current, if the ceramic body is reheated, as in an accumulator cell. W. Soyck[263] demonstrated a similar effect in Steatite if two plates are pressed together and treated under a potential of 1500 volts at 900°C. and then cooled. Initially, the loss factor of this insulator material is (for $\nu = 1000$ cycles) $= 10 \times 10^{-4}$, but after the thermal treatment in the static field on the anode side $= 8 \times 10^{-4}$, on the cathode side $= 13 \times 10^{-4}$, thus the resistance is considerably increased on the anode, reduced on the cathode side, indicating a strong ion impoverishment on the anode.

158. The surface conductance of reduced lead glasses, with metallic lead deposited in its structure, was studied by V. I. Shelyubskiĭ[264] (cf. Vol. I, C. 86, ¶ 116) by measuring the conductance, combined with photometric extinction and reflectance (for $\lambda = 533$ mμ). In the first stage of the reduction, the surface conductance is strongly increased, but after 20 seconds reaches a maximum and falls to a rather constant

[262] *Voprosy Petrog. i Mineral., Akad. Nauk S.S.S.R.*, **2**, 1953, 211-219; see C. Wagner, *Z. physik. Chem. (Leipzig)*, **21**, B, 1953, 25-41, on tarnishing of silver and copper.

[263] *Tonind.-Ztg. u. Keram. Rundschau*, **77**, 1953, 234.

[264] *Doklady Akad. Nauk S.S.S.R.*, **96**, 1954, 745-747; *Stroenie Stekla, Inst. Khim. Silikatov Akad. Nauk S.S.S.R., Trudy Soveshchaniya, Leningrad*, 1953 (Pub. 1955), pp. 219-221, on electron microscopy of black reduced lead glass.

value for more than 40 seconds. Approximately the same type of curves was found for reflectivity. The reduced glass, in the first stage of reduction is a typical semiconductor, until a coalescence of the metal particles occurs, with an increase in diffuse reflection and a maximum after 120 seconds (Fig. B. 93). A volatilization of lead from the surface enriches the glass with silica in a surface layer. H. H. Funk[265] reduced lead glass with hydrogen at 300° to 500°C. and studied especially the ef-

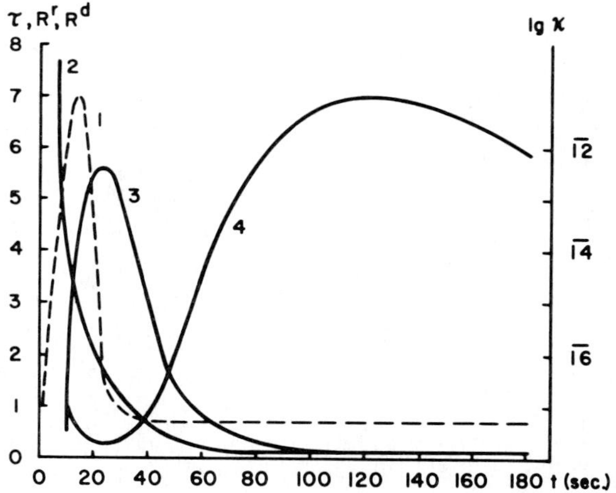

FIG. B. 93. Changes in surface conductance, $\varkappa$ (curve 1), the coefficient $\tau \times 10$ of light transmittance for $\lambda = 533$ mμ (curve 2), of reflectivity, $R^r \times 0.05$ (curve 3), and of diffuse scattering, $R^d \times 100$ (curve 4), of the surface layer of reduced lead glass, as a function of time. (Shelyubskiĭ).

fects of B_2O_3 and Bi_2O_3 added to the glass composition. The bismuth borate glasses showed a remarkably low-ohmic resistance behavior. The temperature coefficient of the conductance of hydrogen-treated lead silicate glasses is much lower than in the same glasses not treated. The surface resistance of higher-ohmic bismuth borate glasses is a linear function of voltage with a negative slope, whereas lower-ohmic glasses of the same type show increasing resistance with increasing voltage on a curve of a higher order. Reduced lead glasses do not show any functional relation to voltage and surface resistance up to 10,000 volts. No amorphous semiconductor effect or Hall effect, nor any changes in the resistance in the magnetic field were observed.

[265] *Glastech. Ber.*, **31**, 1958, 269-272. On the system B_2O_3–Bi_2O_3 see E. M. Levin and Cl. L. McDaniel, *J. Am. Ceram. Soc.*, **45**, 1962, 355-360, and on the dielectric properties of glasses in the systems CdO–Bi_2O_3–SiO_2, CdO–Bi_2O_3–B_2O_3, and CdO–Bi_2O_3–GeO_2 studies of Bh. V. J. Rao, *ibid.*, pp. 555-563, furthermore Ch. Hirayama and E. C. Subbarao, *Phys. & Chem. of Glasses*, **3**, 1962, 111-115.

159. The *non-ohmic* behavior of some glasses at low temperatures and under the influence of very high electrostatic field strengths which was theoretically discussed by H. H. Poole (1921), is confirmed in the investigations of D. A. Stuart and O. L. Anderson,[266] and explained in the light of the Eyring theory of conductivity as a rate process (cf. A. ¶ 137), due to the similarity of conductivity and fluidity.

CHEMICAL RESISTIVITY OF GLASS

160. We mentioned above how *radioactive tracer* methods could be applied in investigations of the ion exchange on the surface of glass electrodes. The excellent methods developed by A. O. Long and J. E. Willard[267] illustrate the great advances made, in comparison with the O. Hahn emanation emission method which was used by M. Heckter (1934) for determination of the absolute active surface of glass grits for an examination of glass leaching in water or other dilute reagents. The chief advantage of the methods of Long and Willard becomes evident when Na^{24} is produced in the glass surface in situ by neutron bombardment in a nuclear reactor and not alone by tracing the exterior solution reacting on the glass surface. If two different Na isotopes are used simultaneously, say Na^{24} and Na^{22} (half-life 15 hours and 3 years respectively), the result differentiates sorption and desorption processes in detail. It thus became evident that in sorption the rate-controlling process is not the ion exchange reaction; a dissolution of the glass in water may be even more rapid than an exchange process. A characteristic example for the experimental method used by Long and Willard is illustrated in Fig. B. 94. Just prior to immersion of the glass samples in a solution labeled with Na^{22}, they were immersed for a certain time (up to 5 hours) in a 0.023 molar sodium salt solution of the same concentration but containing Na^{24}. The amount of Na^{24} sorbed prior to immersion in the Na^{22} solution was determined by drying and counting. The sum of the Na^{22} sorbed and of Na^{24} remaining after immersion in the Na^{22} solution was determined from the

[266] *J. Am. Ceram. Soc.*, **36**, 1953, 27-30. On the range of validity of the non-Ohmian exponential law for electric conductivity at high field strengths see also A. ¶ 137 and V. A. Presnov, *Stroenie Stekla, Inst. Khim. Silikatov Akad. Nauk S.S.S.R., Trudy Soveshchaniya, Leningrad*, 1953 (Pub. 1955), 267-269, with interesting observations of oxidation layers on molybdenum and manganese in contacts with sealing glasses, and recent data of V. A. Presnov, V. I. Gaman, and L. M. Krasil'nikova, *Structure of Glass. Proc. All-Union Conf. Glassy State, 3rd, Leningrad* (English Translation), 1959/60, 220-222. See also P. W. McMillan, "Advances in Glass Technology," Plenum Press, New York, 1962, 333-47.

[267] *Ind. Eng. Chem.*, **44**, 1952, 916-920; see also E. Trebge and R. Fischer, *Silikat Tech.*, **10**, 1959, 351, 385-389. An excellent report on the general usefulness of radioactivity metodos in glass technological investigations was given by S. F. Cox and K. M. Laing, *Glass Ind.*, **35**, 1954, 183-184, 222. Cs^{134} (1.7 years half-life time) is also an element very suitable for desorption experiments.

total counts due to both isotopes when each sample was removed from the Na^{22} solution, and the Na^{22} was determined from the residual count after the Na^{24} had decayed.

161. The results of such tests for different times of immersion, temperatures, and different *pH* show that in each case the initial rate of desorption at 72°C. appears to be greater than the corresponding rate of sorption. After this initial interval the sum of the monolayer represented by the sorbing and desorbing curves remains approximately constant which, however, does not indicate that Na^+ ions enter and leave the surface at equal rates. The great conclusiveness of such experiments is also valid for leaching tests of the neutron-irradiated glass samples (say, with a thermal neutron flux of 10^{12} n/cm.^{2}/sec. for 4 hours). Activity is leached from the

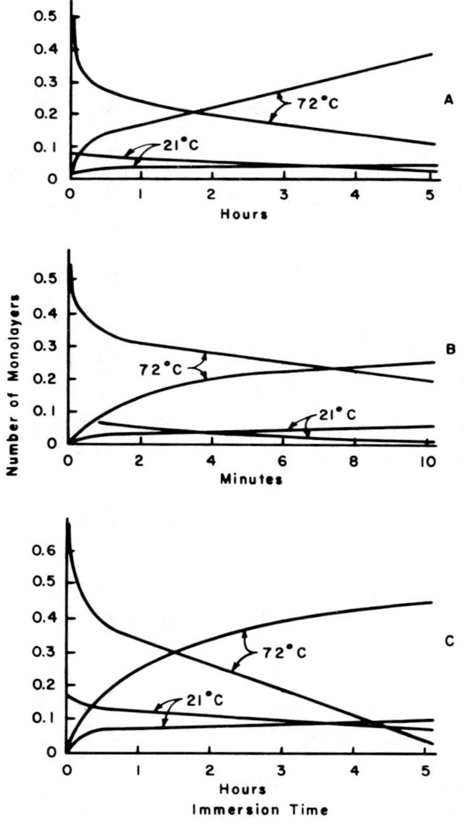

pH = 10 for B and C; 7.5 for A.

FIG. B. 94. Simultaneous sorption and desorption of sodium ions on glass surface. (Long and Willard).

bombarded glass samples (slides) much more rapidly during the first interval of immersion in 0.1 *N* HNO_3 solution than during successive intervals but leaching continues at a lower but apparently constant rate during several immersions following the initial immersion (at 23° and 80°C.). Water is about as effective a leaching agent under the experimental conditions as is the 0.1 *N* HNO_3 solution. Glass heated nearly to fusion prior to bombardment is leached somewhat less readily than that which is only moderately flame-exposed.

162. Cr^{51} in a bichromate cleaning solution for glass surfaces was applied by E. B. Butler and W. H. Johnston[268] for different kinds of commercial instrument glass to determine the retention of Cr ions, by soaking for 24 hours with the solution, and rinsing in running water for 1 minute followed by counting. Later rinsing or rubbing did not change the countings. Pyrex borosilicate glass evidently retained more Cr^{51} than did plate glass or even silica glass (not quartz as written). Higher retention on greasy glass suggests that both, Cr^{3+} and Cr^{6+} ions were retained, indicating an oxidation of grease substance and a deposition of Cr^{3+} ions in reduction products. The coverage may reach 2×10^{-10} moles/cm.^{2} Cr atoms (or 3 per cent monolayer) for clean Pyrex glass (10^{-15} cm.^{2} area covered by 1 atom) as derived from countings with the scintillation spectrometer. If cleaning is incomplete and the grease layer not completely removed retentions of Cr^{3+} up to 0.5 monolayer may occur.

163. The highly critical conditions of modern silicate analysis are the reason for studies of W. Geilmann and G. Tölg[269] on possibilities of eliminating glass containers entirely from the analytical operations, as in the colorimetric determination of iron as perchlorate. The glass in these determinations is to be replaced by organo-plastics, say by "*Hostaflon*" (a trifluorochloroethylene polymerizate), which is useful even for crucibles because of its high softening temperature (above 200°C.) and particularly its complete resistivity to mineral acids, including HF in hot solutions, making Hostaflon superior to all of the common laboratory glasses, even to silica glass and porcelains. In addition the surface of Hostaflon is water-repellent and silica evaporated from HCl solutions does not adhere to the walls of dishes.

164. From the practical viewpoint of measuring the surface durability of glass, H. E. Simpson[270] exposed small glass samples to alternate cycles of "fogging" and "clearing" by temperature drops in a *haze test* oven, from 55° to about 50°C., and recovery to 55° again, with a relative humidity maintained at about 100 per cent. The increasing haze (scattered light) on the glass surface is easily measured by a photoelectric cell. The relative amount of scattering is then a direct measure of glass

[268] *Science*, **120**, 1954, 543-544.

[269] *Glastech. Ber.*, **30**, 1957, 335-341.

[270] *Am. Ceram. Soc. Bull.*, **30**, 1951, 41-45; *J. Am. Ceram. Soc.*, **36**, 1953, 143-146; especially for a series of optical glasses cf. *J. Soc. Glass Technol.*, **37**, 1953, 249-255 (cf. B. ¶ 178).

deterioration and the per cent haze amount plotted versus the time of exposure gives characteristic curves for different composition types of commercial glasses. The age of the samples and previous conditions of storage of the glass are important factors influencing results of this simple and reliable routine method. The results for ground, polished, or fire-polished surface specimens, and of a corrosive treatment by dilute acids and alkaline solutions are equally illustrated in the characteristic curves of practical significance.

165. A *continuous flow* method is recommended by I. R. Beattie[271] which is based on the use of an extraction bulb of silica glass, with refluxing cooling, and a filter bottom crucible as the container of the glass grit samples, The determination of alkalinity of the leachates by titration with 0.001 N HCl solution, and bromothymol as indicator is the analytical standard method in grit extraction methods of this kind. The migration of the Na^+ ions through the glass is the rate-controlling process for leaching by water alone, and thus the integral diffusion coefficients and apparent activation energies can be calculated. The influence of the chemical composition of different glasses is an important factor controlling the Na^+ migration rates. There is a considerable decrease in the diffusion coefficient as Si^{4+} is replaced by Al^{3+} or B^{3+}. Also the replacement of mobile sodium ions by the comparatively immobile calcium ions also caused a marked decrease in the measured diffusion coefficients.

166. A similar Soxhlet principle extraction method of glasses was used recently by L. Žagar and A. Schillmöller,[272] as well for glass fracture-surfaces of grits as for fire-polished beads. The specific surface of the specimens was determined by the krypton adsorption method with an electronic vacuum balance. Žagar and Schillmöller distinguish two different reaction mechanisms in glass leaching, namely the pure alkali diffusion (chiefly below 80°C.), and the hydrolysis of the glass framework (above 80°C.). The extraction process as a function of time follows a parabolic law for most oxide components, but not for silica. In their order of magnitude, the diffusion coefficients of sodium and potassium ions as calculated agree well with those previously determined from radioactive diffusion methods, by H. H. Blau, R. H. Bristow, and J. R. Johnson (1951), based on the self-diffusion of sodium ions labeled with Na^{24} in glass. Between the free activation enthalpies determined for fractured and fire-polished surfaces of the same glass, there are systematic differences which

[271] *J. Soc. Glass Technol.*, **37**, 1953, 240-248 T; *Trans. Faraday Soc.*, **49**, 1953, 1059-1065. See more recently, M. A. Rana and R. W. Douglas, *Phys. & Chem. of Glasses*, **2**, 1961, 179-195, 196-205.

[272] *Glastech. Ber.*, **33**, 1960, 109-116. How critical one must be in the performance of the standard grit tests of chemical durability, was recently demonstrated by E. Wiegel, *Ibid.*, **34**, 1961, [illegible]59-268, who found notable influences of traces of heavy metals (e.g. from the bronze sieves [illegible]ed in grain fractionation) on the results of the hot-water leaching method. Cu, Zn, Sn lower the data, Co, Ni, Mg increase these.

make evident the higher water corrosion durability of the fire-polished samples. The additional energy barrier caused by the higher compaction of the glass framework structure on the boundaries of fire-polished glass is about 1600 cal./mole.

167. The *products* of the leaching reaction between a glass with 14 per cent Na_2O, 12 per cent CaO, and 74 SiO_2, and water at 90°C. were examined by Fr. Fu-Yen Wang and F. V. Tooley[273] for reaction periods up to 4 hours. The principal product is "metasilicic acid," sodium, and calcium metasilicates as determined by analytical methods. It is the constitution of the glass, and not so much the composition of the reactant solution which determines the course of the reactions. Sodium hydroxide is only an intermediate, not a major reaction product. Wang and Tooley also used the reaction products themselves as reactants with the glass surface and found that Na_2SiO_3 reacts with a greater attack than did water alone, whereas a solution of "metasilicic acid" showed a lower degree of reactivity. The rate of leaching was accelerated by increased hydroxyl ion concentrations. The leaching of calcium ions indicates the process of final breakdown of the glass framework structure.

168. Basing his studies on the previous systematic work of J. Enss[274] on the relative corrosivity of glasses by water as a function of chemical composition, Yu. A. Gastev[275] especially studied the composition of the alkali-impoverished surface layer of leached glasses, and emphasized the differences in contents in alkaline earths in these layers when leaching was made simply by water or by acids. The fundamental assumptions of Gastev that quite distinct compounds exist in the composition of the binary sodium silicate glasses is sharply criticized by Yu. A. Shmidt.[276] The coefficient $\alpha = n_{SiO_2}/m \cdot n_{Na_2O}$ for a sodium silicate glass of the composition $Na_2O,mSiO_2$, is a sensitive indicator for the fraction of SiO_2 passing from the surface layer into the solution, if n_{SiO_2} and n_{Na_2O} indicate in moles/cm.^{2} the amount of both oxides going into solution. If $\alpha = 1$ the whole sodium silicate goes into solution, if $\alpha = 0$ the surface layer retains all of the silica. As a function of temperature, α shows the regularities indicated in Fig. B. 95. It is evident that the interval between $\alpha = 1$ and $\alpha = 0$ is relatively narrow in all cases, that is, the transition from complete dissolution to alkali leaching is rather sudden, and there is no stoichiometric ratio which could correspond to one or the other relatively stable compound. On the other hand, it is very characteristic that with increasing temperature the transition curves are

[273] *J. Am. Ceram. Soc.*, **41**, 1958, 467-469; 521-524.

[274] *Glastech. Ber.*, **5**, 1927/28, 449-476, 509-520; see also A. Weberbauer, *Ibid.*, **10**, 1932, 361-374, 426-435, on acid stability and weathering of glass.

[275] *Stroenie Stekla, Inst. Khim. Silikatov Akad. Nauk S.S.S.R., Trudy Soveshchaniya, Leningrad*, 1953 (Pub. 1955), 187-189. On chemical dealkalinization see also J. Asolo and St. de Lajarte, "Advances in Glass Technology," Plenum Press, New York, 1962, 287-306.

[276] *Stroenie Stekla, Inst. Khim. Silikatov Akad. Nauk S.S.S.R., Trudy Soveshchaniya, Leningrad*, 1953 (Pub. 1955), 319-321. See later publications of S. K. Dubrovo and Yu. A. Shmidt.

rapidly shifted from low to highly siliceous compositions. The diagram also makes clear that with acidic leaching α is rapidly shifted to 0, but is raised to complete dissolution of the silicate ($\alpha = 1$) by alkaline leaching (cf. Vol. IV, Section A).

169. It is possible to determine the weights of leached alkalies even in relatively resistive glasses by applying an ultra-sensitive method based on the principle of the "Cartesian diver," using a pressure-controlled hydraulic balance as W. Haller and Gl. L. Calcamuccio[277] demonstrated. The pressure-sensitive glass float (cf. Figs. B. 96 and 97) carrying the sample is observed while pressure on the float is adjusted to

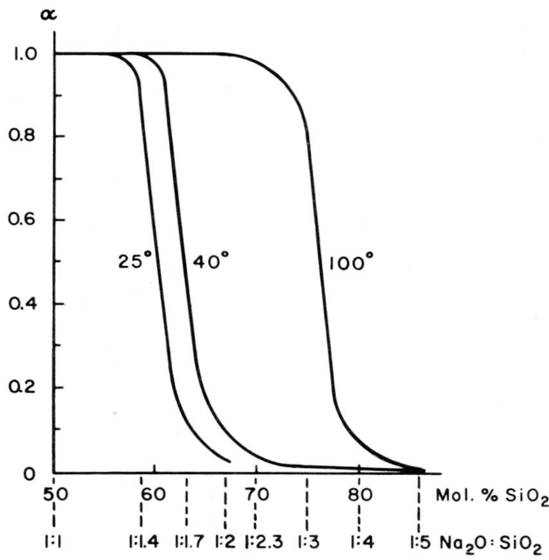

FIG. B. 95. Portion of silica, a, which goes into solution after 1 hour exposure of sodium silicate glasses with water, as a function of composition of the glasses, and of temperature. (Shmidt).

establishment of the equilibrium of the forces involved, and until the movement of the diver ceases. The pressure response of the float is determined by calibration and thus the magnitude of the driving forces, or to simplify, the changes in weight of the system (float + sample) can be calculated from the pressure readings. This method of ultra-high sensitivity is equally suitable for problems, such as the leaching away of alkalies, including every rate process in which density-volume-and concentration parameters are involved. The accuracy is $\pm 10^{-6}$ gm. for a sample weight of 3 gm. if the compressibility of the float system is taken into account. The temperature of the immersion liquid must be controlled within $\pm 0.001°C$.

[277] *Phys. Rev.*, **96**, 1954, 1454; *Rev. Sci. Instr.*, **26**, 1955, 1064-1068.

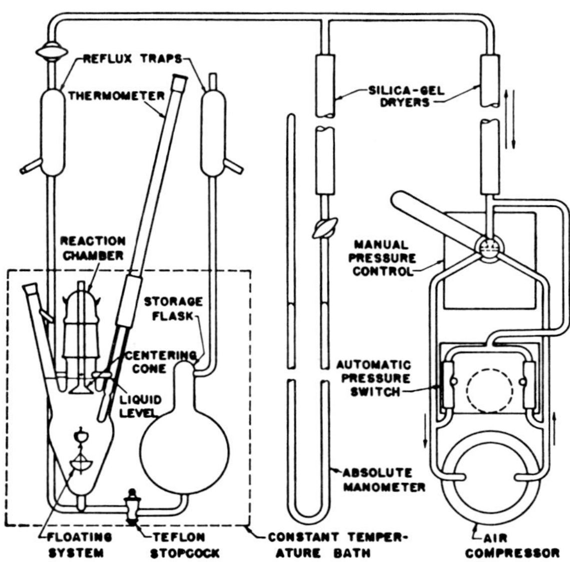

FIG. B. 96. Apparatus for measurement of extremely small changes in weight, volume, and density, by leaching glass samples. Cartesian diver principle. (Haller and Calcamuccio).

170. For stability problems of alkali-rich Jena electrode glasses, H. Wessel[278] gave valuable details on temperature effects of water corrosion, the rate of this attack, delaying and other factors. Since dissolution of the glass surface quite rapidly changes the composition of the corroding liquid, the ratio between both phases, solution and glass surface, changes continuously in a difficultly controlled manner, especially if reaction products are accumulated in the solution and the glass surface is changed by polymerization phenomena forming a more resistant film.

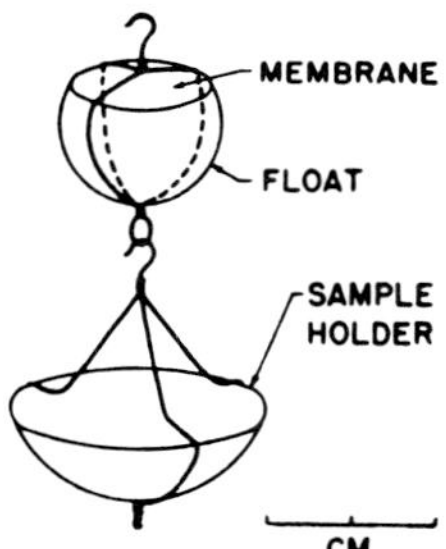

FIG. B. 97. Diagram of the float system for the Cartesian diver apparatus. (Haller and Calcamuccio).

[278] *Silikat Tech.*, 3, 1952, 483-486.

171. Among the wide-spread literature on the influence of chemical composition on the durability of glasses, we first mention the studies of S. Sen and T. V. Tooley[279] on the ratio Na_2O/K_2O in its effect in alkali-lime silicate glasses. Such mixed alkali-lime glasses are attacked by water and it is possible to construct functional relations of the corrosion to the ratio Na_2O/K_2O from analytical data of direct determinations of Na^+, K^+, Ca^{2+}, and Si^{4+} ions, starting from the basic glass 72 per cent silica, 10 CaO, and 18 alkalies. The characteristic minimum in the leaching (for maximum durability) is found for $Na_2O/K_2O = 2.6 : 1.0$ (by weight). For determination of the relative degree of ion exchange and of the framework-breakdown modes of the reactions an empirical index, N, was derived, of the form N =(Number of the Cations in Leachate)/(Number of the Si^{4+} Ions in Leachate) — (Number of Cations in Glass)/(Number of Si^{4+} Ions in Glass). This index gives the amount of ion exchange reaction, which is not yet accompanied by a framework-breakdown, under given experimental conditions. The ion exchange (cf. Fig. B. 98) is more predominant

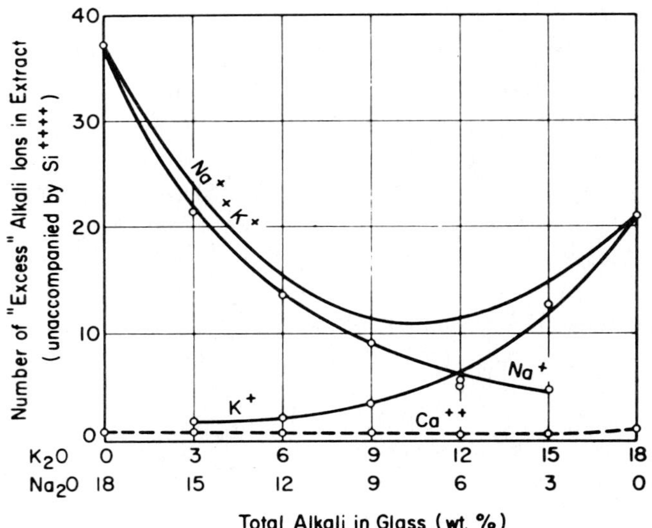

FIG. B. 98. "Excess" alkali ions in leachates from mixed-alkali silicate glasses, indicating the relative tendency of the glasses to ion exchange. (Sen and Tooley).

than the structure breakdown for the end members than for the mixed alkali members of the glass series. The least number of total "excess" alkali ions (= 11) occurs for a glass containing 11 per cent K_2O, and 7 per cent Na_2O.

[279] *J. Am. Ceram. Soc.*, **38**, 1955, 175-177. A detailed analysis of the reaction mechanism on the glass surface with aggressive solutions was recently presented by S. M. Budd and J. Frackiewicz, *Phys. & Chem. of Glasses*, **2**, 1961, 111-114, 114-118; **3**, 1962, 116-120.

172. The very important influence of *alumina* on chemical durability was investigated systematically by T. Uno, H. Shimizu, and H. Yoshikawa,[280] chiefly making evident the additivity relation in improvement of the glasses of soda-lime, or soda-lime-magnesia silicate compositions from 0.01 to 0.55 moles Al_2O_3 added. S. K. Dubrovo and Yu. A. Shmidt,[281] in a series of investigations, started from simple sodium silicate glasses to aluminosilicates, containing 0.5 moles alumina, in their leachability by water and dilute acids (0.1 N HCl, HNO_3, H_2CO_3). In water the amount of Na_2O going into solution continuously decreases with increasing alumina contents. In acid solution there is a sharp decrease in leachability of alkali for 0.3 mole Al_2O_3, but if 0.4 moles are present in the glass a rapid increase in dissolution of alumina also is observed. For 0.5 moles Al_2O_3, the ratio of the oxides going into solution corresponds to their initial ratio in the glass composition. The sodium aluminosilicate glasses are more intensely decomposed by sulfuric acid solutions than by dilute HCl and HNO_3 of the same normality. No indications for the specific formation of any binary or ternary compounds in the glass can be concluded from the leachability data. Natural nepheline was examined in the same way. It is apparently more easily decomposed than the glass $Na_2O,Al_2O_3,2SiO_2$. Figs. B. 99 and 100 give one of the most characteristic diagrams of Dubrovo and Shmidt, in which the numbers n_{Na_2O} of molecules Na_2O, and n_{SiO_2} of SiO_2 are plotted in relation to the number, m, of moles SiO_2 and of the number, k, moles of Al_2O_3, for the extractability in water and in HCl solutions. Neutral solutions, as of NaCl, were examined in their specific action on the Na^+ migration by leaching effects. NaCl decidedly lowers the migration of soda, but enhances that of silica. $CaCl_2$ undergoes side reactions if CO_2 (from the atmosphere) is present, with a precipitation of $CaCO_3$ (as calcite and aragonite). In the presence of $AlCl_3$ there is a remarkable *stabilization* of the glass framework, whereas $MgCl_2$ and $BaCl_2$ do not show such an effect. In every case, Ca^{2+} ions are not able to enter deeply into the glass structure, not even if they are driven by electrolysis, and also do not affect the galvanic potentials of glass electrodes. Al^{3+} ions, on the other hand, are rapidly bound to Na_2O and SiO_2 in the siliceous surface layer formed by normal hydrolysis in the first stage of the

[280] *Asaki Garasu Kenkyu Hokoku.* **1**, 1950, 1-7. On aluminophosphate glasses, in analogy with borophosphate, silicophosphate, and binary phosphate glasses, see K. Takahashi "Advances in Glass Technology," Plenum Press, New York, 1962, 366-376, and specifically on the resistance of phosphate glasses to aqueous HF solutions, K. Takahashi, K. Iida, Y. Watanabe, and M. Oguchi, *Bull. Fac. Engg. Yokohama National Univ.*, **4**, 1955, 69-76.

[281] *Steklo i Keram.*, **11**, 1954 (6) 3-7; *Izvest. Akad. Nauk S.S.S.R.*, Otdel. Khim. Nauk, 1953, 597-606; 1954, 236-243; 244-252; 770-777; 778-783; 1955, 603-610; *Stroenie Stekla, Inst. Khim. Silikatov Akad. Nauk S.S.S.R., Trudy Soveshchaniya, Leningrad*, 1953 (Pub. 1955), 181-184; *Bull. acad. sci. U.R.S.S.*, 1955, 359-364. On interesting galliosilicate glasses (of relatively lower chemical durability) in comparison with aluminosilicate glasses see recently S. K. Dubrovo, *Structure of Glass. Proc. All-Union Conf. Glassy State, 3rd, Leningrad* (English Translation), 1959/60, 373-377.

reaction of $AlCl_3$ with the glass. The aluminosilicate gels precipitated have the characteristics of base exchangers.

173. In *borosilicate* glasses, J. Yamamoto[282] made systematic studies of the influence of lime, magnesia, barium oxide, and alumina on chemical durability, determined by standard leaching methods. BaO, CaO, and MgO did not exert any favorable effect in this direction, whereas alumina increases the stability to hydrolysis if 0.5 to 5 per cent silica are replaced by alumina. "*Nonex*"-type borosilicate glasses varied by addition of ZnO and PbO have some specific improvement effects according to further determinations of chemical durability.[283] Extensive and de-

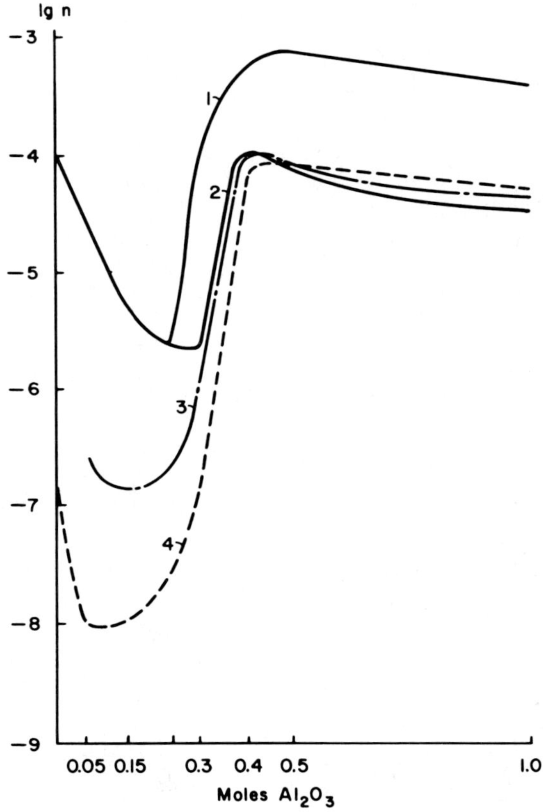

FIG. B. 99. Extractability of sodium aluminosilicate glasses by dilute hydrochloric acid solutions, with numbers, n, of moles Na_2O extracted, and the moles SiO_2 and Al_2O_3, in ratio to the numbers of their molar contents in the initial glass composition. (Dubrovo). (1) log n_{Na_2O} (1 N HCl). (2) n_{Na_2O} (0.1 N HCl). (3) log ($n_{Al_2O_3}/k$) (0.1 N HCl). (4) log (n_{SiO_2}/m) (0.1 N HCl).

[282] *Yogyo Kyokai Shi*, **61**, 1953, 491-494, 495-499; **62**, 1954, 125-128.

[283] *Yogyo Kyokai Shi*, **61**, 1953, 257-265.

tailed investigations of borosilicate glasses, with systematic additions of alumina, were made by Y. Tamura[284] concerning their alkali leachability, in quenched and annealed states, with the remarkable result that alumina-free borosilicate glasses when slowly cooled showed a *higher* alkalinity than the quenched samples. The chemical durability is considerably improved by the introduction of alumina. Tamura's results mean a considerable extension of our knowledge of the immiscibility of qua-

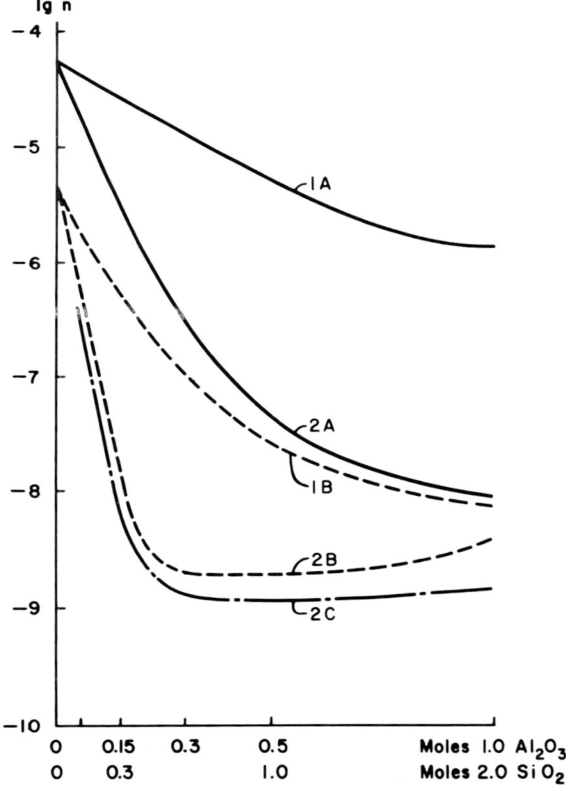

FIG. B. 100. Extractability of sodium aluminosilicate glasses as given in Fig. B. 99, but leached by water. (Dubrovo). (*A*) log (n_{Na_2O}), (*B*) log ($n_{Al_2O_3}/k$), (*C*) log (n_{SiO_2}/m). Index 1 refers to alumina-free sodium silicate glass; index 2 to aluminosilicates, as given in Fig. B. 99, starting from Na_2O, $2SiO_2$.

[284] *Osaka Kogyo Gijutsu Shikenjo Hokoku,* 1953, No. 299, 70 pp. These studies were chiefly made to investigate unmixing reactions to detect the optimum suitability for making glasses of the Vycor type (cf. Vol. V, Section B). Very instructive three-dimensional-perspectivic diagrams of the alkali solubility of different borosilicate glasses are given in the monograph mentioned on pp. 61-67 (Fig. 43 to 48), in a projection method previously used by G. Keppeler and H. Ippach (1927) to show the deep minima in leachability for special compositions.

ternary and more complex borosilicate glasses of the type Na_2O—B_2O_3—Al_2O_3—SiO_2, corresponding to highly resistant industrial types. Decreasing alumina contents will always result in poorer chemical durability. If alumina is very low, the introduction of ZnO and PbO brings about an improvement in this property. For highly siliceous glasses, alumina is the only oxide of strong durability-increasing action.

174. An industrially highly instructive comparison of soda-lime silicate and borosilicate bottle glasses, examined in their alkali leachabilities with dilute solutions of HCl, was performed by Fr. C. Raggon and Fr. R. Bacon[285] in the range from 0.0002 to 6.0 normality of the acid, at a constant temperature of 70°C. and for 7 days exposure. The leachates were analyzed in their concentrations in Na_2O and SiO_2. With increasing acid concentration, extraction of both oxides from borosilicate glass bottles also increased, whereas it remained practically constant for soda-lime silicate glasses. The leaching of silica increased in both types of glasses, although the increase was more pronounced in borosilicate glasses. In 6.0 *N* HCl solution, the leached amount of silica was equal for borosilicate and soda-lime silicate glasses, and the ratio SiO_2/Na_2O (in weight) of the leachates was nearly the same as in the initial glass composition for the borosilicates, which means all of the glass components were passing into solution. Borosilicate glasses were more resistant to attack if the HCl concentration was only 0.0002 to 0.001 *N*, and they were also superior to soda-lime silicate glasses with respect to Na_2O dissolved (cf. Fig. B. 101).

175. *Titanium* containing silicate glasses as substitutes for lead crystal glass, were previously mentioned in B. ¶ 137. In simple alkali-lime silicate glasses a systematic substitution of silica by TiO_2 (up to 45 per cent) was studied by M. A. Bezborodov and I. I. Kisel'.[286] The glasses with only 2 per cent TiO_2 show first a reduction of the chemical durability of modified compositions to water and soda solutions. Glasses higher in TiO_2, however, are more resistant than the titania-free compositions. On the other hand, HCl corrodes intensely the glasses high in TiO_2, although in concentrations of only 8 to 10 per cent TiO_2 there was a very low leaching weight loss. Below 24 per cent, evidently TiO_2 is only a glass framework-modifier constituent in silicate glasses, above 24 per cent it is a framework-forming ingredient, Ti^{4+} replacing Si^{4+} in the tetrahedral structural units. Glasses of the system Na_2O—SiO_2—TiO_2 were studied by V. V. Vargin,[287] especially those starting from the composition 79.5 per cent SiO_2 and 20.5 per cent Na_2O, with systematic additions of TiO_2. The experimental glasses were examined by a leaching test in 20 per cent

[285] *Am. Ceram. Soc. Bull.*, **33**, 1954, 267-271.

[286] *Doklady Akad. Nauk S.S.S.R.*, **103**, 1955, 1073-1076. On conventional and invert titania-containing glasses see H. J. L. Trap and J. M. Stevels, *Phys. & Chem. of Glasses*, **1**, 1960, 107-118, 181-188.

[287] *Doklady Akad. Nauk S.S.S.R.*, **111**, 1956, 848-850.

HCl solution for 3 hours at room temperature. There is a characteristic broad minimum of leachability between 12 and 30 per cent TiO_2, with the lowest amount leached at 21 per cent TiO_2.

176. M. A. Bezborodov and I. S. Kachan[288] made an interesting comparison of glass composition series with TiO_2 and ZrO_2, starting from a glass 75 SiO_2, 10 CaO, and 15 Na_2O, with a simultaneous substitution of SiO_2 by TiO_2 and ZrO_2. The chemical durabilities were determined as well in boiling water as in 20.2 per cent HCl solution (azeotropic mixture), and 0.1 *N* solution of Na_2CO_3, for 5 hours. The

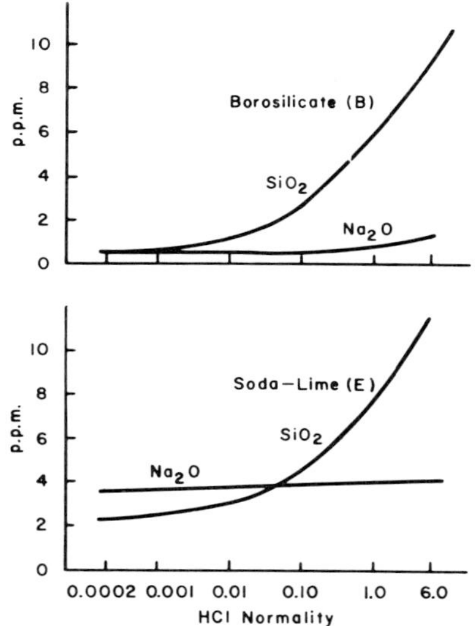

FIG. B. 101. Average concentration of silica and soda in leachates from borosilicate (*B*), and soda-lime silicate (*E*), bottle glasses, as a function of the normality concentration of dilute hydrochloric acid. (Raggon and Bacon).

weight losses in water leaching are very low in zirconium silicate glasses (0.03 to 0.10 per cent). For all glasses they are more or less rapidly decreased in HCl solution with increasing ZrO_2 contents. Particularly sensitive to HCl leaching was the series with 45 SiO_2, $(30 - x)$ TiO_2, and $xZrO_2$ ($x = 0$ to 17 per cent). In 0.1 *N* Na_2CO_3 solution the highest leaching losses were shown by glasses of the series 55 SiO_2, $(20 - x)$ TiO_2, $xZrO_2$, and those of 60 SiO_2, $(15 - x)$ TiO_2, and $xZrO_2$. The curves for losses in Na_2CO_3 solution showed a rather steep slope downward with increasing ZrO_2. If CaO and Na_2O are kept constant, the soda leachability in every case is

[288] *Doklady Akad. Nauk S.S.S.R.*, **115**, 1957, 1148-1151. B. Löcsei, *Silikat Tech.*, **12**, 1961, 64-66, recommends the replacement of feldspar in container glass batches by zircon sand, with a distinct improvement of chemical durability.

lowered, the higher the zirconia content, but likewise the higher that in titania relative to silica. Glasses with 15 or 16 per cent ZrO_2 are, therefore, the compositions most stable to water, HCl, and soda solutions. From the practical viewpoint, the optimum for many purposes is the glass type of very satisfactory chemical durability with 55 SiO_2, 12 or 13 TiO_2, 7 or 8 ZrO_2, 10 CaO, and 15 Na_2O.

177. We mention in this connection also previous investigations of O. K. Botvinkin and B. V. Tarasov,[289] chiefly concerning glasses of the system $Na_2O—ZrO_2—SiO_2$, after an *autoclave* treatment to determine the weight losses by leaching. For all of these glasses the formation of a mechanically very stable hydrated surface film is characteristic, giving off its water content only at 450° to 500°C. The leachability is largest for Na_2O which is immediately extracted, but silica also follows the alkali into the solution phase, whereas zirconia is enriched in the residual surface layer, which, however, still retains some silica and alkali. The shifting of the ratios $Na_2O : ZrO_2 : SiO_2$ from that of the initial glass composition to that of the extraction residues is very complex. The high stability of the zirconium hydroxide films is a characteristic of such glass types as were investigated. Botvinkin and Tarasov interpreted the formation of the film as an adsorption process of zirconium hydroxide (perhaps H_2ZrO_3) on the siliceous reaction layer after the leaching away of most of the alkali and the final formation of highly complex zirconosilicates in the protective film.

178. The systematic study of *weathering* phenomena, especially on hygroscopicity-sensitive *optical* glasses, had the important result two decades ago of finding how one may protect the glass from deterioration by planning. This can simply be attempted by systematic changes in the composition of the glass, but this elementary method is much restricted in its limits of applicability by indispensable requirements of the optical properties of the glass products. But there may also be applied artificial protecting layers of foreign ingredients on the glass surface to improve the transmittance and reduce the reflectivity, while they stabilize the glass surface. Before we discuss these measures we want to emphasize the importance of the studies of H. E. Simpson[290] on the chemical durability of optical glasses as determined by the hazemeter test discussed in B. ¶ 164. The fogging-clearing cycles in about a 100 per cent relative humidity atmosphere constitute a very sharp method for detecting changes in total transmissivity and in the amount of scattered light of weathered samples of Crown Glass, Borosilicate Crown, dense Barium Crown, Light Flint Glass, Heavy Flint, Extraheavy Flint, as the most typical glasses of this kind, all with their specific haze curves. Glasses with highly polarizable ions (lead) appear to be the best.

[289] *Steklo i Keram.*, 11, 1954 (6) 12-14.

[290] *J. Soc. Glass Technol.*, 37, 1953, 249-255. The photoelectric hazemeter is accepted in the ASTM Designation D 672-45 T, of Committee D-20, 1944, 1153-1155, 1945 *Supplement to ASTM Standards* Part III, *Nonmetallic Materials, General*, pp. 362-363.

Less effective are barium glasses, and these are better than Crown Glasses. For the theory of the constitution problems in connection with chemical durability, the concepts of K. Fajans and N. J. Kreidl (1948) proved again to be of great significance, and further those of W. A. Weyl, M. K. Roman and E. C. Marboe (1948) who demonstrated the influence of heavy metal ions on the hygroscopicity of a glass surface.

179. Simpson[291] also applied the electron-microscopic method of *replica techniques* for a detailed study of the texture of "weathered" surfaces on different glasses, such as an orienting macroscopic examination of corroded container surfaces, using a Vinylite liquid plastic dissolved in butyl ketone or acetone. As a general rule, the inside of containers examined was deteriorated more rapidly than outside walls, and, therefore, showed deeper corrosion grooves, striae, and the like. Real electron-microscopic replicas, excellently revealed details of weathering attack on the glass surface and multiple forms of corrosion,[292] but they also made evident surface defects dependent on previous treatments of the glass which are important sources of vulnerabilities enhancing the weathering process from case to case. Not much could be concluded, however, for identification of the observed weathering products, not even from electron-microscopic details. However, it is probable that alkali hydroxides and carbonates participate together with calcium carbonate and silica hydrogel in the formation of weathering crusts. These results are also in agreement with prior electron-microscopic descriptions given by Fr. Oberlies,[293] combined with platinum or platinum-iridium shadow-casting contrasts. Oberlies identified calcite (with typical twins) and gave electron-diffraction data for its identification. This is in agreement with the texture of the swelling layers of fire-polished soda-lime silicate glasses exposed to moisture for 3 days in which crystals of sodium silicate and carbonate hydrates could be observed (of a diameter of 100 to 300 Å.). Pure silica glass shows a uniform swelling layer, whereas weathering of alkali containing glasses as a rule starts from discrete spots.

180. W. Geilmann and G. Tölg[294] applied the replica techniques also for light-microscopic examinations of glass surfaces, as in antique glasses which had undergone a long-time corrosion and deterioration, and further in comparison with deliberately etched glass surfaces. The chief advantage of the replica method for these purposes is evident for heterogeneous, deeply colored, or entirely opaque glass samples, and for the examination of inside walls. In antique glasses, the development

[291] *J. Am. Ceram. Soc.*, **38**, 1955, 81-85; *Glass Ind.*, **36**, 1955, 515-519, 534, 536, 540, for bottle glasses. See also C. Brosset and M. Hagstedt, *Phys. & Chem. of Glasses*, **2**, 1961, 141-144.

[292] Especially *J. Am. Ceram. Soc.*, **41**, 1958, 43-49.

[293] *Glastech. Ber.*, **29**, 1956, 109-120, with many instructive electron micrographs.

[294] *Glastech. Ber.*, **28**, 1955, 299-307; **29**, 1956, 145-168, also with many characteristic micrographs.

of agate-like rhythmic-banded and concentric structures in their details is especially striking (cf. Vol. I, C. ¶ 180). The decomposition in soil is a typical topochemical reaction, developing a colloidal gel phase. Hydrated decomposition layers are simply identified by a slow heating to 400° or 500°C. with the formation of characteristic shrinkage cracks.

181. For the process of spontaneous weathering of window or plate glass in storage in a moist atmosphere the investigations of H. Schroeder[295] are of particular importance in which the corrosiveness of glass by solutions of a *pH* near 7 is examined. For this purpose were used as corrosive agents solutions of Na_2HAsO_4 with an addition of 0.5 molar $BaCl_2$, or sodium acetate solution with 0.5 molar $CrCl_3$ or $FeCl_3$, at 85°C. Methods for measuring the depth to which the corrosion had advanced to form typical "weathering layers" were combined with the study of two different reaction rates, namely first of the leaching process which forms a highly porous layer, and second an etching with a removal of such layers. A pure solution with $pH = 7.3$ has an ability to form "progressive" layers if with increasing concentration a critical minimum rate of corrosion is reached which is equal for the single solutions but a function of temperature. Below this critical rate, only a "regressive" layer formation is observed. A preexisting layer grows to the depth, but simultaneously it is resorbed, even with a higher speed, and finally disappears. The activation energy required for the solutions to form layers are the more reduced the higher the electrostatic charges of the *anions* in the solution. In the range of *pH* from 6 to 14, the measured *pH* effects in the corrosion show a strong maximum for "progressive" layer formation. The position of this maximum depends on the composition of the salt electrolyte and is, say for NaCl, the molarity = 1, for Na_2HAsO_4 the molarity = 0.3. If "activating" agents are added to the electrolyte solution, such as Al^{3+} in concentrations of 10^{-4} to 10^{-3} moles/liter, the removal of the layers is suppressed in favor of the first reaction type mentioned above. The limit between progressive and regressive layer formation is shifted by nearly one order of magnitude to the lower corrosiveness. Many indications are observed by Schroeder from which he concluded that the leaching is determined by the presence of colloid-dispersed particles of a definite size. For aluminum acetate they were estimated to be about 50 mμ in diameter as the maximum.

182. Whereas the phenomenon of "salt corrosion" observed by Schroeder often develops *iridescent* surface reflection phenomena, optical glasses of the heavy Baryte Crown type sometimes show a heavy "frost" which was studied by Sh. Tsuchihashi, E. Sekido, and Y. Nakatani[296] by methods of phase-contrast and electron microscopy combined with electron diffraction. The surface corrosion was initiated either with

[295] *Glastech. Ber.*, **26**, 1953, 91-97.

[296] *Yogyo Kyokai Shi*, **66**, 1958, 233-243. On a corresponding examination of Flint glasses cf. *Ibid.*, **67**, 1959, 269-275.

water alone, or with wet CO_2 and/or SO_2 gas, by HCl vapors, or by solutions of HCl, H_3PO_4, Na_2CO_3, and NH_4Cl. In moist atmosphere, doubtless $Ba(OH)_2.8H_2O$, and $BaCO_3$ were identified among the reaction products, $BaSO_4$ formed from SO_2, and $BaCl_2$ from HCl corrosion. In every case a considerable haze is developed and sometimes interference colors of iridescent layers are also observed. Typical stains (consisting of low-temperature quartz) develop if the layer is highly siliceous, after a more or less complete leaching away of the metal cations. Interference films were observed by F. H. Nicoll[297] on glasses of low refractive index etched by H_2SiF_6 vapors, or silica-enriched H_2SiF_6 solution, with an abnormally *high reflectivity* power. This surprising effect is explained as a superposition effect of common siliceous corrosion layers with deposits formed by precipitations from a silicofluoride solution filling up the skeleton of the layers. This phenomenon is evidently the reversion of the reflection-reducing and transmissivity-increasing action of the improving layers of magnesium fluoride or cryolite, or other substances, applied from vapor phase on the surface of optical glass lenses, which is now generally adopted use for objective constructions,[298] exemplified by cryolite condensation according to Knudsen's transfer method. K. G. Günther[299] discussed the measurement of the vapor pressure of MgF_2, Na_3AlF_6, and other fluorides, and their speeds of evaporation in vacuo, with a functional formulation of the amount, G_0, of material evaporated per time and surface unit, at constant temperature, and the vapor pressure, $p(T)$: $G_0 = p(T) \cdot \sqrt{RTM/2\pi} \cdot C$ (C a geometric constant, M the molecular weight). For MgF_2 and Na_3AlF_6 an improved method of A. Herlet and G. Reich[300] (cf. Vol. IV, Section A) was applied, measuring the weight increase of a condensation surface suspended on a sensitive microbalance. The measured range of vapor pressures was between 10^{-1} and 10^{-4} mm. Hg up to 250°C. Günther modified the Herlet-Reich method for temperatures up to 1400°C. by using a molybdenum radiation furnace and an electrodynamic microbalance to determine the vapor pressures of the fluorides mentioned above, as well as B_2O_3 and SiO.

183. Since E. Berger's *interferometric* investigations (1934) on the laws of reaction rate and mechanism of glass deterioration, Sc. Anderson and D. D. Kimpton[301]

[297] *J. Opt. Soc. Am.*, **42**, 1952, 241-242. Cf. S. M. Thomsen and F. H. Nicoll, U. S. Patent 2,505,629 (April 25, 1950).

[298] An excellent report on actual standard of practice of surface improvement by fluoride films applied on objective glass is that of D. Kossel, "Vergütete Optik," Special Paper of the E. Leitz Optische Werke, Wetzlar, 1955, 12 pp.

[299] *Glastech. Ber.*, **31**, 1958, 9-15.

[300] *Z. angew. Phys.*, **9**, 1957, 14-23.

[301] *J. Am. Ceram. Soc.*, **34**, 1951, 141-145; **36**, 1953, 175-179. The A. H. Pfund method mentioned above is described in *J. Opt. Soc. Am.*, **31**, 1941, 679-682; see on multiple layer reflection A. Vašiček, *Ibid.*, **37**, 1947, 623-634. On the mathematical traetment of glass corrosion data see D. K. Hoganson and J. H. Healy, *J. Am. Ceram. Soc.*, **45**, 1962, 178-181.

made further use of this highly sensitive and simple method to determine the thickness of the leached layers as a function of time and temperature, as in a Crown silicate glass high in BaO, in 3 *N* HCl solution, and in a PbO-rich (60 per cent PbO) X-ray shielding glass, in 0.1 *N* HCl solution, at 50°C. The change in thickness follows the elementary formula $\tau = \alpha t - \beta t^2$ (t the time period of leaching); the constant α is highly sensitive to temperature. So-called "stains" on leached glass correspond to a region of low refractive index moving systematically into the glass. Infrared spectral-photometric curves showed that acid leaching removes from the glass structure those ingredients which damp the tetrahedral and octahedral vibrations, and destroys the $[BaO_6]$ coordinations much more rapidly than the silica framework proper in the structure. At higher temperatures no interference colors appear because the destruction is so complete that parts not directly affected can no longer cling to the base glass. The refractive index of the leached film (measured by the Brewster angle in Pfund's method) at first decreases and then is increased with leaching time making evident a complex multilayer reflection phenomenon.[302]

184. The study of lead-rich X-ray protection glass shows analogous release of lead ions from the glass surface, mostly from the framework-modifying positions. Much more sluggishly are released the far fewer lead cations in network-forming positions. The rate of "stain" formation is complex; it first follows a straight-linear function of time, then another process with a time law of stain growth proportional to the square root of time. Also for PbO-rich glass the refractive index is first rapidly

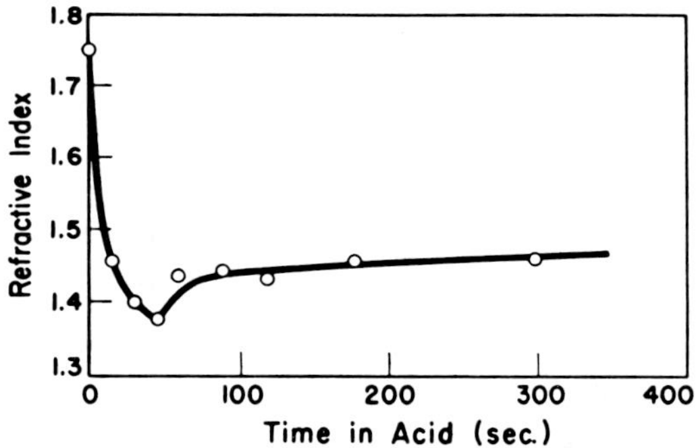

FIG. B. 102. Refractive index of stain layers formed on X-ray shielding plate glass, after etching in 0.1 *N* HCl solution at 50°C., as a function of time. (Anderson and Kimpton).

reduced (from 1.75 to 1.38), then slowly increased (Fig. B. 102) to an end value of 1.45 to 1.46. Because of its high porosity the stain cannot consist entirely of silica alone which should in the present case be considerably lower than 1.46 (about

[302] Cf. A. H. Pfund, *J. Opt. Soc. Am.*, **36**, 1946, 95-99.

1.29) according to Pfund using the ultraviolet "channelled spectrum" method. The variation in thickness of the stain layer shows a better agreement with the cube of thickness versus time than a direct proportionality (cf. Fig. B. 103). The analysis of infrared reflection spectra of the unetched glass in comparison with silica glass, and those of etched glass in different periods of acid exposure (Fig. B. 104) makes evident how the slightest bit of leaching resolves the smooth spectral curve (with a flat maximum at 1000 cm.$^{-1}$ for the original glass), into two peaks at 950 and 1070 cm.$^{-1}$ with a higher reflectivity than that of the non-leached glass. The longer the

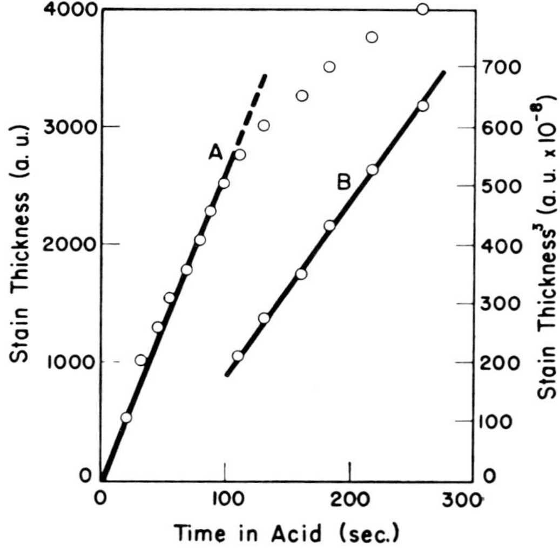

FIG. B. 103. Variation of thickness of stain layer on X-ray shielding glass, by etching with 0.1 *N* HCl solution at 50°C., (*A*) under the assumption of direct proportionality with time, (*B*) for the actual thickness cubed, versus time. (Anderson and Kimpton).

leaching period, the more the 1070 cm.$^{-1}$ peak becomes similar in appearance to that of SiO_2 at about 1110 cm.$^{-1}$. Heating shifts the 1070 cm.$^{-1}$ peak towards higher wave numbers and becomes stronger, whereas the 950 cm.$^{-1}$ peak decreases in intensity. The leaching reaction destroys the [PbO_6] coordinations of the modifying cations, preferably uncoupling tetrahedral and octahedral units which then vibrate undamped.[303] The interference color film on the glass is still composed of [SiO_4] and [PbO_6] coordination units, but with pores occupied before by lead cations of modifier quality and with pore-holes of previous octahedral configurations released. In every case, the layer is *not* a simple silica film; the refractive index of 1.46 is only incidentally coinciding with that of (compact) silica.

[303] Cf. on infrared reflection spectra of glasses as a function of structure: Sc. Anderson, *J. Am. Ceram. Soc.*, 33, 1950, 45-57.

185. Interferometric determinations of chemical durability of crystalline quartz, silica glass, flint, obsidian glass, opal, enamels, and common glasses were also made by R. G. Pike and D. Hubbard,[304] the corrosion agent varying from concentrated sulfuric acid to a 5 per cent solution of sodium hydroxide, the exposure time varying from 1 minute to several weeks, in most cases at constant temperature of 80°C. The shifting of interference fringes as seen in the Pulfrich interferometer was observed in helium light with reference to the untreated surface of the same. Crystalline quartz has a much higher alkaline solution resistance than silica glass. Results are

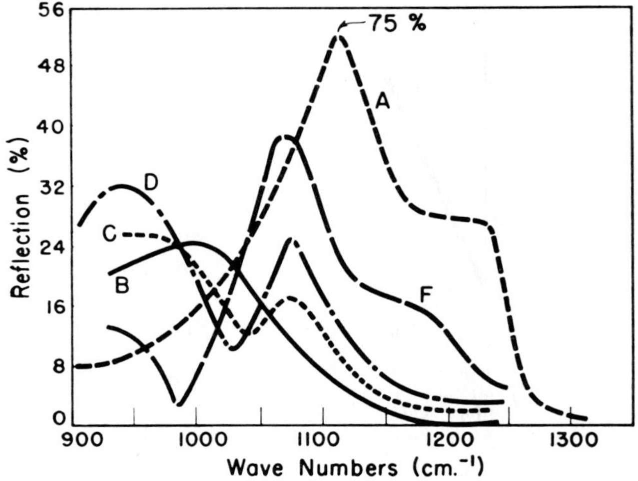

FIG. B. 104. Infrared reflection spectra of silica glass (*A*), nonetched lead-rich X-ray shielding plate glass (*B*), and the same glasses etched (*C*) for 3 seconds, (*D*) for 150 seconds, (*E*) for 542 seconds, in 0.1 *N* HCl solution at 50°C. (Anderson and Kimpton).

given for the hygroscopicity of the different glasses, and other samples in mg. adsorbed water/cm.^{3} of the samples. The unevenness of the distribution of the mobile ions of the exterior solution and the siliceous reaction layer was controlled by a treatment of glass grit at 80°C. with a solution of $pH = 4.1$, followed by a treatment with saturated solution of $Ag(NH_3)Br$ and a potentiometric titration of the excess of Ag^+ and Br^- ions. In this way a detailed distinction of different kinds of siliceous layer reactions could be made to examine the application of the Donnan membrane equilibria theory.

[304] *J. Research Natl. Bur. Standards,* **50**, 1953, 87-93; *Glass Ind.,* **35**, 1954, 82, 106. On the method see D. Hubbard and E. H. Hamilton, *J. Research Natl. Bur. Standards,* **27**, 1941, 143-157; on solubility of quartz and silica glass in comparison see also V. S. Molchanov and N. E. Prikhid'ko, *Izvest. Akad. Nauk S.S.S.R.,* Otdel. Khim. Nauk, 1957, 1151-1157.

186. For the attack of alkaline solutions on fire-polished glass surfaces, L. Schumacher and H. E. Schwiete[305] used the *interference microscope* (in Tl light) as a sensitive method for studying dissolution grooves combined with phase-contrast inspection. There is an extensive analogy between grooves etched by alkalies (1 *N* KOH), or dilute HF solution, at 40°, 60°, 80°C. for 96 and 189 hours. The grooves are larger on the fire-polished surface of Libbey-Owens sheet glass than they are on the rolled side, with craters of 4 μ in diameter and 0.13 μ depth produced in 6 minutes by a 5 per cent HF solution, but in 100 hours in 4 per cent (1 *N*) NaOH solution. The activation energy calculated from the rate constant (in mg./cm.^{2}/hour) is 17,000 cal./mole, the order of the reaction zero for a glass surface constant during the experiment, but changed to 2/3 if free glass grit is corroded, the surface of which is continuously decreased during the reaction. The products formed on the glass surface are $CaCO_3$ and $Mg(OH)_2$ from dissolved alkaline earths, but only a portion of aluminum and little silica.

187. The interference-microscopic etching method was developed with great mastership by J. Löffler[306] to become an indispensable tool for control of chemical discrepancies in the composition of glass heterogeneities in one and the same sample, especially in the analysis of silica and/or alumina-enriched striae (cords) derived from dissolution of refractories in the glass melt in the tank (cf. B. ¶ 104 ff.). Striae in which silica is higher than in the normal glass composition are etched by HF as

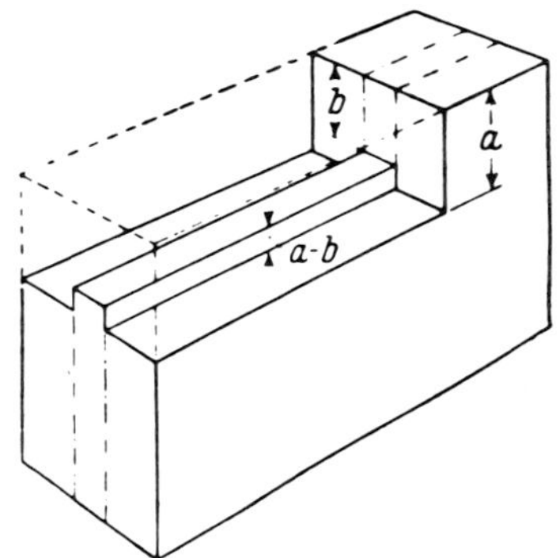

FIG. B. 105. Scheme of etchability of a glass block containing a layer of different chemical composition (excess silica). (Löffler). (*a*) Etchability of the normal glass, (*b*) that of the silica-rich layer (stria).

[305] *Glastech. Ber.*, **33**, 1960, 1-7. One may compare results with those determined by weight losses in the studies of K. H. Lintermanns, *Tonind.-Ztg. u. Keram. Rundschau*, **79**, 1955, 231, in unstirred and stirred 0.1 *N* NaOH solution.

[306] *Glastech. Ber.*, **27**, 1954, 381-392; **30**, 1957, 457-463. See recently an important report of the Physical Properties Committee of the British Society of Glass Technology, *Glass Technol.*, **2**, 1961, 192-198.

crests, those with lower silica as grooves (cf. Figs. B. 105 to 108). The level differences are measured in the interferometer or interference microscope. An increase by 0.9 per cent alumina has about the same effect as one by 2.5 per cent in silica. The relative increases in silica, of course, can be caused also by losses in alkalies by vola-

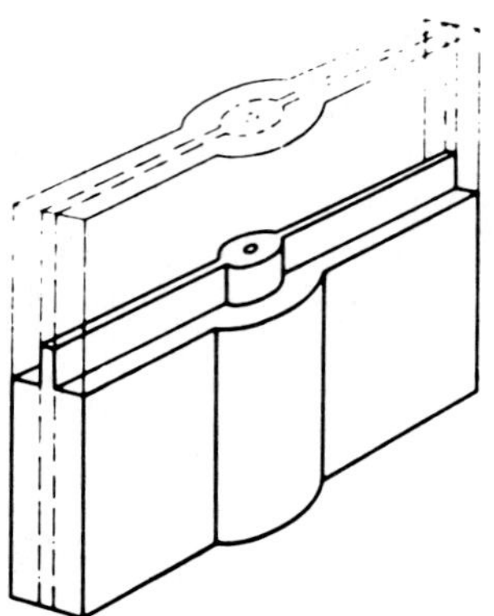

FIG. B. 106. Profile scheme of a fireclay dissolution cord in plate glass. (Löffler).

tilization (cf. B. ¶ 98) from the melt. The rate of dissolution of the glass in 1 per cent HF solution, called "etchability," is a characteristic material constant as a function of the composition of the glass.It is measured by the interferometric method as the decrease in level below the unetched surface and expressed in $\mu/6$ minutes.

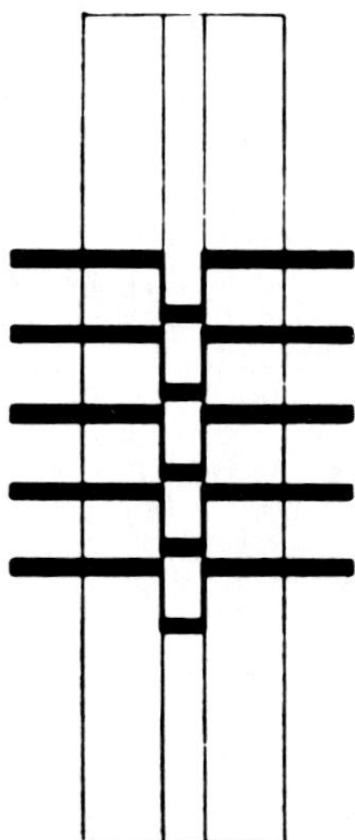

FIG. B. 107. Interference scheme of a silica dissolution cord in plate glass. (Löffler).

Quenching causes considerable increases in etchability. If HBF_4 is used in place of HF, the silica striae form positive, the alumina striae negative reliefs, especially if alumina is low. For striae high in alumina the etchability is not much different than that of the surrounding glass. Löffler also attempted to differentiate silica and alumina striae by etching in molten HPO_3 (at 250°C.), a 10 per cent NaOH solution (at 75°C.), and a mix of HCl + HBF_3.OH (at 70°C.). The interference-microscopic examination detects very small differences, say of its striae in the textural and chemical composition distribution across sections of a given glass sample. The accuracy of a standardized etching method is high, especially in the range from 3.0 μ to 0.0275 μ for the interferometric measurements. Practically in most cases it is sufficient to determine the *relative* etchability by exposing the glass sample simultaneously with a standard glass and to base the calculation on relative profile heights. Qualitative etching with NaOH solution is highly characteristic of alumina cords containing zirconia from a ZAC tank block.

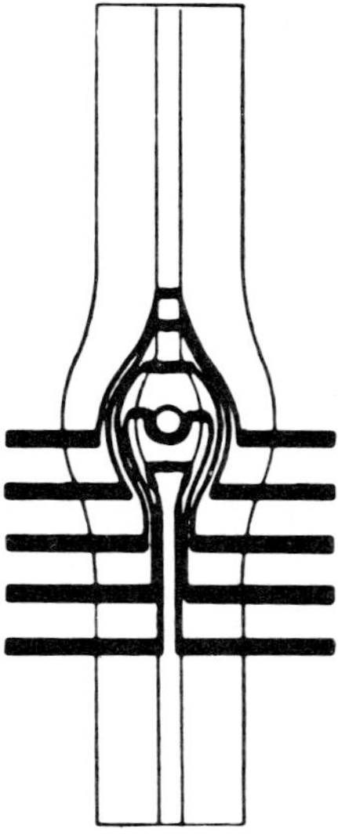

FIG. B. 108. Interference scheme of a fireclay dissolution cord in plate glass. (Löffler).

188. *Stopping* phenomena or so-called *poisoning effects* exerted on the course of corrosion reactions by products of the glass decomposition, or by foreign additions, have often been described, such as by I. R. Beattie[307] in his classical application of the continuous flow method (cf. B. ¶ 165). He also gave a critical discussion of previous observations of the stopping effect of alkali silicates deliberately added

[307] *J. Soc. Glass Technol.*, **37**, 1953, 240-243, especially pp. 242 ff.

to the corrosion solution and their "blocking" reactions. Sh. Furuuchi and T. Uno[308] observed with interferometric methods in the corrosion of sheet glass, by a solution of Na_2CO_3, or sodium silicate with the ratio $SiO_2/Na_2O = 3.02$, the strong inhibiting effect of the silicate anions, especially in initially dilute solutions of the salts. The formation of an iridescent and porous protection film on the glass surface was most evident. If the ratio SiO_2/Na_2O in the attacking sodium silicate solution was varied, the retardation effect was strongest for low ratios SiO_2/Na_2O. The most intense initial corrosion is observed in a solution of sodium silicate with $SiO_2/Na_2O = 1.6$, even higher than for a NaOH solution of the same concentration in Na_2O. If the Na_2O concentration in the attacking solution is raised, the SiO_2/Na_2O ratio causing the strongest retardation systematically shifts toward lower ratios SiO_2/Na_2O (cf.

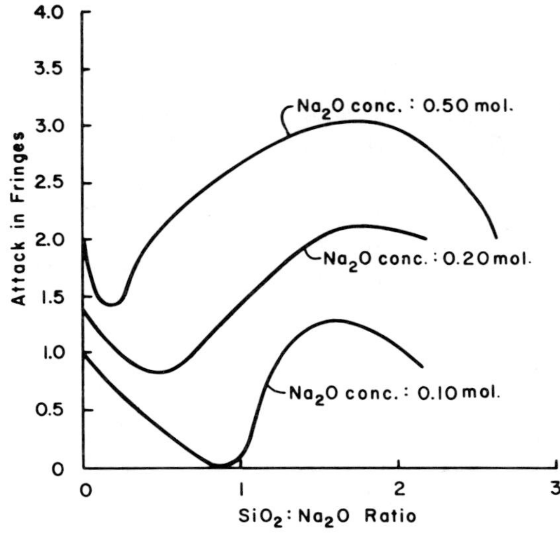

FIG. B. 109. Attack of sheet glass by sodium silicate solutions of variable ratios SiO_2/Na_2O, in three different concentrations. (Furuuchi and Uno).

Fig. B. 109). On the other hand, this ratio for highest attack rate did not alter very much with varying concentration in Na_2O. Very pronounced retardation effects of alkaline attack in soda-lime silicate glasses were observed by G. A. Hudson and Fr. R. Bacon[309] if salts of beryllium, zinc, or aluminum are added to a 3 per cent NaOH solution as the corroding agent. The concentration of the "modifying" salts was 2 to 10 mg. atoms/kg. of the solution. For two different container glass compo-

[308] *Asahi Garasu Kenkyu Hokoku*, **3**, 1953, 1-8.

[309] *Am. Ceram. Soc. Bull.*, **37**, 1958, 185-188; cf. W. Geffcken, *Kolloid-Z.*, **86**, 1939, 11-15; compare also U.S. Patent 2,245,907 (1947).

sitions, beryllium salts proved to be of the highest inhibiting efficiency which reduced the corrosion by as much as 97 per cent of that observed without the inhibiting agent. Zinc salts reduce by up to 90 per cent in higher concentrations. Strontium, antimony, barium, aluminum, and lead in higher concentration ranges reduce by 75 to 45 per cent. Only small inhibiting effects were observed for bismuth, divalent tin, zirconium, and copper salts, magnesium, tetravalent tin, and silicon being practically inert. Of great importance is the fact that *pyrogallol* in amounts of mg./mole was more inhibiting than zinc, but less so than beryllium salts.

189. The specific action of *organic* inhibitors of the group of pyrogallol and its derivatives, which was the subject of some patented inventions[310] is explained by W. A. Weyl[311] by an adsorption on the glass surface and formation of compounds soluble with difficulty whereas the contrasting formation of a soluble complex anion, as with phosphates, brings about an *acceleration* of the alkali attack on glass. Fr. R. Bacon and Fr. C. Raggon[312] observed a remarkable promotion of attack on soda-lime silicate and borosilicate glasses and of free silica (in the form of crystalline, amorphous, and glassy SiO_2) by *citrates* and other organic anions in a *neutral* solution, with a severity which is similar to that of substantially alkaline solutions. The effect is perceptible at $pH = 5$ and is rapidly increased with pH raising to 7.6. There is no doubt that complexes are formed analogous to those of silicon with pyrocatechol of the type

$$\left[\mathrm{Si}\left[\,\bigcirc\,\right]_3\right]^{2-}$$

described in 1931, by A. Rosenheim, B. Raibmann, and G. Schendel[313] (cf. A. ¶ 333). The action of citric acid in digestion with silica is very specific, explaining the stronger reactivity with silica glass than with container glasses containing lime and alumina. In this respect citric acid is even more specific than fluorine anions. It is practically important that SO_2-treated glass bottles show a *greater* sensitivity to citrate corrosion effects than bottles which were not treated with SO_2,[314] although they have very good general chemical durability. The best resistance to a 4 per cent sodium citrate solution was shown by an alkali-resistant flask of a high-in-ZrO_2 glass and other bottles of borosilicate glass with a high content in alumina, but there

[310] U.S. Patent No. 2,248,187 (1947).

[311] *Glass Ind.*, **28**, 1947, 408-412, 428-432.

[312] *J. Am. Ceram. Soc.*, **42**, 1959, 199-205.

[313] *Z. anorg. u. allgem. Chem.*, **196**, 1931, 160-176.

[314] Specifications of the examination of high class medical containers of glass of the U.S.A. Pharmacopeia (15th edition, 1955, 1178 pp., especially pp. 925) must be considered with the fact that citrate solution is prescribed as an anticoagulent agent for human blood storage. On *chelating* effects of Versene, pyrocatechol, and the like, see also F. M. Ernsberger, *J. Am. Ceram. Soc.*, **42**, 1959, 373-375, and Volume IV, Section A. ¶ 494.

is no correspondence between resistance to water and to sodium citrate solution. With the exception of sodium oxalate and Versene (salt of the ethylene diamine tetraacetic acid) all the examined salts of β-hydroxyacids (succinate, ascorbate, malate, tartrate, gluconate, and citrate) showed a specific solubility of silica, four of these with a dissolving power of more than 15 p.p.m. SiO_2 (citrate 170 p.p.m.).

190. The *impoverishment* of the glass surface in *alkalies* by volatilization in its action on the chemical durability was discussed by H. S. Williams and W. A. Weyl,[315] often called the "improvement effect" by alkali removal. We will not discuss here the dealkalization as a solid state reaction per se, but only under the viewpoint of the diffusion process of alkali into a surrounding acceptor, exemplified by high-temperature diffusion into a bed of meta-clay material densely applied to the glass surface (at temperatures up to 600°C.). This method illustrates very well the alkali removal in its distinct effects on changes in chemical durability, and on other characteristic surface reactions of practical interest, such as the application of silver mirrors. The lowest temperature type of the removal reactions is the formation (up to about 300°C.) of the proton-laden surface, in distinction from those glasses from which the alkali was acid-leached or electrodialysed, when both latter processes attack not only the alkali ions but also those of calcium and magnesium. The diffusion into the meta-clay body in alkali removal is, therefore, not able to produce a hydrogen-glass surface of reflection-reducing qualities. It is, on the other hand, able to react with methyl chlorosilanes to produce a water-repellent glass surface (cf. A. ¶ 322). Weyl also discussed the influence of dealkalinization on the mechanical strength of glass and its fatigue endurance, as seen from stress diagrams and tension-compression distribution schemes of dealkalized glass as an improved technological product.

191. The efficiency of glass surface dealkalinization was tested with Na^{24} as a radioactive tracer isotope by J. V. Fitzgerald[316] who also used a solid acceptor (meta-clay, or Cr_2O_3) for the alkali ions. The radioactivity of the sodium ions in the glass was excited by deuteron bombardment in a Cyclotron, using the nuclear reaction $Na^{23}(d,p) \rightarrow Na^{24}$. The activation energy for the acceptor reaction was in meta-clay = 12,000 cal./mole, for Cr_2O_3 = 3,100 cal./mole, the latter material therefore being particularly well suited for dealkalinization studies. During a heating to 537°C. for 30 minutes the amount of Na_2O removed per 1 cm.^{2} was 6 μg.

192. The more technological significance of the dealkalinization process for the glass surface is illustrated by experiments of B. Simmingsköld[317] using ammonium

[315] *Glass Ind.*, **26**, 1945, 275-277, 290-292, 301; 324-326, 339, 341-342, 344; 369-371, 386, 388, 390. Improvement of bottle glass by treatment with different agents is recently discussed by H. R. Persson, *Glass Technol.*, **3**, 1962, 17-35.

[316] *Glass Ind.*, **30**, 1949, 259-261; see also H. Muth, *Glastech. Ber.*, **27**, 1954, 248-255, especially pp. 252 f.

[317] *Proc. Intern. Symposium on Reactivity of Solids. Gothenburg*, 1952, Vol. II, 1954, 797-800.

salts for this purpose. By a treatment in NH_4Cl vapor at 370°C. the water-resistance of common glass is considerably raised, to a degree which is entirely comparable if not superior to the well-known effect of $SO_2 + O_2$ ("lehr gases"). R. M. Barrer, in a discussion remark referring to this interesting observation emphasized the complete analogy of glass dealkalinization with ion exchange reactions on zeolites in NH_4Cl vapor under analogous conditions, an effect which was previously described by F. W. Clarke and G. Steiger in 1902[318] (cf. Vol. IV, Section B. ¶ 69, footnote 129).

193. J. Löffler[319] applied his interference-microscopic etching method (see above, B. ¶ 187) to the study of dealkalinization of glass during welding of one sample to another one in the open flame, on a section perpendicular to the suture, and etching with $HCl + HBF_3.OH$. The welding process itself is normally not sufficient to develop a strong dealkalinization, but as soon as there are acceptors, such as fireclay refractories in the neighborhood, these eagerly resorb the Na_2O vapors evolved from the welding glass surfaces, and the rapid diffusion from inside the glass to the surface comes into action. The alkali volatilization is combined with that of silica in fluorine containing glass. The alkali losses from apparatus glass during open flame operations are in every case remarkable.

194. The classical alkali volatilization from SO_3-vapor treated glass surfaces is discussed by I. Hilgenfeldt and H. Jebsen-Marwedel[320] in the interaction of "ambulant" SO_3 from the furnace atmosphere (flame gases) with the alkali migrating to the open surface of the glass and the formation of Na_2SO_4. Experiments of "$SO_2 + O_2$ leachings" on plate glass grit at 100° to 700°C. show the characteristic acceleration of the Na_2O diffusion to the surface, parallel with its removal as sulfate which was determined by semi-microanalytical methods as a function of the temperature. Calcium ions do not easily diffuse to the surface and the ratio Na_2O/CaO which was in the glass examined at room temperature = 3 : 1, was raised to 12.45 : 1 at 700°C. The thickness of the layer into which SO_3 is able to penetrate is negligible in comparison with the Na_2O diffusion layer. In the exterior zones of the glass the SO_3 is very easily leached out, as from a glass treated at 200°C. with SO_3 for only 5 minutes. Practically, it must be considered from the experiments that considerably more Na_2O is leachable from a glass treated with SO_3 than from the same, non-treated glass. The "improvement effect" is evidently caused by the often mentioned, considerably intensified Na_2O diffusion to the surface from the interior of the glass. In Fig. B. 110 it is schematically demonstrated how one may understand the difference in behavior of normal and of SO_3-treated glass surfaces, chiefly by the different ac-

[318] *U. S. Geol. Survey. Bull.*, **207**, 1902, 57 pp., particularly pp. 29-36.

[319] *Glastech. Ber.*, **30**, 1957, 88-94.

[320] *Glastech. Ber.*, **31**, 1958, 161-170. On consequences of experiments concerning surface tension conditions see B. ¶ 77. On ion exchange and diffusion reactions in glass, see A. Sendt, "Advances in Glass Technology," Plenum Press, New York, 1962, 307-332.

tion of normal hydrolysis alone, or by the intensified Na_2O diffusion to the surface which causes observed increase in leached amounts of Na_2O. The alkali remaining in the surface layer is considerably lower than in the normal glass.

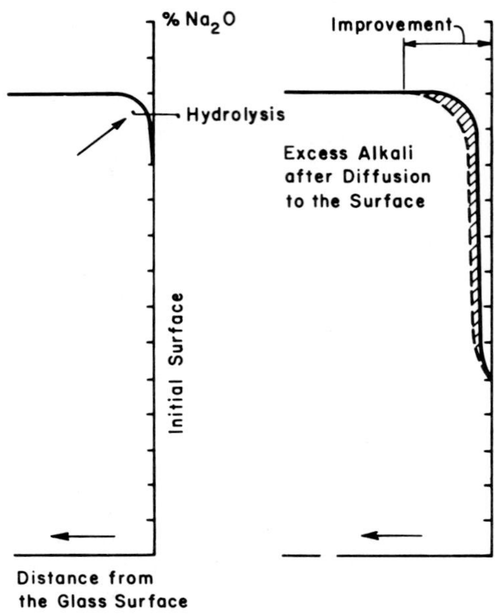

FIG. B. 110. Schematic sketch to illustrate the difference in alkali content of normal (not hydrolyzed) glass surfaces, and those exposed to SO_3-dealkalinization, with improvement effect. (Hilgenfeldt and Jebsen-Marwedel).

195. In lead-rich glasses, and particularly in lead-majolica *glazes* spots resulting from the crystallization of anhydrite and anglesite, $PbSO_4$ are not uncommon since caused by the reaction of SO_3-containing flame gases with the lead silicate melts. Z. A. Nosova and M. E. Yakovleva[321] described such crystallizations of anhydrite in prismatic forms of 2 to 50 μ in diameter, in dull spots occurring in a 45 per cent PbO glaze. They were identified by microscopic-optic examination and by electron diffraction data.

196. The problems of chemical resistivity of *porcelain enamels* was investigated by J. E. Cox[322] for boric acid containing compositions, under alkaline attack con-

[321] *Doklady Akad. Nauk S.S.S.R.*, **100**, 1955, 529-531.

[322] *J. Am. Ceram. Soc.*, **41**, 1958, 336-343. On the resistance of glass to industrial detergents and detergent additions see W. W. Fletcher, E. S. Keir, P. G. Johnson, and B. Slingsby, *Glass Technol.*, **3**, 1962, 195-200. On the electron- microscopic microstructure of weathered Na–Ca silicate glass surface, see R. M. Tichane and G. B. Carrier, *J. Am. Ceram. Soc.*, **44**, 1961, 606-610.

ditions, with boiling 5 per cent NaOH solution, or sodium pyrophosphate solution, and some commercial detergents (bleaches). The use of NaOH as a testing solvent sometimes gives misleading results in comparison with those of typical detergent solutions. The basic reaction mechanism of NaOH is apparently different from that of $Na_4P_2O_7$ or other detergents. If the NaOH solution is so dilute that it has a pH of only 10, the same alkalinity as these agents, the comparison with these has better results, but the conditions of maintaining $pH = 10$ are very critical in the leaching process. As the most appropriate testing agent Cox chose a 1 per cent $Na_4P_2O_7$ solution which behaves in its detergent efficiency similar to commercial detergents for porcelain enamels. The pH of $Na_4P_2O_7$ solutions is reliably buffered as for other detergents and the curves for pH as a function of concentration coincide for both groups of reagents. It is very important to control in such leaching tests the formation of deposits and surface residues on the enamel, which strongly affect the intensity of the attack. Especially titania-opacified enamels have the tendency to develop such deposits of titanosilicates. Sodium hypochlorite bleaches do not affect the results of the alkali tests. Phosphates in their specific accelerating action of attack on sheet glass in caustic solutions were also studied by Sh. Furuuchi and T. Uno[323] by interferometry, in comparison with the effects of zincate, aluminate, and borate solutions which inhibit the attack by alkalies, whereas stannate, plumbate, and vanadate anions are relatively ineffective. The phosphate ion action is at maximum for the ratio $P_2O_5 : Na_2O$ = about 0.1, but rapidly drops to zero at $P_2O_5 : Na_2O = 0.75$ to 1.0, as a function of the concentration in Na_2O.

197. The repeatedly emphasized and very specific *chelating* effects of *o*-dihydroxybenzene (pyrocatechol) and Versene ("EDTA" = ethylenediaminotetraacetic acid salts) as attacking agents on silicates (cf. Vol. IV, Section A) was discussed very recently by F. M. Ernsberger[324] referring to "sequestering" of normal protective cations Ca^{2+}, Mg^{2+} and Al^{3+} in glasses, with a maximum at a pH value which depends on the nature of the anions present, as seen in Fig. B. 111. The character of the attacked surface is very similar to that of glass etched by HF solutions. Chelating effects of organic agents are particularly striking, even the biological action of microorganisms (so-called "fungus" formation on optical glass) for which Fr. Oberlies and G. Pohlmann[325] gave instructive electron micrographs from replicas. Even feldspar cleavage surfaces are etched by such agents, as in soils. Animalic tissues may also affect and corrode glass surface by chelating intercellular metabolism products ("mineralysis").

198. For practical purposes the *autoclave test* of glass corrosion was reexamined

[323] *Asahi Garasu Kenkyu Hokoku*, **3**, 1953, 111-119.

[324] *J. Am. Ceram. Soc.*, **42**, 1959, 373-375.

[325] *Naturwissenschaften*, **45**, 1958, 487, 513-514; **47**, 1960, 58.

by M. Boffé[326] either with glass powder, or with glass strips. In the latter case, deterioration develops layerwise. In the range of steam pressures up to 60 kg./cm.^{2}, only a siliceous layer is formed; between 60 and 120 kg./cm.^{2} the transparent layer is replaced by a white powder, more or less adherent to the glass surface which is covered with fine cracks. In the range from 120 to to 150 kg./cm.^{2}, a dense, consistent white skin is formed which can be removed from the underlying unchanged glass surface only in scales of a characteristic opal-like aspect. Above 150 kg./cm.^{2} steam pressure the glass is subdivided to lumps of an amorphous product low in soda, but

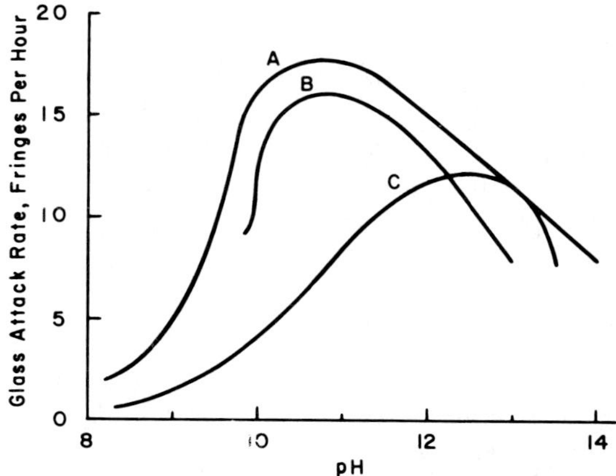

Fig. B. 111. Rate of attack of chelating agents on sheet glass, as a function of pH. (Ernsberger). (*A*) for 0.5 molar sodium phosphate solution + 0.2 per cent EDTA, and 0.4 per cent pyrocatechol, (*B*) for 0.5 molar Na_2CO_3 solution + 0.2 per cent EDTA, 0.4 per cent pyrocatechol, (*C*) for 0.2 per cent EDTA and 0.4 per cent pyrocatechol alone.

only a little silica is dissolved, and lime, magnesia, and alumina are enriched in the white layers. Only below 100 kg./cm.^{2} the attack is more rapid and intense in the liquid phase than by steam. Salts added to the corroding liquid (in 0.1 *N* concentrations) influence the reaction according to their cationic and anionic characteristics, chiefly the electrostatic field strength, z/a^2 of the cation being decisive. While 0.001 and 0.01 *N* NaOH solutions do not behave much different from distilled water, a 0.1 *N* NaOH solution is about 7 times, a 1 *N* solution 25 times, and a 10 *N* solution 50 times more corrosive, as determined by the weight losses of glass samples. The hydroxides of calcium, magnesium and aluminum retard the reaction; their cations are intensely adsorbed on the glass surface. Copper and manganese sulfate exert a specific accelerating effect on the decomposition of glass, whereas $ZnSO_4$ reduces it anomalously. Salts also affect the scaling and adhesion of the white layers on the unchanged glass.

[326] *Verres et réfractaires,* **9**, 1955, 251-263.

199. The very important phenomena of a *surface swelling* of glasses, glazes, and other silicate products under the influence of water vapor resorption, either under natural low-temperature conditions by hygroscopicity, or in the autoclave treatment, as a considerably accelerated reaction of steam with silicates, were frequently discussed in the ceramic literature because of their great practical significance. The functional relations existing between spontaneous "*crazing*" of chinaware glazes were described by H. Harkort[327] as a consequence of the expansion by moisture absorption in the glaze, together with thermal expansion differences between the surface layers of the glazed ware. The moisture stability of a glaze can be improved by additions of chalk to the composition of the glaze batch. The standard autoclave test concerns water absorption, the Steger differential expansion test, or a quenching test deals with the thermal strains developing between glaze and ceramic body. The tensions created in the glaze by moisture expansion are determined by the relation $\sigma_z =$ moisture expansion $\times$ modulus of elasticity/100 (in kg./mm.^{2}). The tensions due to thermal expansion follow a similar equation $\sigma_k = \Delta\alpha \cdot \Delta T \cdot E$ kg./mm.^{2} (E is about 6 kg./mm.^{2} as derived from Winkelmann-Schott additivity rule). The tensile strength of 6.6 kg./cm.^{2} is critical as the limit of glaze mechanical stability, varying somewhat with its composition and thermal treatment factors. Aftereffects of fatigue and *aging* phenomena also influence the stability.

200. F. Sandford[328] gave valuable experimental material on sorption of water in glazes of different compositions, and especially on aging phenomena which are characteristically influenced by the gas atmosphere of the kiln, as if water vapor, CO_2, and air simultaneously act. More recently, P. H. Dal, W. A. Zuleger, and W. H. J. Berden[329] investigated the moisture expansion of ceramics in a detailed monograph, especially for the swelling-decreasing effects of lime additions, and the swelling-increasing observed in feldspar containing glaze compositions, systematically varied in experimental batches. The principal expansion effect is ascribed to the reaction of the moisture (or the steam in the autoclave test) with amorphous silica and γ-alumina, or with the glassy phases and metastable cristobalite. Significant also are differences shown between feldspar powder as such, and products of hydrolysis by percolating water.

201. Equally intensive studies of the moisture swelling of glasses and ceramic products (those used as roof tiles) were performed by Australian investigators, including J. S. Hosking and H. V. Hueber,[330] W. F. Cole with Th. Demediuk,[331]

[327] *Ber. deut. keram. Ges.*, **30**, 1953, 43-47.

[328] *Trans. Intern. Ceram. Congr.*, **3**, Paris, 1952, 8 pp.; *J. Am. Ceram. Soc.*, **34**, 1951, 179-183.

[329] *Klei*, **8**, 1956, 331-476; *Ceram. Abstr.*, 1957, 13-14.

[330] *Nature*, **182**, 1958, 1142-1144; **184**, 1959, 1373-1375; *Australia, Commonwealth Sci. Ind. Research Organization Div. Bldg. Research Tech. Paper*. No. 6, 1959, Melbourne, 56 pp. (with E. H. Waters, and R. E. Lewis).

and A. A. Milne[332] resolving tne corresponding effects on fired meta-kaolin when autoclaved, and with different additions (cf. Vol. V, Section B). Without discussing here the technologically important details of the nature of different clays used in Australia as raw materials, and details of firing curing processes, it is essential to conclude from abundant experimental material how the rate of expansion of lightburnt products is much faster in early stages of the reaction than that of harderburnt products. The lighter-burnt clay products reach their total expansion in a shorter time period. The expansion curves observed follow an equation of the form $E_t = E_T\,(1 - e^{-kt})$ in which E_t is the per cent expansion at time of t hours, E_T the per cent total expansion, k a material constant. The relation given here is similar to a first order chemical reaction in which E_T is a function of the concentration of the components which give rise to expansion and hydration. The deviations from this theory shown in light-burnt tiles indicate that not yet completed solid state reactions develop meta-clay products of large surface area and an open structure whereas well-burnt tiles are more glassy, and, therefore, show a good agreement with the theory (cf. Fig. B. 112).

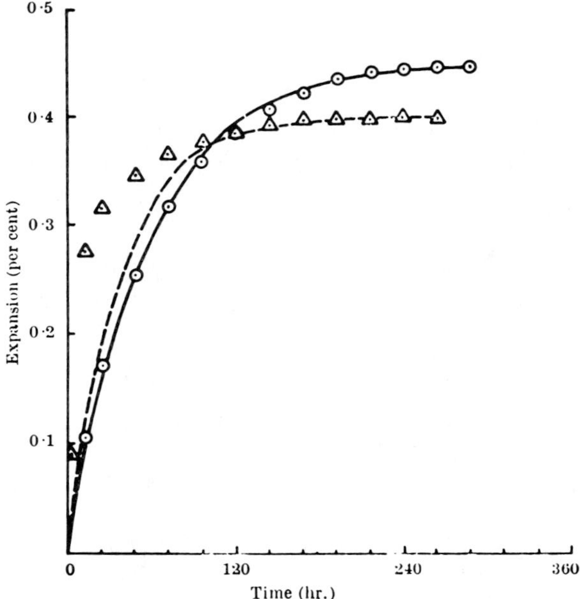

FIG. B. 112. Experimental and derived moisture expansion for a light-burnt, and a hard-burnt brick from one kiln charge. (Hosking and Hueber). Experimental curve, derived curve for hard-burnt material (full), and derived curve for light-burnt material (dotted).

[331] *Nature*, **182**, 1958, 1223-1224; *J. Am. Ceram. Soc.*, **43**, 1960, 359-367, especially on rehydration of γ-alumina, amorphous silica, and mixed alumina-silica gels.

[332] *Trans. Brit. Ceram. Soc.*, **57**, 1958, 148-160.

202. The experiments of Hosking, V. Hueber *et al.* make evident that chemical processes involving hydration of such constituents as amorphous aluminosilicates, amorphous silica, and glasses are responsible for permanent volume increases. The reversible physical swelling effects as observed in under-burnt materials play a relatively minor part in the whole phenomenon. It may be sufficient also to recall the observations of Cl. S. Ross and R. L. Smith (Vol. IV, Section A)[333] on analogous water-resorption in volcanic glasses (perlites, obsidians, and the like). The reactive aluminosilicates responsible for permanent expansion phenomena were compared by Cole and Demediuk with zeolites and permutites (in the general understanding of base-exchanging glasses); but there was *no* evidence of such an identity found. The amorphous sodium aluminosilicates are the most active phases, which develop in moisture strong permanent volume changes, and not so much the glass phases formed in the firing range from 950° to 1200°C. with much disordered cristobalite. The amorphous aluminosilicates are reduced in amount by higher firing temperatures and, therefore, an intensified vitrification improves the moisture stability. The fact that expansions can take place at high firing temperatures suggests that chemically bound water is probably more important in moisture expansion than physically adsorbed water.

DEVITRIFICATION OF GLASS[334]

203. For practical determination of the devitrification and/or the refractory-dissolution "*stones*" in industrial glass, we dispose of several excellent instructive monographs which equally treat the origin, and the petrographic-microscopic identification of the crystal phases of stones. In the first place, we mention here the monograph of H. E. Taylor and D. K. Hill,[335] then the valuable laboratory and plant instructions given by H. R. Swift[336] and by V. D. Fréchette and W. E. Hauth.[337] I. I. Kitaĭgorodskiĭ[338] started in a more theoretical discussion of devitrification phe-

[333] *Am. Mineralogist*, **40**, 1955, 1071-1089.

[334] For this industrially very important phenomenon, including the formation of stones originating by a reaction of glass with refractories, see the excellent Section V in H. Jebsen-Marwedel's book "Glastechnische Fabrikationsfehler," second edition, Springer-Verlag, Berlin, Göttingen, Heidelberg 1959, 524 pp., especially pp. 182-192 (by L. Riedel and H. Petermöller).

[335] "Identification of Stones in Glass by Physical Methods," Monograph of the Glass Delegacy of the University of Sheffield, Sheffield, England, 1952, 70 pp.

[336] *Glass Ind.*, **34**, 1953, 191-194, 220.

[337] *J. Can. Ceram. Soc.*, **15**, 1946, 76-86, with instructive micrographs, and extensive literature. See more extensively in the book of H. Insley and V. D. Fréchette, "Microscopy of Ceramics and Cements," Academic Press, Inc., New York, 1955, 286 pp., especially pp. 161-175.

[338] *Voprosy Petrog. i Mineral., Akad. Nauk S.S.S.R.*, **2**, 1953, 321-329. See also I. I. Kitaĭgorodskiĭ, "Technologie des Glases," German translation from the Russian, by J. Winkler, VEB

nomena from the crystallite hypothesis of glass constitution. Crystallization in large glass tanks is particularly favored by local heterogeneities of the composition in the glass bath, especially if temperature gradients occur, as near the bridges, on which the devitrification can take place over relatively long time periods. D. S. Belyankin and V. V. Lapin[339] developed accurate methods for identification of glass stones, as in special glasses for radio vacuum tubes, with BaO, and other special oxide compositions. Interesting is a description of uniaxial cordierite in such glasses which was previously erroneously identified as nepheline, further discussion of "β-alumina," corundum, mullite, spinel, and even of apatite as a residue from mineral raw materials. In principle, *new* kinds of glass stones derived from zircon, or from zirconia containing refractories of the ZAC group, were described by Lapin in the characteristic "rice-grain" shaped inclusions of baddeleyite,[340] or dendritic growths of zirconium dioxide as devitrification products in the reaction layer and in dissolution cords from ZAC blocks, together with mullite, nepheline, and others (on decomposition of zircon, $ZrSiO_4$, into baddeleyite and glass cf. Vol. V, Section B. ¶ 166 ff.).

204. A very peculiar form of devitrification is that of red-colored spherulites in aventurine glass, as described by M. A. Bezborodov and V. V. Lapin,[341] built-up from crystal skeletons of $Na_2O.2CaO.3SiO_2$ in a green-colored copper silicate glass. The red color of the spherulites is caused by minute crystals of cuprite, Cu_2O, interspersed through this glass in which the spherulites are rather unusual three-phase crystallization products. Typical dissolution stones, derived from the refractories, often occur in lead crystal glass, with a core of sillimanite or disthene, as described by N. A. Shmeleva.[342] Striae and waves protrude from such stones into the glass. Soda-lime silicate glasses dissolve such aluminum silicate stones much easier than potash-lime and lead crystal glasses. In high-in-potash lead crystal glass reacting with fireclay pot walls typical stones develop with corundum, "β-alumina," mullite, cristobalite, carnegieite, and leucite. P. Emer[343] especially studied the formation of stones from the raw materials (minerals and rocks) of refractories and batches. He classified feldspars, disthene, corundum, cordierite and beryl as typical stone-

Verlag Technik, Berlin, 1959, 720 pp., especially pp. 185-193. See also M. V. Okhotin, R. S. Levina, and E. K. Vinogradova, *Steklo i Keram.*, **11**, 1954 (6) 8-11.

[339] *Doklady Akad. Nauk S.S.S.R.*, **59**, 1948, 1594-1601.

[340] *Doklady Akad. Nauk S.S.S.R.*, **84**, 1952, 567-569; *Voprosy Petrog. i Mineral., Akad. Nauk S.S.S.R.*, **2**, 1953, 367-373.

[341] *Doklady Akad. Nauk S.S.S.R.*, **92**, 1953, 389-391.

[342] *Steklo i Keram.*, **10**, 1953 (6) 30-31; *Doklady Akad. Nauk S.S.S.R.*, **88**, 1953, 895-898.

[343] *Glastech. Ber.*, **29**, 1956, 356-366; on diopside in window glass, cf. B. M. Belova and L. G. Gol'denberg, *Steklo i Keram.*, **11**, 1954 (7) 27. This crystallization may be more incidental than a normal devitrification phenomenon. Devitrification of green glass in glass-blowing machining processes was recently described by B. Löcsei, *Epitöanyag*, 1961 (1) 1-7. On devitrification phenomena in borosilicate glasses, see R. Katzschmann, *Silikattechnik*, **13**, 1962, 96-101.

forming ingredients. Coarse-granular quartz, occurring in eruptive rocks, impedes the glass melting and homogenization by the stubborn formation of stones, whereas in phonolites no such tendencies are recalled. Wollastonite phonolite is one of the best examples of the usefulness of effusive rocks as a raw material for manufacturing window glass, nearly without any stone or devitrification troubles, whereas diopside is one of the most typical devitrification stone-formers which may be derived from basic silicate rock melts (basalts, diorites).

205. Stones in optical glass melts were studied by W. Vogel.[344] It is sometimes very difficult to decide whether such formations are real devitrification products or possibly foreign material dropped from the roof of the pot furnace in which such glasses are commonly molten. Usually the latter type is enriched in silica from crusts deposited on the silica brick roof, where they are built up from fly dust of the batches. Other stones, most probably from devitrification, are high in PbO (37.7 to 51 per cent) and K_2O (5.7 to 6.9 per cent), but low in CaO and Na_2O. Tridymite in different stages of its inversion to cristobalite is abundant in the roof-dropped materials, with a characteristic network of cracks, different types of dendritic cristobalite, "fish-scale" or spheroidal "mosaics." The distinction of tridymite and cristobalite is sometimes very difficult but Taylor's and Hill's criteria (given in their monograph mentioned above) are generally helpful for microscopic identifications. The X-ray diagrams identify the stones of cristobalite in the glass, whereas in roof deposits more tridymite is found. Transition forms are typical fern-like secondary crystallizations of tridymite in the latter case. The inversion into cristobalite is mostly metastable, namely below the equilibrium temperature of 1470°C., evidently by mineralizing effects of some constituents of the glass melt.

206. The identification of tridymite and cristobalite appears in an entirely new light by the structural theories developed by O. W. Flörke[345] (cf. Vol. III, B. ¶ 11 ff.), by disordered states in the high-temperature modifications of both crystal phases and their relations to the $\alpha \rightleftharpoons \beta$ inversion temperatures observed. The thermal expansion anomalies of the silica modifications in general, as a function of the states of crystalline order and of the degree of "stuffing" by foreign cations in tridymite, indicate a highly complex behavior of such superstructures and one-dimensionally defective structures. There are, however, also characteristic differences between those dendritic cristobalites which slowly crystallized from the melts and aggregates of cristobalite formed by solid-state reactions. For these latter formations, the so-called "ball-cracks" in the microcrystalline aggregates are highly characteristic. The growth of tabular tridymite from cristobalite dendrites is described as a phe-

[344] *Silikat Tech.*, **6**, 1955, 469-472.

[345] *Ber. deut. keram. Ges.*, 32, 1955, 369-381; *Glastech. Ber.*, **32**, 1959, 1-9. On the devitrification of porous sintered glass bodies see H. Schönborn, *Silikattechnik*, **13**, 1962, 419-424.

nomenon of lattice defects acting as nucleation centers for a renewed growth of defective structures with more or less random alternations of three-rhythmic cristobalite and two-rhythmic tridymite layering of silica units, with $(111)_{\text{cristob.}}$ parallel to $(0001)_{\text{tridym.}}$ A combination of microscopic and X-ray diffraction criteria for the distinction of these transition forms is indispensable for such studies. N. Köppen and O. W. Flörke[346] developed a relatively simple routine method for thin sections applied to a normal powder-diffraction camera. This method is equally suitable for rapid identification of nepheline, "β-alumina," mullite, nosean and the like in stones of a size down to 0.1 mm. in diameter.

207. The indispensable use of X-ray methods for identification of rarer crystallization products, formed either by devitrification, by undissolved batch ingredients, or by incomplete resorption of refractories, is made evident by H. Rooksby.[347] For example, in a glass containing PbO and As_2O_3, crystals of apatite type appear similar to pyromorphite or mimetesite, or simply kalsilite, leucite, etc., or fluorides in opal glasses (cf. B. ¶ 217 ff.).

208. Very interesting studies by W. Vogel[348] concern crystallization in *phosphate* glasses, with excellent isometric crystals of $Al(PO_3)_3$ as a devitrification product in optical glasses of this group. These crystals are in many variable combinations of the regular forms $\{111\}$, $\{1\bar{1}1\}$ and $\{100\}$. At the same time they are sensitive indicators of changes in chemical composition and constitution of the glass from which they originated.

209. In the system Al_2O_3—SiO_2 on the silica-rich side, the crystallization of cristobalite was compared by S. D. Brown and S. S. Kistler[349] with that from purest silica glass (samples of Corning, or "Cab-O-Sil" glass, with less than 0.02 per cent Na_2O, and 0.01 per cent Al_2O_3, besides other minor contaminations). In synthetic glasses with alumina contents up to the molecular ratio $Al^{3+}/Si^{4+} \times 10^2 = 1.097$, the crystal growth of cristobalite is definitely more rapid than that from ultrapure silica glass at a given temperature, although their fluidities (reciprocal viscosities) are lower. The rate of crystal growth, $\dot{g}$, by devitrification follows a zero-order reaction equation

[346] *Glastech. Ber.*, **30**, 1957, 182-186.

[347] *Analyst*, **77**, 1952, 759-764.

[348] *Silikat Tech.*, **6**, 1955, 510-517.

[349] *J. Am. Ceram. Soc.*, **42**, 1959, 263-270. Purest Corning silica glass contains about five times more water than "Cab-O-Sil" glass, as seen from the infrared absorption peaks at 2.8 μ. Sh. D. Brown, *Ibid.*, **43**, 1960, 116-117, found recently very good agreement between the experimental data of H. R. Swift on devitrification rates of crystal growth in soda-lime silicate glasses with an empirical equation for

$$\dot{g} = A.T.\, e^{-B/T\{1-e^{-\alpha(T_L/T)[T_L/T]^{\beta}}-1\}}$$

in the relation of the given temperature T to the Liquidus Temperature, T_L.

of the form $\ln(\dot{g}/T) = \ln A - B/T$, for the experimental temperature range between 1561° and 1730°K. Practically one may determine crystallization rates from increase in the width, w, of the cristobalite zone extending from the periphery of the alumina-silica glass samples into the glass, as a function of time, in the approximation relation $\dot{g} = (dw/dt) \cdot 1/\sqrt{3}$. Both, the viscosity of the glasses (determined by a fiber traction method) and of the rate of growth have a distinct maximum for the ratio $Al^{3+}/Si^{4+} \times 10^2 = 0.225$ (cf. Figs. B. 113 and 114).

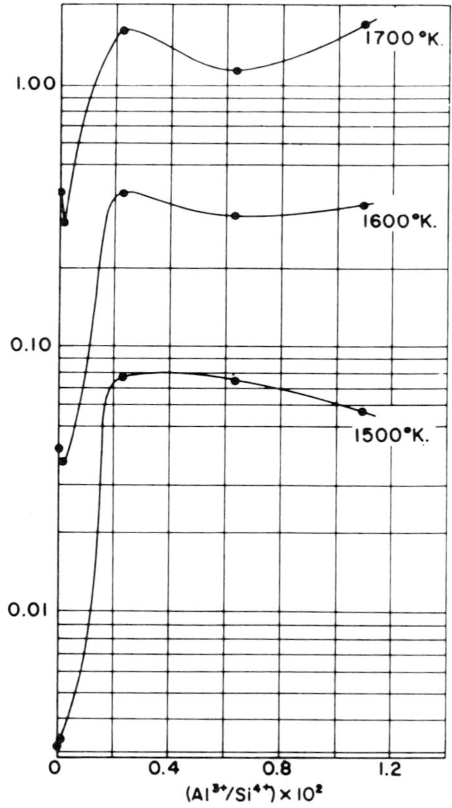

FIG. B. 113. Isotherms of the rate of crystal growth of highly siliceous glasses of system SiO_2—Al_2O_3. (Brown and Kistler).

210. Additional information on the influence of atomic ratios of cations in glass composition on devitrification characteristics was given by I. I. Kitaĭgorodskiĭ, T. N. Keshishyan, and B. G. Varshal[350] for window glass, as a function of the ratios

[350] *Steklo i Keram.*, 12, 1955 (2) 4-5. See studies of G. G. Sentyurin, *Trudy Moskov. Khim.*

Al^{3+}/Ca^{2+} and Al^{3+}/Mg^{2+}. If 1 per cent Al_2O_3 is present in such glasses, the ratio $Al^{3+}/Mg^{2+} = 0.330$ to 0.500 corresponds to a minimum of the rate of crystal growth. At the same time, the temperature interval of crystallization is very narrow. K. P. Azarov, G. V. Berdova, and S. B. Grechanova[351] developed a simple routine device for direct observation of devitrification (also of bubbling processes in glass melts,

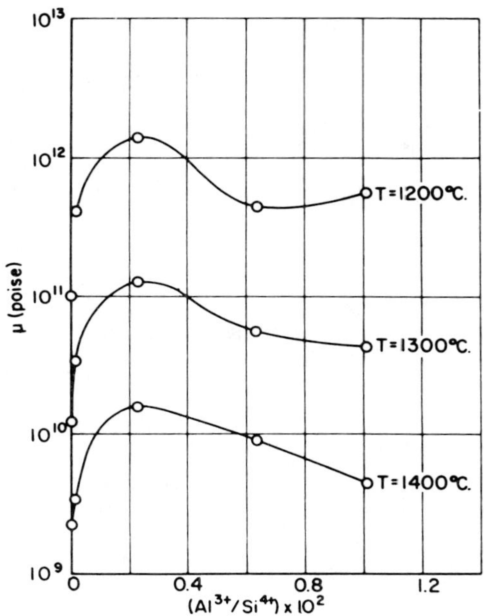

FIG. B. 114. Isotherms of viscosity of siliceous glasses of system SiO_2—Al_2O_3, as a function of the ratio Al^{3+}/Si^{4+}. (Brown and Kistler).

reactions with refractory material, and other results). A thin film of the glass is melted on a slit or on holes in a metal sheet resistance-heated to the fusion temperature of the glass (Fig. B. 115) and then observed with a long-focus microscope objective. For determination of the temperatures a hole is drilled into the junction of a thermocouple flattened to a circular platelet; the whole is mounted similar to the classical meldometer. One may also observe the devitrification in a series of such holes in the heating device, similar to the method of O. H. Grauer and E. H. Hamilton (1950) to control crystallizations in a series of samples at one time. The temper-

Tekhnol. Inst. im. D.I. Mendeleeva, 1949, No. 16, Stroitel'n. Material. Sbornik No. 2, pp. 102-109; further those of T. Uno, H. Shimizu and H. Yoshikawa, *Asahi Garasu Kenkyu Hokoku*, **1**, 1950, 1-7; M. V. Okhotin and R. S. Levina, *Steklo i Keram.*, **11**, 1954 (11) 6-9.

351 *Silikat Tech.*, **10**, 1959, 187-189.

ature gradients are controlled by contact thermocouples. The usefulness of this very simple instrument was demonstrated for boric acid containing and boron-free enamel frits at 730° to 800°C. (cf. B. ¶ 232 f.).

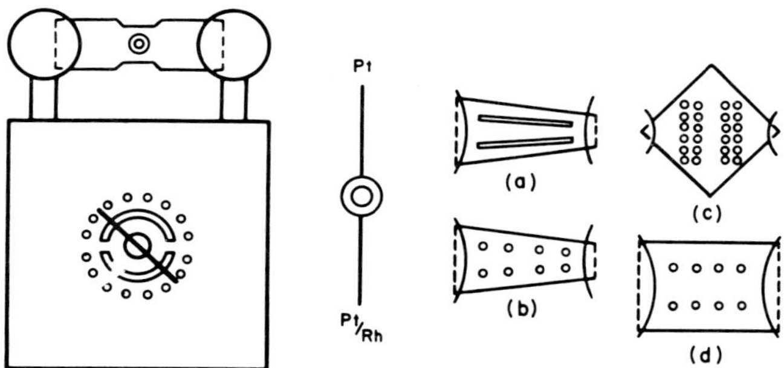

FIG. B. 115. Device for visual observation of devitrification of glass samples. (Azarov, Berdova, and Grechanova).

211. By interferometric-microscopic examination of glass samples with spherulitic crystallizations of devitrite J. Löffler[352] demonstrated how the devitrification centers were grown from surrounding silica-enriched mother glass (with silica contents higher by only 1.4 per cent). Particularly interesting are rhythmic devitrification clusters occurring in the drawing chambers of Fourcault glass machines in which the usual enrichment in silica or in other glass components is absent. In these rhythms devitrite and cristobalite were identified by X-ray diffraction. They may represent a eutectic mixture. The glass in the region of such rhythmic crystallizations is close to the boundary curve between silica modifications and devitrite in the phase diagram of the system Na_2O—CaO—SiO_2.

212. To find a quantitative measure for devitrification phenomena in toto, A. Dietzel and H. Wickert[353] defined a "*degree of vitrification*" in every glass-forming melt which is the reciprocal value of the maximum rate of crystallization. They examined the usefulness of this measure in melts of the system Na_2O—SiO_2 in the concentration range from 0 to 42 molecular per cent Na_2O in which the previously given curve for the degree of vitrification was entirely confirmed and interpreted as the result of combined effects of the glass framework destruction and of the concentration of structural units acting as nuclei of crystallization in the melt. The high-in-silica melts fused in a molybdenum resistor furnace at temperatures between 1450° and 1800°C. show high losses in alkali with rapid quenching (about 30 per cent of the

[352] *Glastech. Ber.*, **33**, 1960, 117-120.

[353] *Glastech. Ber.*, **29**, 1956, 1-4. Cf. previous theoretical discussions of A. Dietzel, *Ibid.*, **22**, 1948/49, pp. 218-219.

alkali introduced). The rate of crystallization was determined by exposure to controlled temperatures; highly siliceous glasses devitrified sometimes in a few seconds. Cristobalite, $Na_2O.2SiO_2$, and $Na_2O.SiO_2$ were the crystal phases for which the rates of crystallizations were measured. There are pronounced minima for $Na_2O.2SiO_2$ and $Na_2O.SiO_2$ in curves for the logarithms of vitrification as a function of the concentration in Na_2O. The eutectics correspond to sharp maxima in the same curves. The melt of $Na_2O.2SiO_2$ is, therefore, more "glassy" than that of $Na_2O.SiO_2$ and the disilicate has a much lower tendency to crystallize. Another wide minimum of vitrification appears in the range from 0.5 to 12 per cent Na_2O, and between 0 and 0.5 per cent the effect of the decrease in viscosity and of framework destruction are particularly evident. Cristobalite crystallized from such melts shows a distinct effect of the crystallization temperature on that of the $\alpha \rightleftharpoons \beta$ inversion. The higher the crystallization temperature and the higher the contents in silica in the melts, the higher is the temperature of this inversion, in agreement with Flörke's postulates.

213. I. F. Ponomarev's classical *temperature gradient method* for determination of devitrifications was rediscussed by this author[354] extensively, as based on G. Tammann's fundamental crystallization diagrams. Ponomarev demonstrates his improved methods in studies on crystallization of glasses, namely of 97 per cent B_2O_3, 3 per cent Na_2O, and similar highly "stable" glasses, such as borax with additions of B_2O_3, $3CaO.P_2O_5$, Al_2O_3, MgO, ZnO, and PbO, and for low-melting (at 400°C. only) fluorine-free types of enamels, based on borax + alumina, or mixes with magnesia. But also low-melting (at 740°C.) silicate glasses could be developed from temperature gradient studies, as when based on feldspar-magnesia-$Na_2O.2SiO_2$-$5Na_2O.Fe_2O_3.8SiO_2$ mixes. A modified and improved temperature gradient method for ceramic studies in firing clay bodies or glazes, such as for studying crazing, bloating, and other defects was recommended by R. L. Stone.[355]

214. Another modern form of the temperature gradient method was described by A. J. Milne,[356] as seen in Fig. B. 116. A vertical Kanthal heating coil furnace, constructed with the temperatures increasing in the axial direction toward the bottom, contains in its cover 12 impervious porcelain tubes of different lengths (20 to 5.25 cm., in steps of 1.25 cm.) each containing a Pt/Pt-Rh thermocouple for controlling

[354] *Voprosy Petrog. i Mineral., Akad. Nauk S.S.S.R.*, **2**, 1953, 306-320; see also T. N. Keshishyan, *Doklady Akad. Nauk S.S.S.R.*, **83**, 1952, 601-602, on the use of the temperature gradient method for industrial tests. I. I. Kitaĭgorodskiĭ, T. N. Keshishyan, and B. G. Varshal, *Steklo i Keram.*, **12**, 1955 (2) 4-5, used this method in their studies on the Al^{3+}/Mg^{2+} ratio in window glass devitrification characteristics (see above).

[355] *J. Am. Ceram. Soc.*, **36**, 1953, 140-142; cf. previous methods of W. Steger, *Ber. deut. keram. Ges.*, **17**, 1937, 177-182.

[356] *J. Soc. Glass Technol.*, **36**, 1952, 275-286, as a modification of an industrial laboratory method of E. Plumat.

the temperature of the samples. These are glass lumps, about 0.05 $cm.^3$ in volume, introduced into the porcelain tubes and supported by a platinum cone as seen in the Figure in exactly the determined position and temperature. After the thermal exposure period the samples are quenched and the crystallization is examined as fixed in its first state. Milne also used for comparison a simple microscope stage furnace for visual observation of crystallization processes. Milne's temperature gradient method proved to be particularly accurate and advantageous in the study of the devitrification of opal glasses.

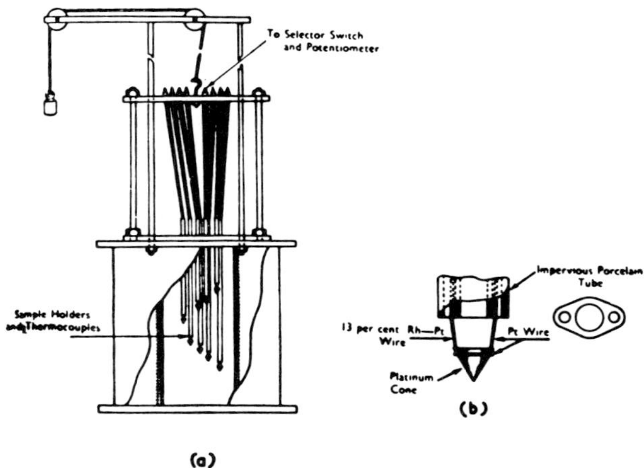

Fig. B. 116. Temperature gradient furnace for devitrification studies, with details of the sample exposure device. (Milne).

215. The crystallization of *feldspars*, and of *feldspathoids* (nepheline, carnegieite, kalsilite, leucite, and others) from glass melts, is described by N. Köppen,[357] particularly that of *analbite* in a reaction zone between siliceous fireclay refractories melting window glass, formed at the debiteuse of a Fourcault tank at 950° to 1000°C. over a period of only four weeks. "Stones," in a sulfate containing glass with reaction products of partially dissolved fireclay material, contain besides nepheline a typical *nosean*, as N. Köppen[358] further demonstrated. In its regular dendrites, nosean may easily be misinterpreted as cristobalite, but nosean shows the very typical forms of {110} (dodecahedron) and the microchemical test for SO_4^{2-} is unambiguous. A confusion with leucite is excluded by its higher refractive index. It may occur that no-

[357] *Glastech. Ber.*, 27, 1954, 169-170, more in detail in *Neues Jahrb. Mineral. Monatsh.*, 1954, 243-249, and *Veröffentl. Max-Planck-Inst. Silikatforsch.*, 14, 1954, 7-13. The crystallization of albite was previously mentioned by D. S. Belyankin, B. V. Ivanov, and V. V. Lapin, in their book "Petrography of Industrial Refractories," Acad. Sci. U.S.S.R., Moscow 1952, 564 pp.

[358] *Glastech. Ber.*, 29, 1956, 16-17.

sean in glass stones may be derived as a residue from phonolite used as a raw material, as in bottle glass batches. More recent observations of T. Fukui[359] confirm the crystallization of the high-temperature modifications of albite-rich plagioclases (above 70 mol. per cent Ab molecule) with negative optical character and low angles 2 V. Fukui ascribes the crystallization of analbite to the mineralizing action of an incidental contamination of the glass by copper oxide. Leucite and carnegieite were observed by N. A. Shmeleva[360] even in high-potash lead crystal glass.

216. Finally, we wish to emphasize the role as nucleation centers which the ultramicroscopic-dispersed silver or gold particles play in photosensitive glasses of the type developed by S D. Stookey,[361] for crystallization of microcrystalline lithium silicate precipitates in glass (easily soluble in HF). Such crystallizations are the essential point in the fascinating process of "chemically machined" three-dimensional photographic image reproductions in opaline glass, as discussed before in Vol. I, C. ¶ 90. Fine-dispersed nuclei of very thin crystals of hematite type, namely of crystalline solutions of Cr_2O_3 in Fe_2O_3, are the characteristic pigment material in special aventurine glasses developed by P. Beyersdorfer,[362] analogous to hematite-colored crystal glazes to be applied on porcelain. The optimum glass composition is in this case a lead borosilicate, with 6.6 per cent Fe_2O_3, and 2.4 per cent Cr_2O_3. The observation in reflected light shows intense Newton interference colors.

OPALESCENCE AND OPACITY OF GLASSES; GLAZES AND ENAMELS

217. In analogy to the classical investigations of H. H. Blau, A. Silverman, and V. Hicks (1934) on the constitution of fluoride-opacified soda-lime silicate glasses in which the X-ray diffraction method was extensively used for identification of crystal phases formed under the mineralizing influences of the fluorine anions, Fr. Jochmann[363] studied from technological viewpoints the so-called "*fluoride gall*" or

[359] *Asahi Garasu Kenkyu Hokoku*, **6**, 1956, 12-19, and especially *Ibid.*, **9**, 1959, 103-127, with detailed studies of nepheline, diopside, aegirine (containing the molecules of diopside and acmite), celsian and analbite.

[360] *Doklady Akad. Nauk S.S.S.R.*, **88**, 1953, 895-898.

[361] *Ind. Eng. Chem.*, **45**, 1953, 115-118. See also F. W. Schuler, *Chem. Eng. Progr.*, **52**, 1956, May, 210-212, specifically on nucleation and crystal growth in photosensitive glasses. A particularly interesting new type of ultra-high strength glass by ion exchange and surface crystallization is described by S. D. Stookey, J. S. Olcott, H. M. Garfinkel, and D. L. Rothermel, "Advances in Glass Technology," Plenum Press, New York, 1962, 397-411.

[362] *Glastech. Ber.*, **30**, 1957, 319-320.

[363] *Glas-Email-Keramo-Tech.*, **5**, 1954, 85-88, 399-403; **6**, 1955, 160-162; **7**, 1956, 157-160. A typical fluorine gall was also described by P. Beyersdorfer, *Ibid.*, **6**, 1955, 305-307, in a glass

immiscibility phenomena which are observed in some opal glasses. In the series of decreasing intensities of the opacifying effects, precipitation and recrystallization is observed for NaF, CaF_2, MgF_2, KF, and BaF_2. It is important to learn from these systematic experiments of Jochmann that ZnO, PbO, and Al_2O_3 do *not* form opacifying particles. The viscosity of the glass melts, however, is increased by these oxides, which, therefore, exert indirectly a stabilizing effect on the degree of recrystallization and opacification in the same basic glass. Scum floating on the glass and the typical gall contained fluorine enriched to 19 per cent whereas the original batch contained only 6 per cent of it. Such a gall is remarkable since it does not contain any sulfates, although the exterior appearance of the unmixed melt is very similar to that of true sulfate galls, forming "oily patches" on the blank glass melt surface. The tendency to form fluoride galls increases with the alumina contents of batches, regardless of whether the fluoride anion is introduced as CaF_2, AlF_3, Na_3AlF_6 (synthetic cryolite), or Na_2SiF_6. Evidently, in melting fluoride opal glasses, much fluorine is volatilized, chiefly as SiF_4, but also as BF_3, some alkali fluoride, or aluminofluoride. Even in clear glass formed at the highest melting temperatures, as in fluorine containing opal glass cullet, still 1 per cent F remains in the glass in every case. The fireclay refractories holding such glass melts undergo serious corrosion.

218. That free fluorides of the alkalies and alkaline earths are the principal crystal phases in fluoride-opacified glasses was also established by G. Rothwell[364] in a more recent series of experiments. As an important general result one must conclude that in the constitution of the basic glass of opal type systems those cations which do not participate in the framework are most inclined to segregate by crystallization as fluorides, that is first the alkali and alkaline earth cations. Interesting also in opal glasses is the occurrence of double-salts, among which the compound $NaF.MgF_2$ is particularly characteristic, which was previously identified by R. J. Callow.[365] Widely distributed in opal glasses of the common type is CaF_2 which also forms crystalline solutions with SrF_2, whereas BaF_2 is somewhat less easily segregated. It is striking that MgF_2 as such is almost never observed, most probably because of the ability of the Mg^{2+} cation to form $[MgO_4]$ tetrahedral coordinations in the framework. ZnO containing opal glasses show a particular tendency of the Zn^{2+} cation to enter into the glassy phase of opal glass, displacing other elements as fluorides. In this way the precipitation of $NaF.MgF_2$ is enhanced from MgO—ZnO fluoride opal glasses.

opacified with CaF_2. On the volatilization of fluorine compounds from opaque glass melts see F. Jochmann, *Sprechsaal*, **90**, 1957, 210-212.

[364] *J. Am. Ceram. Soc.*, **39**, 1956, 407-414.

[365] *J. Soc. Glass Technol.*, **36**, 1952, 261-269. See recently a very complete bibliography (273 references) on F opal glasses, by G. Paoletti, *Vetro e Silicati*, **4**, 1960, (20), 5-14; (21), 5-13; (22), 5-9.

219. In recent Russian literature there is a rich series of investigations on the opalescence phenomenon due to the *unmixing of borosilicate glass* compositions (cf. A. ¶ 207; Vol. I, C. ¶ 59). Especially S. P. Zhdanov, D. I. Levin, and E. A. Poraĭ-Koshits[366] contributed to this important field of glass constitution, which is also intimately connected with methods of preparing porous glasses of the Vycor types (cf. Vol. V, Section B) to which we refer here particularly. The observations of these authors concern especially three stages of changes in the following temperature ranges: (1) from room temperature up to 550°C., in which the intensities slightly decrease with increasing temperature; (2) from 550° to 700°C. with a strong opalescence; (3) above 700°C. in which the opalescence rather suddenly disappears. Although we may not discuss here the colloid-physical side of these phenomena, it is important to know that the Russian authors assume the partial crystallization of the finely dispersed borate glass phase in the heterogeneous glass mixture, especially the increased opalescence in the second stage, with the fusion of the crystalline borate explaining its sudden disappearance in the transition from the second to the third stage, and the depolarization degree as confirmed by measurements with the photoelectric spectrophotometer. The coefficient of depolarization of non-opalescent glass has a maximum at about 500°C., or in the first range, whereas opalescent glasses do not change this coefficient as a function of temperature. The intensity of diffracted light at constant temperatures is a characteristic function of time. With increasing intensity of diffraction, however, the coefficient of depolarization shows a distinct minimum for the intensity ratio log $(I/I_0) = 2.4$, when I_0 is that of ethyl ether as a reference basis.

220. Typical unmixing phenomena in *borate* melts as model systems of borosilicate *glazes* were discussed by L. Shartsis, H. F. Shermer, and A. B. Bestul[367] in connection with measurements of surface tension parameters. The minimum concentrations of the cations in phase-separated binary Ca, Sr, Ba, Mg, Cd, Zn-borates were determined in which a single liquid phase is formed at 1300°C. As additional "auxiliary" cations for the homogenization effect, those of the alkalies and alkaline earths, and also Pb^{2+}, Cd^{2+}, and Al^{3+} were studied. The minimum concentration required for the homogenization effect is increased with decreasing cation radii of either the "auxiliary" cations or of the principal cations. The upper layer of the separated melts is always nearly pure B_2O_3 melt, and therefore, the cations present do not perceptibly affect the surface tension in it. For layered, and for homogeneous ternary borate melts

[366] See especially important reports in *Stroenie Stekla, Inst. Khim. Silikatov Akad. Nauk S.S.S.R., Trudy Soveshchaniya, Leningrad,* 1953 (Pub. 1955), pp. 162-175, 321 (Zhdanov); *Bull. acad. sci. U.R.S.S.,* 1955, 27-33; 353-358; *Izvest. Akad. Nauk S.S.S.R.,* Otdel. Khim. Nauk, 1955, 31-39, 197-207, 395-402; 1956, 267-293 (with N. S. Andreev); further D. P. Dobychin, *Stroenie Stekla, Inst. Khim. Silikatov Akad. Nauk S.S.S.R., Trudy Soveshchaniya, Leningrad,* 1953 (Pub. 1955), 176-180, 318-319.

[367] *J. Am. Ceram. Soc.,* **41**, 1958, 507-516.

containing monovalent ions as auxiliary cations, the surface tension increases with increasing total concentration in alkalies, at rates decreasing with the ionic radii of these auxiliary cations, specifically as a function of their field potentials. Unmixed binary borate melts, therefore, contain the cations more freely in the B_2O_3-enriched layer than is the case in corresponding binary melts (cf. A. ¶ 245). If *silica* is added to borate melts with Ba^{2+}, Pb^{2+}, or Cd^{2+}, the ratio RO/B_2O_3 influences the miscibility limits in a manner illustrated by Fig. B. 117. Compositions on the left side of the

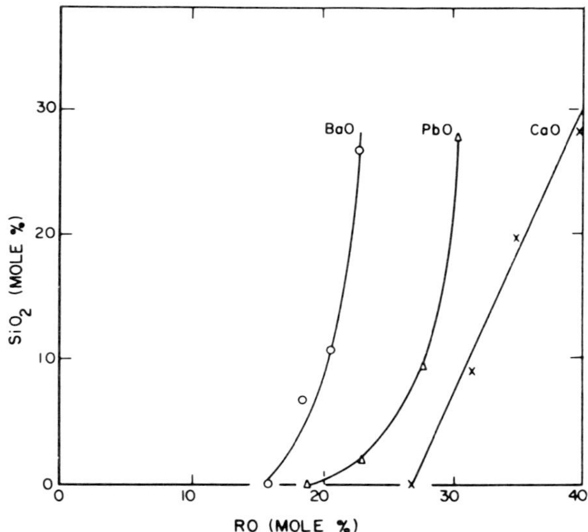

Fig. B. 117. Minimum mole percentages of silica required for homogeneity in $RO—B_2O_3—SiO_2$ systems (R = Ca, Ba, Pb), as a function of the ratio RO/B_2O_3. (Shartsis, Shermer, and Bestul).

curves here indicated form unmixed glasses in the system concerned, on the right side homogeneous melts. At RO/B_2O_3 ratios not too far above the miscibility limits of the binary borate melts, the addition of silica to the homogeneous binary melt brings about phase separation.

221. Silicate immiscibility in the system $Li_2O—TiO_2—SiO_2$ was recently studied by F. A. Hummel, T. Y. Tien, and K. H. Kim.[368] The white opaque glasses of the systems $Li_2O—TiO_2—SiO_2$ and $CaO—TiO_2—SiO_2$ were especially examined with the light and electron microscope which show very fine-dispersed droplets suspended in the glass matrix, with average sizes of 100 to 500 mμ in diameter. The crystallization even enhances the degree of opacity at 700° to 1000°C. If lithium is replaced by sodium, or calcium by barium, the type of opacity is not changed in principle but

[368] *J. Am. Ceram. Soc.*, **43**, 1960, 192-197.

the fusion of the glasses is improved. On the other hand, titanium cannot be replaced by zirconium without a complete change to transparency. The more complex titanosilicate melts include many compositions of industrial significance as glazes or enamel frits when they contain lithium, sodium, calcium, and barium combined.

222. Also *phosphate*-turbid glasses are typical unmixing dispersoids, as H. Schönborn[369] demonstrated. For a given concentration in turbidity agent the total amounts of unmixed material is solely a function of time. Bone ash (in additions of about 4.5 per cent) was used in a common sodium borosilicate glass melt above 1400°C. which is entirely clear after quenching and is easily controlled in its turbidity degree by an adequate thermal treatment. The particles of 1 to 3 μ in diameter which are visible in the light microscope with an immersion objective, show a strong coalescence not below 1200°C. (corresponding to a viscosity $\log \eta = 2.8$ to 3.2) chiefly with the crystallization of $3CaO.P_2O_5$. Slowly cooled industrial mat phosphate type glasses often contain particles up to 20 μ in diameter. The optimum nucleation temperature for different Jena "milk glasses" is 1050° to 1085°C. B_2O_3 and Al_2O_3 promote the degree of turbidity, alkalies and alkaline earths diminish it. CaO and MgO have a particularly strong effect to keep glasses transparent, whereas PbO is rather indifferent. Industrial opaque phosphate glasses are usually manufactured not with bone ash but with a mixture of Na_2HPO_4 and $CaCO_3$. The ratio P_2O_5/CaO is very important for the sharp point at which the curve of translucency drops to very low amounts, CaO containing glasses having the lowest limit of transparency. In lead containing phosphate glasses Schönborn interprets the crystalline phase to be either $3CaO.P_2O_5$ and $3PbO.P_2O_5$, a fluorine or hydroxyl apatite, $2CaO.P_2O_5$ with its characteristic inversion point at 1150°C. in mat glasses, or $CaO.P_2O_5$ in normal phosphate glasses.

223. Much study was given by H. Wondratschek and L. Merker[370] to precision X-ray diffraction methods for the crystalline phases formed in lead *phosphate*- or *arsenate*-opacified *enamels* and glazes (as used say for watch dials) (cf. Vol. III, B. ¶ 446 f.). Apatite-like structure types are predominant in these phases, but some of these were without fluorine or hydroxyl ions, which commonly are thought to be indispensable for the stability of apatite structures. The detailed investigations show first of all that the general assumption of $3PbO.P_2O_5$ or $3PbO.As_2O_5$, as opacifying pigments, is incorrect. The widely varying possibilities of isomorphous sub-

[369] *Silikat Tech.*, **10**, 1959, 390-400.

[370] *Naturwissenschaften*, **43**, 1956, 494-495; *Silikat Tech.*, **8**, 1957, 373-378; *Z. Krist.*, **109**, 1957, 110-114; *Mitt. Ver. deut. Emailfachleute*, **6**, 1958 (2) 13-15; *Veröffentl. Max-Planck-Inst. Silikatforsch.*, **18**, 1958, 159-160. On oxy-pyromorphites and oxy-mimetesites see also H. P. Rooksby, *Analyst*, **77**, 1952, 759-764. On $Pb_3(AsO_4)_2$, $Pb_3(PO_4)_2$, and $Pb_3(VO_4)_2$ see recently R. von Hodenberg, *Ber. deut. keram. Ges.*, **39**, 1962, 69-71, and L. Merker and H. Wondratschek, *Z. anorg. Chem.*, **306**, 1960, 39-43.

stitution in apatite-like structures concern especially the possibility of introducing not only [AsO_4] groups in place of [PO_4], but also of [SiO_4], [GeO_4], [VO_4], and the like in statistical distribution, whereas antimony is in principle different in its crystallochemical behavior. Besides apatite-like phases, also appear α-K_2SO_4(α-Na_2BeF_4)-type complex crystalline solutions of $NaPbAsO_4$, $(Na_{0.5}K_{0.5})PbAsO_4$, and $x NaPbAsO_4 \cdot Pb_2SiO_4$ (with $x = 4$ or 9) and corresponding phosphate complexes. No oxy- or hydroxyl pyromorphites or mimetesites could be identified, nor lead silicoantimonates. To the apatite group belong finally a group of synthetic alkali-lead phosphates of the crystallochemical type $Pb_4A(YO_4)_3$ (with A = Na, K, Rb, Cs, Tl; Y = P, As, V). The compounds $Pb_4Li(PO_4)_3$ and $Pb_4Ag(PO_4)_3$, however, show entirely different structural principles (and a tabular crystal habitus).

224. *Fluoride*-opacified *enamels* which are, of course, in nothing principally different from corresponding fluoride-opacified glasses and glazes mentioned above, were investigated by R. J. Callow and G. Lawson[371] in the process of fluoride precipitation and in time effects during fusion and ripening of the turbidity. The X-ray examination made evident that NaF and KF are the most common precipitation products in those enamels. Callow observed that there are definite equilibrium concentrations in enamel and opal glass melts beyond which the fluoride excess is volatilized. On the other hand, the transmission of white light shows, as a function of the fluorine contents, a rather sharp drop at a definite critical F concentration, such as between 2.5 and 4 per cent for a given sample, with a corresponding transition from a clear glass to the dense opal type. *Fluorosilicate* enamels are particularly important for coatings on aluminum metal, and E. C. S. Rao[372] described such compositions based on the ternary fluoride system $NaF—CaF_2—AlF_3$, modified by the substitution of some lithium for sodium and by introduction of silica, boron and antimony trioxide. The fluoride melts have an excellent adhesion to the metal surface over a reaction layer of the fluorides with the oxidation skin, at a firing temperature of only 565°C. One of the fundamental systems concerning the preparation of fluoride-containing enamel suspending slips was studied by B. W. King[373] in the system $NaF—Na_2O—B_2O_3—H_2O$. The typical "mill liquids" of the technological process of enamel manufacturing by such slips contain moderate amounts of NaF and sodium borates dissolved in well-buffered suspensions with *pH* of about 10, and with a low temperature coefficient of solubility of the salts concerned. The primary crystallization product from the slips is always simply NaF, and no complex

[371] *J. Soc. Glass Technol.*, **36**, 1952, 266-269, 270-274.

[372] *Finish*, **10**, 1953 (5) 31-33, 72-73; see also M. K. Blanchard and A. I. Andrews, *J. Am. Ceram. Soc.*, **27**, 1944, 17-24, 25-31, and on the eutectic system $NaF—Na_2B_4O_7$: A. G. Bergman and I. N. Nikonova, *Zhur. Obshcheĭ Khim.*, **12**, 1942, 449-459. On the system $NaF–AlF_3$ and its solid state reactions, see V. Gottardi, M. Tornati, and B. Locardi, *Vetro e Silicati*, **4**, 1960 (21) 14-21.

[373] *J. Am. Ceram. Soc.*, **37**, 1954, 238-242.

salts are formed by any means. The solubility diagrams of the systems $NaBO_2$—NaF—H_2O and $Na_2B_4O_7$—NaF—H_2O show a very simple character, with the hydrates $NaBO_2.2H_2O$ and $NaBO_2.4H_2O$, on one side, and $Na_2B_4O_7.5H_2O$ and $Na_2B_4O_7.10H_2O$, on the other side, + NaF. Also the fusion diagram NaF—$Na_2B_4O_7$ is very simply eutectic, with the equilibrium temperature of the eutectic at 680°C. and about 20 per cent Na_2F_2.

225. The electron microscope examination of fractured fluoride-opaque glasses and enamels shows irregular clusters of cubic NaF crystals in replica techniques, after shadowing with WoO_3 according to Fr. Kerkhof, R. Seeliger, and W. Westphal.[374] Glasses of this type are "ideally brittle" materials without any plastic flow under the deformation action of mechanical stresses (cf. A. ¶ 44). *Cryolite* as a raw material for enamels was extensively discussed by R. Märker[375] with emphasis given to its excellent suitability as a flux and to its favorable influence on the chemical durability of enamels.

226. Interesting attempts to introduce *vanadium pentoxide* into lead borosilicate glazes as an opacifier are described by A. K. Smalley, B. W. King, and W. H. Duckworth.[376] The concentration in V_2O_5 was varied between 3 and 8 per cent, and maturing temperatures of the easily fusing compositions were 605° to 650°C., as a cover of very satisfactory acid resistivity and adherence to glass or aluminum which can even be improved by the addition of titania, whereas additions of ZrO_2 improve the durability to alkaline solutions. Far more common is the use of *cerium* and *zirconium dioxides* as opacifying agents in enamel and glaze melts. Systematic investigations of the role of CeO_2 in white opaque glazes were made by E. L. Johnson and A. L. Friedberg.[377] They have a good acid resistivity, partially pleasant luster and high reflection, with excellent color stability. The X-ray diffraction examination showed cubic CeO_2 (with fluorite-type structure) as the opacifying crystal phase, often in single crystals up to 1 μ in size, as was seen also in the electron microscope. CeO_2-opacified enamels were thoroughly equivalent or even better than titania-opacified enamels.

227. C. W. F. Jacobs[378] investigated the opacifying crystalline phases in ZrO_2-type glazes by X-ray indentification methods, applied at high temperatures during the maturing (crystallization). Zircon, $ZrSiO_4$, was the principal opacifier in such glazes. This observation is highly remarkable in view of the common experience of glass

[374] *Glastech. Ber.*, **28**, 1955, 262-264.

[375] *Glas-Email-Keramo-Tech.*, **8**, 1957, 117-121, 178-182.

[376] *J. Am. Ceram. Soc.*, **40**, 1957, 253-255.

[377] *J. Am. Ceram. Soc.*, **42**, 1959, 60-69.

[378] *J. Am. Ceram. Soc.*, **37**, 1954, 216-220, 258-266 (with W. J. Baldwin). On the technology modes of application of natural zircon as raw material of zirconium-opacified porcelain glazes see Fr. Zapp, *Keram. Z.*, **9**, 1957, 584-592. The coloring power of manganese compounds added is enhanced by the use of zircon, also the hardness and the abrasion resistivity are increased.

technologists that zirconia (as the monoclinic modification baddeleyite) is highly chemically inert in glass melts and not easily recrystallized to form the silicate. Jacobs also emphasizes that for the crystallization of zircon from such glazes it is rather irrelevant in which form the ZrO_2 was introduced into the batch of the melts. A series of typical X-ray diffraction patterns of a glaze (with 18.5 per cent zircon

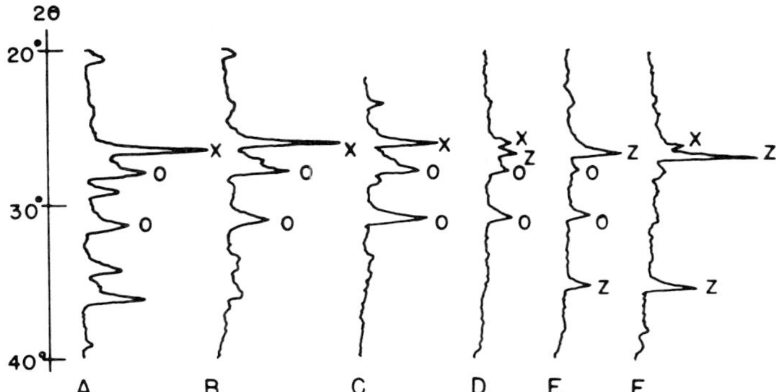

FIG. B. 118. X-ray diffraction patterns of a zirconia-opacified glaze, (*A*) at room temperature, (*B*) heated to 750°C., (*C*) heated to 850°C., (*D*) heated to 950°C., (*E*) heated to 1000°C., (*F*) heated to 1050°C. (Jacobs). The interferences of zircon are indicated by letter *Z*, those of zirconium dioxide by O.

in the original feldspar-borosilicate frit batch) is seen in Fig. B. 118, indicating the resorption of zirconium dioxide and its change into zircon with increasing temperatures. Only when the silica contents are low (35 to 45 per cent), the dioxide is, and remains, the predominant crystal phase as the opacifier agent. Zircon is also the stable product of reactions in normal siliceous glazes, if synthetic double-salts of $ZrSiO_4$ with barium, magnesium, calcium or zinc silicates are introduced, also when zirconium spinel is added to the mill batches as an opacifier ingredient. Jacobs and Baldwin examined further the influence of alkaline earths and zinc silicates on zircon if added to the mill. There is a definitely improving effect on the consistency of the slips and the physical properties of resulting glazes, superior to those prepared from zircon or ZrO_2 alone in the batches, but varying much with the special compositions and the maturing temperatures over the ranges from 1005° to 1050° (Orton cones 06 to 04), and 1250° to 1285°C. (cones 9 and 11), respectively, for two typical groups of glaze compositions.

228. Opaque zinc-zirconia glazes sometimes show crystalline spots which were studied by V. S. Fadeeva and M. E. Yakovleva[379] in the electron microscope with replica methods combined with chromium metal-shadowing. The micrographs show

[379] *Voprosy Petrog. i Mineral., Akad. Nauk S.S.S.R.*, **2**, 1953, 353-357.

triangular, often twinned, crystals of 10 mm. to 6 μ in diameter, which were identified as gahnite (zinc spinel), in addition prismatic zircon (up to 15 μ in length), identified by their optical constants, evidently recrystallized from the undercooled glaze melt (below 1300°C.). An andesine-type plagioclase is an interesting accessory formation in such glazes. Z. A. Nosova and M. E. Yakovleva[380] compared the zirconia-opaque glazes with the common tin dioxide-opacified porcelain glazes. Baddeleyite and zircon were identified either alone, or associated. Baddeleyite glazes apparently have a higher relative opacity and a better covering power than those with zircon because of the higher refractive indices of ZrO_2. In both cases the zirconium compounds are first dissolved and then crystallized from the undercooled melts. The opacity is not satisfactory if the dissolution of the zirconium compounds from the batch frits was incomplete. Zircon chiefly appears from batches with about 54 per cent silica and not more than 16 to 20 per cent alkalies. If silica is below 52 per cent, and the alkalies are about 13 to 24 per cent, the opacifier is ZrO_2 alone. Highly siliceous batch melts remain transparent on cooling. An addition of nepheline syenite to the batch is favorable because it reduces the firing temperatures and makes possible a one-step application of zirconium glazes on faience and half-porcelains.

229. *Tin dioxide*-opacified glazes also require a primary dissolution of the tin compounds and a recrystallization from the undercooled melts for an optimum opacity effect. Soda and barium oxide retard the crystallization of cassiterite, alumina accelerates it; lead oxide (up to 15 per cent) also does not impede the precipitation of the dioxide. If about 9 per cent CaO, and 10 per cent ZnO are added, anorthite or a plagioclase crystallize, in addition to cassiterite, but no gahnite is formed, which, however, appears abundantly if CaO is below 5 per cent. Gahnite increases the opacity effect, especially in alumina-enriched tin glazes. Reducing furnace atmospheres are highly detrimental for the tin glazes because SnO_2 is easily reduced to SnO which is immediately dissolved in the glass melt and does not crystallize on cooling from it. Both for zirconium and tin glazes, the viscosity range between about 400 or 500, and 2500 to 3000 poises is the most unfavorable interval causing bubbling and permanent blister grooves remaining on the surface of the glazes.

230. The chemical interaction of a glaze on a ceramic body at 1000° to 1250°C. was studied by M. E. Yakovleva[381] for a normal alkaline silicate composition, a lead silicate with barium and zinc oxide additions, and also an alkali borosilicate glaze, applied on faience bodies fired at 1000° to 1250°C. Added to the physical conditions of "fitting" by comparable coefficients of thermal expansion, the reactivity of the glaze melt with the ceramic basis was studied by petrographic methods, including

[380] *Trudy Nauch. Issledovatel. Inst. Stroitel. Keram.*, 1955, No. 10, 82-130; *Doklady Akad. Nauk S.S.S.R.*, **91**, 1953, 137-139. Crystallization of titanite, $CaO.TiO_2.SiO_2$, from titania containing glazes may be particularly noted.

[381] *Steklo i Keram.*, **15**, 1958 (6) 30-36.

the examination of the volatilization of Na_2O, B_2O_3, and PbO. The reaction layer may have a positive or a negative influence on the character of the bond between glaze and ceramic body.

231. As a problem intermediate between the use of common ceramic glazes on silicate bodies and enamels on metal sheets, we may here mention the ideas of A. I. Borisenko[382] on the production of coats for metals, but also for corundum (e.g., for cermets), not from traditional *slurries* but from *solutions* such as those of the nitrates of calcium, aluminum, or chromium, in a semicolloidal suspension with silica hydrogel. Aging effects and spontaneous gel precipitation limit the time period for a suitable handling of such suspensions during spraying on metals at temperatures of only 280° to 410°C. Borate containing suspensions have a longer aging period than boron-free compositions. As a rule, the application on metals should be at temperatures by 130° to 180°C. below those used in the slurry enameling process. Borisenko gives examples for the composition of the semicolloidal suspensions for steel coating, porcelain glazing, or corundum glazing. Particularly interesting are compositions to be applied on molybdenum metal, either high in lime and free of B_2O_3, or containing B_2O_3, Na_2O, and ZnO. The layers produced are gas-tight and, therefore, do not require any impregnation with a pore-closing agent.

232. The *viscosity* of fused *enamels* was extensively studied by A. Petzold,[383] expecially in their hyperbolic function of temperature, in the practically important range from 750° to 1000°C. The curves are entirely analogous to those of common glasses. As a function of composition, the viscosity varies in wide limits; if heat-exposed over longer time periods, viscosity data are slightly increased, evidently by losses in volatiles (especially of fluorine compounds), as discussed above for glazes. Petzold's studies include as well the common ground coat enamels (with borax, MnO_2, and Co_3O_4 in their batches) as cover enamels for steel sheet. The viscosity of ground enamels is reduced by FeO, Fe_2O_3, and Fe_3O_4 in the temperature range mentioned before. Whereas FeO causes a progressive decrease in viscosity up to 20 per cent, FeO-containing enamels with Fe_2O_3 show a maximum at 10 per cent Fe_2O_3 at 815°C. FeO and Fe_3O_4 reduce viscosity much more than Fe_2O_3. The fluidity of ground-coat enamels during application cannot be kept constant; it is considerably increased by progressive dissolution of FeO and Fe_2O_3 in the borosilicate melts. This effect is as important as is the wetting behavior and the adaptation of the thermal expansion coefficients of the enamel to that of the metal by the same dissolution reaction. These experimental facts are also in agreement with determinations of the iron oxides dissolved in normally applied ground coat enamels made by E. Zschimmer and K. Meures (1932). K. P. Azarov and V. V. Balandina[384] made systematic

[382] *Doklady Akad. Nauk S.S.S.R.*, **119**, 1958, 339-341.

[383] *Silikat Tech.*, 4, 1953, 391-396, 485-488.

[384] *Doklady Akad. Nauk S.S.S.R.*, **115**, 1957, 1146-1147.

investigations of the solubility of iron oxides in boron-free, in borosilicate, and in titania ground enamels, applying the refractive index method developed by A. I. Andrews and H. R. Swift.[385] The temperatures of heat exposure were 850°C. (for 8 hours), and 950°C. (for 4 hours), to reach equilibrium of dissolution. Boron-containing enamel fluxes have the highest solution capacity for Fe_2O_3 (21 to 23 per cent). Boron-free fluxes showed a solution power of only one third or one half that of borosilicate fluxes, titania enamel being intermediate. Boron-free enamels behave like colloidal suspensions with increased viscosity; the removal of gases on the contact enamel/metal sheet is impeded, therefore boron-free enamels show an undesirable "boiling." They also have a higher surface tension and do not wet the metal surface sufficiently (see below).

233. K. P. Azarov and S. B. Grechanova[386] continuing previous studies on boron-free and borosilicate enamels, investigated the effects of additions of Fe_2O_3 on viscosity. Below 2 per cent added, Fe_2O_3 increases the viscosity of boron-free enamels, whereas in boron containing enamels it is decreased in the softening range. Evidently, the higher basicity of the boron-free enamels favors the formation of $[FeO_4]$ coordinations which participate in the framework constitution of the melts and therefore consolidate the structure. The more acidic borosilicate enamels favor the formation of $[FeO_6]$ groups as framework-modifiers which reduce viscosity. At higher temperatures, however, both kinds of enamels show decreasing viscosity with increasing Fe_2O_3 contents. The boiling phenomenon in boron-free enamels as a consequence of their higher viscosity is confirmed. Characteristic "craters" are left by the outbursting gases which are not leveled by flow of enamel melt and cause brown spots ("copper nails") on the steel sheet surface.

234. Together with the flow characteristics of steel sheet enamels, the problems of their *adherence* are of primary importance for the practitioner. Before we discuss the recent evolution of theories developed to explain the enamel/metal adherence, we mention among the wealth of empirical proposals for best-fitting compositions and formulae the use of *lithium* oxide as an addition to frits for sheet iron ground enamels, acid-resisting antimony-cover coats, acid-resisting cast white, and others, as recommended by S. Hallsworth.[387] The addition of Li_2O reduces the fusion temperature of enamels with widely differing compositions and increases the fluidity with a desirable increase in "gloss." A disadvantage of such enamel fluxes, however, is the increased sensitivity for titania-discoloring to yellow tints.

[385] *J. Am. Ceram. Soc.*, **27**, 1944, 46-50.

[386] *Doklady Akad. Nauk S.S.S.R.*, **118**, 1958, 348-350. On the device used for visual observation of boiling phenomenon cf. K. P. Azarov, G. V. Berdova, and S. B. Grechanova, *Silikat Tech.*, **10**, 1959, 187-189 (cf. B. ¶ 210).

[387] *Inst. Vitreous Enamellers Ltd. Bull.*, **3**, 1951 (1) 18-22; *Sheet Metal Inds.*, **28**, 1951, 189-192.

235. The qualitative aspects of the *adhesion phenomenon* as they are described by electron microscopic replica structures are discussed by I. Buck,[388] directing special attention to the ground coat/iron interfaces and etching (cf. Vol. I, C. ¶ 124 ff., on replica technique). A satisfactory interfacial layer interlaces into both enamel and metal, but no dendritic formations of significant sizes were observed. This normal type of adherence layers is formed in the presence of cobalt, nickel, and manganese oxides in the composition of the ground coat. Underfired frit brings about a heterogeneous appearance of an "immature" layer in large lumps, darkened areas, and so on. The normal layer of glassy characteristics is rich in cobalt, its thickness is 15 to 20 μ. Acicular crystals of relatively high alkali resistance were identified as tridymite by X-ray analysis. Its crystals are particularly abundant in "bubble" structures. H. L. Conaway[389] made practical and very valuable prescriptions for use of common metal microscope techniques in studying the adherence phenomena of enamels on steel sheet.

236. Concerning the extensive discussions on the *theories* of enamel adherence to metal surfaces, we first refer to the fascinating Symposium report by L. S. O'Bannon,[390] chiefly treating of the so-called oxide layer hypothesis, the dendrite hypothesis, and the epitaxis of oxides on iron metal, the role of cobalt and nickel in the ground coat, and the galvanic corrosion theory chiefly based on A. Dietzel's measurements and opinions (1934). B. W. King, H. P. Tripp, and W. H. Duckworth[391] recently recapitulated the essentials of the different theories which are simply illustrated in the schematic sketches Figs. B. 119 and 120, showing Dietzel's presentation of galvanic corrosion in cobalt bearing ground coats on the metal (with local

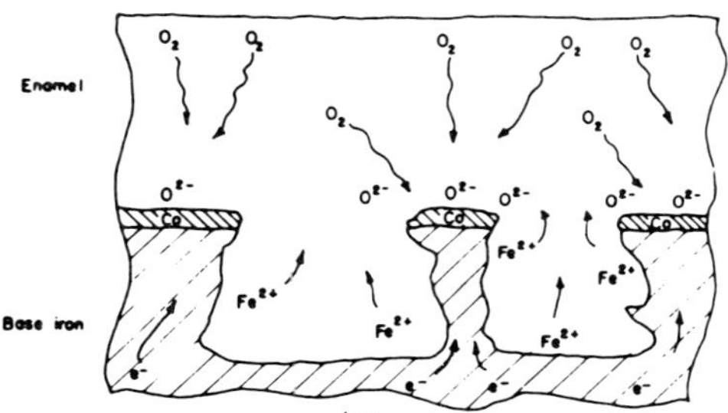

FIG. B. 119. Schematic representation of galvanic corrosion in cobalt-containing ground coat enamel, according to Dietzel's theory. (King, Tripp, and Duckworth).

[388] *Am. Ceram. Soc. Bull.*, 32, 1953, 114-117.

[389] *Proc. Porcelain Enamel Inst. Forum*, 20, 1958, Nos. 5, 6, 7, 54-69.

[390] "Proceedings of the Symposium on Enamel Adherence," held at the New York State College of Ceramics, Alfred University, Alfred, N.Y., on September 24, 1951, 36 pp. (edited by V. D. Fréchette).

[391] *J. Am. Ceram. Soc.*, 42, 1959, 509-525.

galvanic elements), an ideal "dovetail" metal/glass joint, and a more generalized dovetail indentation pattern as adopted by oxide-layer theories. King *et al.* emphasize that really good adherence must result from metal to metal bonds between the atoms in the metal base and proper metallic ions in the enamel. The requirements for this mechanism are first a saturation of the enamel at the interface with the metal oxide

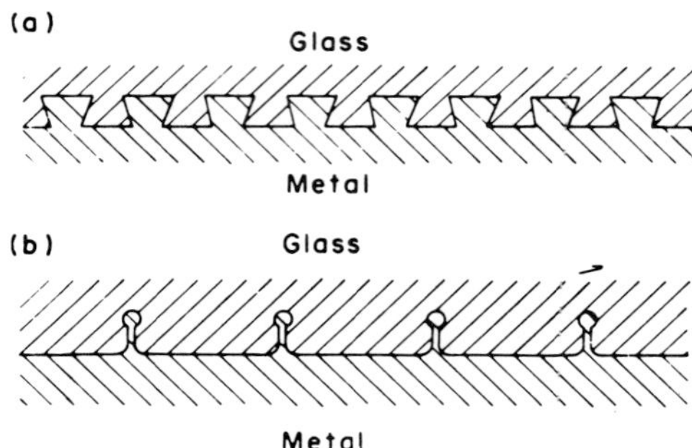

FIG. B. 120. Metal/enamel glass interface promoting mechanical bonding, (*A*) ideal "dovetail" joints, (*B*) generalized dovetail joints. (King, Tripp, and Duckworth).

and second that this oxide dissolved in the glass is not reduced by the metal. The characteristic oxide of iron for this mechanism is FeO. The precipitation of metal flakes or of other particles in the enamel, although they are related to the adherence phenomenon, are in King's opinion *not* the essential factor. The study of the complex reactions coupled with adherence postulates is made much more difficult by firing of enamel in air.

237. In the early stage of firing, a heavy scale is developed, which must be resorbed before the adherence can normally take place. The use of adherence promoting oxides (chiefly of cobalt and nickel) is required in order to maintain these conditions of concentration and oxygen potential in the surrounding of the enamel/metal interface during the period indispensable for the industrial enameling process. Surface *roughness* is doubtless one factor helpful for the development of good adherence (cf. Fig. B. 121) and was demonstrated as such by J. C. Richmond, D. G. Moore, H. B. Kirkpatrick, and W. N. Harrison[392] in impressive photomicrographs of a metallographic section of porcelain enamel ground coat containing 0.8 per cent Co_3O_4, on sandblasted enameling iron or on nickel-plated iron. The analogy between

[392] *J. Am. Ceram. Soc.*, **36**, 1953, 410-416. The definitions of the "adherence index" are those specified by the Porcelain Enamel Institute, Washington, D.C., 1951, as a "Test for Adherence," and "Tentative Standard Test," in terms of amount of metal exposed by a standard deformation treatment, as a percentage of the total deformed area. The "interface ratio" is that of interface length to the length of the reference line $A — A_1$ for microscopic measurement, as seen in Fig. B. 121.

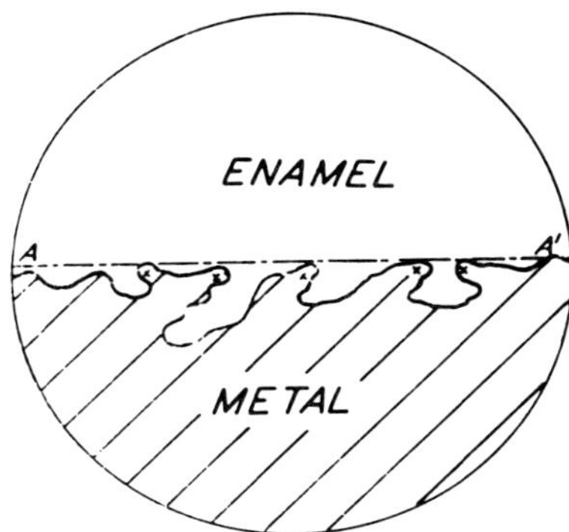

FIG. B. 121. Schematic section of metal/enamel interface, showing anchor points (undercuts marked by x). (Richmond, Moore, Kirkpatrick, and Harrison).

the microscopically determined anchor points (cf. Fig. B. 122), along the straight line $A — A_1$ with the adherence index and the interface ratio, is convincing in this respect. There is a definite maximum of adherence near about 1 per cent cobalt oxide. But also the latter authors conclude that roughness alone is *not* a sufficient condition for adherence, although it is a necessary factor for the efficiency of good enameling.

238. Returning to the conclusions of King *et al.*, we see those factors confirmed by previous authors. Raised roughness of the metal surface must improve the apparent adherence when the bond between enamel and metal was relatively weak.

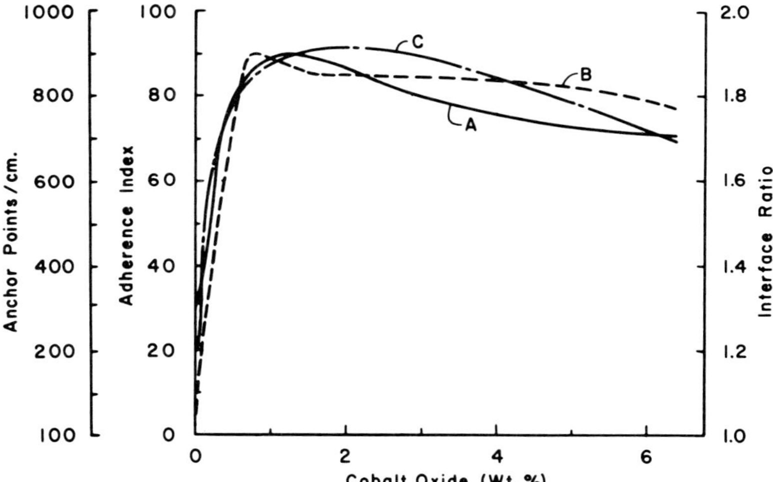

FIG. B. 122. Adherence index, anchor points per cm., and interface ratio, as a function of cobalt contents of ground coat enamels on iron sheet. (Richmond, Moore, Kirkpatrick, and Harrison).

The galvanic-electrolytic theory of Dietzel is also based on anchor point concepts. Against the galvanic theory of adherence, King *et al.* made valid the fact that some metals which exhibit a higher galvanic potential difference in contact with iron than does cobalt, say antimony, do not cause a more extensive attack by electrolytic corrosion. As a matter of fact most of the observations of W. N. Harrison *et al.* speak in favor of the galvanic theory, especially the tests with radioactive tracers for the role of cobalt in porcelain enamels.[393] Any type of chemical bonding as postulated by King must have a particularly high strength at the metal/enamel interfaces. King assumes, for the creation of these optimum conditions of intimate adherence, a state of dynamic equilibrium with a continuous exchange of Fe^{2+} cations at the interface. At lower temperatures, when the atoms and ions are largely fixed to their

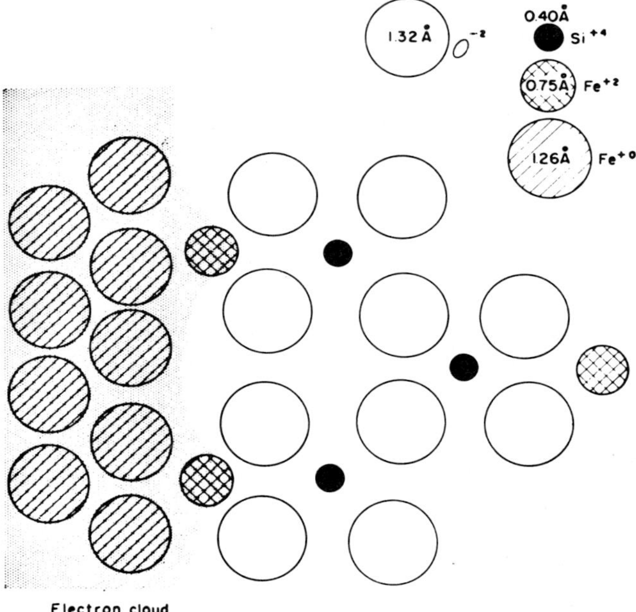

Fig. B. 123. Schematic representation of the iron/enamel glass interface. (King, Tripp, and Duckworth).

positions, there may exist only an exchange of electrons in the scheme Fig. B. 123, the electron cloud partly extending from the metal toward the Fe^{2+} cations. Only when these cations are forced to remain permanently in positions intermediate between the enamel glass structure and the iron metal, they are subjected to strong forces in both phases, causing the linkage-producing adherence.

[393] *J. Am. Ceram. Soc.*, **35**, 1952, 113-120 (with J. C. Richmond, J. W. Pitts, and S. G. Benner).

239. The galvanic corrosion theory for enamel/metal adhesion was defended anew by A. Dietzel,[394] and doubtless is strongly supported by the more recent investigations of D. G. Moore, J. W. Pitts, J. C. Richmond, and W. N. Harrison,[395] in a general examination of the bonding mechanisms between metals and ceramics. Also these authors base their concepts on a *mechanical* anchoring effect of the enamel glass into the pits and grooves formed by the galvanic corrosion in the steel sheet surface. Particular emphasis, however, is given to the galvanic corrosion of the metal base which takes place during the short firing periods in industrial enameling processing.

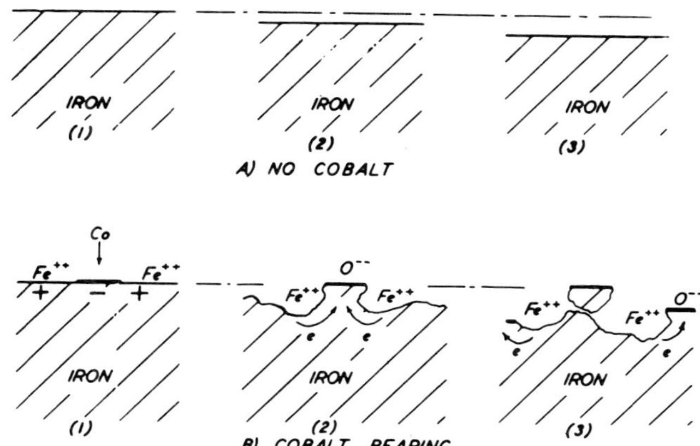

FIG. B. 124. Schematic drawing to show the differences in the corrosion attack by cobalt-free (*A*), and cobalt-containing (*B*) ground coat enamels, with increasing time periods of firing [from (1) to (3)]. (Moore, Pitts, Richmond, and Harrison).

Beyond that, Harrison *et al.* found inconsistencies which may indicate that the mechanism of galvanic attack followed by mechanical anchoring is also *not* the only important factor influencing bond strengths between enamel glass and metal. The influence of cobalt in the enamel is most evident in the principally different type of the corrosion attacks on iron by cobalt-free and cobalt containing ground coats (Fig. B. 124), as a consequence of the cobalt-precipitating reaction $Co^{2+} + Fe$

[394] *Ceram. Age*, **62**, 1953 (6) 17-21, 24, 38-39, as a translation from *Sprechsaal*, **78**, 1945 (5/8) 19-20, (9/12) 34-36, and with reference to the "Symposium on Enamel Adherence," Alfred University, Sept. 1951, mentioned above; a discussion of different theories of adherence is given. See also discussions on the chemical kinetics in porcelain enamel reactions, by R. M. Spriggs, *Am. Ceram. Soc. Bull.*, **35**, 1956, 280-285, and in enamel/metal systems, by R. M. Spriggs and A. L. Friedberg, *J. Am. Ceram. Soc.*, **43**, 1960, 252-262.

[395] *J. Am. Ceram. Soc.*, **37**, 1954, 1-6. See also similar results of R. F. Patrick, E. G. Forst, and G. H. Spencer-Strong, *Ibid.*, **36**, 1953, 305-313, especially concerning the precipitation of cobalt and nickel onto the surface of base metal combined with a roughness effect. Cobalt is precipitated more easily than nickel.

(metal) → Fe^{2+} + Co (metal). For the galvanic cell to function there is need of oxygen anions formed at the cathode as there are Fe^{2+} cations at the anode of the local cells. If the supply of oxygen at the cathode should be shut off for any reason, the flow of electrons could not continue, and the corrosion process would come to a stop. The depth of corrosion of the metal that occurs during firing was determined by inserting a gold alloy as a reference level for the original surface. This depth increasing with time is definitely accelerated by the cobalt oxide present (0.8 per cent). Nickel and arsenic oxides behave similarly, whereas copper oxide and antimony do not increase the corrosion depth. Copper oxide fails to increase appreciably the surface roughness except in the case of sandblasted specimens. Also Sb_2O_3 shows inconsistence with the elementary galvanic theory as mentioned shortly above in King's criticisms.

240. The often used practical measures of "*nickel dipping*," in other words a nickel metal precipitation on steel sheets by submersion in nickel sulfate solution at 65.5° to 75.5°C., was studied by D. G. Moore, J. W. Pitts, and W. N. Harrison[396] in its effect on the adherence with cobalt-free and cobalt containing enamels in common coating processes for iron and special titanium-low carbon steel sheets. Nickel plating decidedly improved adherence measured by the adherence index (see above) during only 2 minutes of firing at 845°C., but considerably decreasing with longer firing periods. The optimum nickel deposition is 0.056 to 0.133 mg./cm.^{2} surface, depending on the type of enamel and on the mode of cleaning the metal surface. The metallographic examination of the interface of nickel-coated specimens made evident that nickel-dipping caused a typical roughening during firing as an indication of distinct galvanic corrosion by local cells in the meaning of Dietzel's theory. Nickel remaining on the steel loses its effectiveness, galvanic-wise, as firing is continued; diffusion of the nickel into the iron sheet is a probable cause for this behavior. If the nickel were *electroplated* on the iron sheet as a continuous layer, no galvanic action could occur because no iron would be exposed to the molten enamel electrolyte, and poor adherence would result, as the experiments confirmed. If the sheet to be enameled is a titanium steel, nickel-dipping causes a definitely larger adherence index than for common enameling iron.

241. The positive effects of *copper oxide* in promoting the adherence of vitreous coatings to a special *stainless steel* was the subject of studies by D. G. Moore and A. G. Eubanks,[397] as well on acid-"pickled" (etched with HCl solution containing some

[396] *J. Am. Ceram. Soc.*, **37**, 1954, 363-369. On successful application of Dietzel's galvanic theory of enamel adherence see also the article of A. Petzold, *Silikat Tech.*, **6**, 1955, 398-402, with particular emphasis given to conclusions of investigators of the National Bureau of Standards (Harrison *et al.*).

[397] *J. Am. Ceram. Soc.*, **39**, 1956, 357-361. See more recently on pickling rates and adherence problems, H. N. Rahn, E. H. Mayer, and J. W. Frame, *J. Am. Ceram. Soc.*, **45**, 1962, 581-585.

$FeCl_3$), as on sandblasted steel surfaces. The effect of copper oxide, however, is decreased with increased firing temperature and time, if the specimens are pickled. By X-ray diffraction methods it was made evident that CuO is reduced to copper metal near the surface during the firing, but no selective corrosion of the stainless steel surface could be observed in the metal microscope, in contrast with the distinct roughness by a cobalt containing coating on iron. If the CuO content of the enamel coating is plotted as a function of the adherence index, the curve obtained is equivalent in principle to that for cobalt enamels on ingot iron.

242. A thorough confirmation of Dietzel's galvanic adherence theory must also be noted in the experimental studies of J. Berk and J. de Jong[398] who used, for their examination of the interfaces of porcelain enamel/metal sheet, light reflection and electron microscopy, combined with light-spectrographic and X-ray-spectrometric methods. Also Berk and de Jong confirm the roughness increase of the metal surface by electrolytic corrosion in contact with cobalt or nickel oxide containing enamel melts during firing. They distinguish two different electrochemical reactions in the interfaces, namely potentials developed between the oxide (say FeO) and the glass phase (as enamel), and between glass/metal. A certain conductivity is possible in the oxides at temperatures as low as 400°C. If the molten enamel contains cobalt, nickel, or copper oxides the metals are plated out first onto the oxide layer. If there are two kinds of ions in the solution, such as Co^{2+} and Fe^{2+} in ground coat enamels the concentrations of both ions are ruled by the equation

$$E_0(\mathrm{Co}^{2+}) - E_0(\mathrm{Fe}^{2+}) = RT/2F \cdot \ln\left(\frac{C_{(\mathrm{Fe}^{2+})}}{C_{(\mathrm{Co}^{2+})}}\right) = \text{about } 0.16 \text{ volts}$$

or a ratio of the ion concentrations Fe^{2+}/Co^{2+} of about 25 : 1.

243. By secondary reactions, Berk and de Jong observed the crystallization of fayalite and magnetite in the interface layer, due to a secondary oxidation of iron metal. By another electrochemical reaction, in a galvanic cell of iron/enamel/Co or Ni-Fe alloy, a cathodic precipitation of alloys onto the preexisting alloy particles can proceed, combined with a pitting on the anodic areas, or a roughening, exemplified along the boundaries of the ferrite crystals in the sheet structure, as made visible by etching. The enamel melt becomes gradually impoverished in Fe^{2+} ions after 16 minutes of firing treatment at 800°C. and the glass melt originally containing 0.5 per cent cobalt oxide becomes much more transmittent for light of 1 μ than another enamel without cobalt oxide. The reason of this phenomenon is seen in the cathodic deposition of iron metal and oxidation of Fe^{2+} to Fe^{3+} cations in the anodic area.

[398] *J. Am. Ceram. Soc.*, **41**, 1958, 287-293.

244. Also A. Petzold and M. Krüger[399] investigated the adherence of ground coat enamels on steel sheet as a function of the addition of adherence promoting agents, for comparison by the conventional impact test, determination of the torsion strength, and reflectance of the de-enamelized steel as a measure of the roughness effects (MPI test). The adherence strength is increased by increasing contents in Co_3O_4 up to 15 per cent as a limit value. The same general tendency is observed for Ni_2O_3 (up to 2.5 per cent) as an addition to ground coat enamel compositions. For As_2S_3 there is an optimum effect at 0.7 per cent as an adhesion agent. No difference was observed in the results whether Co_3O_4 was molten with the enamel, or added to the milling charge, or if it was partially milled with the batch, and partially molten in the enamel. For nickel-dipped steel sheets, in distinction from common steel, there is a maximum of adherence strength if 0.13 mg.nickel/cm.^{2} was deposited (cf. above the data given by Harrison *et al.*).

245. A. Petzold and H. Betzer[400] recommended using in the place of cobalt or nickel oxide common mineral antimonite (Sb_2S_3) as an adherence promoting agent for ground coat enamels, following previous recommendations of K. P. Azarov and N. S. Kharchenkova (1938). Enamel batches with 0.25 to 1.0 per cent Sb_2S_3 were fused at 850° to 900°C. and have an excellent adherence. The ground coat was dark-gray or deep-greenish black. The use of Na_3SbS_3 in place of Sb_2S_3 proved to be not suitable since the resulting enamel shows black spots in the surface of the cover coats.

246. The *oxidation* of iron sheets in different industrial enamel processes, such as H_2SO_4-"pickling" (etching), H_3PO_4-etching (phosphate coating), or nickel-"flashing" (0.040 mg. Ni/cm.^{2}) was studied by L. E. Fussell and R. L. Hadley[401] in the range from 705° to 845°C. It follows the parabolic law for growth in thickness of the reaction layers $w^2 = kt + C$, as a function of time, and temperature dependence follows an Arrhenius equation $k = A \cdot e^{-Q/RT}$. All three iron oxides, wüstite, hematite, and magnetite, were found to be intact after cooling, regardless of surface treatments. Nickel-flashing suppresses the oxidation of the iron because the nickel layer reduces the rate of diffusion of iron outward from the metal surface. Some nickel, on the other hand, diffuses into the iron sheet and some NiO remains in wüstite. Phosphate coating does not influence the rates of oxidation of iron sheet but distinctly roughens the surface.

[399] *Silikat Tech.*, **11**, 1960, 267-270; on the reflectance test according to L. Stuckert, cf. *Emailwaren-Ind.*, **21**, 1944, 43-46, and A. Dietzel and E. Wegner, *Ber. deut. keram. Ges.*, **26**, 1949, 132-137; *Mitt. Ver. deut. Emailfachleute*, **2**, 1954, 33-34 (with R. Bauer).

[400] *Silikat Tech.*, **4**, 1953, 297-298; cf. K. P. Azarov and N. S. Kharchenkova, *Zhur. Priklad. Khim.*, **12**, 1938, 1598-1600.

[401] *J. Am. Ceram. Soc.*, **41**, 1958, 81-88. On the oxidation of iron through an enamel coat layer, see V. Gottardi and G. Chioatto, *Vetro e Silicati*, **2**, 1957 (4), 10-14, and on oxidation of metal alloys, *ibid.*, (6), 16-20.

247. H. Crystal and G. Bullock[402] studied in a similar way the influence of iron deposition and nickel-flashing on the adherence of oxides to enameling steel sheet. In every case, nickel-flashing enhances the adherence of oxides. For a directly applied (titania-opacified) cover coat enamel a correlation was observed between adherence of this coat and oxide adherence to the metal. A *direct* application of cover coat on steel is enhanced by HNO_3-pickling with, or without, nickel-flashing. The common HCl pickling with an addition of $FeCl_3$ to the bath, influences the amount of hydrogen absorbed by the steel during the pickling, and the amount of nickel deposited on the steel during flashing. The experiments of Crystal and Bullock are interesting contributions to the possibilities of direct application of cover coat enamels on steel sheet when preoxidized, without the use of cobalt containing ground coat enamels. Attempts to introduce suitable amounts of iron oxide into the enamel and thus to promote adherence by coating the steel surface with a layer of deposited and active iron were not successful, whereas adhesion of the enamel on HNO_3-pickled steel was extremely good. The role of nickel flashing is evidently to achieve adherence of directly applied enamels to an industrially acceptable standard.

248. The enamelling of steels containing more than 0.5 per cent *chromium* presents many difficulties in adherence of the ground coat enamel, as A. Dietzel and F. Schneider[403] observed, whereas even highly alloyed stainless steels (such as "Nichrotherm" with 16 per cent Cr, 30 per cent Ni, and 3 per cent Ti) show an excellent ground coat enamelling, evidently by the presence of nickel which compensates for the unfavorable effects of chromium. In Cr containing cast iron and sheets Ni is usually absent and then the difficulties mentioned by Dietzel and Schneider become evident, as A. Petzold[404] demonstrated. This latter author described the successful application of normal cobalt containing ground coats on stainless steel. The titanium containing nichrotherm alloy behaves similarly to common titanium special enamelling steels of the type "Ti-Namel" for ground coats containing TiO_2 and ZrO_2.

249. Acid-pickling of enamelling steel surfaces brings about the tremendous problems involved in the evolution of *hydrogen* gas and its absorption and release in the metal combined with the renowned phenomena called "*reboiling*" and "*fish-scaling*." It is the unanimous opinion of all investigators that hydrogen evolution is the reason for the detrimental volume effects of the gas bubbles observed on the interface between enamel and metal sheet, which also starts the reboiling. P. K. Chu, J. H. Keeler, and H. M. Davis[405] extensively analyzed these gas bubbles by microchemical methods from individual "blisters." More than 84 per cent of the gas proved to be hydrogen. Where this hydrogen originated is an important problem.

[402] *J. Am. Ceram. Soc.*, **42**, 1959, 30-39, with literature on oxidation of iron-nickel alloys.
[403] *Ber. deut. keram. Ges.*, **28**, 1951, 81-82.
[404] *Silikat Tech.*, **4**, 1953, 346.
[405] *J. Am. Ceram. Soc.*, **36**, 1953, 48-59.

Chu, Keeler, and Davis discuss the possibility of the reaction between water vapor and iron metal during the enamel firing, as an ever-present source of the gas. Water is evidently supplied from the enamel slip, not only as structural water of clay ingredients with a relatively high temperature of dehydroxylation, but also as a part of the glass structure of the frit, as from highly hydrated salts like soda and borax (both as dekahydrates). In every case the steel/enamel interface absorbs at 500°C. much of hydrogen evolved by water-iron reaction. A 3 per cent Sb_2O_3 containing enamel absorbs hydrogen very readily at 500° to 700°C. Details of the water-iron reaction could be systematically studied. Neither CO nor CO_2 were found in gases evolved from the metal/enamel interface, only nitrogen and air-oxygen in low amounts.

250. The enamel itself acts as a kind of reservoir of hydroxyl ions which always are available to react with iron to regenerate hydrogen when reheated with steel. The iron-water reaction as such is not avoidable. Hydrogen may be evolved by simple heating of steel chips with "dried" enamel frit, and retarded blistering (by "splotches" and "shines") is observed throughout. It is, therefore, the special art of impeding detrimental degrees of this "hydrogen-illness" of the steel interface to avoid disastrous fish-scaling. Particularly important as a powerful source of troubles by hydrogen evolution and fish-scaling is the acid-pickling process in enamelling iron sheets. This fact was made most evident by experiments of E. E. Bryant, B. J. Sweo, G. E. Miller, and M. L. Simmons[406] who attempted to classify different ground coat enamels in their resistance to fish scaling by determination of time required to observe the first signs of fish-scaling. This resistance is definitely decreased even for enamels with a normally excellent adherence behavior when ground coats were overfired. Hydrogen is definitely able to diffuse through the iron sheet as manometric measurements made evident, but it is difficult to find direct correlations between the rate of hydrogen acceptance and diffusion or with its amounts of resistance to fish-scaling.

251. In every case, the nature of the steel sheet quality, besides pickling and firing conditions, is of primordial importance for ease of fish-scaling. It is only possible for one and the same steel quality to characterize empirically the tendency to fish-scaling for different enamels. Also D. G. Moore and M. A. Mason[407] examined gas

[406] *Am. Ceram. Soc. Bull.*, **32**, 1953, 248-252. On the action of moisture in the furnace atmosphere during the firing of ground coat enamels, see J. F. Benzel, J. F. Uher, F. G. Allenbaugh, and B. J. Sweo, *J. Am. Ceram. Soc.*, **44**, 1961, 1-6. On the importance of glass-lined equipment in chemical plants see, e.g., H. London, *Glass Technol.*, **2**, 1961, 248-253.

[407] *J. Am. Ceram. Soc.*, **36**, 1953, 241-249. On microanalysis of gases in bubbles of primary boiling cf. S. E. Freeman and V. W. Meloche, *Ibid.*, **18**, 1935, 123-125. See also the extensive studies of D. Horstmann, *Stahl u. Eisen*, **81**, 1961, 629-640, in which the defects in enameled steel sheets are interpreted by comparison with the phase equilibria in the $\alpha + \gamma$ field, and on the boundary curve between this field and the γ- field of the system Fe-C.

evolution from porcelain enamels during firing by microscopic observation of changes of bubbling during firing for normal and *anhydrous* enamel compositions, combined with a mass-spectrometric analysis of the gases trapped in the bubbles after variable firing time periods. Especially important is the question whether the *carbon* content of the steel may interfere with the gas composition. In this latter respect, the radioactive tracer C^{14} could be successfully used. The composition of the gases was principally hydrogen, CO, and CO_2; 77 mol. per cent of the gases were CO and CO_2 in the primary boiling stage, but after 6 minutes of firing both dropped to 26 mol. per cent. The first stages of blistering (primary boiling) were observed when the enamels were applied to a low-in carbon steel, chiefly as a consequence of an oxidation of the carbon in the metal. Hydrogen is evolved by the typical reaction of water with the hot metal. The diffusion into the metal is associated with some diffusion of hydrogen into the enamel as firing continues. Rapid cooling expels the hydrogen from the iron, and the typical accumulation of hydrogen bubbles at the interface enamel/metal occurs. Moore and Mason ascribe a considerable portion of bubbling to contaminations, as by organic substance in the clay mill additions. It is, therefore, most important to control the suspensions of clay in the slips, pretreated before firing. If the metal sheet was decarbonized, no primary boiling occurred, and also not if a layer of electrodeposited (electroplated) nickel was applied to the metal sheet.

252. The appearance of hydrogen blisters on cooling is explained by the characteristic lower solubility of this gas in iron (and also in nickel) at lower, rather than at

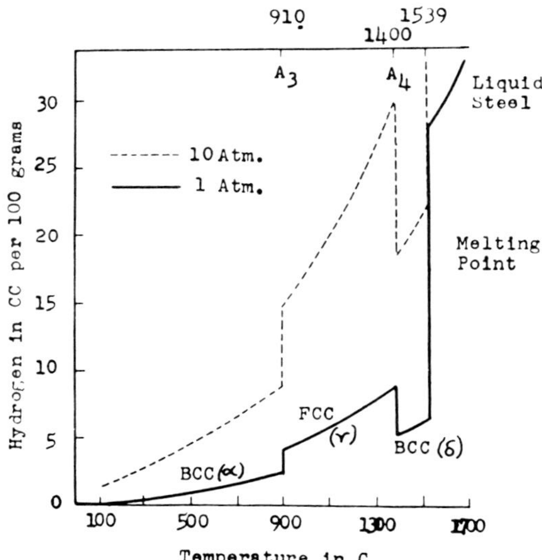

FIG. B. 125. Solubility of hydrogen in pure iron metal, as a function of temperature, and for two pressures.(Sieverts and Martin).

higher temperatures, first demonstrated by A. Sieverts,[408] and was most recently emphasized by G. P .K. Chu,[409] as seen in Fig. B. 125. The suitability of different iron sheets concerning their hydrogen diffusion behavior was studied by this author in a simple electrolytic cell shown in Fig. B. 126, with two iron sheet cathodes and

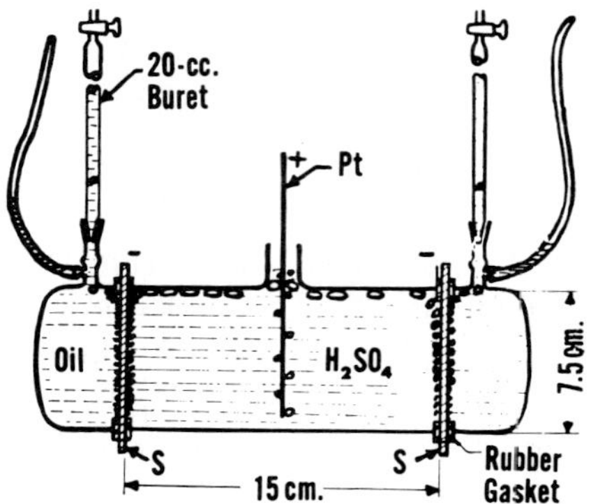

FIG. B. 126. Electrolysis cell for the determination of hydrogen diffusion through steel sheets, *S*, by cathodic charging. (Chu).

a central platinum anode. One observed in a few minutes how the hydrogen absorbed in statu nascendi diffuses through the metal to the remote surface of the sheets and can be collected in a gas burette. If the remote side of the steel cathodes is enamel-coated, the coat will chip after a relatively short time, thus demonstrating the entire fish-scaling phenomena. "Mild" steels are particularly susceptible to hydrogen diffusion, whereas nickel- or chromium-treated (say nickel sulfate-dipped) steels show diffusion considerably reduced. Fish-scaling can also be alleviated by alloying to the steel sheets elements like Ti, V, Ta, Nb, which form hydrides and keep hydrogen from being expelled during cooling.[410]

253. The recent observations of Chu sufficiently explain the fish-scaling phenomenon in its complex behavior, and also the results of previous experiments of J. H.

[408] *Z. physik. Chem. (Leipzig)*, **77**, 1911, 591-613; **183**, A, 1938, 19-37 (with G. Zapf and H. Moritz). See further E. Martin, *Arch. Eisenhüttenw.* **3**, 1929/30, 407-416.

[409] *Ind. Eng. Chem.*, **50**, 1958, 59A-60A, from general viewpoints of the importance of hydrogen behavior for chemical equipment construction, further *Inst. Vitreous Enamellers Ltd. Bull.*, **10**, 1960, 235-251; *Sheet Metal Inds.*, **35**, 1958, 585-619. See also J. D. Sullivan, D. H. Nelson, and F. W. Nelson, *J. Am. Ceram. Soc.*, **45**, 1963, 509-512, referring to glass-coated steel and its hydrogen defects.

[410] Moore and Mason, in their investigations mentioned above assume that in Ti containing steel the carbon bound in the carbide TiC is much less reactive than in common steels. TiC does not interfere with the mobility of hydrogen diffusing through the metal. On the behavior of hydrogen in steels higher in carbon contents, cf. P. Bardenheuer and G. Thanheiser, *Mitt. Kaiser-Wilhelm-Inst. Eisenforsch. Düsseldorf*, **10**, 1928, 323-342.

Keeler, G. P. K. Chu, and H. M. Davis,[411] further those of R. M. Hudson, J. K. Magor, and G. L. Stragand,[412] who spent much care in determining hydrogen evolved by heating at 150°–160°C. after a H_2SO_4 etching (at 38°C.) for enamelling steels, in a gas burette as shown in Fig. B. 127, and in the charging curve as given in

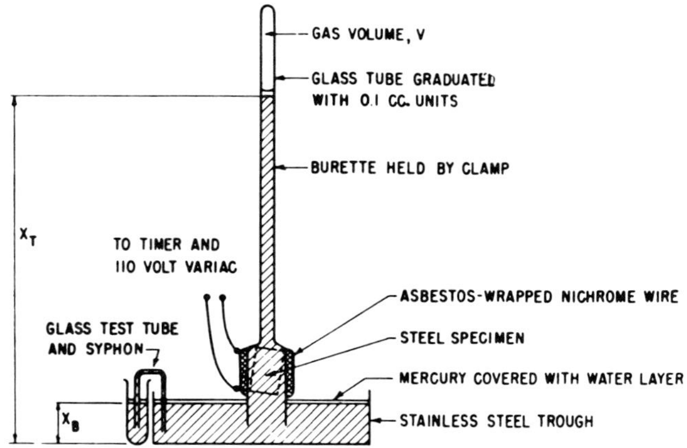

FIG. B. 127. Hydrogen extraction burette for examination of enamelling steels. (Hudson, Magor, and Stragand).

Fig. B. 128. For low-in-carbon steels of similar composition it was observed that the seriousness of reboiling depends on lower or higher hydrogen solubilities in the metal. Although a given level of hydrogen solubility is not necessarily related to the same degree of reboiling for different steel grades, the hydrogen acid-charging method is a useful routine way of evaluating different steels in their suitability for enamelling, such as rimmed and commercial steels. A. Dietzel and E. Wegner[413] recapitulated the experience concerning hydrogen evolution, fish-scaling, blistering, and other effects, concluding with the practical postulates that in every case an acid-pickling period too long should be avoided, that the ground coat frit should be smelted in an oxidizing atmosphere, and that as little clay as possible should be added. A portion of the clay may be replaced by bentonite with small additions of fluorspar and $NaNO_2$ to degas the enamel rapidly during the firing period. The firing range should be extended by an addition of quartz to the mill, in order to give hydrogen and water vapor time to escape; dry air should be introduced if closed muffles are used in the kiln.

[411] *Sheet Metal Inds.*, **31**, 1954, 233-240.

[412] *J. Am. Ceram. Soc.*, **41**, 1958, 23-27. See also R. M. Hudson, M. Kotyk, and G. L. Stragand, *ibid.*, **43**, 1960, 564-570, and R. M. Hudson and K. J. Riedy, *Ibid.*, **44**, 1961, 467-471.

[413] *Mitt. Ver. deut. Emailfachleute*, **5**, 1957, (2) 5-10; (3) 11-17; *Veröffentl. Max-Planck-Inst. Silikatforsch.*, **17**, 1957, 223-234.

254. The particular importance of viscous flow of enamels containing not only common fluxes like CaF_2 but also adherence-promoting additions like cobalt and nickel oxide, MnO_2, TiO_2, Fe_2O_3, and others, is evident from the studies of Tin Boo Yee, J. S. Machin, and A. I. Andrews[414] in the range of industrial firing temperatures. The viscosimeter best suitable for such purposes is an oscillation cylinder type as described by J. S. Machin and D. L. Hanna (1945). Calcium fluoride is a most efficient agent for modifying the viscosity of ground coat enamels, causing in small additions a drastic lowering of viscosity. Also small amounts of Fe_2O_3 dissolved in such enamels reduce the viscosity profitably in the firing range. The typical coloring oxides have a slight although significant effect on fluidity of the frits. Titania in amounts up to 11 weight per cent causes an additional major increase in fluidity. The limits of application of all these fluxes by the imminent rise of crystallization are evident in the example of CaF_2 for a typical frit, as seen in Fig. B. 129. Lithium fluoride is another powerful agent for controlling fluidity of porcelain enamels, as recommended by P. A. Huppert,[415] especially in respect to the favorable influence exerted on the thermal expansion properties of cover coats. Very recently J. E. Cox, D. M. Gerstner, and Helen M. O'Rourke[416] used the commercial Brookfield

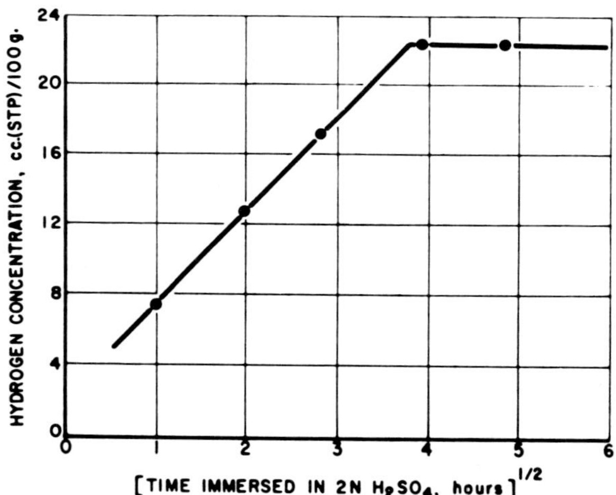

FIG. B. 128. Typical hydrogen charging rate curve for enamelling steel. (Hudson, Magor, and Stragand).

[414] *J. Am. Ceram. Soc.*, **38**, 1955, 378-382. For effects of titania see also N. S. C. Millar and C. H. Buck, *Metal Finishing J. (London)*, **2**, 1956, 151-156, 160. On changes in viscosity by additions of Fe_2O_3 see K. P. Azarov and S. B. Grechanova, *Doklady Akad. Nauk S.S.S.R.*, **118**, 1958, 348-350, for boron-free and borosilicate enamel compositions.

[415] *Am. Ceram. Soc. Bull.*, **38**, 1959, 57-60.

[416] *Am. Ceram. Soc. Bull.*, **39**, 1960. 78-81

"Synchro-Lectric" viscosimeter for routine measurements of viscosity of industrial ground coat frits, especially in the range of their optimum firing and flow temperatures, namely for $\eta = 3{,}000$ to 5,000 poises. Low-temperature frits may have a viscosity of 25,000 poises under their optimum operation conditions.

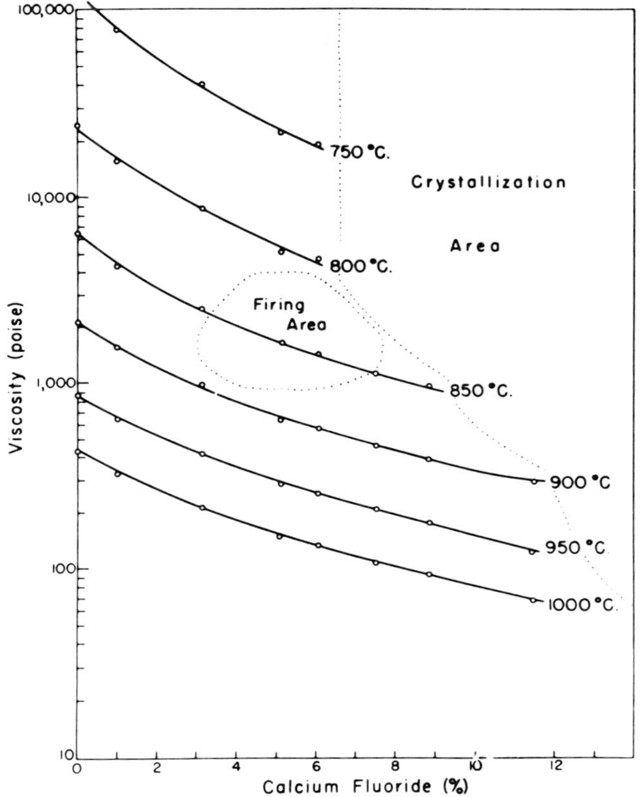

FIG. B. 129. Effect of calcium fluoride additions on the viscosity of an enamel frit, as a function of temperature. (T. B. Yee, Machin, and Andrews).

255. Of fundamental importance for adhesion problems are also the *wettability* characteristics of metals by silicate and enamel-like melts of which we discussed already in B. ¶ 232 ff., industrial questions concerning the possibility of developing gas-tight and durable metal/glass sealings. From a viewpoint very close to enamel-technological viewpoints, A. G. Pincus[417] described the metallographic examination of such seals, using a taper technique (cf. Fig. B. 130) for preparing very thin sample

[417] *J. Am. Ceram. Soc.*, **36**, 1953, 152-158.

sections, with all the finer transitions, as from nickel-iron metal wire to copper plating layers, an eutectic copper-silver solder, a manganese-molybdenum alloy, to the glass and ceramic body of the sealing, as a typical complex case of such multiple transitions which are necessary for metallized ceramics in modern vacuum tube constructions. Another example is a seal for forsterite/forsterite ceramics by manganese-molybdenum metallizing and a silver-copper solder, or a special titanium hydride technique for vacuum tube applications, of a molybdenum-alumina adhesion process, also described by A. G. Pincus.[418] We further refer to W. Düsing's instructive report on vacuum-tight metal/glass sealings[419] (cf. B. ¶ 151). The viscous flow in such seals was studied by H. E. Hagy and H. N. Ritland[420] for simple geometries (sandwich or wire samples), by measurement of the birefringence distribution during cooling at different rates, down to 125°C. below the strain point of the sealing glass.

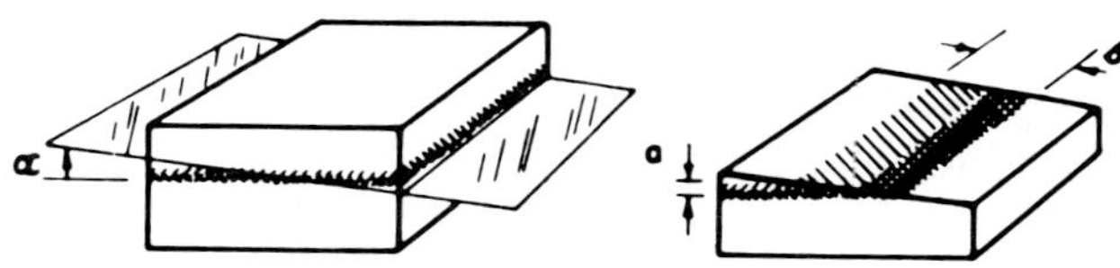

FIG. B. 130. Geometry of taper-section method for microscopic examination of metal/ceramics sealings. (Pincus). If $a = 5^\circ 43'$, the ratio b/a is $= 10 : 1$.

A typical titanium steel and a stabilized chromium-iron alloy were used as metal base, with suitable glass compositions (say a sodium borosilicate glass). Typical *solder glasses* with low-temperature flow-characteristics, especially lead borosilicates with more than 70 per cent PbO were investigated by J. Gallup and A. G. F. Dingwall[421] for the electronic tube industry. E. P. Denton and H. Rawson[422] pursued this same problem on the basis of zinc-borate-vanadate compositions, as solder glasses for the Fernico (iron-nickel-cobalt alloy) sealing glass. The corresponding use of V_2O_5 enamel frits for an adherence on nickel-chronium (stainless) steel was recommended by M. Tashiro, S. Sakka, and H. Teranishi.[423] Such frits are even superior to MoO_3 in their application.

[418] Report RL-976 Gen. Electric Research Laboratory, Schenectady, N.Y. December 1953, 24 pp.

[419] *Glastech. Ber.*, **26**, 1953, 232-238.

[420] *J. Am. Ceram. Soc.*, **40**, 1957, 58-62.

[421] *Am. Ceram. Soc. Bull.*, **36**, 1957, 47-51.

[422] *J. Soc. Glass Technol.*, **40**, 1956, 252-259 T. See also many special solder glasses described by O. Knapp, *Silikat Tech.*, **9**, 1958, 153-155.

[423] *Yogyo Kyokai Shi*, **61**, 1953, 537-540. A complete thermodynamic calculation of reactions occuring on the contacts of tantalum metal and sodium silicate glass, with formation of $Na_2Ta_2O_6$ as the interface material in a vacuum at 1000°C., was given by St. P. Mitoff, *J. Am. Ceram. Soc.*, **40**, 1957, 118-120.

256. The wettability of steel sheet by borosilicate ground coat enamels was extensively studied by K. P. Azarov,[424] especially in connection with the old problem of the substitution of an increased content of Na_2O for boric acid in the melts. The results of such a measure were always unsatisfactory because the *surface tension* of the melt on the metal surface was considerably raised and at the same time the chemical durability of the coat decreased. For specific effects of B_2O_3 low surface tension and the slope of viscosity as a function of the temperature gradient are characteristic. Boron-free enamels could be improved to a certain degree by introduction of lithium oxide and iron. A lead silicate glass with 88 per cent PbO had a favorable effect decreasing the softening temperature, even below that of boron containing enamels. BaO, MoO_3, CaO, and MgO, but particularly SrO, raise softening temperatures, whereas ZnO, Na_2O, K_2O, and NaF have relatively subordinate effects. Scaling by iron oxidation is stronger for boron-free enamels than for boron containing compositions.

257. Nickel-plating is the best measure to keep oxidation under control. With additions of iron oxide to the frit charge an increased boiling was always observed. The surface tension on the enamel/sheet contacts affects the rate of removal of gases from the interface and promotes a more or less smooth flow of the melt on the sheet. The surface tension can also be reduced by addition of sulfides or selenides of iron, copper or antimony (see below). Another attempt to improve boron-free enamels was made by A. Petzold and H. Betzer[425] who recommended the addition of BaO in place of B_2O_3. Such a special enamel is particularly useful if B_2O_3 must be excluded thoroughly, such as for high-temperature enamels in atomic reactor construction, because of the high absorption power of boron for thermal neutrons. Soda- and titania-enriched boron-free enamels are not satisfactory in this case because of their incomplete wetting characteristics. Barium may be introduced into the frits especially as the carbonate, nitrate, or sulfate, the latter mixed with carbon (in the ratio $BaSO_4 : C = 1 : 2$). Barium sulfide formed by reduction is an excellent wetting ingredient of such improved enamel coats, which may also contain 1 per cent of Sb_2S_3, added in the mill, and common cobalt or nickel oxide additions.

258. R. M. King and R. L. Cook[426] made instructive measurements of the *contact angles* of enamel drops on metal sheets, as a function of composition and time, in a closed tube furnace and in controlled atmospheres. The influence of dry and moist furnace atmospheres is most striking in the sessile drops of the enamel melts in their continuously changing shapes, as seen for one example in Fig. B. 131, and as a function of time in the curves of Fig. B. 132. The presence of oxides formed on the steel

[424] *Zhur. Priklad. Khim.*, **27**, 1954, 33-42; *Silikat Tech.*, **5**, 1954, 347-348.
[425] *Silikat Tech.*, **8**, 1957, 434-436.
[426] *Am. Ceram. Soc. Bull.*, **36**, 1957, 293-296.

sheets was qualitatively confirmed by the fact that complete wetting by the enamel melt occurred only if oxides acted as an interlayer, whereas on nonoxidized iron the melt drops were repelled. The presence of only 0.6 to 1.6 per cent moisture made a definite improvement in the wetting characteristics. The quantity of oxide formed in the interlayer on the metal could be directly determined by the cathodic cleaning method in a 10 per cent ammonium citrate solution, as recommended by H. H. Uhlig.[427] Very similar contact angle measurements were made by J. C. Williams

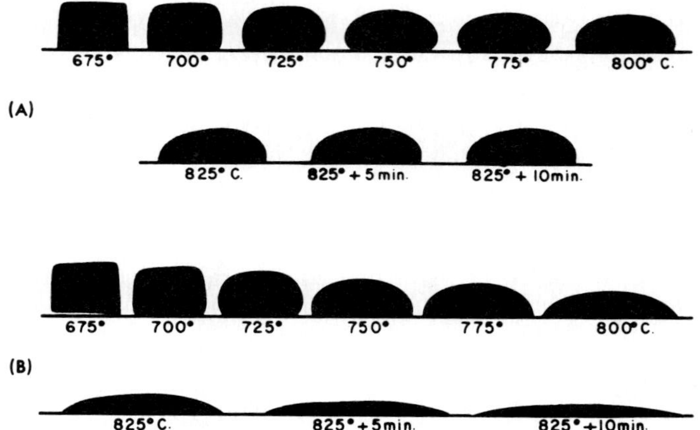

FIG. B. 131. Change of contact angle of an enamel melt drop on iron sheet (pickled), with increasing temperature (*A*), in dry furnace atmosphere, and a slight oxide interlayer, (*B*) in moist furnace atmosphere, with metal surface oxide interlayer. (King and Cook).

and J. W. Nielsen[428] for high-in-alumina ceramic surfaces wetted either in the original or in a molybdenum-metallized state by molten brazing solders (metals and alloys), quasi in the reversal phenomenon of normal enameling. Also in this case the typical decrease of the contact angle with time was observed for the system alumina-fused silver, copper, or eutectic alloys of silver-copper and gold-nickel at reduced pressures. In all cases, however, the metal layer, on cooling, separated from the ceramic body taking an adherent layer of the ceramic material with it. This indicates a destructive misfit between the thermal expansion of the ceramic and the metal layer, as well as a nonductile behavior of the metal films in the metallizing treatment.

259. More recently, Y. Godron[429] discussed the physical factors ruling surface energy on ceramics or glass/metal interfaces and wettability, with premises for the

[427] "Corrosion Handbook," John Wiley & Sons, New York, 1948, 1192 pp., especially p. 1082.

[428] *J. Am. Ceram. Soc.*, **42**, 1959, 229-235.

[429] *Silicates inds.*, **24**, 1959, 539-549; cf. F. Zackay, D. W. Mitchell, St. P. Mitoff, and J. A. Pask, *J. Am. Ceram. Soc.*, **36**, 1953, 84-89 (cf. A. ¶ 94). See also Fr. D. Gaidos and J. A. Pask, "Advances in Glass Technology," Plenum Press, New York, 1962, 548-565.

contact angle method, especially concerning noble metals and their alloys, as well as for molybdenum, wolfram, and tantalum. The systems of oxides and carbides with interfaces to metal are particularly characteristic of *cermets*, of the type WoC/Co or TiC/Ni, or/nickel with 10 per cent molybdenum, including modern high-temperature

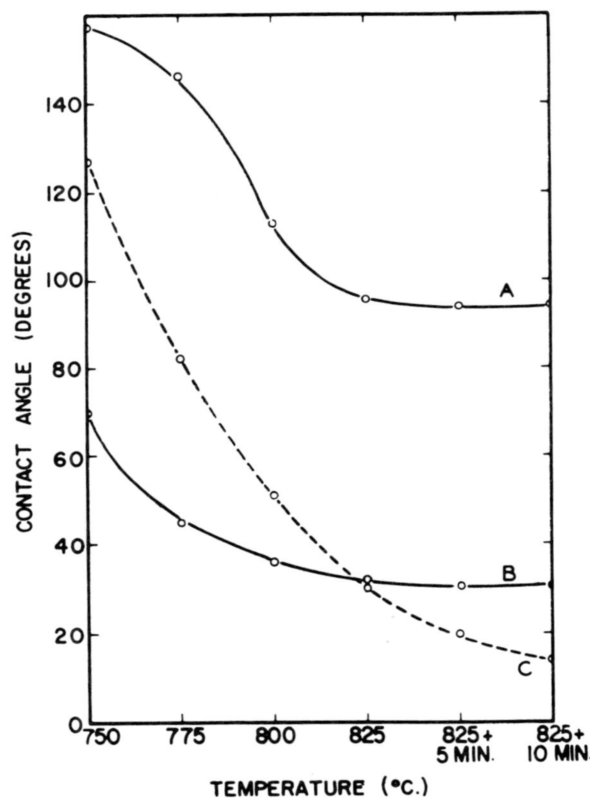

FIG. B. 132. Change of contact angle with temperature and time. (King and Cook). (*A*) 0.050 mg./cm.^{2} oxide, in dry nitrogen atmosphere, (*B*) 0.256 mg./cm.^{2} oxide in dry nitrogen, (*C*) 0.273 mg./cm.^{2} oxide, in moist atmosphere.

enamelling for protection from oxidation. In this respect it is *not* self-evident that one and the same enamel should have equally good protection qualities at different temperatures. An enamel containing V_2O_5 applied on a metal surface may be an excellent protection layer at lower temperatures, whereas it acts as a corrosion agent, contrary to what was expected, at higher temperatures.

260. The particularly plausible application of *sulfides* as surface tension-reducing agents in enamel compositions was previously discussed by A. Dietzel (1942), and was further experimentally studied by A. Dietzel and E. Wegner[430] with the interesting result that in boron-free, or low-in-B_2O_3 enamels, often more or less unintentionally, sulfides had been used as improving agents, as by introduction of metallurgical slags of the desulfuration process. Measurements of surface tension with the bubble-pressure method, however, showed that this reduction in slag-containing coats is not caused by the sulfide S^- anion but by that of the sulfates, SO_4^{2-} (cf. A. ¶ 114). Sulfide-coating enamels, in their firing range, are easily oxidized in the thin film of the molten coat applied onto the metal during its exposure to the oxidizing muffle atmosphere. Sulfate anions thus formed act intensely in favor of better wetting characteristics. Only when the scaling layer on the sheet is dissolved in the coat melt, the iron metal reduces again sulfate to sulfide anions, which then form deep-brown coloring complex sulfoferrites characteristic of the final aspect of the ground enamel.

261. *The volatilization* of *lead* oxide from the surface of enamels (and the same from glazes) as a phenomenon related to surface energy actions was studied by R. L. Hallse and R. L. Cook[431] by measuring weight losses at temperatures from 800° to 1300°C. from the surface of various lead silicate melts. The differences for free lead oxide and lead silicate melts (called mono- and di-silicates) corresponding about to the eutectics with 84.6 per cent PbO, 15.4 per cent SiO_2, and 70.3 PbO, 29.3 SiO_2, in the presence of one per cent of alumina, in the system PbO—Al_2O_3—SiO_2, according to E. N. Bunting and R. F. Geller (1943), melting at 716° and 730°C., respectively, are seen from Fig. B. 133. Here are shown the volatilization curves as a function of the logarithm of time (in hours), and in Fig. B. 134, the same as a function of temperature. Additions of lime or magnesia have a nearly negligible effect on the volatility, whereas alumina decidedly slows down the rate of volatilization because it strongly increases the viscosity of the melts. The experiments of Hallse and Cook demonstrate how profitable it is to reduce the volatilization losses from lead-rich glazes and enamel melts by adding PbO not as litharge but in a silicate-bound form. That PbO volatilization from borate melts is much lower than from lead silicate melts was further shown by Ch. Hirayama.[432] Experimental melts with a PbO content above 71 per cent demonstrated that volatilization from liquid surfaces is controlled by diffusion of the oxide through the melt towards the surface. The approximate vapor pressure of PbO at 900°C. is 0.2 mm. Hg, whereas that of B_2O_3 at the same temperature is of the order of magnitude of 8×10^{-6} mm. Hg.

[430] *Mitt. Ver. deut. Emailfachleute*, **2**, 1954 (3) 13-14; *Veröffentl. Max-Planck Inst. Silikatforsch.*, **14**, 1954, 204-205. On the metallurgical side of these problems concerning wetting and surface tension between metal and slag baths cf. S. I. Popel', O. A. Esin, G. F. Konovalov, and N. S. Smirnov, *Doklady Akad. Nauk S.S.S.R.*, **112**, 1957, 104-106 (cf. C. ¶ 2).

[431] *J. Am. Ceram. Soc.*, **41**, 1958,331-336.

[432] *J. Am. Ceram. Soc.*, **43**, 1960, 505-509.

262. The *enameling of cast iron*, the latter as an iron-carbon alloy with variable contents in C (up to 5 per cent of it), with graphite and cementite, Fe_3C, as the characteristic phases, was in principle already discussed in B. ¶ 251, concerning the problem of the presence of CO and CO_2 in gas bubbles evolved at the interfaces. Whereas P. K. Chu, J. H. Keeler, and H. M. Davis[433] emphasized the predominance

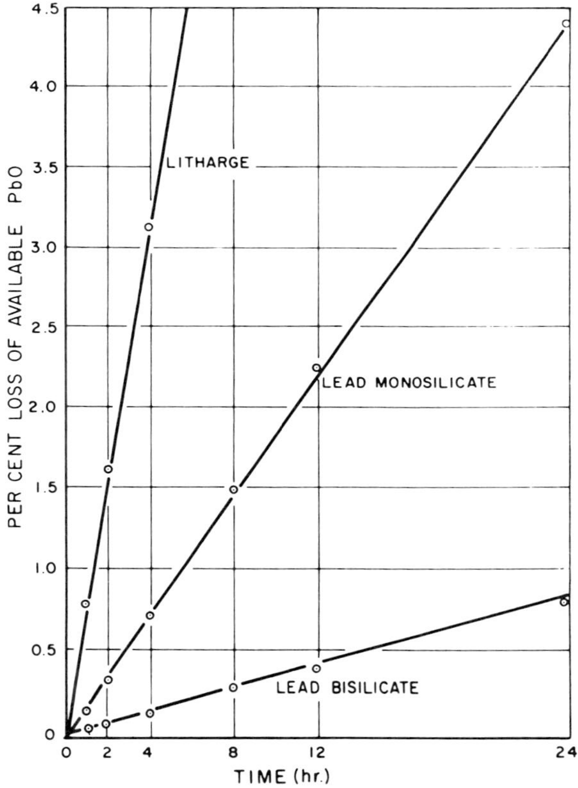

FIG. B. 133. Weight loss of pure PbO (litharge), and mono- and disilicate melts, as a function of logarithm of time, at 1100°C. (Hallse and Cook).

of hydrogen in these gases evolved in blisters by the water-iron reaction, Chu, Davis, and J. K. Magor[434] recognized by metallographic-microscopic examination of different cast iron qualities the importance of structural factors in Fe—C alloys. There is no reason to regard the total carbon amounts or the total volume of the

[433] *J. Am. Ceram. Soc.*, 36, 1953, 48-59. Cf. D. G. Moore and M. A. Mason, *Ibid.*, 36, 1953, 241-249.

[434] *J. Am. Ceram. Soc.*, 37, 1954, 391-401.

CO and CO_2 evolved under standard experimental conditions as a reliable measure to control the enamelling behavior, since the formation and distribution of the cementite phase is by far more a factor of critical significance. For the accurate study of gas evolution in cast iron samples, a special gas-extraction furnace was constructed (Fig. B. 135) with a magnetic elevator device working in the vacuum of the furnace heating chamber. In every case, a well-distributed cementite in pearlite tends to cause only minor defects by gas evolution at high temperatures from the carbon content of the iron, namely by fine "specks" or half-concealed bubbles. Large clusters of cementite, especially in the metal surface, on the other hand, cause large gas bubbles in the enamel coat, producing blisters, pinholes, and even black specks, whereas a skin-decarburized iron of relatively high carbon content in the bulk can be enamelled without any gas-associated defects.

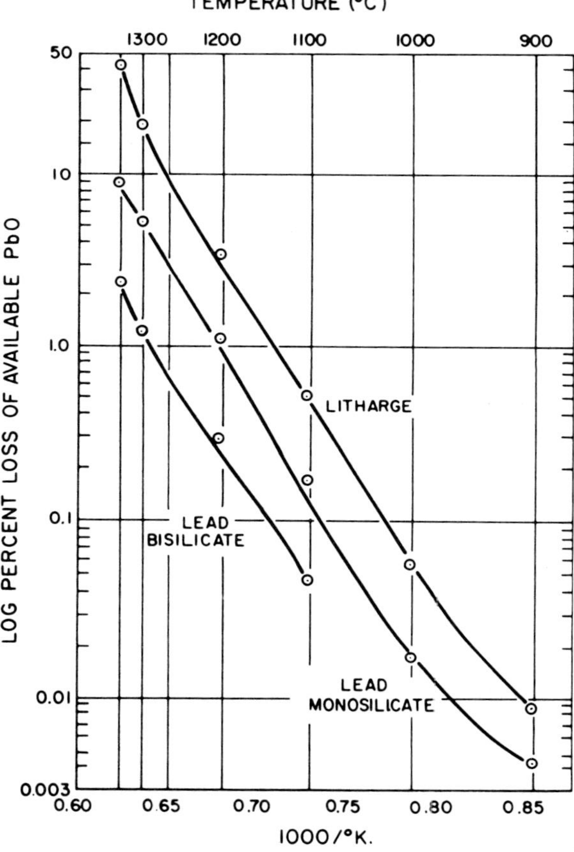

FIG. B. 134. Weight loss of pure PbO (litharge), mono- and disilicate melts, as a function of temperature. (Hallse and Cook).

263. The metallographic characteristics of cast iron in its suitability for enamelling is also emphasized by A. Dietzel and E. Wegner,[435] with a maximum permissible copper content of 0.20 per cent only. As a practically important prescription, it is

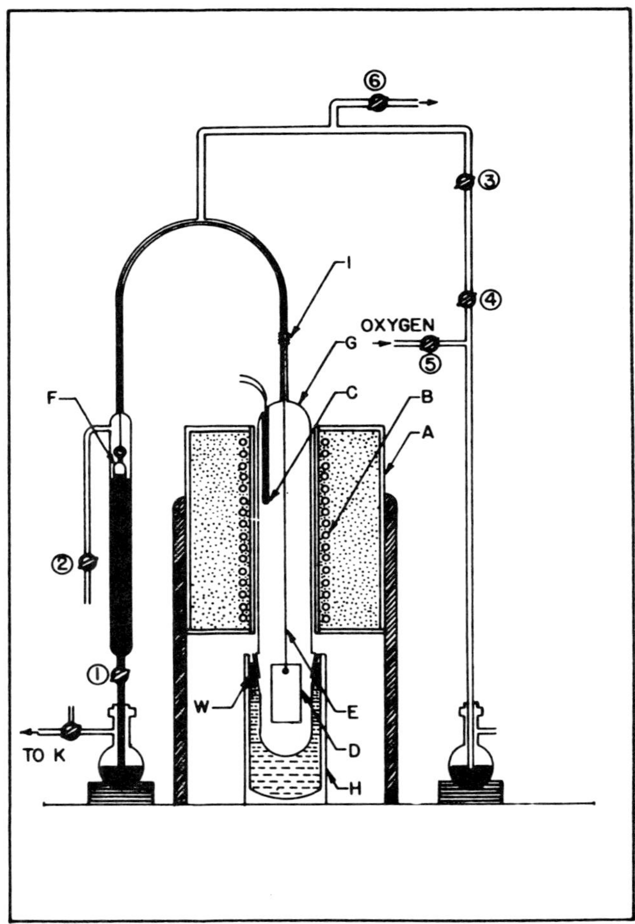

Fig. B. 135. Gas extraction furnace, and vacuum sample elevator device for the study of steel, or cast iron enamelling reactions. (Chu, Magor, and Davis).

postulated that grinding of cast iron with silicon carbide wheels before enamelling must strictly be avoided. It is highly detrimental because a strong gas evolution in enamel by oxidation of minute carbide particles will ensue. The cast iron shall be

[435] *Mitt. Ver. deut. Emailfachleute*, **6**, 1958 (2) 11-13; *Veröffentl. Max-Planck-Inst. Silikatforsch.*, **18**, 1958, 139-141.

low in gas content, low in pores, and as dense as possible to avoid boiling from cavities. D. Kamran[436] studied the adherence of ground coats on cast iron, as a consequence of an oxide interlayer between metal and enamel, with a good wetting from both sides. This oxide layer consisting of magnetite, some wüstite, but only hematite near the surface, must be preserved and must not be dissolved, whereas in steel sheet enamelling the original scale layer must be removed by dissolution. The influence of moisture in the furnace atmosphere during firing (at a constant temperature of 800°C.) on formation of the oxide layer is very important for a rapid formation of a stable oxide interlayer. The growth in thickness of this layer as a function of time follows the Jander parabolic law.

264. A fascinating Symposium on the structure and properties of ceramic glazes, also to be extended to that of enamels, was given by the British Ceramic Society in 1956[437] in which H. Moore first discussed the structural effects on chemical durability as a function of composition, chiefly for B_2O_3 containing types, with additions of ZnO, Al_2O_3, and BeO which form framework-participating [RO_4] coordinations. The more problematic role of PbO in borosilicate compositions is interpreted as favoring [BO_4] coordinations if the molecular ratio PbO : B_2O_3 is 2 : 5, with a corresponding strong reduction of softening temperatures. Also [PO_4] will improve the durability of glazes. W. James and A. W. Norris emphasize physical properties of glazes in general, and viscosity in particular, in the working range from 600° to 800°C. for lead-free compositions, in comparison with industrial sheet glass, as measured with a special low-temperature viscosimeter with cathetometer observation of the deformation (sag) as a function of temperature of horizontally-held bar specimens (in a chuck contrivance of nonscaling metal). For high temperatures a penetrating-buoyancy platinum rod device is used for viscosimetry, the principle of which is shown in Fig. B. 136. A glaze opacified with 5 per cent SnO_2 behaved as a true clear Newtonian liquid under the shear conditions of the penetration test, in contrast with the behavior of an anomalous organic liquid (cf. Fig. B. 137), as seen in the curves representing x^2, the square of the displacement of the rod, as a function of time. In every case, the opacifying agent did not increase viscosity at a constant temperature of 1090°C.

265. Bloor discussed especially glazes for electrotechnical ceramics (on the basis of common porcelains, Steatite, zircon, cordierite, and titania special bodies) in the practically important correlation between the maturing temperature with the

[436] *Mitt. Ver. deut. Emailfachleute*, **6**, 1958 (10) 85-91; *Veröffentl. Max-Planck-Inst. Silikatforsch.*, **18**, 1958, 151-157.

[437] "Symposium on the Structure and Properties of Glazes," *Trans. Brit. Ceram. Soc.*, **55**, 1956, 589-600 (H. Moore); 601-630 (W. James and A. W. Norris); 631-660 (E. C. Bloor); 674-688 (A. W. Norris). See also A. S. Watts, *J. Am. Ceram. Soc.*, **38**, 1955, 343-352; **41**, 1958, 249-253; **44**, 1961, 272-276, on eutectic compositions in multicomponent glaze systems.

ratios of the total oxygen atoms to the sum of the structure-forming constituents (chiefly Si, Al, B), and the ratio of oxygen atoms to the sum of the framework-modifying components. In complete analogy with the corresponding regularities for common glasses, the glazes and enamels show the same correlation of composition

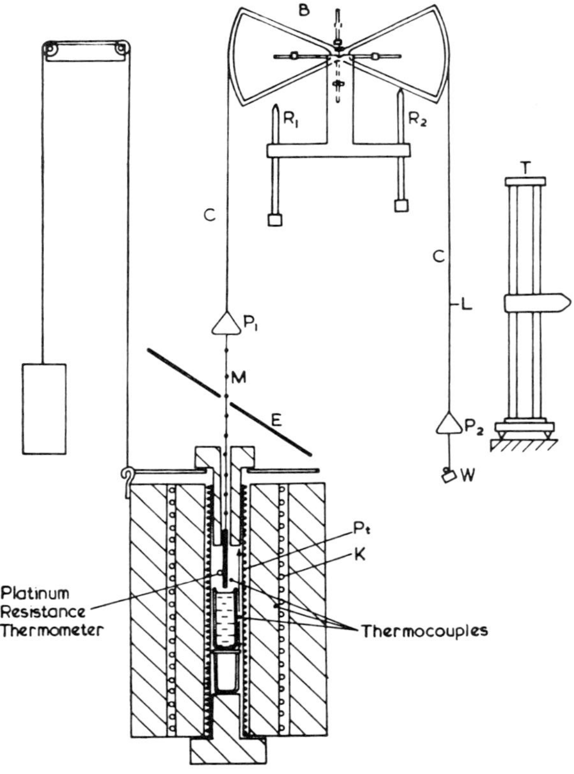

FIG. B. 136. Penetrating rod viscosimeter for determination of viscosity of glaze melts at high temperature. (James and Norris).

factors to viscosity, surface tension, and thermal expansion, if calculated on a molar basis (cf. Figs. B. 138 to 140). Volatilization from the free surface of a molten glaze film and heterogeneities originating from a reaction with the ceramic body exert a more or less profound effect on properties of the fired product, as demonstrated by A. W. Norris, such as in the measurement data of refractive indices throughout the thickness of a glaze applied to identical "biscuit" samples (prefired without the glaze) after different firings. Also A. Dietzel, L. Merker, B. Haas, and O. Schmitt[438]

[438] *Mitt. Ver. deut. Emailfachleute*, **6**, 1958 (5) 40-43; *Veröffentl. Max-Planck-Inst. Silikatforsch.*, **18**, 1958, 143-146.

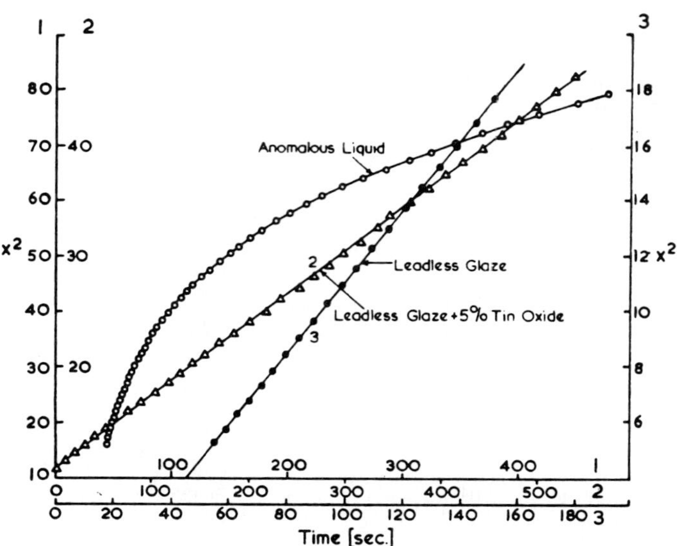

FIG. B. 137. Square of deformation (x^2) curves for a clear glaze, an opaque glaze, and an anomalous organic liquid, for comparison, as a function of time. (James and Norris).

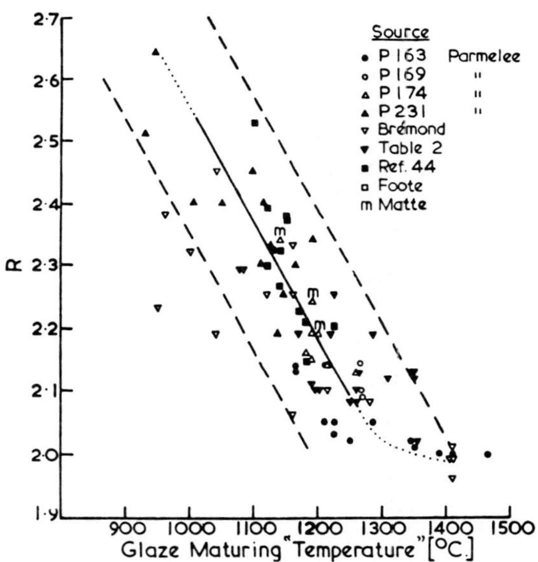

FIG. B. 138. Glaze maturing temperatures, as a function of ratio of the sum of oxygen atoms to sum of aluminum and silicon atoms in glaze composition. (Bloor).

discussed chemical durability (weathering stability) and the technologically important physical properties of majolica enamels as a function of composition, especially concerning the most adequate contents in alkalies, B_2O_3, and BaO, and those measures to improve the (usually not too good) durability, by systematic additions especially of alumina, CaO, ZnO, TiO_2 which also improve the flow characteristics. Compromises between a satisfactory fluidity and chemical durability cannot be avoided.

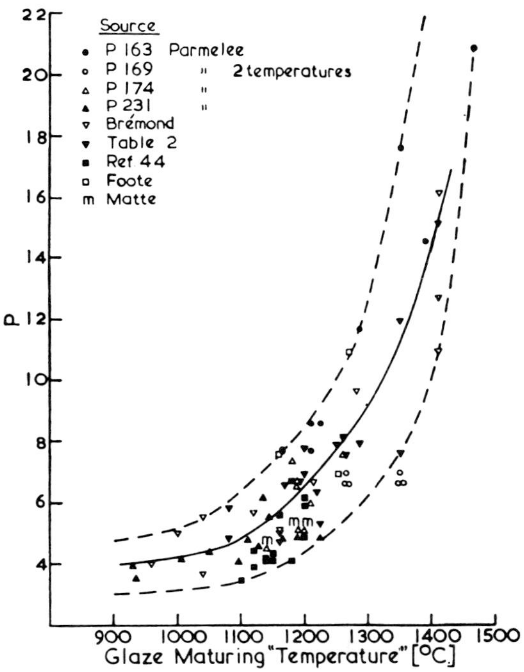

FIG. B. 139. Glaze maturing temperatures, as a function of ratio of the sum of oxygen atoms to sum of Na,K,Mg,Ca,Sr,Ba,Zn,Pb, in glaze composition. (Bloor).

266. The *opacity* of glazes and enamels as their most striking property is quantitatively measured commonly by the accomplished photoelectric reflectance meters or "Leucometers" of which the C. Zeiss instrument with this trade mark is a good example. W. Kerstan[439] gave a complete discussion of optical conditions for the white-opacifying effects of different dioxides (of Sn, Ce, Zr, Ti) and of Sb_2O_3 in the increas-

[439] *Ber. deut. keram. Ges.*, **31**, 1954, 404-409; **32**, 1955, 390-391, with critical remarks; *Euro-Ceram.*, **4**, 1954, 134-135; cf. on emulsion turbidities: H. Bollenbach, *Ber. deut. keram. Ges.*, **9**, 1928, 495-499.

ing series of refractive indices, on the basis of Fresnel's formula for the intensity of reflectance:

$$I_r = I_e \cdot \left(\frac{(n_2 - n_1)^2}{(n_2 + n_1)^2} \right)$$

with I_e the intensity of the incident light, I_r that of the diffusely reflected light, n_2 the refractive index of the opacifier agent, n_1 that of the glaze or enamel glass phase. Gas turbidity shows an efficiency (determined by the difference $n_2 - n_1$) higher than that of SnO_2 but lower than that of CeO_2. The optimum effect is observed for SnO_2 at a particle size of 0.2 μ, for ZrO_2 in every case also below 2 μ. Great in-

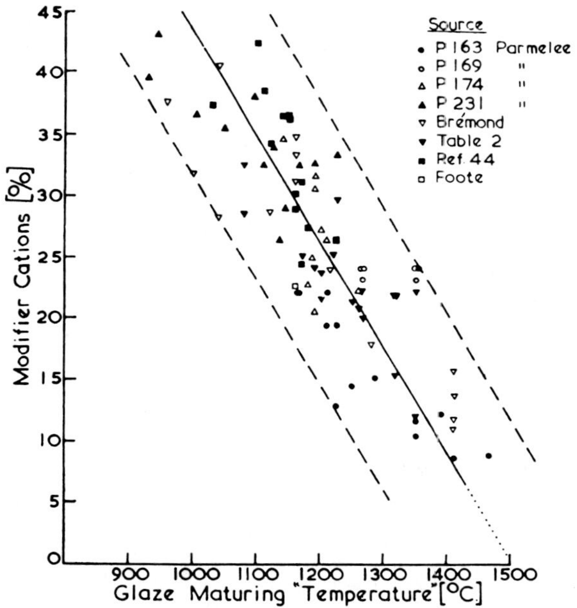

FIG. B. 140. Glaze maturing temperature, as a function of ratio of sum of Li,Na,K,Mg,Ca, Sr,Ba,Pb, atoms to the sum of these atoms + silicon and aluminum (Bloor).

terest is found for so-called *unmixing* turbidities (often erroneously called calcium or boron turbidity), caused by the partial immiscibility phenomenon as in the system $CaO—B_2O_3—SiO_2$, and characterized by formation of an emulsion-like suspension of different glass phases. Another valuable comparison of SnO_2 and ZrO_2 glazes was made by Z. A. Nosova and M. E. Yakovleva[440] (cf. B. ¶ 228, where the question of crystallization of $ZrSiO_4$ and/or baddeleyite is especially discussed). These authors are of the opinion that the leucometric-photoelectric determination of the albedo

[440] *Trudy Nauch. Issledovatel. Inst. Stroitel. Keram.*, 1955, 82-130.

(whiteness) of tin- or zirconium-opacified glazes (and enamels) is not sufficient for the characterization of such products. It may, however, be used as a relative basis of comparison for different materials and as a plant-routine method for opacity effects.

267. In these practical aspects, comparative studies of E. Kato, Sh. Morita, and I. Choshi[441] may also be mentioned in which the observation of a *tin-titanite* (sphen), $CaO.SnO_2.TiO_2$, is most remarkable, as a crystal phase identified together with cassiterite in SnO_2-opacified products. Also a *tin-perovskite*, $CaSnO_3$, which favors crystallization of the calcium titanostannate was observed. Sometimes this latter phase crystallizes so rapidly that no more cassiterite is observed. Evidently the temperature of the melting process plays a role in crystallization of all three tin compounds, cassiterite being formed preferably in compositions fired at higher temperatures.

268. Actually, the use of SnO_2 as an opacifying agent in enamels is in most cases abandoned and replaced by that of *titania*. A wealth of literature on the TiO_2-opacifying effects is available in which the crystallization of rutile and anatase forms the true pigment of the opacity phenomenon. The titania is introduced to the batches usually as anatase. W. F. Sullivan and S. S. Cole[442] first studied the colloid-chemical processes in the hydrolysis of titanium sulfate solutions to form a dispersed, but rapidly flocculating, titania hydrosol from which anatase crystallizes in minute crystals. Water and SO_4^{2-} are intensely adsorbed on the surface of the particles and in capillary cavities of the precipitates. The slow process of dehydration starts at 150°C., an evolution of SO_3 from the residue at 650°C., and a growth of anatase particles at 600°C. with ensuing slow inversion into rutile in the range from 700° 950°C. as a function of the thermal treatment methods and of the presence of contaminations. The kinetics of this inversion anatase $\rightarrow$ rutile is of the first order, with a distinct induction period dependent on the mechanism of the spontaneous nucleation, an activation energy of about 100,000 cal./mole and a frequency of 10^{20} to 10^{22} per hour. Precision X-ray investigations made evident that both the anatase structure below the inversion range and the rutile structure formed above it have a typical defectiveness with a widening of the unit cell dimensions.

269. Crystal growth as a function of temperature shows typical anisotropy effects for the interference lines of (200), (101), and (004) as determined from the formula

$$d(hkl) = \frac{0.9\ \lambda}{\beta^{1/2} \cdot \cos \delta}$$

($\beta^{1/2}$ the half-maximum line breadth in radians, cf. Vol. I, C. ¶ 96 f.). The average

[441] *Nagoya Kogyo Gijutsu Shikensho Hokoku*, **7**, 1958 (6) 81-91; **8**, 1959 (5) 44-50 (with Sh. Kanaoka); *Ceram. Abstr.*, 1959, 270, i, c.

[442] *J. Am. Ceram. Soc.*, **42**, 1959, 127-133.

dimension $\overline{D} = (D_{200}.D_{101}.D_{004})^{1/3}$ is shown as a function of temperature in Fig. B. 141. Also R. E. Cowan[443] made precision studies of the TiO_2 modifications in opacified enamels under the influence of additions of MgO and P_2O_5, in the temperature range from 700° to 820°C. During firing of such enamels about 50 per cent of the TiO_2 present in the frit crystallizes; with increasing firing temperature, anatase rapidly disappears and rutile increases in amount. Particularly remarkable is the observation of Cowan that quartz also crystallizes as an independent phase from compositions

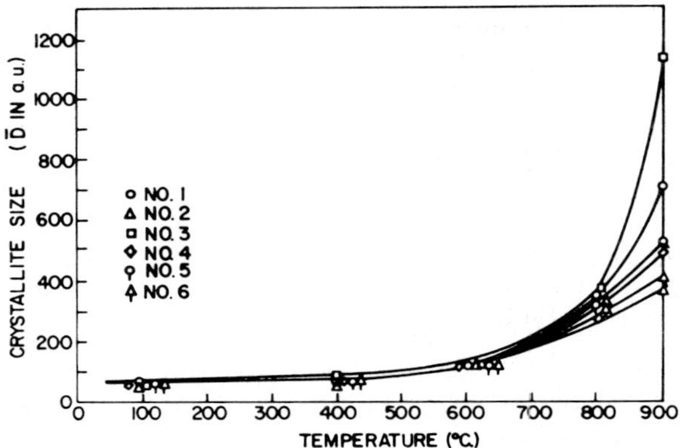

Fig. B. 141. Change of average crystallite dimensions of rutile, as a function of the calcination temperature. (Sullivan and Cole).

with only 1 per cent MgO. This oxide does not affect the amount of anatase observed but decidedly retains the crystallization of rutile. The X-ray diffraction method is in every case, a most valuable help in the quantitative determination of the opacifying crystalline phases. The particular effect of P_2O_5 additions with respect to the reflectance stability of Ti-enamels is not solely interpreted by the amount of the crystalline opacifiers. Additions of Li_2O, on the other hand, in enamels fired at 650° to 700°C. distinctly accelerate the inversion of anatase into rutile as seen from semiquantitative X-ray diffraction measurements made by M. D. Houston.[444] The devitrification tendency of lithium-rich compositions establishes technological limits to the use of Li_2O as a corrective agent for enamel compositions; silica then easily crystallizes as α-cristobalite. Cryolite as a raw material for enamel compositions is also a valuable accelerating agent for the crystallization of rutile, as R. Märker[445] observed. At the same time, chemical durability (acid resistance) is increased by the introduction of alumina.

[443] *Am. Ceram. Soc. Bull.*, **35**, 1956, 53-56.
[444] *Am. Ceram. Soc. Bull.*, **36**, 1957, 139-141.
[445] *Glas-Email-Keramo-Tech.*, **8**, 1957, 117-121, 178-182.

270. Tin-Boo Yee and A. I. Andrews[446] extensively investigated spontaneous *nucleation* and *crystal growth* laws of titania as rutile and anatase, as a function of time in TiO_2-opacified enamels during the firing period, on the basis of G. Tammann's classical laws. The results may be seen in Fig. B. 142 which shows particularly well the temperature difference of the maxima of the rate of crystal growth at about

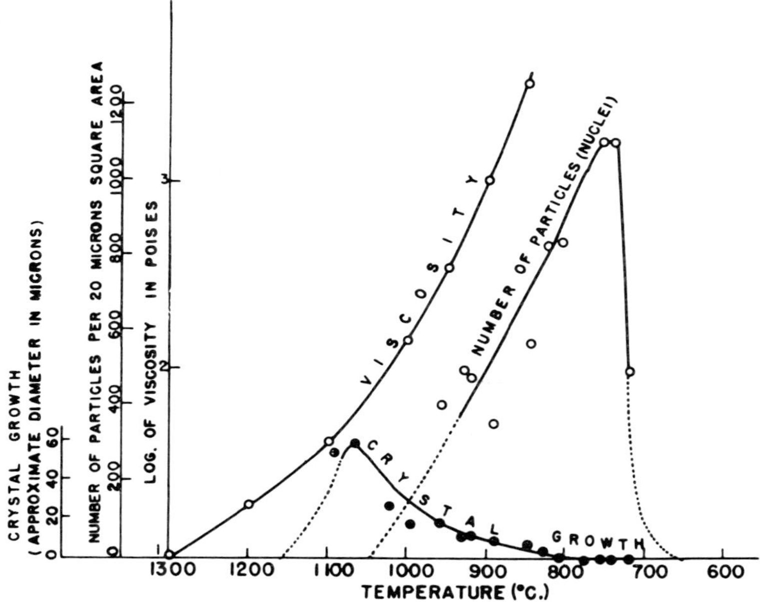

FIG. B. 142. Relation of viscosity, number of spontaneous nuclei, and crystal growth in titania-opacified enamel. Firing for 2 minutes, as a function of temperature. (Yee and Andrews).

1080°C., and that of spontaneous nucleation at about 730°C. The experimental studies concerned the range from 650° to 1300°C. Remarkable color changes of the systems were observed in reflected light, from bluish-white to white to a cream-tint white. Very thin films of the enamel blown from the melts were used in light- and electron-microscopic examinations. The crystals had sizes varying from about 0.1 μ (only anatase, at 720° to 750°C.), 0.3 to 2.0 μ (at 775°C.), and 1 $^1/_2$ μ (at 800°C.).

271. Problems of the crystallization of titania either as anatase and/or as rutile in TiO_2-opacified porcelain enamels were also studied by R. D. Shannon[447] who

[446] *J. Am. Ceram. Soc.*, **39**, 1956, 188-195. Differential-thermal curves are also given by F. Imoto and K. Hirao, *Yogyo Kyokai Shi*, **63**, 1955, 198-202, in agreement with those of Yee and Andrews, characterized by a distinct exothermic effect in the range from 550° to 750°C., corresponding to nucleation of anatase and rutile.

[447] *Am. Ceram. Soc. Bull.*, **38**, 1959, 110-116.

discussed the interesting hypothesis that anatase crystallized from such complex mixtures is not a pure TiO_2 modification but a "stuffed" crystal phase comparable, in this respect, to tridymite and "β-alumina." The constant contents in SO_3 trapped from the raw material (hydrolyzed titanium sulfate) or of P_2O_5 in anatase is an important fact in this connection. Shannon further extensively discussed the nucleation mechanisms in the general meaning of Tammann's theory and of Yee's and Andrews' confirming experiments. The specific influence of contaminations on nucleation is most evident by the fact that sulfates and phosphates prevailingly favor the precipitation of anatase, whereas in the presence of halides rutile forms directly, a fact well known from the pigment manufacturing industry. Shannon is inclined to ascribe the blue-white reflectance type in fired enamels to crystallization of anatase, in agreement with the spectrophotometric results of A. L. Friedberg, R. B. Fischer, and A. E. Petersen,[448] who based their deductions on the theory of the Rayleigh light diffraction (scattering) phenomenon. The diffraction by rutile is slightly larger than that of anatase, if the same particle sizes (0.1 to 0.3 μ) are studied, as S. S. Cole (1952) had previously observed. Whereas anatase appears to be more stable and consistent in size and color effects, rutile is higher in phase stability. The properties of enamels are normally better with anatase and more desirable from the practical viewpoint. The rule that higher firing temperatures, in the presence of agents to increase the fluidity of glazes and porcelain enamels, promote the crystallization of rutile is confirmed by K. H. Styhr and M. D. Beals.[449] Practically important is the influence of alumina, which increases the whiteness of the products. The fritting temperatures must in this case be raised to 1350° to 1550°C. although glazing temperatures are still in the range from 840° to 1110°C.

272. Concerning *chemical stability* of Ti-enamels, V. V. Vargin[450] made interesting contributions to the technological possibilities of substituting corrosion-resistant metals with enamels for simple *water* corrosion. The presence of three- and four-valent metal cations (particularly in TiO_2 and ZrO_2) considerably increases the durability; 10 to 12 per cent B_2O_3 reduces the leaching losses, but higher B_2O_3 amounts increase them again. Enamels with an excellent *acid* resistivity were developed on the basis of Ti, Al, and B contents, and also on slight additions of fluorine. The attack by *alkaline* solutions is usually much more serious since no protecting silica hydrogel layer is formed in the corrosion reaction. Nevertheless, zirconia containing compositions show improved alkali resistance, whereas titania promotes alkali at-

[448] *J. Am. Ceram. Soc.*, **31**, 1948, 246-253. Shannon writes the formula of anatase $[Ti_{1-x}(S,P_x]O_2$. The effect of P_2O_5 on whiteness of Ti-enamels was further studied by V. V. Vargin and R. I. Rylova, *Zhur. Priklad. Khim.*, **29**, 1956, 293-295, indicating preferred crystallization of anatase, and retardation of rutile.

[449] *Am. Ceram. Soc. Bull.*, **37**, 1958, 480-485. Cf. previous investigations of A. Dietzel and R. Boncke (1942).

[450] *Khim. Nauka i Prom.*, **3**, 1958, 76-82.

tack. Concerning opacity, V. V. Vargin and V. Ya. Senderovich[451] emphasize the possibility of applying titania-white coats in *one* operation on steel sheets since a 0.1 mm. thick layer, with an index of reflectance of 70 to 75 per cent, is sufficient for this purpose. Concerning the leachability of B_2O_3 containing Ti enamels, Vargin and Senderovich find for one composition type a minimum at about 12 per cent TiO_2 content, with a sudden increase in reflectance from 15 to 70 per cent during the crystallization of rutile. Another enamel type, in which titania replaces silica, did not show a minimum of leachability but an accelerated acid-leaching above 12 per cent TiO_2, and a discontinuity of reflectance at 10 per cent titania replacing silica. The problems of an improved alkali resistance of porcelain enamels were also studied by J. E. Cox[452] (cf. B. ¶ 196), with NaOH, or sodium pyrophosphate solution, and commercial detergents, reporting the important result that conventional tests with a 5 per cent NaOH solution are often misleading whereas a pyrophosphate solution test has the advantage of an excellent buffering (in a 1 per cent solution).

273. The last-mentioned practical problem of a *one-coat* application of Ti-enamels directly on steel sheet is most attractive and was, therefore, newly discussed by E. D. Lynch and A. L. Friedberg.[453] According to their results it is the crystallization of wüstite at the interface enamel/metal which contributes principally to the adherence of the enamel. No good adherence was observed if wüstite was absent. By X-ray diffraction methods, the following crystal phases were identified in the interface region: wüstite, magnetite, ilmenite ($FeTiO_3$), anatase and rutile, in addition to metal crystallites. An unidentified phase with a characteristic interference line $d = 2.59$ Å. is also mentioned.[454] As a stabilizing process for wüstite, the nickel-flashing of the metal sheet which acts by the reaction $Ni + Fe + Fe_3O_4 \rightarrow Ni + 4\ FeO$ is interpreted. A very characteristic phase distribution diagram based on quantitative X-ray determinations is shown in Fig. B. 143, in which case the abundant crystallization of wüstite is very obvious. Magnetite could not be distinguished safely enough, because of the well-known coincidence of magnetite and ilmenite lines in the X-ray diffraction diagrams. In every case, the X-ray examination techniques from samples systematically taken from the surface of the enamel down to that of the steel sheet, combined with metallographic examinations, is a method superior to common acid etching of the metal. Acid-etching always removes much of the iron oxides, also the iron metal in the coating, and even the nickel-iron alloy formed.

[451] *Doklady Akad. Nauk S.S.S.R.*, **84**, 1952, 1213-1216; *Silikat Tech.*, **4**, 1953, 201-202; *Trudy Leningrad. Tekhnol. Inst. im. Lensoveta*, 1955, No. 34, 38-44, 45-47.

[452] *J. Am. Ceram. Soc.*, **41**, 1958, 336-343, with discussion remarks of J. T. Roberts, ibidem p. 241, and a reply of Cox pp. 342-343.

[453] *J. Am. Ceram. Soc.*, **38**, 1955, 257-263.

[454] This phase was previously observed by G. S. Douglas and M. Zander, *J. Am. Ceram. Soc.*, **35**, 1952, 5-11.

For metallographic study, the taper technique (cf. A. G. Pincus, ¶ 255) under an angle of 6° to the enamel surface is most profitable. Extensive studies on titania-white enamelling in a one-coat operation of Ti-steel (Ti-Namel) based on a lithium-Ti-enamel composition were further made by A. Petzold.[455]

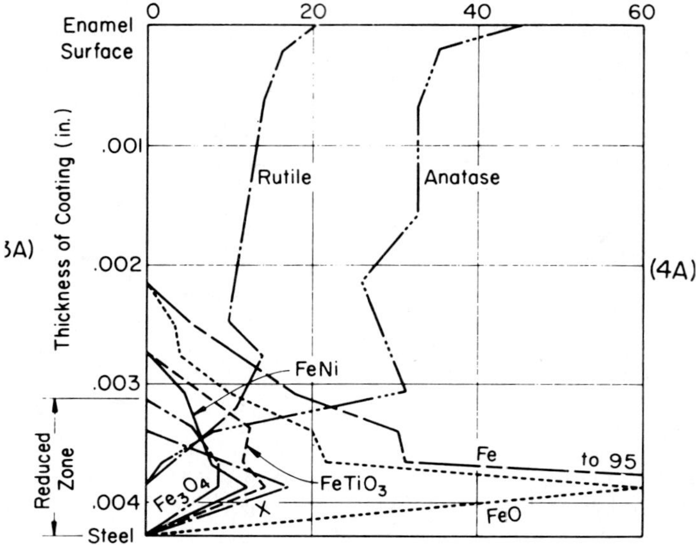

FIG. B. 143. Quantitative X-ray analysis of samples taken from different levels of one-operation-applied titania enamel, on nickel-flashed steel sheet. (Lynch and Friedberg).

274. The excellent properties of *cerium dioxide*-opacified enamels which are striking as revealed by the investigations of E. L. Johnson and A. L. Friedberg[456] (cf. B. ¶ 226) are unfortunately hindered by the high price of this oxide. Much more economical, and also of great efficiency in opacity is *antimony oxide* (Sb_2O_3 as isometric senarmontite, and orthorhombic valentinite) which was compared with titania by O. W. Flörke[457] who confirmed previous observations of B. W. King and

[455] *Sprechsaal,* **89,** 1956, 541-545; **90,** 1957, 87-90, 152-155, 207-210 (with I. Lange and H. Betzer). Steel used for direct-coatings should be low in carbon, preferably below 0.10 per cent C. A special metal surface treatment for this requirement (decarburization) is feasible. Sand-blasting is not sufficient; best results were shown with HNO_3-pickling.

[456] *J. Am. Ceram. Soc.,* **42,** 1959, 60-69.

[457] *Mitt. Ver. deut. Emailfachleute,* **6,** 1958, 49-52; *Veröffentl. Max-Planck-Inst. Silikatforsch.,* **18,** 1958, 147-150. Data for refractive indices of Sb_2O_3 in the literature are usually given much too high, e.g. = 2.6 by L. Stuckert (1941), and by A. Petzold, "Email," VEB Verlag Technik, Berlin, 1955. For valentinite it is $\gamma = 2.358$, for $\alpha = 2.18$, for senarmontite $n = 2.087$; cf. E. Kordes, "Optische Daten zur Bestimmung anorganischer Substanzen mit dem Polarisationsmikroskop," Verlag Chemie, Weinheim a.d. Bergstrasse, 1960, pp. 38 and 102.

A. I. Andrews.[458] These latter authors emphasized that Sb_2O_3 as such is not the opacifying phase in such enamels but that the Sb_2O_3 on heating in air takes up additional oxygen to form higher oxides, especially if the oxygen potential is kept high in the melts by addition of nitrates into the batch composition. Flörke confirms the formation of Sb_2O_5 and excludes the mixed oxide $Sb^{III}Sb^{V}O_4$ which is cubic (n = 2.087), or orthorhombic (as mineral cervantite). In no case could Sb_2O_3 in one or the other modification be observed in antimony enamels, if the oxygen potential is normally high. If one starts from synthetic valentinite or senarmontite and slowly melts the enamel mixes one observed the difference in stability by the more rapid transition of senarmontite into Sb_2O_4, whereas valentinite remained volatile for a longer time causing larger losses in antimony from the enamel melts.

[458] *J. Am. Ceram. Soc.*, **23**, 1940, 225-228, 228-232. King and Andrews speak of the best opacity when the calcinate has the ratio $CaO : CaF_2 : Sb_2O_{3-5} = 1 : 2 : 3$, with the same structure as Sb_2O_5.

Section C

Industrial Slags

CONSTITUTION OF SLAGS

1. Since the classical investigations of J. H. L. Vogt (1884) the evident analogy between the occurrence of natural magmatic *sulfide-silicate systems* with metallurgical slags often has been emphasized and studied. Also the modern literature in this field[1] contains many investigations on the nature of magmatic sulfide ore deposits. We mention as two illustrative examples, the occurrence of primary pyrite in the Leona rhyolite of Alameda County, California, described by G. D. Robinson,[2] and the differentiated ore magmas of the Tsuchikura and Besshi Mines in Japan, as described by A. Matsubara.[3] For the rhyolite containing pyrite the disseminated appearance of the sulfide in the effusive rock is characteristic and the existence of iron disulfide is not in disagreement with the fundamental data given by E. T. Allen, J. L. Crenshaw, and J. Johnston,[4] in their investigation of the system Fe—S, especially in relation to the rather low temperature of the rhyolite effusion. The type of the pyrite crystals is that of primary phenocrysts of rounded and embayed forms, indicating some later reaction with the melt (magmatic corrosion). The considerably larger ore bodies of the Japanese mines mentioned (chiefly with chalcopyrite, pyrite, some scarce sphalerite and pyrrhotite) show an origin of magmatic flow by liquation-unmixing of incomplete equilibrium in the system silicate—FeS_2—$CuFeS_2$, the copper-rich compositions being very scarce. The flows must have taken place

[1] A valuable compilation of literature up to 1950 was given by W. Fischer and S. Wolf in the book "Schwefel in Schlacke und Schlackenwolle," E. Schweizerbartsche Verlagsbuchhandlung, Stuttgart 1951, 231 pp. We will repeatedly refer to this important source of information.

[2] *Am. Mineralogist,* **38,** 1953, 1204-1217, especially pp. 1212 f.

[3] *J. Geol. Soc. Japan,* **59,** 1953 (3) 79-87. Massive pyrite with sphalerite and chalcopyrite of the same deposit, however, is hydrothermal in its origin, and not correlated to primary pyrite in the rhyolite.

[4] *Am. J. Sci.,* [4] **33,** 1912, 169-236; *Z. anorg. u. allgem. Chem.,* **76,** 1912, 201-273.

as a "shoot" intrusion, in the consistency of a thick jelly paste, with a high surface tension of the sulfide versus silicate phases. In the Besshi Mine the occurrence of much quartz is remarkable, with three kinds of differentiates, namely (most abundant) pyrite ore, silicic ore in the "banded ore," of lenticular irregular shapes, and chalcopyrite in "banded ore," with sharp boundaries. The temperature of the "shoot intrusion" may have been so low that it could not have been much above the critical point of water, namely above 375°C., in a state of high viscosity of a typical "sulfide matte."

2. Not only because of immiscibility phenomena is the presence of sulfides in metallurgical slags of great significance but equally for the strong influence of sulfur on the *phase boundary tensions* between metal melts and fused slags. S. I. Popel', O. A. Esin, G. F. Konovalov, and N. S. Smirnov[5] studied the behavior of slag inclusions in steel ingot casts concerning specifically the surface tension conditions on the phase boundary metal/slag in the desulfurization process of steel. Sulfur is known to decrease considerably surface tension as also in enamel/steel sheet systems (cf. B. ¶ 260), and to promote the wetting characteristics on solid surfaces. For the study of physical conditions on such phase boundaries, Popel' and Esin used an X-ray shadow-casting method,[6] as the components of the investigated systems pure iron, iron sulfide, FeS, and a synthetic slag composition of steel slag or blast furnace slag type (with 35, and 47 per cent CaO; 27.2 and 1.0 per cent FeO, respectively), fused in magnesia and alumina crucibles. A dissolution of the crucible material in the slag melt did not appreciably influence phase boundary phenomena. The metal droplets in the slag melts are considerably flattened if the sulfur content of the metal phase is increased, the surface tension decreased. At 1530° to 1570°C. the surface tension of pure iron to slag is 1015 and 700 ergs/cm.^{2}, respectively, and decreased to about 500 and 300 ergs/cm.^{2}, respectively, if 2 per cent sulfur was added to the steel melt. This result is particularly important for an understanding of the transition of sulfur from the metal melt into the slags and explains also the great stability of non-metal inclusions in steel, if they contain much sulfur and ferrous oxide at the same time. Such an emulsion-like system does not easily coalesce because the electrokinetic potential impeding this process is considerably changed by the action of sulfur.[7]

3. A very interesting study on the *uranium* contents of α-$CaSiO_3$ slags produced in phosphorus distillation from phosphorite ores was made by S. Young and Z. I. Altschuler,[8] combined with an excellent petrographic and X-ray examination of

[5] *Doklady Akad. Nauk S.S.S.R.*, **112**, 1957, 104-106.

[6] On this method see S. I. Popel', O. A. Esin, and P. V. Gel'd, *Doklady Akad. Nauk S.S.S.R.*, **74**, 1950, 1097-1100; **83**, 1952, 253-255.

[7] Cf. B. V. Deryagin, *Bull. acad. sci. U.R.S.S.*, 1937, 1153-1164.

[8] *Ind. Eng. Chem.*, **50**, 1958, 793-796.

the α-$CaSiO_3$ crystallization from such slags. The uranium is easily made evident by radioactive contact prints on nuclear photographic emulsion, which indicates a distinct enrichment of uranium in these slags from the original phosphate ore (Florida phosphorite as the raw material). The characteristic α-tracks in the prints indicate a specific enrichment in the fibrous end-phases (with some alumina content of the slag), in comparison with pseudowollastonite crystals proper. Another anomalous concentration of uranium is observed in spherules of iron phosphide (Fe_2P) which unmixes from the silicate melt but is also distributed in these minute inclusions amid crystals of α-$CaSiO_3$.

4. The metallurgically highly important contents of sedimentary iron ores in *phosphorus*, *vanadium*, and *arsenic* is particularly evident and well studied in the so-called "tobacco" ores of the Kerch Peninsula, namely by I. A. Shamraĭ and V. I. Sorochinskaya.[9] They are roe-like, dark-brown colored marine ores derived from a primary dark-green oolithic ore often containing vivianite, siderite, rhodochrosite, and manganospherite, the ooliths chiefly consisting of goethite. Chamosite also occurs in the primary horizons. For such and related investigations, the infrared spectroscopic method is a valuable help in slag examination as W. Pepperhoff[10] demonstrated for the range from 6 to 14 μ in the investigation of the basic systems SiO_2—Al_2O_3, CaO—P_2O_5, CaO—P_2O_5—SiO_2, with the crystal phases anorthite, gehlenite, Ca_2SiO_4, $CaO.P_2O_5$, $2CaO.P_2O_5$, $3CaO.P_2O_5$, $4CaO.P_2O_5$, and the apatite phases with OH^-, Cl^-, and F^- anions, and also lead apatite and silicocarnotite (cf. Vol. III, B. ¶ 435).

5. The highly dispersed distribution of metals in silicate melts of the so-called "metal cloud" type has its analogue in the systems of heavy metals (lead, zinc, antimony, bismuth, cadmium) dispersed in their fused halides which were described by J. D. Corbett, S. von Winbush, and Fr. C. Albers,[11] with the formation of slightly stable "subhalides" many of which form diamagnetic liquid solutes and some of which are highly volatile (gaseous).

6. On the mineralogical constitution of *blast furnace slags* new experimental experience was collected, especially for crystallization in the system CaO—Al_2O_3—SiO_2. R. Sersale and Elvira Gregorio[12] determined the glass contents of tempered blast furnace slags, either by direct microscopic-planimetric methods, or by differential-thermal analysis. They also determined caloric effects in devitrification and examined for Ca_2SiO_4 by ultraviolet luminescence tests. In the differential-

[9] *Doklady Akad. Nauk S.S.S.R.*, **120**, 1958, 875-878.

[10] *Arch. Eisenhüttenw.* **29**, 1958, 151-158.

[11] *J. Am. Chem. Soc.*, **79**, 1957, 3020-3024.

[12] *Ricerca sci.*, **21**, 1951, 2152-2166. For critical remarks on the use of differential-thermal analysis for the characterization of blast furnace slags, see W. Schrämli, *Zement-Kalk-Gips*, **16**, 1963, 140-147; cf. Vol. V, C. ¶ 7, footnote 17.

thermal curves, the exothermic effect of crystallization of melilite (gehlenite- åkermanite crystalline solutions, usually of complex composition) is particularly characteristic for alkali-poor slags. If the alkali content of the slags is relatively high, mostly caused by sea-water granulation, another exothermic effect indicates the crystallization of the ternary compound $8CaO.Na_2O.3Al_2O_3$ (cf. Vol. V, Section C.) which is also identified by X-ray diffraction diagrams. The zinc melilite (hardystonite, $2CaO.ZnO.2SiO_2$) with its congruent fusion equilibrium (cf. Vol. III, B. ¶ 103) is a characteristic crystal phase in metallurgical slags from lead blast furnace slags associated with rankinite, pseudowollastonite, willemite, and the like, in the ternary system $CaO—ZnO—SiO_2$ as determined by E. R. Segnit.[13] Such slags may be enriched in ZnO up to 20 per cent.

7. From a more general viewpoint, the influence of active surfaces on the crystallization from fused slags was investigated by M. A. and V. A. Matveev,[14] as from a cupola slag which first was consolidated as a glass and then systematically devitrified. The crystallization is ruled by the number, N, of crystal centers as a function of the number of nuclei formed in a unit volume per unit of time, c_v, and of the volume of the melt, in the equation $N = c_v \cdot v \cdot t \cdot f$, in which the parameter f means the active surface of crystallization. Very extensively, C. Brisi and A. Molinari[15] studied, with classical microscopic and X-ray methods the crystallization of two blast furnace slags, one with 45.7 per cent CaO and 3.9 per cent MgO, the other with only 21.4 per cent CaO, and 32.4 per cent MgO, from the glassy state and by thermal treatments changed to crystal phases. For increasing temperatures, the isotherms of devitrification show at lowest temperatures a typical sigmoid course, indicating an incubation interval of nucleation before the rapid crystal growth starts (cf. Vol. III, A. ¶ 30 ff.). The crystallization is approximately a reaction of the first order with $K = 8.3 \times 10^{-3}$, perhaps also of the order $n = 1.1$. The growing crystals were identified as gehlenite in the magnesia-poor slag (with $a_0 = 7.73$ Å.; $c_0 = 5.01$ Å.), for that high in magnesia crystalline solution phases of the series forsterite-monticellite. Interesting is the early indication of $8CaO.Na_2O.3Al_2O_3$ above 700°C. which, however, is unstable and, therefore, later resorbed, with the alkali content entering the melilite structure. If the manganous oxide content of a blast furnace slag is high, additional phases may occur in crystallization. L. I. Karyakin and L. A. Tseĭtlin[16] observed in a steel melt of 250 kg. weight fused at 1530° to 1550°C. in a silica crucible (heated in an induction furnace), in addition to characteristic reaction products on the surface of the crucible (galaxite, $MnO.Al_2O_3$, pseudowollastonite, hematite, and pseudomorphs of cristobalite after quartz), a glass with galaxite octa-

[13] *J. Am. Ceram. Soc.*, **37**, 1954, 273-277.

[14] *Trudy Moskov. Khim. Tekhnol. Inst. im. D. I. Mendeleeva*, 1956, No. 21, 74-79.

[15] *Silicates inds.*, **23**, 1958, 502-207.

[16] *Doklady Akad. Nauk S.S.S.R.*, **130**, 1960, 1091-1094.

hedra, an iron-tephroite, and prismatic manganese cordierite, $2MnO.2Al_2O_3.5SiO_2$, incrustations with manganese anorthite, derived from the addition of ferromanganese and aluminum for deoxidation to the metal bath.

8. There are many analogies between the industrial production of artificial pavement stones from cast blast furnace slags and natural basalts, and related effusive rocks as castable compositions. It is, therefore, plausible to treat both branches of a modern silicate brick production together. For the theory of the crystallization processes from such complex melts as basalts and diabases are, N. N. Ormont[17] developed so-called "asymmetrical" phase equilibrium diagrams, such as that for diopside—albite—anorthite for representation of every given ternary complex, not only from the mineral components but in oxidic composition. The ternary system is thus referred to the fundamental quinary system Na_2O—CaO—MgO—Al_2O_3—SiO_2, for which special polynary graphic constructions and numerical calculation methods are applied. From the discussion of about sixty basalt analyses, Ormont demonstrates distinct groupings in the midst and by the sides of the eutectic curve between the primary crystallization fields of diopside (pyroxene) and plagioclase feldspars. These groups correspond to compositions of *different* suitability as castable raw materials for pavement stones. The best types for this purpose are those of the Rhineland and from Southern Russia (Don District) with a remarkably low viscosity, and characterized by a primary crystallization of pyroxenes. Rather unsuitable are diabases with primary plagioclase which have a high viscosity and form much glass. Intermediate qualities show basalts with both pyroxene and plagioclase simultaneously crystallized on, or near, the eutectic curve. Very distinct effects are exerted by the FeO and TiO_2 contents. In slowly cooled melts magnetite and ilmenite appear, whereas in rapidly cooled melts both iron and titanium enter the pyroxene phases. The asymmetric diagrams of Ormont are a great help in finding the most-suitable compositions of basalts; the common grouping according to the silica contents is inaccurate and often misleading.

9. Experiments corresponding to Ormont's principles were made systematically by P. G. Verbitskiĭ[18] with basic rocks from the Central Dnieper Districts, chiefly ophitic diabases, quartz diabases, and their metamorphic derivatives (with uralitized augite, albitized or scapolitized plagioclase, and the like). The fusion temperatures vary between 1180° and 1300°C. depending on the degree of metamorphic changes and the viscosity. Crystallization sequence in most cases is not in agreement with the classical Rosenbusch sequence rules, fusion temperature also not a simple function of the silica contents, but depends greatly on the sum of silica and alumina on one hand, FeO, MgO, CaO and alkalies on the other hand. The chief requirements

[17] *Vestnik Moskov. Univ. Ser. Fiz. Mat. i Estestven. Nauk*, 1950 (3) 117-133; 1951 (7) 107-121.
[18] *Zapiski Vsesoyuz. Mineral. Obshchestva*, **83**, 1954, 153-157.

for the suitability of basaltic rocks in "petrurgy" (the technology of castable rocks) are discussed by J. Voldán,[19] for materials castable at temperatures below 1300°C. Melts of basic melilite and nepheline basalts (with about 42.2 per cent silica) are highly corrosive to refractory linings of the fusion kilns, and easily crack on cooling because of inner strains between the glass matrix and olivine phenocrysts. In this respect basanites (with 43 to 46 per cent silica) have better qualities, higher viscosity, and better thermal and chemical durability. Plagioclase basalts are lowest in suitability because of their very sluggish crystallization and tendency to remain glassy.

10. A detailed study of Austrian castable basalts for petrurgical purposes was made by E. Eipeltauer and F. Thenner,[20] especially for high-quality wear- and abrasion-resistent gutter grooves, and the like. A coarse-crystalline texture is highly unfavorable for such purposes but an addition of magnetite and some soda makes the texture fine-granular. A thermal after-treatment is then not needed, the mechanical strength being considerably increased. Such cast products are free from the renowned shrink holes of common casts. For satisfactory improvement of cast products the suppression of plagioclase crystallization is essential, whereas magnetite, olivine, and pyroxenes are desirable constituents, which were identified by X-ray diagrams and convincing micrographs of polished sections in reflected light.

11. In many respects, *combustion* slags (better: *ashes*) from hard and soft coals may be compared in their chemical and mineralogical composition to blast furnace slags. A. Nicol and Marthe Dominé-Bergès[21] examined, in this respect, different normal blast furnace slags, a producer slag, and fly ash collected from a Cottrell dust separator, and determined their characteristic crystal phases. Mullite occurred only in the fly ash. Characteristic of all these slags was their hydrolysis behavior and the *pH* value attained. The electrolytic conductance in the supernatant liquid solutions (from powder suspensions) after an exposure to the action of direct current of the intensity of 0.3 milliamperes and 1800 to 5500 volts, was applied over a time up to 1 month, until exhaustion of the soluble. After this treatment the Ca^{2+} in solution was determined. For glassy blast furnace slags, the electrolysis curves rise steeply with time to extraction amounts of 88, 92, and 100 per cent in maximum of available CaO amounts. The *pH* was 9.5 to 10.0, whereas the producer and fly ash slags reached a *pH* of only 7 to 8. Devitrified slags showed a lower extractibility for calcium than the glassy samples. Characteristic also is the dilatometric shrinkage

[19] *Silikat Tech.*, **7**, 1956, 48-53; see also experiments with rocks of Bohemia, in *Sklář a keram.*, **5**, 1955, No. 1 and 2; "Advances in Glass Technology," Plenum Press. New York, 1962, 382-396. on chemical classification of castable basalts see also P. Beyersdorfer, *Silikat Tech.*, **5**, 1954, 381-384, especially for electric arc welding powders.

[20] *Radex Rundschau*, 1958, 224-234. See also extensive investigations on melt-cast products of this class by B. Löcsei, *Epitöanyag*, 1958, 408-414, 446-459; 1959, 247-256.

[21] *Silicates inds.*, **20**, 1955, 410-418.

amount of slag powder compacts in cylindrical samples after a calcination at 500° to 950°C. The highest shrinkage at 700° to 900°C. is distinctly determined by the devitrification whereas the final shrinkage of complete crystallization at highest temperatures is relatively low.

12. J. Endell[22] studied the influence of clayish components in *combustion ashes* derived from a primary kaolinite or hydromica content of the coal sediments. Complicated reactions occur in soft coal ashes, if the natural alkaline earth humates react with the contents in CaO and SO_3, or other sulfur compounds, to form $CaSO_4$, which in its way reacts with silica and alumina in meta-kaolinite to form calcium aluminates and silicates, with a volatilization of the sulfur as SO_2 or SO_3. If clay substance is absent, soft coal ashes often contain calcium sulfide (oldhamite). In this case they become undesirably adherent and form lumpy-sticky combustion ashes. By a high content in iron compounds, soft coal ashes easily develop a calcium ferrite type of slags with very fluid melts. Such ashes are important in their fusion behavior for the control of modern melt chamber furnaces (with fusion pockets), as so-called "short" slags, whereas "long" slags, chiefly from hard coals, show a long range of temperature between softening and consolidation. In practice often coals of both types must be blended to produce a well-running slag of a distinct viscosity. Soft coal (lignite) ashes alone are mostly too high in fluidity, highly clayish coals and their ashes more normal for Cyclone firings.

13. Returning to the mineralogical composition of *blast furnace slags*, we mention the interesting discussion of H. M. Richardson, F. Ball, and G. R. Rigby[23] who compare the occurrence of $3CaO.Al_2O_3$ in blast furnace stack samples, formed when alkalies and CaO react with the firebrick lining, as well as the occasional formation of hedenbergite and ferrosilite. They compared also the compound $Al_2O_3.Fe_2O_3$ with the structure of $\varkappa$-alumina,[24] and iron cordierite, $2FeO.2Al_2O_3.5SiO_2$, especially in the brickwork of the blast furnace hearth. Willemite in the upper cooler portions of the furnace, on the other hand, is made responsible for failure of firebrick linings by the catalytic effect, exerted on the formation of free carbon in the decomposition of CO, and ensuing bursting (cf. Vol. V, Section B). B. G. Baldwin[25] emphasizes the prevailing importance of the system $CaO—MgO—Al_2O_3—SiO_2$ (see below) for the constitution of blast furnace slags because more than 95 per cent of these belong to it. Baldwin determined the Liquidus temperatures of more than 150 of these as

[22] *Ber. deut. keram. Ges.*, **32**, 1955, 69-71; *Silicates inds.*, **22**, 1957, 156-159. Experience with ashes of borax-impregnated wood shavings, which intensely corroded firebrick linings and roofings in McCashney incinerators, is presented by E. R. Segnit, R. M. Liversidge, and T. Gelb, *Australian J. Appl. Sci.*, **14**, 1963, 155-167; see also Vol. V, B. ¶ 109, footnote 241.

[23] *Trans. Intern. Ceram. Congr.*, **3**, Paris 1952, 173-181.

[24] Cf. A. Muan and C. L. Gee, *J. Am. Ceram. Soc.*, **39**, 1956, especially pp. 211 ff.

[25] *J. Iron Steel Inst. (London)*, **186**, 1957, 388-395.

a function of the composition, with a direct observation of their crystallization. The mean Liquidus temperatures are about 1500°C. in maximum, but periodical fluctuations in temperature occur in practice. To be free-flowing the slags must be above their Liquidus temperatures. The viscosity is chiefly determined by the basicity, specifically by the ratio $(CaO + MgO)/(SiO_2 + Al_2O_3)$. A thorough comparison of physico-chemical properties of the slags as calculated form their composition in the quaternary system $CaO—MgO—Al_2O_3—SiO_2$ with slags from the basic process and for foundry iron was made by N. L. Zhilo, A. V. Rudneva, and G. A. Sokolov,[26] with special reference to the behavior of potash in basic and acidic slags. The practically undesirable effects of alkalies on the properties of the slags are caused by the formation of the molecules of kaliophilite and leucite, both with framework structures of a high stability and, therefore, with high fusion points (see also C. ¶ 73 f.).

14. In a fundamental study of phase equilibria diagrams of the system $CaO—MgO—Al_2O_3—SiO_2$, built up from the system $CaO—MgO—SiO_2$ by systematic stepwise additions of 5 per cent alumina up to 35 per cent of this oxide (cf. Vol. III, B. ¶ 283 f.), E. F. Osborn, R. C. DeVries, K. G. Gee, and H. M. Kraner[27] provided rich numerical and graphic data for solution of the practically important problem of determination of the *optimum* composition of blast furnace slags. Such an optimum must correspond to a slag which is entirely liquid at service temperatures, has a minimum in viscosity, and in addition a maximum in desulfurization potential (cf. C. ¶¶ 102,

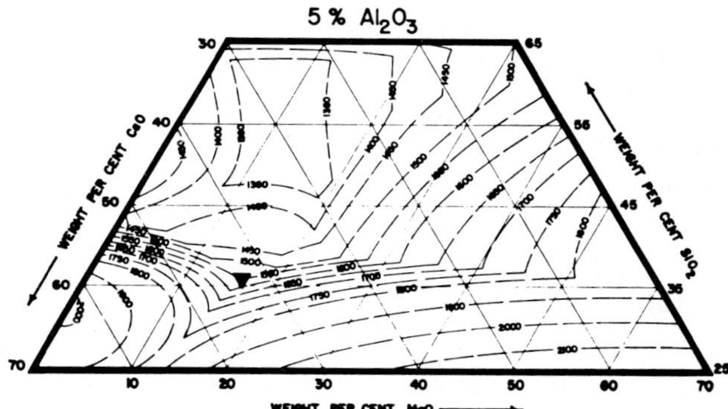

FIG. C. 1. Diagram of a part of the Liquidus surface of the 5 per cent alumina plane showing the position of optimum compositions for blast furnace slags (indicated by the full triangle). (Osborn, De Vries, Gee, and Kraner).

[26] *Izvest. Akad. Nauk S.S.S.R.*, Otdel. Tekhn. Nauk, 1957 (6) 37-42.

[27] *J. Metals*, **6**, 1954 (1) 33-45. See also studies of A. T. Prince, *J. Am. Ceram. Soc.*, **37**, 1954, 402-408, on the Liquidus relationships of the 10 per cent MgO plane in the same quaternary system, with reference to slag problems (cf. Vol. III, B. ¶ 275), especially for their use in hydraulic binders with Portland Cement (Vol. V, Section C).

112). In the three-dimensional representation, optimum compositions are found in an inclined-cylindrical volume intersecting the planes of constant alumina contents, and systematically shifted towards the side plane $CaO—Al_2O_3—SiO_2$ with increasing alumina contents (cf. Figs. C. 1-4, indicated by small triangles, and Fig. C. 5 for an

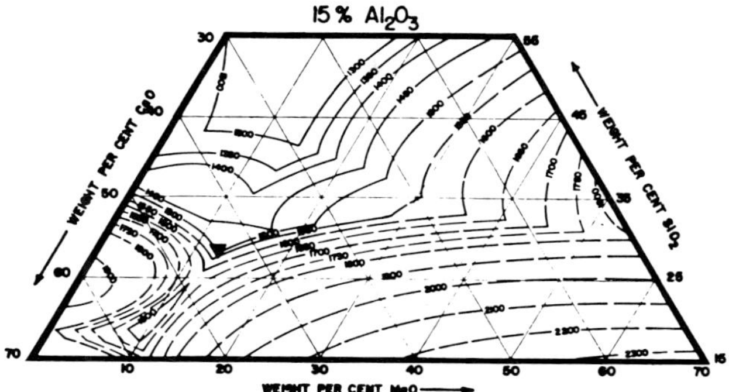

FIG. C. 2. Same as Fig. C. 1, for the 15 per cent alumina plane. (Osborn, De Vries, Gee, and Kraner).

example taken from practice). The primary crystallization volume of monticellite extends upward from the base system $CaO—MgO—SiO_2$ as a four-sided prism terminating against the spinel volume before reaching the 15 per cent Al_2O_3 plane. The merwinite volume extends in a similar manner from the base upward through the

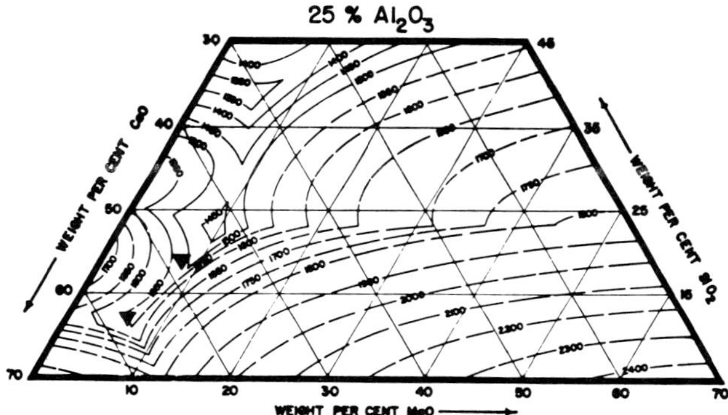

FIG. C. 3. Same as Fig. C. 1, for the 25 per cent alumina plane. (Osborn, De Vries, Gee, and Kraner).

20 per cent Al_2O_3 plane before it is truncated. As seen from a three-dimensional model constructed for the 10 per cent Al_2O_3 compositions in an evolution of the isotherms to contour lines on the Liquidus-surfaces, the steep rise of the Ca_2SiO_4 and

MgO L.-surfaces in contrast with the plateau-like form of monticellite, merwinite, and melilite (åkermanite-gehlenite) surfaces is particularly notable.

15. There is a re-entering angle of the merwinite plateau into the steeply rising surfaces of Ca_2SiO_4 and MgO, which is the key to the position of optimum compo-

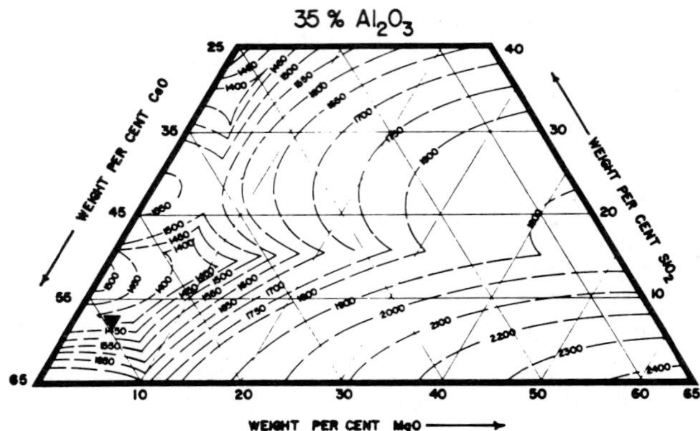

FIG. C. 4. Same as Fig. C. 1, for the 35 per cent alumina plane. (Osborn, De Vries, Gee, and Kraner).

sitions for blast furnace slags. These optimum compositions lie in the upper, narrow portion of the plateau as it approaches those steep fronts of the Ca_2SiO_4 and MgO volumes at lowest silica contents. The Liquidus temperatures are lower than actual

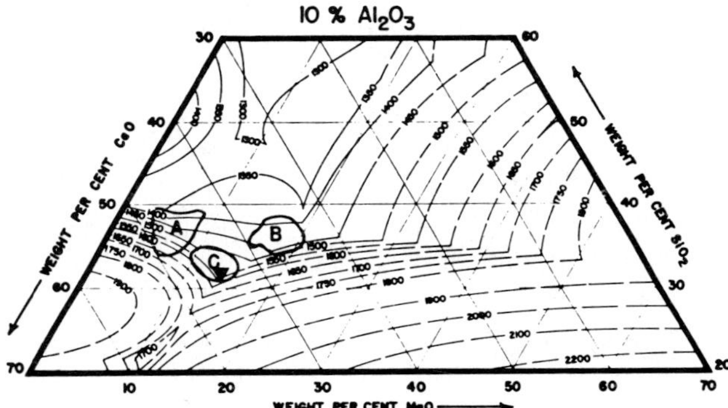

FIG. C. 5. Diagram of a part of the 10 per cent alumina plane, showing the position of representative blast furnace slags, recalculated to 100 per cent ($CaO + MgO + Al_2O_3 + SiO_2$). The areas delimited by heavy lines (*A*, *B*, *C*) include daily average analyses of slags from three furnaces running with 10 ± 1 per cent alumina. Slag compositions *C* are excellent, they correspond to the optimum region. Slags *A* and *B* are not well-efficient desulfurizers. (Osborn, De Vries, Gee, and Kraner).

temperatures determined for slags in the blast furnace. The plateau permits a rather considerable latitude in variation of the slag compositions, and there is no operational difficulty in control of the blast furnace concerning the slag. Only if the slag compositions come close to those steep fronts, trouble may result when the slag becomes somewhat low in silica and moves into the high-Liquidus regions. Then crystallization may occur, which immediately raises the viscosity and makes the slag "gummy" (sticky). For the commonly used range of alumina contents between 8 and 15 per cent, the variation in CaO and MgO for optimum slag compositions varies only slightly, centering around 14 per cent MgO, and 44 per cent CaO. Silica is about complementary to alumina, being 32 per cent at 10 per cent Al_2O_3 and decreasing as alumina is increased.

16. P. D. S. St. Pierre[28] criticized somewhat the conclusiveness of Osborn's and other authors' results, namely by emphasizing the too large interpolations needed to determine temperatures in the vicinity of real blast furnace slag compositions. Interpolations of only 10 per cent can lead to serious errors in Liquidus temperatures. The reason for these discrepancies are seen by St. Pierre to lie in the presence of accessorial oxides in real slags in contrast to the pure system $CaO—MgO—Al_2O_3—SiO_2$. Especially if a slight content in FeO is neglected, the predictions from the pure system may be badly misleading, as was demonstrated by experimental studies of the fluxing action of systematic additions of MgO and CaO (corresponding to the use of dolomitic limestones) on the fusion temperature of 1 : 4 and 1 : 3 alumina : silica ratios of the "gangue," containing the equivalent to 3 per cent FeO in the final slag. The conclusion is made that it is *not* possible to generalize on the behavior of MgO in slags, even if the discussion is restricted to observations of slags with 40 to 55 per cent of fluxing material (typical of modern practice).

17. If the concentration in FeO in slags becomes high, as is the case in open hearth slags, or in special reaction products of slag type, such as between oxidized portions of a steel melt with refractories, or with foundry sands as mold materials, the conditions of crystallization of complex iron containing silicates are valid, as E. F. Osborn[29] outlined in a general discussion of such equilibria diagrams and slag crystallizations (these special facts will be discussed in Vol. III, B. ¶ 283 f.). A particularly instructive example of reactions of this kind was demonstrated by W. Eitel and N. Köppen[30] in the reaction slags of Portland cement-bonded foundry sand with a steel cast body in which the oxidized, fayalite, and magnetite-rich portions of the steel melt easily react in contact with the highly calcareous silicates of the

[28] *Can. Mining Met. Bull.*, **59**, 1956, 224-231.

[29] Pttsburgh Regional Technical Meeting, American Iron and Steel Institute, October 14, 1954, 33 pp. See also E. F. Osborn and A. Muan, on steelplant refractories, in "Electrical Furnace Steelmaking," Interscience Publishers, New York, 1963, Vol. II, pp. 215-281.

[30] *Giesserei*, 1954, 701-707; 1955, 761-765.

cement matrix. They form crystalline solutions of the series $CaSiO_3$—$FeSiO_3$ (cf. Vol. III, B. ¶ 333), also pseudowollastonite in characteristic crystal forms, as observed in polished sections, and finally tridymite in the siliceous reaction glass. Such partly iron-rich slag melts are rather fluid and may adhere firmly to the steel castings as a "burnt" layer. The conditions of cement-bonded foundry sands for steel casting have some analogy with the siliceous (acidic) fettling sinters in rammed soles of open hearth furnaces as described by I. S. Kaĭnarskiĭ and L. I. Karyakin.[31] There is a distinct zoning, first on the hottest side an entirely glassy product with about 97 per cent silica, of obsidian-like aspect, followed by a densely sintered less siliceous glass zone with cristobalite as devitrification product, and an "exhausted" sand sinter richly impregnated with fayalite, magnetite, some pseudowollastonite, bustamite, and the like. The porosity of the second glassy layer is important for penetration of the aggressive melts into the rammed sole. The silica content of the iron-calcium silicate zone is about 71 per cent. Residual quartz and occasional tridymite in characteristic twins are normal in it. Cristobalite appears in a scaly or acicular habitus, even as irregularly aggregated meta-cristobalite, and excellent paramorphs of cristobalite after quartz and tridymite are observed.

18. The conditions of such formations are also similar to those discussed by W. F. Ford and J. White[32] when highly acidic and basic slag compositions (e.g., from magnesite and dolomite) come into contact. The differences in viscosity are then high and equilibria of the system CaO—MgO—FeO—Fe_2O_3—SiO_2 are concerned. Ford and White, for the simplified conditions of the system CaO—MgO—Fe_2O_3—SiO_2 report tabular calculations of amounts of the liquid phase formed, especially on the working faces of refractories, by slag absorption. These reactions in the open hearth linings were further discussed by H. M. Kraner[33] especially demonstrating the field near the silica apex of the system CaO—FeO—SiO_2 at 1285°C., CaO—MgO—SiO_2 at 1250°C., and the solubility of magnesia in CaO—FeO—SiO_2 slags at 1600°C. (cf. Fig. C. 6) according to K. L. Fetters and J. Chipman.[34] They discussed related problems influencing the service life of an open hearth, with practical instructions. For this decisive problem the detrimental effect of alumina, or of small contents in alkalies, on silica brick refractories (cf. Vol. V, Section B) is not less important. Kraner also emphasizes the tremendous effects of the oxygen partial pressure conditions which bring about a considerably stronger corrosion on silica under reducing conditions of the furnace atmosphere, and, therefore, an accelerated wear of the

[31] *Voprosy Petrog. i Mineral., Akad. Nauk S.S.S.R.*, **2**, 1953, 330-341.

[32] *Refractories J.*, **34**, 1958 (4) 171-174.

[33] *Am. Ceram. Soc. Bull.*, **37**, 1958, 312-316. See also H. M. Kraner and A. Muan, on reactions in ferromanganese blast furnace hearth refractories, in *Trans. Met. Soc., Am. Inst. Mining Met. Engrs.*, **224**, 1962, 763-770.

[34] *Am. Inst. Mining Met. Engrs. Trans.*, **145**, 1941, 95-112; *Am. Inst. Mining Met. Eng. Tech. Publ.*, No. 1316, 1941, 13 pp.

roof portions, as is seen from the equilibrium diagrams of the system Fe—Si—O according to A. Muan (cf. Vol. III, B. ¶ 351 f., and C. ¶ 94).[35] In the regenerative chambers and checkerwork of a basic open hearth furnace, I. A. Pitskhelauri[36] studied the zoning of silica brick, with α-$CaSiO_3$, tridymite in an iron oxide-impregnated melt (refractive index of the glass 1.600), as the reaction product with Fe_2O_3 dust, and another pale-yellow glass (n = 1.560 to 1.590), with hematite and charac-

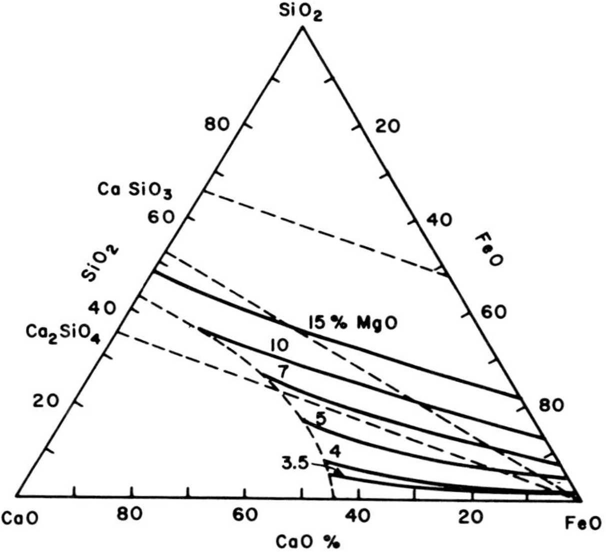

Fig. C. 6. Solubility of magnesia in slags of the system SiO_2—CaO—FeO, at 1600°C. (Fetters and Chipman).

teristic iron-titania pyroxenes. In fireclay brick the reaction zones contain some anorthite in characteristic polysynthetic twins. The surface of such brick was covered with a glass containing a similar pyroxene of complex composition, and also crusts with some alkalies, ZnO, PbO, CuO. In steel-metallurgical kilns running at 1350° to 1450°C. exposed in their lining to iron scale dusts, L. I. Karyakin and O. M. Margulis[37] observed in such zones also the unusual short-prismatic ("isometric"), 2.5 per cent Fe_2O_3 containing mullite (mentioned in Vol. III, B. ¶ 128 f.), in a glass of n = 1.540 to 1.550, besides normal fine-acicular mullite.

19. The important problems of steel ingot melts under the influence of reaction slags with refractory materials were investigated by O. M. Margulis and A. G.

[35] *Am. Inst. Mining Met. Engrs. Trans.*, 203, 1955, 965-976; *J. Metals*, 7, 1955, 965-976.
[36] *Ogneupory*, 21, 1956, 207-211.
[37] *Ogneupory*, 22, 1957, 123-126.

Karaulov[38] by the method of radioactive tracing, namely by introduction of Ca^{45} into the lining materials. Previous experiments of V. A. Grigoryan and A. M. Samarin[39] had shown that calcium does not migrate into the pans but only into a thin layer which is removed after the first melt. The fired brick marked with Ca^{45} were built into the lining of pans, siphons, and other structures, to come into direct contact with the steel bath. The exogenic inclusions in the metal were controlled in their amounts and radioactivity after isolation by electrolysis (cf. C. ¶ 90). The resulting data make evident that the amount of refractory material dissolved in the inclusions was relatively small, and depended on their chemical nature. It was say 4.6 per cent for fireclay lining, 2.5 per cent for kaolin brick, and 1.5 per cent for a highly aluminous body. The relative amount of oxide in the inclusions, which is a measure for contamination of the steel, was 1.82 for fireclay, 1.62 for kaolin brick, and 1.56 for "superduty" material, in agreement with the practical experience that runner channels after service look worst by steel corrosion of fireclay linings, best for graphite-fireclay bodies. More recently also M. I. Tsekhanskiĭ, K. B. Khusnoyarov, and G. D. Susloparov[40] used Ca^{45} for tracing refractories of runner brick and for finding the origin of silicate inclusions from these in steel castings. In slag floating on the surface of the metal bath the content in refractory material was about 20 to 24 per cent; 2 to 3 per cent of these were caused by dissolution of the ladle, 18 to 22 per cent from the runner brick. The total content of refractory inclusions and their distribution in the ingot, however, was largely a matter of chance.

20. For constitution of many metallurgical slags the orthosilicate portions of the system $CaO—FeO—SiO_2$ are of high importance. W. C. Allen and R. B. Snow[41] (cf. Vol. III, B. ¶ 320) investigated this portion extensively in the fields between fayalite, Ca_2SiO_4, wüstite, and CaO. Characteristic of this system is the absence of a ternary eutectic, crystallization of olivine-type crystalline solutions, and the appearance of a cross- and a parallel-twinned modification of Ca_2SiO_4, of $3CaO.SiO_2$, free lime, and calcium-wüstite crystalline solution phases from corresponding slag compositions. The lime phase may contain up to 10 per cent FeO in crystalline solution, whereas the wüstite phase proper can take up 28 per cent of CaO. Particularly important is the crystallization of $3CaO.SiO_2$ in open hearth slags and also in hearth bottom refractories. Even the presence of only 0.3 to 0.4 per cent sulfur in the melts on the section Ca_2SiO_4—wüstite can considerably shift the equilibria temperatures to as low as 1340°C., and especially increase the amount of the liquid phase, in con-

[38] *Ogneupory*, **21**, 1956, 253-258. On the method for the introduction of Ca^{45} into refractories in ceramic production see Z. V. Levintovich, A. N. Lyulichev, O. M. Margulis, and D. M. Shakhtin, *Ibid.*, **22**, 1957 (1) 29-31. See also *Doklady Akad. Nauk S.S.S.R.*, **109**, 1956, 821-823.

[39] *Izvest. Akad. Nauk S.S.S.R.*, Otdel. Tekhnich. Nauk 1954 (3) 91-101.

[40] *Ogneupory*, **23**, 1958 (2) 82-87.

[41] *J. Am. Ceram. Soc.*, **38**, 1955, 264-280.

trast with the fusion behavior of the thoroughly pure ingredients which show sharp equilibria temperature effects. Characteristic also is the latent tendency of melts in the system $CaO.FeO.SiO_2$—wüstite to unmix as seen from the S-shaped flexure on the Liquidus-curve in this system, with a pseudoeutectic of only 1200°C. These phenomena also illustrate the importance of iron oxide contents in such melts. In the section fayalite—wüstite a similar flexure occurs and a pseudoeutectic of 1177°C., with an equal influence of the presence of Fe_2O_3. A nonidentified "yellow" phase also was observed in polished sections of mill scale, as an interstitial material between crystals of wüstite. It may be a sulfide (in every case it is not magnetite).

21. The appearance of $3CaO.SiO_2$ and of "parallel-twinned" Ca_2SiO_4 is particularly important for slags in the bottoms of open hearths. This kind of Ca_2SiO_4 does not dust and is evidently stabilized by FeO contents. Fayalite is not easily reduced to metallic iron in the blast furnace until it has become liquid. CaO is made more resistant to atmospheric hydration by its limited crystalline solubility with FeO. Important is the observation that manganese-free steel production slags contain a high amount of undissolved Ca_2SiO_4 even when the ratio $CaO : SiO_2$ is 3 : 1. At higher ratios, however, the melting may be more complete owing to the low-temperature valley along the boundary of the $3CaO.SiO_2$-field which separates the high-melting fields of Ca_2SiO_4 and CaO (cf. Vol. III, B. ¶ 332).

22. F. Z. Dolkart and L. A. Kuz'mina[42] observed interesting intermediate forms of the inversion of β into γ-Ca_2SiO_4 from reaction products of slaked slags on the hearth of a steel furnace. Ca_2SiO_4 is here connected with anisotropic ferrites (chiefly $2CaO.Fe_2O_3$ in crystalline solution with $4CaO.Al_2O_3.Fe_2O_3$) and with periclase. The inversion $\beta \rightarrow \gamma$-Ca_2SiO_4 is stabilized by FeO over a period of more than three years, in agreement with previous observations of this phenomenon by P. P. Budnikov and R. D. Azelitskaya.[43]

23. The great significance of the $CaO : SiO_2$ ratio in blast furnace slags for their suitability in large highway building projects was emphasized by L. Blondiau[44] in his extensive report on practical observations of their use as aggregate (in the place of high-grade porphyry) for roads and airport runways in Belgium. Especially the porosity of the slag which must be very low (only 1 per cent and less) is influenced by the $CaO : SiO_2$ ratio. Highly porous slags are to be rejected and Ca_2SiO_4 containing slags should be avoided. For production of blast furnace slag cements similar strict

[42] *Ogneupory*, **23**, 1958 (1) 41-42; on similar phenomena in reaction products of aluminum silicate refractories as a lining in lime shaft kilns see also: Fr. Massazza, *Ceramica* (*Milan*), **14**, 1958 (8) 47-54 (cf. Vol. V, Section B).

[43] *Doklady Akad. Nauk S.S.S.R.*, **108**, 1956, 515-517.

[44] *Silicates inds.*, **20**, 1955, 97-105; 147-161, 205-211; *Rev. matériaux construct. et trav. publ.*, **487**, 1956, 83-94; **488**, 1956, 109-126.

specifications for the ratios $CaO : SiO_2$ and $CaO : Al_2O_3$ are developed (cf. Vol. V, Section C), and special microscopic tests in polished sections are recommended for the examination of slags for the presence of detrimental constituents. For production of *foam slags* important as heat insulation material in building construction, W. Ruopp[45] gave analogous recommendations, especially for slag melts foamed with high-pressure steam for an adequate thermal treatment to remove inner strains.

24. *Pyroxenes* of the crystalline solution series $FeSiO_3$—$MgSiO_3$ were described by N. N. Kurtseva[46] from electrometallurgical slags of nickel processing, with well-developed magnesium-ferrosilite crystals of tabular or columnar habitus. This special slag contained 70 per cent fayalite, 10 per cent cristobalite, 3 to 4 per cent of the ferrosilite, 9 per cent glass, and 7 to 8 per cent FeS, besides droplets of NiS, Cu_2S, and magnetite. The columnar pyroxene contained 58 per cent $FeSiO_3$, 42 per cent $MgSiO_3$, the tabular pyroxene 76 $FeSiO_3$, and 24 $MgSiO_3$, but no NiO. Both pyroxenes are of the clinoenstatite type in their X-ray diagrams. We mentioned already in C. ¶ 6 that *hardystonite*, $2CaO.ZnO.2SiO_2$, is a rather frequent melilite-type phase in zinc slags of the lead blast furnace process. Another blast furnace slag with 3 per cent chromium oxide and corresponding synthetic compositions were examined by Y. Matsuhita[47] with a $CaO : SiO_2$ ratio between 0.8 and 1.2 which after fusion at 1360° to 1400°C. for 2 to 6 hours and cooling *in vacuo* showed a very interesting occurrence of predominantly CrO as the typical chromium compound.

25. *Titanium-enriched* slags with crystal phases partly unknown before came rapidly into the foreground of metallurgical interest in the last decade. F. Z. Dolkart[48] and N. V. Gul'ko described very hard slags, strikingly low in silica (about 3 per cent only), but high in TiO_2 (20 per cent) and alumina (70 per cent) from the alumino-thermic ferrotitanium metallurgical process, with a high content in pleochroic (with blue and greenish colors) corundum, and $CaO.6Al_2O_3$ (cf. Vol. III, B. ¶ 237, ¶ 278). Particularly interesting are the structure-defective rutile phases and lower titanium oxides which if heated in an oxidizing atmosphere above 1450°C. are changed into TiO_2. Characteristic also is the compound $Al_2O_3.TiO_2$ (cf. Vol. V, Section B). Slags of this type (with TiO, Ti_2O_3, and TiN) have a certain practical suitability as additions to refractories, as in magnesite brick batches, because of their high refractoriness (up to 1700° and even 1800°C.).[49]

26. Also A. S. Berezhnoĭ and N. V. Gul'ko[50] described such refractories prepared by the calcination of a mixture of 60 per cent Ti-slag with 40 per cent magnesia,

[45] *Stahl u. Eisen*, **77**, 1957, 36-43.
[46] *Doklady Akad. Nauk S.S.S.R.*, **92**, 1953, 153-155.
[47] *Tetsu to Hagane*, **42**, 1956, 945-950.
[48] *Doklady Akad. Nauk S.S.S.R.*, **98**, 1954, 137-139.
[49] F. Z. Dolkart, *Ogneupory*, **21**, 1956, 300-305.
[50] *Ukrain. Khim. Zhur.*, **21**, 1955, 158-166.

but the shrinkage upon firing is too great. Dolkart started from mixtures of 30 per cent Ti-slag and 70 per cent magnesia, plastified by sulfite cellulose waste brine. He described the firing products as a mixture of 35 per cent periclase, 50 per cent of a crystalline solution (of spinel type) of $MgO.Al_2O_3$, Mg_2TiO_4, little $MgO.Fe_2O_3$, monticellite and some glass. Bricks of this composition are good for linings in Portland cement burning kilns, furnaces for burning metallurgical dolomite, or finally for electrosteel furnaces. Berezhnoĭ and Gul'ko studied the system $MgO—Al_2O_3—TiO_2$ in extenso, with geikielite ($MgTiO_3$) as a typical crystal phase, and the already mentioned spinel crystalline solutions of Mg_2TiO_4 with $Al_2O_3.TiO_2$. Characteristic is the reaction of alumina with $MgTiO_3$, to form crystalline solutions of ($MgO.Al_2O_3$ + Mg_2TiO_4) on one hand, ($Al_2O_3.TiO_2$ + $MgO.2TiO_2$) on the other hand. In the ternary system there are two eutectics, the first with corundum, spinel, and $Al_2O_3.TiO_2$, the second with spinel crystalline solutions, geikielite, and anosovite. The latter eutectic has a fusion temperature of about 1570°C.

27. With the name *anosovite*, V. V. Lapin, N. N. Kurtseva, and O. P. Ostrogorskaya[51] defined a crystal phase isolated from an aluminous Ti-steel slag (70 per cent Al_2O_3, 26 per cent TiO_2), which is associated with perovskite ($CaTiO_3$), titanite, and iron metal, and is an isomorphous complex of $Al_2O_3.TiO_2$, $RO.TiO_2$ ($R = Mg$, Fe^{2+}, Mn^{2+}), $R'_2O_3.TiO_2$ ($R' = Ti^{3+}$, V^{3+}, Cr^{3+}) and some TiO_2. The excess titania is similar to that observed in pseudobrookite. Anosovite is described as an independent phase of weak bireflection, brownish interference colors, and a reflection power of 11 to 12 per cent in polished sections, differing from another more "rose-colored" crystal phase similar to pure Ti_2O_3, with a reflection power of only 7 per cent, but a distinct anisotropy. Both phases differ in their X-ray diagrams and differential-thermal analysis curves during the oxidation to anatase as the end phase. A. V. Rudneva and T. Ya. Malysheva[52] further studied the complex composition of the anosovite phase, starting from the compound $TiO.2TiO_2 = Ti_3O_5$, or $(RO, R'_2O_3).2TiO_2$, with $R = Ti^{2+}, Mg^{2+}, Fe^{2+}$; $R' = Al^{3+}, Ti^{2+}, Fe^{3+}$. Anosovite could be isolated by a treatment of fine-powdered titaniferous slag with 10 per cent HCl solution on the water bath for 2 hours, then with 30 per cent KOH solution to remove silicate glass. The analyses of pure anosovite showed 6 to 7 per cent MgO, 28 to 30 per cent Ti_2O_3, sometimes with excess Ti_2O_3. The degree of oxidation of Ti can vary in some limits with the oxygen partial pressure of the furnace atmosphere. The "rose-colored phase"

[51] *Doklady Akad. Nauk S.S.S.R.*, **109**, 1956, 824-827. On general importance of reflected light examination of polished sections in the metal microscope for such and related slag types see R. B. Snow, *J. Am. Ceram. Soc.*, **36**, 1953, 299-304. It is in this connection highly interesting that A. V. Rudneva and T. Ya. Malysheva, *Doklady Akad. Nauk S.S.S.R.*, **136**, 1961, 191-194, recently described two new slag minerals, called ceftosil and celanite, the first a fluorosilicate of calcium and rare earths, the second a calcium titanate of perovskite type with rare earths, similar to knopite.

[52] *Doklady Akad. Nauk S.S.S.R.*, **115**, 1957, 787-790.

mentioned above is evidently a Ti_2O_3-enriched anosovite. Characteristic is the absence of Ti-augite and spinel in Ti-Al slags as used in these studies. Alumina is also low in the rapidly chilled TiO_2 and Ti_2O_3 containing glass, although it is present in anosovite.

28. Silicate slags high in titania, as the most abundant silicate phase, show a complex Ti-augite, and anosovite, described by A. V. Rudneva, M. S. Model', and T. Ya. Malysheva,[53] associated with crystalline solutions of ilmenite-geikielite-pyrophanite-crichtonite types (series Mg_2TiO_4—Fe_2TiO_4, and RO.TiO_2, with R = Mg^{2+}, Fe^{2+}, Mn^{2+}) and anosovite. Cubic crystals of 2RO.TiO_2 with FeO can be reduced into iron metal at 1100°C. The distinction of different mineral phases was possible in their characteristic reflection and anisotropy effects in polished sections. M. S. Model'[54] made a much detailed study of the different titanium oxide phases of industrial Ti slags and showed that anosovite of the simplest composition TiO.2TiO_2 is especially typical of aluminous and magnesian Ti slags. Another phase of the same composition Ti_3O_5 is observed if rutile is reduced, or in Ti-slags containing CaO. Model' further distinguished a phase Ti_2O_3.4TiO_2(Ti_6O_{11}) which forms crystalline solutions with rutile. It is never observed in magnesian Ti-slags which always contain rutile and anosovite. These highly complex phase relations illustrate the tremendous importance of defect structures (by subtraction or substitution), or semiconductor types in the system TiO_2—TiO. Pure anosovite is probably metastable since it is only found in crystalline solutions with MgO.2TiO_2, FeO.2TiO_2, and Al_2O_3.TiO_2 whereas the second phase Ti_3O_5 always crystallizes from the synthetic melts directly.

29. What Rudneva and Malysheva called "opaque rutile"[55] is a reduction product of pure TiO_2 with the ratio Ti : O = 1 : 1.9 (20 per cent Ti_2O_3). In reflected light such a phase shows a distinct anisotropy and a stronger reflection (16 to 20 per cent) than anosovite (11 to 13 per cent) and pure Ti_2O_3 (14 per cent). This property the "opaque rutile" has in common with Ti_6O_{11}, and all transitions between both are possible in their optical qualities, although the X-ray diagrams are different. Slags

[53] *Doklady Akad. Nauk S.S.S.R.*,, **115**, 1957, 141-144.

[54] *Doklady Akad. Nauk S.S.S.R.*, **124**, 1959, 887-889. We emphasize in this connection the occurrence of dititanates, say FeO.2TiO_2, in the matrix of highly titaniferous alumina (bauxite) slags which are used as abrasives and were described with petrographic methods by H. N. Baumann, *Am. Ceram. Soc. Bull.*, **35**, 1956, 387-390. Such a dititanate matrix may account for toughness of this kind of abrasives.

[55] *Doklady Akad. Nauk S.S.S.R.*,, **125**, 1959, 363-365; cf. M. E. Filonenko, V. I. Kudryavtsev, and I. V. Lavrov, *Ibid.*, **86**, 1952, 561-564, further St. Anderson, B. Collén, U. Kuylenstierna, and A. Magnéli, *Acta Chem. Scand.*, **11**, 1957, 1641-1652, 1653-1657 (X-ray examinations). On crystal structure of anosovite see G. S. Zhdanov and A. A. Rusakov, *Trudy Inst. Krist. Akad. Nauk S.S.S.R.*, **9**, 1954, 165-210; it is similar to that of pseudobrookite. Recent contributions on the phases Ti_2O_3 and Ti_3O_5 see by F. Trojer, *Radex-Rundschau*, 1962, 212-218.

with less than 30 per cent Ti_2O_3 show only one crystal phase of the series TiO_2—Ti_2O_3. If the ratio Ti : O is below 1.80, Ti_3O_5 appears in the modification different from anosovite. Ti_6O_{11} is not formed from complex slags of the system TiO_2—Ti_2O_3—FeO, especially not from those containing constituents soluble in anosovite (MgO, MnO, alumina, silica). Anosovite is particularly frequent in electrosteel furnace slags with 10 to 30 per cent of those oxides. Under the highest reduction potentials opaque rutile disappears completely and Ti_2O_3 is predominant.

30. As "*tagirovite*" A. V. Rudneva[56] described the ilmenite-type Ti_2O_3 occurring in titanium slags. It is also most probably the "rose-colored" phase described by V. V. Lapin *et al.* (see above). A rather constant alumina content of 6 to 7 per cent occurs in it, illustrating the high degree of isomorphic relations between aluminum and titanium ions in corundum-type structures. In Ti-augite, anosovite, and the like, TiO and TiO_2 only appear as contaminations and accessories, if not due to analytical errors. Characteristic of synthetic ilmenite of tagirovite type in microscopic examination are the spindle-shaped streaks with a striking anisotropy and distinct birefringence, in yellowish or pinkish-brown colors. Anosovite is formed at higher temperatures of reduction (1150° to 1250°C) and surrounds, in typical reaction rims, the unchanged ilmenite with tagirovite spindles and iron metal patches. The reaction can proceed to a complete replacement of the rhombohedral ilmenite structure by orthorhombic anosovite.

31. *Baĭkovite* is called by A. V. Rudneva and T. Ya Malysheva[57] a dark-brown titanium mineral in Ti silicate steel slags, associated with Ti-augite and melilite, of the unusual composition $2CaO.3MgO.Ti_2O_3.2TiO_2.2SiO_2$. Evidently it is a product of highly reducing conditions of the slags at 1500° to 1600°C., whereas Ti-augite and melilite crystallized not much above 1300° to 1350°C. The X-ray powder diagram of baĭkovite is similar to, but distinctly different in details from, that of Ti-augite. Isolated crystals have the aspect of low symmetry, either monoclinic or triclinic.

32. The great importance of *spinel* phases for metallurgical processes, chiefly in steel refining in their often complex chemical and mineralogical characters was extensively discussed in a report given by E. F. Osborn,[58] especially dealing with refractories of the magnesite and chromite group. Besides formation of magnetite in the fundamental system Fe—O, and in the system CaO—FeO—MgO—SiO_2, the

[56] *Doklady Akad. Nauk S.S.S.R.*, **125**, 1959, 149-152.

[57] *Doklady Akad. Nauk S.S.S.R.*, **130**, 1960, 1329-1332. On Ti^{3+} containing Ti-augites cf. A. M. Tsvetkov, *Trudy Inst. Geol. Nauk, Petrog. Sektor im. Levinson-Lessinga, Akad. Nauk S.S.S.R.*, Petrog. Ser. No. 41, 1951, and prior observation of E. V. Britske, K. Kh. Tagirov, and I. V. Shmanink, *Bull. acad. sci. U.R.S.S.*, Chem.-Tekhnol. Ser. 1941 (1) 13-21; (2) 9-40, on Ti silicate slags from blast furnace, smelting with Ti-magnetite, ilmenite, and related problems.

[58] American Iron and Steel Inst., Chicago Regional Technical Meeting, October 17, 1956, 16 pp. See also B. ¶ 17, footnote 58 (with A. Muan).

magnetite-aluminum spinels are only stable at high temperatures. They break down below 1380° to 1410°C. into other structures, such as hematite, $FeO.Al_2O_3$, and corundum as previously shown by A. Muan and C. L. Gee (cf. Vol. III, B. ¶ 366). We have to refer further to the system $FeO—Al_2O_3$ with the formation of hematite and hercynite, $FeO.Al_2O_3$, as the characteristic spinel phase, studied by W. A. Fischer and A. Hoffmann[59] (cf. Vol. III, B. ¶ 341). FeO forms with $FeO.Al_2O_3$ an eutectic melt at 1330°C. containing about 5 per cent alumina. Pure $FeO.Al_2O_3$ melts at 1780°C. It forms crystalline solutions with alumina from 1500° to 1750°C. with a highest alumina content of 64.2 per cent at 1750°C. and another eutectic with alumina and this saturated crystalline solution of $FeO.Al_2O_3$ with 39 to 63 molecular per cent Fe_3O_4 (magnetite). This formation of crystalline solutions with magnetite is a consequence of the constant presence of some trivalent iron in equilibrium with divalent iron and oxygen under the partial pressure of the furnace atmosphere.

33. Details of the fundamental system Fe—Al—O comprising these complex phenomena, were recently given by L. M. Atlas and W. K. Sumida[60] also including the equilibria of the gas phase with Fe metal, wüstite, and the complex spinel phases mentioned above. Iron-rich spinels tolerate a considerable cation deficit, but this property decreases sharply with increasing alumina concentrations. Therefore, the spinel field is at maximum in high iron concentrations and converges towards $FeO.Al_2O_3$. The structure of the compound $Al_2O_3.Fe_2O_3$ was determined as one of the so-called $\varkappa$-alumina type. The system $FeO—Al_2O_3$ is not strictly binary. Iron metal and wüstite are in equilibrium with a spinel varying from 83 per cent $FeO.Al_2O_3$ and 17 magnetite at 1000°C. to higher concentrations in aluminum at higher temperatures. Particular equilibrium details are discussed as soon as a liquid phase appears in the system FeO—MgO—O, the liquid varying from oxygen-rich wüstite to a wüstite-spinel eutectic.

34. The *phosphate* slags of the basic open hearth and steel converter (Thomas) process have their particular importance as phosphate fertilizers. G. Trömel[61] gave an excellent report on their suitability for this purpose. Phosphate-poor slags chiefly of $CaSiO_3$ type, from electric furnace smelting of natural phosphate rocks, have recently had a certain interest as light-weight aggregate material for building construction when bloated by steam in the liquid state, as W. C. Scott and W. D.

[59] *Arch. Eisenhüttenw.*, **27**, 1956, 343-346.

[60] *J. Am. Ceram. Soc.*, **41**, 1958, 150-160.

[61] In Ullmann's "Enzyklopädie der Technischen Chemie," 3rd edition, Vol. 6, 1955, especially pp. 122-155, including industrial superphosphate production and specification tests. A most valuable discussion on production of manganese, vanadium, chromium, etc., in basic phosphate slags besides their use as fertilizers was also given by G. Trömel in *Stahl u. Eisen*, **74**, 1954, 1275-1281, with important references.

Sandberg[62] described (C. ¶ 23 on foam slags, and Vol. V, Section B). The citric acid solubility of phosphate-rich fertilizer slag is much unfavorably influenced by dissolution of refractory material during their production in the molten state. Sh. Nagai and J. Ando[63] studied these solubility reductions systematically in experimental melts. Alumina is the most unfavorable agent for devaluation of phosphate slags. It not only raises the chemical resistivity of the glassy phosphate slags but it forms entirely inert spinel and calcium aluminosilicates, and promotes crystallization of fluorapatite, as made evident in thin sections. Phosphate slag fertilizers should not contain more than 6 per cent alumina to have a solubility of 95 to 96 per cent in citric acid. Japanese phosphate slags from natural calcium-magnesium phosphate rocks are sometimes so high in alumina (more than 10 per cent) that their citric acid solubility is decreased to 90 per cent of pure phosphate fertilizer. This fundamental problem of the introduction of alumina contaminations by reaction of phosphate slag melts with fireclay refractories was further studied by the same authors for a clayish phosphate rock (low-grade phosphorite ores) concerning a distinctly favorable effect[64] of a fusion with forsterite (from peridotite rocks). Good phosphate slags of this kind are glassy, with Si^{4+}, Al^{3+}, P^{5+}, and Fe^{3+} as framework formers, Ca^{2+}, Mg^{2+}, Fe^{2+} as modifiers. As the latter ingredients are raised, in like measure the glasses become chemically more resistant.

35. As a consequence of studies with magnesium rock additions, Ando was successful in developing new types of calcium-magnesium phosphate fertilizers, as discussed in Vol. III, B. ¶ 438 ff. He further showed an industrially suitable way to produce at the same time a nickel matte and such glassy Ca-Mg phosphates,[65] the latter of very high citric acid solubility when the crystallization of apatite was avoided (P_2O_5 below 19 per cent). In slags with more than 13 per cent P_2O_5 primary crystallization of forsterite or tridymite was observed. Fusion at 1350°C. produces 90 per cent of the nickel content in garnierite ores as a matte together with 92 per cent of P_2O_5 from the phosphate rocks. J. Ando[66] compared the glassy development of Ca-Mg phosphate slags with corresponding silicate glasses. The degree of "glassiness" is chiefly ruled by the ratio of the number of cations in octahedral coordination with O^{2-} anions to that in tetrahedral coordination. For these investigations Ando used a semi-quantitative X-ray diffraction method in which the shifting of the "center of gravity" of the X-ray halo in the diffraction diagrams (cf. Fig. C. 7) is a function of this ratio

[62] *Chem. Eng.*, **60**, 1953 (7) 182-183.

[63] *Denki Kagaku*, **20**, 1952, 150-154. The influence of silicic acid, alumina and titania as a constituent of Thomas phosphate slags was quite recently studied by K.-H. Obst and H.-J. Ueberhorst, *Stahl u. Eisen*, **81**, 1961, 729-735.

[64] *Denki Kagaku*, **20**, 1952, 274-278.

[65] *Ind. Eng. Chem.*, **51**, 1959, 1267-1270.

[66] *Bull. Chem. Soc. Japan*, **33**, 1960, 685-692.

and the apparent basicity number (in the parameter A.B.N. = CaO + MgO + 0.7 FeO + 0.25 Al_2O_3, usually between 47 and 52 for fertilizer slags). It is highly interesting to learn that in phosphate slag glasses more Mg^{2+}, Al^{3+}, and Fe^{3+} occurs in tetrahedral coordination than in silicate glasses; even a part of the Ca^{2+} and Na^+ ions are apparently coordinated to O^{2-} in this type. A coupled replacement of $3\ Si^{4+} \rightleftharpoons Mg^{2+} + 2\ P^{5+}$ is presumed.

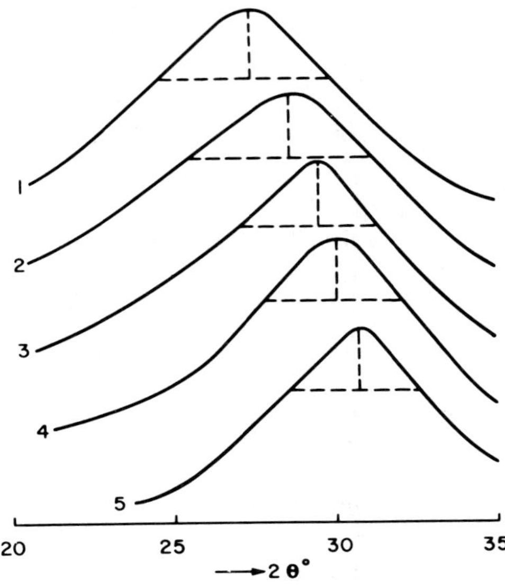

Fig. C. 7. Shifting of the "center of gravity" in the haloes of X-ray diffraction diagrams for glassy calcium-magnesium phosphate slags (1) and (2), a silicophosphate (3), an aluminosilicate composition (4), and a calcium aluminate slag (5). (Ando). From (1) to (5) the apparent basicity is increasing.

36. Concerning the constitution of phosphate slags from basic converter processes the investigation of the fundamental system $3CaO.SiO_2$—$3CaO.P_2O_5$ by N. A. Toropov, A. I. Borisenko, and P. V. Shirokova[67] may be recalled, especially because of its complex nonbinary reactions with the formation of $4CaO.P_2O_5$ and crystalline solutions of silicophosphates (silicocarnotite). In every case, no free lime was observed, as would be expected from the decomposition of $3CaO.SiO_2$. If free CaO is formed in excess, it goes immediately into crystalline solution in silicocarnotite and "nagelschmidtite" phases to form silicophosphates of high basicity.

37. Very interesting studies were made by G. Trömel and Chr. Zaminer[68] on isolated, well-developed crystals of the so-called S-phase (previously identified with the

[67] *Doklady Akad. Nauk S.S.S.R.*, **92**, 1953, 1015-1018.

[68] *Arch. Eisenhüttenw.*, **30**, 1959, 205-209. Results of these authors differ in details from those

compound $5CaO.P_2O_5.SiO_2$ = silicocarnotite cf. Trömel *et al.* 1948), but now interpreted as a crystalline solution. Blue-colored crystals (containing 3.5 per cent V_2O_5, besides 28.8 per cent P_2O_5 and only 8.76 per cent SiO_2) of pseudohexagonal habitus, but of orthorhombic symmetry, are described sometimes in aggregations as shown in Fig. C. 8a, b, c. A close relationship to real hexagonality is seen in the transition

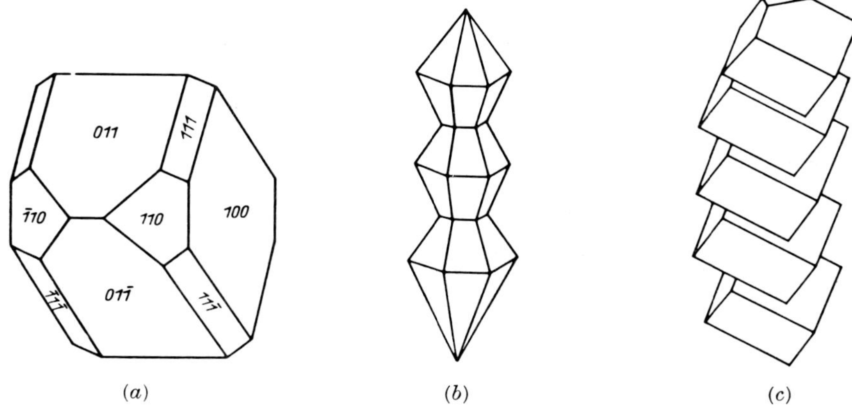

Fig. C. 8. Crystals of the "S" phase in phosphate slags, (*a*) an idealized single crystal, (*b*) a pseudohexagonal composite, (*c*) a column of orthorhombic crystals. (Trömel and Zaminer).

to the hexagonal R-phase above 1300°C. as seen from the section Ca_2SiO_4—$3CaO.P_2O_5$—SiO_2. Trömel and Zaminer described brown crystals of $4CaO.P_2O_5$ with a lamellar aggregation parallel [100] and [010], similar to microcline lamellae of potassium feldspar. Also apatite in brown-colored needles (free from fluorine, but with a considerable isomorphous change of $[SiO_4]$ groups for $[PO_4]$, with a strong pleochroism by a content in FeO and Fe_2O_3, and finally black crystals of $2CaO.Fe_2O_3$ occur, with considerable contents in alumina, titania, chromia, and smaller contents in MnO, P_2O_5, and V_2O_5. There are also interesting crystal skeletons of oxides containing CaO, MgO, MnO, and FeO.

38. The physical chemistry of phosphate slags and their role in metallurgical steel refining processes is discussed and explained by P. T. Carter[69] in a basic attempt to introduce *activities* of the constituents in the kinetics of the reactions of their formation, and especially in lowering the activities by formation of calcium silicates and phosphates. Both positive and negative deviations from Raoult's law may occur, negative deviations by formation of compounds, positive deviations by *separation into two liquid phases.* Such positive deviation phenomena are more marked, as in

of D. R. Riley and E. R. Segnit, *Mineral. Mag.*, **28**, 1948, 496-504, especially concerning the space group of the S-phase.

[69] *Roy. Tech. Coll. Glasgow, Rept. on Research.* **8**, 1955/56, 43-46.

slags of the system $FeO—(Ca, Mg)O—P_2O_5$, than in $FeO—(Ca, Mg)O—SiO_2$, and actually immiscibility was observed. The maximum differences in the free energy curves for the systems $FeO—SiO_2$ and $CaO—SiO_2$ occur near orthosilicate compositions, in the systems $FeO—P_2O_5$ and $CaO—P_2O_5$ near the composition of orthophosphates. For more complex silicate-phosphate slags maximum deviations from ideality should occur at intermediate compositions, say at $CaO/(SiO_2 + P_2O_5)$ ratios between 2 and 3. This is in agreement with FeO activity curves derived by E. T. Turkdogan and J. Pearson,[70] and also with the results of W. Oelsen and H. Maetz (1941) in the system $FeO—CaO—P_2O_5$. The difference of the free energies of formation of $3FeO.P_2O_5$ and $3CaO.P_2O_5$ is greater than the critical value corresponding to separation into two liquid phases, whereas it is less for Fe_2SiO_4 and Ca_2SiO_4. If SiO_2 is added to $FeO—CaO—P_2O_5$ slags showing immiscibility, the free energy difference will be reduced to the critical value, and liquid immiscibility disappears, which is what Oelsen and Maetz (1948) accurately observed. Similarly it would appear that separation into two liquids would occur in the system $CaO—MnO—P_2O_5$. The area of immiscibility, however, would be less extensive than in $CaO—FeO—P_2O_5$.

39. The influence of *fluorspar* additions on the distribution of phosphorus between the metal bath and the CaO—FeO slags was studied by I. Yu. Kozhevnikov, O. V. Travin, and E. N. Yarkho,[71] with extensive thermochemical calculations, referring to P. Herasymenko and G. E. Speight[72] who assumed that the addition of CaF_2 considerably changes equilibrium constants in the reactions of steel-slag dephosphorization by formation of complex fluorophosphate anions. Because of the different charges of the F^- and O^{2-} anions, these authors expected different actions of F^-, SO_4^{2-}, PO_4^{3-}, and AlO_2^- anions. With introduction of P^{32} as a radioactive tracer in slags of the system $CaO—FeO—SiO_2$ it was possible to determine by thermodynamic calculation the linear functional relation of the logarithm of the distribution coefficient of P^{32} versus 1/T. A systematic replacement of CaO by CaF_2 (for the constant temperature of 1600°C.) brings about a linear decrease of log L_P as a function of CaF_2 concentration. The formation of $3CaO.P_2O_5$ is stabilized and increased even for very low concentrations in P_2O_5 and apatite is formed. This reaction confirms the general metallurgical experience that the most stable molecular configurations are preferred to be formed. It is also in agreement with aspects given by electrochemical investiga-

[70] *J. Iron Steel Inst.* (*London*), **173**, 1953, 217-223. From this investigation comes the important conclusion that in basic and acidic steel-making practice FeO activity is not affected by temperature in the range from 1550° to 1650°C. FeO shows a positive deviation from ideality and is a function of fluoride contents in the slag. Cf. also K. L. Fetters and J. Chipman, *Metals Technol.*, **8**, 1941 (2) 1-13.

[71] *Doklady Akad. Nauk S.S.S.R.*, **122**, 1958, 635-638. On the detrimental effect of CaF_2 in phosphate slags for their citric acid solubility see K.-H. Obst and H.-J. Ueberhorst, *Stahl u. Eisen*, **81**, 1961, especially pp. 732-735.

[72] *J. Iron Steel Inst.* (*London*), **166**, 1950, 169-183.

tions of liquid slags and their ionic equilibria. CaF_2 strongly decreases the solubility of phosphates in citric acid. It is, therefore, detrimental to phosphate fertilizers, as is well known, and inadequate in dephosphorizing of pig iron. CaF_2 should only be added in the "boiling" period of the open hearth process under conditions of total tapping of the furnace.

40. Much more fundamentally, G. Trömel[73] investigated the comprehensive system $CaO—FeO—P_2O_5$, or better said, $CaO—Fe—O—P_2O_5$ which is indispensable for a discussion on the laws ruling the distribution of phosphorus in metal/slag, as a consequence of the special fields of the system in which the composition of the slags is to be projected. Real equilibria are only reached if a stable refractory container (crucible) material is used. The special artifice in using containers made of $3CaO.P_2O_5$, or $4CaO.P_2O_5$, or even of pure CaO, gave deep-going information on dephosphorization reactions, on the basis of definite saturation conditions in slag. The mass action law is then applied to different special equations. The functional relation of P_2O_5 in slags and metal bath cannot be generalized for the whole system uniformly. This means that in projections of these conditions there are different curves, not one general curve relation for defined saturation conditions. Only for every individual curve, thermodynamic calculations can be made which may be a real contribution to a solution of basic dephosphorization problems.

41. Going back to the fundamental phase equilibria of the systems $CaO—P_2O_5$ and $CaO—FeO—P_2O_5$, as partial systems of the basic system $CaO—Fe—O—P_2O_5$, G. Trömel[74] emphasizes particularly the rapidity of reactions at 1600°C. which make the common crucible materials inadequate for equilibrium determination in the complex system. The artifice of using crucibles prepared by ceramic methods of powder-sintering from compounds of solid phases occurring in the system (e.g., CaO, $3CaO.P_2O_5$, and $4CaO.P_2O_5$ as mentioned above) was the key to full success in establishing the total diagram of dephosphorization. On one side there are lime-free slags consisting of FeO_n and $3FeO.P_2O_5$, on the other side slags with highest possible lime contents, namely saturated in CaO for the constant temperature of 1600°C. The iron metal bath in equilibrium contains 0.010 to 0.012 per cent P, specifically lower than in metallurgical melts of common practice as observed in optimo. At the limit of the miscibility gap seen in Fig. C. 9 one may observe phosphorus contents in iron which are not much higher, with slags still far away from lime saturation. In the lime-rich portion of the system a relatively low lime content in

[73] The author of the present book owes many important aspects for complex problems here involved to a valuable personal correspondence with G. Trömel. Facts given above are for a considerable part given in the words of G. Trömel himself to whom the author expresses his sincere gratitude for fascinating discussions granted. On the use and preparation of crucibles made of $3CaO.P_2O_5$ see G. Trömel and W. Oelsen, *Arch. Eisenhüttenw.*, **23**, 1952, 17-20.

[74] *Arch. Einsenhüttenw.*, **26**, 1955, 497-506.

the slag does not affect very much the phosphorus content in the iron. CaO binds P_2O_5 very strongly and no considerable lime excess is required to remove P entirely from the iron.

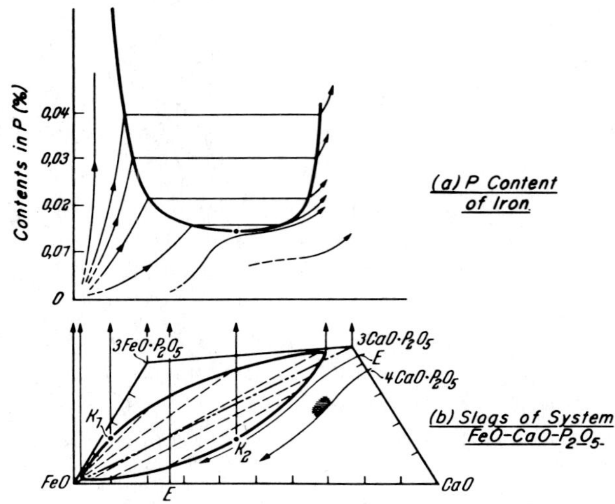

FIG. C. 9. Functional relation of phosphorus contents in iron metal melt at 1600°C., and slag composition in the system P_2O_5—FeO—CaO. (Trömel and Oelsen). The shaded field corresponds to typical lime-rich phosphate slags.

42. In the wide range between 1550° and 1725°C. it is evident that with increasing temperature the phosphorus content in iron is only moderately raised. Slags taken shortly before and at the end of the dephosphorization reactions in a steel converter were examined in polished sections by K. E. Mayer, H. J. Därmann, G. Trömel, and K. H. Obst[75] in an industrial melt. It became evident that often the liquid slags contain lime-rich striae which are entirely independent of superficially enveloped lime lumps. The size of the lime additions (8 to 40 mm.) is not unimportant for the rapidity of the process. Smaller grain sizes, which accelerate dissolution in the melt and lime striae, which later may cause a volume instability (swelling) of the phosphate slag fertilizer, are avoided. It is advisable to bind fine-powdered lime by tar to a crumbly aggregate; perhaps, it may be possible to sinter the lime with calcium ferrite.

43. With H. W. Fritze, G. Trömel[76] gave a new detailed determination of the systems CaO—FeO—P_2O_5 (Fig. C. 10), and CaO—Fe—O—P_2O_5 (Fig. C. 11 and 12),

[75] *Stahl u. Eisen*, **77**, 1957, 1614-1618.

[76] *Arch. Eisenhüttenw.*, **28**, 1957, 489-495; **30**, 1959, 461-472. See recently a comprehensive discussion of equilibria between Fe metal and calcium phosphate slags, by G. Trömel, W. Fix and

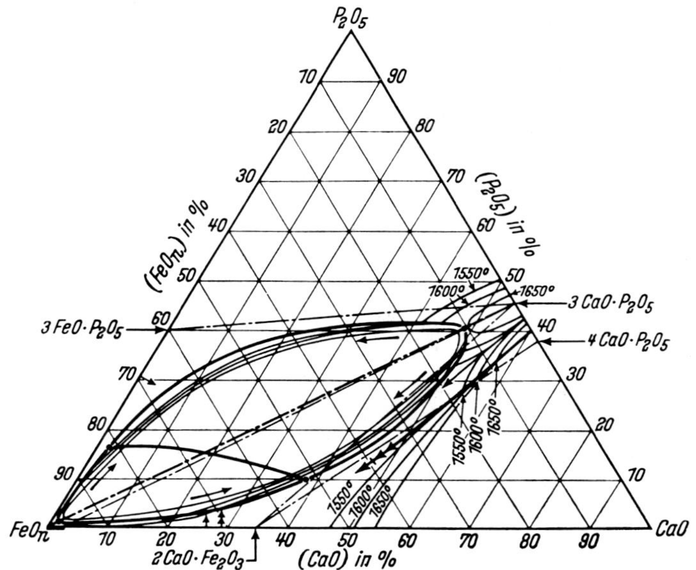

FIG. C. 10. Revised phase equilibrium diagram of system CaO—FeO—P_2O_5. (Trömel and Fritze).

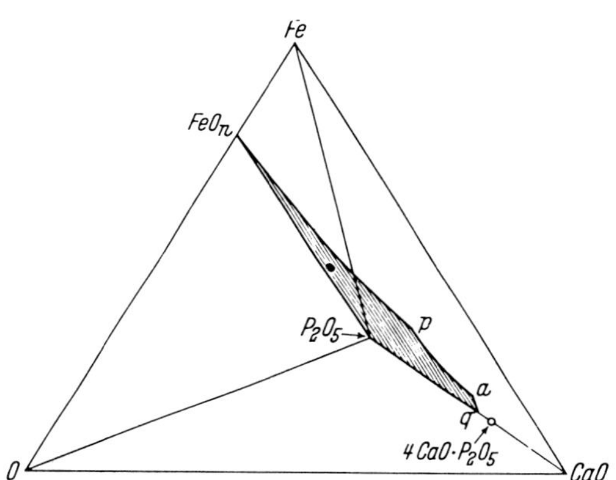

FIG. C. 11. System CaO—P_2O_5—Fe—O, general aspect, with plane (shaded) containing representative points of slags saturated in iron, in point *a*, in addition, saturated in $4CaO.P_2O_5$. (Trömel and Fritze).

H. W. Fritze, *Ibid.*, **32**, 1961, 353-359, furthermore G. Trömel and W. Fix, *Ibid.*, **33**, 1962, 745-755, on these equilibria in presence of SiO_2 and MnO, and G. Trömel and Kl. Schwerdtfeger, *Ibid.*, **34**, 1963, 101-108, with CaO-free phosphate slags.

with the lime-rich portion of the latter (Fig. C. 13), at 1600°C. constant. Below a CaO-saturated slag melt the P content of the iron bath is increased with increasing temperature. If on the other hand, silica is present in the slag melt composition, phosphorus contents are lower but the increase with temperature becomes steeper as seen in Fig. C. 14. Nevertheless, the P contents observed in industrial plant melts of iron are considerably higher than those found in laboratory experiments.

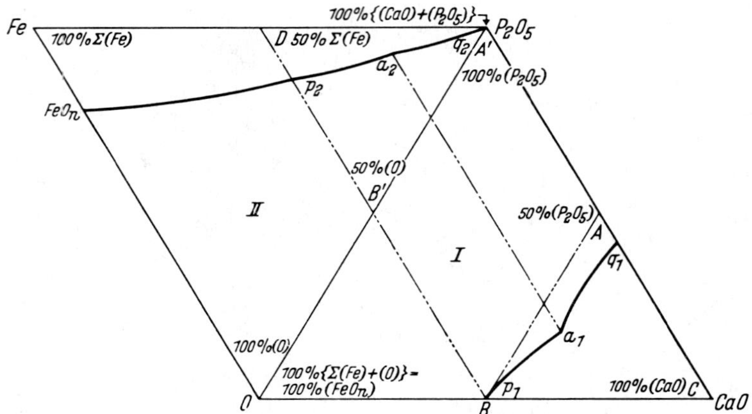

FIG. C. 12. System CaO—P_2O_5—Fe—O, in planar projection. (Trömel and Fritze).

The contents in phosphorus, as those in *oxygen* in the metal bath, in the same way are increased with rising temperature. Also in this case, the laboratory data of oxygen contents are below those observed in practice. The diagrams given by Trömel and Fritze show for 1600°C. corresponding compositions of the metal phases, in a special method of projection which permits direct transfer of determined values for CaO and P_2O_5 and data of the sum of the iron oxides (FeO_n) in the slags. The experiments were also extended to temperatures of 1550°, 1600°, 1650° and 1700°C. (in crucibles of $3CaO.P_2O_5$ and $4CaO.P_2O_5$ mixed with CaO). Isothermal sections for the system CaO—Fe—O—P_2O_5 are given, including the ratio Fe_2O_3 : FeO, which is higher in calcium phosphate slags than in ferrous oxide-rich slags,[77] and this ratio as a function of temperature. For conditions of saturation in CaO and $4CaO.P_2O_5$, the lowest phosphorus and oxygen contents in the iron were found, which in this case at 1600°C. is only 0.009 per cent.

44. Interesting is the occurrence of an *apatite*-type crystal phase in the neighborhood of the primary crystallization surface of $3CaO.P_2O_5$ in the system CaO—FeO—P_2O_5. In expectation that this might be a hydroxyl apatite, which is notor-

[77] E. T. Turkdogan and P. M. Bills, *J. Iron Steel Inst.* (*London*), **188**, 1958, 143-153.

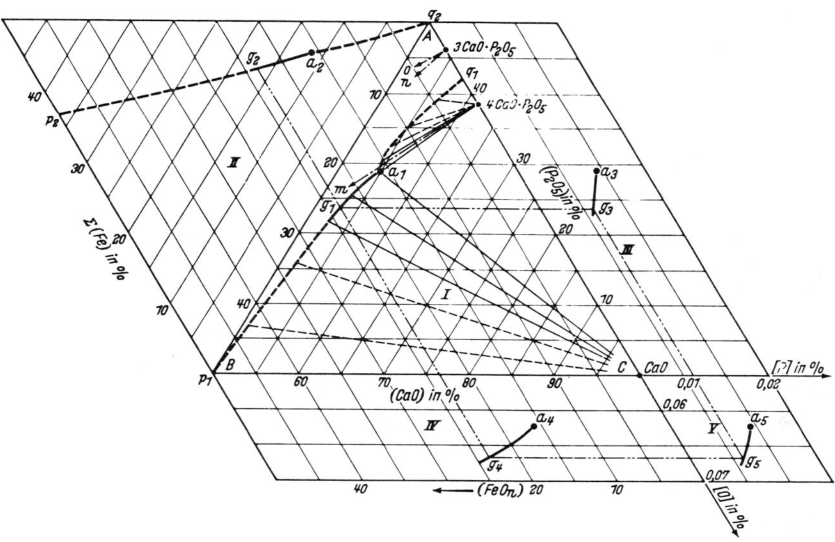

FIG. C. 13. Lime-rich portion of system $CaO—P_2O_5—Fe—O$, at 1600°C. (Trömel and Fritze).

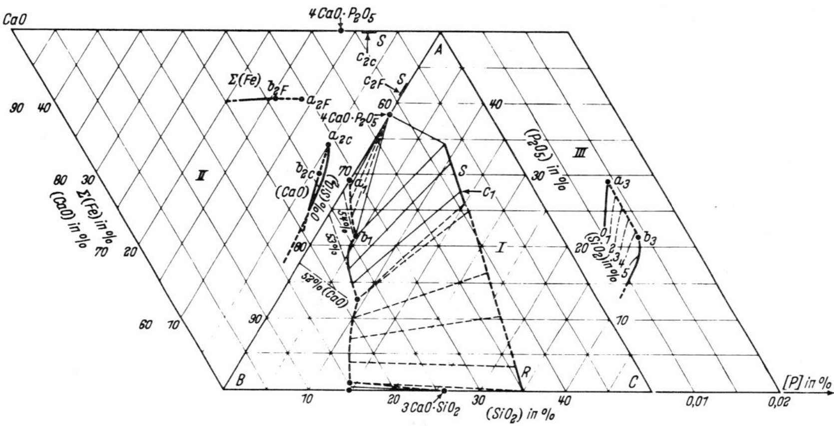

FIG. C. 14. Phase equilibria in system $CaO—P_2O_5—Fe—O$ at 1600°C. (Trömel and Fritze). Diagram shows the phases saturated in CaO and Fe.

iously stable even at high temperatures, a special analytical method was developed by K. H. Obst and H. Malissa[78] for determination of water in slags heated up to 1450°C., based on the evolution of acetylene from calcium carbide in the presence of water, expelled from a real hydrate mineral, or from a hydroxyl containing silicate.

[78] *Arch. Einsenhüttenw.*, **30**, 1959, 601-603.

The results, however, were negative; the apatite observed is, therefore, most probably anhydrous, and it may be presumed that it may be an iron oxide containing apatite form.

VISCOSITY AND CORROSIVITY OF MOLTEN SLAGS

45. For accurate measurement of the viscosity of metallurgical slags M. P. Volarovich and O. I. Yatsunskaya[79] constructed a special rotation viscosimeter based on the principle of a torque pendulum. The melt is fused in a graphite crucible, the rotor being a graphite rod with two loads for increasing the moment of inertia of the oscillating system (cf. A. ¶ 17). The slag is kept liquid about 100° to 150°C. above the softening temperature and then slowly cooled keeping the measuring temperatures constant for about 30 minutes. Chromium containing open hearth slags (with 16.4 per cent, down to 4.9 per cent Cr_2O_3) were used for the measurements.

46. Systematic studies of the influence of the chemical composition of blast furnace slags on their viscosity behavior were previously made by I. P. Semik,[80] especially for the system $CaO—Al_2O_3—SiO_2$, in which the viscosity is the higher the higher the ratio SiO_2/CaO at constant alumina contents, and the higher the latter, at constant ratio SiO_2/CaO. Additions of 4 per cent magnesia lower the viscosity at 1400° to 1500°C. to about half the original data. This lowering effect is the stronger, the lower the temperature. In calculations, the monoxide contents can be considered simply to be CaO without a considerable error at 1500° to 1600°C., but at 1400°C. there would arise a very marked error. In slags high in alumina and with CaO above SiO_2, addition of magnesia can cause an increase in viscosity at lower temperatures, not the normal decrease in viscosity.

47. CaS in small amounts, as a common ingredient in blast furnace slags, has no significant effect on viscosity, but in acidic slags it can considerably increase it whereas for basic slags (low in alumina) above 1500°C. the effect is negligible. Unmixing phenomena are characteristic of slags with CaS in amounts of 6 to 8 per cent and more. Titanium blast furnace slags show, at 1600°C. and a following slow cooling, a relatively high fluidity in a graphite crucible and in an atmosphere of nitrogen and CO, even if titania goes up to 26 per cent, but a typical "thickening" occurs at higher TiO_2 contents. With increasing TiO_2 the fusibility of the slags decreases, and also the viscosity decreases at higher temperature. Qualitatively, the effect of titania additions is similar to that of lime, but quantitatively there are important differences.

[79] *Zavodskaya Lab.*, **4**, 1950, 813-818.

[80] *Bull. acad. sci. U.R.S.S.*, Otdel. Tekhnich. Nauk 1941 (4) 55-66; 1946, 1655-1671; *Stal'*, **7**, 1947, 202-206.

In every case, much titanium occurs in blast furnace slags in the form of the lower oxides (cf. C. ¶ 27 ff.). The ratio CaO/SiO_2 is very important for the viscosity of titanium slags and for their fusibility. A highly titaniferous slag with $CaO/SiO_2 = 0.9$ melts at about 1300°C. These studies of Semik are particularly illustrative for the smelting of Ti-Magnetite ores of the Kusin metallurgical center, with nepheline and miaskite added as fluxes.

48. Synthetic blast furnace slags molten at 1600°C. (with slight additions of CaS, manganese ore, and $MnCO_3$) were used to study the source of serious errors which arise due to the abnormally high viscosity in surface layers of the melt, which then often act like a solid even up to 1600°C. The rotating spindle of the viscosimeter can thus be "blocked" by a surface layer and give viscosity data which are by 2 to 3 times too high. This effect was observed for all basic iron silicates, all cast iron and ferromanganese slags. It was most pronounced in Bessemer slags with 46 to 47 per cent CaO, 10 to 12 per cent alumina, the more striking the more basic the slag composition, and the lower the ratio Si^{4+}/O^{2-}. However, a 40 per cent SiO_2 containing acidic ferrosilicon slag behaved entirely normally. The older data of A. Feild and P. H. Royster (1916), are unreliable for the reason that the occurrence of a surface layer influences the viscosity in the system $CaO—Al_2O_3—SiO_2$, whereas the data of R. S. McCaffery (1931) are confirmed by Semik. Viscosity in connection with *desulfurization* properties of blast furnace slags high in magnesia were studied by Semik for various alumina contents (10 to 15 per cent) and magnesia up to 40 per cent. Such extremely high magnesia amounts make the slags difficult to melt, whereas the fusibility, and at the same time the desulfurization activity, is raised by increasing contents in MgO for slags with $CaO + MgO = 55$ per cent, with variable silica and alumina contents.

49. Especially for acidic blast furnace slags (with 50 to 60 per cent SiO_2), V. I. Loginov[81] investigated extensively viscosity behavior under the aspects of using such slags in coke blast furnaces for pig iron. The viscosity was measured in the same rotation viscosimeter Semik had used before. The composition of some slags projected in the system $CaO—FeO—SiO_2$ was in a broad field from Ca_2SiO_4 to $Ca_3Si_2O_7$ and $CaFeSiO_4$ (iron monticellite), to the primary crystallization of tridymite, or in the system $CaO—Al_2O_3—SiO_2$ with primary $CaSiO_3$, $Ca_3Si_2O_7$, and at the boundary of the anorthite and gehlenite fields. Discontinuities in viscosity versus temperature curves indicate the tendency to crystallization, as for "long" slags at 20 poises, for tap cinder slags at 30 poises. The corresponding temperatures are important for the flow characteristics of different slags. The curves also show the primordial effects of changes in the ratio CaO/SiO_2 of the contents in FeO, MnO, and of the absolute

[81] "Investigations on the Blast Furnace Process," Academy of Sciences, U.S.S.R., Moscow, 1957, 148-166.

silica concentration in the range from 1450° to 1300°C. for common furnace operation in flowing slag. In general, iron-rich slags and pure iron silicates above 1400°C. are more viscous than basic slags. The slopes of the viscosity versus temperature curves are wholly variable and often intersect. The behavior of "shorter" or "longer" slag types is of great importance for the uniform operation of the blast furnace process. Isothermic curves for 1350° and 1300°C. show that even for different slag types one characteristic silica content corresponds to a minimum in viscosity. The higher the FeO contents and the lower the corresponding CaO, the more is the minimum shifted to lower silica contents. The minimum is characteristic of a CaO/SiO_2 ratio of 0.9 to 1.0. The effect of alumina is observed from the isothermic curves which also have viscosity minima at low temperatures. At 1450°C. the viscosity curves for slags of different acidity (40 to 60 per cent SiO_2), and as a function of alumina concentration, show a rapid increase in viscosity with the same sum of $SiO_2 + Al_2O_3$. The lower CaO/SiO_2 ratio, and the higher the alumina content, the more viscous are such tap cinder slags. The studies of Loginov contain many more practical rules for the operation of slag running in blast furnaces which can not be discussed here in detail.

50. *Soft-coal combustion ashes* ("slags") in their viscosity characteristics as a function of their composition were discussed by J. Endell,[82] in comparison to those from hard coals. Ashes containing 50 per cent CaO, 15 to 25 per cent Fe_2O_3, 6 to 12 per cent SiO_2, and 6 to 15 per cent Al_2O_3, are very "short" because of their high contents in calcium ferrite, whereas the average hard coal ashes are "long" in viscosity. The low fusion temperature and high fluidity of ferrite-type ashes are not only troublesome because of their high corrosivity to refractory linings (cf. C. ¶ 12) but also for the continuous running of cyclone heaters and fusion chamber firings in which they may easily "freeze," and suddenly solidify. The viscosity behavior can be improved ("lengthened") by the deliberate addition of clay to soft coal ashes. With the oscillating viscosimeter of G. Heidtkamp and K. Endell (1936) the viscosity values were measured with special reference to the effect of additions of alumina and silica. Not only the temperature of solidification is lowered to 1100°C. and even 1050°C. but viscosity isotherms show a less steep course indicating lengthening effects. Polyeutectics of the system $CaO—Fe_2O_3—Al_2O_3—SiO_2$ are thus reached, indicating more favorable flow characteristics. Hard coal ashes, on the contrary, behave in a first approximation like fired ceramic clays with their typical long fusion interval and corresponding smooth log η versus temperature diagrams.

51. Concerning a direct applicability of the equilibrium diagram of the system $CaO—Al_2O_3—SiO_2$ to the viscosity of slags, W. F. Ford and J. White[83] discussed especially the mutual reaction products of blast furnace slags with silica-alumina

[82] *Silicates inds.*, **20**, 1955, 405-409; **22**, 1957, 156-159.
[83] *Trans. Brit. Ceram. Soc.*, **50**, 1951, 461-505.

refractories. The common use of the determination of softening characteristics by the cone falling (deformation) method brings about a series of systematic discrepancies from conclusions to be made from the equilibrium diagram. They make evident how much the cone falling method as a highly complex rheological test on flow or subsidence under the action of gravity depends in its indications on viscosity and the amount of eutectic liquid phase formed during heating. The physicochemical relations between slag melt and refractory composition are given by sections of a quasi-binary type to show the different primary crystallization phases and liquid phase compositions. In Fig. C. 15 is shown only one of these diagrams as an example of a typical pseudosystem of a normal fireclay (with only 22.5 per cent alumina, 74.4 SiO_2), and a blast furnace slag (with 43.9 per cent CaO, 21.1 per cent Al_2O_3, 27.3 per cent SiO_2, and 2.5 per cent MgO), and in Fig. C. 16 the principle of determination of the amount of liquid phase formed at a constant temperature (indicated by an isothermal curve projected), and the elementary lever rule construction for ternary systems. The eutectic temperature of only 1165°C. is the decisive factor in this case.

52. A systematic study of such theoretical sections in comparison with what the softening test indicates (dotted curve in Fig. C. 15) and with calculations of the liq-

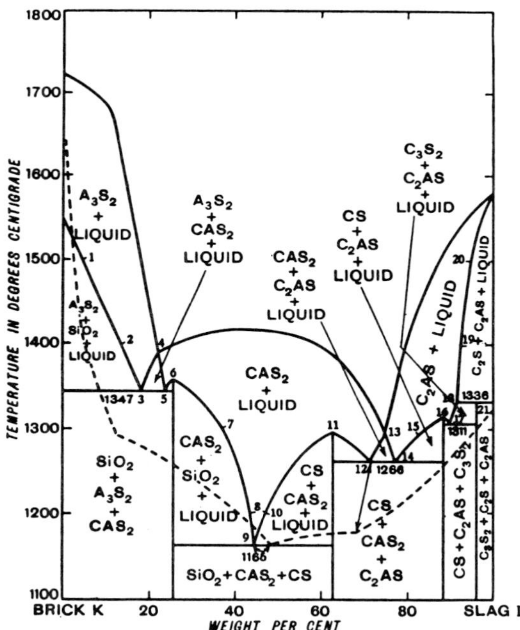

FIG. C. 15. Melting and phase relations in an ideal section through system SiO_2—Al_2O_3—CaO, from composition of a fireclay body (brick *K*), to a normal blast furnace slag. (Ford and White).

uid amounts, show gradual changes of the primary crystallization products with increasing alumina contents of the refractories as a consequence of changes in the composition fields. By nonequilibria, the ideal phase sequences on cooling will not be observed in practice, and especially peritectic reactions would be incomplete or

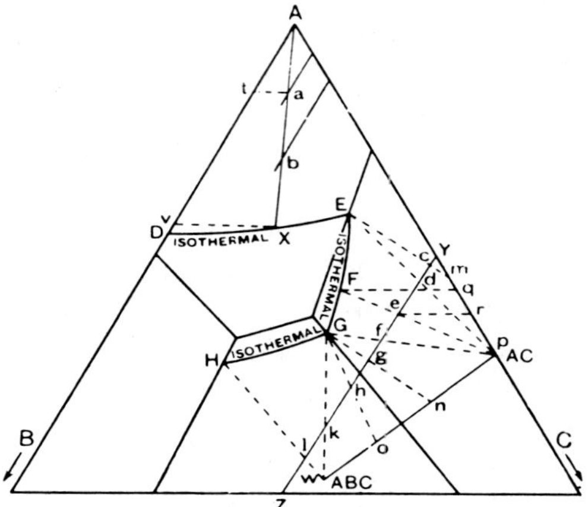

FIG. C. 16. Calculation of the contents of ternary mixtures of system SiO_2—Al_2O_3—CaO, in liquid phase. (Ford and White).

even suppressed. To show effects of changes in alumina contents, Fig. C. 17 gives in (*a*) the theoretical liquid phase contents of fireclay brick reacted with a basic blast furnace slag melt as shown in Fig. C. 15, in (*b*) the corresponding diagram for a superduty (alumina-enriched) brick (with 59.0 per cent alumina, 38.5 per cent silica) and a slag high in alumina (27.9 per cent, 25.3 per cent SiO_2, and 37.2 per cent CaO), derived in both cases from the ideal diagrams. The isotherms in Fig. C.

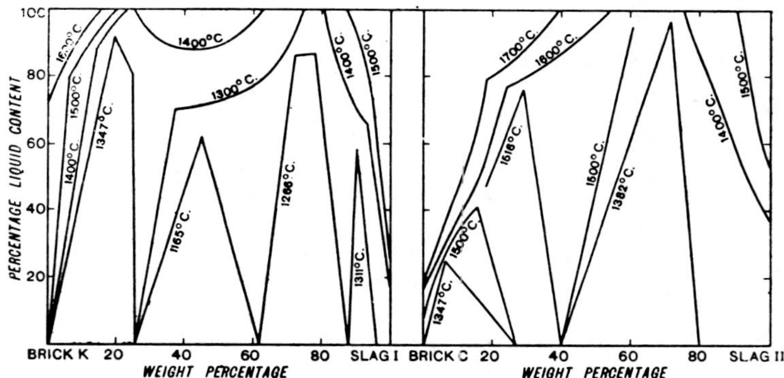

FIG. C. 17. Effects of temperature on theoretical liquid contents of mixtures of siliceous and aluminous fireclay brick, with blast furnace slags of system SiO_2—Al_2O_3—CaO. (Ford and White).

17 illustrate great differences in fluidity and liquid amounts to be expected from interactions of slags with refractories of the types here considered. Calculated fluidity ($1/\eta$) curves as a function of temperature for the ternary eutectics at 1165°, 1266°, and 1347°C. in the system $CaO—Al_2O_3—SiO_2$ are seen in Fig. C. 18.

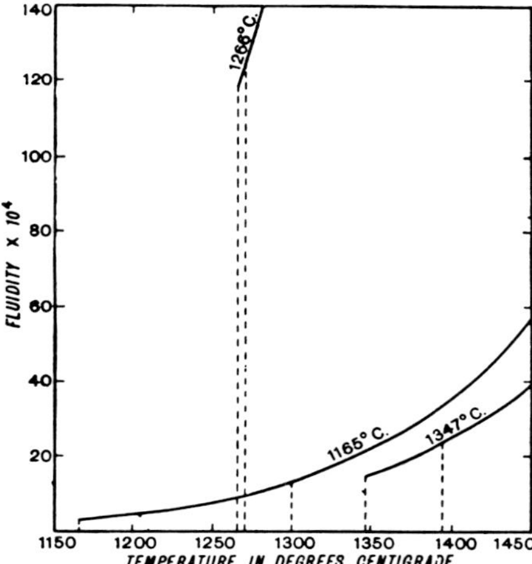

FIG. C. 18. Calculated fluidity curve of ternary eutectics in system $SiO_2—Al_2O_3—CaO$. (Ford and White).

53. For the special conditions of reaction slags of burnt lime in vertical shaft kilns lined with fireclay, L. Halm[84] applied the same principles of discussion for the $CaO—Al_2O_3—SiO_2$ equilibrium diagram combined with petrographic studies (cf. Vol. V, Section B), and viscosity data given by J. S. Machin and Tin-Boo Yee (1948). Special reference is made to the 1165°C. eutectic and the conditions of viscosity in the field between the points of anorthite, gehlenite, and $CaSiO_3$ which are most important for reaction melts occurring in the practice of lime burning. Further, A. Bertrand[85] studied lime-aluminosilicate slags formed in the metallurgical operation of lead and antimony ores (chiefly galenite and cerussite, with antimony sulfide) including their gangues + lime. The ternary 1165°C. eutectic melt does not dissolve any sulfides nor lead and antimony as metals, and a simultaneous, one-operation reduction of both types of ores, sulfides, and oxides, is possible but with very critical viscosity conditions in the temperature range required.

54. For modern blast furnace practice with dolomitic lime applied in the charges, and a *magnesia* containing slag resulting, an extension of viscosity studies to the

[84] *Trans. Intern. Ceram. Congr.*, **4**, 1954, Florence, 303-324, especially pp. 311 ff.

[85] *Chim. & ind. (Paris)*, **77**, 1957, 1281-1287.

quaternary system CaO—MgO—Al_2O_3—SiO_2 was a necessity. P. Kozakevich[86] used a coaxial rotation viscosimeter to study crystallization and liquefaction behavior of basic slags of this type between 1480° and 1540°C. (corresponding ternary slags crystallize at 1600° to 1700°C.). In many cases basic slags proved to be not entirely homogeneous; in this case additions of fluxing agents (lime or BaO) in amounts up to 12 per cent are required to lower the liquidus points. If sulfur is introduced (as S^{2-} anion replacing O^{2-}) an amount up to 2 per cent is without a distinct effect on the viscosity. Particularly strong is the reduction in viscosity (and in the Liquidus points) for quaternary slags with additions of calcium and magnesium fluoride, in the same way these fluxes act in acidic slags.

55. The fundamental investigations of E. F. Osborn, R. C. DeVries, K. H. Gee, and H. M. Kraner[87] on the optimum compositions of quaternary blast furnace slags (cf. C. ¶ 14 ff.), in reference to viscosity characteristics demonstrate the rule that, on a weight per cent basis, CaO and MgO are approximately equivalent in their effects on viscosity. In order to have a well-running slag flow of lowest viscosity, the optimum composition is about the same as that necessary for a maximum desulfurization effect. Perhaps, both properties will not accurately coincide in an optimum composition but they practically are close together. From measurements of J. S. Machin, Tin-Boo Yee (1948), and with D. L. Hanna[88] over the range from 1350° to 1500°C. it is evident that the viscosity of liquid slag compositions are regularly decreasing with increasing CaO. The isokoms (curves of constant viscosity) are nearly parallel to the lines of constant CaO contents, and alumina and silica are approximately equivalent with respect to their action on viscosity. Magnesia may be somewhat more effective even than lime in lowering the viscosity as noted from their numerous diagrams.

56. Very recently, E. T. Turkdogan and P. M. Bills[89] emphasized that for quaternary lime-magnesia-alumina silicate slags the total "acid" concentration cannot be given simply by the sum $SiO_2 + Al_2O_3$ in molecular concentrations. The SiO_2 equivalent of alumina, $N_A = N_{SiO_2,\text{ binary}} - N_{SiO_2,\text{ ternary}}$ varies with the ratio Al_2O_3/CaO (Fig. C. 19) and the concentration in alumina. On a molar basis, on the other hand, CaO and MgO have the same effects on viscosity as Osborn *et al.* concluded. For a constant temperature, a relationship between viscosity and composition is shown by the curve Fig. C. 20, by plotting $N_{SiO_2} + N_A$, which is valid over a wide range of compositions. Below 1350°C. $\log \eta$ is a linear function of $1/T$; at higher temperatures, however, there are considerable deviations from linearity.

[86] *Rev. mét.*, **51**, 1954, 569-587. T. Lindemer, *J. Am. Ceram. Soc.*, **44**, 1961, 466, offered instructive data on the effects of CaF_2 additions on the fusion temperatures of siliceous slags.

[87] *J. Metals*, **6**, 1954, 33-45, especially p. 41.

[88] *J. Am. Ceram. Soc.*, **35**, 1952, 322-325; **37**, 1954, 177-186.

[89] *Am. Ceram. Soc. Bull.*, **39**, 1960, 682-687.

As the heat of activation, ΔH^*, decreases with increasing temperature (Fig. C. 21), it may be assumed that structural changes take place in the melt when the temperature is raised.

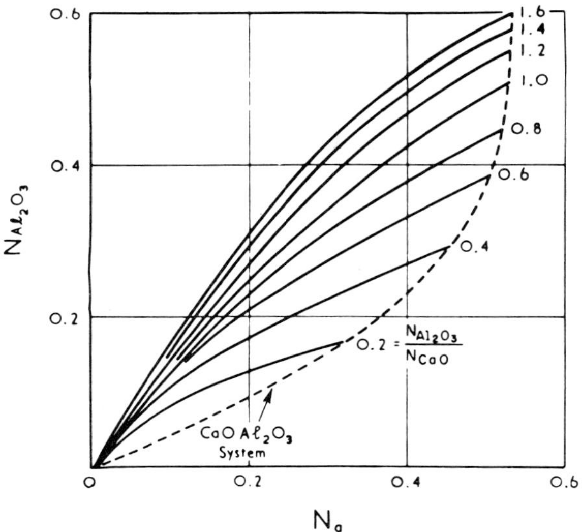

FIG. C. 19. Silica equivalence of alumina, N_A, as a function of molecular alumina concentration, and of molar ratio Al_2O_3/CaO in slags of system SiO_2—Al_2O_3—CaO—MgO. (Turkdogan and Bills).

57. A comparison of the viscous flow of slags in the systems CaO—SiO_2, CaO—MgO—SiO_2, and CaO—MnO—SiO_2 was made by D. T. Livey, H. Towers, H. B. Bell, and R. Hay,[90] based on the modern theories of reaction rates developed chiefly by H. Eyring.[91] The free energies and entropies of activation of flow were calculated and tabulated in relation to Eyring's equation, giving viscosity as a function of free energy and temperature:

$$\eta = \frac{h \cdot N}{V} \cdot e^{\Delta G/RT}$$

For silicate melts the energies of activation of viscous flow are much larger than those for nonpolar liquids (e.g., C_2H_4), hydroxyl-bonded liquids (e.g., H_2O), and common ionic liquids (e.g., molten KCl and NaCl). An increase in the ionic strength

[90] *Trans. Brit. Ceram. Soc.*, **53**, 1954, 741-747.

[91] Cf. S. Glasstone, K. J. Laidler, and H. Eyring, "The Theory of Rate Processes," McGraw-Hill Book Company, Inc., New York and London, 1941, 611 pp., especially important pp. 480-516. On viscosity of different pure liquids of homopolar and heteropolar type see especially A. G. Ward, *Trans. Faraday Soc.*, **33**, 1937, 88-97.

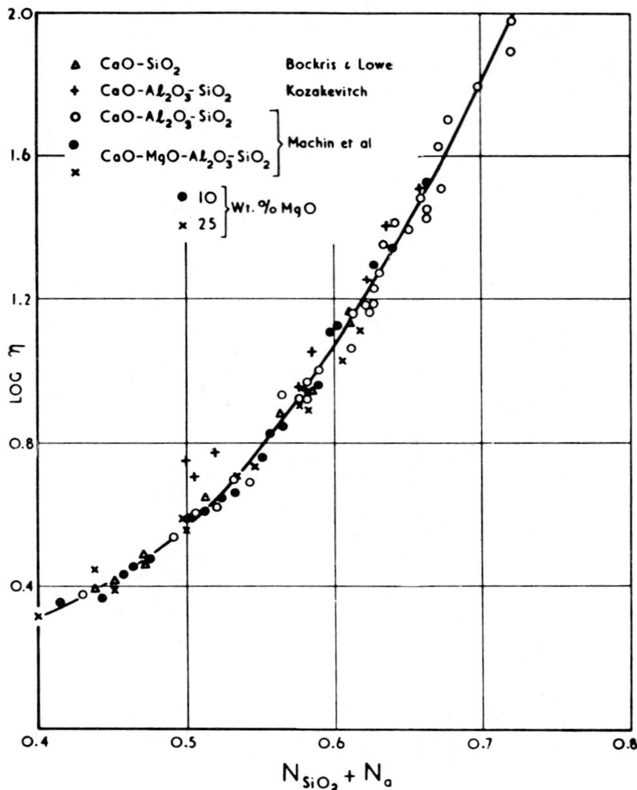

FIG. C. 20. Variation of viscosity of slags of system SiO_2—Al_2O_3—CaO—MgO, at 1500°C., with composition. (Turkdogan and Bills).

of the cation added to a silicate melt increasingly removed the effect of framework structure within the liquid. The magnesium ions alone make a remarkable exception to this rule. The accompanying table impressively shows these differences.

Compound	Heat of fusion ΔH_f* in kcal.	Activation energy of viscous flow E_η in kcal.	$\Delta E_\eta/\Delta H_f$	
C_2H_4		0.739		
H_2O		3.05		
NaCl	7.2	9.1	1.26	
KCl	6.4	7.4	1.15	
$MnO.SiO_2$	8.7	16.8	1.96	
$CaO.MgO.2SiO_2$	18.5	36.5	1.97	(Aver.
$CaO.SiO_2$	19.8	36.7	1.86	1.97)
$MgO.SiO_2$	14.4	30.0	2.10	

The consequences for slag investigations are interesting in that the assumption of E. T. Turkdogan and J. Pearson[92] is not correct when concluding that CaO, MgO, and MnO would have an equivalent effect on the activity of FeO when added to $FeO—SiO_2$ slags.

58. Melts of the system $Na_2O—MnO—SiO_2$ show a remarkable gap of miscibility according to R. Hay, P. T. Carter, and S. K. Kabi[93] amidst the field of primary crys-

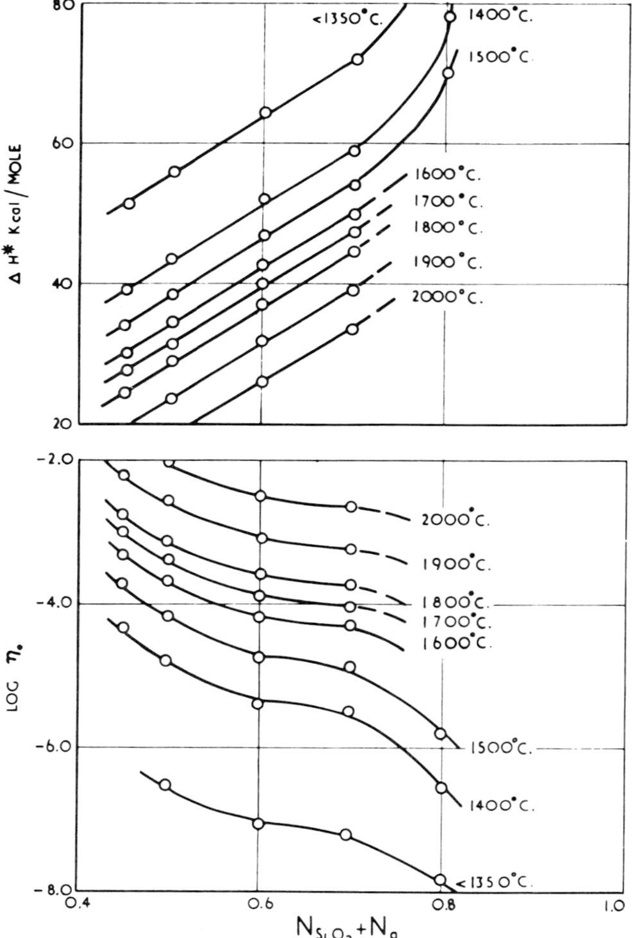

FIG. C. 21. Variation of activation energy, ΔH^*, and of logarithm of viscosity, log η, with temperature and composition. (Turkdogan and Bills).

[92] *J. Iron Steel Inst.* (*London*), 173, 1953, 217-223.
[93] *J. Soc. Glass Technol.*, 40, 1956, 429-444.

tallization of $MnO.SiO_2$ (cf. Vol. III, B. ¶ 306). In a certain degree this is analogous to the immiscibility field near the FeO apex of system $CaO—FeO—P_2O_5$ according to G. Trömel and W. Oelsen (see above), and to another gap in system $MnO—Al_2O_3—SiO_2$, according to H. Towers and G. M. Gworek,[94] when alumina is added to a rhodonite melt, whereas no gaps occur in the apparently analogous systems $Na_2O—FeO—SiO_2$ and $FeO—Al_2O_3—SiO_2$. It is also remarkable that one of the liquid phases which separate in the immiscibility relation may fairly closely correspond to a compound, namely FeO, $MnO.SiO_2$, or silica. Such "limit compounds" are in a relation with the cationic radii, as shown in the accompanying table.

Apparent compound	Percentage of the basic oxide	Cation radius in Å.
$SrO.4SiO_2$	30	1.27
$CaO.2SiO_2$	32	1.08
$MnO.2SiO_2$	37	0.91
$ZnO.2SiO_2$	40	0.83
$2FeO.3SiO_2$	44	0.83
$2MgO.3SiO_2$	31	0.78

The acidity of these "limit compounds" is increasing with the radius of the metal cation but Mg^{2+} is an exception in this case.

59. For the system $FeO—Al_2O_3—SiO_2$, P. Röntgen, H. Winterhager, and R. Kammel[95] determined the viscosity of synthetic slags of this composition with a discussion of presumable molecular states of the liquid and crystalline phases, that is, of the action of principally the same binding forces in both. K. Mori[96] challenged the same problem by a study of electrical conductivity as a function of composition and constitution in ferrous titanate slags, fused in an iron crucible (under nitrogen atmosphere) in the temperature range from 1200° to 1450°C. The specific conductance was between 30 and 300 Ω^{-1} cm.$^{-1}$ slightly variable with temperature. An anomalous discontinuity occurred in the range of the temperature of solidification in slags which are high in TiO_2. If TiO_2 was below 40 molecular per cent conductance decreases with increasing titania contents, but above 40 per cent it is reversed. This complex behavior is interpreted as indicating a semiconductor effect by Fe^{2+} ions and of electrons derived from titanium oxides, with a participation also of Ti^{4+} and O^{2-} ions in the transference phenomenon.

[94] *J. West Scot. Iron Steel Inst.*, **51**, 1943/44, 123-132.

[95] *Z. Erzbergbau u. Metallhüttenw.*, **9**, 1956, 207-214.

[96] *Tetsu to Hagane*, **42**, 1956, 1024-1029. On acidic and basic slags in reaction to titanium containing steel baths in electric induction furnaces see P. Bardenheuer and W. A. Fischer, *Arch. Eisenhüttenw.*, **25**, 1954, 515-521. By dissolution of titanium oxides glassy slags become more fluid and are most reactive at 15 to 20 per cent TiO_2.

60. Especially for *welding slags*, and fluxes for electric arc welding, often of highly complex composition, the action of fluxing agents, which reduce the viscosity to a high fluidity of the melts, is of fundamental significance. The addition of fluorides for this purpose is common, and reactions in the system $CaO—SiO_2—CaF_2$ (cf. Vol. IV, Section A) and related systems explain the presence of cuspidine, $3CaO.2SiO_2.CaF_2$, and of villaumite, NaF, in the constitution of solidified slags as described by V. V. Lapin,[97] specifically cuspidine with its characteristic polysynthetic twin lamellae. N. V. Zaimskikh and O. A. Esin[98] studied for different welding slag compositions as chipped from a welded seam, or the protecting slags, either high in TiO_2, FeO, and MnO, or containing CaF_2, and determined the viscosity and electric conductance at the same time by a torsion rotation viscosimeter and a Thompson alternating-current bridge. The highest conductance and lowest viscosity data (cf. Figs. C. 22 and 23) are observed in slags high in FeO and MnO, whereas with

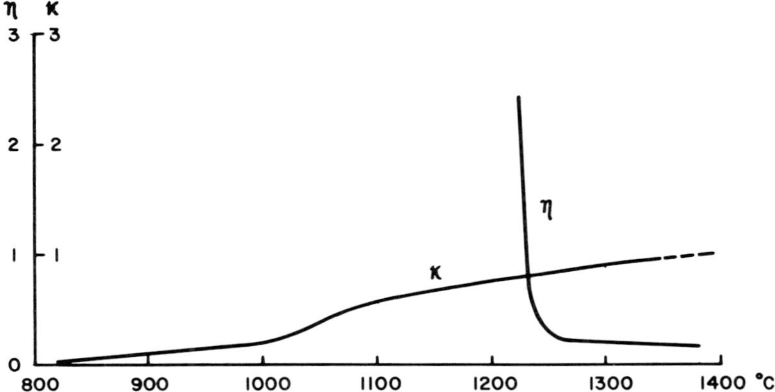

FIG. C. 22. Viscosity and specific electric conductance of a welding slag high in FeO,MnO, and TiO_2, as a function of temperature. Viscosity (η) in logarithmic scale. (Zaimskikh and Esin).

increasing contents in calcium aluminosilicates both parameters are lowered, as is common experience in acidic slags. If a primary crystallization occurs in the slag melts, a sharp break point appears in the viscosity versus temperature curves. Acidic slags have a strong tendency to solidify as glasses, and for these the viscosity changes

[97] *Doklady Akad. Nauk S.S.S.R.*, 31, 1947, 694-696. The attempt to explain the strong fluxing effect of CaF_2 on slag melts of the system $CaO—Al_2O_3—SiO_2$ was made by Tr. Baak, in "The Physical Chemistry of Iron and Steelmaking," editor J. F. Elliott, Technology Press, Massachusetts Institute of Technology, Wiley, 1958, pp. 84-86, who studied possibilities of a solvation of the calcium silicate by the fluoride, and activity coefficients in the system $CaSiO_3—CaF_2$ (if water reaction and evolution of SiF_4 is strictly excluded).

[98] *Zhur. Priklad. Khim.*, 26, 1953 (1) 70-76.

smoothly with temperature. Also Zaimskikh and Esin attempted to explain viscosity-conductance behavior of such melts from the viewpoint of a theory of ionic positions (cf. A. ¶ 163).

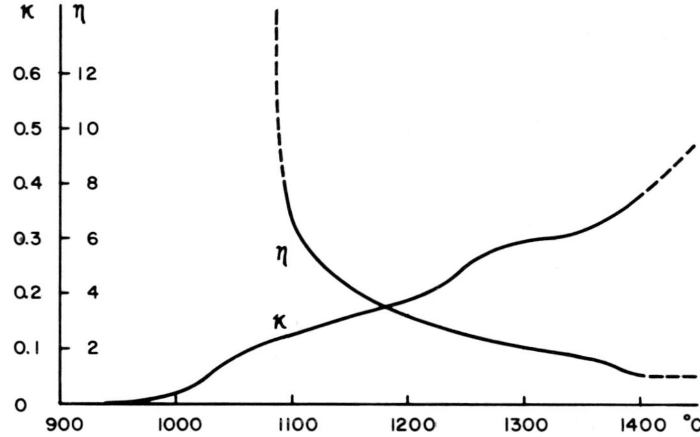

FIG. C. 23. Viscosity and specific electric conductance of a slag high in $MnO.SiO_2$, containing in addition 8 per cent CaF_2. Viscosity (η) in logarithmic scale. (Zaimskikh and Esin).

61. The action of soda in iron silicate melts was studied by P. T. Carter and M. Ibrahim[99] (cf. Vol. III, B. ¶ 320 f.), with a strong tendency to form sodium silicates rather than iron silicates. Even small percentages of any melt richer in silica than the disilicate composition always result in the decomposition of fayalite to form sodium disilicate and wüstite. The high stability of $Na_2O.2SiO_2$ is undoubtedly due to the ease with which the sodium ions fit into the interstices of the silicate framework structure in the melts with a pronounced tendency to form glassy phases. Another fact of importance in the metallurgy of iron is the surprising effect of small additions of soda to ferrosilicate melts near the composition $FeO.SiO_2$. Only 1 per cent Na_2O lowers the melting temperature of a typical acidic slag composition (with 50 to 55 per cent SiO_2) by several 100°C. Such compositions are normally rather viscous at steel-making temperatures but there is no indication observed of a corresponding effect of soda additions on the viscosity of these slags (cf. C. ¶ 105).

62. The effects of zinc oxide and sulfide on the viscosity of molten blast furnace slags were studied by F. Johannsen and K. Holler[100] for the systems $CaSiO_3$—FeO—SiO_2, and CaO—FeO—Al_2O_3—SiO_2. Up to 20 per cent ZnO the viscosity is lowered, and the solidification temperature is reduced by 50° to 100°C., but alumina contents

[99] *J. Soc. Glass Technol.*, **36**, 1952, 142-163.
[100] *Z. Erzbergbau u. Metallhüttenw.*, **9**, 1956, 511-522.

of 5 to 10 per cent increase the viscosity of ZnO containing slags considerably, although the solidification temperature may be lowered below 1000°C. Zinc sulfide reacts with ferrous silicate melts in slags to form iron sulfide and dissolved zinc silicate; the reaction constant at 1200°C. is 3.0 ± 0.5. The influence of zinc sulfide on the viscosity is low, but on cooling crystallization occurs, which increases the viscosity sharply with a break point on the viscosity versus temperature curve. The temperature of it depends on the concentration in ZnS; it is 1150°C. for about 0.3 per cent ZnS, and 1320°C. for 4.3 per cent ZnS.

63. Welding slags for metal and slag baths in electric arc welding operations were extensively studied by W. M. Conn.[101] The particularity of their action, in contrast with the conditions given in the open hearth melts, is characterized by a very short (for a few seconds) contact between the liquid steel and the slag melt thus making the problem of equilibrium of reactions highly critical. A low viscosity of the slags is indispensable for a rapid removal of gases from the baths. Therefore, also very special conditions of surface tension and crystallization rates are ruling. Conn comes to the conclusion that during the welding equilibria in the liquids are approximately reached if the fluxing agents promote a sufficiently high fusion rate, a low viscosity and surface tension. On the other hand, the heat conductivity and thermal diffusion should be low. For these multiple requirements, mixtures of the system $Al_2O_3—SiO_2$ on the aluminous side are suitable, with the addition of fluorides. It is further possible to use mixtures of low-melting and refractory ingredients (say aluminum silicates) in place of uniformly melting compositions. The refractory ingredients would then act (in a suitable grain size distribution) as nuclei for the removal of gases from the liquid bath.

64. Determinations of the viscosity of *manganese*-alumina silicate slags by Sh. M. Mikiashvili, A. M. Samarin, and L. M. Tsylev,[102] in the range of their lowest fusion temperatures, were made by a rotation viscosimeter over a wide range of viscosity, for alumina contents between 5 and 30 per cent, at 1250° to 1290°C. The viscosity is increased with the silica contents, dependent on the size of the complex silicate anions. Lowest viscosity was observed for slags with the ratio $MnO/Al_2O_3 = 6$ and 20 to 35 per cent silica. The viscosity is increased with decreasing MnO/SiO_2 ratio, the fluidity is highest for this ratio = 2.6, and up to 22 per cent alumina. Melts in the crystallization field of tephroite are least viscous (cf. Vol. III, B. ¶ 310 ff.). By the elementary ceramic test of cone deformation temperature observation,

[101] *Schweissen u. Schneiden,* **10**, 1958 (5) 174-181, with extensive bibliography. See further the book of W. M. Conn, "Die Technische Physik der Lichtbogenschweissung einschliesslich der Schweissmittel," Springer-Verlag, Berlin, Göttingen, Heidelberg, and J. Bergmann, München, 1959, 386 pp. especialy pp. 181-227.

[102] *Izvest. Akad. Nauk S.S.S.R.,* Otdel. Tekhnich. Nauk 1957 (1) 115-122; on the viscosimeter see *Zavodskaya Lab.,* 1951, No. 5.

F. Ball, H. M. Richardson, and G. R. Rigby[103] investigated the action of MnO and of mixtures of MnO and FeO in the molecular ratio 1 : 1 on the viscosity behavior of aluminum silicate refractories in nitrogen atmosphere. The crystallization of manganese anorthite and manganese cordierite was confirmed by X-ray analysis. The deformation temperatures as a function of composition show minima in the neighborhood of the eutectics of the system $MnO—Al_2O_3—SiO_2$. Manganous oxide reduces the deformation temperature of aluminum silicate refractories somewhat more than FeO. Highly aluminous alumina-silica bricks are more affected by both oxides in their thermal durability than bricks high in silica.

65. Starting from the special conditions of electric arc welding (see above), P. Beyersdorfer[104] discussed the chemical *acidity-basicity* problems of silicate melts of the slag type. The amphoteric nature of alumina in such compositions is of great significance for aspects of the chemical character of slag melts. The weight ratio of oxygen in the acidic portions to that in the basic parts is often used for establishing formulae similar to the classical Seger symbols for ceramic bodies. Thus Beyersdorfer develops a modified symbolic formula system based on the "chemical characteristics" $C = p + 3\,q$ in which is $p = (R_2O + RO)/\text{acids}$, and $q = R_2O_3/\text{acids}$, both in gram-moles, determining the distribution of acidic and basic constituents in silicate slag melts. Aluminates are formed if less silica is available than required for the formation of orthosilicates (with C above 2). In slag calculations titania is to be added to SiO_2, CaF_2 to be neglected. Practical experience in arc welding slags is that iron oxides go into the slag, reduced silicon into the metal phase. The higher silica in the slag, the more silicon is alloyed in the metal phase. The complex composition of most welding slags makes it difficult to characterize their mineralogical constitution, but with a sufficient probability it is concluded that for C above 2 either corundum, or spinels, or both may crystallize, associated with $CaSiO_3$, $CaO.Al_2O_3$, and gehlenite.

66. K. Werner[105] challenged the basicity problem of fused slags by determination of the electrolytic dissociation to replace analytical methods to determine the ratio CaO/SiO_2 (in weight per cent). In general, increasing amounts of CaO added to iron silicate melts reduce conductance, and it becomes more and more electronic. For complex systems of lime with silica, FeO, MnO, and other oxides, no distinct relation exists between basicity and conductance. Also a measurement of the electromotive force of galvanic cells like $Pt/(Al_2O_3 + FeO)/(Al_2O_3)/Pt$ at 1500°C. does not give a simple correlation between the FeO concentration and the EMF, nor for similar cells with iron silicates as electrolytes to which lime is added. No reproducibility at all

[103] *Trans. Brit. Ceram. Soc.*, **56**, 1957, 200-216.

[104] *Silikat Tech.*, **5**, 1954, 381-384.

[105] *Metallurgie (Berlin)*, **4**, 1954, 379-384. See also electrochemical data given by Ph. Visseriat, *Silicates inds.*, **20**, 1955, 200-204, on migration of Mn^{2+}, Cr^{3+}, V^{5+} and Mo^{6+}, ions from the metal (steel bath) into refractories under action of galvanic potentials in corrosion cells.

was observed. Therefore, it is not yet possible to give a conclusive electrochemical determination of the basicity characteristics of slags.

67. An extensive attempt to give accurate physicochemical aspects on the important *corrosivity* of slags on refractories in metallurgical furnaces was made by I. G. van Gijn[106] concerning the diffusion processes here ruling and possibilities for the development of more or less effective protection layers on the hot surface. The application of the Fick law made evident that the decrease in thickness, Δd, of the refractory layer (in horizontal position) is a function of viscosity (η) of the corrosive slag melt, of temperature, (T), and of the square root of time, (t), duration of the reaction: $\Delta d = A\sqrt{T/\eta}\cdot\sqrt{t}$ (A a constant). Starting from the model experiment with exposure of an alumina-silica refractory brick in a rather basic slag of the system $CaO—Al_2O_3—SiO_2$ (with 50 per cent CaO, 10 per cent Al_2O_3), isotherms of solubility were made, which show a sharp maximum for the 20 per cent alumina material. Isotherms of log η as a function of alumina content in the refractory body also show maxima in the same concentration range. The more important effects of viscosity bring about a complex function of $A/\sqrt{\eta}$ versus alumina contents, the isotherms of which show for 1300° and 1400°C. a rapid climax at 70 to 80 per cent alumina. These experimental factors are, of course, only valid for constant conditions of the glass matrix constitution and porosity of given refractory material, migration of alkalies in the body following the temperature gradient (for factors discussed in Vol. V, Section B). The viscosity and its temperature coefficient ("length") of molten corrosive slag determine the rate of penetration into the pores, static pressure and differences in surface tension (in dynes/cm.^{2}). Particularly "long" are the acidic Bessemer converter slags, "short" are cupola furnace slags, those of the open hearth, and nonferrous metallurgical slags. The latter type of slag is highly corrosive to common fireclay refractories. The rate of corrosion may become so high that no penetration occurs and a rapid eating away from the surface is observed. Different diffusion rates characterize the well-known zoning phenomenon, with typically colored fronts of diffusion for Fe^{2+}, V^{3+}, and Ti^{3+} ions in a level more shallow than that of alkalies and alkaline earths. Galvanic potentials are built up by diffusion processes.

68. Serious problems arising from corrosivity of most metallurgical slags on refractory lining materials, especially in iron and steel making, have caused tremendous efforts to establish clear evidence and testing methods for this fundamental phenomenon. We mention here the analogy with processes to be discussed in Vol. V, Section B, such as corrosion by fly dusts[107] of soft coal ashes with the often repeated

[106] *Silicates inds.*, **23**, 1958, 63-66, 137-139. A thorough study of fundamental physico-chemical problems of scorification of steel plant ladle bricks on the basis of fireclay, see by H. Neises and H. E. Schwiete, *Sprechsaal*, **94**, 1961, 1-5, 24-29, 49.

[107] N.N., in *Silikat Tech.*, **5**, 1954, 473-475.

advice to replace common siliceous fireclay refractories with basic bricks. G. R. Rigby, H. M. Richardson, and G. H. B. Lovell[108] discussed corrosion effects observed by ferrugineous slags also on basic refractories in which magnesia and lime behave much differently, since in magnesite brick only occurs the exchange of Mg^{2+} and Fe^{2+} cations, whereas in lime-rich brick material FeO and magnetite react to form calcium ferrites and metallic iron is deposited. Further Fe_2O_3 reacts with forsterite brick to form magnesium ferrite and clinoenstatite with a very marked reduction of the refractoriness of forsterite brick. Stabilized dolomite brick are corroded by ferrugineous slags, as was shown by R. Kiyoura and T. Sata,[109] using synthetic alumina-iron oxide silicate melts, with molecular ratio $Al_2O_3/Fe_2O_3 = 1$, in test crucibles made of different dolomite composites (cf. Vol. V, Section B). To imitate practical operating conditions and to accelerate the corrosion of slags with refractories, the same authors recommend a special shaking (oscillating) test machine[110] for 25 c. p. m., useful up to 1600°C.

69. Steel foundry slags in steel ingot casting were studied by G. van Gijn and K. Kooij[111] under aspects of the determining role of the ratio Mn : Si in steel. Those from steel with Mn : Si below 1.0 are usually not corrosive, but manganese-rich steels bring about a strong scorification, at a casting temperature of 1640°C. Plant experiments are described with a microscopic examination of the refractories and of reaction zones of steel/slag contacts which always show a cellular structure caused by the turbulent flow of the steel and local disruptions of the metal from the surface. The Reynolds number of flowing steel is calculated to be about 60,000, i.e., far above the critical limit of the transition from lamellar to turbulence flow phenomena.

70. The characteristic reaction of manganese in steel with silica in the refractories and slag compositions to free silicon according to the equation $2\,Mn + SiO_2 \rightarrow 2\,MnO + Si$ is also extensively discussed by Ph. Visseriat.[112] MnO is bound in the slag to silicate and aluminosilicate of manganese, of course. Concerning the turbulent flow *erosion* exerted by steel whirling, as in nozzles and channels, around stopper and collar, during casting, Visseriat recommends laboratory test methods imitating these mechanical effects in a rotary model furnace. The flow of steel in channels can be replaced by a vertical up-and-down movement of the channel bricks. A similar

[108] *Trans. Intern. Ceram. Congr.*, **3**, Paris 1952, 193-203.

[109] *Sekko to Sekkai*, 1958 (36) 31-38.

[110] *Yogyo Kyokai Shi*, **62**, 1954, 768-772. We mention in this connection also the testing method based on the principle of a constant impact flow of corrosive slag melt onto refractory surface as developed by M. S. Kamenichnyĭ, M. P. Nazarov, I. Ya. Zalkind, and T. V. Bursian, *Ogneupory*, **17**, 1952, 414-420; *Silikat Tech.*, **4**, 1953, 225. The name of the first author is here written erroneously M. S. Kamenitschin (misprint).

[111] *Silicates inds.*, **19**, 1954, 317-325.

[112] *Silicates inds.*, **20**, 1955, 200-204.

viewpoint of combining corrosion and erosion effects in slag-refractories-metal bath systems of testing is emphasized by J. Massieye and L. Lécrivain[113] in a rotary laboratory kiln at a temperature of 1600° to 1650°C., to replace the unsatisfactory common pill and dipping tests. At a service temperature of 1700°C. in a medium-frequency induction furnace, A. Cocco[114] observed the reaction of chromium-nickel steel, stainless steel, and molybdenum or wolfram-chromium-vanadium alloyed steels to refractories and slag formed with interesting microscopic identifications and X-ray diffraction analyses. The migration of alloyed elements into slag and refractory bodies could thus be followed, especially into foundry sand used for the tamped moulding bed (with a boric acid binder). Besides cristobalite and tridymite as the common high-temperature modifications inverted from quartz sand, characteristic spinels are described as reaction products, chiefly of the type RFe_2O_4 and RCr_2O_4 (subordinate also RAl_2O_4), corundum, and mullite.

71. The important role of *surface tension* effects, measured in the contact angle, Θ, of slags on different solid surfaces (refractories, or metal baths) in metallurgical processes is made evident by experimental investigations of H. Towers.[115] Optimum conditions for reducing slag corrosion on refractories are to create a maximum possible contact angle, combined with the action of a high interfacial free energy between both phases (the refractory body being assumed as a homogeneous phase unit). Towers developed a method for continuously observing and measuring the contact angles at high temperatures (cf. A. ¶ 100), and numerical data are given for angles on the contact between slag melts and platinum sheet, or on alumina. The influence of furnace atmosphere (air or nitrogen) is highly important in such measurements. For example, changes in oxygen partial pressure in kiln atmosphere bring about a typical time-dependent change of Θ for contacts of molten slags (of the system $CaO—Al_2O_3—SiO_2$) with graphite. For these changes a thermodynamic calculation of heterogeneous gas reactions acting is possible, including specific effects of reaction products on surface energies. Fig. C. 24 gives the logarithm of the partial pressure of CO as a function of temperature for surface contact slag/graphite, calculated from activities of silica = 1.0, 0.3, and 0.1, if the simple reactions $SiO_2 + 2\,C \rightarrow Si + 2\,CO$ and $SiO_2 + 3\,C \rightarrow SiC + 2\,CO$ are assumed to rule the interaction of slag with graphite. One must, however, not omit the contingency that vapor of SiO might be formed and participate in heterogeneous reactions. In every case, continuous diffusion and evaporation from contacts occur.

[113] *Bull. soc. franç. céram.*, 1955 (28) 13-22.

[114] *Fonderia ital.*, **8**, 1959 (5) 175-179.

[115] *Trans. Brit. Ceram. Soc.*, **53**, 1954, 180-202. On effects of furnace atmosphere in metallurgical processes on refractories, and, therefore, also on slags cf. H. M. Kraner, *Am. Ceram. Soc. Bull.*, **34**, 1955, 173-176, especially concerning the system Fe—O, and particular action of zinc containing vapors, catalytic effects on the break-down of CO with deposition of carbon, bursting, and the like, cf. Vol. V, Section B.

72. The change in contact angle Θ with time, however, is not able to give any specific information on the progress of reactions presumed nor actual data of surface energies. Some conclusions may be made concerning a solubility of SiO in the slag, thus lowering surface energy of the slag with respect to both gas phase and graphite.

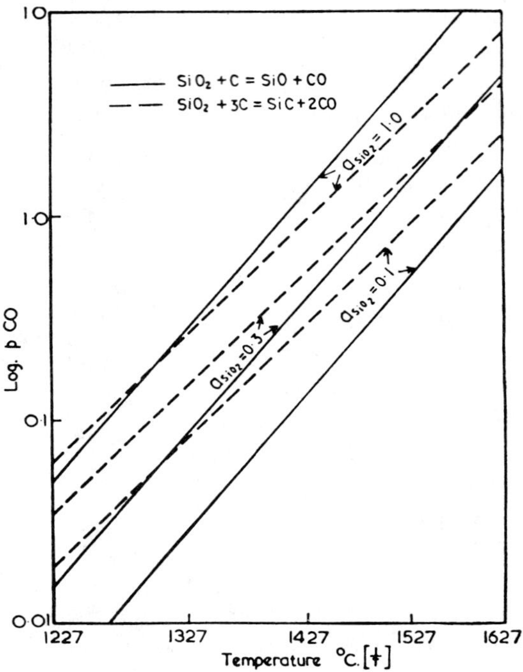

FIG. C. 24. Functional relation of CO partial pressure in furnace atmosphere, for system slag melt/graphite, for variable activities of silica, and temperatures. (Towers).

The change in shape of the slag droplet proceeds in steps as seen from the curves of cos Θ versus time. For each SiO concentration an equilibrium state may be possible following the Dupré equation

$$\cos \Theta = \frac{\gamma_{SL} - \gamma_{SG}}{\gamma_{LG}}$$

If the effect of SiO on surface energies is greatest at lowest concentrations, and gradually decreases as the concentration increases, the gradual approach to a finite end equilibrium for cos Θ would be understood. It is empirically found to be reached at the final contact angle of 20° for the experimental system and for the presumed saturation of the slag in SiO. The reality of the formation of SiO in liquid slags is made evident by electronmicroscopic studies of P. V. Gel'd, N. N. Buĭnov, and

R. M. Lerinman[116] (cf. Vol. IV, Section A), especially in flue deposits. In electric arc steel furnaces, strongly reducing conditions with acidic slags are most favorable to the volatilization of SiO in a typical aerogel dispersion with spherical particles of less than 1 μ in diameter.

73. *Volatile alkalies* in upper portions of blast furnaces are the cause of particular wear phenomena on refractory lining, as described by O. M. Margulis and E. A. Gin'yar.[117] In one of the inspected furnaces, slag layers on the lining brick consisted principally of nepheline, alkali silicates, and wollastonite, forming a slag of low refractoriness (softening at 1190° to 1300°C.). The alkalies had penetrated 5 to 6 cm. deep into the brick, at some places even to 10 cm., and densified the structure, but no carbon deposits were observed, probably because of the very low iron content of the brick. Liquid alkaline slag was observed in the boshes, in contrast with resorption of alkali vapors in the shaft portions of the furnace. If alkalies, particularly potash, enter the blast furnace slag, together with FeO, viscosity is considerably lowered, and crystallization temperature of acidic and basic alumina-free slags decreased, as N. L. Zhilo, A. V. Rudneva, G. A. Sokolov, and L. M. Tsylev[118] observed. Also in acidic slags with 5 to 10 per cent alumina, viscosity and crystallization temperature are lowered by potash resorption, whereas in basic slags both parameters are *increased*. An optimum slag composition of such alkali-enriched slags would be 0 to 17 per cent K_2O, 35 to 52 per cent CaO, and 42 to 48 per cent silica. FeO has a fluxing effect similar to potash, but more pronounced in acidic than in basic slags. If the ratio CaO/SiO_2 is = 0.6 to 1.16, addition of 3 to 16 per cent alkalies lowered viscosity. The alkali displaces FeO in the molten slags and accelerates, in this way, reduction to metallic iron. The potash forms, in primary alkali-enriched blast furnace slags, molecules of kaliophilite (kalsilite) and leucite.

74. D. S. Belyankin, B. V. Ivanov, and V. T. Basov[119] contributed much to the knowledge of interactions between slags and inner wall lining refractories of blast furnaces by their masterly microscopic analyses of thin sections of reaction products, as a function of composition (especially alkali contents) of rock materials and wastes or coke ashes in the charges. The role of an hydrolysis of alkali halides to form highly reactive hydroxides, and, in a secondary reaction, carbonates and cyanides

[116] *Izvest. Akad. Nauk S.S.S.R.*, Ser. Fiz. **15**, 1951, 366-372.

[117] *Ogneupory*, **22**, 1957, 549-556.

[118] *Izvest, Akad. Nauk S.S.S.R.*, Otdel. Tekhnich. Nauk 1957 (2) 27-35; (6) 37-42. See also F. Trojer on so-called alkali bursting of refractories, in a microscopic interpretation of volume changes by replacement reactions; cf. Vol. V, Section B, and *Radex-Rundschau*, 1956, especially, pp. 191-195, and *Ber. deut. keram. Ges.*, **38**, 1961, 557-566.

[119] "Investigations on the Blast Furnace Process," Academy of Sciences of the U.S.S.R., Moscow 1957, pp. 196-214. See also book of D. S. Belyankin, V. V. Lapin, and B. V. Ivanov, "Technische Petrographie," translated by J. Winkler, VEB Verlag Technik, Berlin, 1960, 455 pp., especially pp. 90-95.

is also made evident, with their most detrimental actions on the lining and a striking local enrichment. Vapors of K_2CO_3 react even with basic or semi-acidic refractories more intensely than with common fireclay brick. Alumina-enriched brick behaved somewhat more favorably to KCl vapors and no warping was observed. Besides the usual formation of kaliophilite, leucite, and the like, Belyankin *et al.* observed thermonatrite, $Na_2CO_3.H_2O$, as an interesting formation in blast furnace brick linings. They described typical replacement structures of alkali aluminosilicates over grains of the fireclay bodies, with a crystal growth coarser, nearer to the hot surface. Also $5CaO.3Al_2O_3$ was then observed in the completely "metamorphic" structure of the refractory body, anorthite, mullite and leucite in the deeper portions, with a glass of refractive index $n = 1.610$.

75. The specific reactions between *potash* and alumina-silica refractories of blast furnace linings were recently studied by S. E. McCune, T. P. Greaney, W. C. Allen, and R. B. Snow.[120] In general, their observations confirm results of Zhilo, Belyankin, et al. In addition, laboratory tests were made with mixtures of KCl and K_2CO_3 applied to different alumina-enriched fireclay bricks (up to 70 per cent alumina) at 870°C. Whereas this mixture did not cause peeling or alkali-bursting, typical cracking occurred by treating with fused KCN at 815°C. under particularly reducing conditions. The formation of kaliophilite and leucite was made evident by X-ray diffraction analysis in the same way that these minerals are wide-spread in blast furnace linings of the low-temperature region (cf. Vol. V, Section B) after service.

76. A typical resorption of silica-alumina refractories in basic blast furnace slags (with 47.3 per cent lime, 10.3 per cent alumina) was studied by D. N. Poluboyarinov and L. V. Smirnova,[121] at 1400° to 1450°C. With increasing alumina contents of the refractories the extent and rate of dissolution in the slag was decreased consistently. The rate of dissolution was determined in initial stages of the resorption reactions for a refractory with 30 to 45 per cent alumina in the order of magnitude of 10^{-2} gm./minute/°C., and 7×10^{-3} gm./min./°C. for a refractory with 70 per cent alumina. Refractories of iron and steel making also must often stand particularly strong corrosion effects from the combustion gas atmosphere of metallurgical furnaces, combined with the action of droplets and splashes of corrosive slags, metal, and metal oxide vapors. Especially strong corrosion effects of combustion gases on burners from natural gas, enriched in methane, are observed by M. Inzigneri,[122] and E. Biagiotti and G. Grungo,[123] with cracking reactions of hydrocarbons to form unsaturated compounds and even free radicals, with a serious volatilization of silica as SiO, and ensuing changes of mullite into corundum. Slag splashes of fayalite-rich

[120] *J. Am. Ceram. Soc.*, **40**, 1957, 187-195.
[121] *Ogneupory*, **17**, 1952, 71-81.
[122] *Ceramica* (*Milan*), **8**, 1953 (6) 39-45.
[123] *Ceramica* (*Milan*), **8**, 1953 (7) 39-61.

composition are particularly detrimental because such FeO compounds formed on refractories act as catalysts for cracking reactions with local depositions of carbon.

77. *Alkali sulfides* in contact with blast furnace slags and refractories may bring about an interesting reaction to volatilize silica in the form of SiS, namely by a reaction of the type $Na_2S + SiO_2 \rightarrow Na_2O.SiO_2 + SiS$, as R. Rasch[124] described it. Commonly, the sulfide will have been formed by reduction of sulfates first to sulfites, and then by disintegration of the latter to sulfide, as it is typically observed in glass tanks running with a reducing atmosphere, or in the batch (cf. B. ¶ 27).

78. Particular conditions are observed if refractories stand in low-temperature coke furnaces and are exposed to salt coal ashes. F. Kögl[125] described similar damages on refractories in service of boiler-combustion chambers and industrial hearths. He tested to determine their corrosion durability by blowing molten slag and ash material under an angle of 30° to 45° in the flame current towards the surface of the refractory sample. Low-temperature coke, the ash of which is high in $CaSO_4$ and Fe_2O_3 (cf. C. ¶ 12), is particularly detrimental to silica-alumina refractories. An ash with 13.3 per cent alkalies (alkali salt coal ash) melting at 1200° to 1300°C., was also used for such tests in which superduty fireclay brick stood somewhat better than common fireclay material, corundum, silicon carbide, chrome dolomite, forsterite, and with magnesite best, but in the three last-named bodies a high degree of slag infiltration was observed.

79. In view of the variable porosity of refractories, H. E. Schwiete and L. Žagar[126] emphasized capillary suction and impregnation in refractories. Starting from the ideal case of capillary suction, the Hagen-Poiseuille law can be used as the basis for calculations of practical applicability. If V_t is the amount of liquid sucked up in time t, A is the cross section through which the suction and impregnation takes place, ε is the "open" porosity, r the efficient pore radius, σ the surface energy in ergs/cm.^{2} of the liquid phase, η its viscosity, and Θ the contact angle of liquid/solid phase, then

$$V_t = A \cdot \sqrt{\frac{1}{2}\varepsilon^2 r} \cdot \sqrt{\sigma \cdot \cos\Theta/\eta} \cdot \sqrt{t}$$

The slope of the linear relation of V_t/A versus t is the impregnation coefficient

$$f = \sqrt{\frac{1}{2}\varepsilon^2 r} \cdot \sqrt{\sigma \cdot \cos\Theta/\eta}$$

[124] *Silikat Tech.*, 2, 1951, 303-304, 309.

[125] *Silikat Tech.*, 6, 1955, 113-116.

[126] *Naturwissenschaften*, 42, 1955, 533-534; *Tonind.-Ztg. u. Keram. Rundschau*, 83, 1959, 115-118. On L. Žagar's investigations of the physical factors governing the interaction of silicate (slag) melts on refractories see his Habilitationsschrift, Technische Hochschule Aachen 1960, and *Veröffentl. Inst. Gesteinshüttenkunde Rhein.-Westfäl. Tech. Hochschule Aachen*, 3, 1960, 37-92.

Experiment No.	F_{texture}	F_{mobility}	$F_{\text{calculated}}$	$F_{\text{experimental}}$	Ratio $F_{\text{calc./exptl.}}$
6	0.00235	8.70	0.0205	0.0220	0.93
7	0.01100	8.70	0.0957	0.0286	(3.35)
8	0.00289	8.70	0.0251	0.0370	0.68

81. The stability of different refractory oxides (of aluminum, beryllium, magnesium, thorium, titanium, and zirconium) in contact with metals at high temperatures was experimentally tested because of its importance for nuclear energy reactors and supersonic flight purposes, by G. Economos[127] in a medium-frequency induction furnace in high vacuum especially at 1400°, 1600°, and 1800°C. The reaction with silicon at 1400°C. is slight for Al_2O_3, MgO, and TiO_2; with no perceptible reaction observed for ZrO_2. At 1600°C. the reaction is violent with all oxides mentioned above and at 1800°C. heavy volatilization occurred. As reaction products mullite, forsterite, and zircon, respectively were observed. Molten aluminum metal wettens and impregnates all common fireclay refractories, K. J. Brondyke[128] observed. The pickup of silicon by the reaction of $4\ Al + 3\ SiO_2 \rightarrow 2\ Al_2O_3 + 3\ Si$ in the metal melt is a serious problem of aluminum metallurgy, especially for alloys with low silicon limits. Penetration is so intense that the refractory bottom portion of the furnaces can be literally dragged out as a monolithic impregnated mass. Surprisingly enough, porosity of the refractory body has little effect on the degree of penetration. The rate of reaction is ruled by the rate of diffusion of aluminum through the penetration zone to the nonreacted refractory material.

82. In reverberatory furnaces for ZnO production chrome magnesite bricks are often destroyed by a reaction with acidic iron and calcium silicates acting on the spinel-forsterite binder of the basic brick, especially in combination with ZnO. As A. M. Davidson, P. A. Polvoĭ and G. A. Rashin[129] observed, iron migrates from the scum with low-melting silicates and spinellides. Orthosilicates of the bond are changed to calcium-magnesium-iron pyroxenes, whereas chromite grains are more resistant and the structure of the refractory body becomes highly irregular and nonuniform. In contrast with such refractories for zinc furnaces, zinc spinel gahnite, $ZnAl_2O_4$, is not only highly refractory (up to 1910°C.) when in a well-sintered state, but also resistant to molten aluminum, zinc, tin, and lead, further to NiO, CaO, and MgO, and to silicate slags up to 1600°C. I. S. Kaĭnarskiĭ and N. A. Sidorov,[130]

[127] *Ind. Eng. Chem.*, **45**, 1953, 458-459.

[128] *J. Am. Ceram. Soc.*, **36**, 1953, 171-174.

[129] *Ogneupory*, **22**, 1957, 417-425.

[130] *Ogneupory*, **23**, 1958, 19-23. See also *Trudy Khar'kov. Politekhn. Inst. im. V. I. Lenina, Ser. Khim.-Tekhnol.*, **13**, 1957 (4), 165-171, and *Ibid.*, pp. 149-151, on the thermal decay of gahnite at high temperatures.

therefore, considered and recommend the use of gahnite sinters as refractories. Gahnite, however, reacts with free silica, lime, and manganese oxides, also especially violently with alkalies under phenomena of strong volume increase (swelling). In an oxidizing atmosphere a service temperature of 1500°C. is most suitable, such as for gahnite saggers for firing refractories as crucibles for fusion of aluminum, zinc, tin, for lining zinc distillation retorts, and similar uses.

83. D. M. Chizhikov, Z. F. Gulyanitskaya, V. P. Schastlivyĭ, and R. N. Petrova[131] for metallurgical slag compositions, studied especially the system $CaO—FeO—SiO_2$, with a substitution of FeO by ZnO. For this purpose electrical conductance was measured (in pure nitrogen atmosphere) from 1150° to 1450°C., magnetic susceptibility (by Gouy's method) from 700° to 1400°C. in a field of 4000 Oersted units, heat capacity from 100° to 1300°C. The melts had a constant ratio $SiO_2 : FeO = 0.9$. Substitution of SiO_2 by ZnO brings about an increase in conductance, but it is decreased if FeO is substituted by ZnO. Heat capacity is reduced by an exchange of ZnO for FeO. Measurements were made for the ratio $SiO_2/CaO = 0.8$, 1.0, and 1.6, with a constant sum FeO + ZnO (fusion points between 1130° and 1230°C.), with a maximum fusion temperature at $SiO_2/CaO = 1.0$, and 7 per cent ZnO. Conductance is reduced for $SiO_2/CaO = 1.6$ by additions of 20 per cent ZnO in maximum from 0.7 to 0.26 Ω^{-1} cm.$^{-1}$. Also increasing temperatures diminished specific conductance. The magnetic susceptibility is relatively independent of temperature, but depends on the FeO contents. All of the melts are paramagnetic, with a Curie temperature of about 700°C. The effect of ZnO substituting for FeO, on heat capacity is rather insignificant, with a flat minimum for $SiO_2/CaO = 1.0$. Crystallization of calcium silicates from the melts causes a distinct exothermic effect at 800° to 1000°C. Specific conductance for 1300°C. has a maximum if 10 per cent ZnO is substituted for FeO, probably by formation of a compound. In an analogous series of experiments, D. M. Chizhikov, Z. F. Gulyanitskaya, and V. P. Schastlivyi[132] measured specific electrical conductance of liquid melts of the system $FeO—SiO_2$ (with some Fe_2O_3), with 52 to 100 per cent FeO at 1150° to 1450°C. If alkaline earth oxides (lime, magnesia, BaO) are added, the specific conductance is increased for the constant ratio $SiO_2/FeO = 0.9$. It is also increased if SiO_2 is replaced by these monoxides. If FeO is replaced by lime or magnesia, specific conductance is initially increased, but with continued substitution the effect is reversed, whereas BaO causes a continuous decrease in specific conductance.

84. The often discussed problems of corrosion of refractories by *vanadium* containing ashes from natural oils as fuels in industrial firing were extensively studied by M. C. K. Jones, and R. L. Hardy.[133] Although the absolute content of ashes in crude

[131] *Doklady Akad. Nauk S.S.S.R.*, **129**, 1959, 174-176.

[132] *Zhur. Neorg. Khim.*, **4**, 1959, 2067-2071.

[133] *Ind. Eng. Chem.* **44**, 1952, 2615-2619. See furthermore J. R. McLaren and H. M. Richardson,

mineral oils is only between 0.001 and about 0.05 per cent, this nonvolatile material is accumulated in heavy fuel oils and their ashes. Contents in V_2O_5 are destructive for fireclay brick on direct contact at high temperatures by combination with alkalies (chiefly Na) and alkaline earths. The most aggressive ingredient is vanadium pentoxide. Whereas without vanadium, usually only a hard film of glaze is formed on the brick surface which may result in failure by spalling, fluxing action of V_2O_5-rich slag is much more intense and causes deep-going damage. More recently A. J. Hammer[134] gave a highly instructive bibliography on the mode of occurrence of vanadium in clays, either as sulfide (patronite, VS_4), or in complex organic compounds of the porphyrin group, wide-spread in oil-shales, bituminous rocks, mineral oils, waxes, coals, and the like. W. R. Foster, M. H. Leipold, and T. S. Shevlin[135] studied especially the system $Na_2O—SO_3—V_2O_5$, as a basis of comparison for corrosion studies in the laboratory to make them consistent with observations of refractory damage caused by similar ashes. It is important that sodium sulfate and V_2O_5 as such, are not compatible but react on heating, evolving SO_3, and forming $NaVO_3$, or complex vanadates (but not Na_3VO_4, or $Na_4V_2O_7$). V_2O_5, as such, is a relatively harmless corrosion agent and also does not occur in fuel ashes, but $NaVO_3$ causes the most serious attack of any of the possible corrodents, because of its association with SO_3 in statu nascendi by the reaction mentioned above. Only when sodium sulfate is in large excess over V_2O_5 in the ashes it may act independently as a corrosive material.

EQUILIBRIA OF SLAGS WITH METAL BATHS

85. In his fundamental investigations on the physical chemistry of silicate melts in general, and metallurgical slag melts in particular (cf. A. ¶ 169), O. A. Esin[136] with Yu. P. Nikitin recognized the particular importance of *exchange currents* between the liquid slags and liquid metal (baths). By alternating current measurements (with frequencies between 50 and 17,000 c. p. s.) of electrical conductance and capacity at 1480° to 1560°C. in magnesia crucibles (graphite and wolfram electrodes being introduced in cylindrical bore-holes) it was found that capacity of the systems is practically independent of frequency, indicating a low concentration in ions determining contact potentials. The capacity was say 4.6 microfarads on contact of an iron melt (containing 4.3 per cent carbon) with a boric acid melt containing 1.54

Trans. Brit. Ceram. Soc. **58**, 1959, 188-197, G. R. Rigby and R. Hutton, *J. Am. Ceram. Soc.*, **45**, 1962, 68-73; V. Cirilli, A. Burdese, and C. Brisi, *Atti Accad. Sci. Torino*, **95**, 1960/61, 32 pp., on the corrosion of refractory alloys.

[134] *Ceram. Age*, **35**, 1955 (11) 24-27.

[135] *Corrosion*, **12**, 1956, 539-548.

[136] *Doklady Akad. Nauk S.S.S.R.*, **116**, 1957, 63-65; see also *Ibid.*, **111**, 1955, 133-135.

per cent FeO, as the "model" slag, under a current intensity of 1 milliampere/cm.^{2}. However, when a sodium borate melt was used as the slag (with 1.24 per cent FeO) a double-layer was found on the contact, raising the capacity to 15.3 microfarads/cm.^{2} and an exchange current flow of 120 milliampere. The reaction resistance $R_p = RT/nF \cdot 1/i_0$ was determined exclusively by concentration in FeO, not by the presence of silica, alumina or lime in variable concentrations in the slag. Further experiments concerned melts of ferrosilicon (with 44.9 per cent Si) with the boric acid slag mentioned above (with 50 milliamperes/cm.^{2}), and iron-phosphorus (19.8 per cent P), with a calcium phosphate melt (containing 65 per cent P_2O_5 and 28.4 per cent CaO) (260 milliampere/cm.^{2}). An anomalously high intensity of the exchange current (880 milliampere/cm.^{2}) was observed on the contact of iron with a slag 28 per cent FeO, 13 per cent CaO, 35 per cent SiO_2, and 24 per cent Al_2O_3. High intensities were also observed when *sulfur* was introduced into the metal bath (such as 4.2 per cent carbon, 0.13 per cent sulfur) in contact with a lime-alumina silicate slag containing 2.25 per cent CaS, 1.72 per cent FeO (145 milliampere/cm.^{2}). Analogously, the concentration in MnO in the slag electrolyte determines current intensity on contact with baths of Mn-C or Mn-Si alloys. A slag with 36 per cent MnO gave an intensity of 1100 milliampere/cm.^{2}. Valence interchanges of the type $R^{3+} \rightleftharpoons R^{2+}$ in the melt solution were very important for rapid electron exchange of ions on contacts. There is also a regular interrelation between the intensity of exchange currents and surface tensions, σ_n, of the slag melt, and that on contacts with the metal phase, σ_m. For iron alloys with carbon, silicon, and phosphorous, and alloys of manganese with carbon and silicon, this functional relation is simply linear.

86. *Kinetics* of the ions exchange between slag and metal bath was studied by Nikitin and Esin[137] for a slag with 54 per cent silica, 15 alumina, and 31 lime. The diffusion coefficient of Fe^{2+} at 1500°C. in it was determined to be 2.4 to 3.1 $\times 10^{-6}$ cm.^{2}/sec., in good agreement with measurements using radioactive isotope tracers. For a model slag of sodium borate (with 15 per cent Na_2O) the diffusion coefficient of Ag^+ at 840°C. was 0.6 $\times 10^{-7}$, and at 940°C. 1.4 $\times 10^{-7}$ cm.^{2}/sec., the activation energy = 23,000 cal./gm. atom. The ion exchange current is a linear function of iron concentration in iron-carbon alloys, versus slags containing lime, alumina, silica, and soda, boric and phosphoric acid ions. The addition of Na_2O always raises the density of negative charges on the electrode in the systems Fe—C and Fe—P, and diminishes surface tension on phase boundaries. Cast iron (with 4.3 per cent C) versus the lime-aluminosilicate slag mentioned above at 1350°C for an Fe^{2+} concentration of 0.36 showed a current intensity i_0 = 22 milliampere/cm.^{2}, and at 1550°C. a concentration in Fe^{2+} of 0.52 and i_0 = 50 milliampere/cm.^{2}. The activation energy E_1 = 23,500 cal./gm. atom, or of the reaction Fe^{2+} (slag) + 2 $e^- \rightarrow$ Fe (alloy), and E_2 = 13,000 cal./gm. atom, for the reversed reaction. The difference between these activa-

[137] *Doklady Akad. Nauk S.S.S.R.*, 122, 1958, 106-108.

tion energies is explained by the fact that carbon in the metal melt creates reducing conditions abruptly on the phase boundary. For the analogous reaction $Ag^+ + e^- \rightarrow Ag$ (metal) activation energy $E_1 = 12{,}600$; $E_2 = 22{,}800$ cal./gm. atom. This anomaly is probably caused by formation of complexes of silver with borate anions in the melt. The bonding of Fe^{2+} and Ag^+ ions in aqueous solutions, for a comparison, is less firm than it is in slags. Silicon, carbon, and phosphorus are enriched on the phase boundaries, the metal is relatively impoverished in these elements. Remarkable is the behavior of a calcium carbide (10.8 per cent) containing lime-magnesia aluminosilicate slag which exchanges with cast iron (4.3 per cent C) in the following way: 2 C (metal) + 2 $e^- \rightarrow C^{2-}$ (slag), with a high intensity i_0. The capillary activity of CaC_2 on the phase boundaries is evident from the reduction of surface tension.

87. Nikitin and Esin[138] investigated the *electrocapillary phenomenon* of synthetic slags on contacts with metal sulfides like Cu_2S, Ni_3S_2, and a manganese-carbon alloy (7 per cent C, 0.14 per cent Si, and 0.37 per cent P) at 1350° to 1400°C. with interfacial tensions, σ, for the sulfides, at 1480° to 1500°C. with the Mn—C alloy. The interfacial tension of Ni_3S_2 was affected at the interval of potential differences, ζ, from + 0.3 to — 0.4 volt, for Cu_2S from + 0.5 to — 0.3 volt, and for Mn—C from + 0.55 to — 0.6 volt (for current densities not above 50 milliampere/cm.^{2}). The alloy surface is negatively charged. Specific charge density, ε, was found by extrapolation at zero polarizing current to be 15×10^{-6} coulomb/cm.^{2} for Ni_3S_2, 12×10^{-6} for Cu_2S, and 9×10^{-6} for the Mn—C alloy. O. A. Esin and V. V. Khlynov[139] more recently studied metal losses by the appearance of small metal droplets in the slag as a consequence of electrocapillary effects in the two-phase system under electrostatic potentials. In a highly basic (52 per cent CaO) slag melt at 1370° to 1500°C. the surface transference of droplets (of 0.5 to 3 mm. in diameter) of copper, nickel, manganese, and silver, and of Ni_3S_2 was observed in a crucible made of pure magnesia, with a direct current potential of 5 to 6 volts/cm. and an intensity of 0.5 to 5 amp. The metals migrate to the cathode, Ni_3S_2 to the anode.

88. The electrocapillary effect is caused by double-layer conditions on the contacts between slag/metal phases, with equilibria of the type $M \rightarrow M^{2+} + 2\,e^-$. If the sulfur content in the system Ni—Ni_3S_2 is varied, mobility and direction of the migration of droplets to the anode is not changed for the wide range between 26, and 0.4 per cent S, although surface tension of nickel (1756 ergs/cm.^{2}) is considerably reduced by Ni_3S_2 (with 452 ergs/cm.^{2}). But for only 0.15 per cent S mobility of droplets becomes zero, and with still lower sulfur contents direction of the migration is reversed towards the cathode. The reducing or oxidizing character of furnace atmosphere is

[138] *Doklady Akad. Nauk S.S.S.R.*, **107**, 1956, 847-849.

[139] *Doklady Akad. Nauk S.S.S.R.*, **120**, 1958, 134-136. Similar problems on ferrotitanium droplets settling in molten aluminothermic slags were recently discussed by Yu. L. Pliner and S. I. Suchil'nikov, *Ibid.*, **135**, 1960, 1187-1189, to calculate the rates of sedimentation.

an important factor in electrocapillary phenomena: with increasing reduction of oxides by hydrogen on the surface of the slag the double-layer on the droplets is changed in its electrocapillary character, migration rate for nickel is reduced to zero, for copper and silver even reverted: M^{2+} (slag) + 2 $e^- \rightarrow$ M. If oxygen is present the reaction is O (metal) + 2 $e^- \rightarrow O^{2-}$ (slag), causing a positive charge on the surface of the droplets. Applying an equation of A. N. Frumkin and V. Levich,[140] migration rates, v, are determined by the relation $v = U/Er$ in which energy factor U includes viscosities of the liquid phases, specific conductance, $\varkappa$, of the slag, specific charge, ε, of the double-layer, and its capacity, C. In good agreement with experimental data for small droplets, migration speeds are determined by the quotient $\varepsilon/2\eta$ (η the viscosity).

89. Khlynov and Esin[141] finally demonstrated how a complete removal of sulfide inclusions (droplets) from molten slags is possible in an electrostatic field of 5 to 7 volt intensity, with silicon carbide electrodes, say at 1400°C., from a lime-alumino-silicate slag containing Ni and Co sulfides. These inclusion droplets migrate under the influence of the direct current field to the cathode with a rate of 7 to 10 $\times$ 10^{-3} cm./sec. (mobility 1 to 1.5 $\times$ 10^{-2} cm./sec./volt). Under these conditions the direction of migration is inverse to that observed for Ni_3S_2 in experiments described above, and mobility is also much lower. The negative charge of the sulfide droplets suspended in a slag free from FeO is decreased if increasing additions of FeO are made, goes through the zero value, and becomes positive for a FeO content above 8 per cent. If FeO content is kept constant at 23 per cent in the slag, lime, and silica, however, are varied in its composition, mobility is not affected even if the viscosity is increased by four times because viscosity effects are compensated by a reduction of the boundary diffusion current and of depolarization coefficient. Experiments show that, practically spoken, it should be possible to remove every sulfide inclusion from metallurgical slags by using electrocapillary effects. Droplets collected near the cathode are much larger than those suspended in the slag melt.

90. The study of liquid droplet *inclusions in the metal bath*, especially of those caused by more or less violent reactions of the deoxidation process of steel-making, has been much advanced by modern analytical methods for their isolation from the metal and their spectral-analytical determination. We present here on this important field of steel metallurgy only a brief hint on investigations of W. Koch.[142] For an adequate and mildly acting isolation, electrolytical dissolution of metal is nowadays a fully adopted laboratory method, combined with chlorination methods of anode residues for the vacuum sublimation of metal chlorides formed. Many characteristic

140 *Zhur. Fiz. Khim.*, **19**, 1945, 573-600; *Acta Physicochim. U.R.S.S.*, **20**, 1945, 769-808.

141 *Doklady Akad. Nauk S.S.S.R.*, **123**, 1958, 330-332.

142 *Stahl u. Eisen*, **68**, 1948, 321-333 (with P. Klinger); **72**, 1952, 1056-1063; **74**, 1954, 264-271 (with Fr. Wever).

examples of such inclusions are described, together with comparison data of the sensitivity of chemical- and spectral-analytical methods. Since deoxidation inclusions are among the most troublesome phenomena which are detrimental to steel manufacturing, further studies after the nowadays established variability of their composition concentrate on conditions of phase segregation of liquid slag droplets from the metal bath, factors affecting their rise to the upper surface of the molten metal, and anomalies occurring on this way upward. As Koch and Wever observed, it is not only desirable that inclusion droplets have the lowest possible viscosity. On the contrary, the most fluid inclusions may have the most undesirable tendency to spread over primary grain boundaries of the crystallites of consolidating metal and thus become frozen-in as glass films. It is the practitioner's task to find the best possibilities for a reduction of amounts of inclusions and to influence their physical characteristics without a change in composition of the steel. In every case, inclusions are by no means merely an incidental phenomenon. They have their regular rules of formation and character which provide a possibility of solving the practical problems involved with success.

91. If the inclusions are derived from wear of refractory lining, ladles and other parts such as those found below the crust of rimming steel ingots, they may have the original alumina/silica ratio of fireclays used, with much glass, and crystals of mullite and fayalite. Such inclusions show a characteristic increase in number from the beginning to the end of pouring, owing to increase in viscosity of the metal which retards the rise of the particles, as V. A. Bron and D. P. Zegzhda[143] observed. Large and dispersed inclusions of this kind differ in chemical composition with considerable shifts in the ratios Al_2O_3/SiO_2 in favor of alumina enrichments. P. S. Mamykin and S. G. Zlashkin[144] describe the wear of runner brick in steel casting causing inclusions in addition to those formed from deoxidation reactions. Especially for runners, penetration of a lower viscosity reaction slag is not undesirable, because the surface layer enriched in FeO and MnO with a characteristic "comb" structure stands steel erosion action by peeling away better than fireclay material itself.

92. The important *wetting* properties of slags in the electrosteel process were extensively discussed by A. I. Kholodov, S. I. Suchil'nikov, and I. P. Malkin[145] for three-phase contacts gas/slag melt/metal bath, in which oxygen is driven to surface layers by its capillary activity, the slag being positively, the metal phase negatively charged. Sodium ions are more surface-active than those of calcium and displace these from the double-layer. Capillary wetting phenomena therefore have a strong influence on removal of silicate inclusions from the metal since they also determine the degassing process, especially in electrosteel metallurgy. The contact angle, Θ,

[143] *Ogneupory*, **16**, 1951, 518-524.

[144] *Voprosy Petrog. i Mineral., Akad. Nauk S.S.S.R.*, **2**, 1953, 271-280.

[145] *Doklady Akad. Nauk S.S.S.R.*, **101**, 1955, 1093-1096.

of synthetic slags to cast iron melt (with 4 per cent carbon, 1.5 per cent silicon) was optically measured, say for calcium aluminosilicate slags with additions of 5, 15, or 20 per cent calcium fluoride, or with 0, 0.27, and 2.31 per cent calcium carbide. The

$$\cos \Theta = \frac{\sigma_{MG} - \sigma_{MS}}{\sigma_{GS}}$$

for all slags shows increasing values with increasing temperature, the curves being normally concave to the temperature axis (convex only for one slag with 5 per cent CaF_2). The cos Θ is decreased with the ratio CaO/SiO_2 in a linear relation increasing from 2.5 to 3.1. If the carbon content in the steel bath is increased from 0.2 to 4

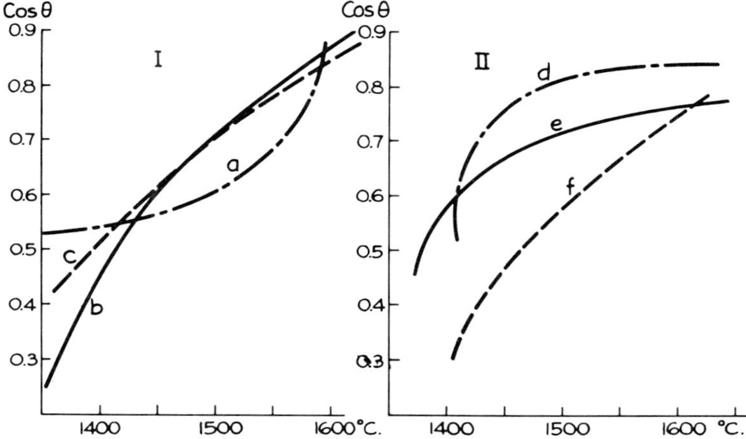

FIG. C. 27. Wetting characteristics of cast iron, by synthetic (*I*), and industrial (*II*) slags, with (*a*) 5 per cent CaF_2, (*b*) with 20 per cent CaF_2, (*c*) with 15 per cent CaF_2, (*d*) with 2.3 per cent CaC_2, (*e*) with 0.27 per cent CaC_2, (*f*) without CaC_2. (Kholodov, Suchil'nikov, and Malkin).

per cent, surface tension is decreased, silicon acts in a similar way (see Figs. C. 27, 28). Typical deoxidation slag melts of the system $MnO—Al_2O_3—SiO_2$ in their surface tension relation to an iron metal bath were studied by Sh. M. Mikiashvili, A. M. Samarin, and L. M. Tsylev.[146] The contact angle Θ increases with increasing silica and lower MnO content. Surface tension depends slightly on temperature in the range from 1450° to 1500°C. Industrially pure iron (with 0.15 per cent contaminations) has a surface tension of 1306 to 1310 dynes/cm. at 1500° to 1595°C. In any case, MnO is the active agent in slags in contact with the iron bath. As MnO is increased, surface tension is uniformly decreased.

[146] *Izvest. Akad. Nauk S.S.S.R.*, Otdel. Tekhn. Nauk 1957 (4) 54-62.

93. P. M. Shurygin and O. A. Esin[147] showed how iron can be dissolved in liquid metallurgical slags. A drop of gold in contact with $FeO—Fe_2O_3$-rich slag melts, say in an iron crucible, is rapidly enriched in iron; in 12 minutes an alloy with 60 per cent iron was formed. This transfer of iron from the slag occurs even if the crucible material is pure magnesia; but in this case the limit concentration in the gold-iron alloy is only about 2 per cent Fe. From a fayalite slag at 1250°C., iron is also transferred into molten copper if a crucible of magnesia is used, much more rapidly,

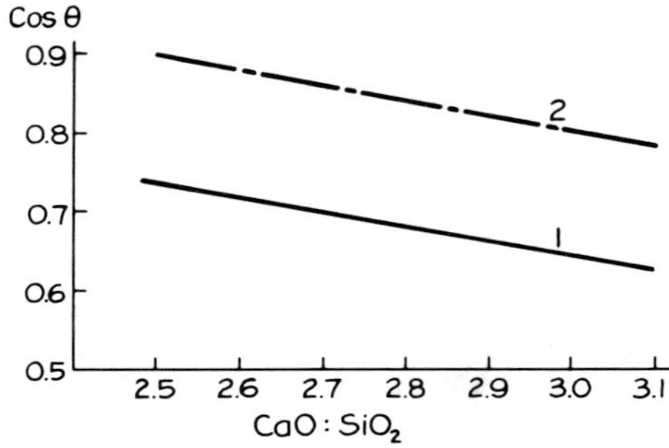

FIG. C. 28. Functional relation of wettability of cast iron by slags, dependent on the ratio CaO/SiO_2, (1) for 1400° to 1430°C., (2) for 1580°/1620°C. (Kholodov, Suchil'nikov, and Malkin).

however, in an iron crucible. Evidently, decomposition of FeO into iron and Fe_2O_3 at 572°C. is important for these reactions. Gold takes up iron even when there is no direct contact of a $FeO—Fe_2O_3$-containing slag with metallic iron. Shurygin and Esin, therefore, concluded that iron must be an ingredient of the slags themselves, not in a colloidal suspension, but in true solution analogous to metal clouds as defined by R. Lorenz (1901). An equilibrium reaction $2\ Fe^{3+}$ (slag) + Fe (slag) $\rightleftharpoons$ $3\ Fe^{2+}$ (slag) may be established as well for the slag melt as for the metal (if present), combined with distribution equilibrium of iron in the melt and in gold. At given temperatures the iron concentration in the gold melt decreases with increasing oxidation potential in the melt, in good agreement with experimental results and following the mass action law. Because of the process 2 Au (metal) + Fe^{2+} (slag) $\rightarrow$ $2\ Au^{2+}$ (slag) + Fe (slag), the activity coefficient of iron dissolved in gold cannot be equal to activity of the free iron metal. The distribution coefficient of iron in the slag and in the gold is not known, and concentration of the dissolved iron in the slag cannot be calculated. Gold is dissolved the more in the slag, the lower the oxidation potential of the slag, or the ratio Fe_2O_3 : FeO. Surface tension of gold/air is 1180 ergs/cm.^{2}, that of gold/iron 805 ergs/cm.^{2}; that of the alloy 97 per cent gold, 3 per cent iron/fayalite melt at 1300°C. = 720 ergs/cm.^{2} Gold is evidently dissolved as atoms.

[147] *Doklady Akad. Nauk S.S.S.R.*, **95**, 1954, 1043-1045.

94. The fundamental role of phase equilibria in the systems Fe—O, and Fe—O—Si, the latter including systems of the "iron oxides," FeO_n, with silica, is made evident by extensive investigations of L. S. Darken and R. W. Gurry,[148] and of A. Muan and E. F. Osborn,[149] to be discussed in Vol. III, B. ¶ 314 f., and Vol. IV, Section A concerning heterogeneous gas reactions influencing equilibria as a function of partial pressure of oxygen. The same criteria were applied to the important system MgO—FeO_n—SiO_2 by Muan and Osborn (cf. III, B. ¶ 360 ff.). We do not wish to discuss here details of the phase equilibrium diagrams which give the results of those investigations covering a wide range of partial pressures of oxygen (from 10^{-9} to 10^0 atm.) but to emphasize their primordial importance for modern aspects of the physical-chemical behavior of steel slags, and also for the life of refractories in steel-plant practice. It is particularly essential that the practitioner may learn from the diagrams and experimental data of Muan and Osborn how refractories react to changes in partial pressure of oxygen in metallurgical furnace processes, that is, how they "breathe" as the amount of oxygen in the furnace atmosphere fluctuates. Spalling, peeling, or even general disintegration of roof and wall refractories of basic (magnesite) type can be expected if oxygen partial pressure at the high temperature of operation is alternately increased and decreased. Fluctuations in temperature may have a similar effect, inasmuch as decrease in temperature enhances structural changes in the crystalline phases of the same nature as those caused by an increase in partial pressure of oxygen.

95. Concerning the behavior of forsterite brick, it was an important result of Muan and Osborn that refractoriness of such bricks is retained until considerable amounts of Fe_2O_3 (about 60 per cent) are added. At higher oxygen partial pressures, as in air, crystalline phases take up iron oxide in relatively small amounts. A critical limit concentration is soon reached, above which magnesium ferrite and pyroxenes appear as additional phases. A liquid may also appear in such mixtures near 1374°C. The refractoriness of forsterite brick is then rapidly lost. When enough magnesia is added, in excess, then liquid phases may be developed not below 1675°C. even if 60 per cent Fe_2O_3 could be formed in air atmosphere, and refractoriness in air would be better than under strongly reducing conditions. If, on the other hand, silica were in excess over the forsterite ratio $SiO_2/MgO = 0.5$, refractoriness in air is again very low. Excess silica is very detrimental to forsterite brick and much more so in air than under reducing conditions.

96. As an important supplement we also mention studies of I. A. Ostrovskiĭ and Ya. I. Ol'chanskiĭ[150] on the system fayalite—magnetite with its eutectic melt at only 1142°C. and a composition of 17 per cent Fe_3O_4 and 83 per cent Fe_2SiO_4, as

[148] *J. Am. Chem. Soc.*, **67**, 1945; 1398-1412; **68**, 1946, 798-816.

[149] *J. Metals*, 7, 1955, 965-976; *J. Am. Ceram. Soc.*, **39**, 1956, 121-140.

[150] *Doklady Akad. Nauk S.S.S.R.*, **107**, 1956, 881-883.

determined in platinum crucibles in evacuated silica glass tubes, and a visual observation of primary crystallizations from melts with less than 20 per cent magnetite.

97. The tremendous problem of the constitution of *sulfide* containing slags in steel production (cf. C. ¶ 1 ff.) is extensively discussed in the book of W. Fischer and S. Wolf,[151] together with the fascinating problem of the existence of sulfosilicates. On aluminum sulfides and the question of the existence of complex compounds of the system Al—Si—S, E. Kohlmeyer and H. W. Retzlaff[152] gave valuable information on their experiments on sulfurization of liquid aluminum-silicon alloys. There is a remarkable contrast in the relatively high volatility of SiS and SiS_2, and non-volatility of AlS and Al_2S_3 up to more than 1600°C. No sulfoaluminates or aluminum sulfosilicates were observed.

98. Referring to the complex problems of *deoxidation* of steels and their products we want first to mention observations of a reversed reduction of silica by the carbon content of pig iron melts, which is the rule if crucibles of loamy sand are used. Such reactions are not only troublesome for increasing silicon contamination in the metal melts but they affect considerably physicochemical measurements such as those made by W. A. Fischer and R. Schäfer[153] on electric resistance and galvanic cell electromotive forces. Reductions of a similar kind even in crucibles of pure alumina (sintered corundum) or magnesite are most obstructive to the discussion of these galvanic effects in metal baths in contact with such oxides (cf. A. ¶ 170).

99. Reduction effects exerted on silica in refractories are a big problem for changes observed in running brick for steel casting, as L. Halm[154] described them. The manganese content of steel determines the reaction $2\,Mn + SiO_2 \rightarrow 2\,MnO + Si$ as first studied by F. Koerber and W. Oelsen,[155] and brings about manganese aluminosilicate inclusions in steel ingots if aluminum is present at the same time. The silicon content of steel can easily be increased to 0.8 per cent. Moisture in the running brick may also cause a troublesome and stubborn evolution of hydrogen in the steel bath. The MnO-enriched and silica-impoverished vitrified glaze observed on running channels and stoppers, in the same way, is a reaction product of manganese with silica of the refractory material. Halm's results are in good agreement also with those of G. van Gijn and K. Kooij[156] on steel ingot castings and the strong scorification by reaction of manganese with refractories used in a foundry at a casting temperature

[151] "Schwefel in Schlacken und Schlackenwolle," E. Schweizerbartsche Verlagsbuchhandlung, Stuttgart 1951, 231 pp., especially pp. 113-119, on compounds of silicon containing sulfides.

[152] *Z. anorg. u. allgem. Chem.*, **261**, 1950, 248-260.

[153] *Arch. Eisenhüttenw.*, **24**, 1953, 105-111.

[154] *Silicates inds.*, **19**, 1954, 189-206.

[155] *Mitt. Kaiser-Wilhelm-Inst. Eisenforsch. Düsseldorf*, **15**, 1933, 271-309.

[156] *Silicates inds.*, **19**, 1954, 317-325.

of 1640°C. (cf. C. ¶ 69 f.). They also agree with observations of Ph. Visseriat[157] on ladle brick.

100. Reaction products on contacts between liquid steel and fireclay lining brick were studied with microscopic and chemical analysis by F. Wahlberg and L. Fredholm[158] who found manganese aluminosilicates as well as mullite. Extensive studies of deoxidation inclusions in steel ingots, their isolation and analytical-microscopic character must again be mentioned here from C. ¶ 90 when we outlined modern methods used by W. Koch, P. Klinger, and Fr. Wever,[159] plus the recent studies of W. A. Fischer and M. Wahlster[160] on the rate of separation of steel inclusions from the metal, and deoxidation products with elementary silicon. We wish only to call the attention of the reader to the multiple forms of the more or less siliceous, mostly glassy inclusions, with many crystalline separations of wüstite, cristobalite, corundum, and spherical aggregations of manganese and iron aluminosilicates of the more siliceous portions in the system $MnO—Al_2O_3—SiO_2$ (cf. Vol. III, B. ¶ 310 ff.). Very remarkable also is the result of W. Koch, that in highly silicified steels (up to 1.14 per cent Si) in the presence of silicate inclusions with 19 to 55 per cent SiO_2, aluminum metal is observed dissolved in the steel in amounts of some 10^{-3} per cent if the preferably glassy inclusions contain up to 30 per cent alumina, but several 10^{-2} per cent if the alumina content is up to 70 per cent in the mostly crystalline inclusions. Reduction of silica or of silicates by aluminum would correspond to the equilibrium constant

$$K_{\text{Si,Al}} = \frac{(\text{SiO}_2)^3\,[\text{Al}]^4}{(\text{Al}_2\text{O}_3)^2\,[\text{Si}]^3}$$

Fischer and Wahlster determined especially the equilibrium constant for deoxidation by silicon:

$$K_{\text{Si}} = \frac{[\text{Si}]\,[\text{O}]^2}{a_{\text{SiO}_2}}$$

[157] *Silicates inds.*, **20**, 1955, 200-204.

[158] *Jernkontorets Ann.*, **137**, 1953 (1) 1-26.

[159] *Stahl u. Eisen*, **68**, 1948, 321-333; **72**, 1952, 1056-1063; **74**, 1954, 264-271. See also more recently C. A. Müller, and E. Plöckinger, *Radex Rundschau*, 1957, 738-753, and E. Plöckinger and A. Randak, *Ibid.*, 1957, 754-768, with interesting examples of hot-deformed silicate glass inclusions by rolling, in different degrees of plasticity. Highly fluid silicates show a poor segregation from the steel melt, and the metal then becomes high in total oxygen. Alumina-richer silicates have good separation characteristics, with a low total oxygen content in steel. Less favorable is the separation with calcium-silicon alloys as deoxidation additions. W. Koch, *Stahl u. Eisen*, **81**, 1961, 1592-1598, emphasizes the effects of the volatilization of calcium from such deoxidation alloys and ensuing distribution of Ca in slag inclusions in steel ingots. On favorable effects of aluminum deoxidation by influence on wettability and surface tension of separating particles in contact with the steel melt see R. Rosegger, *Radex Rundschau*, 1958, 292-305, to control speed of the rise of inclusions following Stokes law.

[160] *Arch. Eisenhüttenw.*, **28**, 1957, 601-609; **29**, 1958, 1-9. On slag reaction of aluminum melts with alumina-silica refractories changing these to corundum see also K. J. Brondyke, *J. Am. Ceram. Soc.*, **36**, 1953, 171-174 (cf. C. ¶ 81).

in analogy to the constant $K_{Al} = [Al]^2 \cdot [O]^3 = 4 \times 10^{-4}$ for deoxidation by aluminum as determined at 1600°C. by N. A. Gokcen and J. Chipman.[161] There are characteristic differences in deoxidation curves as a function of time in silica, corundum, lime, and CaO + 15 per cent CaF_2 crucibles in experiments of silicon addition to Armco iron melts, fused in a coreless medium-frequency induction furnace. For constant experimental conditions, separation time of deoxidation products was by far (by about 10 times) shorter in a crucible of lime + CaF_2 than in a silica crucible, with intermediate speeds observed in alumina, magnesia, and pure CaO crucibles. For sampling by suction from open melts the use of silica glass (Rotosil) tubes proved to be a safe measure, to avoid any oxidation during sampling and cooling for a subsequent metallographic examination, and an isolation of inclusions by electrolysis. Inclusions taken before deoxidation by silicon show wüstite prevailing, those after deoxidation highly siliceous glasses of the miscibility gap of the system FeO—SiO_2. No crystalline silica was observed.

101. Inclusion problems were also described in detail by M. Šicha[162] who attributed a great role to formation of volatile silicon monoxide by reduction of silica with magnesium or aluminum, and also by hydrogen. Thermodynamics of the evolution of SiO by reaction of silica with hydrogen were established by experiments of N. C. Tombs and A. J. E. Welch[163] (cf. Vol. IV, Section A). In the temperature range of steel-making, the formation of SiO is thoroughly normal and to be expected by this reaction. This is also confirmed by calculations of K. Sanbongi and M. Otani[164] who worked in a CaO crucible at 1550° to 1650°C. with a determination of activity of silica in a molten binary calcium silicate slag (saturated in Ca_2SiO_4):

$$a_{SiO_2} = \frac{K^{S}_{Si\text{-}H}\,[Si]}{P^2_{(H_2)}/P_{(H_2O)}}$$

and the activity of CaO from

$$\log\left(\frac{a_{CaO}}{N_{CaO}}\right) - \log\left(\frac{a'_{CaO}}{N'_{CaO}}\right) = -\int_{N_{SiO_2}}^{N'_{SiO_2}}\left(\frac{N_{SiO_2}}{N_{CaO}}\right) d\log k\left(\frac{a_{SiO_2}}{N_{SiO_2}}\right)$$

(a' and N' referring to the standard state).

102. Since titanium has a greater affinity to oxygen than silicon there must be a relatively high oxidizing loss in titanium steels (alloyed from ferro-titanium prepared separately by aluminothermic method, and later added to the steel bath), by a

[161] *J. Metals,* **5**, 1953, 173-178. On sampling with silica glass suction tubes cf. W. A. Fischer and H. vom Ende, *Arch. Eisenhüttenw.* **23**, 1952, 21-33.

[162] *Hutnické listy,* **8**, 1953, 506-512, 566-577; *Chem. Abstr.,* **49**, 1955, 2275-2276.

[163] *J. Iron Steel Inst.* (*London*), **172**, 1952, 69-78.

[164] *Tetsu to Hagane,* **37**, 1951, 462-488.

prompt reaction with silica in furnace linings and slags, as P. Bardenheuer and W. A. Fischer[165] observed. In a basic induction furnace lining (with a slag of 30 per cent CaO, 70 per cent TiO_2) a strong reaction with slag zone refractories occurred (with resorption of magnesia in the slag melt making it highly viscous), but no titanium losses occurred when the furnace was closed. A basic TiO_2 containing slag which also easily resorbs magnesia in large amounts has a strongly corrosive effect on the basic lining, stronger than an approximately equivalent silicate slag. If an aluminum containing ferrotitanium is used as deoxidation agent, first aluminum reacts and brings about alumina-rich inclusions, which are often of high dispersity and have a very slow tendency to rise in the steel melt. In order totally to avoid titanium losses by reaction with silica, the use of lime-corundum (60 : 40) slags is to be recommended in processing with an electric arc furnace or open hearth. Additions of ferrosilicon are to be avoided to keep silica content of the slags very low.

103. Lime chromite slags produced in the metallurgical ferrochrome process were described by D. S. Belyankin and V. V. Lapin[166] from converter slags. The ratio $CaO : Cr_2O_3$ varies between 4.7 and 2.2. The slags contain α-$CaO.Cr_2O_3$, a green chrome spinel, perhaps monticellite, and calcium fluoride. Another slag high in Fe_2O_3 shows crystalline solutions of calcium chromite with ferrite, of strong pleochroism with yellow and red-brown colors, and β-Ca_2SiO_4. Free lime is efficiently stabilized by iron and manganese oxides.

104. *Radioactive tracer* methods have been repeatedly applied to problems of the interaction of refractories with metallurgical slags, as we noted in C. ¶ 19 in experiments of Z. V. Levintovich, A. N. Lyulichev, O. M. Margulis, and D. M. Shakhtin,[167] especially to trace with Ca^{45} the distribution of calcium in refractories. V. A. Grigoryan and A. M. Samarin[168] applied this method to study the origin of silicate inclusions in steel ingot castings (for ball-bearings), or with electrosteel with acidic and basic linings. When siphoning the steel, the lining may furnish up to 9 per cent of nonmetallic inclusions and up to 18 per cent may be derived from the destruction of ladle linings. Also O. M. Margulis and A. G. Karaulov[169] "indexed" with Ca^{45} the linings of ladles, siphon channels, spouts, and other parts, which come into direct contact with flowing steel. Inclusions were isolated by electrolysis, and examined by radiographic prints, but even the tracer method could not sufficiently answer problems of the influence of refractoriness of linings on quality of steel.

[165] *Arch. Eisenhüttenw.*, **25**, 1954, 515-521. On problems of distribution of manganese between steel melts and FeO-MnO slags see W. A. Fischer and H. J. Fleischer, *Ibid.*, **32**, 1961, 1-10.

[166] *Doklady Akad. Nauk S.S.S.R.*, **91**, 1953, 911-914; *Silikat Tech.*, **4**, 1953, 549-550.

[167] *Ogneupory*, **22**, 1957 (1) 29-31.

[168] *Izvest. Akad. Nauk S.S.S.R.*, Otdel. Tekhn. Nauk 1954 (3) 91-101.

[169] *Ogneupory*, **21**, 1956, 253-258; cf. also M. I. Tsekhanskiĭ, H. B. Khusnoyarov, and G. D. Susloparov, *Ibid.*, **23**, 1958, 82-87.

105. Interesting reactions of *desulfurization* of iron melts by fused *soda* (soda slag process) were studied by C. E. A. Shanahan[170] with a high degree of efficiency. Practically important also is the fact that corrosion exerted on refractory linings of the siphons, ladles, mixers, and the like, in which the process is operated, is not so high that it would be intolerable. It is only advisable to remove fume, and to skim off the soda slag, and no greater practical troubles are found in the process than is normal in Bessemer converters. The usefulness of soda wastes for glass-making is generally accepted,[171] especially for manufacturing dark bottles and for black enamels, in which considerable reduction of surface tension by alkali sulfides is most desirable for a good flow on the steel sheet. In its general importance for the basic understanding of the soda slag desulfurization process, studies of P. T. Carter and A. Ibrahim[172] on the system Na_2O—FeO—SiO_2 (cf. Vol. III, B. ¶ 320 f.) may be emphasized.

106. The analogous *soda dephosphorization* process of pig iron was discussed by W. R. Maddocks and E. T. Turkdogan.[173] At 1400°C. a reduction of original phosphorus contents by 50 to 90 per cent is possible, for carbon only 25 to 60 per cent, with a relatively basic slag composition (basicity ratio 1.2 to 2.5). If the slag is more siliceous (55 to 60 per cent silica), higher temperatures are required to dissociate fayalite. Not especially for soda dephosphorization, but in a more general understanding, studies of O. A. Esin and V. N. Shikhov[174] are interesting, wherein they introduced into the iron bath ferrophosphorus, indexed by P^{32} with a complete kinetics of distribution of the radioactive tracer element in metal and slag at 1550°C. as a function of time. The result is highly remarkable in that not so much does the chemical reaction rates determine output of the dephosphorization process, as does the rate of diffusion of the phosphorus. Equilibrium concentrations are proportional to original phosphorus contents in the pig iron.

107. In their fundamental investigations on optimum compositions of blast furnace slags in the system CaO—MgO—Al_2O_3—SiO_2, E. F. Osborn, R. C. DeVries, K. G. Gee, and H. M. Kraner[175] (cf. Vol. III, B. ¶ 283 f.; C. ¶ 14 ff.) repeatedly emphasized the analogy between favorable viscosity and desulfurization conditions in such slags, including the possibilities of using K_2O or Na_2O as desulfurization agents after tapping. They also present a diagram (Fig. C. 29) based on older data of W. F.

[170] *Iron & Steel (London)*, **29**, 1956 (1) 9-14.

[171] Cf. H. Anders, *Tonind.-Ztg. u. Keram. Rundschau*, **76**, 1952, 165; M. Í. Nekrich and I. Borovikov, *Steklo i Keram.*, **8**, 1951 (9) 10; 9, 1952 (4) 9, with Ya. P. Mostovoĭ.

[172] *J. Soc. Glass Technol.*, **36**, 1952, 142-163.

[173] *J. Iron Steel Inst. (London)*, **162**, 1949, 249-264.

[174] *Doklady Akad. Nauk S.S.S.R.*, **102**, 1955, 327-330; *Izvest. Akad. Nauk S.S.S.R.*, Otdel. Tekhn. Nauk 1955 (3) 79-89.

[175] *J. Metals*, **6**, 1954 (1) 33-45, especially pp. 38-41.

Holbrook and T. L. Joseph[176] for a constant alumina content of 10 per cent with characteristic curves of constant desulfurization capacities. O. A. Esin and V. N. Shikhov[177] studied the distribution between liquid iron (of a relatively high purity) and slag melts at 1550° to 1700°C. for slags of different basicities, with systematic additions of iron sulfide marked with S^{35} and Fe^{59}. In thin layers cut from the metal

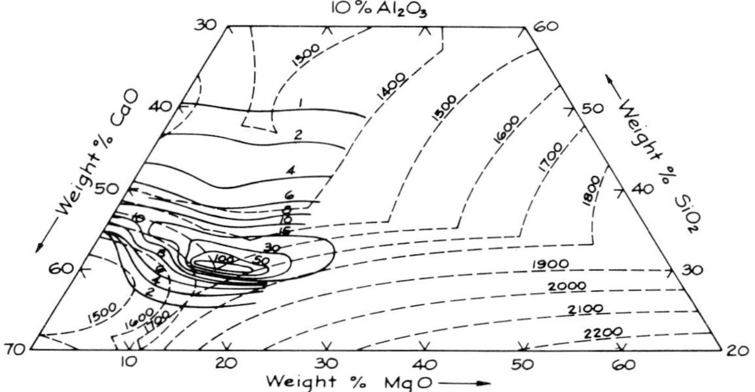

Fig. C. 29. Curves of constant capacity of desulfurization of melts in a portion of system SiO_2—Al_2O_3—CaO—MgO, with isotherms (dashed curves are the Liquidus isotherms). (Osborn, De Vries, Gee, and Kraner). Data are from Holbrook and Joseph.

and slag the isotope contents could thus easily be determined with high accuracy of distribution kinetics. With increasing surface ratios of metal to slag up to 1 : 7 to 8, the rate of desulfurization is increased as it was previously determined in experiments with pig iron. The higher the initial sulfur contents of the iron (from 0.005 to 0.1 per cent) the more rapid is the transition of sulfur into the slag.

108. The process is of the first order for basic and neutral slag compositions, whereas for siliceous (acidic) slags the order is 0.75. Temperature coefficient of transition is large for siliceous slags, less for neutral and basic slags. At a constant temperature of 1580°C. the concentration increases of S^{35} and Fe^{59} in the slags are a function of time. The kinetic equation includes ionization and electrostatic charge exchange reactions in the metal/slag system. The reversed transition from slag into metal follows a simple linear relation of the first order. In spite of exterior analogies there is a distinct difference in desulfurization of pig iron melts high in carbon and steels low in carbon. In the first case, the transition of iron into the slag is impeded. The

[176] *Am. Inst. Mining Met. Engrs. Trans.*, **120**, 1936, 99-117; *Am. Inst. Mining Met. Eng. Tech. Publ.*, **960**, 1936, 19 pp.; **131**, 1938, 127-138.

[177] *Doklady Akad. Nauk S.S.S.R.*, **102**, 1955, 583-586; *Izvest. Akad. Nauk S.S.S.R.*, Otdel. Tekhn. Nauk 1955 (2) 105-112; (3) 79-89.

following equation illustrates desulfurization of cast iron:

$$S(\text{metal}) + O^{2-}(\text{slag}) \rightarrow S^{2-}(\text{slag}) + O(\text{metal})$$

and for steel melts:

$$S(\text{metal}) + Fe(\text{metal}) \rightarrow S^{2-}(\text{slag}) + Fe^{2+}(\text{slag})$$

ruling the kinetics of the process. The fact that a temperature increase by 60° to 100°C. increases the rate of desulfurization for basic slags only by 1.5 to 2 times, but for acidic slags by more than 10 times, indicates that in basic slags the convection diffusion of sulfur controls the process, whereas in acidic slags the reaction itself determines the rates.

109. Interesting equilibrium studies between calcium aluminate slags and a furnace atmosphere of CO, CO_2, and SO_2, were made by P. T. Carter and T. G. Macfarlane[178] to determine activities of CaO and Al_2O_3. The thermodynamic discussion concerned especially the free energies for the formation of $3CaO.Al_2O_3$, $12CaO.7Al_2O_3$,

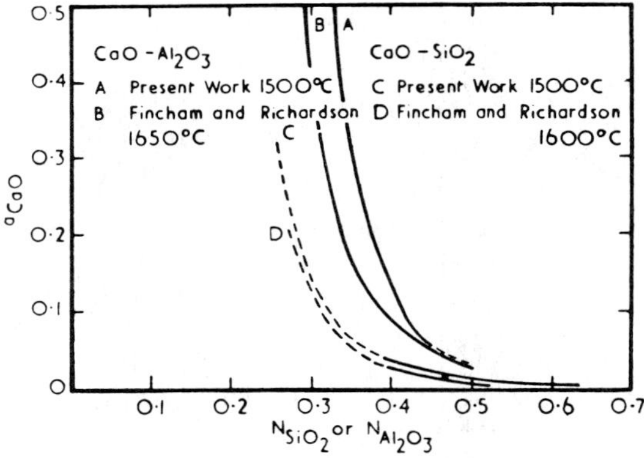

FIG. C. 30. Comparison of the activities of CaO in slags of the binary systems SiO_2—CaO and Al_2O_3—CaO. (Carter and Macfarlane).

and $CaO.Al_2O_3$ from the oxides. The sulfur absorbed by different slags depends on the ratio $(p_{SO_2}/p_{O_2})^2$ which decreases with increasing temperature in gas compositions used in the experiments. Corresponding studies with slags of the system CaO—SiO_2 at 1500°C. were used to derive activities of CaO and SiO_2 in the melts and free energies of the formation of $CaO.SiO_2$ and Ca_2SiO_4. It is easily understood that reducing conditions favor formation of CaS in the slags, but calcium sulfate at high oxi-

[178] *J. Iron Steel Inst.* (*London*), **185**, 1957 (1) 54-66. See previously C. J. B. Fincham and F. D. Richardson, *Proc. Roy. Soc.* (*London*), *A*, **223**, 1954, 40-62; *J. Iron. Steel Inst.* (*London*), **178**, 1954, 4-15.

dation potentials. If oxygen pressure is lower than 10^{-8} atm., the sulfur transfer takes place almost entirely by formation of sulfide, whereas sulfate prevails for oxygen pressures above 10^{-3} atms. A comparison of activities of CaO in calcium aluminate and silicate slags (cf. Fig. C. 30) immediately illustrates that a_{CaO} is small for all compositions more acidic than Ca_2SiO_4, but increases rapidly as N_{CaO} is increased beyond that corresponding to Ca_2SiO_4. Temperature effects are apparently rather small.

110. Results of Carter and Macfarlane are in agreement with the common assumption of slag behavior in steel-making that formation of Ca_2SiO_4 determines the limit in calculation of basic or oxygen activities. At lower basicities than that corresponding to Ca_2SiO_4 desulfurization and dephosphorization powers of basic steel making are very poor. The corresponding rapid increase in a_{Ca} in calcium aluminate slags occurs, however, at much lower basicities and the value of a_{CaO} is much more dependent on temperature. Similar characteristics may be expected for complex slag compositions of steel metallurgy practice, and it is obvious how dubious is the value of assessing the effect of alumina on base or oxygen ion activity by assuming 3CaO. Al_2O_3 formation, or anions of the type AlO_2^-.

111. Distribution equilibrium of sulfur between metal bath and slag in the open hearth process was extensively discussed by L. A. Shvartsman, A. I. Osipov, V. F. Surov, M. L. Sazanov, S. A. Telesov, and A. M. Opengengen,[179] especially under the impression of considerable deviations of practical data from the theoretical thermodynamic distribution coefficient of sulfur in the baths, evidently caused by a very sluggish mass transfer beyond phase boundaries. If we define, with J. Grant and J. Chipman,[180] the basicity of the slag as the difference, Δ, of the moles of basic and acidic oxides (FeO assumed to be neutral), the distribution coefficient, L_S, is a linear function of this difference. The activity a_{FeS} is a function of the slag composition including FeO and MnO; the activity of FeO, however, cannot be neglected if it replaces silica. Increasing contents in FeO in open hearth slags bring about a decreasing viscosity (at constant temperature), the mass transfer is accelerated, and with it also the desulfurization process. If unusually high sulfur concentrations appear in the slag during the fusion period, this may be a consequence of absorption of sulfur from flame gases (see above), and agrees with observations by radioactive tracers. Establishment of equilibrium with the metal bath is a pronouncedly slow process; the difference Δ may even become negative in the fusion period. The slag is then more acidic, L_S is low, and no desulfurization occurs at all. Later, Δ increases, becomes positive, and the slag starts its desulfurizing action with increasing L_S for the distribution coefficients. A large portion of the finishing and the whole period

[179] *Doklady Akad. Nauk S.S.S.R.*, **120**, 1958, 599-601.

[180] *Am. Inst. Mining Met. Eng. Tech. Publ.*, No. 1988, 1946, 16 pp.; *Metals Technol.*, 1946, No. 3.

of fusion is practically without any important desulfurization effect because of sluggishness of the distribution equilibrium on the phase boundaries metal/slag bath. The use of accelerating additions is, therefore, most desirable.

112. An interesting practical attempt to determine by elementary calculations and nomograms the oxidation capacities and desulfurization potentials of metallurgical slags as a function of temperature and chemical composition was made by Tr. Tr. Negrescu,[181] based on the presence of cation groups of rather unusual types like 4 MgO_2, 5 SiO_2; 3 FeO, 8 SiO_2; 3 MnO, 10 SiO_2; 3 CaO, 5 SiO_2; BaO, 3 SiO_2, near the eutectic compositions occurring in binary systems. Such groupings appear associated with electrostatic charges in the constitution of the liquid melts and are stable over wide ranges of temperature, likewise in variable slag composition fields,

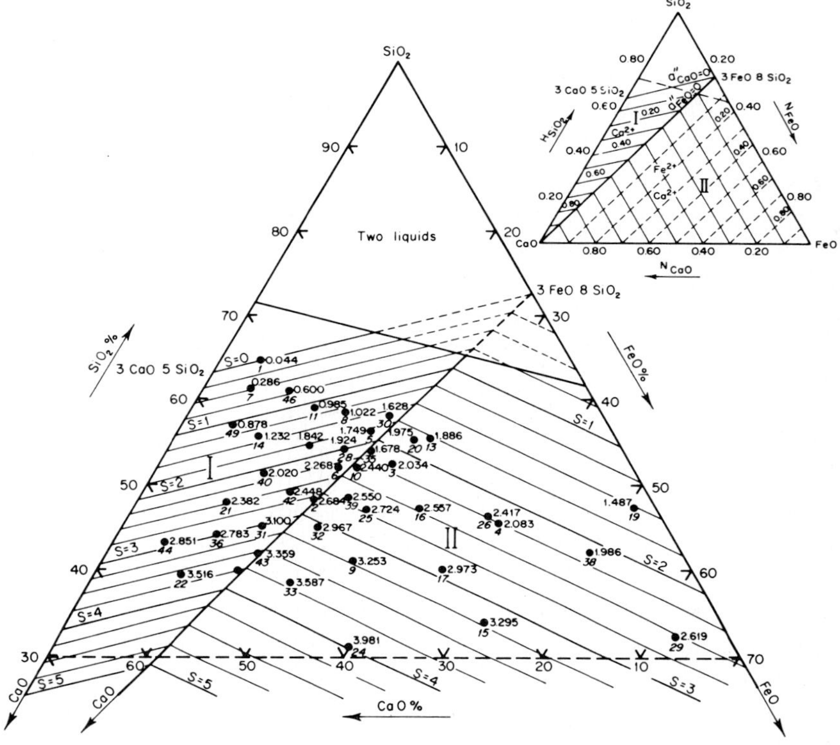

Fig. C. 31. Nomogram illustrating the absorption capacity for sulfur. (Fincham and Robertson), of slags in the system SiO_2—CaO—FeO, with the characteristic groupings 3CaO,5SiO_2 and 3FeO,8SiO_2. (Negrescu).

[181] *Ind. chim. belge*, **24**, 1959, 1349-1356; *Rev. mét. Acad. rép. populaire Roumaine*, **1**, 1956, 5-32; **2**, 1957, 5-49; **3**, 1958, 5-40.

say of the systems, $CaO—SiO_2$, $CaO—MgO—SiO_2$, CaO—MgO—FeO, CaO—MnO—SiO_2, $CaO—BaO—SiO_2$, and $CaO—Al_2O_3—SiO_2$, for 1425° to 1575°C. A very similar relation exists between the nomograms given for desulfurization capacities of slags and for oxidizing power of FeO and Fe_2O_3-containing slags of refining furnaces. Ions which do not belong to the association groups mentioned are properly reactive constituents of the slags. As an example of Negrescu's diagrams we give in Fig. C. 31 the nomogram for the system $CaO—FeO—SiO_2$ at 1500°C. indicating capacities of different ternary compositions for the absorption of sulfur, precisely as Finch and Richardson defined this property. It corresponds to the general formulation of the desulfurization coefficient $N_S = k_T[N_{CaO} - 2/5\ (N_{SiO_2} - 5/4\ N_{MgO} - 8/3\ N_{FeO} - 10/3\ N_{MnO})]$. In Fig. B. 31, slags of field I have calcium ions disposable for absorbing S^{2-} anions, but not O^{2-} anions, whereas in field II Fe^{2+} cations are disposable besides Ca^{2+} ions, and both S^{2-} and O^{2-} anions can be "captured."

113. Concerning the fascinating problems of how physicochemical experience and theory of metal/slag systems may be applied in *geochemical* and *cosmogonic* questions, we have to refer to studies of Ya. I. Ol'shanskiĭ[182] on the ternary system Fe—FeO—FeS, and the quaternary system $Fe—FeO—FeS—SiO_2$, as model systems of modern aspects of material constitution of the interior of the earth in its different zones corresponding to the theories of V. M. Goldschmidt (1922) and H. S. Washington (1925). Fe—FeO—FeS for isothermal sections at 1140°, 1290°, and 1400°C. shows the well-known large ternary miscibility gap. The high vapor pressure of sulfur, and reaction of the sulfide melt with the oxide to evolve SO_2 make the equilibrium studies rather inaccurate and restricted (cf. Fig. C. 32), but it is remarkable that the lowest isotherm for 1140°C. and the lower boundary curve of the miscibility gap overlap to a certain extent the line FeS—FeO, indicating by negative coordinates for Fe the presence of excess oxygen, in other words, presence of some trivalent iron (in magnetite) in the melt. The immiscibility phenomenon in the quaternary system with silica makes even the appearance of three liquid phases in equilibrium possible. There is also a constant transition from the pure iron melt to silicate liquid, and a gradual transition of sulfur from Fe to FeS in which FeO is easily dissolved. Even silica can be taken up in this sulfide-oxide liquid phase to a certain degree. In the fields where only one liquid appears, a continuous transition from all to all of the end compositions is possible. If pure fayalite melt is added, a disintegration of the silicate was observed into iron and a complex $FeO + Fe_2O_3$ silicate melt. If FeS is present, the reaction $FeS + Fe_2SiO_4 \rightarrow SiO_2 + (FeO + FeS)$ prevails, thus all of the silica in fayalite is split off. It is, therefore, characteristic of the quaternary system that finally FeS, Fe, and silica (besides some residual fayalite) occur. Melts high in FeS cannot simultaneously contain much silica.

[182] *Doklady Akad. Nauk S.S.S.R.*, **80**, 1951, 893-896; *Izvest. Akad. Nauk S.S.S.R.*, Ser. Geol., 1951 (6) 128-152.

114. Temperature conditions in the innermost parts of the earth were discussed by F. W. Preston[183] from the viewpoints of heat radiation equilibria and radiative transfer, comparable to that in homogeneous glass melts in a large tank at high temperatures. We considered such conditions in B. ¶ 110 ff. from the practical viewpoint of glass technology.

115. The system Fe—S—O was discussed by G. Kullerud[184] for phase relations, especially the changes occurring in the narrow range from about 700° to 743°C. Below 675°C. pyrite, FeS_2, and pyrrhotite, FeS, coexist with a liquid phase, and hematite, Fe_2O_3, is not stable in the presence of FeS, but at 675°C. the reaction of pyrite with magnetite to form pyrrhotite and hematite starts, and paragenesis of

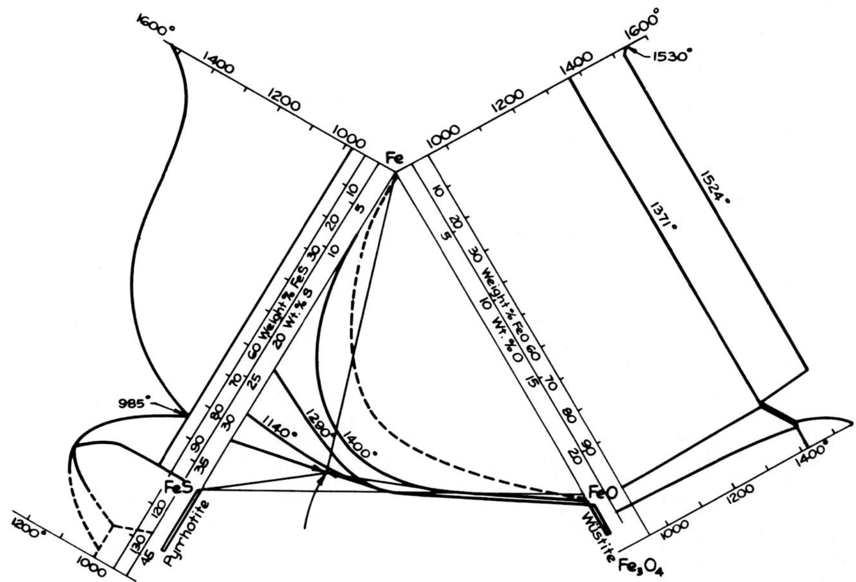

FIG. C. 32. Ternary miscibility gap in the liquid phases of the system Fe—FeO—FeS, with isotherms. (Ol'shanskiĭ).

hematite with FeS and magnetite may become stable, whereas magnetite and pyrite are incompatible. At 700°C. the reaction of pyrite with hematite to form FeS and SO_2 occurs, hematite and pyrite, as well as magnetite and pyrite, being incompatible. Above 743°C. pyrite is no longer stable at all, regardless of the amounts of sulfur

[183] *Am. J. Sci.*, **254**, 1956, 754-757; see also S. P. Clark, *Am. Mineralogist*, **42**, 1957, 732-742, especially on radiative conductivity pp. 737 ff.

[184] *Carnegie Inst. Wash. Year Books*, **56**, 1956/57, 198-200.

present, but hematite and magnetite + pyrrhotite are a mineral association which is stable in an almost ubiquitous occurrence in nature, not only in terrestrial rocks, but also in meteoric stones.

116. For deep-seated telluric silicate rocks the eclogite problem is still the key to a fundamentally important understanding for geochemistry and geophysics. We mention in this connection interesting observations of eclogite in veins from an Alpine orthogneiss, as described by St. Zucchetti,[185] evidently of eruptive nature with interspersed pyrrhotite. The presumed eclogite magma is a basic system erupted under ultrahigh pressures from deep-seated hearths in the interior of the earth. A classical case of convergence of eruptive and highly metamorphic eclogitization is the occurrence described from Archaean complexes of crystalline schists in connection with the kimberlite pipes from the Siberian platform, described by A. P. Bobrievich and V. S. Sobolev[186] as derived from primary norites, under an estimated pressure of 15,000 to 17,000 atm., in very deep zones from which only xenoliths were transported to the surface by the kimberlite pipe explosions.

[185] *Periodico Mineral.* (*Rome*), **24**, 1956, 349-359.
[186] *Zapiski Vsesoyuz. Mineral. Obshchestva*, **86**, 1957, 3-17.

AUTHOR INDEX

C

D

E

F

G

H

I

M

Q

R

S

W

Y

Z

SUBJECT INDEX

A

B

C

D

E

G

UNIV. COLL. N.W.
BANGOR
LIBRARY

M

P

S

T

U

V

W

COMPOUND INDEX

Chemical Campounds

Organic Compounds

MINERAL INDEX

E

F

G

H

I

K

L

M

N

V

W

Z